Pemberton

Mathematics for Cambridge IGCSE®

Second edition

For the updated syllabus

Extended

Sue Pemberton

Oxford excellence for Cambridge IGCSE®

OXFORD

Great Clarendon Street, Oxford OX2 6DP, United Kingdom

Oxford University Press is a department of the University of Oxford. It furthers the University's objective of excellence in research, scholarship, and education by publishing worldwide. Oxford is a registered trade mark of Oxford University Press in the UK and in certain other countries

First published in 2012

Second edition 2016

British Library Cataloguing in Publication Data

Data available

ISBN: 978-0-19-837840-2
10 9 8 7 6 5 4 3 2 1

Printed in China by Golden Cup

Acknowledgments
® IGCSE is the registered trademark of Cambridge International Examinations.

All examination-style questions and answers within this publication have been written by the author. In examination, the way marks are awarded may be different.

The publishers would like to thank the following for permission to reproduce photographs:

P90: Kuttly/Shutterstock; **P257**: R. Formidable; **P271**: Andreas Meyer/Shutterstock; **P271**: Jorgen Mcleman/Shutterstock.

Cover Image: Shutterstock

Contents

Access your support website for extra exam revision material and presentations www.oxfordsecondary.com/9780198378402

Syllabus topic		Page numbers in student book
E1: Number		
E1.1	Identify and use natural numbers, integers (positive, negative and zero), prime numbers, square numbers, common factors and common multiples, rational and irrational numbers, real numbers.	3–9, 378–379
E1.2	Use language, notation and Venn diagrams to describe sets and represent relationships between sets. Definition of sets e.g. $A = \{x: x \text{ is a natural number}\}$, $B = \{(x,y): y = mx + c\}$, $C = \{x: a \leq x \leq b\}$, $D = \{a, b, c, \ldots\}$	272–283
E1.3	Calculate squares, square roots, cubes and cube roots of numbers.	Embedded throughout book
E1.4	Use directed numbers in practical situations.	3
E1.5	Use the language and notation of simple vulgar and decimal fractions and percentages in appropriate contexts. Recognise equivalence and convert between these forms.	10–13
E1.6	Order quantities by magnitude and demonstrate familiarity with the symbols $=, \neq, >, <, \geq, \leq$	62–63
E1.7	Understand the meaning and rules of indices. Use the standard form $A \times 10^n$ where n is a positive or negative integer, and $1 \leq A < 10$.	58–61, 284–287
E1.8	Use the four rules for calculations with whole numbers, decimals and vulgar (and mixed) fractions, including correct ordering of operations and use of brackets.	2, 10–13
E1.9	Make estimates of numbers, quantities and lengths, give approximations to specified numbers of significant figures and decimal places and round off answers to reasonable accuracy in the context of a given problem.	13, 14–16, 17–18
E1.10	Give appropriate upper and lower bounds for data given to a specified accuracy. Obtain appropriate upper and lower bounds to solutions of simple problems given data to a specified accuracy.	422–4257
E1.11	Demonstrate an understanding of ratio and proportion. Use common measures of rate. Calculate average speed.	20–23, 25–26, 35–38, 268–271

E1.12	Calculate a given percentage of a quantity. Express one quantity as a percentage of another. Calculate percentage increase or decrease. Carry out calculations involving reverse percentages.	50–57, 168–169, 264–267
E1.13	Use a calculator efficiently. Apply appropriate checks of accuracy.	Embedded throughout the book
E1.14	Calculate times in terms of the 24-hour and 12-hour clock. Read clocks, dials and timetables.	Embedded throughout the book
E1.15	Calculate using money and convert from one currency to another.	220–221
E1.16	Use given data to solve problems on personal and household finance involving earnings, simple interest and compound interest. Extract data from tables and charts.	266–267
E1.17	Use exponential growth and decay in relation to population and finance.	380–383
E2: Algebra and graphs		
E2.1	Use letters to express generalised numbers and express basic arithmetic processes algebraically. Substitute numbers for words and letters in formulae. Construct and transform complicated formulae and equations.	17–27, 118–119, 388–389
E2.2	Manipulate directed numbers. Use brackets and extract common factors. Expand products of algebraic expressions. Factorise where possible expressions of the form: $ax + bx + kay + kby$, $a^2x^2 - b^2y^2$, $a^2 + 2ab + b^2$, $ax^2 + bx + c$.	3–4, 18–21, 114–117, 230–233
E2.3	Manipulate algebraic fractions. Factorise and simplify rational expressions.	64–67, 388–389
E2.4	Use and interpret positive, negative and zero indices. Use and interpret fractional indices. Use the rules of indices.	58–61, 284–287
E2.5	Solve simple linear equations in one unknown. Solve simultaneous linear equations in two unknowns. Solve quadratic equations by factorisation, completing the square or by use of the formula. Solve simple linear inequalities.	20–23, 62–63, 108–113, 288–291, 384–387, 426–427, 432–435
E2.6	Represent inequalities graphically and use this representation in the solution of simple linear programming problems.	72–75, 428–431
E2.7	Continue a given number sequence. Recognise patterns in sequences and relationships between different sequences. Find the nth term of sequences (including quadratic and cubic sequences, exponential sequences and simple combinations of these).	340–349

About this book

This revised edition has been specially prepared to help you achieve your highest potential on the Cambridge IGCSE® Mathematics 0580, extended syllabus. The book is fully up to date and covers the latest syllabus in depth.

The author is an experienced mathematics teacher and an examiner. Packed with carefully chosen examples, helpful tips and plenty of exercises to give you confidence in your abilities, the author's many years of experience ensure that the book is carefully designed to help you succeed.

The contents are organised into nine units and within each unit the sections are grouped into the broad topics: number, algebra, shape and space, and probability and statistics. Each section concludes with a selection of examination-style questions. A unit is intended to be covered in around 20 hours; this should leave ample time for revision and exam practice if the syllabus is being taught over two years.

In all the examination papers it is permissible to use an electronic calculator provided that it is not an algebraic or graphical calculator. Four figure trigonometric tables are also permitted. Opportunities to practise using a calculator arise throughout the book, though in practice many questions will not require the use of a calculator and candidates should be able to use mental and written methods. Indeed certain questions may require evidence of a written method being used.

Note that because this book is written for the extended syllabus a number of core topics – basic arithmetic, measures, time, money, etc. – are not covered explicitly but are woven into the treatment of more advanced topics.

The support website contains extra exam practice material and eighty-six presentations, one for each section of the book, that provide many more fully worked examples. All this material can be found online at www.oxfordsecondary.com/9780198378402

Order of operations

THIS SECTION WILL SHOW YOU HOW TO
- Perform operations in the correct order

When a calculation involves more than one operation it is important to do the operations in the correct order.

1 Work out the **B**rackets first.
2 Work out the **I**ndices next.
3 Work out the **D**ivisions and **M**ultiplications next.
4 Work out the **A**dditions and **S**ubtractions last.

Memory aid

B I D M A S
- **B**rackets
- **I**ndices
- **D**ivision
- **M**ultiplication
- **A**ddition
- **S**ubtraction

EXAMPLE

Calculate $150 - 3^2 \times (8 + 4) + 6 \div 2$

$150 - 3^2 \times (8 + 4) + 6 \div 2$ *brackets first*
$= 150 - 3^2 \times 12 + 6 \div 2$ *indices next, 3^2 means $3 \times 3 = 9$*
$= 150 - 9 \times 12 + 6 \div 2$ *division and multiplication next*
$= 150 - 108 + 3 = 45$ *addition and subtraction last*

EXERCISE 1.1

Work out:

1 $2 + 3 \times 5$ **2** $8 \div 2 + 4$ **3** $3 \times 4 - 5 \times 2$ **4** $5 \times 6 - 8 \div 2$

5 $7 - 2 \times 3 + 5$ **6** $10 - 2 \times 3$ **7** $6 + (3 \times 5) - 2$ **8** $10 + 3^2 \times 4$

9 $(5 + 4) - 3 \times 2$ **10** $(5 \times 4) - (3 \times 2)$ **11** $(4 \times 3) \times 2^2$ **12** $5 \times (4 - 3) \times 2$

13 $5 \times (16 - 3 \times 2)$ **14** $38 \div 2 - 4 \times 3$ **15** $56 - (2^3 + 4)$ **16** $4 + (3 \times 2)^2$

17 $(4 \times 3 + 2)^2$ **18** $(2 \times 3^2) + 4$ **19** $4^3 - 5 \times 6$ **20** $5 \times (2^3 + 3^2)$

EXERCISE 1.2

Copy these and use brackets (where necessary) to make the statements true.

1 $2 + 3 \times 4 + 5 = 25$ **2** $2 \times 3 + 4 \times 5 = 26$

3 $2 + 3 \times 4 + 5 = 29$ **4** $2 + 3 \times 4^2 = 80$

5 $2 \times 3 + 4 \times 5 = 46$ **6** $5 + 4 \times 3 - 2 = 15$

7 $2 \times 3 + 4 \times 5 = 50$ **8** $2 \times 3 + 4 \times 5 = 70$

9 $2 + 3 \times 4 + 5 = 45$ **10** $2 + 3 \times 4^2 = 50$

Directed numbers

THIS SECTION WILL SHOW YOU HOW TO
- Perform the four rules on directed numbers

The **positive** and **negative** whole numbers are called **integers**.
They can be shown on a number line.

NEGATIVE NUMBERS POSITIVE NUMBERS

−6 −5 −4 −3 −2 −1 0 1 2 3 4 5 6

The number line can be used in practical situations.

To find the difference between a temperature of 5 °C and a temperature of −3 °C, you find the gap between these two numbers on the number line.
The difference is 8 °C.

Adding and subtracting directed numbers

The rules for adding and subtracting directed numbers are:

Change	$-2 + +5$	to	$-2 + 5 = 3$
Change	$-2 + -5$	to	$-2 - 5 = -7$
Change	$-2 - +5$	to	$-2 - 5 = -7$
Change	$-2 - -5$	to	$-2 + 5 = 3$

EXAMPLE

Work out **a** $(-9) - (-3)$ **b** $(+8) - (+15)$ **c** $(29) + (-12)$

a $-9 - -3 = -9 + 3 = -6$ change − − to +
b $+8 - +15 = +8 - 15 = -7$ change − + to −
c $29 + -12 = 29 - 12 = 17$ change + − to −

EXERCISE 1.3

Work out:

1 $(-3) + (-5)$ **2** $6 - (-4)$ **3** $8 + (-10)$

4 $2 + (-5)$ **5** $(-4) - (+2)$ **6** $(-15) + (-3)$

7 $36 + (-8)$ **8** $29 - (+1)$ **9** $(-52) - (-38)$

10 $(-54) + (-3)$ **11** $(-16) + (-2)$ **12** $(-20) - (-20)$

13 $(-57) + (+5)$ **14** $41 + (-16)$ **15** $52 - (-3)$

16 $(-5) - (+10)$ **17** $(-7) - (-14)$ **18** $(-42) + (-5)$

19 $(-8) + (-2) + (-5)$ **20** $(-6) - (+2) - (-3)$ **21** $7 - (-2) + (-3)$

22 $(+9) - (-6) + (-6)$ **23** $7 - (+9) + (-3)$ **24** $46 - (-12) + (-5)$

Multiplying and dividing directed numbers

The rules for multiplying and dividing directed numbers are:

Multiplication	Division
+ × + = +	+ ÷ + = +
+ × − = −	+ ÷ − = −
− × + = −	− ÷ + = −
− × − = +	− ÷ − = +

If the two signs are the same, the answer will be positive.
If the two signs are different, the answer will be negative.

EXAMPLE

Work out **a** $(-6) \div (-2)$ **b** $5 \times (-8)$ **c** $(-4)^3$

a $(-6) \div (-2) = 3$ *the two signs are the same so the answer is positive*

b $5 \times (-8) = -40$ *the two signs are different so the answer is negative*

c $(-4)^3 = -4 \times -4 \times -4$ *first multiply −4 by −4*
$= 16 \times -4$ *then multiply by −4 again*
$= -64$

EXERCISE 1.4

Work out:

1 $(-12) \times (-5)$ **2** $(-8) \times (+4)$ **3** $(+16) \times (-2)$

4 $(-52) \div (-13)$ **5** $(-55) \div (+5)$ **6** $(-145) \div (-5)$

7 $(+20) \div (-2)$ **8** $(-95) \div (-19)$ **9** $(-11) \times (-11)$

10 $(-3) \times (-4) \times (-5)$ **11** $(-2) \times (+8) \times (-4)$ **12** $(+6) \times (-3) \times (-7)$

13 $(-2) \times (-5) \times (+6)$ **14** $(-9)^2$ **15** $(-15)^2$

16 $(-5)^3$ **17** $(-60)^3$ **18** $(-4)^3 \times (-1)^3$

19 $(-2)^5 \times (-10)^2$ **20** $(-1)^{13}$ **21** $(-1)^{15} \times (-1)^{24}$

22 $\frac{-6}{-3}$ **23** $\frac{(-10) \times (+3)}{-15}$ **24** $\frac{(-12) \times (-5)}{(-2) \times (+10)}$

25 Check your answers to questions **1** to **18** using a calculator.

Find the missing numbers:

26 $\frac{\square^2 \times (-12)}{8 \div (-2)} = 48$ **27** $\frac{(-5) \times \square}{10} = 3$ **28** $\frac{\square \div 3}{(-2) \times (-6)} = -1$

KEY WORDS
positive
negative
integer

Multiples, factors, primes, squares and cubes

THIS SECTION WILL SHOW YOU HOW TO

- Identify and use factors, multiples, primes, squares and cubes

Factors

The whole numbers that divide exactly into 15 are called **factors** of 15.
The factors of 15 are 1, 3, 5 and 15.

EXAMPLE

List all the factors of 24.

$24 = 1 \times 24$ — *write 24 as the product of two factors*
$24 = 2 \times 12$ — *repeat until all pairs have been found*
$24 = 3 \times 8$
$24 = 4 \times 6$
Factors of 24 = 1, 2, 3, 4, 6, 8, 12 and 24.

EXAMPLE

Find the **highest common factor (HCF)** of 20 and 36.

Factors of 20 = (1), (2), (4), 5, 10, 20 — *list the factors of both 20 and 36*
Factors of 36 = (1), (2), 3, (4), 6, 9, 12, 18, 36
Common factors of 20 and 36 are 1, 2, and 4. — *find the numbers that are in both lists*
Highest common factor of 20 and 36 is 4. — *select the highest number*

Multiples

The **multiples** of 6 are the numbers 6, 12, 18, 24, 30 …

EXAMPLE

Find the **lowest common multiple (LCM)** of 12 and 9.

Multiples of 12 = 12, 24, (36), 48, 60, (72), 84 … — *list the multiples of 12 and 9*
Multiples of 9 = 9, 18, 27, (36), 45, 54, 63, (72), 81 …
Common multiples of 12 and 9 are 36, 72 … — *find the numbers that are in both lists*
Lowest common multiple of 12 and 9 is 36. — *select the lowest number*

EXERCISE 1.5

1 Write down all the factors of:

a	10	b	15	c	9	d	17	e	60
f	80	g	100	h	64	i	125	j	90

2 Find the common factors of:

a	6 and 8	b	10 and 15	c	9 and 18
d	16 and 20	e	20 and 25	f	12 and 30
g	80 and 100	h	42 and 48	i	6, 12 and 42

3 Find the highest common factor (HCF) of:

a	6 and 8	b	10 and 15	c	90 and 18
d	36 and 45	e	23 and 46	f	20 and 24
g	30 and 45	h	42 and 48	i	8, 32 and 44

4 List the first six multiples of each of the following numbers.

a	10	b	6	c	9	d	18	e	25
f	40	g	100	h	12	i	27	j	121

5 Find the lowest common multiple (LCM) of:

a	6 and 8	b	5 and 15	c	6 and 9
d	7 and 8	e	4 and 6	f	12 and 8
g	14 and 21	h	11 and 5	i	8, 10 and 12

6 A piece of rope can be cut into an exact number of 6 m lengths.
The rope could also be cut into an exact number of 8 m lengths.
What is the shortest possible length of the rope?

7 A light flashes every 15 minutes.
A second light flashes every 18 minutes.
Both lights flash together at 2 a.m.
What will be the time when they next flash together?

8 A bell rings every 20 seconds. A second bell rings every 25 seconds.
A third bell rings every 30 seconds. They all ring together at 8 p.m.
How long will it be before all three bells ring together again?

Primes

A **prime** number is a number that has exactly two factors.
5 is a prime number because it has exactly two factors (1 and 5)

EXAMPLE

List the first ten prime numbers

The prime numbers are: 2, 3, 5, 7, 11, 13, 17, 19, 23 and 29

A **prime factor** is a factor that is also a prime number.

EXAMPLE

List the prime factors of 30.

The factors of 30 are: 1, (2), (3), (5), 6, 10, 15 and 30 *select the factors that are prime*
The prime factors of 30 are: 2, 3 and 5.

Numbers can be written as the **product of prime factors**.

For example $120 = 2 \times 60$
$= 2 \times 2 \times 30$
$= 2 \times 2 \times 2 \times 15$
$= 2 \times 2 \times 2 \times 3 \times 5$
$= 2^3 \times 3 \times 5$

The next example shows how a factor tree can be used.

EXAMPLE

Write 84 as the product of prime factors.

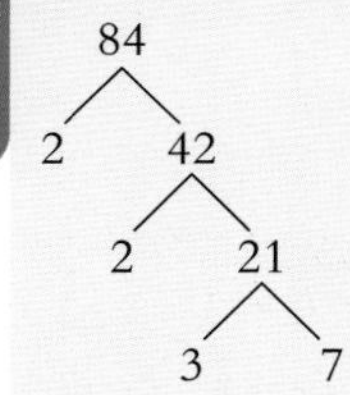

$84 = 2 \times 2 \times 3 \times 7 = 2^2 \times 3 \times 7$

Expressing numbers as the product of prime factors can help you to find highest common factors (HCF) and lowest common multiples (LCM).

EXAMPLE

Find the HCF and the LCM of 270 and 420.

First write 270 and 420 as the product of prime factors.
$270 = 2 \times 3 \times 3 \times 3 \times 5$ and $420 = 2 \times 2 \times 3 \times 5 \times 7$
Write the prime factors on a diagram.

Prime factors of 270: 3 3 | intersection: 2 3 5 | Prime factors of 420: 2 7

The HCF is the product of the numbers in the intersection $= 2 \times 3 \times 5 = 30$.
The LCM is the product of all the numbers in the diagram $= 3 \times 3 \times 2 \times 3 \times 5 \times 2 \times 7 = 3780$.

UNIT 1

EXERCISE 1.6

1 Which of the following numbers are prime numbers?
11, 17, 21, 35, 47, 69, 72, 81.

2 Which of the following numbers are prime numbers?
3, 13, 23, 33, 43, 53, 63, 73.

3 Calculate the value of the following.

a $2 \times 5 \times 5$ **b** $2 \times 3 \times 11$ **c** $3 \times 3 \times 3 \times 7$ **d** $2 \times 5 \times 7$
e $2 \times 3 \times 3 \times 13$ **f** $2 \times 2 \times 3 \times 11$ **g** $3^3 \times 5^2 \times 7$ **h** $2^5 \times 3^2 \times 5 \times 11$

4 Write each of the numbers as the product of prime factors.

a 10 **b** 150 **c** 81 **d** 60
e 74 **f** 100 **g** 98 **h** 250
i 1110 **j** 275 **k** 2004 **l** 2210

5 3, 7, 10, 15, 19, 21, 35

a Which of these numbers are prime numbers?
b Which of these numbers are multiples of 3?
c Which of these numbers are multiples of 5?
d Which of these numbers are factors of 30?
e Which of these numbers are factors of 380?

6 List the prime numbers between 80 and 100.

7 Sanjit thinks that 713 is a prime number. Explain why he is wrong.

8 The number 12 can be written as the sum of two prime numbers.

$12 = 5 + 7$

Write each of the following numbers as the sum of two prime numbers.

a 10 **b** 14 **c** 25 **d** 49 **e** 30
f 20 **g** 38 **h** 82 **i** 36 **j** 48

9 Find the HCF and LCM of:

a 24 and 42 **b** 45 and 105 **c** 70 and 42 **d** 40 and 75

10 Find the HCF and LCM of:

a 30, 36 and 42 **b** 28, 35 and 56 **c** 60, 90 and 210

Square and cube numbers

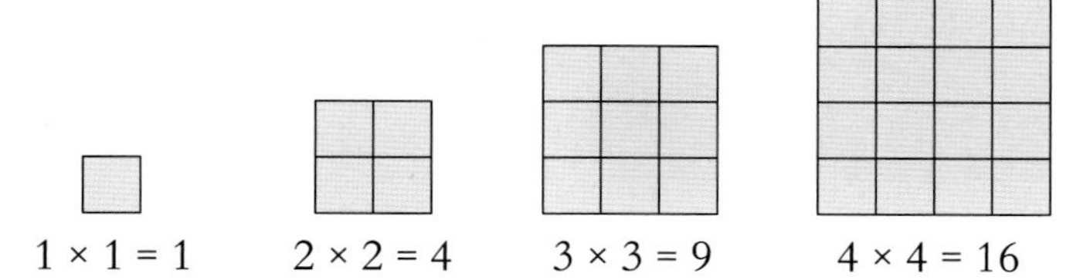

The numbers 1, 4, 9 and 16 are called **square numbers**.
The number 169 is also a square number because $13 \times 13 = 169$
13×13 can be written as 13^2

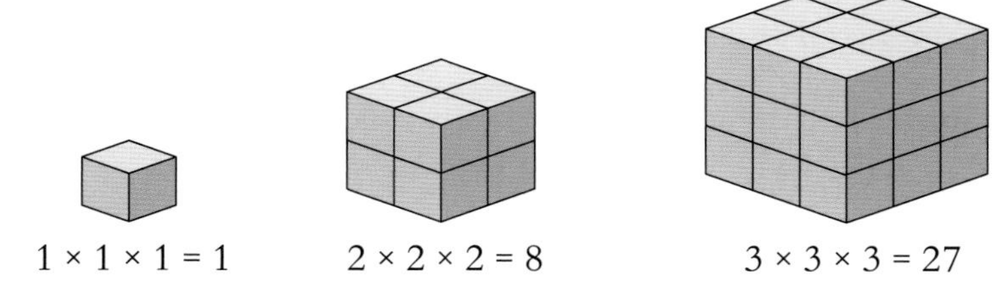

The numbers 1, 8 and 27 are called **cube numbers**.
The number 8000 is a cube number because $20 \times 20 \times 20 = 8000$.
$20 \times 20 \times 20$ can also be written as 20^3.

EXERCISE 1.7

1 Which is biggest 2^3 or 3^2?

2 1, 5, 9, 20, 27, 56, 48, 49, 52, 64, 275, 289, 343, 436, 512.
 a Write down the numbers that are square numbers.
 b Write down the numbers that are cube numbers.

3 The number 64 is a square number because $8 \times 8 = 64$.
It is also a cube number because $4 \times 4 \times 4 = 64$.
Can you find another number (bigger than 1) that is both a square and a cube number?

KEY WORDS
factor
highest common factor (HCF)
multiple
lowest common multiple (LCM)
prime
prime factor
product of prime factors
square numbers
cube numbers

Four rules for fractions

THIS SECTION WILL SHOW YOU HOW TO

- Use the language of fractions
- Apply the four rules for fractions

The language of fractions

In the fraction $\frac{3}{7}$, the top part is called the **numerator** and the bottom part is called the **denominator**.

The fraction $\frac{8}{3}$ is called an **improper fraction** because the numerator is bigger than the denominator.

The number $2\frac{2}{3}$ is called a **mixed number** because it is made up of a whole number and a fraction.

To simplify a fraction you divide the numerator and denominator by a common factor.

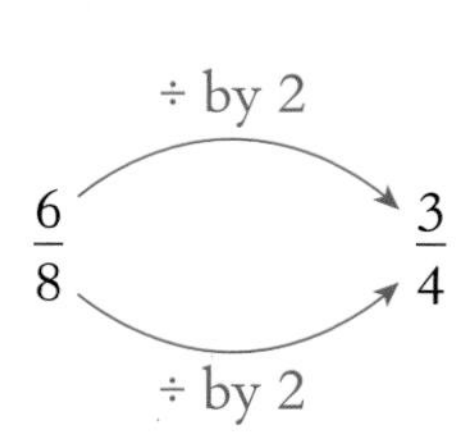

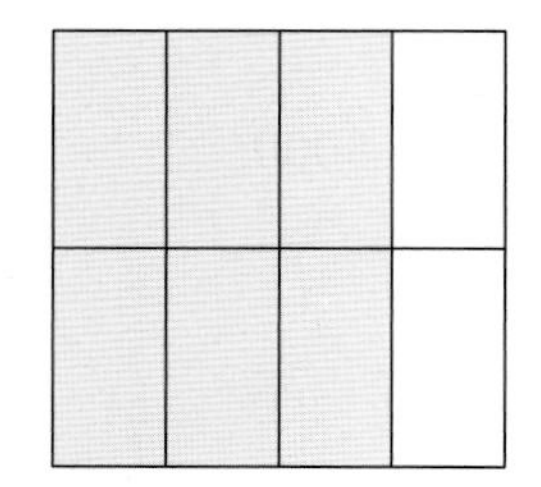

=

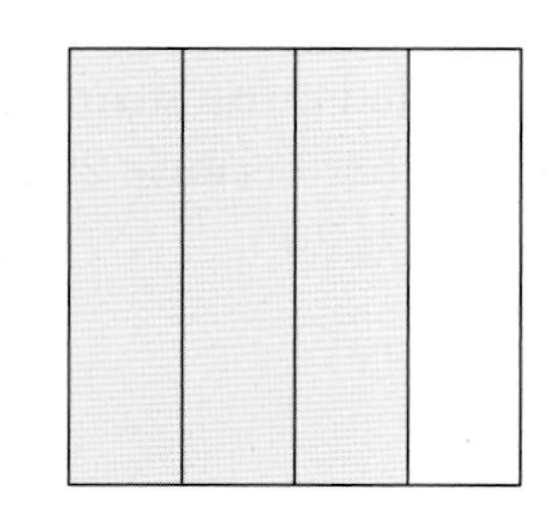

EXERCISE 1.8

Simplify:

1 $\frac{6}{12}$ **2** $\frac{35}{49}$ **3** $\frac{18}{24}$ **4** $\frac{6}{15}$ **5** $\frac{80}{128}$ **6** $\frac{150}{175}$

7 $\frac{36}{63}$ **8** $\frac{60}{75}$ **9** $\frac{42}{48}$ **10** $\frac{36}{132}$ **11** $\frac{51}{60}$ **12** $\frac{84}{144}$

Change to mixed numbers:

13 $\frac{8}{5}$ **14** $\frac{7}{3}$ **15** $\frac{15}{7}$ **16** $\frac{20}{17}$ **17** $\frac{13}{2}$ **18** $\frac{19}{4}$

19 $\frac{17}{12}$ **20** $\frac{47}{15}$ **21** $\frac{15}{8}$ **22** $\frac{22}{21}$ **23** $\frac{17}{5}$ **24** $\frac{30}{9}$

EXERCISE 1.11

1 Round each of these correct to 1 decimal place (1 d.p.).

a	5.347	b	8.75	c	5.99	d	30.021	e	0.06
f	8.88	g	4.2067	h	16.255	i	643.991	j	7.006

2 Round each of these correct to 2 decimal places (2 d.p.).

a	6.352	b	8.388	c	16.1555	d	4.2081	e	5.6999
f	0.0384	g	0.0666	h	2.994	i	3.056	j	7.1025

3 Round each of these correct to 3 decimal places (3 d.p.).

a	7.2222	b	6.1616	c	35.6855	d	82.0089	e	24.7894
f	3.00002	g	6.0091	h	6.66666	i	102.1028	j	5.0096

4 Round each of these correct to 1 significant figure (1 s.f.).

a	2.56	b	8.09	c	9.7	d	48.26	e	352
f	26700	g	51000	h	0.0022	i	0.087	j	0.308

5 Round each of these correct to 2 significant figures (2 s.f.).

a	1.7328	b	3.094	c	5680	d	61300	e	15.93
f	20.07	g	103.24	h	0.00633	i	0.0577	j	0.000399

6 Round each of these correct to 3 significant figures (3 s.f.).

a	27.348	b	6.5137	c	2587	d	148.8	e	16.745
f	0.3492	g	0.07175	h	0.008076	i	10.066	j	39992

7 In 2010 the population of the world was estimated to be 6 831 000 000.
Round this number to 1 s.f.

8 The total area of the Atlantic Ocean is about 106 400 000 square kilometres.
Round this number to

a 1 s.f. b 2 s.f. c 3 s.f.

9 Mount Everest is the highest mountain in the world.
The summit is 8848 metres above sea level. Round this number to

a 1 s.f. b 2 s.f. c 3 s.f.

10 An egg has mass of 0.06395 kg. Round this number to

a 1 s.f. b 2 s.f. c 3 s.f.
d 1 d.p. e 2 d.p. f 3 d.p.

11 The length of the Earth's equator is 40 008.629 km.
Round this number to

a 1 s.f. b 2 s.f. c 3 s.f.
d 4 s.f e 1 d.p. f 2 d.p.

Estimating

It is important that you know how to check if your answer to a calculation is a sensible answer. To do this you need to **estimate** the answer.
To estimate the answer to a calculation use the following steps.

1 Round each of the numbers to 1 significant figure.
2 Do the calculation using your rounded numbers.

EXAMPLE

Estimate the answers to these calculations.

a $5.94591 \div 2.0738$ **b** $\dfrac{(19.719)^2 - 2.9513 \times 22.628}{685 \div (5.16 + 1.7694)}$

a $5.94591 \div 2.0738$ *first round each of the numbers to 1 s.f.*

$\approx 6 \div 2$

$= 3$ *[the answer using a calculator is 2.86715....]*

b $\dfrac{(19.719)^2 - 2.9513 \times 22.628}{685 \div (5.16 + 1.7694)}$ *first round each of the numbers to 1 s.f.*

$\approx \dfrac{(20)^2 - 3 \times 20}{700 \div (5 + 2)}$ *remember to do the operations in the correct order (BIDMAS)*

$= \dfrac{400 - 60}{700 \div 7} = \dfrac{340}{100} = 3.4$ *[the answer using a calculator is 3.257899....]*

EXERCISE 1.12

1 Estimate the answers to these calculations.

a 6.4×9.8 **b** 2.16×7.79 **c** 289×12.3
d 49.3×52.1 **e** 218×372.9 **f** $17.94 \div 3.15$
g $22.8 \div 3.89$ **h** $188.3 \div 42.76$ **i** $5.76 + 4.73 \times 2.28$

2 Estimate the answers to these calculations.

a $\dfrac{84.3 + 17.8}{7.16 + 2.947}$ **b** $\dfrac{8.26 - 1.99}{1.207 + 1.806}$ **c** $\dfrac{5.16 + (1.94)^2}{7.26 - 3.78}$

3 Use a calculator to find the answers to questions **1** and **2**, give your answers correct to four significant figures. Comment on whether your estimate is either an over- or an under-estimate.

4 Estimate the cost of 2.16 kg of bananas at \$3.84 per kg.

5 A theatre sells tickets for a show. They sell 103 tickets at \$18.99 per ticket, 321 tickets at \$21.50 per ticket and 48 tickets at \$27.99 per ticket. Estimate the total amount of money obtained from the sale of the tickets.

KEY WORDS
round
decimal places (d.p.)
significant figures (s.f.)
estimate

Simplifying algebraic expressions

THIS SECTION WILL SHOW YOU HOW TO

- Collect like terms
- Expand single brackets

You can simplify **expressions** by collecting **like terms**.

EXAMPLE

Simplify these expressions.

a $9x + 5y - 3x - 8y$ **b** $4x^2 - 3x - 7 - 2x^2 - 5x + 1$ **c** $7x^2y - 3xy - 7xy - 2x^2y$

a $9x + 5y - 3x - 8y$ *find the like terms*

$= 9x + 5y - 3x - 8y$ *put the like terms next to each other*

$= 9x - 3x + 5y - 8y$ *9 – 3 = 6 and +5 – 8 = –3*

$= 6x - 3y$

b $4x^2 - 3x - 7 - 2x^2 - 5x + 1$ *find the like terms*

$= 4x^2 - 3x - 7 - 2x^2 - 5x + 1$ *put the like terms next to each other*

$= 4x^2 - 2x^2 - 3x - 5x - 7 + 1$ *4 – 2 = 2 –3 – 5 = –8 –7 + 1 = –6*

$= 2x^2 - 8x - 6$

c $7x^2y - 3xy - 7xy - 2x^2y$ *find the like terms*

$= 7x^2y - 3xy - 7xy - 2x^2y$ *put the like terms next to each other*

$= 7x^2y - 2x^2y - 3xy - 7xy$ *7 – 2 = 5 and –3 – 7 = –10*

$= 5x^2y - 10xy$

EXERCISE 1.13

Simplify these expressions.

1 $5x - 8x + 2x$

2 $3y - 4y - 2y$

3 $8xy - 2xy$

4 $6xy - 9yx$

5 $6x - 4y + x - 5y$

6 $3p + 9q + 2p - 4q$

7 $3xy + 5x + 2xy - 8x$

8 $5x^2 - 9x - 6x + 2x^2$

9 $9 - 4x - 2 + 8x$

10 $4xy - 3yx + 2xy - 10yx$

11 $x^2 - 6 - 4x^2 + 15$

12 $7y^2 - 2y^2 + 3y - 5y - 11y^2$

13 $6xy - x + 2xy$

14 $a + 8ab - 12ab$

15 $7ab - 4bc + 3ab - 2bc$

16 $x^2 + x^4 + x^2 - 5 + 3x^2$

17 $6x^2 + 5x^3 - 7x^2 + x$

18 $7fg + 2gh - 5fg - 6gh - 4fg$

19 $6a^2b - 4ab^2 - 12ab^2 - 2a^2b$

20 $8cd^2 - 5c^2d - 4c^2d - 3cd^2$

21 $x + \frac{3}{x} + 5x + \frac{7}{x}$

22 $\frac{8}{x} - \frac{2}{y} + \frac{3}{x} - \frac{5}{y} - \frac{2}{x}$

23 $3xy^2 + 7xy - 6x^2y - 5y^2x - 4x^2y + x^2 + 5xy - y^2$

24 $7xy - 5x^2 + 3y^2 - 2x^2 - 9xy + 7x^2 - 15y^2 + 3 + 2xy + 12y^2$

Expressions involving brackets

$2(x + 3)$ means the same as $2 \times (x + 3)$
The diagram below helps to show that $2(x + 3) = 2x + 6$

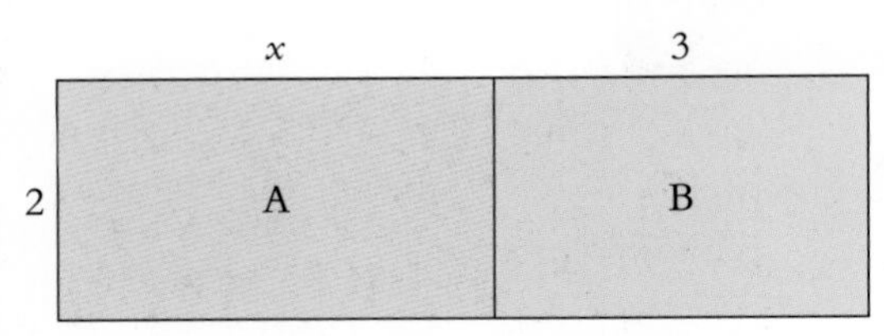

The area of rectangle A is $2x$
The area of rectangle B is 6
The area of the whole rectangle is $2x + 6$

When you multiply out the brackets you must multiply each term inside the bracket by the term outside the bracket.

$5(2x + 3y)$ means $5 \times 2x + 5 \times 3y = 10x + 15y$

EXAMPLE

Expand **a** $7(2a - 5b)$ **b** $9x(3x - 4y + 8)$

a $7(2a - 5b) = 7 \times 2a - 7 \times 5b$
$= 14a - 35b$

b $9x(3x - 4y + 8) = 9x \times 3x - 9x \times 4y + 9x \times 8$
$= 27x^2 - 36xy + 72x$

You need to be very careful when multiplying by a negative number
To expand $-2(x - 6)$ you multiply both terms in the bracket by -2.

$$-2(x - 6) = (-2 \times x) + (-2 \times -6) = -2x + 12$$

EXAMPLE

Expand **a** $-5(3x + 6)$ **b** $-8(7 - 5y)$

a $-5(3x + 6) = (-5 \times 3x) + (-5 \times 6)$
$= -15x + -30$
$= -15x - 30$

b $-8(7 - 5y) = (-8 \times 7) + (-8 \times -5y)$
$= -56 + 40y$

EXAMPLE

Expand and simplify
a $6(2x - 3) + 5(3x - 1)$ **b** $7(2x - 3) - 5(3x - 2)$

a $6(2x - 3) = 12x - 18$ and $+5(3x - 1) = +15x - 5$ *expand each set of brackets*
$6(2x - 3) + 5(3x - 1) = 12x - 18 + 15x - 5$ *collect like terms*
$= 27x - 23$

b $7(2x - 3) = 14x - 21$ and $-5(3x - 2) = -15x + 10$ *expand each set of brackets*
$7(2x - 3) - 5(3x - 2) = 14x - 21 - 15x + 10$ *collect like terms*
$= -x - 11$

EXERCISE 1.14

1 Freda says that $7(2x + 3) = 14x + 3$.
Explain why she is wrong.

2 Expand these expressions.

a $4(x + 3)$ b $6(y + 2)$ c $4(x + 5)$
d $2(3 - a)$ e $7(y - 4)$ f $8(x - 9)$
g $5(2x + 3)$ h $6(3y + 4)$ i $7(5a + 6)$
j $3(2x + 2y - 1)$ k $5(4a + 5b - 3)$ l $8(3p - 4q + 7r)$
m $\frac{1}{2}(8x - 2)$ n $\frac{1}{4}(12x + 8)$ o $\frac{1}{3}(15y - 6)$

3 Expand these expressions.

a $-3(x + 4)$ b $-2(x - 6)$ c $-5(6 - 4x)$
d $-8(2x + 5)$ e $-7(3x - 8)$ f $-8(x - 9)$
g $-(5x + 4)$ h $-(2x - 7)$ i $-(3x + 8)$
j $-6(3x - 3y + 5)$ k $-4(3p + 4q - 5)$ l $-3(9x^2 + 3x - 2)$

4 Expand these expressions.

a $x(x + 3)$ b $y(y - 5)$ c $a(7 - a)$
d $2x(5 - 3x)$ e $5y(y + 8)$ f $3x(2x + 4y)$

5 Expand and simplify.

a $2(x + 3) + 3(x + 8)$ b $5(x - 2) + 6(x + 7)$ c $6(y + 1) + 3(y - 2)$
d $4(x - 8) + 7(x - 6)$ e $6(5x - 2y) + 2(3x + 4y)$ f $7(2x^2 - 3x) + 2(5x - 6x^2)$
g $x(x + 4) + x(x + 2)$ h $x(x - 3) + x(x + 7)$ i $2x(3x + 4) + 5x(6 - 3x)$
j $2x(3y - 4) + 3y(2x - 5)$ k $2x(3x - 5y) + 3y(2x + 7y) + 5x(3y - 2x)$

6 Joe says that $5(2x - 3) - 4(x - 6) = 10x - 15 - 4x - 24 = 6x - 39$.
Explain why he is wrong.

7 Expand and simplify.

a $3(x - 4) - 2(x - 6)$ b $5(y + 2) - 3(y + 7)$ c $7(a + 2) - 8(3 - 2a)$
d $6(7x - 2) - 2(3x + 4)$ e $3(x + 4) - (x - 5)$ f $7(2 - y) - (y + 3)$
g $4(2h - 3g) - 3(3h - 2g)$ h $3x(2x + 1) - x(5x + 8)$ i $5p(p - 6) - 2p(7 - 3p)$
j $2a(3b - 5a + 7) - 5a(7a + 3b - 5) - 3a(2a - 5b - 8)$

8 Expand and simplify.

a $10 + 2(x + 3)$ b $15 - 3(x - 4)$
c $20 - 2(3x - 4y + 6)$ d $9 - 5(y + 2)$
e $8 - 7(x - 2)$ f $17x - 3(x + 5)$
g $10y - (2y - x)$ h $5x - 2x(x - 3)$

KEY WORDS

expression
like terms
expand

Solving linear equations

THIS SECTION WILL SHOW YOU HOW TO

- Set up simple equations
- Solve linear equations

EXAMPLE

Solve **a** $6x - 5 = 11$ **b** $9 - 5x = 2$

a $6x - 5 = 11$

$6x - 5 + 5 = 11 + 5$ *add 5 to both sides*

$6x = 16$

$\frac{6x}{6} = \frac{16}{6}$ *divide both sides by 6*

$x = 2\frac{2}{3}$ *change to a mixed number*

b $9 - 5x = 2$

$9 - 5x + 5x = 2 + 5x$ *add 5x to both sides*

$9 = 2 + 5x$

$9 - 2 = 2 + 5x - 2$ *subtract 2 from both sides*

$7 = 5x$ *divide both sides by 5*

$\frac{7}{5} = \frac{5x}{5}$ *divide both sides by 5*

$x = 1\frac{2}{5}$ *change to a mixed number*

EXAMPLE

Solve **a** $3(x - 4) = 5(2x + 3)$ **b** $5 - 6y = 2 - 4y$

a $3(x - 4) = 5(2x + 3)$ *expand the brackets*

$3x - 12 = 10x + 15$

$3x - 12 - 3x = 10x + 15 - 3x$ *take 3x from both sides*

$-12 = 7x + 15$

$-12 - 15 = 7x + 15 - 15$ *take 15 from both sides*

$-27 = 7x$ *divide both sides by 7*

$\frac{-27}{7} = \frac{7x}{7}$

$x = -3\frac{6}{7}$ *change to a mixed number*

b $5 - 6y = 2 - 4y$

$5 - 6y + 6y = 2 - 4y + 6y$ *add 6y to both sides*

$5 = 2 + 2y$

$5 - 2 = 2 + 2y - 2$ *take 2 from both sides*

$3 = 2y$

$\frac{3}{2} = \frac{2y}{2}$ *divide both sides by 2*

$y = 1\frac{1}{2}$ *change to a mixed number*

EXERCISE 1.15

1 Solve these equations.

a $2x + 3 = 15$ b $7x - 1 = 62$ c $5x + 2 = 37$ d $6x - 2 = 5$
e $8x + 3 = 15$ f $4x - 1 = 52$ g $18 - 2x = 12$ h $6 - 3x = 8$
i $10 - 4x = 7$ j $30 - 8x = 15$ k $10 - 2x = 15$ l $6 - 5x = 20$
m $3x + 4 = 27$ n $4(y - 5) = 14$ o $2(x + 3) = 7$ p $3(2x + 5) = 26$
q $4(3x - 1) = 80$ r $5(3x + 8) = 4$ s $2(13 - x) = 22$ t $5(10 - 3x) = 8$

2 Solve these equations.

a $12x + 5 = 8x + 13$ b $7x + 2 = 3x + 22$ c $6x - 3 = 5x + 4$ d $2x - 3 = 6x - 17$
e $3x + 1 = 2 - 5x$ f $7x - 9 = 10 + 2x$ g $3x + 7 = 8 - 2x$ h $7x - 3 = 9x + 10$
i $3 - 2x = 8 - x$ j $10 - 3x = 5 + 2x$ k $6 - 7x = 3 - 5x$ l $5 - 2x = 10 - 7x$

3 Solve these equations.

a $2(x - 3) + 4(2x - 7) = 6$ b $3x + 11 = 5(x + 1)$ c $7 - x = 3(5 - x)$
d $5(2x + 2) = 6(x + 5)$ e $3(2x + 5) - 2(x - 5) = 51$ f $3(3x - 2) - 7(x - 2) = 0$
g $4(1 - 2x) - 2(3 - x) = 0$ h $6(2x - 5) - 4(x - 2) = 14$ i $37 - 2(3x + 1) = 30 - 3x$

4 The sum of three consecutive numbers is 171.
Find the numbers.

TOP TIP

Let the numbers be x, $x + 1$ and $x + 2$.

5 AB is a straight line. Find the value of x.

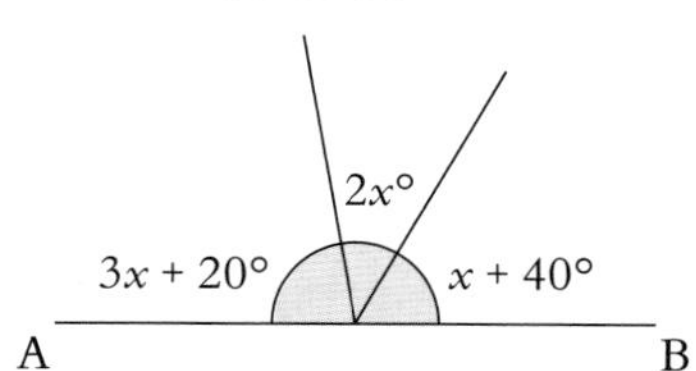

TOP TIP

The angles on a straight line add up to 180°.

6 ABCD is a rectangle.
Find the value of x.

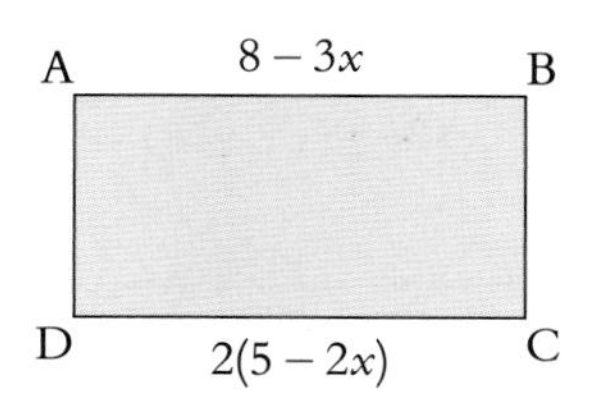

TOP TIP

Opposite sides of a rectangle are equal.

7 The perimeter of the triangle is 10 cm.
Find the value of x.

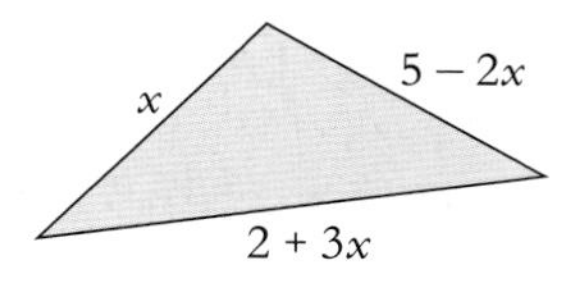

8 Expression A is 18 more than expression B.
Find the value of x.

A
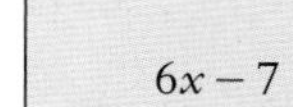

B
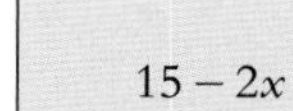

Solving linear equations involving fractions

EXAMPLE

Solve **a** $\frac{x}{2} + 4 = 12$ **b** $\frac{x-3}{4} = 5$ **c** $\frac{3}{x} - 2 = 8$

a $\frac{x}{2} + 4 = 12$ *subtract 4 from both sides*

$\frac{x}{2} = 8$ *multiply both sides by 2*

$2 \times \frac{x}{2} = 2 \times 8$ ➡ **NOTE:** $2 \times \frac{x}{2} = x$

$x = 16$

b $\frac{x-3}{4} = 5$ *multiply both sides by 4*

$4 \times \frac{(x-3)}{4} = 4 \times 5$ ➡ **NOTE:** $4 \times \frac{(x-3)}{4} = x - 3$

$x - 3 = 20$ *add 3 to both sides*

$x = 23$

c $\frac{3}{x} - 2 = 8$ *add 2 to both sides*

$\frac{3}{x} = 10$ *multiply both sides by* x

$x \times \frac{3}{x} = 10 \times x$ ➡ **NOTE:** $x \times \frac{3}{x} = 3$

$3 = 10x$ *divide both sides by 10*

$x = \frac{3}{10}$

The next example shows how to solve an equation when there is a single fraction on both sides of the equation.
You multiply both fractions by the lowest common multiple (LCM) of the two denominators.

EXAMPLE

Solve $\frac{x-4}{5} = \frac{x-2}{6}$

$\frac{x-4}{5} = \frac{x-2}{6}$ *multiply both sides by 30*

$30 \times \frac{(x-4)}{5} = 30 \times \frac{(x-2)}{6}$ *30 ÷ 5 = 6 and 30 ÷ 6 = 5*

$6(x - 4) = 5(x - 2)$ *expand the brackets*

$6x - 24 = 5x - 10$ *add 24 to both sides*

$6x = 5x + 14$ *subtract 5x from both sides*

$x = 14$

TOP TIP

You can simplify an equation with a single fraction on each side by 'cross-multiplying'.

$\frac{x-4}{5} = \frac{x-2}{6}$

$6(x - 4) = 5(x - 2)$

EXERCISE 1.16

1 Solve these equations.

a $\frac{x}{4} = 5$ b $\frac{x}{3} = 20$ c $\frac{2x}{5} = 30$ d $\frac{3x}{4} = 5$

e $\frac{x}{3} + 2 = 17$ f $\frac{x}{4} - 1 = -8$ g $\frac{2x}{5} - 3 = 7$ h $\frac{3x}{7} + 2 = 6$

i $\frac{x}{2} + 10 = 3$ j $16 - \frac{x}{2} = 5$ k $18 - \frac{x}{3} = 10$ l $19 - \frac{4x}{5} = 3$

2 Solve these equations.

a $\frac{2x+1}{3} = 5$ b $\frac{3x+2}{5} = 7$ c $\frac{4x-1}{3} = 4$ d $\frac{6-2x}{3} = 1$

e $\frac{10-4x}{5} = 4$ f $\frac{2(x+4)}{3} = 6$ g $\frac{3(2x-1)}{4} = 6$ h $\frac{5(2-3x)}{4} = -3$

3 Solve these equations.

a $\frac{24}{x} = 8$ b $24 = \frac{12}{x}$ c $4 = \frac{7}{x}$ d $5 = \frac{3}{2x}$

e $\frac{5}{x} - 2 = 8$ f $\frac{2}{x} + 4 = 5$ g $\frac{30}{x} - 2 = 4$ h $\frac{8}{x} - 7 = -11$

i $5 - \frac{3}{x} = 3$ j $24 - \frac{12}{x} = 30$ k $18 - \frac{15}{2x} = 15$ l $9 - \frac{36}{x} = 18$

m $\frac{8}{x+1} = 2$ n $\frac{6}{x-1} = 3$ o $\frac{10}{3x+2} = 2$ p $\frac{8}{2-5x} = 4$

4 Solve these equations.

a $\frac{3-x}{3} = \frac{2+x}{2}$ b $\frac{x+1}{4} = \frac{8-x}{5}$ c $\frac{2x+2}{5} = \frac{8-x}{2}$ d $\frac{3x-1}{4} = \frac{x}{2}$

e $\frac{5x+2}{3} = \frac{3x}{4}$ f $\frac{x+1}{3} = \frac{x-1}{4}$ g $\frac{x+5}{4} = \frac{3x+1}{5}$ h $\frac{6x-3}{5} = \frac{4x+5}{2}$

5 Explain why you cannot solve the equation $\frac{6x+2}{3} = \frac{4x-5}{2}$

6 The perimeter of the equilateral triangle is 15 cm. Find the value of x.

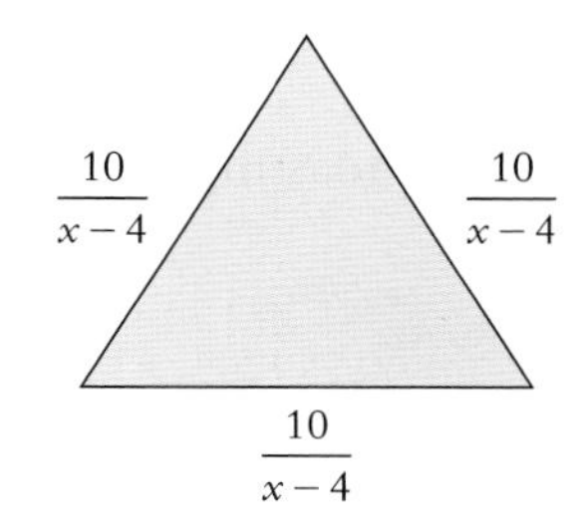

KEY WORDS
equation
solve

UNIT 1

Constructing formulae

THIS SECTION WILL SHOW YOU HOW TO

- Construct formulae

A **formula** is a rule that shows how **variables** are connected.

EXAMPLE

A chocolate has a mass of 15 grams and a sweet has a mass of 22 grams.
I have x chocolates and $x + 7$ sweets.
Write down a formula for the total mass M, in grams, of the chocolates and sweets.

mass of x chocolates $= 15x$
mass of $x + 7$ sweets $= 22(x + 7)$
$M = 15x + 22(x + 7)$ *expand brackets*
$M = 15x + 22x + 154$ *collect like terms*
$M = 37x + 154$

➡ **NOTE:** the two variables in this question are x and M.

EXERCISE 1.17

1 A theatre sells 135 tickets at \$10 each and n tickets at \$13 each
Write down a formula for the total cost P, in \$, of the tickets.

2 I have \$100 to spend. I buy n books at \$9 each.
Write down a formula for the amount of money Q, in \$, that I have left after buying the books.

3 Light bulbs cost \$$b$ each and candles cost \$$c$ each.
Write down a formula for the total cost C, in \$, of 4 light bulbs and 7 candles.

4 Rulers cost r cents each, pencils cost p cents each and files cost \$$f$ each.
Write down a formula for the total cost T, in \$, of 5 rulers, 6 pencils and 4 files.

5 The cost of hiring a tent is made up of two parts. There is a fixed charge of \$30 and then an extra charge of \$4 for each day that the tent is rented.
Write down a formula for the total cost T, in \$, when a tent is rented for n days.

6 The instructions for calculating the roasting time for a chicken are given below.

Allow 45 minutes per kg plus 20 minutes.

Write down a formula for the total cooking time T, in minutes, to cook a chicken that has a mass of m kg.

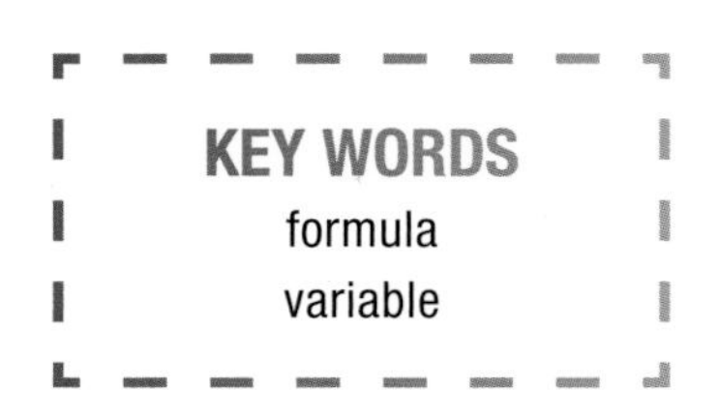

EXERCISE 1.20

1

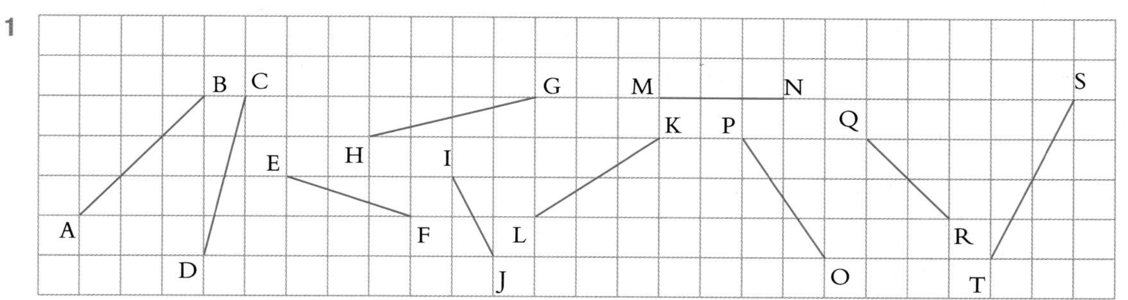

Find the gradient of the lines.

a AB b CD c EF d GH e IJ
f KL g MN h OP i QR j ST

2 Use a graph to help you work out the gradient of the line joining these pairs of points.

a $(5, 3), (6, 4)$ b $(-2, 4), (6, -5)$ c $(-1, 1), (3, 7)$
d $(2, -4), (5, -3)$ e $(1, 9), (4, -3)$ f $(5, 6), (-2, -3)$
g $(6, 4), (8, 4)$ h $(6, -3), (5, 8)$ i $(2, -4), (5, 8)$
j $(14, 0), (10, 4)$ k $(-7, -8), (-2, -6)$ l $(0, -6), (4, -5)$

3 Calculate the gradient of the line joining these pairs of points using the formula: gradient $= \dfrac{y_2 - y_1}{x_2 - x_1}$.

a $(2, 3), (5, 6)$ b $(-2, 4), (1, 10)$ c $(4, 2), (6, 3)$
d $(2, -3), (6, 5)$ e $(2, 5), (6, 3)$ f $(1, 4), (1, -2)$
g $(-3, -1), (-6, -4)$ h $(6, -3), (5, -3)$

4 The line joining $(2, 1)$ to $(7, a)$ has a gradient of 1. Find the value of a.

5 The line joining $(4, 2)$ to $(6, b)$ has a gradient of 3. Find the value of b.

6 The line joining $(0, 3)$ to $(6, c)$ has a gradient of $\frac{2}{3}$. Find the value of c.

7 The line joining $(3, 6)$ to $(7, d)$ has a gradient of $\frac{1}{2}$. Find the value of d.

8 The line joining $(-4, -1)$ to $(-1, e)$ has a gradient of 2. Find the value of e.

9 The line joining $(-2, 1)$ to $(6, 3)$ is parallel to the line joining $(2, -4)$ to $(6, f)$.
Find the value of f.

10 The line joining $(8, 6)$ to $(2g, -g)$ has a gradient of 2.
Find the value of g.

11 A steep road has gradient $\frac{4}{15}$.
The road rises by 3 meters. Calculate the horizontal distance x.

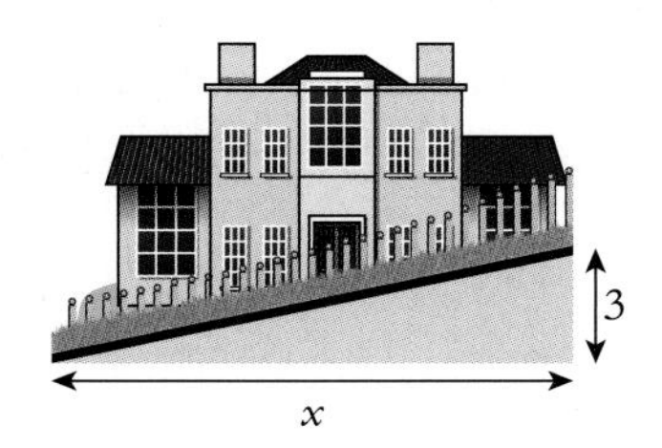

The graphs of $x = c$, $y = c$, and $y = x$

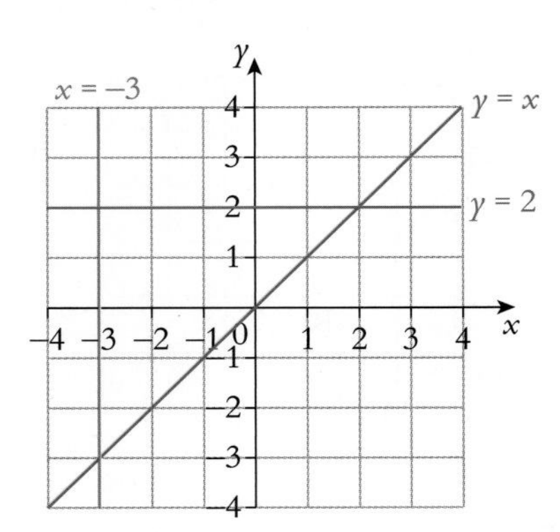

The x and y coordinates are the same at every point on the line $y = x$
eg (1, 1), (−3, −3), (2, 2) …

The y coordinate is 2 at every point on the line $y = 2$
eg (4, 2), (0, 2), (−3, 2) …

The x coordinate is -3 at every point on the line $x = -3$
eg (−3, 2), (−3, −1), (-3, 4) …

➡ **NOTE:** these are important graphs that you should remember

Straight line graphs of the form $y = ax + b$

EXAMPLE

Plot the graph of $y = \frac{1}{2}x + 1$

Choose 3 values of x and work out the corresponding y values.

when $x = 0$, $y = \left(\frac{1}{2} \times 0\right) + 1 = 1$

when $x = 2$, $y = \left(\frac{1}{2} \times 2\right) + 1 = 2$

when $x = 4$, $y = \left(\frac{1}{2} \times 4\right) + 1 = 3$

x	0	2	4
y	1	2	3

➡ **NOTE:** to find each y value you halve the x value and then add 1

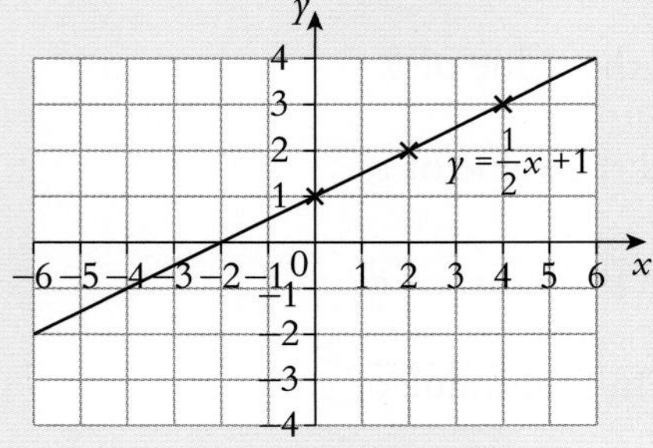

EXAMPLE

The line $y = ax - 5$ passes through the point (4, 11)
Find the value of a.

$y = ax - 5$ *substitute $x = 4$ and $y = 11$ into the equation*
$11 = 4a - 5$ *add 5 to both sides*
$4a = 16$ *divide both sides by 4*
$a = 4$

EXERCISE 1.21

1 Write down the equations of each of these lines.

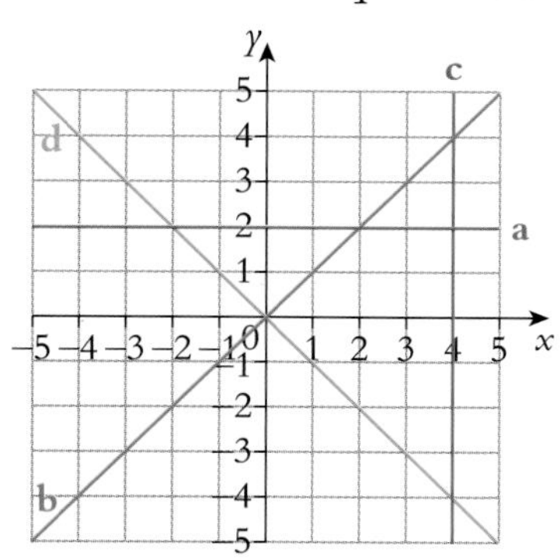

2 For each of the following copy and complete the table of values and then draw a graph of the function. Use coordinate axes with values from −6 to 6.

a $y = x + 2$

x	0	1	2
y			

b $y = x - 5$

x	0	1	2
y			

c $y = 2x - 2$

x	0	1	2
y			

d $y = 3x - 4$

x	0	1	2
y			

e $y = \frac{1}{2}x - 1$

x	−2	0	2
y			

f $y = \frac{1}{3}x - 4$

x	−3	0	3
y			

g $y = 3 - 2x$

x	0	1	2
y			

h $y = 5 - 3x$

x	0	1	2
y			

3 Plot these graphs on the same set of axes.

$y = 3x$ $\quad y = 3x + 2$ $\quad y = 3x - 5$

What do you notice about the graphs?

4 Plot these graphs on the same set of axes.

$y = 3x - 2$ $\quad y = 2x + 1$

Write down the coordinates of the point where the two lines inersect.

5 Plot these graphs on the same set of axes.

$y = 2x - 3$ $\quad y = \frac{1}{2}x + 3$

Write down the coordinates of the point where the two lines inersect.

6 The graph $y = ax + 4$ passes through the point (2, 8). Find the value of a.

7 The graph $y = bx - 3$ passes through the point (−2, 6). Find the value of b.

8 The graph $y = 3x + c$ passes through the point (3, −5). Find the value of c.

UNIT 1

Graphs of the form $ax + by = c$

An easy method to draw or sketch a graph of this form is to find the axis crossing points.
Putting $x = 0$ into the equation will tell you where the graph crosses the y-axis.
Putting $y = 0$ into the equation will tell you where the graph crosses the x-axis.

EXAMPLE

Sketch the graph of $3x + 4y = 12$.

when $x = 0$, $4y = 12$

$y = 3$ *the graph crosses the y-axis at (0, 3)*

when $y = 0$, $3x = 12$

$x = 4$ *the graph crosses the x-axis at (4, 0)*

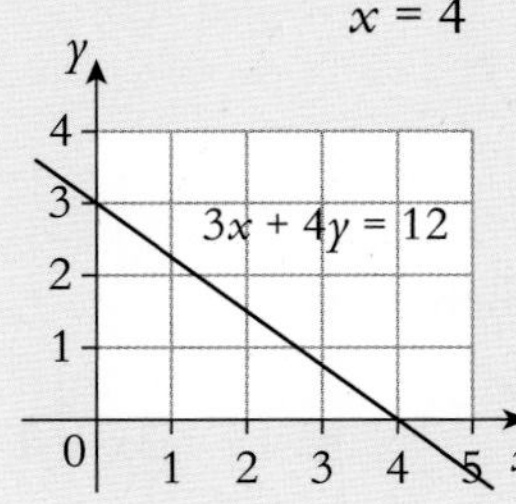

EXERCISE 1.22

Copy and complete the table for each of the following functions and sketch the graph.

1 $2x + y = 4$

x	0	
y		0

2 $3x + 2y = 6$

x	0	
y		0

3 $2x + 5y = 10$

x	0	
y		0

4 $y + 3x = 9$

x	0	
y		0

5 $2x + 3y = -12$

x	0	
y		0

6 $6x + 5y = -30$

x	0	
y		0

7 $x - 2y = 6$

x	0	
y		0

8 $4x - 5y = 20$

x	0	
y		0

9 $2x - 3y = 12$

x	0	
y		0

10 $y - 3x = 15$

x	0	
y		0

11 $5x - 3y = -30$

x	0	
y		0

12 $2x - \frac{1}{3}y = -4$

x	0	
y		0

KEY WORDS

gradient

2 For each shape write down the number of lines of symmetry and the order of rotational symmetry.

a b c

d e f

g h i

j k l

3 For each pattern write down the number of lines of symmetry and the order of rotational symmetry.

a 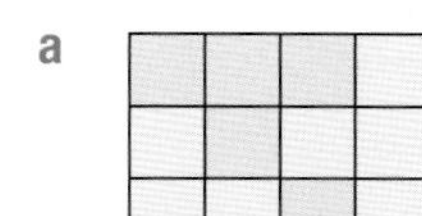 b 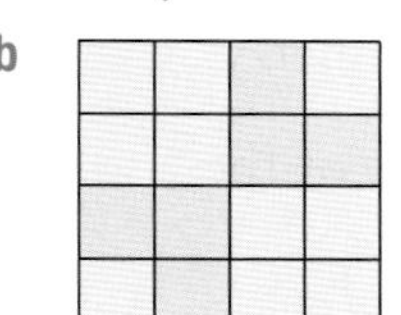 c 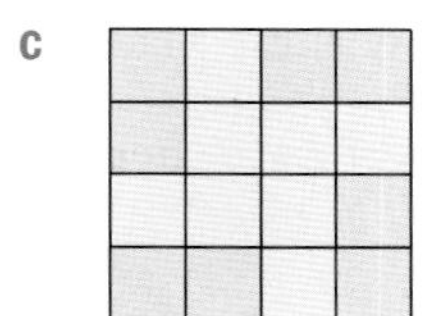

4 Make three copies of this grid. Colour each grid so that it has

a One line of symmetry and rotational symmetry of order 1
b Two lines of symmetry and rotational symmetry of order 2
c No lines of symmetry and rotational symmetry of order 2

5 For each of the following curves write down the number of lines of symmetry and the order of rotational symmetry.

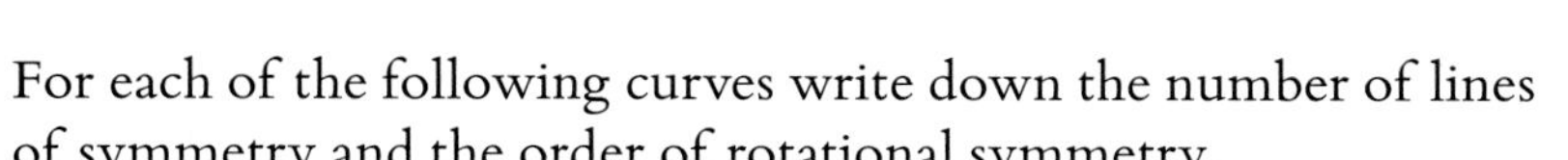

a 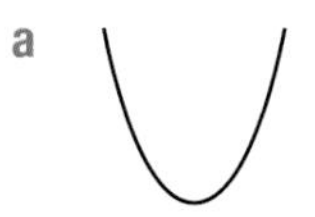 b 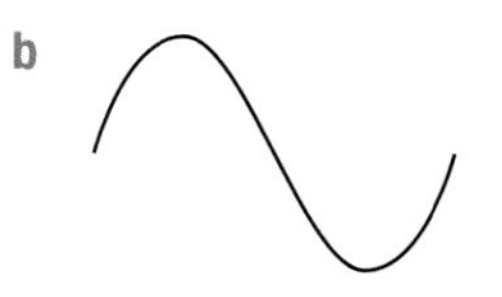 c 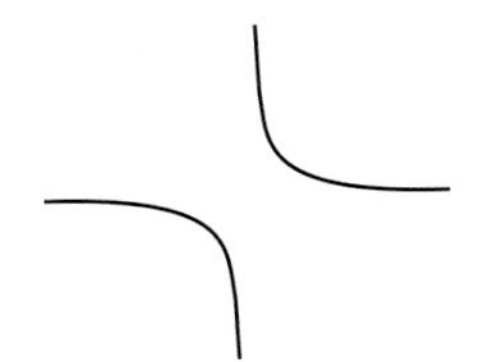 d 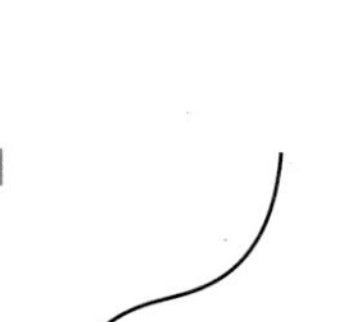 

6 For an unknown quadrilateral you are told how many lines of symmetry it has, its order of rotational symmetry, whether its diagonals cross at right angles and whether the diagonals bisect one another. Given this information can you always identify the quadrilateral.

TOP TIP

Start by constructing a table of properties of quadrilaterals.

Planes of symmetry

This cuboid has 3 **planes of symmetry**.

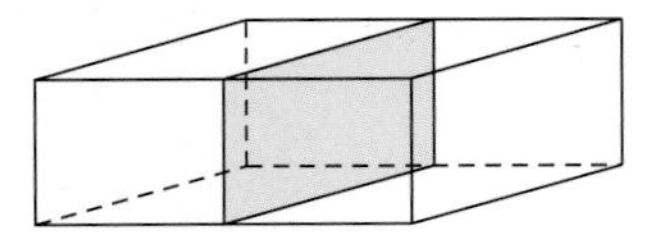
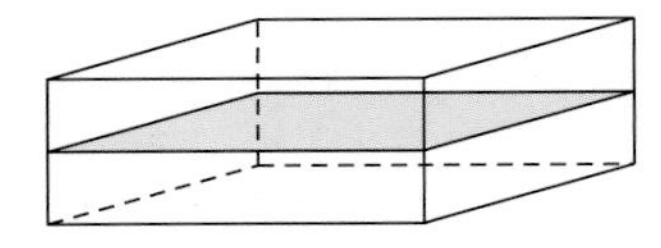
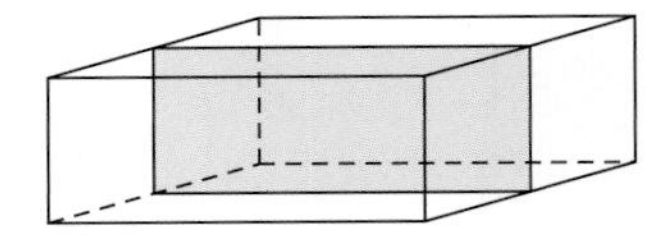

A plane of symmetry cuts a solid into two equal parts.
Each part is a mirror image of the other.

EXAMPLE

How many planes of symmetry does a square based pyramid have?

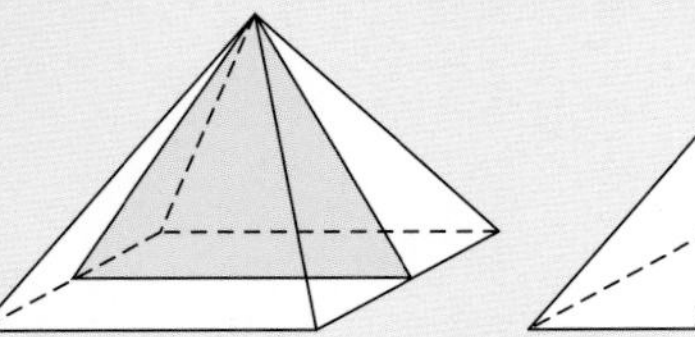
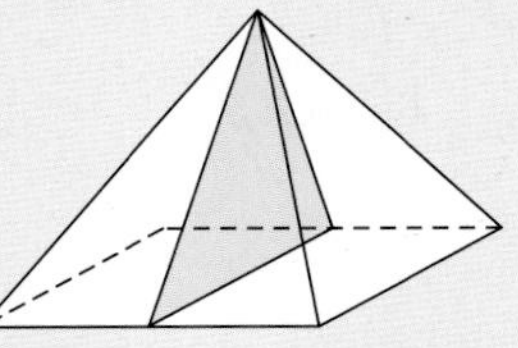
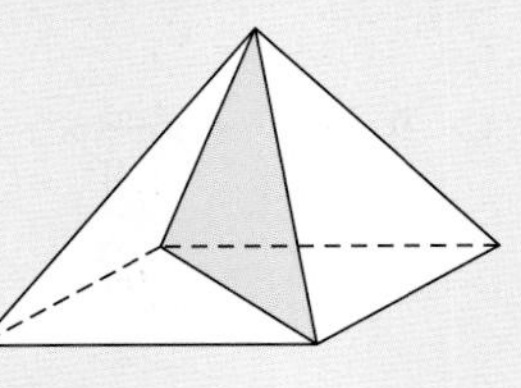
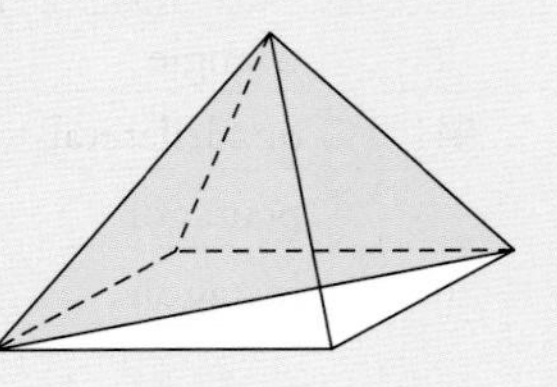

A square based pyramid has 4 planes of symmetry.

EXERCISE 1.25

1 The end face of this prism is an equilateral triangle.
A plane of symmetry is shown.
The prism has three more planes of symmetry.
Draw three diagrams to show these planes of symmetry.

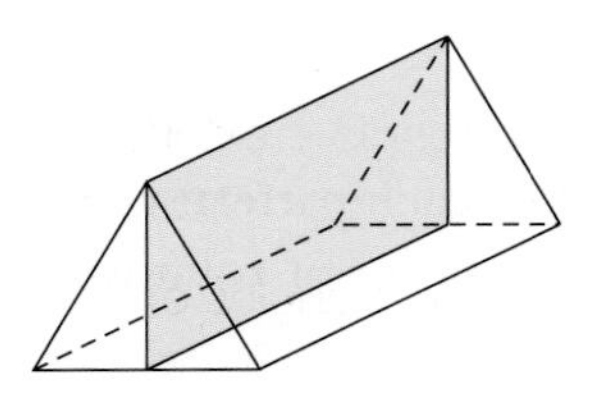

2 How many planes of symmetry are there for each of these solid objects.

a 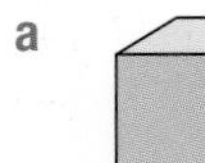
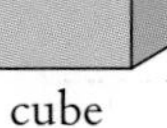
cube

b 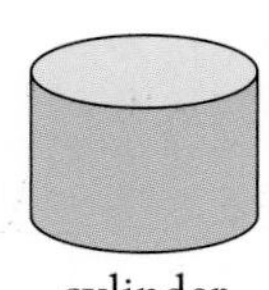
cylinder

c 
sphere

d 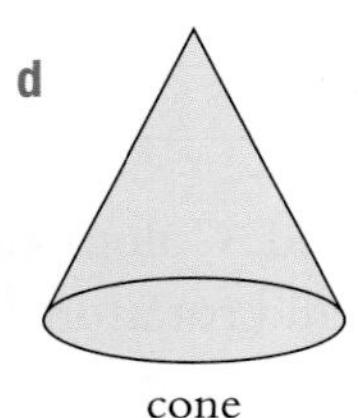
cone

e 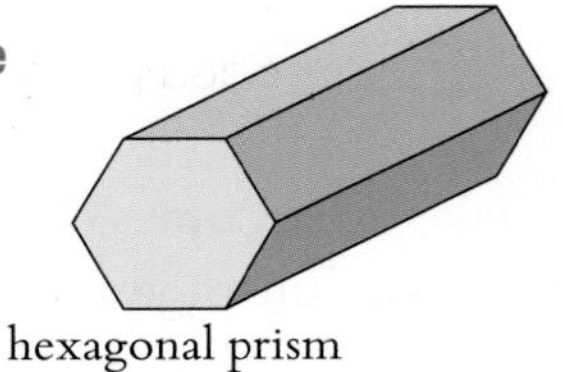
hexagonal prism

3 These solids are made from small cubes.
How many planes of symmetry does each solid have?

a 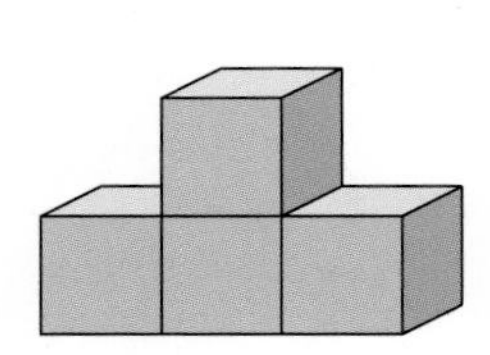

b 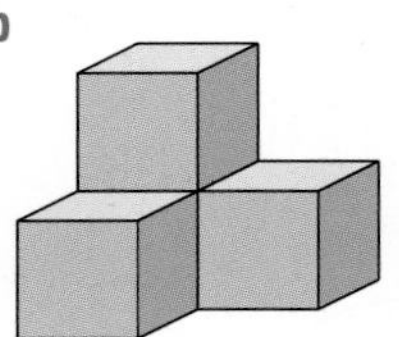

c 

d 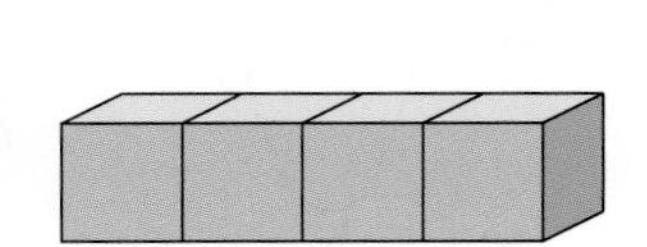

4 Draw a prism whose cross section is an isosceles trapezium.
How many planes of symmetry does the prism have?

KEY WORDS
line of symmetry
order of rotational symmetry
plane of symmetry

Polygons

THIS SECTION WILL SHOW YOU HOW TO
- Find the sum of the interior angles in a polygon
- Find the size of exterior and interior angles in a regular polygon

A **polygon** is a shape enclosed by straight lines.
An n-sided polygon can be divided into $(n - 2)$ triangles.

Number of sides	Name of polygon	Sum of interior angles
3	triangle	$(3 - 2) \times 180° = 180°$
4	quadrilateral	$(4 - 2) \times 180° = 360°$
5	pentagon	$(5 - 2) \times 180° = 540°$
6	hexagon	$(6 - 2) \times 180° = 720°$

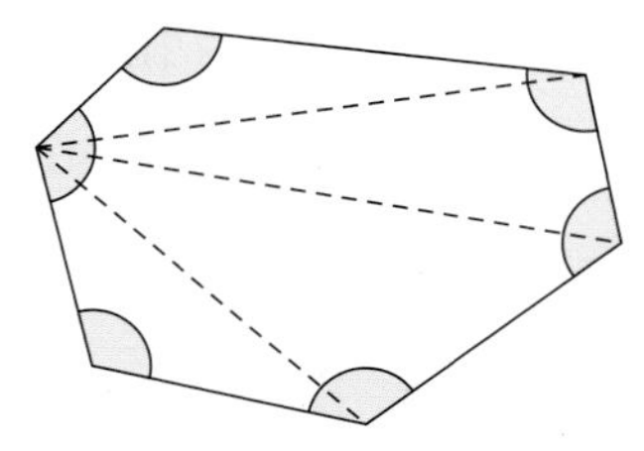

the sum of the **interior angles** in an n-sided polygon = $(n - 2) \times 180°$

interior angle + exterior angle = 180°

the exterior angles of any polygon always add up to 360°

A **regular polygon** has all sides equal and all angles the same.

In a regular n-sided polygon

exterior angle = $\frac{360°}{n°}$

interior angle = $180° - \frac{360°}{n°}$

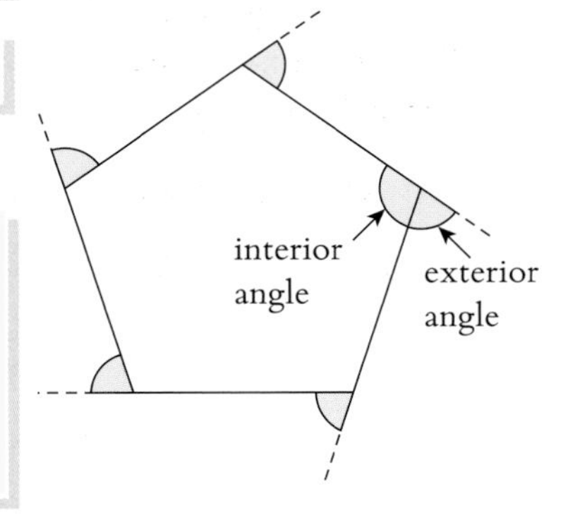

EXAMPLE

ABCDE is a regular pentagon.
The lines BA and DE are extended to meet at F.
Calculate the size of angle AFE.

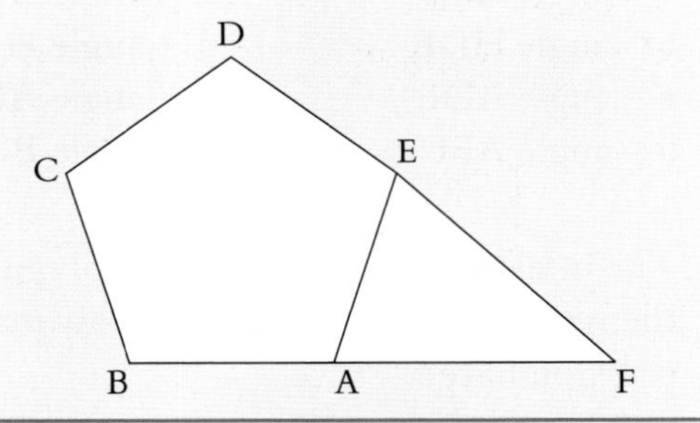

Angle FAE = $\frac{360}{5}$ = 72° — *angle FAE is an exterior angle of the pentagon*

Angle FEA is also 72°

Angle AFE = 180 − (72 + 72) — *angles in a triangle add up to 180°*

= 180 − 144 = 36°

EXERCISE 1.26

1 Jay says that the sum of the interior angles of a six-sided polygon is $6 \times 180° = 1080°$. Is he correct? Explain your answer.

2 Four of the angles of a pentagon are 123°, 84°, 113° and 96°. Calculate the fifth angle.

3 A regular polygon has nine sides. Calculate the size of the interior angle.

4 The exterior angle of a regular polygon is 30°. How many sides does the polygon have?

5 The interior angle of a regular polygon is 162°. How many sides does it have?

6 A polygon has 10 sides. Find the sum of the interior angles.

7 Calculate the size of x for each of the diagrams.

a
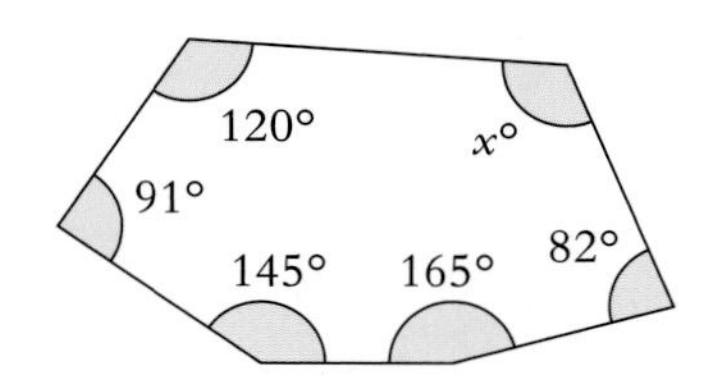

b
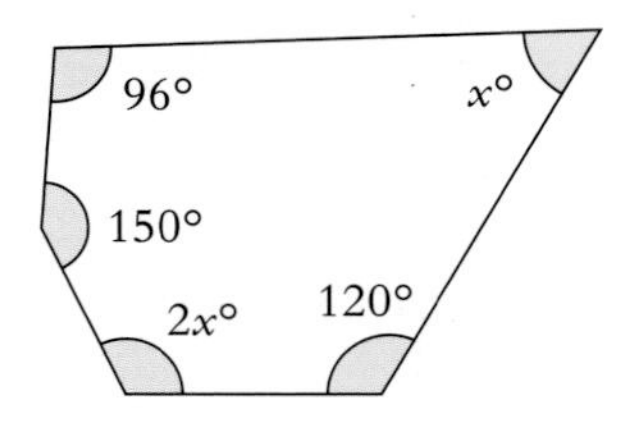

c
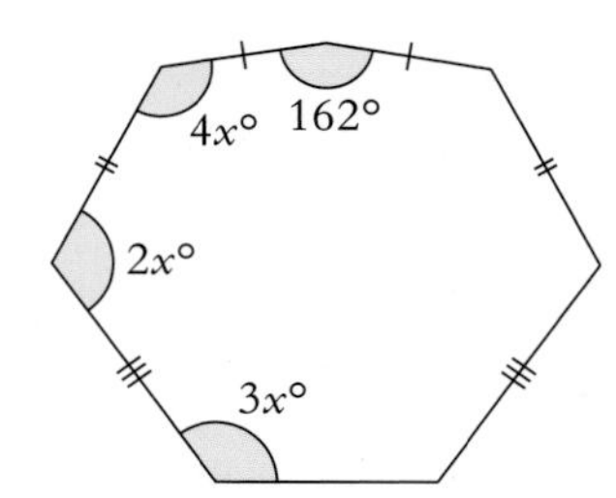

d
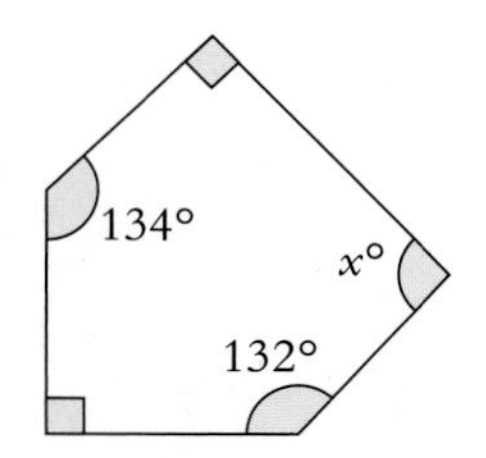

e
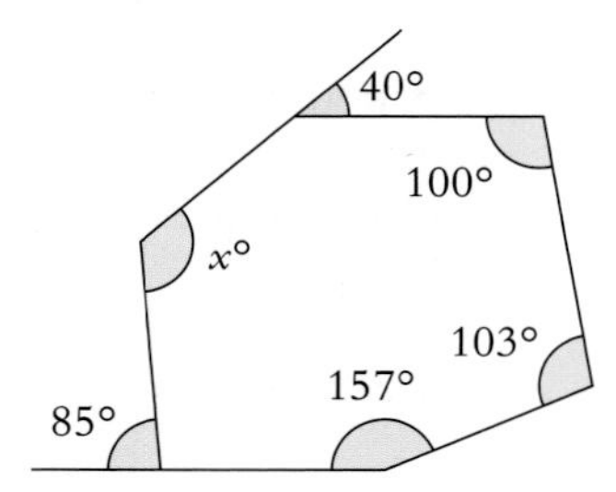

f
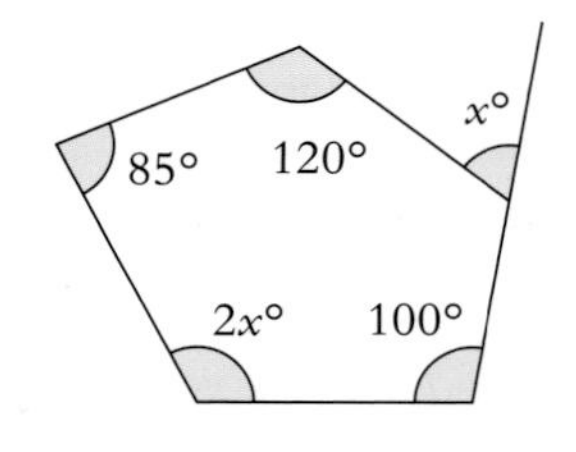

8 The diagram shows a regular octagon and a regular pentagon. Calculate:

a angle AED b angle DEF
c angle HDE d angle HDG
e angle ADE f angle ADG
g angle AEF h angle BDA

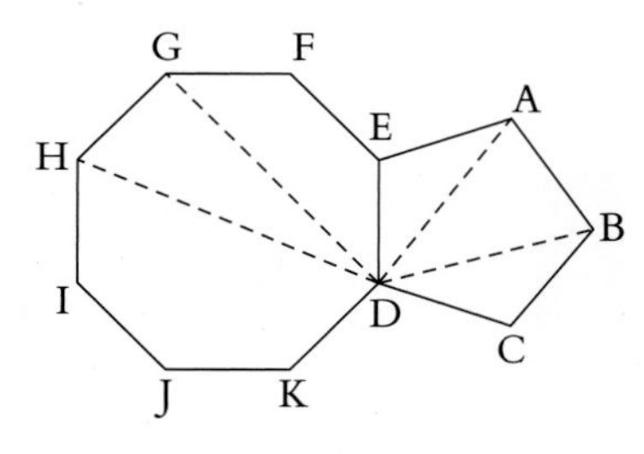

9 The interior angle of a regular polygon is eight times the size of the exterior angle. How many sides does the polygon have?

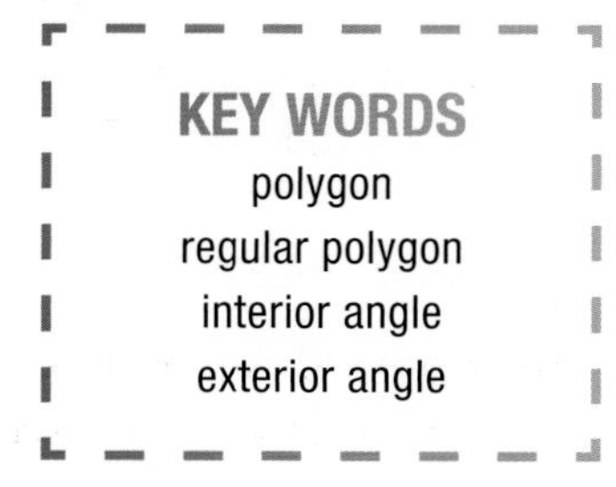

EXERCISE 1.28

1 a Gina says the mode for the number of fish is 6.
Is she correct? Explain your answer.

b Gregor says that the median number of fish is 2.
Is he correct? Explain your answer.

Number of fish	Frequency
0	5
1	6
2	5
3	3
4	1

2 The table shows the number of people in each car passing a checkpoint.

Number of people	Frequency
1	52
2	38
3	8
4	2

a Find the mode for the number of people in a car.

b Find the median number of people in a car.

c Copy and complete the table.
Calculate the mean number of people in a car.

Number of people (x)	Frequency (f)	$f \times x$
1	52	$1 \times 52 =$
2	38	
3	8	
4	2	
Total		

3 Twenty children were asked how many pets they own. The table shows the results.

Number of pets	0	1	2	3	4
Frequency	5	9	4	1	1

Find the mode, median and mean number of pets.

4 The table shows the number of suitcases belonging to each of 100 air passengers.

Number of suitcases	0	1	2	3
Frequency	3	55	38	4

Find the mode, median and mean number of suitcases.

5 The table shows the scores in a quiz for 40 students.

Quiz score	5	6	7	8	9
Frequency	3	7	17	12	1

Find the mode, median and mean quiz score.

6 The table shows the test marks for n students. The mean test mark is 4.

Test mark	1	2	3	4	5
Frequency	3	5	6	2	x

a Find the values of x and n.

b Find the median test mark.

KEY WORDS
frequency table

Unit 1 Examination-style questions

1 Use brackets to make the following statement correct.

$$15 - 2 \times 3 + 4 \div 2 = 2.5$$ [1]

2 The lowest recorded temperature in Alaska is −62.2 °C.
The highest recorded temperature in Alaska is 37.8 °C.

Find the difference between these temperatures. [1]

3 Find the highest common factor (HCF) of 240 and 336. [2]

4 **(a)** Write 140 as a product of prime factors. [2]
(b) Find the lowest common multiple (LCM) of 84 and 140. [2]

5 Write down the prime numbers between 50 and 60. [1]

6 **Without using a calculator**, work out $5\frac{2}{7} - 3\frac{5}{6}$.
You must show all your working and give your answer as a mixed number in its simplest form. [3]

7 **Without using a calculator**, work out $2\frac{5}{6} \div 1\frac{1}{3}$.
You must show all your working and give your answer as a mixed number in its simplest form. [3]

8 **(a)** Write the number 302.658 correct to 2 decimal places, [1]
(b) Write the number 302.658 correct to 2 significant figures. [1]

9 Write each number correct to 1 significant figure and estimate the value of the calculation. You must show your working.

$$\frac{27.8^2}{6.85 + 2.34}$$ [2]

10 Expand the brackets and simplify.

$$3(2x - 1) - 2(x + 4)$$ [2]

11 Solve.

$$2 - \frac{x}{5} = 6$$ [2]

UNIT 1

12 Solve.

(a) $5(2x - 1) = 7$ [3]

(b) $\frac{12 - 5x}{6} = 3 - 2x$ [3]

13 Solve.

(a) $10 - 2(x - 4) = 5(3x + 2)$ [3]

(b) $\frac{2x - 3}{5} = \frac{1 - 4x}{3}$ [3]

14 Ali hires a car.
There is a fixed charge of $170 and then an extra charge of $40 for each day that the car is hired.

Write down a formula for the total cost T, in dollars, when the car is hired for n days. [2]

15 Peaches cost 80 cents each and apples cost 65 cents each.
Anna buys x peaches and 14 apples and spends a total of $17.90

(a) Write down an equation, in terms of x, for the total cost of this fruit. [2]
(b) Solve your equation in **part (a)**. [2]

16 $e = 2f + 5g^3$

Find the value of e when $f = 4$ and $g = -5$. [2]

17 The equation of a straight line is $3x + 2y = 6$.

(a) Write down the gradient of this line, [1]
(b) Write down the coordinates of the point where the line crosses the y-axis. [1]

18 A is the point $(5, -3)$ and B is the point $(-9, 4)$.

Find the gradient of the line AB. [2]

19 P is the point $(-4, 5)$ and Q is the point $(2a, -3a)$.
The gradient of the line PQ is -2.

Find the value of a. [2]

20 P is the point $(p, 15)$ and Q is the point $(-4, q)$.
The line PQ has gradient -2.

Find an expression for q in terms of p. [3]

21 The line $y = mx + 4$ passes through the point $(5, -6)$.

Find the value of m. [2]

22

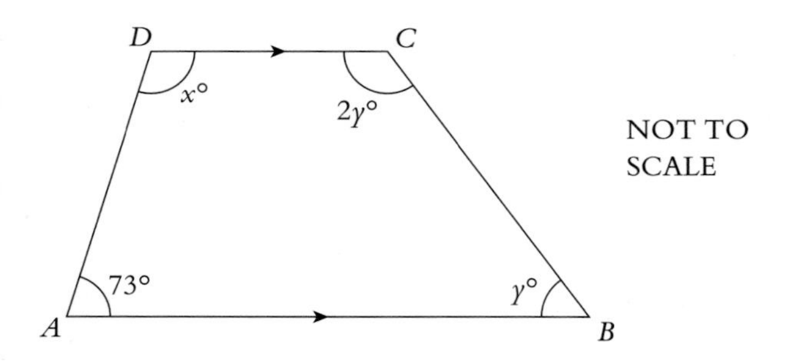

$ABCD$ is a trapezium with AB parallel to CD.

Find the value of x and the value of y. [2]

23

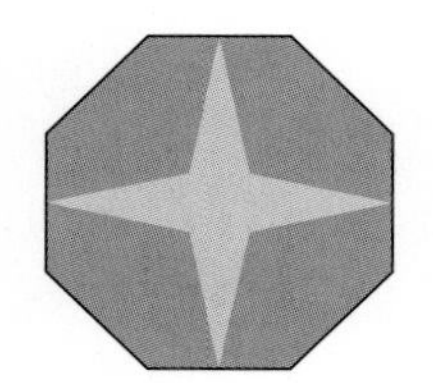

For the diagram above, write down

(a) the order of rotational symmetry, [1]
(b) the number of lines of symmetry. [1]

24

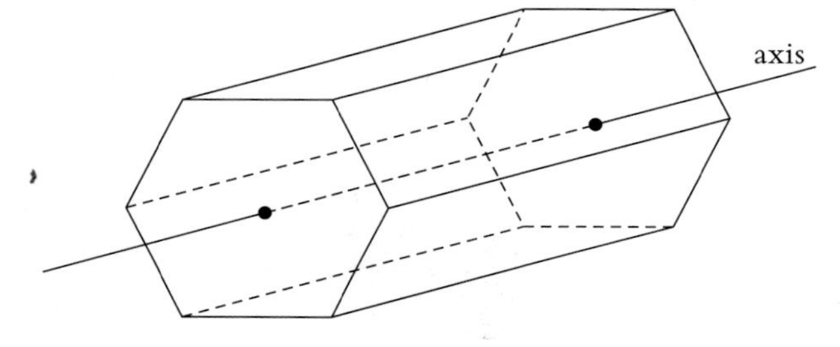

The diagram shows a prism.
The cross-section of the prism is a regular hexagon.

(a) Write down the order of rotational symmetry of this prism about the axis shown. [1]
(b) Write down the number of planes of symmetry for this prism. [1]

25 The interior angle of a regular polygon is 162°.

Find the number of sides for this polygon. [2]

26 The interior angle of a regular polygon is seven times the size of the exterior angle.

How many sides does this polygon have? [3]

27

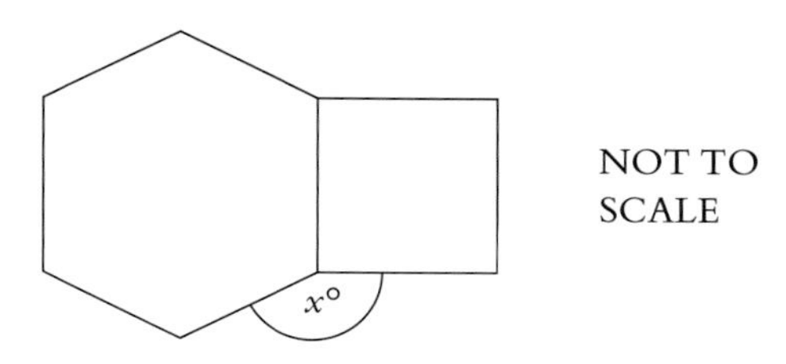

The diagram shows a regular hexagon joined to a square.

Find the value of x. [3]

28 Alan counts the number of sweets in each of 8 packets.
The results are shown below.

22 28 25 29 24 28 26 23

(a) **(i)** Find the range. [1]
(ii) Write down the mode. [1]
(iii) Find the median. [2]
(iv) Calculate the mean. [2]

(b) Alan buys another n packets of sweets.
The mean for these n packets is 26.

Find, in terms of n, an expression for the mean number of sweets in the $(n + 8)$ packets. [2]

29 A six-sided dice, numbered 1 to 6, is rolled 40 times.

The frequency table shows the results.

Score	1	2	3	4	5	6
Frequency	10	3	5	7	8	7

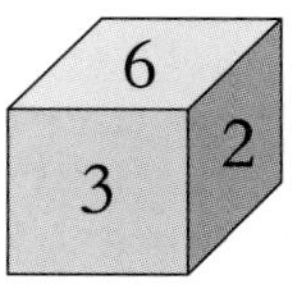

(a) Write down the modal score. [1]
(b) Find the median score. [1]
(c) Calculate the mean score. [3]
(d) The dice is then rolled another 10 times.
The mean score for these 10 rolls is 3.4

Calculate the mean score for all 50 rolls. [3]

30 The table shows information about the number of text messages received, in one hour, by each student in a class.

Number of text messages	0	1	2	3	4	7
Frequency	7	12	15	x	2	1

The mean number of text messages is 1.8

Calculate the value of x. [3]

Percentages 1

THIS SECTION WILL SHOW YOU HOW TO

- Find percentages of a quantity
- Increase and decrease by a given percentage

Percentage means parts of 100 so $37\% = \frac{37}{100}$ (= 0.37)

Percentages are often used to compare quantities.

You should know how to convert fractions and decimals to percentages.

To change a fraction or decimal to a percentage, multiply by 100

$\frac{1}{4} = \frac{1}{4} \times 100\% = 25\%$ $\quad$ $0.03 = 0.03 \times 100\% = 3\%$

To change a percentage to a fraction or a decimal, divide by 100.

$29\% = 29 \div 100 = \frac{29}{100}$ or 0.29

EXAMPLE

a Find 27% of 540 m **b** Find 8% of \$350

a $27\% \text{ of } 540\,\text{m} = \frac{27}{100} \times 540$

$= 145.8\,\text{m}$

Alternative method:

0.27 × 540 = 145.8 m

b $8\% \text{ of } \$350 = \frac{8}{100} \times 350$

$= \$28$

Alternative method:

0.08 × 350 = \$28

EXAMPLE

An airline increases its ticket prices by 15%.
Calculate the new price of a ticket costing \$1300.

Increase in price = 15% of \$1300

$= \frac{15}{100} \times 1300$

$= \$195$

New price = \$1300 + \$195

= \$1495

Quicker method:

multiplying factor is 1.15 (= 1 + 0.15)

1.15 × 1300 = \$1495

EXAMPLE

A coat costing \$180 is reduced by 12% in a sale.
Calculate the price of the coat in the sale.

Decrease in price = 12% of \$180

$= \frac{12}{100} \times 180$

$= \$21.60$

New price = \$180 − \$21.60 = \$158.40

Quicker method:

multiplying factor is 0.88 (= 1 − 0.12)

0.88 × 180 = \$158.40

EXERCISE 2.1

1 Write these percentages as fractions in their simplest form.

a 24% b 80% c 58% d 65% e 30%

2 Write these fractions as percentages.

a $\frac{3}{5}$ b $\frac{7}{20}$ c $\frac{18}{25}$ d $\frac{37}{50}$ e $\frac{7}{8}$

3 Write these percentages as decimals.

a 18% b 70% c 6% d 47% e 2.5%

4 Write these decimals as percentages.

a 0.28 b 0.8 c 0.08 d 1.2 e 0.755

5 Calculate the following.

a 15% of $80 b 52% of 600 km c 27% of 500 kg

6 Which is bigger: 35% of $50 or 30% of $60?

7 a Increase $58 by 12% b Increase $72 by 24%
c Increase $136 by 40% d Increase $230 by 18%
e Increase $440 by 2.5% f Increase $90 by 120%

8 a Decrease $64 by 15% b Decrease $400 by 32%
c Decrease $250 by 8% d Decrease $128 by 12.5%
e Decrease $500 by 3.5% f Decrease $360 by 0.2%

9 Nadia is paid $700 dollars a week. Her pay is increased by 5%.
Calculate her new weekly pay.

10 A holiday is advertised as costing $950.
Raju is given a 7% discount for booking online.
Calculate how much he pays for the holiday.

11 A new car costs $1750.
The car depreciates in value by 27% in its first year.
Find the value of the car after one year.

12 A rare book is bought for $3250. It increases in value by 8%.
Calculate the new value of the book.

13 Rukhusana and Shahida invested $300 each in company shares.
Rukhusana's investment increased in value by 5.2% in the first year.
Shahida's investment decreased in value by 0.8% in the first year.
At the end of the first year, find how much more Rukhusana's investment is worth than Shahida's.

Ratio

THIS SECTION WILL SHOW YOU HOW TO

- Simplify ratios
- Divide a quantity in a given ratio
- Solve problems using ratios

Simplifying ratios

You can use **ratios** to compare one quantity with another quantity. If there are 4 teachers and 18 students on a school trip then the ratio of teachers to students can be written as

$$\text{teachers} : \text{students} = 4 : 18$$

Ratios are simplified in a similar way to fractions.

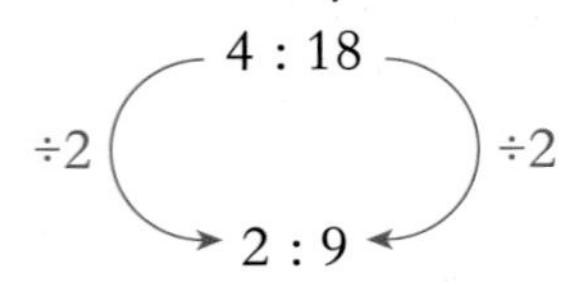

➡ **NOTE:** divide both sides of the ratio by the common factor 2.

This means that for every 2 teachers there are 9 students.

A ratio in its simplest form (lowest terms) has integer values that cannot be cancelled further.

EXAMPLE

Write these ratios in their simplest form.

a $44 : 36$ **b** $1.7 : 0.4$ **c** $\frac{1}{2} : \frac{2}{3}$

a $44 : 36$ (÷4) $= 11 : 9$

b $1.7 : 0.4$ (×10) $= 17 : 4$

c $\frac{1}{2} : \frac{2}{3}$ (×6) $= 3 : 4$

Writing a ratio in the form 1 : *n*

Map scales are often written as ratios in the form $1 : n$

EXAMPLE

Write the ratio 2 cm : 5 km in the form $1 : n$.

2 cm : 5 km	*change 5 km to m*
= 2 cm : 5000 m	*change 5000 m to cm*
= 2 cm : 500 000 cm	*divide both sides of the ratio by 2*
= 1 : 250 000	

EXERCISE 2.5

1 Sarah says that $3^x \times 3^y = 9^{x+y}$. Is she correct? Explain your answer.

2 Simplify these expressions.

a $x^2 \times x^7$ b $b^4 \times b^4$ c $y^8 \times y^2$ d $c^3 \times c^9$

e $a^4 \times a^2 \times a^5$ f $p^2 \times p^2 \times p^4$ g $y^3 \times y^5 \times y$ h $a^2 \times a \times a^4$

TOP TIP

$a = a^1$

3 a $5x^2 \times 3x^4$ b $2b \times 4b^3$ c $7y^8 \times 2y^6$ d $5c^3 \times 5c^3$

e $2a^2 \times 3a$ f $7p^4 \times 2p^3 \times 5p$ g $3y \times 4y^2 \times 5y^3$ h $3a^2 \times 3a^2 \times 3a^2$

UNIT 2

4 a $2x^2 \times 3y^2 \times x^4$ b $3x^4 \times 2y \times 5y^2$ c $5y^2 \times x^2 \times 2y$ d $x^3 \times 2y^2 \times 3x$

e $5xy^2 \times 3x^2y$ f $7xy \times 5x^2$ g $4x^3y \times xy$ h $3a^5b^2 \times 2a^3b^2$

5 a $x^8 \div x^3$ b $b^7 \div b^6$ c $a^6 \div a^3$ d $x^6 \div x$

e $6x^5 \div x^2$ f $8y^7 \div 4y^5$ g $15b^9 \div 3b^2$ h $21c^8 \div 7c^7$

6 a $\frac{5y^4}{y^2}$ b $\frac{12a^2b}{a}$ c $\frac{2x^2}{8x}$ d $\frac{5y^2}{10y}$

e $\frac{6a^4}{3a^3}$ f $\frac{16a^2b^2c^2}{4abc}$ g $\frac{a^3b^4c^2}{ab^2c}$ h $\frac{\left(5x^2\right)^2}{5x^2}$

i $\frac{x^3 \times x^6}{x^4}$ j $\frac{y^8 \times y^2}{y^3}$ k $\frac{3a^4 \times 2a^5}{a^2}$ l $\frac{6x^4 \times 2x^4}{4x^2}$

TOP TIP

Simplify the numerator first.

7 a $(x^3)^2$ b $(a^2)^5$ c $(b^4)^4$ d $(y^7)^3$

e $3(a^2)^3$ f $5(x^5)^4$ g $(2x^3)^3$ h $(3y^2)^4$

i $(2a^2)^3$ j $(3y^4)^2$ k $2y^2(3x^2)^2$ l $5x(2x^2)^3$

8 a $(3xy^2)^3$ b $(2a^2b^3)^5$ c $(5x^4y^3)^2$ d $(10x^5y^7)^3$

9 a $\frac{3x^2 + 5x^2}{2x^2}$ b $\frac{15x^{10}}{3x^4} + 8x^6$ c $\frac{\left(2x^2\right)^3 \times \left(3x\right)^2}{2x^5}$ d $\frac{5a^2b^4}{3c} \times \frac{abc}{15a^2b^2}$

e $\frac{4a^5b}{cd} \div \frac{a}{c^2d^3}$ f $\frac{\left(4x^8y^3\right)^3}{2x^5y^2}$ g $\frac{\left(2x^2y\right)^6}{\left(2xy^2\right)^3}$ h $\left(\left(2xy^2\right)^2\right)^2$

Negative powers: $4^3 \div 4^5 = \frac{\cancel{4} \times \cancel{4} \times \cancel{4}}{4 \times 4 \times \cancel{4} \times \cancel{4} \times \cancel{4}} = \frac{1}{4^2}$

and $4^3 \div 4^5 = 4^{3-5} = 4^{-2}$

so $4^{-2} = \frac{1}{4^2}$

RULE 4 $a^{-m} = \frac{1}{a^m}$

Zero powers: $5^4 \div 5^4 = \frac{\cancel{5} \times \cancel{5} \times \cancel{5} \times \cancel{5}}{\cancel{5} \times \cancel{5} \times \cancel{5} \times \cancel{5}} = 1$

and $5^4 \div 5^4 = 5^{4-4} = 5^0$

so $5^0 = 1$

RULE 5 $a^0 = 1$

Any number raised to the power zero is equal to 1.

EXAMPLE

Work out the value of **a** 5^{-2} **b** 8^0 **c** $(-3)^{-5}$ **d** $\left(\frac{3}{5}\right)^{-1}$

a $5^{-2} = \frac{1}{5^2} = \frac{1}{5 \times 5} = \frac{1}{25}$

b $8^0 = 1$

c $(-3)^{-5} = \frac{1}{(-3)^5} = \frac{1}{(-3)\times(-3)\times(-3)\times(-3)\times(-3)} = \frac{1}{-243} = -\frac{1}{243}$

d $\left(\frac{3}{5}\right)^{-1} = \frac{1}{\left(\frac{3}{5}\right)^1} = 1 \div \frac{3}{5} = 1 \times \frac{5}{3} = \frac{5}{3}$

➡ **NOTE:** it is very useful to know that $\left(\frac{3}{5}\right)^{-1}$ is the same as $\left(\frac{5}{3}\right)$.

EXAMPLE

Simplify

a $x^9 \times x^{-3} \times x^{-4}$ **b** $5x^{-5} \times 4x^2$ **c** $x^6 \div x^{-8}$ **d** $\left(x^{-2}\right)^3$ **e** $\left(5x^2\right)^{-3}$

a $x^9 \times x^{-3} \times x^{-4} = x^{9+-3+-4}$ *add the powers*

$= x^{9-3-4}$

$= x^2$

b $5x^{-5} \times 4x^2 = 5 \times 4 \times x^{-5} \times x^2$

$= 20 \times x^{-5+2}$

$= 20x^{-3}$

c $x^6 \div x^{-8} = x^{6--8}$ *subtract the powers*

$= x^{6+8}$

$= x^{14}$

d $\left(x^{-2}\right)^3 = x^{-2\times3}$ *multiply the powers*

$= x^{-6}$

e $(5x^2)^{-3} = \frac{1}{(5x^2)^3} = \frac{1}{5x^2 \times 5x^2 \times 5x^2}$

$= \frac{1}{5 \times 5 \times 5 \times x^2 \times x^2 \times x^2}$

$= \frac{1}{125x^6}$ or $\frac{1}{125}x^{-6}$

or $\left(5x^2\right)^{-3} = (5)^{-3}\left(x^2\right)^{-3}$

$= 5^{-3}x^{-6}$

$= \frac{1}{125}x^{-6}$ or $\frac{1}{125x^6}$

EXERCISE 2.6

Work out the value of these.

1 a 3^{-1} b 5^0 c 2^{-3} d 3^{-3} e 7^0 f 10^{-3}
g 4^{-2} h 7^{-1} i 10^{-4} j 5^{-3} k 2^{-5} l 3^{-4}

2 a $(-3)^{-2}$ b $(-2)^{-3}$ c $(-6)^{-2}$ d $(-9)^{-1}$ e $(-2)^{-4}$ f $(-10)^{-3}$

3 a $\left(\frac{2}{3}\right)^{-1}$ b $\left(\frac{2}{5}\right)^{-1}$ c $\left(\frac{3}{4}\right)^{-1}$ d $\left(\frac{5}{6}\right)^{-1}$ e $\left(\frac{2}{7}\right)^{-1}$ f $\left(\frac{4}{5}\right)^{-1}$

4 a $\left(\frac{2}{3}\right)^{-2}$ b $\left(\frac{2}{5}\right)^{-2}$ c $\left(\frac{3}{4}\right)^{-2}$ d $\left(\frac{5}{6}\right)^{-2}$
e $\left(\frac{2}{7}\right)^{-2}$ f $\left(\frac{4}{5}\right)^{-2}$ g $\left(\frac{1}{2}\right)^{-3}$ h $\left(\frac{1}{10}\right)^{-3}$

TOP TIP

$\left(\frac{2}{3}\right)^{-2} = \left(\frac{3}{2}\right)^{2}$

Simplify the following expressions. Give your answers in index form.

5 a $a^6 \times a^{-2}$ b $b^{-5} \times b^{-4}$ c $c^{-3} \times c$ d $d^{-6} \times d^{-7}$
e $d^{-3} \times d^{-4} \times d^{10}$ f $x^2 \times x^5 \times x^{-4}$ g $y^4 \times y^2 \times y^{-2}$ h $d^{-3} \times d^{-4} \times d^{10}$
i $a^{-3} \times a^3 \div a^{-5}$ j $b^6 \times b^{-3} \div b^2$ k $c^4 \times c^{-7} \div c^{-3}$ l $d^{-3} \times d^{-2} \div d^{-1}$

6 a $\dfrac{x^2 \times x^4}{x^8}$ b $\dfrac{x^{-3} \times x^5}{x^{-3}}$ c $\dfrac{4x^5 \times 3x^2}{2x^{-2}}$ d $\dfrac{8x^2}{4x^3 \times 2x^4}$

7 a $2x^{-2} \times 4x^3$ b $5x^{-1} \times 3x^{-2}$ c $2x^{-2} \times 3x^5$ d $7x^{-1} \times 4x^{-4}$

8 a $(x^2)^{-1}$ b $(y^3)^{-2}$ c $(a^{-4})^5$ d $(b^{-2})^4$
e $(a^{-3})^{-2}$ f $(y^{-1})^{-2}$ g $(a^{-2})^{-5}$ h $(a^{-1})^{-1}$

9 a $(3a^{-1})^2$ b $(5b^{-1})^3$ c $(2c^{-2})^4$ d $(4d^{-2})^2$
e $(2x^2)^{-2}$ f $(3y^2)^{-3}$ g $(5x^3)^{-2}$ h $(2x^{-2})^{-2}$

10 a $x^{-1} \div \dfrac{1}{x^2}$ b $x^3 \div \dfrac{2}{x^4}$ c $y \div \dfrac{1}{y^{-2}}$ d $\dfrac{4}{x^2} + 3x^{-2}$

11 Find the odd one out in these three expressions.

$5a^3b^{-4} \times a^{-1}b$	$2a^2 + 3b^{-3}$	$\dfrac{5a^{-2}b^{-2}}{a^{-4}b}$

Solving linear inequalities

THIS SECTION WILL SHOW YOU HOW TO

- Solve linear inequalities
- Represent solutions to linear inequalities on a number line.

You need to know these **inequality** signs:

$>$ means 'is greater than' $\geq$ means 'is greater than or equal to'
$<$ means 'is less than' $\leq$ means 'is less than or equal to'

You solve a linear inequality in a similar way to solving linear equations.

$2x - 3 < 10$ *add 3 to both sides*
$2x < 13$ *divide both sides by 2*
$x < 6.5$

You must be careful if you multiply or divide an inequality by a negative number. You must reverse the inequality sign.
(For example $-3 < 7$, multiplying both sides by -1 gives $3 > -7$)

EXAMPLE

Solve $7 - 2x > 1$ and show your answer on a number line.

Method 1

$7 - 2x > 1$ *add 2x to both sides*
$7 > 1 + 2x$ *take 1 from both sides*
$6 > 2x$ *divide both sides by 2*
$3 > x$ so $x < 3$

−4 −3 −2 −1 0 1 2 3 4

➡ **NOTE:** a hollow circle is used for < or >.

Method 2

$7 - 2x > 1$ *take 7 from both sides*
$-2x > -6$ *divide both sides by −2*
$x < 3$ *and reverse the sign*

EXAMPLE

Solve $4(x + 1) \leq 4 - 2(x + 3)$ and show your answer on a number line.

$4(x + 1) \leq 4 - 2(x + 3)$ *multiply out the brackets*
$4x + 4 \leq 4 - 2x - 6$
$4x + 4 \leq -2 - 2x$ *add 2x to both sides*
$6x + 4 \leq -2$ *take 4 from both sides*
$6x \leq -6$ *divide both sides by 6*
$x \leq -1$

−4 −3 −2 −1 0 1 2 3 4

➡ **NOTE:** a solid circle is used for ≤ or ≥.

EXAMPLE

Solve $-1 \leq 2x + 1 < 5$ and show your answer on a number line.

$-1 \leq 2x + 1 < 5$ *subtract 1 throughout*
$-2 \leq 2x < 4$ *divide by 2*
$-1 \leq x < 2$

−4 −3 −2 −1 0 1 2 3 4

➡ **NOTE:** if a question asks for the integer values that satisfy the inequality $-1 \leq 2x + 1 < 5$ the final answer will be −1, 0 and 1. (An integer is a whole number.)

EXERCISE 2.7

Write the inequalities shown on these diagrams.

1 $-3\ -2\ -1\ 0\ 1\ 2\ 3$

2 $-3\ -2\ -1\ 0\ 1\ 2\ 3$

3 $-3\ -2\ -1\ 0\ 1\ 2\ 3$

4 $-3\ -2\ -1\ 0\ 1\ 2\ 3$

Solve these inequalities.

5 $2x - 5 \le 7$
6 $3x + 2 \ge 14$
7 $5x + 4 < 39$
8 $6x - 2 > 10$
9 $8x + 3 \ge 15$
10 $4x - 5 < 19$
11 $2x + 6 \le 4$
12 $6 > 3x - 15$
13 $10 - 4x \le 6$
14 $30 - 2x > 15$
15 $4 - 2x \ge 16$
16 $6 - 5x > -4$
17 $27 - 3x \le 6$
18 $1 - 10x > 3$
19 $7 - 2x < 12$
20 $16 \le 8 - 2x$
21 $3(2x + 5) \le 21$
22 $4(y - 5) > 12$
23 $3(2x + 5) \ge 27$
24 $4(3x - 1) < 80$
25 $7 > 2(x + 3)$
26 $-25 \le 5(3x + 4)$
27 $3(2 - x) \ge -6$
28 $2(13 - x) < 22$
29 $5x - 2 \ge 3x + 4$
30 $8x + 4 < 3x - 31$
31 $2x + 1 \le 5x - 26$
32 $5 - 2x > 3 - x$
33 $3(x + 1) \le 4(x - 1)$
34 $2(x + 2) \ge 5(x - 1)$
35 $3(x + 2) - 2(x - 4) \ge 4(x + 2)$
36 $2(3x - 1) - 5(x + 2) > 3(x - 5)$
37 $\frac{x-1}{8} \le 1 - x$
38 $x - 3 \ge \frac{x+12}{2}$
39 $\frac{x-3}{3} > \frac{x-2}{5}$
40 $\frac{x+2}{2} \le \frac{x-1}{7}$
41 $\frac{x-5}{3} > \frac{4-x}{2}$
42 $\frac{x-1}{2} > \frac{x+2}{5}$
43 $\frac{x-3}{5} \ge \frac{x+3}{2}$
44 $\frac{2x-3}{4} < \frac{3x-1}{5}$
45 $5 \le 3 - \frac{2x+1}{5}$

Solve these inequalities and then list the integer solutions.

46 $-6 \le 2x \le 10$
47 $-9 \le 3x < 12$
48 $8 < 4x \le 36$
49 $-1 < \frac{x}{2} \le 2$
50 $0 < \frac{x}{2} - 1 < 1$
51 $-8 \le 2n - 3 < 8$
52 $3 < 2x + 3 \le 11$
53 $-2 \le 2(x - 4) \le 6$
54 $-3 < 3(x - 4) \le 15$
55 Write down the smallest integer that satisfies the inequality $3x + 2 \ge 12$
56 Find the largest integer that satisfies $5(3x - 7) < 28$
57 $3x + 2y \le 6$ and x and y are both positive integers (not including 0).
List all the possible pairs of values for x and y.

KEY WORDS

inequality

Manipulating algebraic fractions

THIS SECTION WILL SHOW YOU HOW TO

- Add and subtract algebraic fractions
- Solve equations involving algebraic fractions

Adding and subtracting algebraic fractions

Reminder

$\frac{2}{3}+\frac{1}{4}$ Similarly $\frac{2x}{3}+\frac{x}{4}$ *the common denominator is $3 \times 4 = 12$*

$=\frac{8}{12}+\frac{3}{12}$ $=\frac{8x}{12}+\frac{3x}{12}$ $\frac{2}{3}=\frac{8}{12}$ *and* $\frac{1}{4}=\frac{3}{12}$

$=\frac{8+3}{12}$ $=\frac{8x+3x}{12}$

$=\frac{11}{12}$ $=\frac{11x}{12}$

EXAMPLE

Simplify $\frac{2x}{3}+\frac{x}{4}-\frac{3x}{5}$.

$\frac{2x}{3}+\frac{x}{4}-\frac{3x}{5}$ *the common denominator is $3 \times 4 \times 5 = 60$*

$=\frac{40x}{60}+\frac{15x}{60}-\frac{36x}{60}$ $\frac{2}{3}=\frac{40}{60}$, $\frac{1}{4}=\frac{15}{60}$ *and* $\frac{3}{5}=\frac{36}{60}$

$=\frac{40x+15x-36x}{60}$

$=\frac{19x}{60}$

EXAMPLE

Simplify $\frac{5(2x-1)}{6}-\frac{2(2x-3)}{5}$.

$\frac{5(2x-1)}{6}-\frac{2(2x-3)}{5}$ *the common denominator is $6 \times 5 = 30$*

$=\frac{25(2x-1)}{30}-\frac{12(2x-3)}{30}$ $\frac{5}{6}=\frac{25}{30}$ *and* $\frac{2}{5}=\frac{12}{30}$

$=\frac{25(2x-1)-12(2x-3)}{30}$ *remove the brackets and be careful with the signs*

$=\frac{50x-25-24x+36}{30}$ *collect like terms in the numerator*

$=\frac{26x+11}{30}$

EXERCISE 2.8

1 Simplify the following algebraic fractions.

a $\frac{3x}{5}+\frac{x}{5}$ b $\frac{2x}{7}+\frac{3x}{7}$ c $\frac{x}{4}+\frac{x}{4}$

d $\frac{x}{8}+\frac{3x}{8}$ e $\frac{5x}{6}-\frac{x}{6}$ f $\frac{5x}{8}-\frac{x}{8}$

g $\frac{7x}{10}-\frac{3x}{10}$ h $\frac{4x}{5}-\frac{2x}{5}$ i $\frac{4x}{7}+\frac{2x}{7}-\frac{x}{7}$

2 Pedro says that $\frac{y}{3}+\frac{2y}{5}=\frac{3y}{8}$.

Explain why he is wrong.

Simplify the following algebraic fractions.

3 a $\frac{x}{2}+\frac{x}{4}$ b $\frac{x}{3}+\frac{x}{4}$ c $\frac{2x}{5}+\frac{x}{3}$

d $\frac{3x}{5}+\frac{x}{4}$ e $\frac{x}{2}-\frac{x}{3}$ f $\frac{2x}{3}-\frac{x}{5}$

g $\frac{9x}{10}-\frac{2x}{5}$ h $\frac{8x}{9}-\frac{3x}{4}$ i $\frac{x}{2}-\frac{3x}{10}+\frac{2x}{5}$

4 a $\frac{x}{3}+\frac{x+2}{4}$ b $\frac{x-3}{5}+\frac{x+1}{6}$ c $\frac{2x-3}{7}+\frac{x+4}{5}$

d $\frac{6x-3}{7}+\frac{x+4}{5}$ e $\frac{3(x+1)}{4}+\frac{2(x+3)}{3}$ f $\frac{8(x-1)}{11}+\frac{2(x+4)}{3}$

5 a $\frac{6x+1}{3}-\frac{x+2}{4}$ b $\frac{3x-2}{5}-\frac{x-3}{8}$ c $\frac{4x+2}{5}-\frac{3x+4}{6}$

d $\frac{2x+3}{9}-\frac{5x-1}{4}$ e $\frac{2(3x+4)}{3}-\frac{3(2x-1)}{5}$ f $\frac{3(x-1)}{4}-\frac{2(x+6)}{9}$

6 $\frac{2x+1}{4}-\frac{3(x-2)}{5}+\frac{2(x+1)}{3}$

7 Find the odd one out.

$\frac{2x}{3}-\frac{x}{4}$	$\frac{x}{4}+\frac{x}{3}-\frac{x}{6}$	$\frac{x}{2}-\frac{x}{12}$	$\frac{3x}{4}-\frac{2x}{3}$	$\frac{x}{4}+\frac{x}{6}$

UNIT 2

UNIT 2

Solving equations

EXAMPLE

Solve $\frac{2x}{5}+\frac{3x}{2}=38$

There are two methods for solving this type of equation. Use the method that you prefer.

Method 1

$\frac{2x}{5}+\frac{3x}{2}=38$ *common denominator is 10*

$\frac{4x}{10}+\frac{15x}{10}=38$

$\frac{19x}{10}=38$ *multiply by 10*

$19x = 380$ *divide by 19*

$x = 20$

Method 2

$\frac{2x}{5}+\frac{3x}{2}=38$ *multiply by 10*

${}^{2}\cancel{10}\times\frac{2x}{\cancel{5}}+{}^{5}\cancel{10}\times\frac{3x}{\cancel{2}}=10\times 38$ *cancel*

$4x + 15x = 380$

$19x = 380$ *divide by 19*

$x = 20$

EXAMPLE

Solve $\frac{2x+8}{3}-\frac{x-1}{2}=4$

Method 1

$\frac{2x+8}{3}-\frac{x-1}{2}=4$ *change fractions to a common denominator of 6*

$\frac{2(2x+8)}{6}-\frac{3(x-1)}{6}=4$

$\frac{2(2x+8)-3(x-1)}{6}=4$ *remove the brackets and be careful with the signs!*

$\frac{4x+16-3x+3}{6}=4$ *collect like terms in the numerator*

$\frac{x+19}{6}=4$ *multiply both sides by 6*

$x + 19 = 24$ *subtract 19 from both sides*

$x = 5$

Method 2

$\frac{2x+8}{3}-\frac{x-1}{2}=4$ *multiply both sides by 6*

${}^{2}\cancel{6}\times\frac{(2x+8)}{\cancel{3}}-{}^{3}\cancel{6}\times\frac{(x+1)}{\cancel{2}}$ *cancel*

$=6\times 4$

$2(2x + 8) - 3(x - 1) = 24$ *remove the brackets and be careful with the signs!*

$4x + 16 - 3x + 3 = 24$ *collect like terms*

$x + 19 = 24$ *take 19 from both sides*

$x = 5$

EXERCISE 2.9

Solve these equations.

1
a $\frac{x}{3}-\frac{x}{4}=3$
b $\frac{2x}{3}-\frac{x}{2}=3$
c $\frac{x}{4}+\frac{x}{5}=18$
d $\frac{2x}{3}+\frac{3x}{2}=26$
e $\frac{3x}{5}+\frac{x}{4}=34$
f $\frac{2x}{9}-\frac{x}{2}=10$
g $\frac{4x}{5}+\frac{3x}{2}=23$
h $\frac{6x}{7}-\frac{x}{3}=22$

2
a $\frac{x}{5}-\frac{x}{6}=\frac{1}{2}$
b $\frac{3x}{4}-\frac{2x}{3}=\frac{1}{3}$
c $\frac{2x}{5}+\frac{x}{4}=2\frac{3}{5}$
d $\frac{2x}{9}-\frac{x}{8}=\frac{7}{12}$
e $\frac{x}{8}-\frac{2x}{5}=1\frac{1}{10}$
f $\frac{3x}{5}-\frac{3x}{8}=\frac{3}{20}$
g $\frac{x}{4}-\frac{x}{7}=\frac{9}{14}$
h $\frac{x}{8}+\frac{3x}{5}=-1\frac{9}{20}$

3
a $\frac{3x+4}{2}+\frac{8-x}{3}=7$
b $\frac{x+1}{4}+\frac{2x+1}{5}=5$
c $\frac{4x}{5}+\frac{3x-7}{2}=8$
d $\frac{x+9}{6}+\frac{5x-3}{3}=6$
e $\frac{11x+5}{4}+\frac{4x-3}{3}=-12$
f $\frac{3x+2}{2}+\frac{x+6}{3}=14$
g $\frac{4x-1}{5}+\frac{5x-16}{2}=5$
h $\frac{x}{4}+\frac{x-5}{3}=3$
i $\frac{3-x}{6}+\frac{x+8}{5}=2$
j $\frac{3x+10}{8}+\frac{4x+11}{6}=1$
k $\frac{2x+1}{2}+\frac{6x-3}{4}=6$
l $\frac{6x-1}{3}+\frac{15x+7}{9}=1\frac{2}{3}$

4
a $\frac{x+8}{5}-\frac{x+1}{4}=1$
b $\frac{19x+5}{6}-\frac{2(x+4)}{5}=2$
c $\frac{5x+6}{3}-\frac{5x-1}{7}=5$
d $\frac{5x-2}{4}-\frac{x+6}{3}=3$
e $\frac{4x-1}{3}-\frac{20-x}{8}=3$
f $\frac{11x+5}{4}-\frac{3x-5}{2}=5$
g $\frac{2(x-2)}{3}-\frac{6x-5}{5}=-3$
h $\frac{3x-5}{4}-\frac{5(x-4)}{3}=-1$
i $\frac{3(3-x)}{2}-\frac{2(x+6)}{3}=7$
j $\frac{3(5-x)}{5}-\frac{4(2-x)}{7}=2$
k $\frac{3(4x+1)}{7}-\frac{14x-5}{8}=1$
l $\frac{21x+11}{6}-\frac{9x-1}{3}=2\frac{1}{3}$

5 $\frac{5x-4}{6}-\frac{2x+1}{5}=\frac{3x-2}{4}-\frac{x+1}{3}$

6 The perimeter of this rectangle is 20 cm.

a Show that $17x + 31 = 150$

b Hence, or otherwise, find the sides of the rectangle.

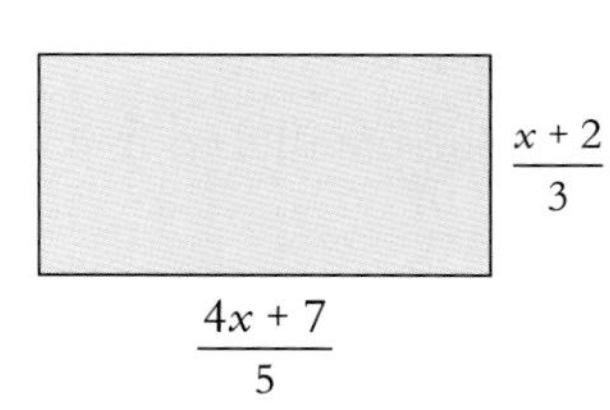

7 The perimeter of this isosceles triangle is 22 cm. Find the value of x.

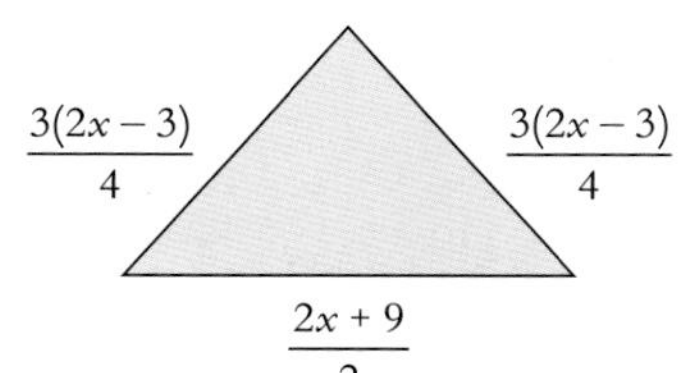

KEY WORDS
algebraic fraction

UNIT 2

The general equation of a straight line

THIS SECTION WILL SHOW YOU HOW TO

- Use the equation $y = mx + c$
- Find the midpoint of a line segment

The diagram shows the graph of $y = 2x + 1$
From the graph you can see that the

- gradient = 2
- **y-intercept** = 1

This important result can be written as

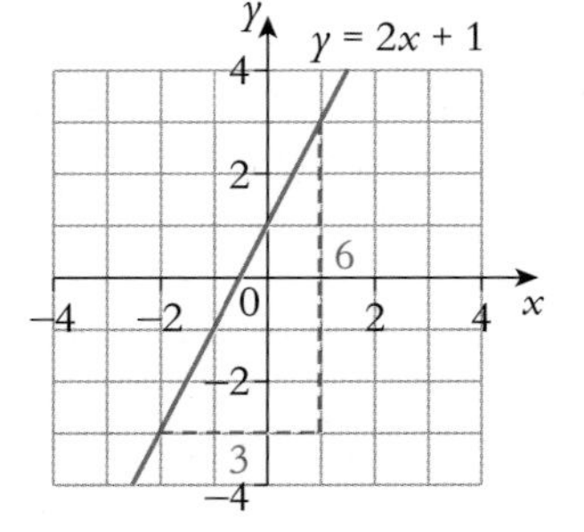

The graph of $y = mx + c$ is a straight line where m is the gradient and c is the y-intercept.

EXAMPLE

Find the gradient and y-intercept for these lines.

a $y = 3x - 5$ **b** $2y - x = 8$

a $y = 3x - 5$ *compare with $y = mx + c$*

$m = 3$ and $c = -5$

gradient = 3 y-intercept = −5

b $2y - x = 8$ *add x to both sides*

$2y = x + 8$ *divide both sides by 2*

$y = \frac{1}{2}x + 4$ *compare with $y = mx + c$*

$m = \frac{1}{2}$ and $c = 4$

gradient = $\frac{1}{2}$ y-intercept = 4

Parallel lines

Parallel lines have the same gradient.

EXAMPLE

A line is parallel to the line $4y = 2x + 8$ and passes through the point (4, 5).
Find the equation of the line.

$4y = 2x + 8$ *divide both sides by 4*

$y = \frac{1}{2}x + 2$ *compare with $y = mx + c$*

This line has a gradient of $\frac{1}{2}$ so the line parallel to it will also have gradient $\frac{1}{2}$

The equation of the parallel line is $y = \frac{1}{2}x + c$

The line passes through (4, 5) so substitute $x = 4$ and $y = 5$ into $y = \frac{1}{2}x + c$

$5 = \frac{1}{2} \times 4 + c$

$5 = 2 + c$

$c = 3$

The equation of the parallel line is $y = \frac{1}{2}x + 3$

EXERCISE 2.10

1 Write down the gradient and y-intercept for these lines.

a $y = 3x + 2$ b $y = 2x - 5$ c $y = \frac{1}{2}x + 3$ d $y = \frac{2}{3}x - 1$

e $y = 3 + 2x$ f $y = 4 - 5x$ g $y = 7 - \frac{1}{2}x$ h $y = \frac{1}{2} + 3x$

2 Find the gradient and y-intercept for these lines.

a $2y = 3x + 4$ b $3y = 6x - 2$ c $5y = 2x - 3$ d $2x + 3y = 5$

e $3x + 2x = 6$ f $x - 3y = 6$ g $4x - 2y = 3$ h $3x + 7y = 2$

3 For each of the lines a, b, c and d write down

i the gradient

ii the y-intercept

iii the equation of the line.

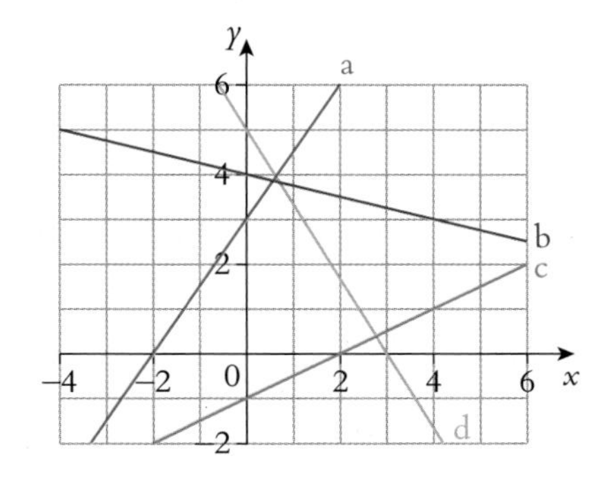

4 Write down the equation of each of these lines.

Line A	Line B	Line C
Gradient = 0	Gradient = 2	Gradient = 3
Passes through (5, −2)	Passes through (0, 0)	Passes through (0, −2)

Line D	Line E	Line F
Parallel to $y = \frac{1}{2}x + 4$	Parallel to $3x + 4y = 12$	Parallel to $2y - x = 6$
Passes through (0, 6)	Passes through (2, 5)	Passes through (−6, 1)

5 Sketch each of these graphs, labelling the point where the graph crosses the y-axis.

a $y = x + 4$ b $y = 3x - 5$ c $y = \frac{1}{2}x - 3$ d $y = 5 - 2x$

e $y + 2x = -4$ f $2x + 3y = 12$ g $2x - y = 3$ h $5x - 2y = 10$

6

A $2y = (x + 8)$	B $y = 3x + 2$	C $x = y + 2$	D $3x + y = 2$	E $x = 2y$

From the list above write down the lines that

a pass through (0, 0)

b are parallel

c pass through the point (−2, −4)

d have negative gradient

e are reflections of each other in the y-axis.

Finding the equation of a line passing through two given points

Some questions give you the y-intercept and another point on the line.

EXAMPLE

Find the equation of the line passing through (0, −2) and (4, 1).

Using a sketch graph:

gradient $= \frac{3}{4}$ so $m = \frac{3}{4}$ or gradient $= \frac{y_2 - y_1}{x_2 - x_1} = \frac{1 - -2}{4 - 0} = \frac{3}{4}$

y-intercept = −2 so $c = -2$

Using $y = mx + c$, $m = \frac{3}{4}$ and $c = -2$

The equation of the line is $y = \frac{3}{4}x - 2$

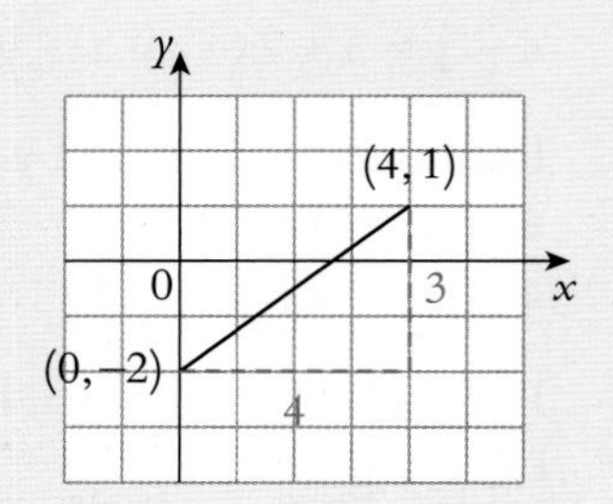

When the two points given do not include the y-intercept it can be more difficult.

EXAMPLE

Find the equation of the line passing through (−3, 5) and (5, −3).

Using a sketch graph: gradient $= -\frac{8}{8} = -1$ so $m = -1$

Using $y = mx + c$ and $m = -1$

$y = -x + c$

The line passes through (−3, 5) so substitute $x = -3$ and $y = 5$ into $y = -x + c$

$5 = 3 + c$

$c = 2$

The equation of the line is $y = -x + 2$.

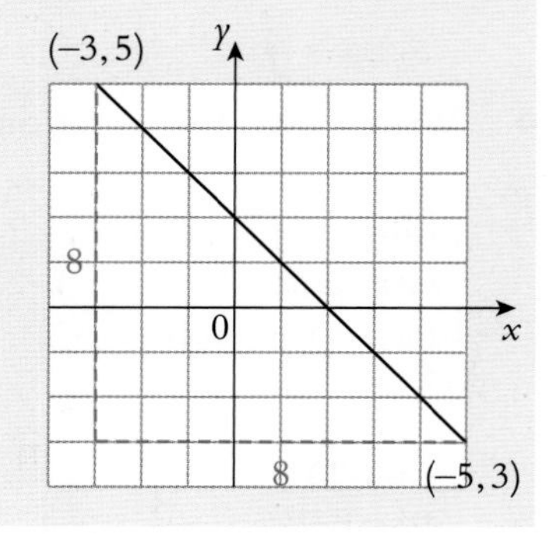

EXERCISE 2.11

Find the equations of the lines passing through these points.

1 a (0, 6) and (8, 10) b (0, −3) and (−2, 1) c (0, 1) and (9, 10)
d (0, −4) and (3, 8) e (0, −2) and (−6,10) f (0, 5) and (9, 11)
g (0, −10) and (5, −13) h (0, 8) and (10, 12) i (0, 5) and (−4, 2)

2 a (2, 3) and (6, 5) b (−1, −1) and (1, −7) c (−3, 2) and (1, 6)
d (1, 0) and (−2, 6) e (4, −1) and (1, −4) f (2, 1) and (5, 3)
g (−4, 4) and (1, 3) h (−4, −6) and (4, −4) i (−5, −4) and (−1, −2)

Finding the midpoint of a line

The **midpoint** of a line segment can be found using a diagram or using the rule midpoint = (mean of x coordinates, mean of y coordinates)

This rule can be written more formally as

> The midpoint of the line segment joining the points (x_1, y_1) and (x_2, y_2) is given by midpoint $= \left(\frac{x_1+x_2}{2}, \frac{y_1+y_2}{2}\right)$

EXAMPLE

Find the midpoint of the line joining the points (1, 2) and (9, 5).

From the graph the midpoint is (5, 3.5).

This answer can also be found using the formula.

First decide which values to use for x_1, y_1, x_2 and y_2.

(1, 2) ↑↑ (x_1, y_1) (9, 5) ↑↑ (x_2, y_2)

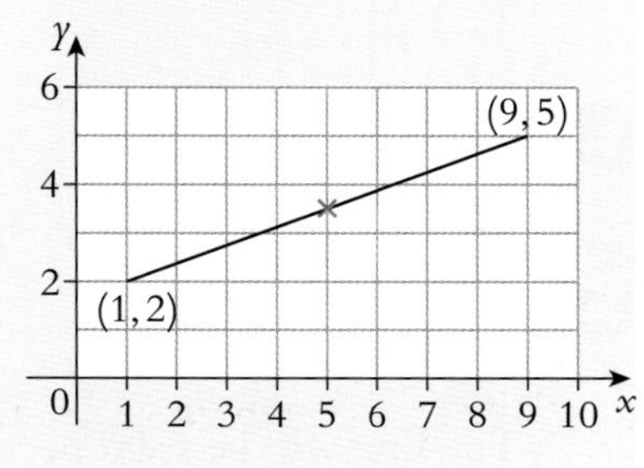

$$\text{midpoint} = \left(\frac{x_1+x_2}{2}, \frac{y_1+y_2}{2}\right) = \left(\frac{1+9}{2}, \frac{2+5}{2}\right) = \left(\frac{10}{2}, \frac{7}{2}\right) = \left(5, 3\frac{1}{2}\right)$$

EXAMPLE

Without drawing the line segment find the midpoint of $(-5, -2)$ and $\left(-2, 2\frac{1}{2}\right)$.

$(-5, -2)$ ↑↑ (x_1, y_1) $\left(-2, 2\frac{1}{2}\right)$ ↑↑ (x_2, y_2)

$$\text{midpoint} = \left(\frac{x_1+x_2}{2}, \frac{y_1+y_2}{2}\right) = \left(\frac{-5+-2}{2}, \frac{-2+2\frac{1}{2}}{2}\right) = \left(\frac{-7}{2}, \frac{\frac{1}{2}}{2}\right) = \left(-3\frac{1}{2}, \frac{1}{4}\right)$$

EXERCISE 2.12

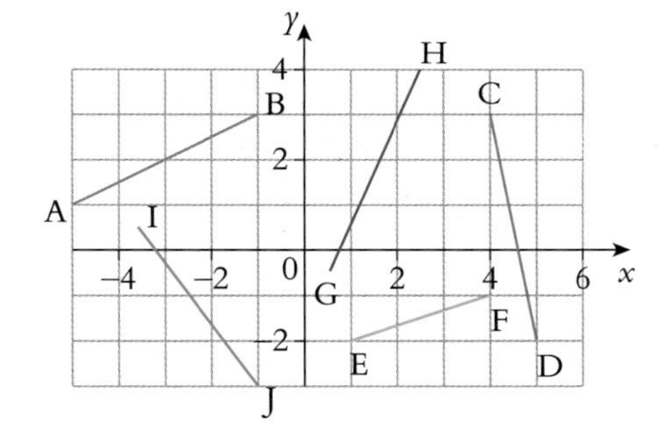

1 Write down the midpoints of the following lines.

a AB b CD c EF

d GH e IJ

2 Use the formula to find the midpoint of the lines joining the following points.

a (2, 3) and (6, 4) b (−4, 2) and (0, −2)

c (1, −4) and (4, 2) d (−6, 4) and (−2, 6)

e (−2, 2) and (−1, 5) f (4, −2) and (5, −6)

g (−6, −6) and (−1, −3) h (−5.5, −2.5) and (−2, −2)

i (1, −1.5) and (4, 3)

KEY WORDS

y-intercept

midpoint

Representing linear inequalities on graphs

THIS SECTION WILL SHOW YOU HOW TO

- Represent linear inequalities on a graph

You can represent an inequality by a **region** on a graph.
There are some important rules that you must follow.

If the inequality is ≥ or ≤ the **boundary line** for the region is shown by a solid line.
If the inequality is > or < the boundary line for the region is shown by a broken (dashed) line.
You are usually expected to shade the unwanted region.

EXAMPLE

Show the region $x + y \geq 3$ on a graph. Leave the required region unshaded.

First draw the boundary line $x + y = 3$.

x	0	3
y	3	0

The line will be solid because of the ≥ symbol.
The line divides the graph into two regions.
Find the required region using a trial point that is not on the line.
Using a trial point of (3, 2), substitute $x = 3$ and $y = 2$ into $x + y \geq 3$ to see if the inequality is true for the trial point.
$3 + 2 \geq 3$ is true. So (3, 2) is in the required region.

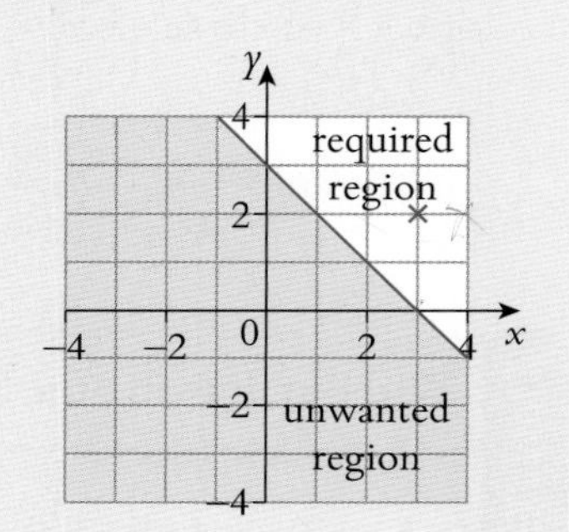

EXAMPLE

Show the region $y < \frac{1}{2}x + 1$ on a graph. Leave the required region unshaded.

x	0	2	4
y	1	2	3

Draw the boundary line $y = \frac{1}{2}x + 1$

The line is broken because of the < symbol.
Using a trial point of (0, 0): $0 < \frac{1}{2} \times 0 + 1$ is true.
So (0, 0) is in the required region.

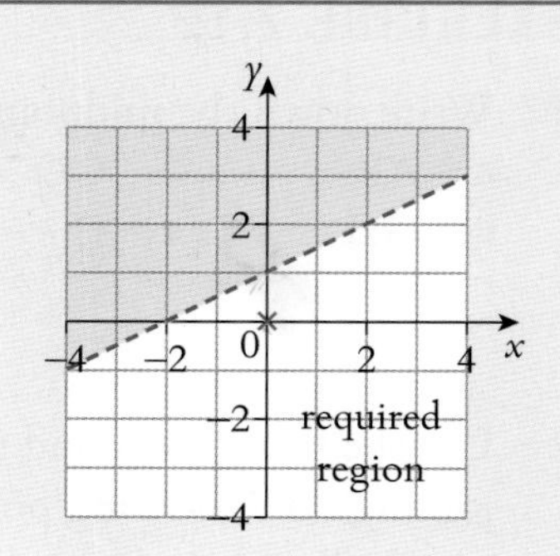

To convert an area in cm^2 to an area in mm^2 you need to remember that:

$1\,cm^2 =$ 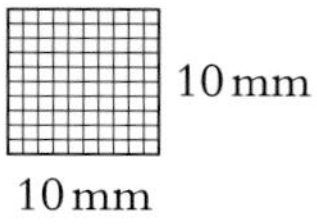

$= 10 \times 10 = 100\,mm^2$

$1\,cm^2 = 100\,mm^2$

$1\,m^2 = 10\,000\,cm^2$

$1\,km^2 = 1\,000\,000\,m^2$

EXERCISE 2.14

1 Find the perimeter and area of these shapes. All lengths are in cm.

a 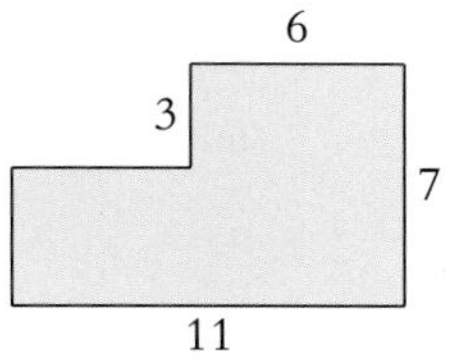

b 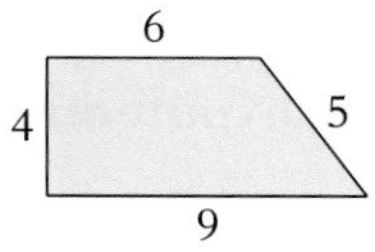

c 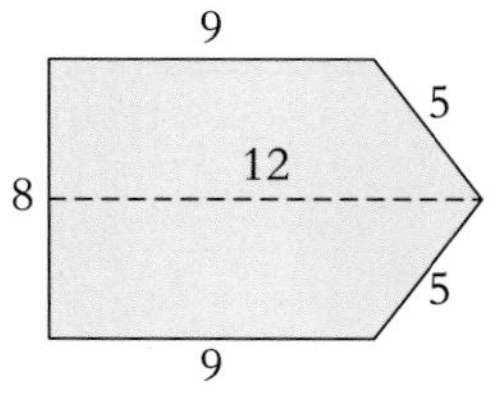

d 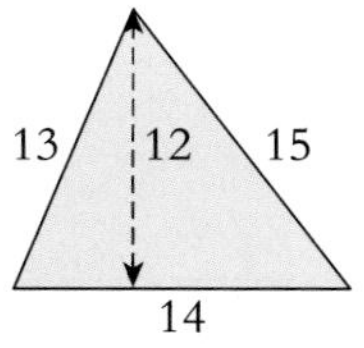

2 Find the area of these parallelograms. All lengths are in cm.

a 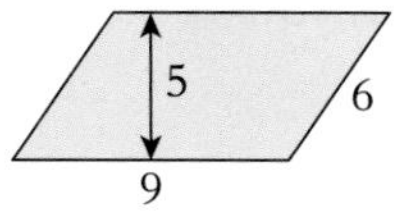

b 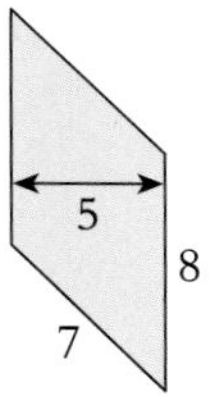

c 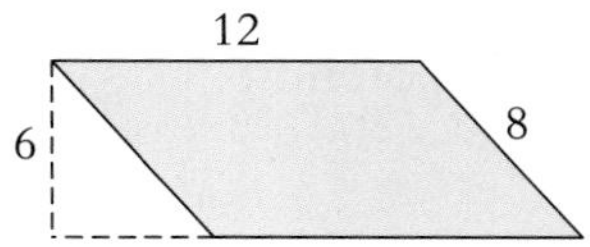

3 Find the area of these trapeziums.

a 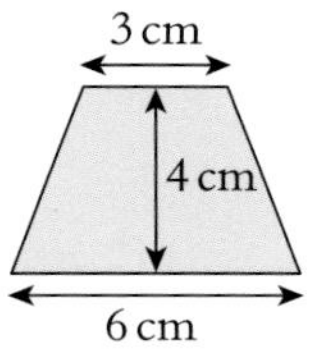

b 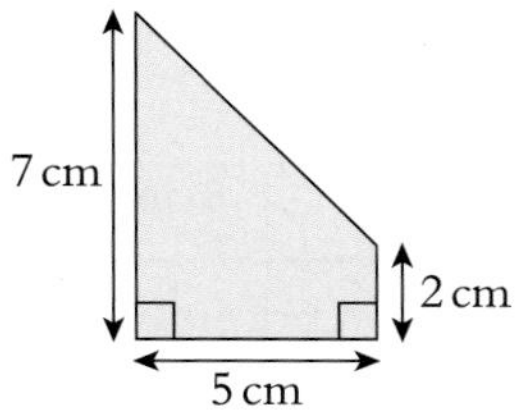

c 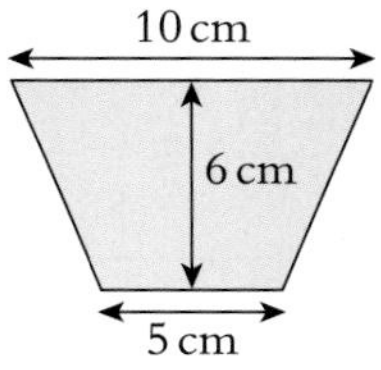

d 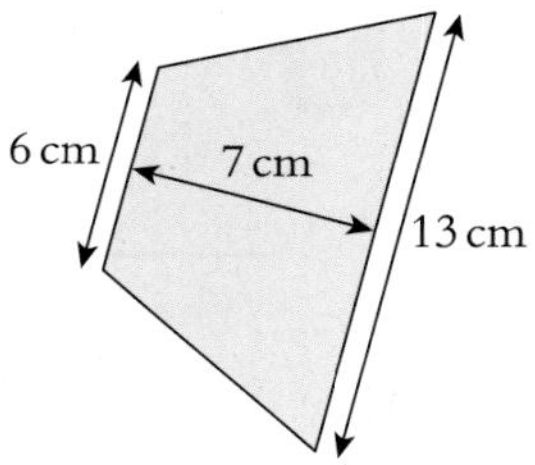

4 Find the area of these shapes. All lengths are in cm.

a 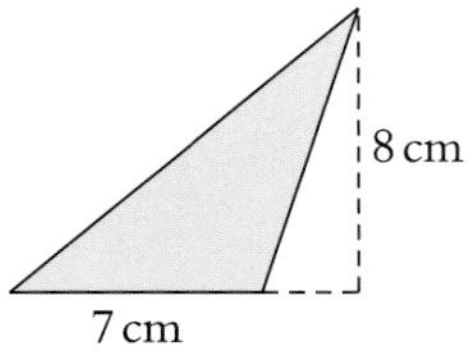

b 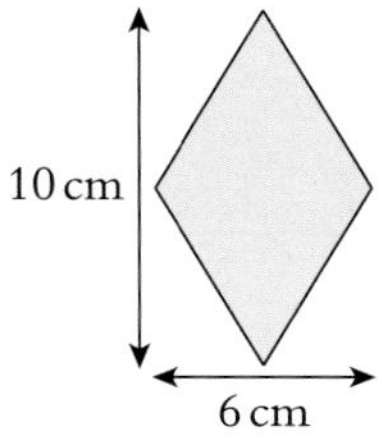

c 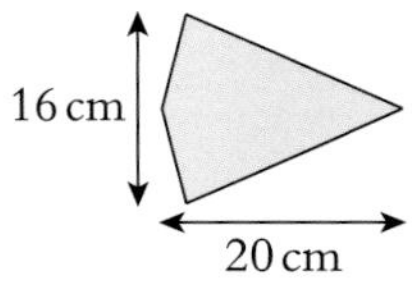

d 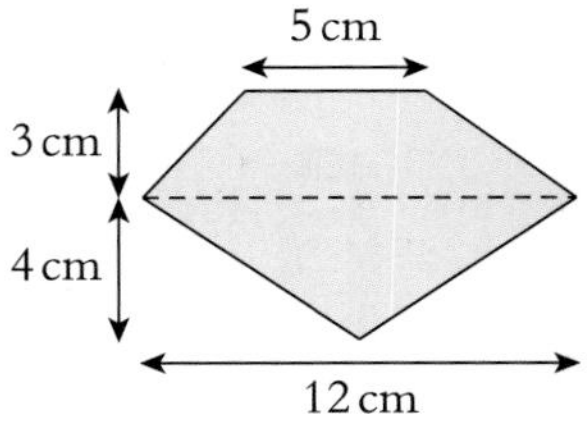

5 A trapezium has an area of $36\,cm^2$. The two parallel sides are 3 cm and 6 cm. Find the distance between the two parallel sides.

6 The trapezium has two parallel sides of length x cm and $3x + 2$ cm.
The distance between the parallel sides is 6 cm.
The area of the trapezium is $108\,cm^2$.
Find the value of x.

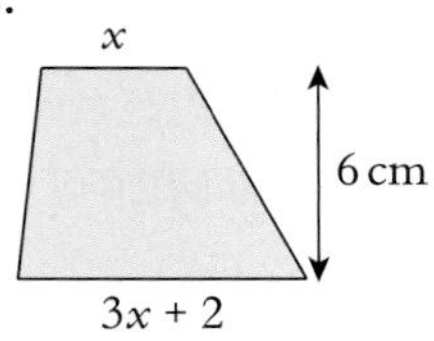

KEY WORDS
area
perimeter

Pythagoras

THIS SECTION WILL SHOW YOU HOW TO

- Apply Pythagoras' theorem to right-angled triangles
- Find the distance between two co-ordinate points

You can use **Pythagoras' theorem** when you know the length of two sides in a right-angled triangle and you want to find the length of the third side.

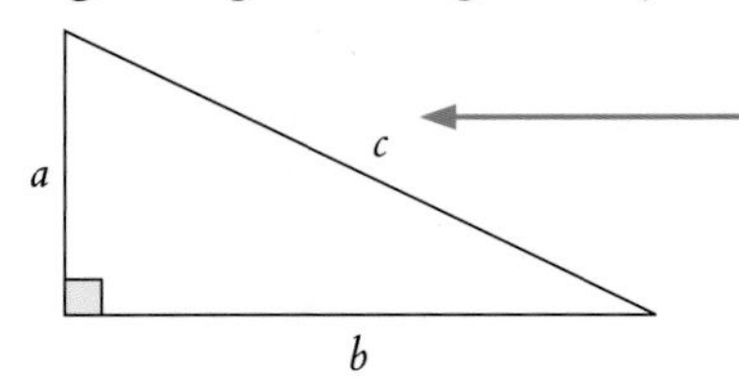

the longest side in a right-angled triangle is called the **hypotenuse**

Pythagoras' theorem for right-angled triangles says that:

The sum of the squares of the lengths of the two shorter sides is equal to the square of the length of the hypotenuse.

$$a^2 + b^2 = c^2$$

EXAMPLE

Find the length of x.

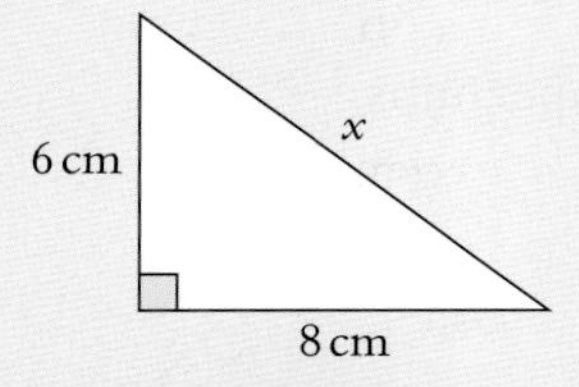

The two shorter sides are 6 cm and 8 cm. The hypotenuse is x cm.

$6^2 + 8^2 = x^2$ *square 6 and 8*

$36 + 64 = x^2$ *$6 \times 6 = 36$ and $8 \times 8 = 64$*

$100 = x^2$ *square root both sides*

$x = \sqrt{100} = 10$ cm

EXAMPLE

Find the length of x.

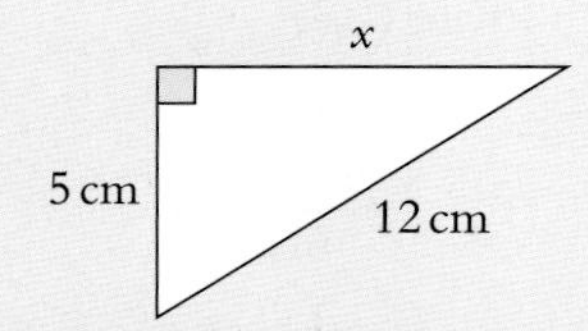

The two shorter sides are x cm and 5 cm. The hypotenuse is 12 cm.

$x^2 + 5^2 = 12^2$ *square 5 and 12*

$x^2 + 25 = 144$ *take 25 from both sides*

$x^2 = 119$ *square root both sides*

$x = \sqrt{119}$

$x = 10.908712\ldots$ *round to a suitable degree of accuracy*

$x = 10.9$ cm (to 3 s.f.)

EXERCISE 2.15

1 Calculate x for each of these triangles.

a
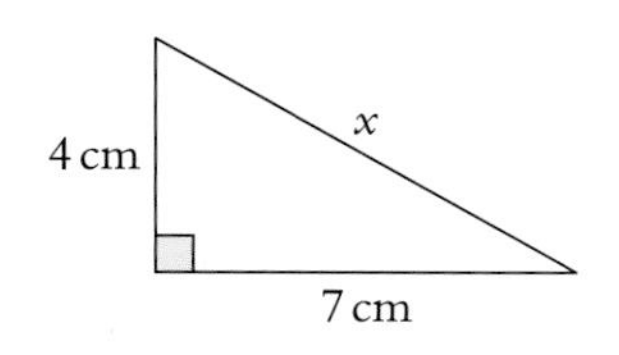

b
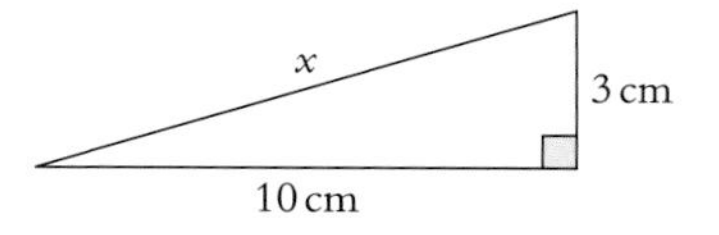

c
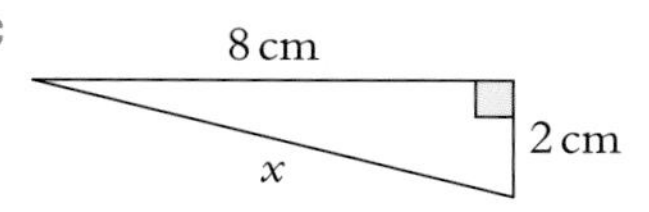

d
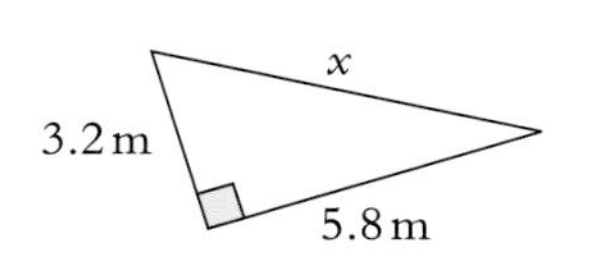

e
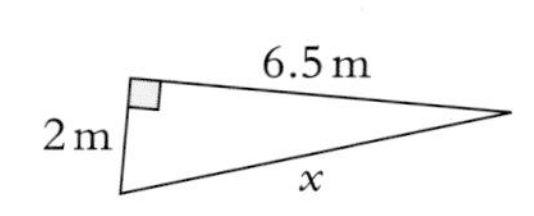

f
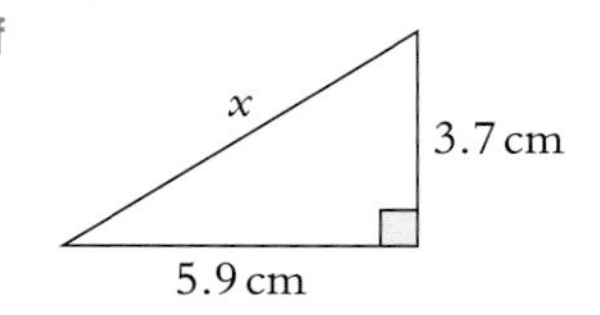

2 Manuel is trying to calculate the value of x for this triangle.
This is what he writes

$6^2 + 4^2 = x^2$

$36 + 16 = x^2$

$x^2 = 52$

$x = \sqrt{52}$

$x = 7.21$ cm

He has made a mistake. Correct his working.

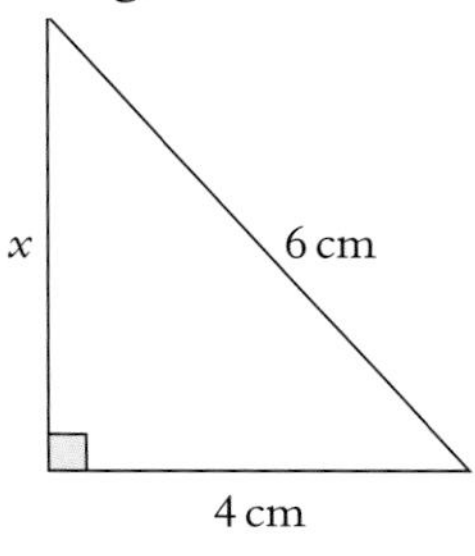

3 Calculate x for each of these triangles.

a
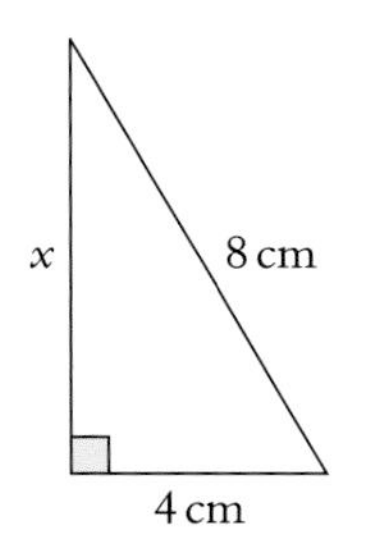

b
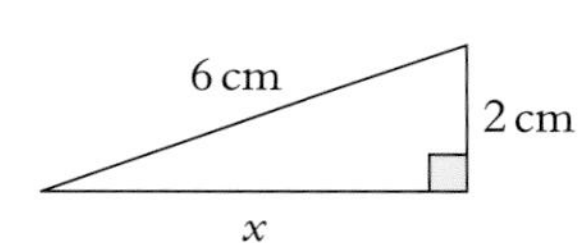

c
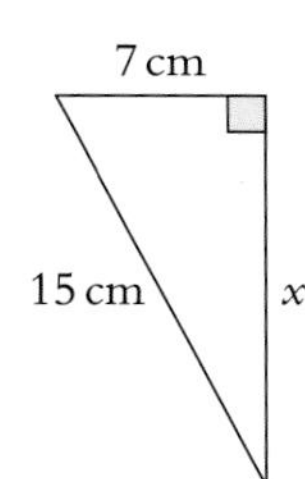

d
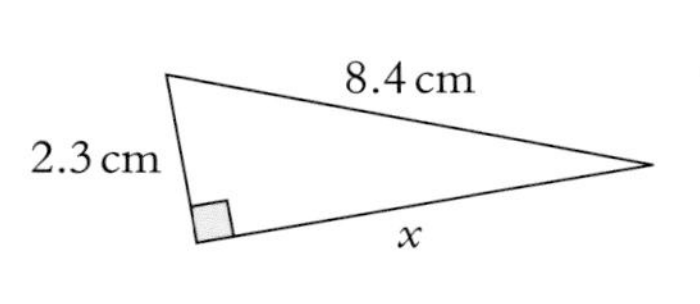

e
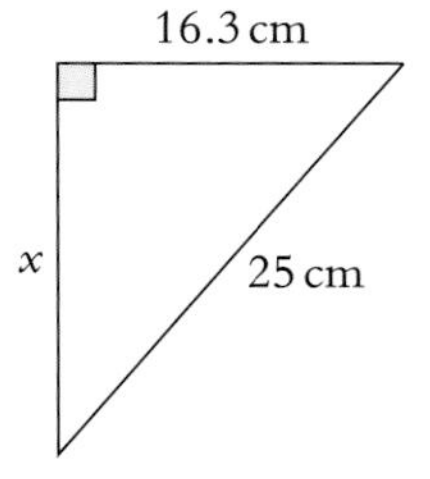

f
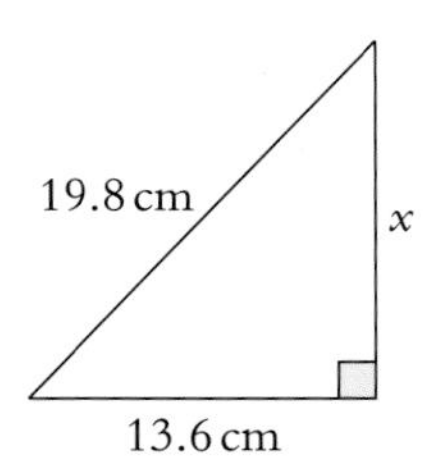

4 A boat sails 15 km due south and then 35 km due east.
How far is the boat from its starting position?

5 A ladder is 5 m long. It rests against a vertical wall with the bottom of the ladder 1.5 m from the base of the wall. How far up the wall does the ladder reach?

6 Calculate x for these right-angled triangles.

a

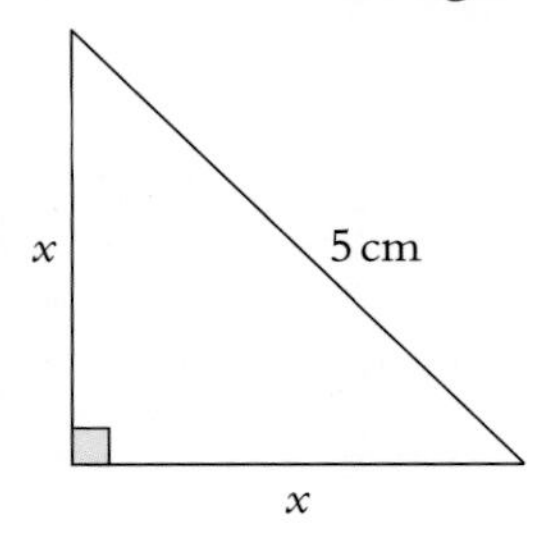

b

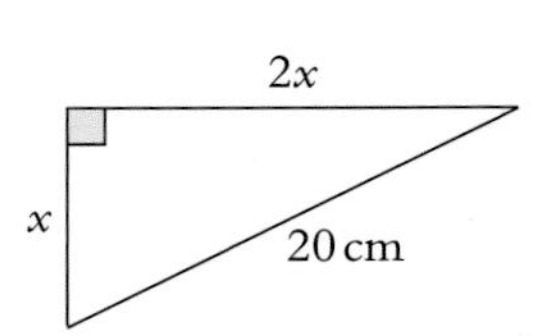

c

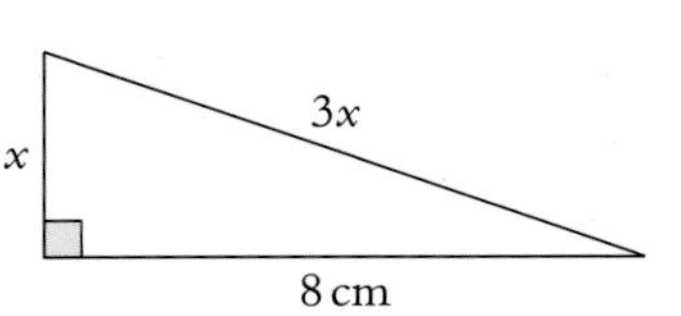

7 Calculate AD.

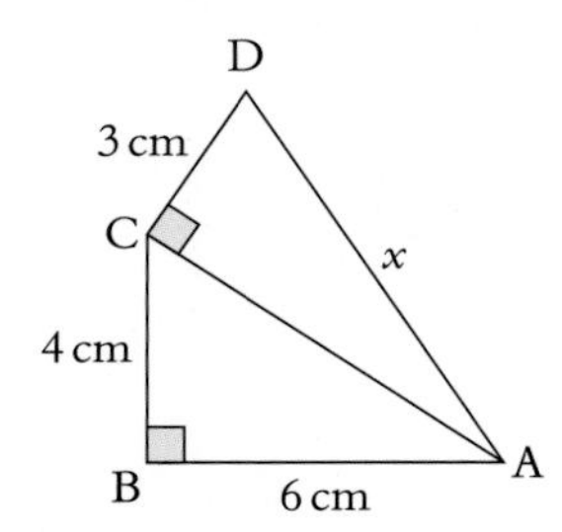

8 Calculate AB.

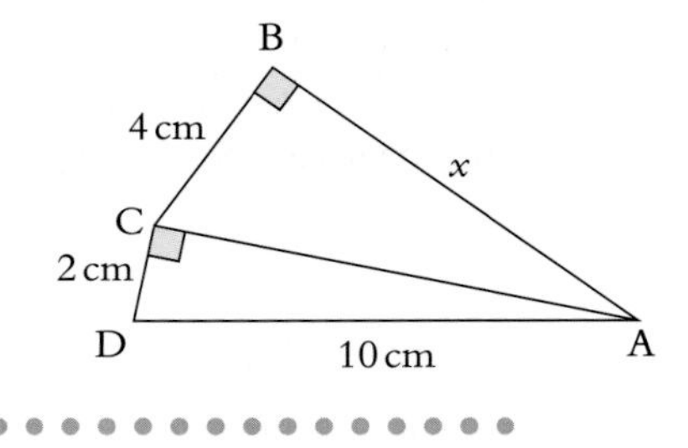

TOP TIP

Calculate the length of AC first.

9 Find the length of DB.

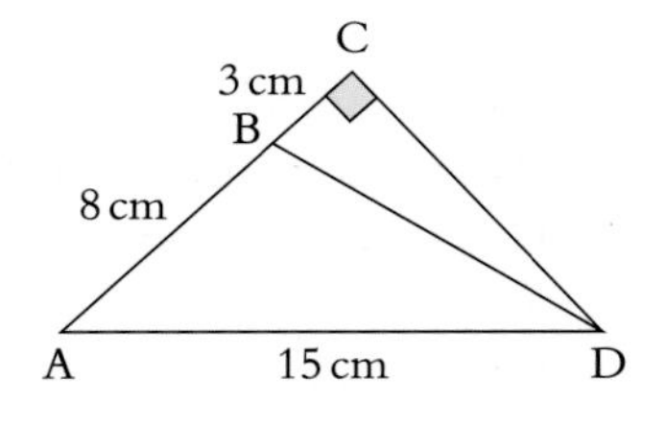

TOP TIP

Calculate the length of CD first.

10 Calculate

a the length of AD

b the perimeter of ABCD

c the area of ABCD

d the area of triangle ADC.

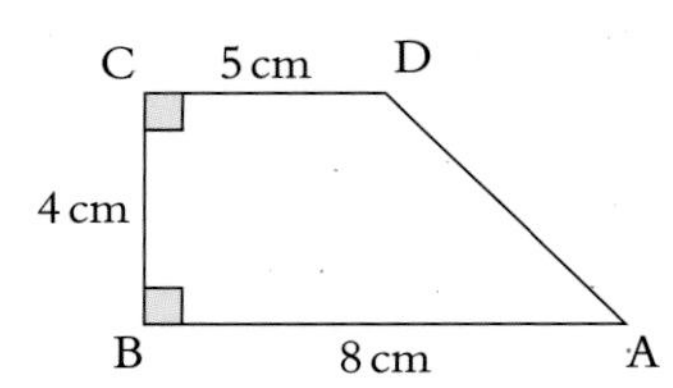

11 An equilateral triangle has sides of length 5 cm.
Calculate

a the height of the triangle

b the area of the triangle.

12 A rhombus has a perimeter of 24 cm. The longest diagonal is 10 cm. Calculate the length of the shortest diagonal.

13 Sungu says this triangle is a right-angled triangle is he correct?

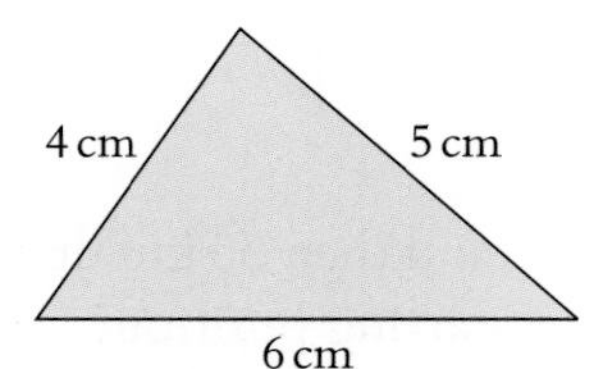

TOP TIP

Is $4^2 + 5^2 = 6^2$?

14 A triangle has sides of length 48 cm, 55 cm and 73 cm.
Is it a right-angled triangle? Explain your answer.

15 Suri travels 5 miles west, then 4 miles north, then 8 miles east.
How far is she from her starting point?

16 The circle has radius 4 cm.
The vertices of the rectangle lie on the circumference of the circle.
The rectangle has width 6 cm.
Calculate the height of the rectangle.

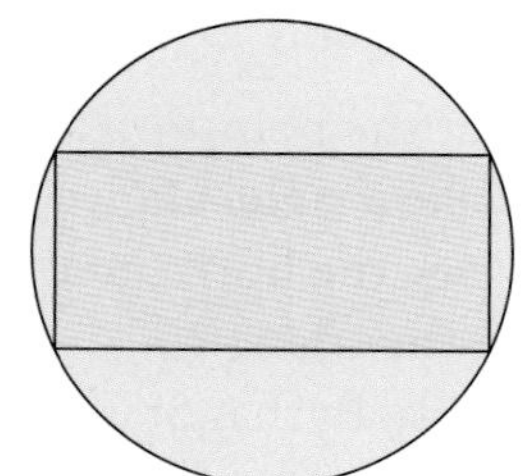

17 A hollow circular cylinder is used for storing thin metal rods.
The cylinder has a height of 15 cm and a base radius 4 cm.
Calculate the length of the longest rod that will fit exactly inside the cylinder.

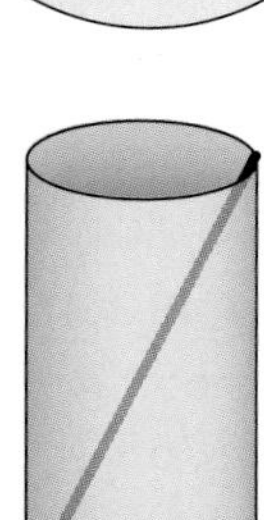

18 The perpendicular height of a circular cone is twice the length of the diameter of the circular base.
The slant height is 10 cm.
Calculate the diameter of the circular base.

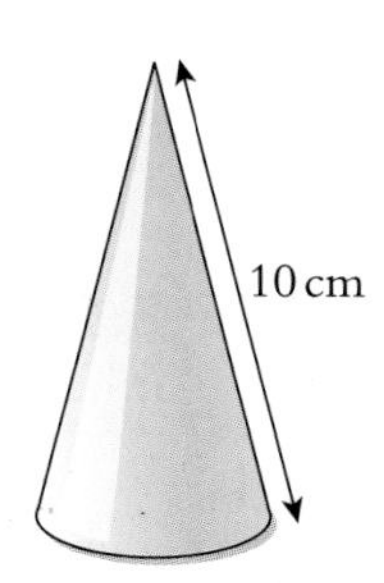

19 The diagram shows a square with sides of length 8 cm.
Inside the square there are four touching circles of radius 2 cm and a smaller circle of radius r cm.
Calculate the radius r.

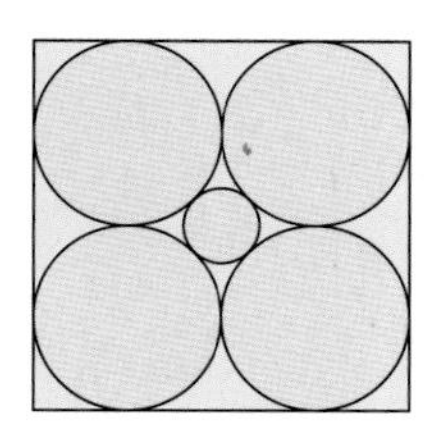

UNIT 2

Distance between two points on a graph

You can use Pythagoras' theorem to find the length of a line joining two points on a graph.

To do this you

- Plot the points on a graph
- Make a right-angled triangle
- Mark the horizontal distance and vertical distance on the triangle
- Apply Pythagoras' theorem to the triangle.

EXAMPLE

Find the length of the line joining A (−3, 2) and B (4, 6).

$AB^2 = 7^2 + 4^2$

$AB^2 = 49 + 16$

$AB^2 = 65$

$AB = \sqrt{65}$

$AB = 8.062...$

$AB = 8.06$ (to 3 s.f.)

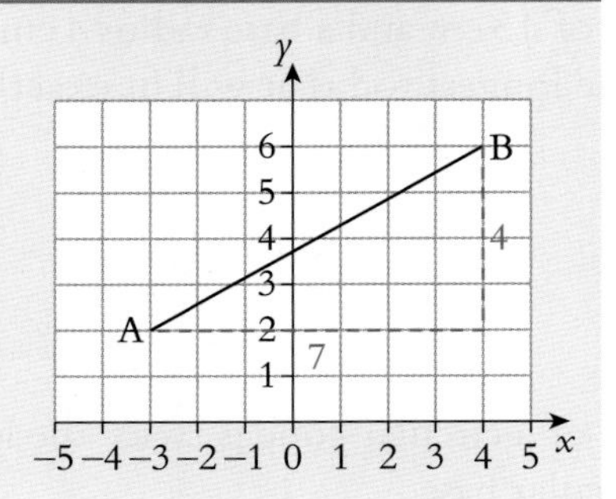

EXERCISE 2.16

1 Work out the length of the following lines.

a AB b CD

c EF d GH

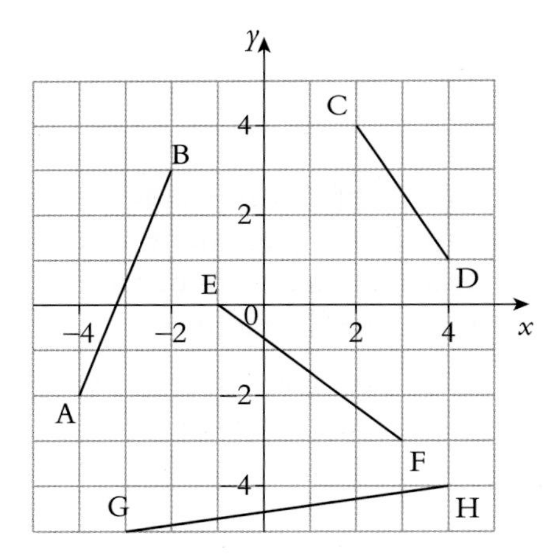

2 Calculate the length of the lines joining the following pairs of points.

a (2, −4) and (6, 3) b (5, 8) and (2, −7)

c (3, −6) and (−8, 5) d (0, −4) and (−3, 4)

e (−9, 2) and (1, −4) f (6, −2) and (−6, 5)

g (5, −1) and (−4, −4) h (−2, −2) and (−5, −1)

i (−2, −4) and (−6, −3)

Pythagoras in 3-D

To find the length of AG you need to look at triangle AGC.

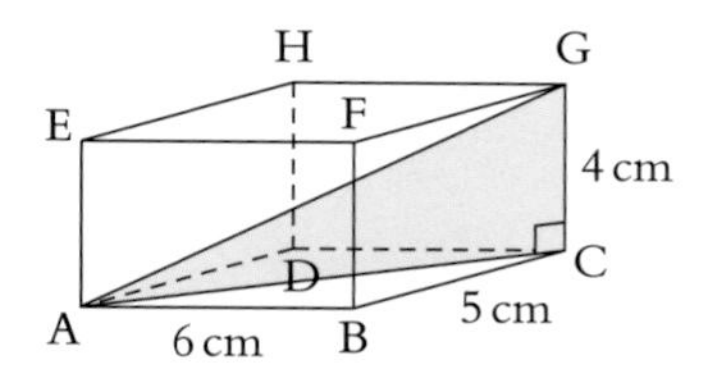

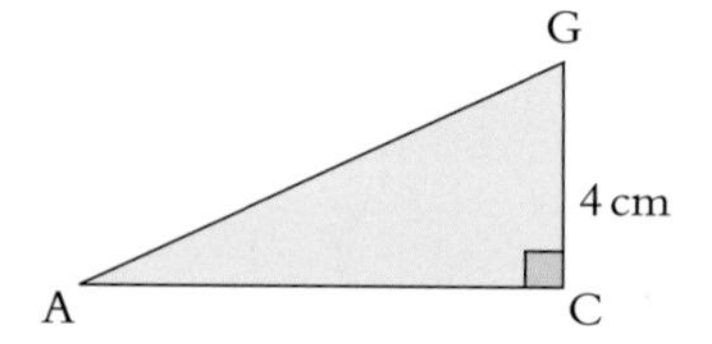

Before you can calculate AG you need to calculate AC using Pythagoras on triangle ABC.

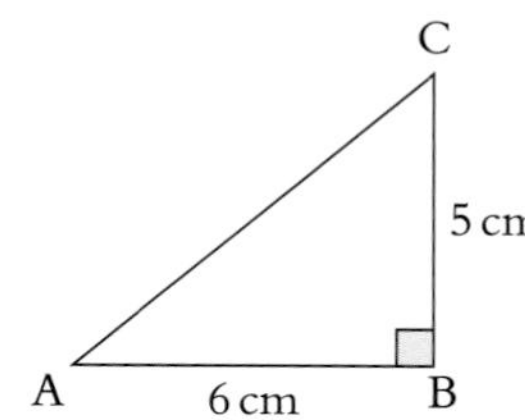

$AC^2 = 6^2 + 5^2$
$AC^2 = 36 + 25$
$AC^2 = 61$
$AC = \sqrt{61}$

Now use Pythagoras on triangle ACG.
$AC^2 + 4^2 = AG^2$
$61 + 16 = AG^2$
$AG^2 = 77$
$AG = 8.77$ cm (to 3 s.f.)

It is useful to remember that AG can be calculated directly using

$AG^2 = 6^2 + 5^2 + 4^2$
$AG^2 = 36 + 25 + 16$
$AG^2 = 77$
$AG = 8.77$ cm

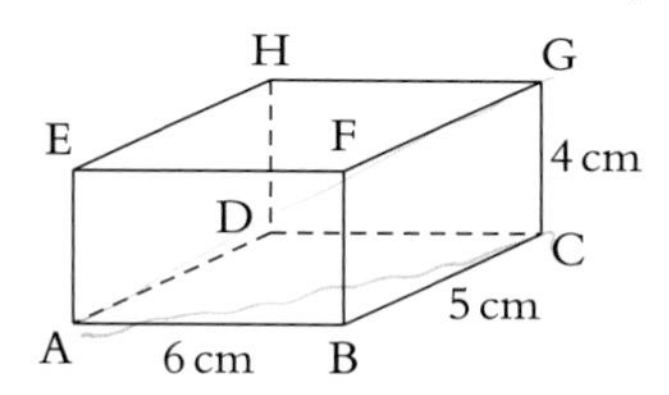

EXERCISE 2.17

1 Find the length of AG for each of these cuboids.

a
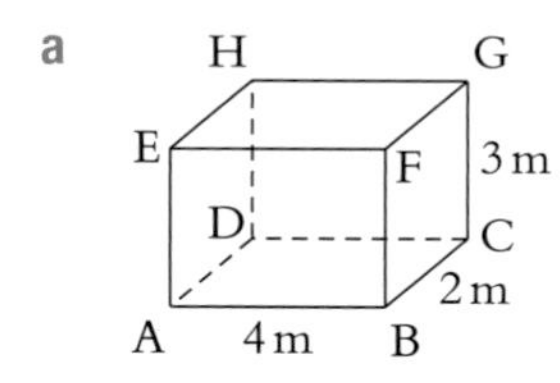

b
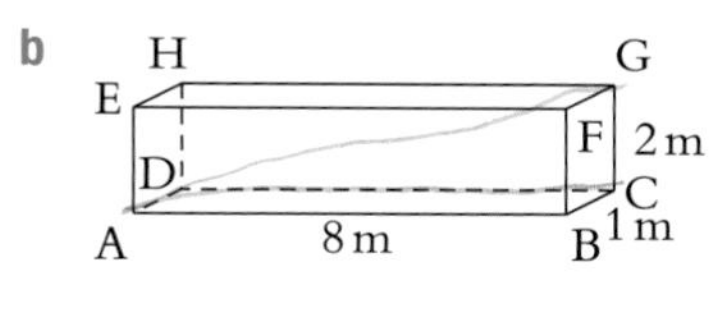

c
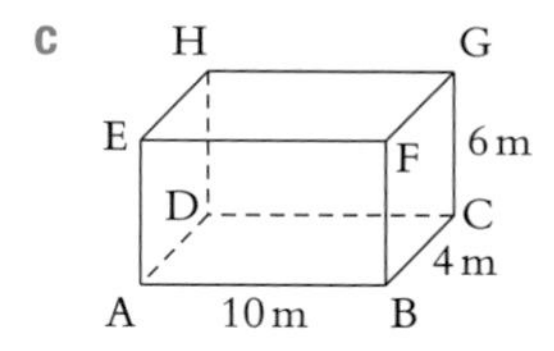

d
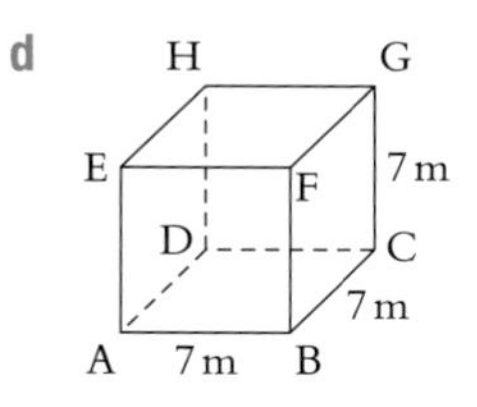

2 Find the length of AC in this triangular prism.

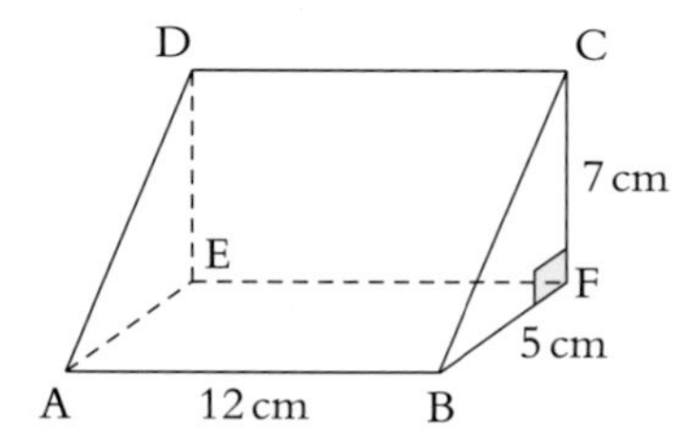

3 ABCDEFGH is a cuboid.
P is the midpoint of GC.
Calculate the length of AP.

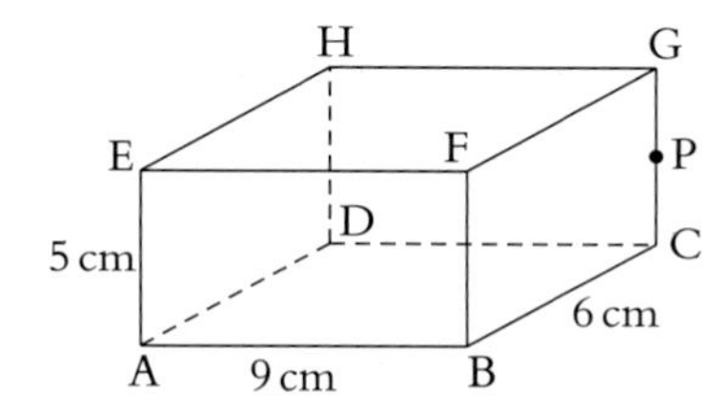

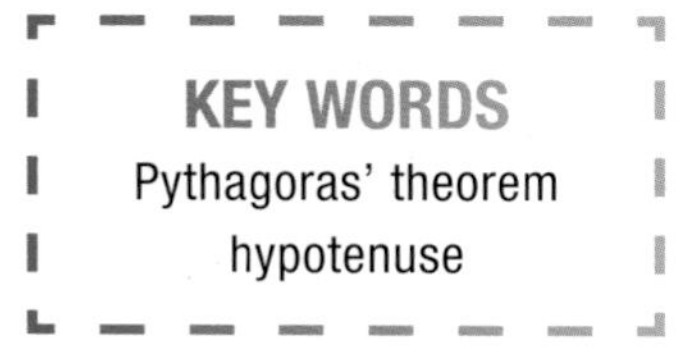

Geometrical constructions

THIS SECTION WILL SHOW YOU HOW TO

- Use a straight edge and compasses to construct a triangle given the lengths of the sides
- Use a straight edge and compasses to construct angle bisectors and perpendicular bisectors

EXAMPLE

Construct triangle ABC in which AB = 6.5 cm, AC = 3.3 cm and BC = 4.5 cm.

- Draw AB = 6.5 cm
- With your compass point on A, draw an arc radius 3.3 cm
- With your compass point on B, draw an arc radius 4.5 cm
- Label the point of intersection of the two arcs as C
- Draw the lines AC and BC

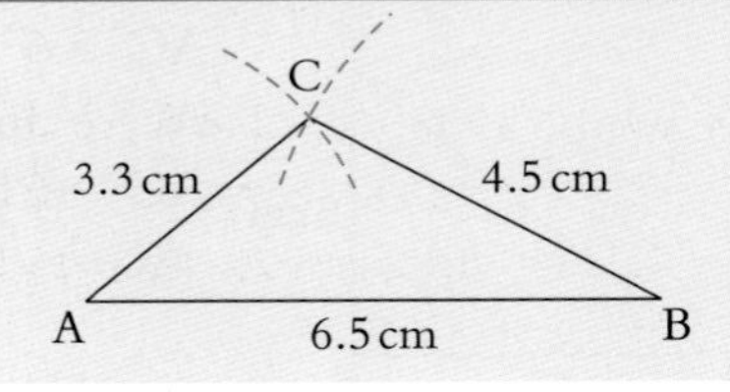

EXAMPLE

Construct an angle of 60° without using a protractor.

- Draw a base line AB
- With your compass point on A, draw an arc to intersect AB at X
- With your compass point on X, draw an arc with the same radius to intersect the first arc at Y
- Draw the line AY Angle BAY = 60°

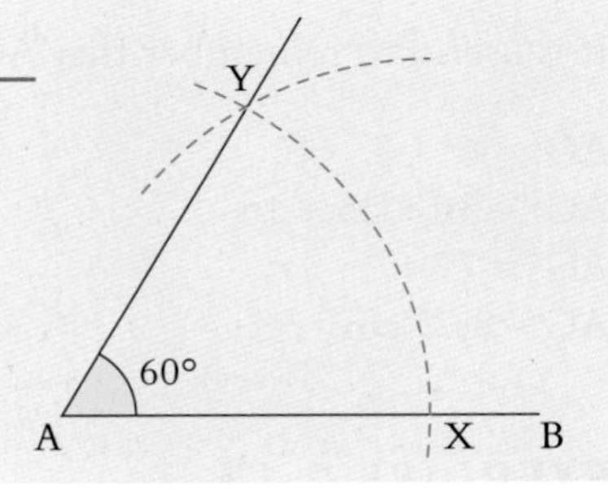

EXAMPLE

Constuct the bisector of angle ABC.

- With your compass point on B, draw an arc to intersect the lines BA and BC at P and Q
- With the same radius, draw arcs from P and Q to intersect each other at R
- Label the point of intersection of the two arcs as R
- Draw the line BR
- Angle ABR = Angle CBR

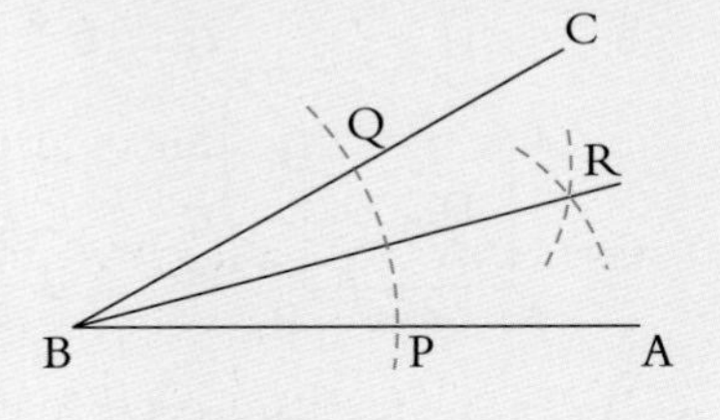

EXAMPLE

Construct the perpendicular bisector of the line AB.

- With your compass point on A, draw arcs (with the same radius) above and below the line AB
- Using the same radius and your compass point on B, draw arcs above and below the line AB, so that they intersect the first arcs (at P and Q)
- Draw the line PQ
- PQ is the perpendicular bisector of AB

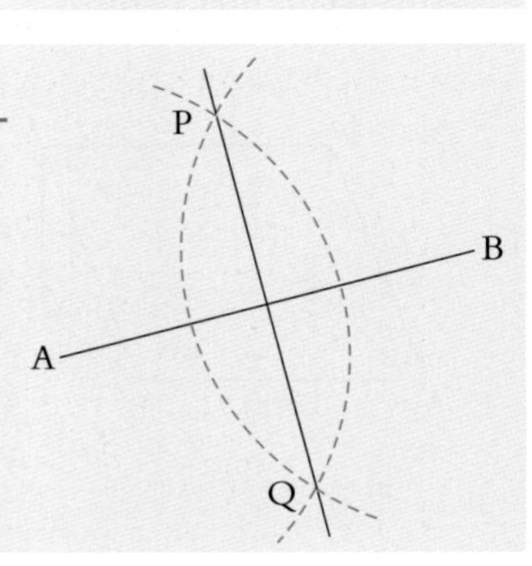

EXERCISE 2.18

Use a straight edge and compasses for each of these constructions.

1 Draw any *acute* angle. Construct the angle bisector.
Check your results with a protractor.

2 Draw any *obtuse* angle. Construct the angle bisector.
Check your results with a protractor.

3 Constuct an angle of 60°. Bisect your 60° angle to make two angles of 30°.

4 Draw a line AB that is 10 cm long.
Construct the perpendicular bisector of your line AB.
Check your results with a protractor and ruler.

5 Plot the points A(2, 10) and B(8, 1) on a pair of axes.
Construct the perpendicular bisector of the line joining A to B.
What are the coordinates of the midpoint of the line AB?

6 Construct an equilateral triangle with sides of length 8 cm.

7 Construct a rhombus whose sides are of length 8 cm
and whose shorter diagonal is 5 cm.
What is the length of the longer diagonal?

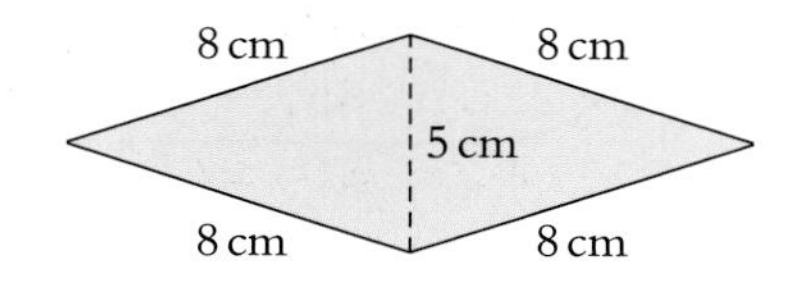

8 Construct a regular hexagon where the length of the line AD is 8 cm.

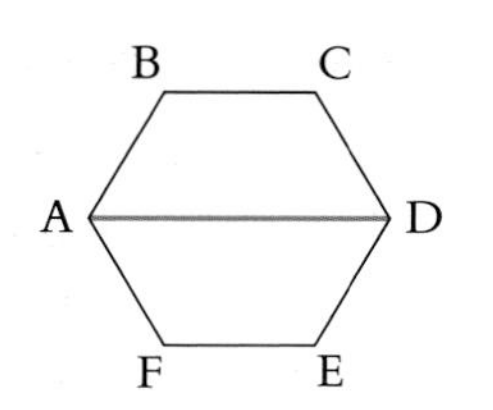

9 Construct a parallelogram with diagonals of length 9 cm
and 7 cm that intersect at 60°.
What is the length of the longest side of the parallelogram?

10 Construct a circle that passes through each of the points A, B and C.
What are the coordinates of the centre of the circle?

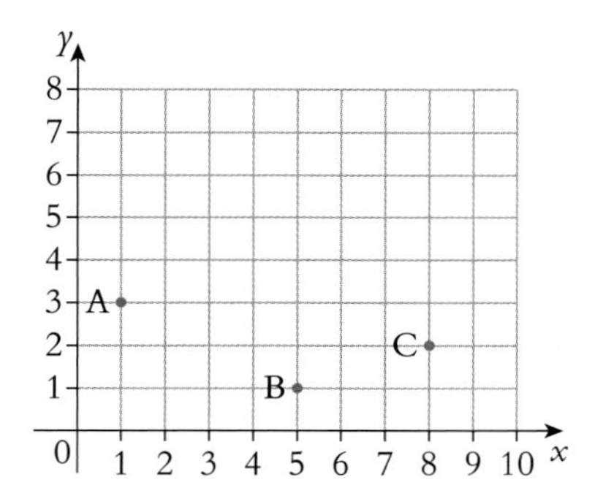

TOP TIP

Construct the perpendicular bisectors of AB and BC.

KEY WORDS

construct

bisector

Loci

THIS SECTION WILL SHOW YOU HOW TO

- Find a set of points that satisfy a given rule

A **locus** is a set of points that satisfy a given rule.
It can be a line, a curve or a region.
Loci is the plural of locus.
These examples show some common loci.

EXAMPLE

Find the locus of points that are 1 cm from the point A.

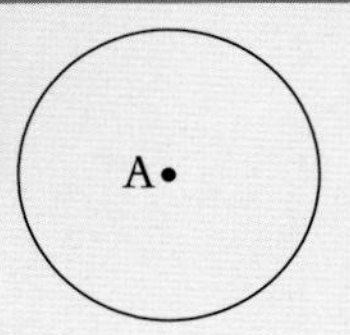

The locus is a circle of radius 1 cm.

EXAMPLE

Find the locus of points that are 1 cm from the line AB.

The locus is made from two straight lines and two semicircles.

EXAMPLE

Find the locus of points that are the same distance from the point A as they are from the point B.

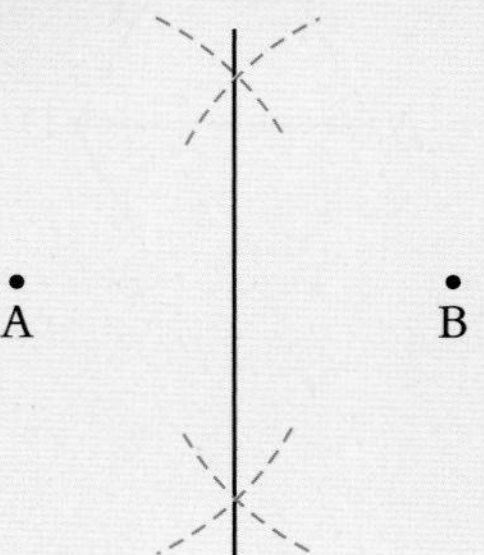

➡ **NOTE:** equidistant can also be used to mean 'the same distance'.

The locus is the perpendicular bisector of the line AB.

EXAMPLE

Find the locus of points that are the same distance from the line AB as they are from the line AC.

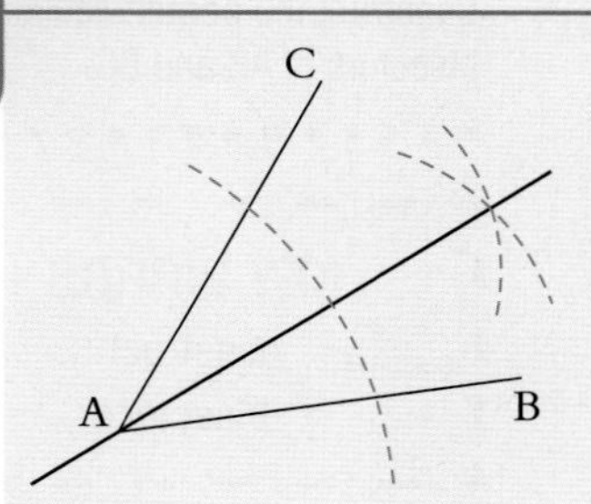

The locus is the bisector of angle BAC.

EXERCISE 2.20

1 Find the circumference and area of these circles. All lengths are in cm.

a
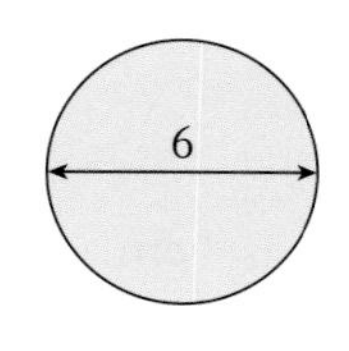

b
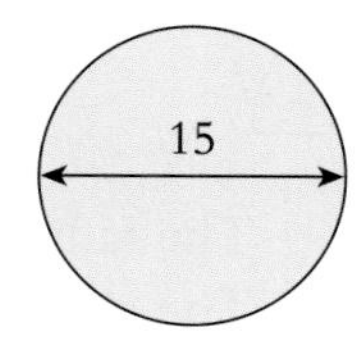

c
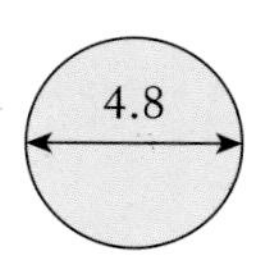

d
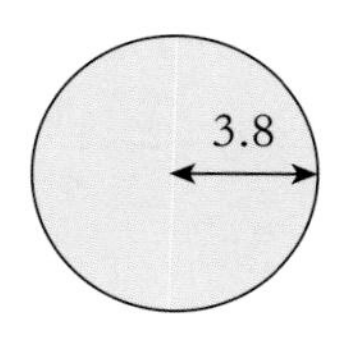

e
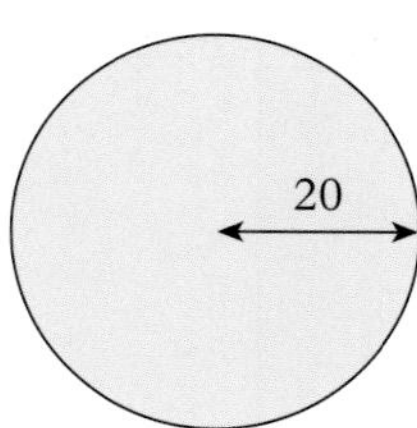

f
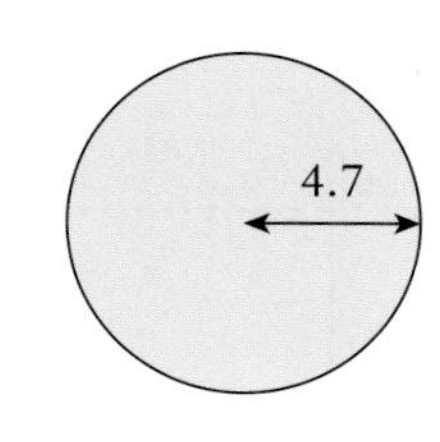

2 The wheels on a car have a diameter of 52 cm.

a Calculate the distance travelled by the car when the wheels rotate once.

b How many km does the car travel if the wheels rotate 5000 times?

3 A circular table has an area of 1.25 m^2.
Calculate **a** the radius **b** the circumference of the table.

4 The length of the sides of each square is 12 cm.
The circles fit tightly inside the square.
Calculate the shaded area for each diagram.

a
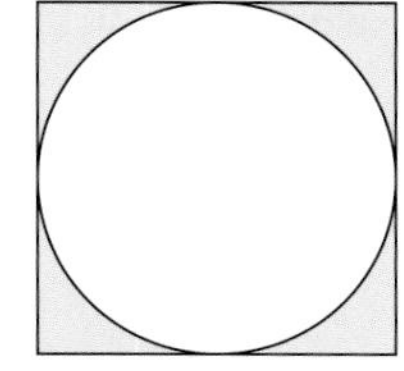

b
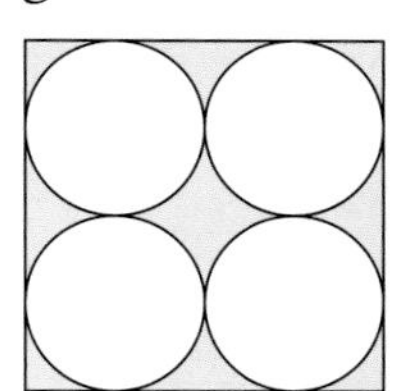

c
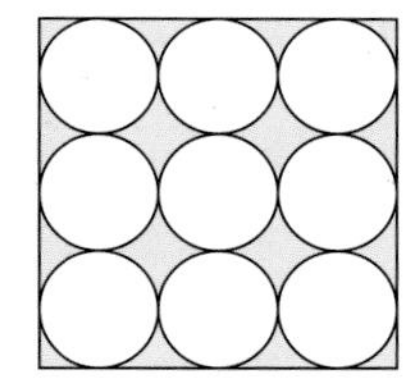

d What do you notice about your answers? Can you explain why?

5 The radius of the large circle is 5 cm.
The radius of the small circle is 3 cm.
Calculate the shaded area.

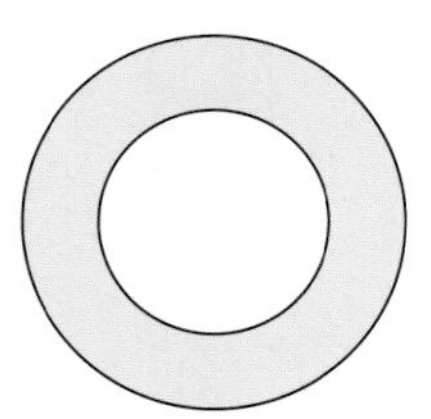

6 The radius of the circle is 10 cm.
The vertices of the square lie on the circumference of the circle.
Calculate the shaded area.

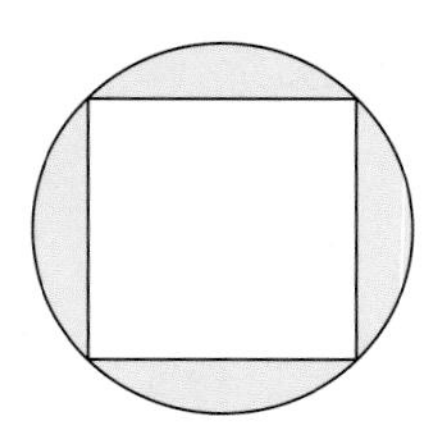

Semicircles and quadrants

A **semicircle** is half a circle. A **quadrant** is quarter of a circle.

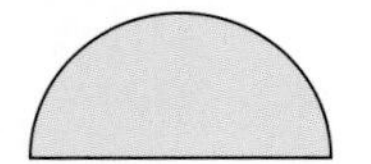

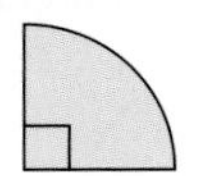

EXAMPLE

Find the perimeter and area of the shape.

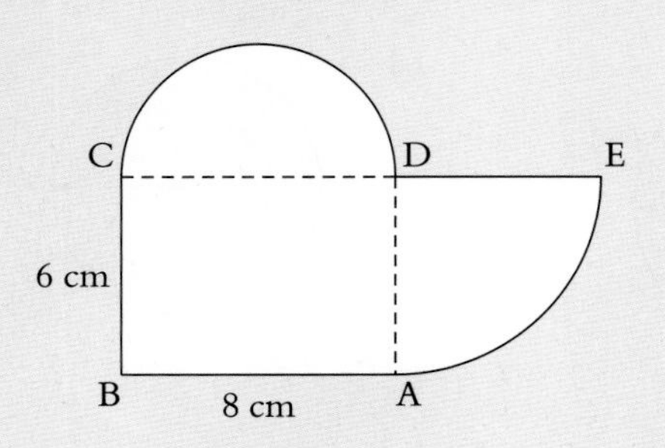

The radius of the semicircle is 4 cm.

The radius of the quadrant is 6 cm.

Perimeter = AB + BC + arc CD + DE + arc EA

$= 8 + 6 + \frac{2 \times \pi \times 4}{2} + 6 + \frac{2 \times \pi \times 6}{4}$

$= 8 + 6 + 4\pi + 6 + 3\pi$

$= 41.99114\ldots$

$= 42.0\,\text{cm}$ (to 3 s.f.)

Area = area of rectangle + area of semicircle + area of quadrant

$= (8 \times 6) + \frac{\pi \times 4^2}{2} + \frac{\pi \times 6^2}{4}$

$= 48 + 8\pi + 9\pi$

$= 101.407\ldots$

$= 101\,\text{cm}^2$ (to 3 s.f.)

EXERCISE 2.21

Calculate the perimeter and area of each of these shaded shapes. All lengths are in cm.

1

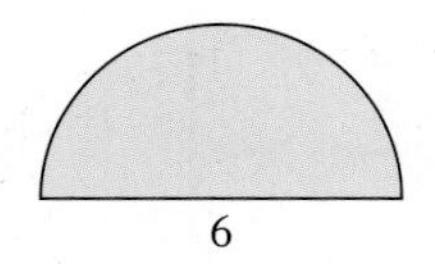

2

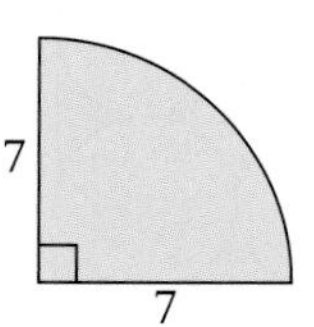

3

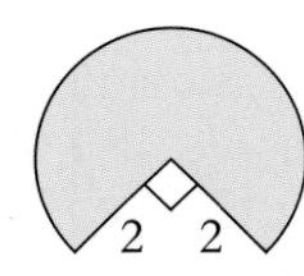

4

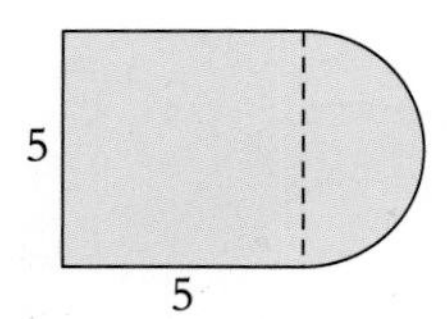

5

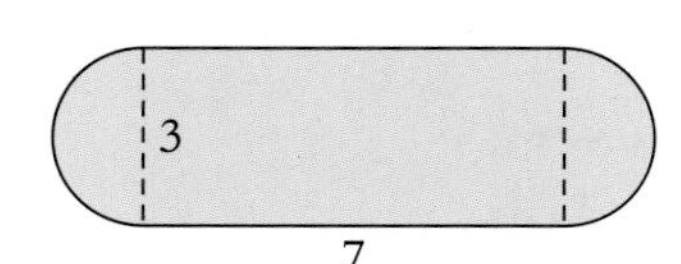

6

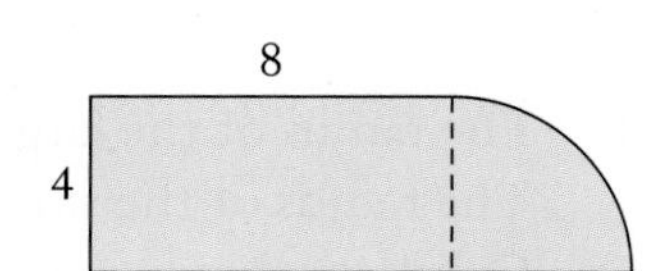

7

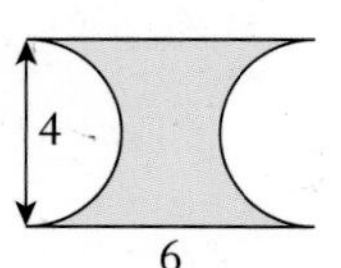

8

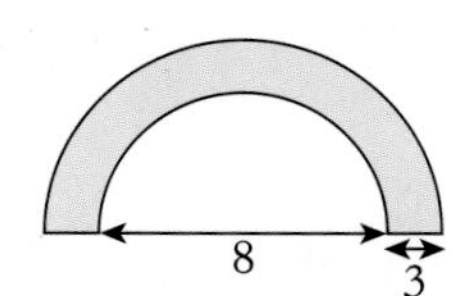

9

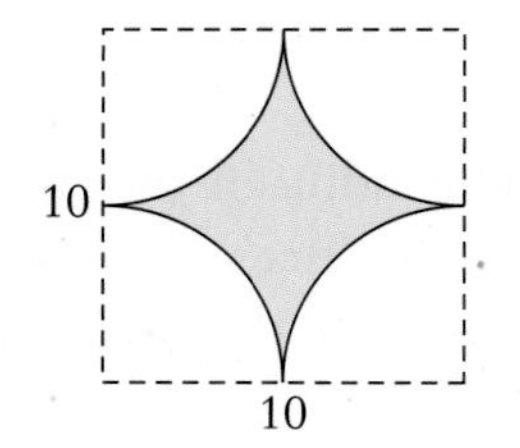

10

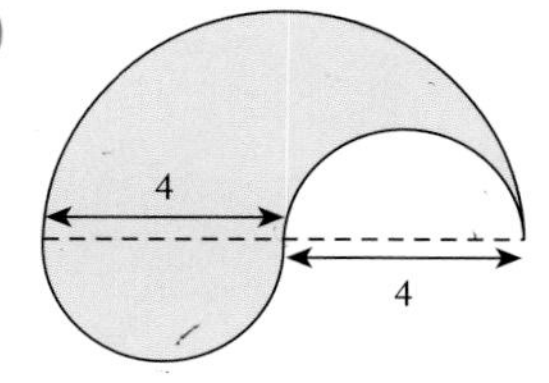

11

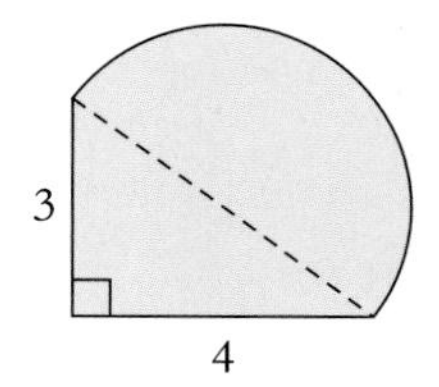

12 The diagram shows four rods of radius 2 cm.
The rods are held tightly together by a piece of wire.
Calculate the length of the wire.

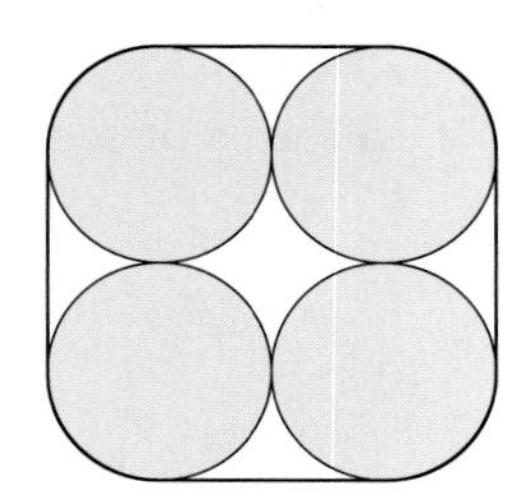

13 **a** The area of a semicircle is 6 cm^2. Find the radius and perimeter.
b The area of a semicircle is 24 cm^2. Find the radius and perimeter.
c Compare your answers to parts **a** and **b**. What do you notice?

14 **a** The area of a quadrant is 5 cm^2. Find the radius and perimeter.
b The area of a quadrant is 45 cm^2. Find the radius and perimeter.
c Compare your answers to parts **a** and **b**. What do you notice?

15 The perimeter of a quadrant is 24 cm.
Calculate the area of the quadrant.

16 This shape is made from two identical quadrants.
The perimeter of the shape is 40 cm.
Calculate the radius of the quadrants.

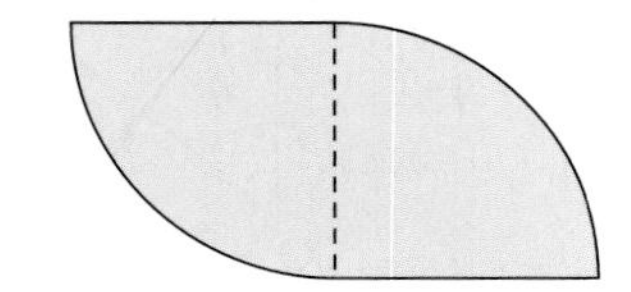

17 A circle has a circumference of x cm and an area of $2x\text{ cm}^2$.
Calculate the radius of the circle.

Arcs and sectors

An **arc** is part of the circumference of a circle.
A **sector** of a circle is a region bounded by an arc and two radii.

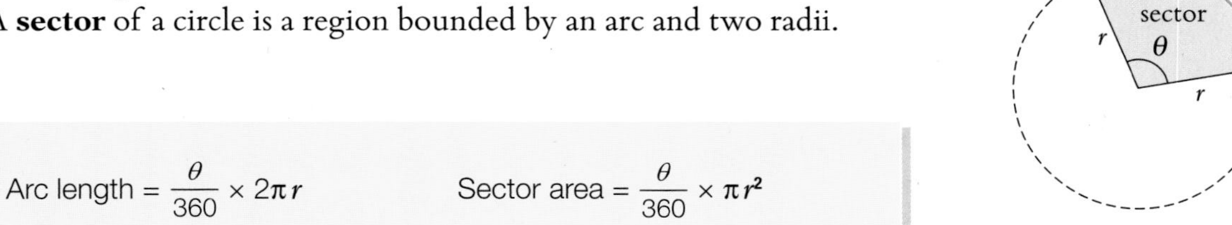

EXAMPLE

Find

a the area of the sector OAB

b the length of the arc AB

c the perimeter of the sector AOB.

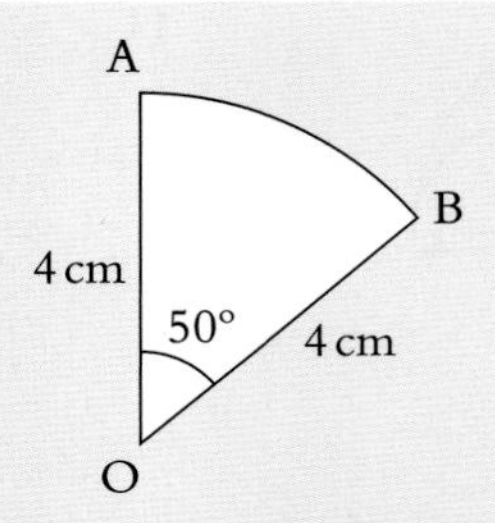

a area of sector OAB $= \frac{50}{360} \times \pi r^2$

$= \frac{50}{360} \times \pi \times 4^2$

$= 6.98\text{ cm}^2$ (to 3 s.f.)

b arc length AB $= \frac{50}{360} \times 2\pi r$

$= \frac{50}{360} \times 2 \times \pi \times 4$

$= 3.49\text{ cm}$ (to 3 s.f.)

c perimeter of sector OAB = OA + OB + arc AB

$= 4 + 4 + 3.49$

$= 11.5\text{ cm}$ (to 3 s.f.)

EXAMPLE

a Arc AB = 7 cm.
Find the value of θ

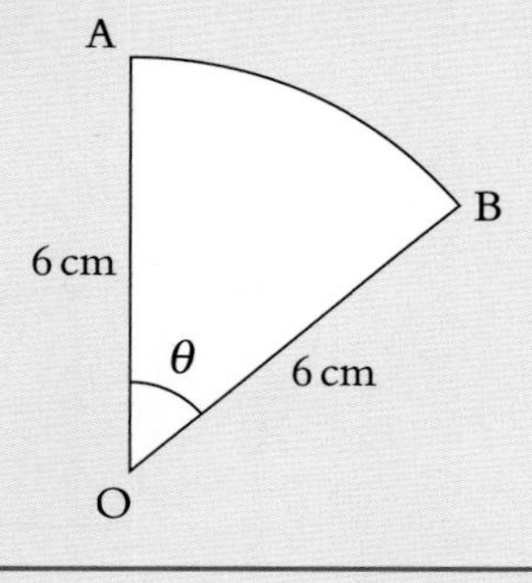

b Area of sector AOB = 50 cm².
Find the value of r.

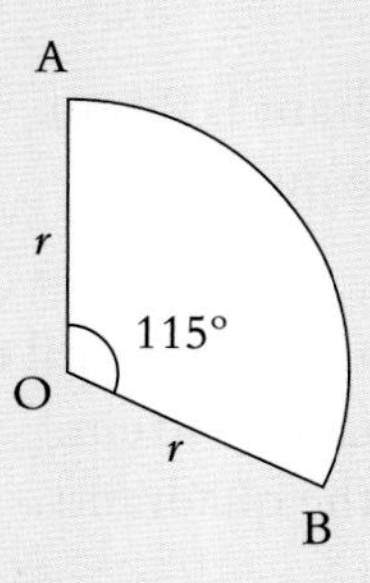

a Arc length $= \frac{\theta}{360} \times 2\pi r$

$7 = \frac{\theta}{360} \times 2 \times \pi \times r$

$7 = \frac{\theta}{360} \times 12\pi$

$\theta = \frac{7 \times 360}{12\pi}$

$\theta = 66.8°$ (to 3 s.f.)

b Area of sector $= \frac{\theta}{360} \times \pi r^2$

$50 = \frac{115}{360} \times \pi r^2$

$50 \times 360 = 115\pi \times r^2$

$r = \sqrt{\frac{50 \times 360}{115\pi}}$

$r = 7.06\text{ cm}$ (to 3 s.f.)

EXERCISE 2.22

1 Find the length of the arc AB for each of these.

a

b

c

d

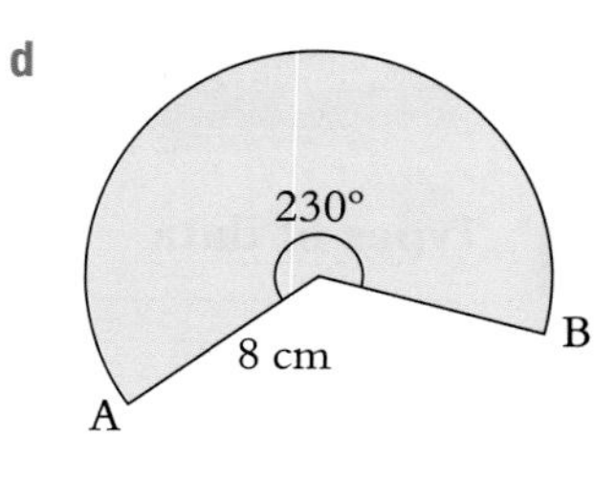

2 Find the area of the sector for each of these.

a

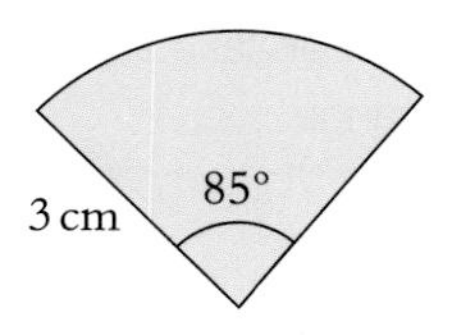

b

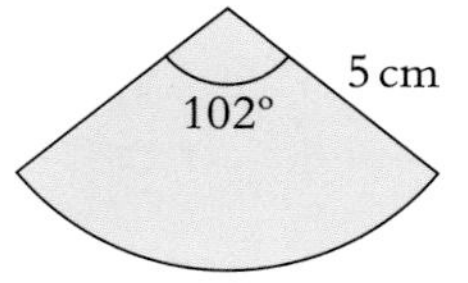

c

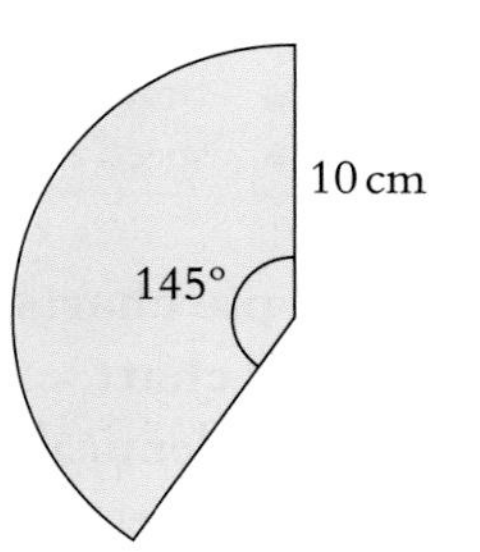

d

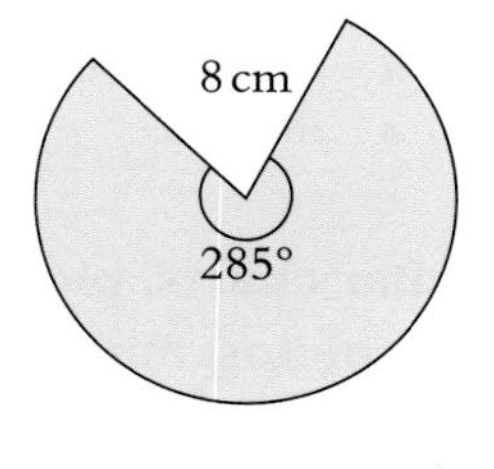

3 Find the perimeter and area of these shapes.

a

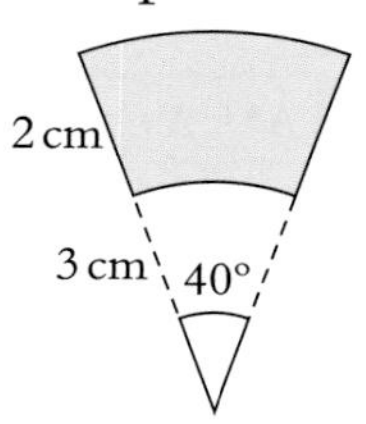

b

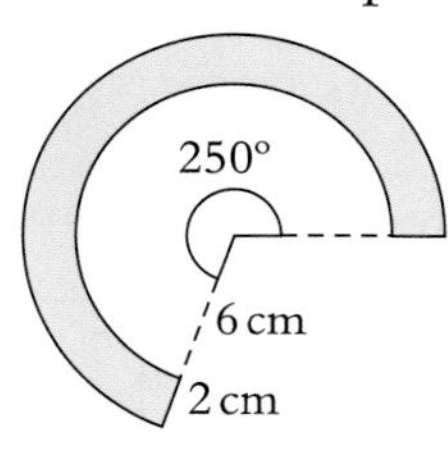

c

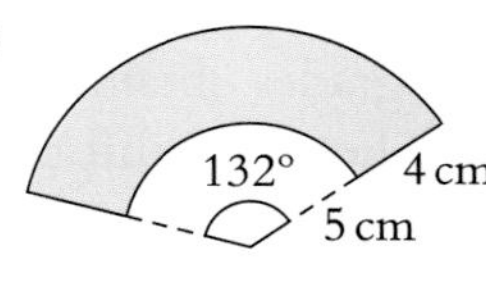

d

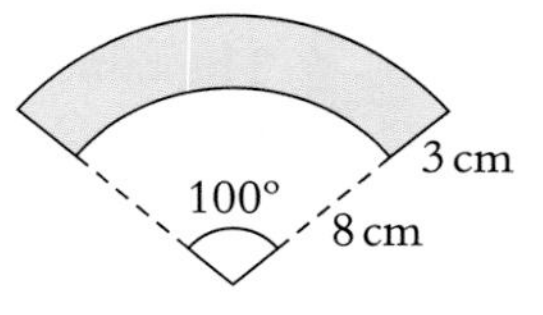

4 The area of the sector is $6\,\text{cm}^2$.
Calculate the value of θ.

5

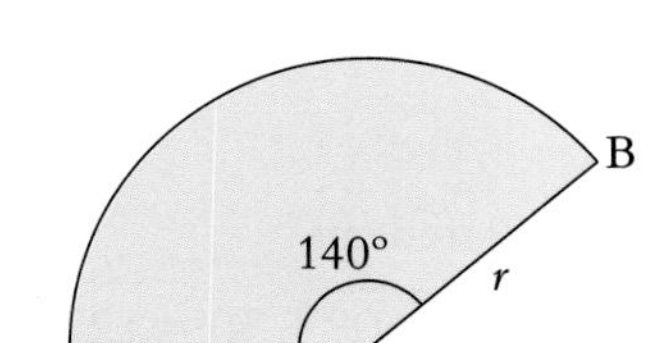

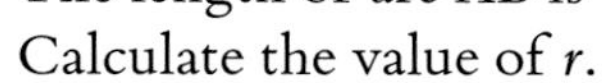

The length of arc AB is 10 cm.
Calculate the value of r.

6 This shape is made from a square of length 5 cm and four identical sectors of a circle of radius 5 cm and angle 60°.
Calculate **a** the perimeter
b the area of the shape.

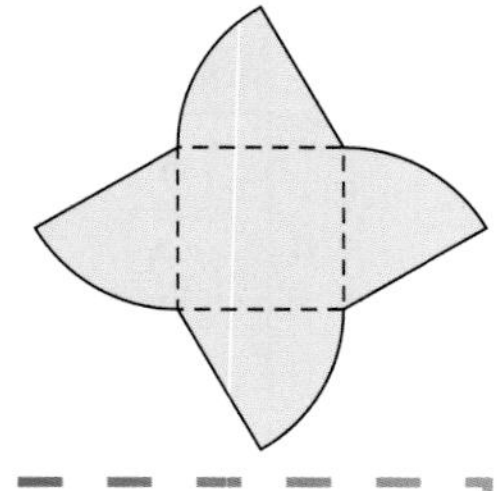

7 The diagram shows three identical coins of radius 1 cm.
Calculate the shaded area in the middle of the diagram.

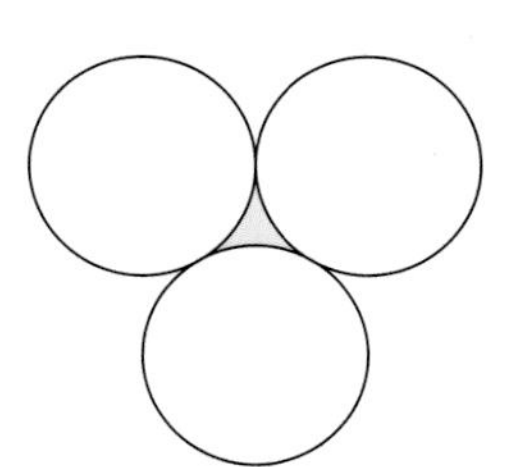

KEY WORDS
circumference
diameter, radius
semicircle, quadrant
arc, sector

Displaying data

THIS SECTION WILL SHOW YOU HOW TO

- Display data and interpret data

Types of data

Discrete Data	Continuous Data
• marks in a test • number of people in a boat • number of coins • shoe size	• height of students • mass of sand • time to complete a puzzle • speed of a car

Discrete data can only have exact numbers.

Continuous data can take any value in a particular range.

Bar charts, pictograms and pie charts

A **pictogram** is similar to a **bar chart**, except that the frequencies are represented by a number of identical pictures.

In a **pie chart** the frequency is represented by the angle of the sector of a circle.

EXAMPLE

The table shows the grades obtained by 30 students in an examination. Show this information on

a a bar chart **b** a pictogram **c** a pie chart.

Grade	A*	A	B	C	D
Number of students	3	12	10	3	2

a

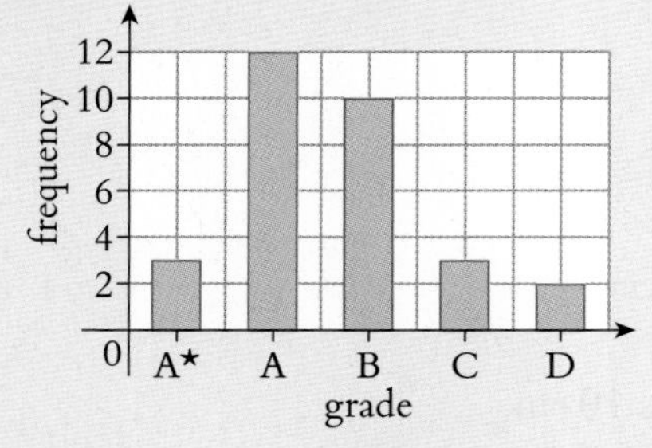

b

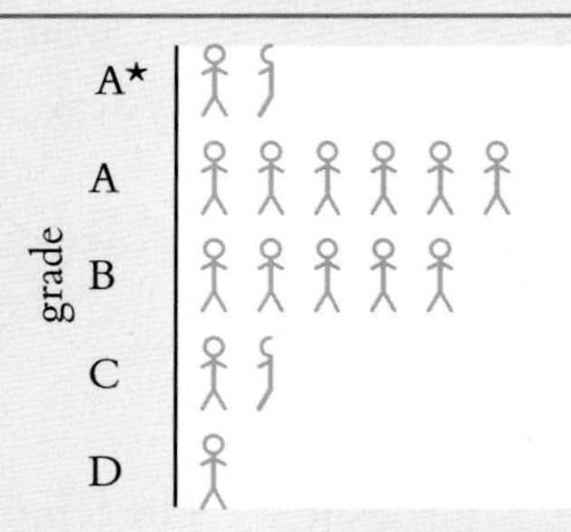

Key: represents 2 students

c The pie chart represents 30 students. Each student is represented by $360° \div 30 = 12°$

Grade	Frequency	Angle
A*	3	$3 \times 12° = 36°$
A	12	$12 \times 12° = 144°$
B	10	$10 \times 12° = 120°$
C	3	$3 \times 12° = 36°$
D	2	$2 \times 12° = 24°$
		360°

Examination grades

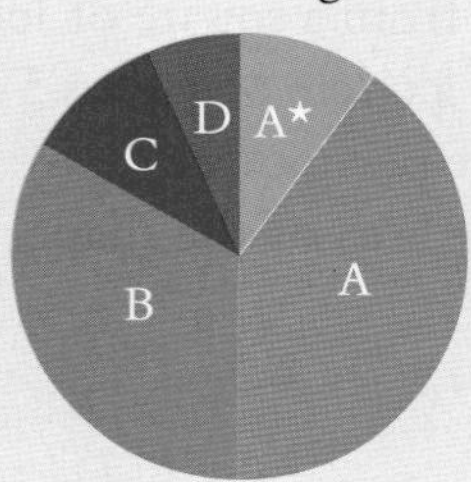

Displaying continuous data

Continuous data is often put into a grouped frequency table with equal class widths. A **frequency diagram** can then be drawn to represent the data.

EXAMPLE

2.83	3.22	3.79	2.95	3.88	3.08	4.23	1.74	3.47	2.77
3.56	2.03	3.14	4.11	2.51	2.38	3.98	3.37	3.66	3.33

The table shows the mass (m), in kilograms of 20 newborn babies.

a Arrange the data into a grouped frequency table using class intervals of $1.5 < m \leq 2.0$, $2.0 < m \leq 2.5$, $2.5 < m \leq 3.0$, $3.0 < m \leq 3.5$, $3.5 < m \leq 4.0$ and $4.0 < m \leq 4.5$.

b Use the grouped frequency table in part **a** to draw a frequency diagram to represent the data.

a

mass (m) kg	tally	frequency
$1.5 < m \leq 2.0$	|	1
$2.0 < m \leq 2.5$	||	2
$2.5 < m \leq 3.0$	||||	4
$3.0 < m \leq 3.5$	𝍸 |	6
$3.5 < m \leq 4.0$	𝍸	5
$4.0 < m \leq 4.5$	||	2

b

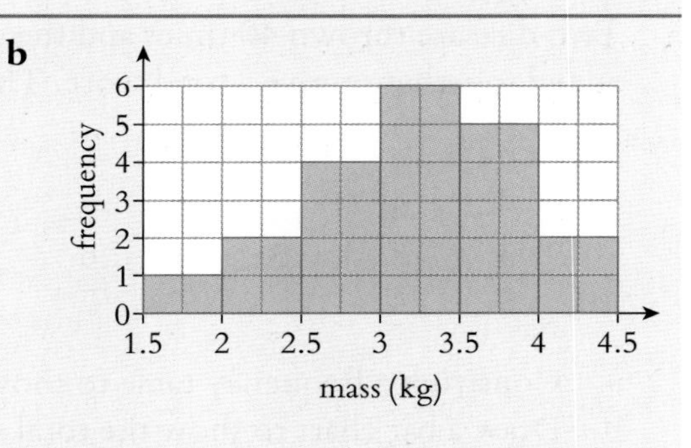

➡ **NOTE:** for continuous data

- there are no gaps between the bars
- the horizontal axis is written as a continuous number line

EXERCISE 2.23

1 The table shows the colour of 60 students' mobile phones.

black	silver	blue	other
25	20	10	5

Show this information on

a a bar chart b a pictogram c a pie chart.

2 The table shows the sales of four different flavour packs of crisps.

plain	vinegar	chicken	cheese
16	8	12	4

Show this information on

a a bar chart b a pictogram c a pie chart.

3 The table shows how 50 adults travel to work.

walk	cycle	bus	car	train
15	5	10	18	2

Show this information on

a a bar chart b a pictogram c a pie chart.

4 The pie chart shows the ingredients used to make a cake. The cake mixture has a total mass of 900 g. Measure each of the angles on the pie chart and then calculate the mass of each of the ingredients used to make the cake.

5 Two dice are thrown 40 times and the scores on each dice are added together to give a total score. These are the results.

5	7	4	8	10	6	7	3	9	6
6	9	8	7	5	2	11	10	11	9
7	3	8	5	12	6	7	4	8	6
10	6	4	11	7	7	8	5	9	4

a Construct a frequency table to show the total scores.

b Draw a bar chart to show the total scores.

6 The table shows information about the ages of people in a swimming pool.

a Draw a frequency diagram to show the information in the table.

b Find the percentage of people that are over 30 years old.

Age (x years)	Number of people
$0 < x \leq 10$	4
$10 < x \leq 20$	20
$20 < x \leq 30$	15
$30 < x \leq 40$	6
$40 < x \leq 50$	4
$50 < x \leq 60$	1

7 The table shows information about the times taken by 32 teachers to travel to work.

a Draw a frequency diagram to show the information in the table.

b Find the percentage of teachers that take more than 5 minutes.

Time (t minutes)	Frequency
$0 < t \leq 5$	8
$5 < t \leq 10$	12
$10 < t \leq 15$	6
$15 < t \leq 20$	4
$20 < t \leq 25$	2

Two-way tables

A **two-way table** shows two or more sets of data at the same time.

EXERCISE 2.24

1 The table shows information about the number of boys and girls in Class 5A who did and did not do their homework.

	Boys	Girls
Homework	13	15
No homework	4	1

a How many boys are in Class 5A?

b How many students did not do their homework?

c How many students are there altogether in Class 5A?

2 In a survey 100 students are asked if they are left-handed or right-handed.
The table shows the results.
Copy and complete the table.

	Boys	Girls	Total
Left-handed	7		**18**
Right-handed		41	
Total			**100**

3 Students in a small school must attend one lunchtime club.
They have a choice of Athletics Club, Drama Club or Music Club.
The table shows information about the number of students attending each club.

	Boys	Girls	Total
Athletics	56	42	
Drama	58	37	
Music	16		
Total			**250**

a Copy and complete the table.

b What percentage of the students attend the Music Club?

4 The table shows information about the times taken by a group of people to solve a puzzle.

		TIME (t minutes)		
		$0 < t \leq 10$	$10 < t \leq 20$	$20 < t \leq 30$
AGE (x years)	$10 < x \leq 20$	5	9	23
	$20 < x \leq 30$	15	10	7
	$30 < x \leq 40$	13	12	15
	$40 < x \leq 50$	10	15	16

a How many people are there altogether?

b How many people took more than 10 minutes?

c What percentage of the people took 10 minutes or less?

KEY WORDS
discrete data
continuous data
bar chart
pictogram
pie chart
frequency diagram
two-way table

Unit 2 Examination-style questions

1 The normal selling price of a car is $18 000.
This price is reduced by 14% in a sale.

Calculate the sale price of the car. [2]

2 At noon, the number of bacteria in a culture was 25 500.
One hour later this number had increased by 8%.

Calculate the number of bacteria at 1 p.m. [2]

3 Raju was paid $800 each week.
His pay is increased by x% to $828 each week.

Find the value of x. [3]

4 Alan and Barbara share $500 in the ratio 13 : 7.

Calculate the amount that each receives. [2]

5 Edward, Fred and George share some money in the ratio 5 : 7 : 3.
George receives $237.

Find the total amount of money that was shared. [2]

6 A map has a scale of 1 : 20 000.
A road is 16 cm long on the map.

Find the actual length, in kilometres, of the road. [2]

7 Simplify.

(a) $x^{10} \div x^2$ [1]
(b) $(x^3)^4$ [1]

8 **(a)** Work out $\left(\frac{2}{3}\right)^{-3}$. [1]

(b) Simplify.

(i) $7x^0$ [1]

(ii) $8x^5 \times 4x^{-3}$ [2]

9 $5^x = \frac{1}{625}$

Find the value of x. [1]

10 Solve the inequality.

$5x - 2 > 7x + 3$ [2]

11 $\frac{2x-5}{3} \leq \frac{x+2}{5}$

(a) Solve the inequality. [3]

(b) Using your answer to **part (a)**, write down the positive integer values of x that satisfy the inequality. [1]

12 Simplify.

$\frac{2x}{3} - \frac{x}{8}$ [2]

13 Solve.

$\frac{5-x}{6} + \frac{2x+3}{4} = 5$ [4]

14 E is the point (5, 7) and F is the point (0, −3).

Find the equation of the line EF. [3]

15 P is the point (−2, 9) and Q is the point (4, −3).

(a) Find the coordinates of the midpoint of the line PQ. [2]

(b) Calculate the length of the line PQ. [3]

(c) Find the equation of the line through P and Q.
Write your answer in the form $y = mx + c$. [3]

(d) Another line is parallel to PQ and passes through the point (4, −9).
Write down the equation of this line. [2]

16 Change 925 000 cm^2 to m^2. [1]

17 In this diagram all lengths are in centimetres.

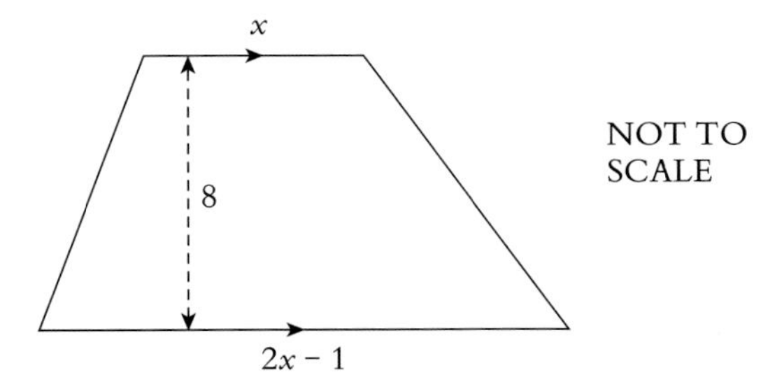

NOT TO SCALE

The trapezium has two parallel sides of length x cm and $(2x - 1)$ cm.
The perpendicular distance between the two parallel sides is 8 cm.
The area of the trapezium is 110 cm^2.

Find the value of x. [3]

18

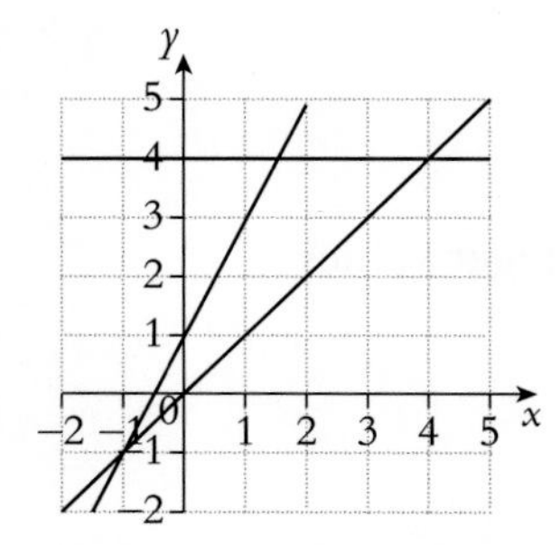

By shading the **unwanted** regions on the grid, find and label the region R that satisfies the following three inequalities.

$y \leq 4$ $\quad$ $y \leq 2x + 1$ $\quad$ $y \leq x$ [3]

19

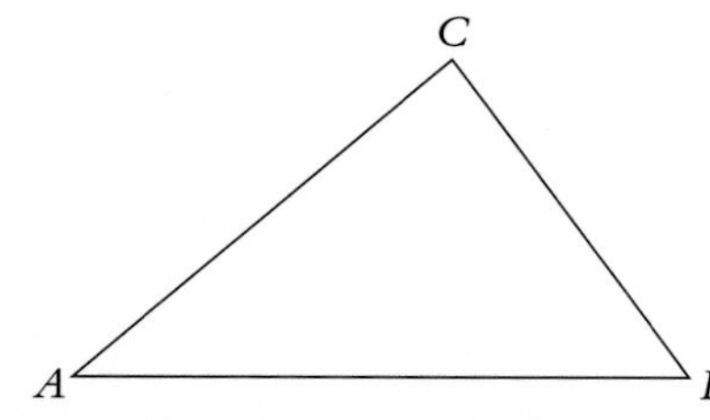

(a) Using a straight edge and compasses only, construct

(i) the bisector of angle ABC, [2]

(ii) the perpendicular bisector of the line BC. [2]

(b) Shade the region inside the triangle that is

- nearer to BC than to AB

and

- nearer to C than to B. [1]

20

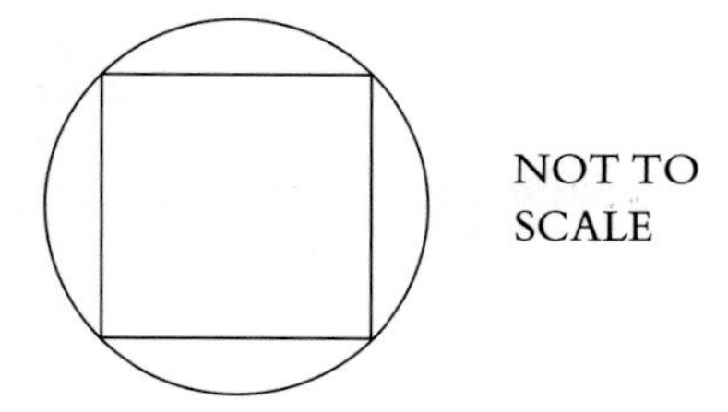

NOT TO SCALE

The circle has radius 5 cm.
The vertices of the square lie on the circumference of the circle.

Calculate the length of the sides of the square. [2]

21

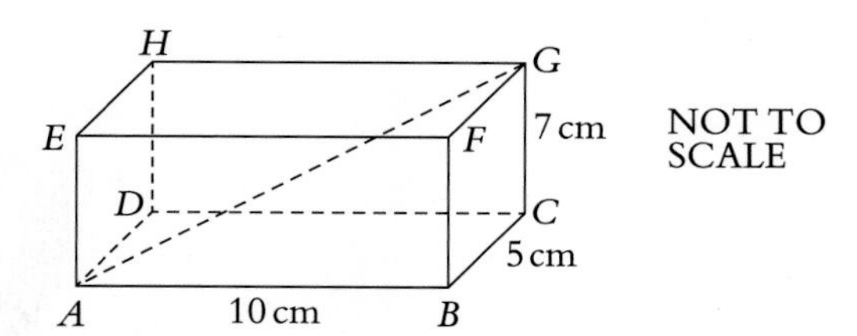

NOT TO SCALE

The diagram shows a cuboid $ABCDEFGH$.
$AB = 10$ cm, $BC = 5$ cm and $CG = 7$ cm.

Calculate the length of the line AG. [4]

22

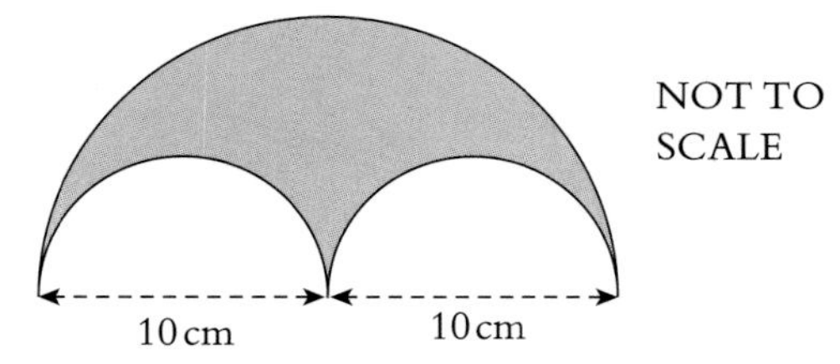

Two semicircles with diameter 10 cm are removed from a semicircle with diameter 20 cm.

(a) Calculate the area of the shaded shape. [3]
(b) Calculate the perimeter of the shaded shape. [3]

23

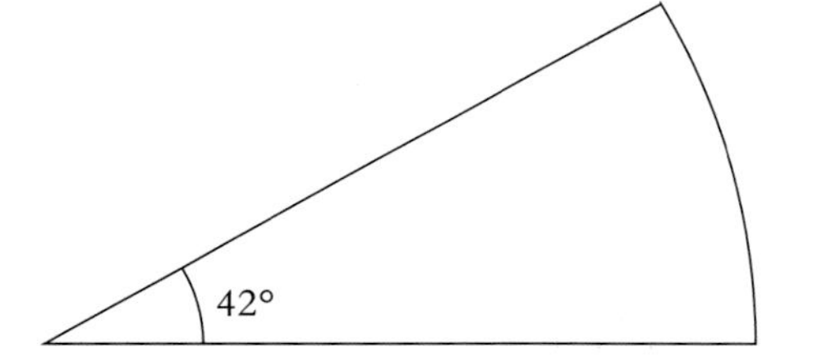

The diagram shows the sector of a circle with sector angle 42°.
The perimeter of the sector is 45 cm.

Calculate the radius of the circle. [3]

24

Grade	A*	A	B	C	D	E
Frequency	5	11	15	8	4	2

The table shows the results of a group of students in an examination.

(a) Write down the modal grade. [1]
(b) Paul draws a pie chart to represent the results.
Calculate the sector angle for grade A. [2]

25 In a survey, 54 students were asked how they travelled to school.
The table shows some of the results.

	Walked	Cycled	Bus	Total
Boys	8	10		25
Girls		7	5	
Total				54

(a) Complete the table. [2]
(b) Find the percentage of students who cycled to school. [1]

Standard form

THIS SECTION WILL SHOW YOU HOW TO

- Write numbers in standard form
- Solve problems and do calculations in standard form

A number in standard form must be written as

$A \times 10^n$ where $1 \leq A < 10$ and n is an integer

It is a very useful way of writing very large and very small numbers.

Large numbers

EXAMPLE

Write these numbers in standard form

a 8 000 000 **b** 30 500

a $8\,000\,000 = 8 \times 10^6$ **b** $30\,500 = 3.05 \times 10^4$

NOTE:

$10^1 = 10$
$10^2 = 100$
$10^3 = 1000$
$10^4 = 10\,000$
$10^5 = 100\,000$
$10^6 = 1\,000\,000$

EXAMPLE

Write these as ordinary numbers

a 1.36×10^5 **b** 7×10^8

a $1.36 \times 10^5 = 136\,000$ **b** $7 \times 10^8 = 700\,000\,000$

EXAMPLE

$a = 4.5 \times 10^8$ and $b = 3 \times 10^5$

Calculate the following giving your answers in standard form.

a $a + b$ **b** $a \times b$ **c** $a \div b$

a $a + b = (4.5 \times 10^8) + (3 \times 10^5)$ *change so they are both* $\times 10^5$
$= (4500 \times 10^5) + (3 \times 10^5)$
$= 4503 \times 10^5$ *this is not in standard form so replace 4503 by* 4.503×10^3
$= 4.503 \times 10^3 \times 10^5$ $10^3 \times 10^5 = 10^{3+5} = 10^8$
$= 4.503 \times 10^8$

b $a \times b = (4.5 \times 10^8) \times (3 \times 10^5)$ *multiply 4.5 by 3 and* 10^8 *by* 10^5
$= 13.5 \times 10^{8+5}$
$= 13.5 \times 10^{13}$ *this is not in standard form so replace 13.5 by* 1.35×10^1
$= 1.35 \times 10^1 \times 10^{13}$
$= 1.35 \times 10^{14}$

c $a \div b = (4.5 \times 10^8) \div (3 \times 10^5)$ *divide 4.5 by 3 and* 10^8 *by* 10^5
$= 1.5 \times 10^{8-5}$
$= 1.5 \times 10^3$

EXERCISE 3.1

1 Which of these numbers are written in standard form?

a 4×10^2 b 15×10^6 c 9.5×10^4 d 10×10^3
e 23.4×10^7 f 1×10^8 g 7.07×10^5 h 50.03×10^8

2 Write these numbers in standard form.

a 54 000 b 200 c 700 000 d 5330
e 82 000 000 f 3020 g 666 000 h 36.36

3 Write these as ordinary numbers.

a 4.8×10^5 b 3.7×10^2 c 5.64×10^8 d 7.06×10^7
e 2×10^4 f 1.1×10^6 g 3.001×10^2 h 9.5×10^9

4 Calculate the following, giving your answer in standard form.

a $(4 \times 10^3) + (2 \times 10^3)$ b $(4 \times 10^5) + (5 \times 10^4)$ c $(8 \times 10^4) - (5 \times 10^4)$
d $(4.3 \times 10^6) - (2 \times 10^5)$ e $(4 \times 10^3) \times (8 \times 10^5)$ f $(2.1 \times 10^5) \times (3 \times 10^2)$
g $(8 \times 10^7) \div (2 \times 10^4)$ h $(7.5 \times 10^9) \div (3 \times 10^2)$ i $(4 \times 10^6)^2$
j $\dfrac{(4 \times 10^8) + (5 \times 10^7)}{9 \times 10^3}$ k $\dfrac{(2.1 \times 10^3) \times (4 \times 10^5)}{2 \times 10^6}$ l $\dfrac{(4.9 \times 10^8) - (3 \times 10^6)}{(7 \times 10^5) + (3 \times 10^5)}$

5 Check your answers to question 4 using a calculator.

6 $p = 6 \times 10^9$ and $q = 2 \times 10^8$
Calculate the following giving your answers in standard form.

a $p + q$ b $p - q$ c $2p - 5q$ d $p \times q$
e $p \div q$ f p^2 g $\dfrac{p+q}{q}$ h $\dfrac{p^2}{2q}$

7 Simplify the ratios.

a $6 \times 10^5 : 2 \times 10^4$ b $2.5 \times 10^6 : 1 \times 10^8$ c $5.6 \times 10^7 : 11.2 \times 10^4$

8 Solve the equations.

a $5x + 3 \times 10^4 = 5.1 \times 10^5$ b $2(x - 6 \times 10^3) = 3 \times 10^4$ c $3(x - 2 \times 10^4) = 2(x + 7 \times 10^5)$

9 Simplify, giving your answer in standard form.

a $\sqrt{4 \times 10^8}$ b $\sqrt{9 \times 10^4}$ c $\sqrt{3.6 \times 10^7}$ d $\sqrt{4.9 \times 10^9}$

10 Write these in order of size, smallest first.

2.2×10^{14} 2.04×10^{14} 5.8×10^{13} 2.13×10^4

11 The distance between the Earth and the Sun is 149 million kilometres.
Write this number in standard form.

12 The area of land on the Earth is 1.5×10^8.
The area of ocean on the Earth is 3.6×10^8.
Write the ratio of land to ocean in its simplest form.

13 The population of China is approximately 1.34×10^9.
The population of Argentina is approximately 4.05×10^7.
Calculate the difference in population, giving your answer in standard form.

Small numbers

EXAMPLE

Write these numbers in standard form

a 0.0007 **b** 0.0000061

a $0.0007 = 7 \times 10^{-4}$ **b** $0.0000061 = 6.1 \times 10^{-6}$

➡ **NOTE:**

$10^{-1} = 0.1$
$10^{-2} = 0.01$
$10^{-3} = 0.001$
$10^{-4} = 0.0001$
$10^{-5} = 0.00001$
$10^{-6} = 0.000001$

EXAMPLE

Write these as ordinary numbers

a 4.23×10^{-2} **b** 9×10^{-4}

a $4.23 \times 10^{-2} = 0.0423$ **b** $9 \times 10^{-4} = 0.0009$

EXAMPLE

$a = 3 \times 10^5$ $b = 4 \times 10^{-3}$ $c = 2 \times 10^{-4}$

Calculate the following giving your answers in standard form.

a $a \times b$ **b** $b \times c$ **c** $a \div c$ **d** $c \div b$ **e** $b - c$

a $a \times b = (3 \times 10^5) \times (4 \times 10^{-3})$ *multiply 3 by 4 and 10^5 by 10^{-3}*
$= 12 \times 10^2$ *this is not in standard form so replace 12 by 1.2×10^1*
$= 1.2 \times 10^1 \times 10^2$ *$10^1 \times 10^2 = 10^{1+2} = 10^3$*
$= 1.2 \times 10^3$

b $b \times c = (4 \times 10^{-3}) \times (2 \times 10^{-4})$ *multiply 4 by 2 and 10^{-3} by 10^{-4}, $10^{-3} \times 10^{-4} = 10^{-3+-4} = 10^{-7}$*
$= 8 \times 10^{-7}$

c $a \div c = (3 \times 10^5) \div (2 \times 10^{-4})$ *divide 3 by 2 and 10^5 by 10^{-4}, $10^5 \div 10^{-4} = 10^{5--4} = 10^9$*
$= 1.5 \times 10^9$

d $c \div b = (2 \times 10^{-4}) \div (4 \times 10^{-3})$ *divide 2 by 4 and 10^{-4} by 10^{-3}, $10^{-4} \div 10^{-3} = 10^{-4--3} = 10^{-1}$*
$= 0.5 \times 10^{-1}$ *this is not in standard form so replace 0.5 by 5×10^{-1}*
$= 5 \times 10^{-1} \times 10^{-1}$ *$10^{-1} \times 10^{-1} = 10^{-1+-1} = 10^{-2}$*
$= 5 \times 10^{-2}$

e **Method 1**
$b - c = (4 \times 10^{-3}) - (2 \times 10^{-4})$
$= (4 \times 10^{-3}) - (0.2 \times 10^{-3})$
$= 3.8 \times 10^{-3}$

Method 2
$b - c = (4 \times 10^{-3}) - (2 \times 10^{-4})$
$= 0.004 - 0.0002$
$= 0.0038$
$= 3.8 \times 10^{-3}$

EXERCISE 3.2

1 Write these numbers in standard form.

a	0.015	b	0.000 67	c	0.0009
d	0.000 606	e	3 200 000	f	0.00052
g	0.44	h	53 080 000	i	0.000 000 057
j	0.000 613	k	700 200	l	0.00089

2 Write these as ordinary numbers.

a	8.8×10^{-5}	**b**	2.3×10^{-7}	**c**	6×10^{-3}
d	5.11×10^{-2}	**e**	9.9×10^{4}	**f**	1.04×10^{-6}
g	6.8×10^{-4}	**h**	6×10^{8}	**i**	3.8×10^{-1}
j	2.2×10^{1}	**k**	4.08×10^{5}	**l**	9.5×10^{-6}

3 Calculate the following, giving your answer in standard form.

a $(4 \times 10^{-2}) + (2 \times 10^{-2})$
b $(7 \times 10^{-5}) + (2 \times 10^{-6})$
c $(8 \times 10^{-8}) - (3 \times 10^{-8})$
d $(6.2 \times 10^{-4}) - (2 \times 10^{-5})$
e $(5 \times 10^{-5}) \times (6 \times 10^{-5})$
f $(3.2 \times 10^{-3}) \times (2 \times 10^{-4})$
g $(9 \times 10^{7}) \times (3 \times 10^{-5})$
h $(4.8 \times 10^{-6}) \div (2 \times 10^{-4})$
i $(3 \times 10^{-5})^2$
j $\dfrac{(5 \times 10^{4}) + (2 \times 10^{4})}{3.5 \times 10^{-2}}$
k $\dfrac{(1.8 \times 10^{-2}) \times (3 \times 10^{8})}{2 \times 10^{3}}$
l $\dfrac{(9 \times 10^{-4}) + (7 \times 10^{-4})}{(4 \times 10^{-5}) \times (2 \times 10^{9})}$

4 Check your answers to question 3 using a calculator.

5 Simplify, giving your answer in standard form.

a $\sqrt{9 \times 10^{-6}}$
b $\sqrt{4 \times 10^{-10}}$
c $\sqrt{6.4 \times 10^{-5}}$
d $\sqrt{8.1 \times 10^{-13}}$

6 $p = 6 \times 10^{-8}$ and $q = 3 \times 10^{-9}$
Calculate the following, giving your answers in standard form.

a $p + q$
b $p - q$
c $3p + 2q$
d $p \times q$
e $p \div q$
f p^2
g q^2
h $\dfrac{5p^2}{4q}$

7 The mass, in grams, of a single hydrogen atom is 1.67×10^{-24} g.
Find the total mass of 500 hydrogen atoms.
Give your answer in standard form.

8 The mass, in grams, of a single oxygen atom is 2.67×10^{-23} g.
Calculate the number of oxygen atoms in 5 grams of oxygen.
Give your answer in standard form.

9 The time taken by light to travel one metre is approximately 3×10^{-9} seconds.
How long does it take light to travel 5 kilometres?
Give your answer in standard form.

10 The mass, in kilograms, of a grain of sand is 3.5×10^{-10} kg.
Find the total mass of 75 000 grains of sand.
Give your answer in standard form.

KEY WORDS
standard form

Simultaneous equations 1

THIS SECTION WILL SHOW YOU HOW TO

- Solve simultaneous equations using a graph
- Solve simultaneous equations using the elimination method

Simultaneous equations are a set of equations that are all satisfied by the same values of the variables. Simultaneous equations can be solved using a graph or by using algebra.

EXAMPLE

Solve these simultaneous equations using a graph.

$x + 2y = 4$

$x - y = 1$

First, make a table of values for each equation.

$x + 2y = 4$

x	0	4
y	2	0

$x - y = 1$

x	0	1
y	−1	0

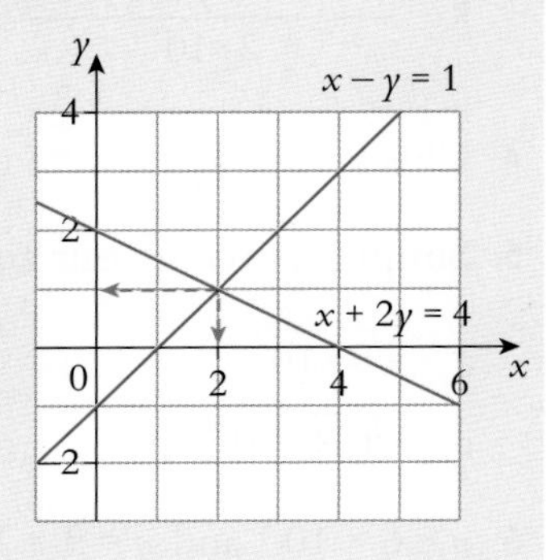

Then draw both lines on one set of axes.

The lines intersect at the point (2, 1).

The solution to the simultaneous equations is $x = 2$, $y = 1$

➡ **CHECK:** substitute the values for x and y into the original equations:

$2 + (2 \times 1) = 4$ ✓

$2 - 1 = 1$ ✓

EXERCISE 3.3

1 Copy and complete these tables for the given equations.

$y = x + 2$

x	0	2	4
y			

$y = 4 - x$

x	0	2	4
y			

Draw both graphs on one set of axes.

Use your graph to solve the simultaneous equations $y = x + 2$ and $y = 4 - x$

2 Copy and complete these tables for the given equations.

$y = \frac{1}{2}x + 1$

x	0	2	4
y			

$5x + 4y = 20$

x	0	
y		0

Draw both graphs on one set of axes.

Use your graph to solve the simultaneous equations $y = \frac{1}{2}x + 1$ and $5x + 4y = 20$

Solving simultaneous equations using a graph takes time and can be inaccurate if the coordinates of the point of intersection are not integers. It is quicker and easier to use an algebraic method.

Elimination method

The elimination method involves adding or subtracting the equations (or multiples of them) to **eliminate** one of the variables.

EXAMPLE

Solve these simultaneous equations. $3x + 2y = 19$
$x + 2y = 13$

$3x + 2y = 19 \quad (1)$
$x + 2y = 13 \quad (2)$

Subtracting equation (2) from equation (1) will eliminate the y terms.

$2x = 6$
$x = 3$

To find y substitute $x = 3$ in equation (1) and then solve.

$9 + 2y = 19$
$2y = 10$
$y = 5$

The solution is $x = 3$, $y = 5$.

➡ CHECK:

$(3 \times 3) + (2 \times 5) = 19$ ✓
$3 + (2 \times 5) = 13$ ✓

EXAMPLE

Solve these simultaneous equations. $4x + 3y = 1$
$2x - 3y = 14$

$4x + 3y = 1 \quad (1)$
$2x - 3y = 14 \quad (2)$

Adding equations (1) and (2) will eliminate the y terms.

$6x = 15$
$x = 2\frac{1}{2}$

Substitute $x = 2\frac{1}{2}$ in equation (1) and then solve to find y:

$10 + 3y = 1$
$3y = -9$
$y = -3$

The solution is $x = 2\frac{1}{2}$, $y = -3$.

➡ CHECK:

$\left(4 \times 2\frac{1}{2}\right) + (3 \times -3) = 10 - 9 = 1$ ✓

$\left(2 \times 2\frac{1}{2}\right) - (3 \times -3) = 5 - -9 = 14$ ✓

UNIT 3

EXERCISE 3.4

Solve these simultaneous equations.

1

a $2x + y = 10$
$x + y = 9$

b $3x + 2y = 19$
$x + 2y = 9$

c $2x + y = 19$
$2x + 3y = 15$

d $4x + 3y = 28$
$4x + y = 20$

e $5x + 2y = 8$
$3x + 2y = 4$

f $7x + y = -17$
$5x + y = -11$

g $4x + 5y = -34$
$4x + 2y = -28$

h $3x + 8y = -1$
$5x + 8y = 9$

2

a $5x - 2y = 2$
$3x + 2y = 14$

b $3x + 5y = 52$
$4x - 5y = 11$

c $4x - 3y = -22$
$5x + 3y = 40$

d $2x - 9y = -13$
$2x + 9y = 41$

e $4x + 9y = 2$
$3x - 9y = 33$

f $5x - 4y = -31$
$2x + 4y = 10$

g $7x + 2y = 39$
$3x - 2y = 31$

h $6x + y = -21$
$10x - y = -19$

3

a $3x - 7y = 4$
$x - 7y = -8$

b $-2x + 5y = 9$
$-2x + 3y = 3$

c $8x - y = 42$
$6x - y = 32$

d $4x - 3y = -3$
$8x - 3y = 3$

e $-3x + 2y = 17$
$-3x + 4y = 22$

f $2x - 3y = 0$
$5x - 3y = 6$

g $8x - 4y = 1$
$3x - 4y = 6$

h $7x - 2y = -35$
$3x - 2y = -19$

4 The sum of two numbers is 15 and the difference is 4. Let the two numbers be x and y. Write down a pair of simultaneous equations involving x and y. Solve the equations.

5 Find two numbers with

a a mean of 27 and a difference of 24

b a mean of 4 and a difference of 22.

6 Find the values of x and y.

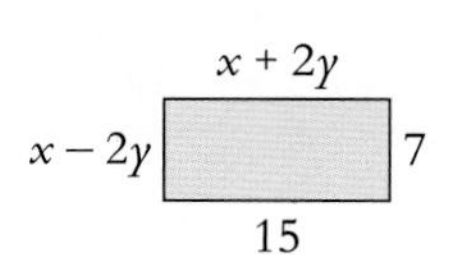

7 Find the values of x and y.

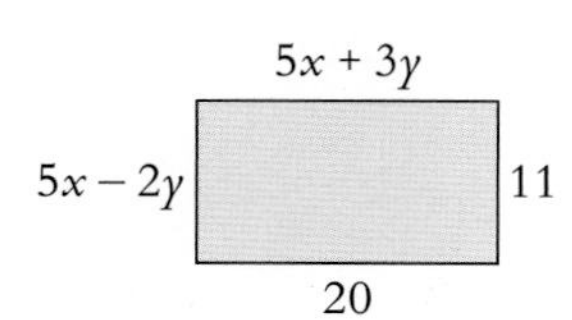

8 Seven pens and two pencils cost \$6.59. Seven pens and 5 pencils cost \$7.55
Find the cost of a pencil and the cost of a pen.

9 The line $y = mx + c$ passes through the points $(-2, 1)$ and $(4, 4)$.
Write down a pair of simultaneous equations involving m and c.
Solve the simultaneous equations.

10 The cost of a taxi journey is made of two parts.
There is a fixed charge of \$$x$ and then an extra charge of \$$y$ per km.
The total cost for a 5 km journey is \$14.60. The total cost for a 12 km journey is \$23.84
Find the total cost of a 24 km journey.

To eliminate one of the variables you might first need to multiply one or both of the equations by a suitable number before adding or subtracting.

EXAMPLE

Solve these simultaneous equations. $8x + 3y = 7$
$3x + 5y = -9$

$8x + 3y = 7$ (1)
$3x + 5y = -9$ (2)

Multiply equation (1) by 5 and equation (2) by 3.

$40x + 15y = 35$
$9x + 15y = -27$

Subtracting these two equations will eliminate the y terms.

$31x = 62$
$x = 2$

Substitute $x = 2$ in equation (1) and then solve to find y:

$16 + 3y = 7$
$3y = -9$
$y = -3.$

The solution is $x = 2$, $y = -3$.

➡ **CHECK:**

$(8 \times 2) + (3 \times -3) = 16 + -9 = 7$ ✓
$(3 \times 2) + (5 \times -3) = 6 + -15 = -9$ ✓

Some simultaneous equations need to be re-arranged before solving them.

EXAMPLE

Solve these simultaneous equations. $5x + 2y - 25 = 0$
$2x - 1 = y$

First re-arrange the equations into the form $ax + by = c$

$5x + 2y = 25$ (1)
$2x - y = 1$ (2)

Multiply equation (2) by 2.

$5x + 2y = 25$
$4x - 2y = 2$

Adding the equations will eliminate the y terms

$9x = 27$
$x = 3$

Substitute $x = 3$ in equation (1) and then solve to find y:

$15 + 2y = 25$
$2y = 10$
$y = 5$

The solution is $x = 3$, $y = 5$.

➡ **CHECK:**

$(5 \times 3) + (2 \times 5) = 25$ ✓
$(2 \times 3) - 5 = 1$ ✓

EXERCISE 3.5

Solve these simultaneous equations.

1 a $2x + 3y = 21$, $5x - y = 10$ b $8x + 2y = 34$, $3x + y = 13$ c $4x + 3y = 36$, $2x - y = 8$ d $3x + 7y = 58$, $x + 2y = 17$

e $5x + 2y = 21$, $x + y = 3$ f $6x + 5y = 45$, $2x - 3y = 29$ g $4x + 2y = 22$, $6x - y = 2$ h $7x + 8y = 1$, $x + 4y = -7$

2 a $3x + 8y = 68$, $5x - 6y = -22$ b $4x + 3y = 18$, $3x - 2y = 5$ c $5x - 2y = 19$, $2x - 3y = 1$ d $4x + 5y = 18$, $6x + 2y = 16$

e $3x - 4y = 0$, $2x + 5y = 46$ f $9x + 7y = 127$, $3x + 2y = 41$ g $2x + 3y = 33$, $6x - 5y = -27$ h $8x + 5y = 43$, $5x + 3y = 26$

3 a $10x - 2y = -13$, $4x - 3y = -14$ b $2x + 7y = -29$, $3x - 4y = 29$ c $5x + 6y = 32$, $3x + 9y = 57$ d $8x - 5y = -14$, $3x - 4y = -1$

e $6x - 7y = -6$, $8x + 2y = 26$ f $2x - 8y = 0$, $3x + 12y = 36$ g $8x + 7y = -7$, $6x + 5y = -4$ h $2x - 3y = -30$, $3x - 2y = -35$

4 a $x = 10 - 3y$, $2x - 3y = 2$ b $y = x - 1$, $3x - y = 5$ c $x + 2y - 8 = 0$, $2x - 3y - 14 = 0$ d $3x + 2y = 11$, $y - 2x = 3$

e $y - 3x = -9$, $x + 2y = -4$ f $y = x + 3$, $3y + x = 5$ g $3y - 5x = 13$, $3x + y + 5 = 0$ h $3y = 2x + 4$, $2y = 7 - 3x$

5 a $\frac{x}{4} + \frac{y}{3} = 8$, $\frac{x}{2} - \frac{y}{5} = 3$ b $\frac{x}{2} + \frac{y}{4} = 6$, $\frac{x}{5} - \frac{y}{2} = 12$ c $\frac{x}{6} - \frac{y}{2} = -8$, $\frac{x}{3} - \frac{y}{5} = -8$ d $\frac{x}{8} + \frac{y}{3} = 1$, $\frac{x}{12} + \frac{y}{2} = -1$

6 Four times one number added to three times another number gives 33. The sum of the two numbers is 10. Find the two numbers.

7 Five times one number take away two times another number gives 15. The sum of the two numbers is 6.5. Find the two numbers.

8 A bag contains 36 coins. The coins are a mixture of one-dollar coins and 50-cent coins. The total value of the coins in the bag is $26.50. How many one-dollar coins are there?

9 Cinema tickets cost $8.50 per child and $14 per adult. One evening the cinema sells 201 tickets for a total of $2550. How many adult tickets were sold?

10 Eight bottles of fruit juice and three bottles of water cost \$17.89
Five bottles of fruit juice and four bottles of water cost \$13.37
Find the cost of a bottle of fruit juice.

11 Four lemons and three oranges cost \$5.45
Seven lemons and four oranges cost \$9.30
Find the cost of an orange.

12 The straight line $ax + by = 15$ passes through the points $(-5, 6)$ and $(15, -6)$.
Find the values of a and b.

13 The straight line $ax + by = 9$ passes through the points $(12, 5)$ and $(-6, -7)$.
Find the values of a and b.

14 For the rectangle ABCD, find
- **a** the values of x and y
- **b** the perimeter of the rectangle
- **c** the area of the rectangle.

C $4x + 2y$ D
$2x + 2y$ $x - y$
B $x - 3y + 4$ A

15 For the square ABCD, find
- **a** the values of x and y
- **b** the perimeter of the square
- **c** the area of the square.

C $2y - 2x$ D
$8x - y - 3$ $y + 1$
B $4x - 1$ A

16 For the equilateral triangle ABC, find
- **a** the values of x and y
- **b** the perimeter of the triangle
- **c** the area of the triangle.

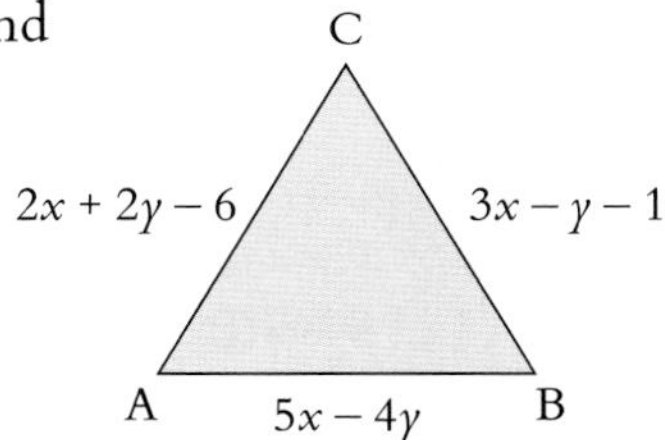

KEY WORDS
simultaneous equations
eliminate

UNIT 3

Factorising 1

THIS SECTION WILL SHOW YOU HOW TO
- Factorise simple expressions
- Use factorisation to simplify fractions

Expanding $3x(5x + 4)$ gives $15x^2 + 12x$.
Factorising is the reverse of expanding.
So factorising $15x^2 + 12x$ gives $3x(5x + 4)$.

EXAMPLE

Factorise $6x + 8$

$6x + 8$
$= 2(3x + 4)$

2 is the highest common factor (HCF) of the two terms
$6x = 2 \times 3x$ and $8 = 2 \times 4$
It is sensible to check your answer by expanding the brackets.
$2(3x + 4) = 2 \times 3x + 2 \times 4 = 6x + 8$ ✓

EXAMPLE

Factorise $5x^2 + 10xy$

$5x^2 + 10xy$
$= 5x(x + 2y)$

$5x$ is the HCF of the two terms
$5x^2 = 5x \times x$ and $10xy = 5x \times 2y$

EXAMPLE

Factorise $xy^2 + x^3y^4$

$xy^2 + x^3y^4$
$= xy^2(1 + x^2y^2)$

xy^2 is the HCF of the two terms
$xy^2 = xy^2 \times 1$ and $x^3y^4 = xy^2 \times x^2y^2$

EXAMPLE

Factorise $x(x + 2) + 5(x + 2)$

$x(x + 2) + 5(x + 2)$
$= (x + 2)(x + 5)$

$x + 2$ is the HCF of the two terms
➡ **NOTE:** $(x + 5)(x + 2)$ is also correct

EXAMPLE

Factorise $xy - 2x + 3y - 6$

$xy - 2x + 3y - 6$
$= x(y - 2) + 3(y - 2)$
$= (y - 2)(x + 3)$

factorise the first two terms and the last two terms
$y - 2$ is the HCF of these two terms
➡ **NOTE:** $(x + 3)(y - 2)$ is also correct

EXERCISE 3.6

1 Abdullah has got his factorising homework wrong. Copy out each question and correct his mistakes.

a) $6x + 12y$	b) $5xy - 10xy^3$	c) $4x + 6y$
$= 2(3x + 6y)$	$= 5x(y - y^2)$	$= 4\left(x + 1\frac{1}{2}y\right)$

Factorise these expressions completely.

2 a $4x + 8$ b $3x - 6$ c $8x + 2$
d $10x - 15$ e $7y - 14$ f $9y + 12$
g $20 - 8x$ h $36 - 12y$ i $42 - 7x$
j $18y - 4$ k $24x + 16$ l $22y - 55$

3 a $x^2 + 5x$ b $x^2 - 8x$ c $y^2 - 4y$
d $a^2 + 9a$ e $4x^2 - 8x$ f $3y^2 - 9y$
g $10x^2 - 4x$ h $8y - 2y^2$ i $15y^2 + 5y$
j $14x - 7x^2$ k $6x - 4x^2$ l $12y^2 - 9y$

4 a $8\pi + 4$ b $15 - 9\pi$
c $\pi r^2 + \pi rl$ d $2\pi r^2 + 2\pi rh$

5 a $6x^2y - 3xy$ b $10x^4y^2 - 5x^4y$ c $8xy^3 - 32x^3y$
d $15x^3y^2 - 12x^2y^4$ e $12x^2y - 15xy$ f $28pq^2 - 35p^2q$
g $6a^2b + 9ab$ h $21x^5y - 28x^2y$ i $56a^4b^7 - 35ab^5$

6 a $3a + 3b - 3c$ b $5x + 5y + 5x^2$ c $2ax - ay - 3ay^2$
d $x^3 + 4x^2 - 2x$ e $y^4 + 6y^3 - 6y^2$ f $p^2q^2r + pqr + pq^2r$
g $20a^2b^2c^2 - 8a^2b^2 + 16abc^2$ h $15x^3y^2 - 27x^2y + 21x^2y^2$
i $8p^7q^5r^3 + 6p^4q^3r^2 - 4p^3q^3r^3$

7 a $x(y + 2) + 3(y + 2)$ b $a(b - 4) + 2(b - 4)$ c $p(q + 3) - 5(q + 3)$
d $x(y - 5) - 3(y - 5)$ e $x(x + y)^2 + y(x + y)^2$ f $(a + b)^3 - (a + b)^2$
g $(x - y)^2 - (x - y)^3$ h $5(x + y) + (x + y)^2$ i $(x - y)^3 - 4(x - y)^2$

8 a $xy + y + 3x + 3$ b $5y - 10 + xy - 2x$ c $4a + 4b + ac + bc$
d $pq + 2pr + 3q + 6r$ e $3x + 12 - xy - 4y$ f $xy - 3x - 4y + 12$

You can use factorisation to simplify algebraic fractions.

EXAMPLE

Simplify $\frac{x^2+4x}{x}$

$\frac{x^2+4x}{x}$ *factorise the numerator*

$= \frac{{}^{1}\cancel{x}(x+4)}{\cancel{x}_{1}}$ *cancel*

$= x + 4$

EXAMPLE

Simplify $\frac{6x-12}{18}$

$\frac{6x-12}{18}$ *factorise the numerator*

$= \frac{{}^{1}\cancel{6}(x-2)}{\cancel{18}_{3}}$ *cancel*

$= \frac{x-2}{3}$

EXAMPLE

Simplify $\frac{x^2-xy}{7x^2-7xy}$

$\frac{x^2-xy}{7x^2-7xy}$ *factorise the numerator and denominator*

$= \frac{{}^{1}\cancel{x}\cancel{(x-y)}^{1}}{7\cancel{x}_{1}\cancel{(x-y)}_{1}}$ *cancel*

$= \frac{1}{7}$

EXAMPLE

Simplify $\frac{3x+3y}{x^2-xy} \times \frac{5x-5y}{2x+2y}$

$= \frac{3(x+y)}{x(x-y)} \times \frac{5(x-y)}{2(x+y)}$ *factorise both numerators and denominators*

$= \frac{3\cancel{(x+y)}^{1}}{x\cancel{(x-y)}_{1}} \times \frac{5\cancel{(x-y)}^{1}}{2\cancel{(x+y)}_{1}}$ *cancel*

$= \frac{15}{2x}$

EXERCISE 3.7

1 Sheeba has got her simplifying algebraic fractions homework wrong. Copy out each question and correct her mistakes.

a) $\frac{4x-8}{4} = 4x - 2$

b) $\frac{x^2+2x}{4x} = \frac{x^2+2}{4}$

c) $\frac{6x+8}{10} = \frac{2(3x+8)}{10} = \frac{(3x+8)}{5}$

Simplify these algebraic fractions.

2 a $\frac{3x+6}{3}$ b $\frac{8y+4}{2}$ c $\frac{5x+10}{5}$ d $\frac{12y+8}{4}$

e $\frac{6x-12}{8}$ f $\frac{9x+6}{12}$ g $\frac{15-20y}{10}$ h $\frac{24-16a}{20}$

3 a $\frac{x^2+9x}{x}$ b $\frac{y^2-8y}{y}$ c $\frac{x^2-3x}{x^2}$ d $\frac{x-x^3}{x}$

e $\frac{8x+4y}{4z}$ f $\frac{5x+5y}{5z}$ g $\frac{x^2+xy}{xy}$ h $\frac{x^2-2x}{xy}$

4 a $\frac{3x+3y}{x+y}$ b $\frac{5a-5b}{a-b}$ c $\frac{x-y}{6x-6y}$ d $\frac{2x+2y}{10x+10y}$

e $\frac{2-2y}{5-5y}$ f $\frac{3x+3xy}{7+7y}$ g $\frac{xy+x}{yz+z}$ h $\frac{10x-10y}{15x-15y}$

5 a $\frac{x^2+x}{x+1}$ b $\frac{5x}{x^2-5x}$ c $\frac{x^2-3x}{5x-15}$ d $\frac{4y-6y^2}{10-15y}$

e $\frac{x^2-x}{(x-1)^2}$ f $\frac{3y-6}{(y-2)^3}$ g $\frac{x^2+3x}{x+3}$ h $\frac{3x^2y-6xy}{6x-12}$

6 a $\frac{4p+4q}{2p-2q} \times \frac{5p-5q}{3p+3q}$ b $\frac{7a+14b}{a-b} \times \frac{2a-2b}{3a+6b}$ c $\frac{x^2-3x}{y} \times \frac{x}{xy-3y}$

d $\frac{x^2+x}{y^2+y} \times \frac{y}{x}$ e $\frac{7x-7y}{xy} \times \frac{x^2+xy}{x-y}$ f $\frac{f^2-6f}{2g} \times \frac{4g}{fg-6g}$

7 a $\frac{2x-4y}{6xy} \div \frac{4x-8y}{5x^2y^2}$ b $\frac{3x^2-6x}{6y-2y^2} \div \frac{5x^2-10x}{9y-3y^2}$

KEY WORDS
factorise

Rearranging formulae 1

THIS SECTION WILL SHOW YOU HOW TO
- Change the subject of a formula

In the formula $P = 2x + 2y$, P is called the **subject** of the formula. The formula can be rearranged to make x the subject as follows:

$P = 2x + 2y$ *take $2y$ from both sides*

$P - 2y = 2x$ *divide both sides by 2*

$\frac{P - 2y}{2} = x$

EXAMPLE

Rearrange the formula $y = \frac{m - n}{2}$ to make m the subject.

$y = \frac{m - n}{2}$ *multiply both sides by 2*

$2y = m - n$ *add n to both sides*

$2y + n = m$

EXAMPLE

Rearrange the formula $x = \frac{y^2}{3} + z$ to make y the subject.

$x = \frac{y^2}{3} + z$ *take z from both sides*

$x - z = \frac{y^2}{3}$ *multiply both sides by 3*

$3(x - z) = y^2$ *square root both sides*

$y = \pm\sqrt{3(x - z)}$

➡ **NOTE:** remember for example that both $5^2 = 25$ and $(-5)^2 = 25$

EXAMPLE

Rearrange the formula $T = 2\pi\sqrt{\frac{l}{g}}$ to make l the subject.

$T = 2\pi\sqrt{\frac{l}{g}}$ *divide both sides by 2π*

$\frac{T}{2\pi} = \sqrt{\frac{l}{g}}$ *square both sides*

$\frac{T^2}{4\pi^2} = \frac{l}{g}$ *multiply both sides by g*

$\frac{T^2 g}{4\pi^2} = l$

EXERCISE 3.8

1 Hassan has got his rearranging formulae homework wrong. Copy out each question and correct his mistakes.

a) $p = 3x - q$
$p - q = 3x$
$\frac{p-q}{3} = 3$

b) $2x + a = 8b$
$x + a = 4b$
$x = 4b - a$

c) $y = \sqrt{x} + b$
$y^2 = x + b^2$
$y^2 - b^2 = x$

Make x the subject of these formulae.

2 a $x + 3 = y$ b $x - 6 = 5y$ c $x + y = 4$ d $x - a = b$
e $y = x - 8$ f $x + 2y - 5 = 0$ g $x - 4y + 8 = 2y$ h $y - x = 10$

3 a $6x = y$ b $9x = 3y$ c $3x = y - 4$ d $ax = 7 - 5y$
e $\frac{x}{2} = p$ f $\frac{x}{5} = 3 + d$ g $\frac{x}{2} = y - 8$ h $\frac{x}{a} = 10 - 4y$
i $\frac{2x}{3} = y$ j $\frac{4x}{5} = 3y$ k $\frac{3x}{7} = a + b$ l $\frac{ax}{b} = c + d$

4 a $2x + c = d$ b $5x - y = 2z$ c $4x + 3y = 6$ d $8x - 3b = c + d$
e $ax + b = d$ f $bx - c = cd$ g $px - q = a + b$ h $y - xy = 8$

5 a $x(y + z) = 2$ b $p = (2q + 3)x$ c $a = 3(2x + y)$ d $p = q(x - r)$

6 a $\frac{x+a}{b} = c$ b $\frac{x-2f}{3g} = h$ c $\frac{ax+by}{c} = d$ d $\frac{p-qx}{r} = m$
e $\frac{p}{x} = q$ f $f = \frac{g}{x}$ g $\frac{p+q}{x} = r$ h $5 = \frac{p-q}{x}$

7 a $ax^2 = b$ b $px^2 = q$ c $abx^2 = c$ d $ax^2 = bc$
e $x^2 + a = b$ f $x^2 - d = e$ g $ax^2 + b = c$ h $px^2 - 5q = 4r$
i $\frac{x^2}{4} - a = b$ j $\frac{x^2}{a} + bc = d$ k $f - \frac{x^2}{g} = 3h$ l $3fg - \frac{x^2}{2} = g$

8 a $\sqrt{x + a} = b$ b $\sqrt{x - p} = qr$ c $\sqrt{2x - y} = 4$ d $2\sqrt{x - y} = z$

9 Sofia and Karl are asked to make h the subject of the formula $A = 2\pi r(r + h)$

Sofia writes:
$A = 2\pi r(r + h)$
$A = 2\pi r^2 + 2\pi rh$
$A - 2\pi r^2 = 2\pi rh$
$\frac{A - 2\pi r^2}{2\pi r} = h$

Karl writes:
$A = 2\pi r(r + h)$
$\frac{A}{2\pi r} = r + h$
$\frac{A}{2\pi r} - r = h$

KEY WORDS
subject

Who is correct? Explain your answer.

UNIT 3

Similar triangles

THIS SECTION WILL SHOW YOU HOW TO

- Recognise congruent and similar shapes
- Use the properties of similar shapes to find missing lengths and angles

Congruency

Two shapes are **congruent** if one of the shapes fits exactly on top of the other shape.

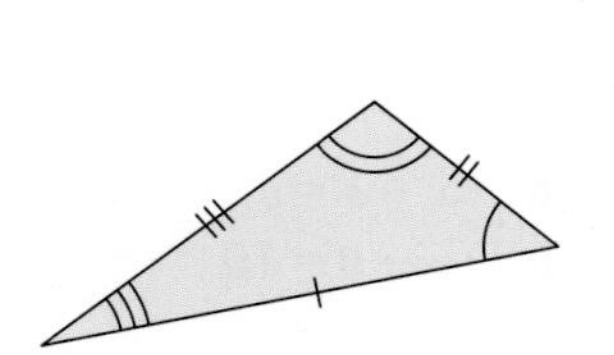

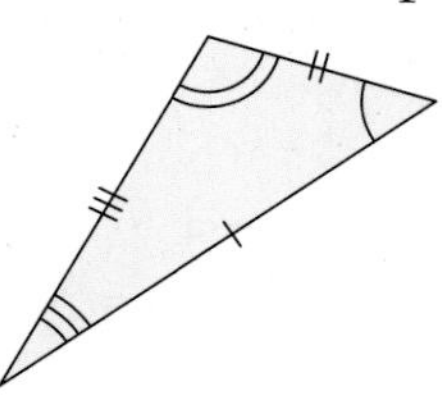

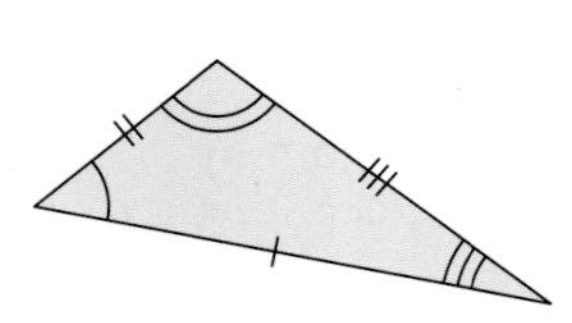

These three triangles are all congruent.

In congruent shapes

- corresponding angles are equal
- corresponding lengths are equal.

To prove that two triangles are congruent you must show that they satisfy one of the following four sets of conditions.

SSS three sides are equal.

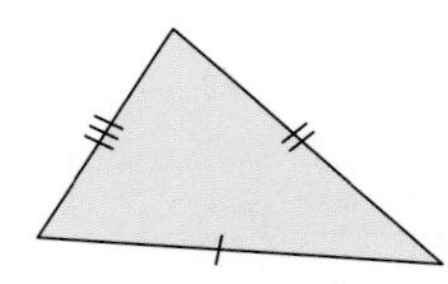

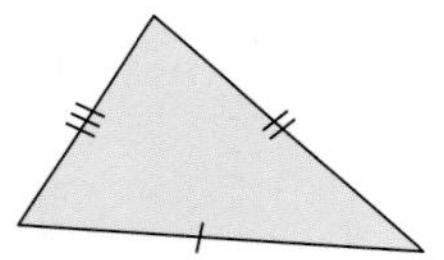

SAS two sides and the included angle are the same.

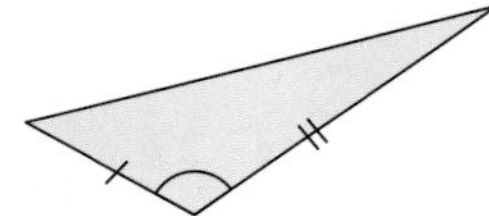

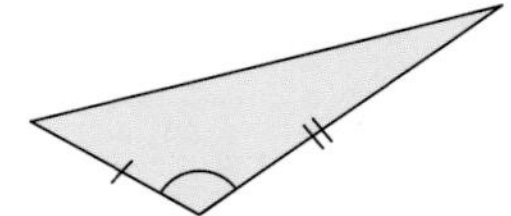

ASA two angles and the included side are the same.

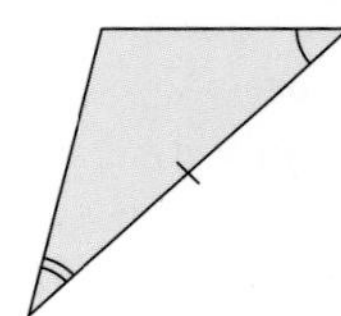

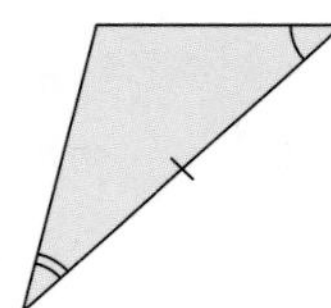

RHS right-angled triangles with hypotenuse and one other side the same.

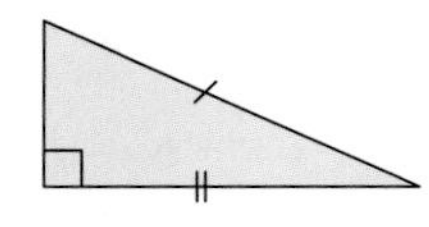

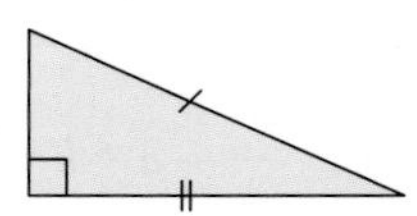

EXAMPLE

Explain why the following pairs of triangles are congruent.

a

6 cm
30°
3 cm

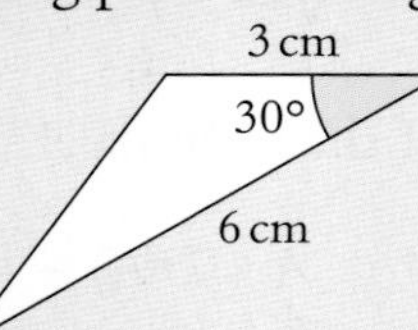

b

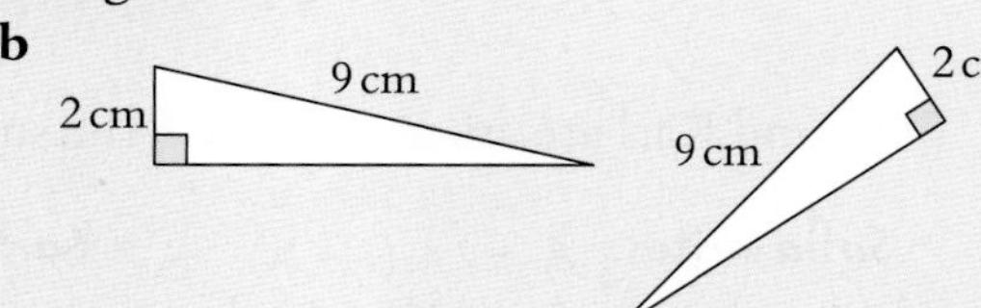

a Two sides and the included angle are the same. (SAS)

b The triangle is right-angled (R), the hypotenuses (H) are the same and one other side (S) is the same. (RHS)

EXERCISE 3.9

1 Using axes from 0 to 10, draw each pair of triangles.
Which pairs of triangles are congruent?

a	Triangle A: (0, 5) (0, 9) (1, 6)	Triangle B: (3, 10) (7, 10) (5, 9)
b	Triangle C: (0, 2) (1, 0) (3, 2)	Triangle D: (1, 7) (1, 10) (3, 8)
c	Triangle E: (4, 4) (6, 6) (3, 7)	Triangle F: (7, 7) (10, 8) (8, 10)
d	Triangle G: (2, 3) (6, 3) (7, 4)	Triangle H: (6, 5) (9, 5) (10, 6)
e	Triangle I: (0, 4) (1, 3) (2, 7)	Triangle J: (9, 0) (10, 1) (8, 3)
f	Triangle K: (3, 9) (6, 8) (7, 6)	Triangle L: (5, 0) (7, 1) (8, 4)

2 Explain why each of the following pairs of triangles are congruent.

a

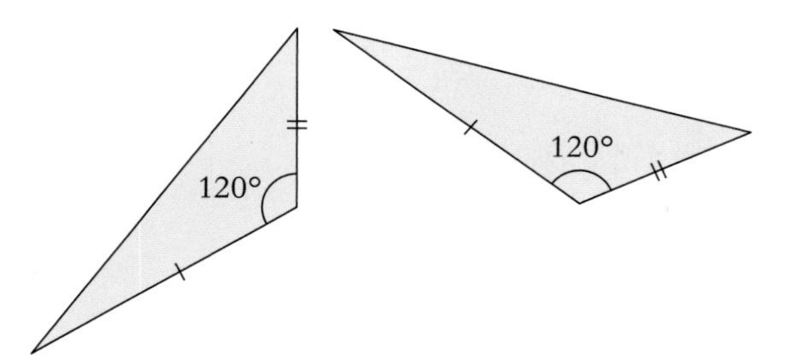

b

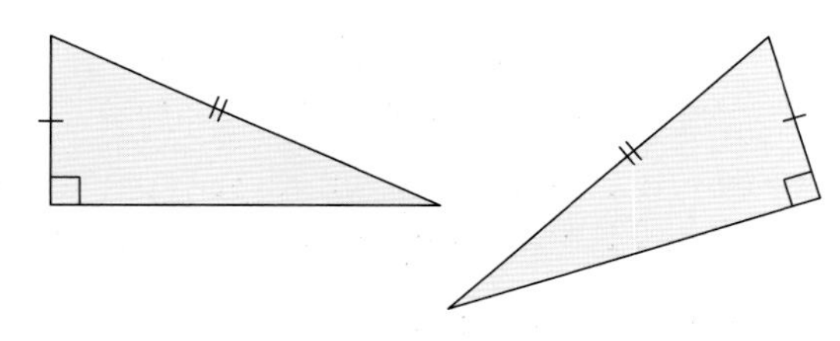

c

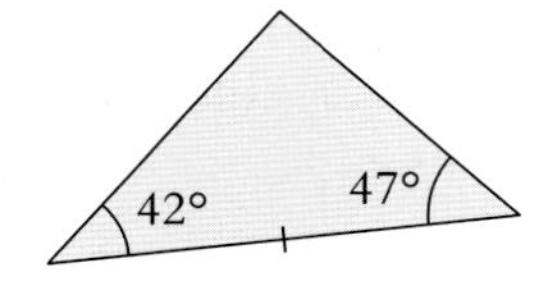

47°
42°

d

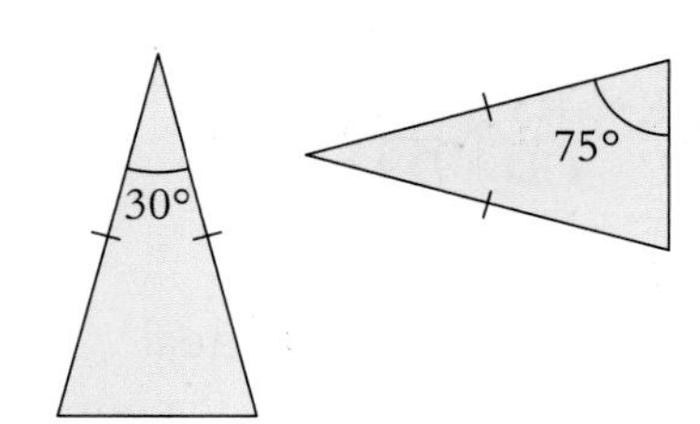

e

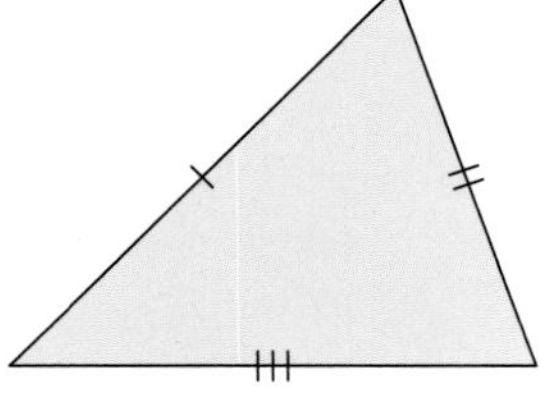

f

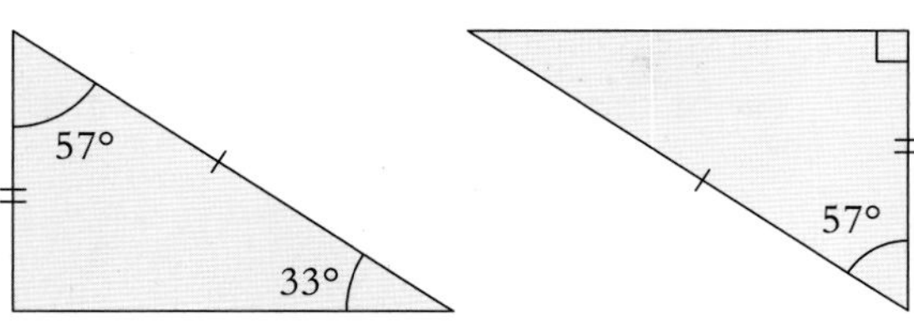

3 ABCD is a parallelogram.
Explain why triangles ABC and ACD are congruent.

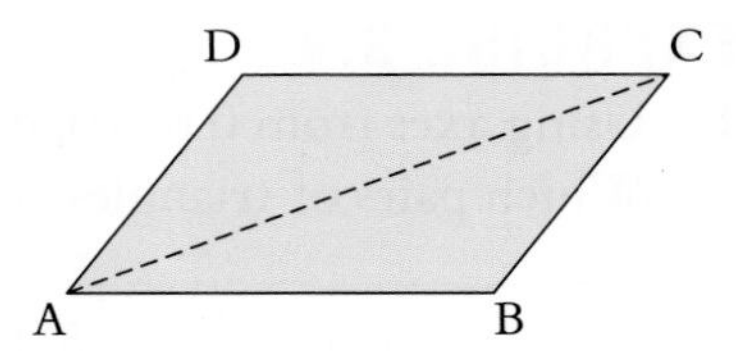

4 PQRS is a rectangle.
Explain why triangles PQS and QSR are congruent.

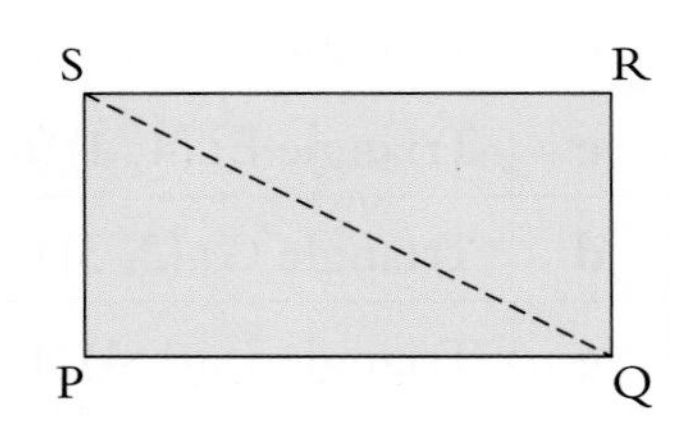

5 Explain why triangles ABC and PQR are congruent.

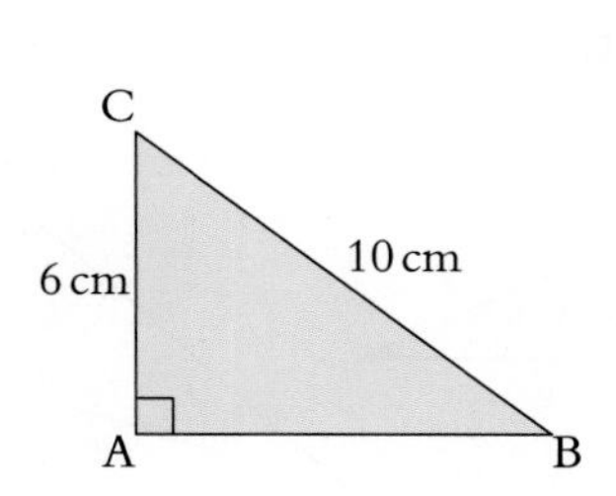

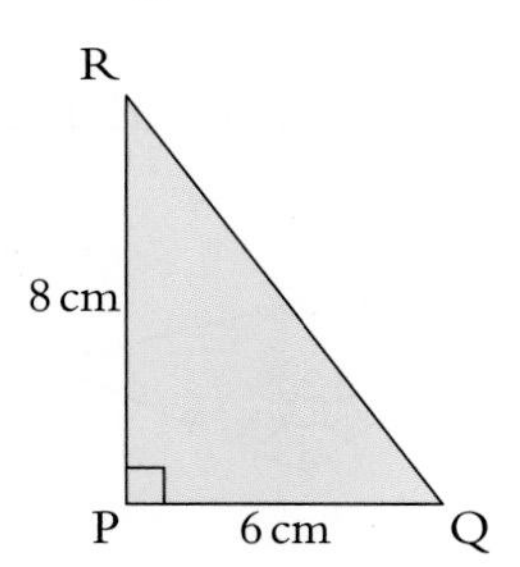

Similar shapes
These two quadrilaterals are **similar**.

$$\frac{PQ}{AB} = \frac{QR}{BC} = \frac{RS}{CD} = \frac{SP}{DA} = 2$$

In similar shapes

- Corresponding angles are equal
- Corresponding sides are in the same ratio.

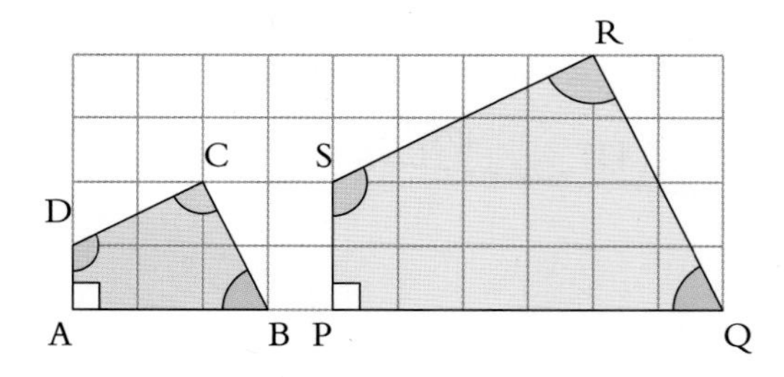

Similar triangles
To show that two triangles are similar it is sufficient to show that just *one* of the above conditions is satisfied.

EXAMPLE

Are these triangles similar?

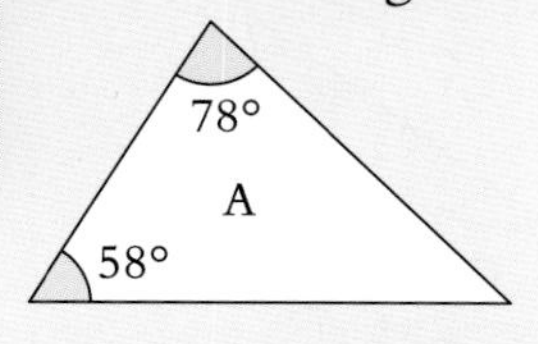

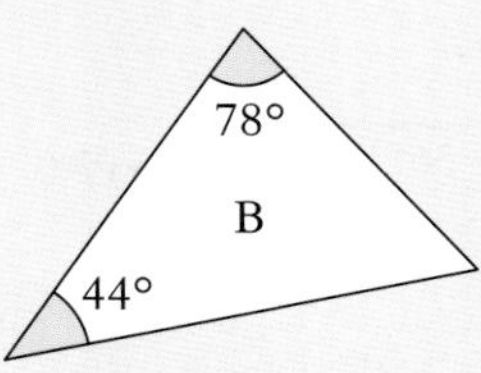

For the triangles to be similar, corresponding angles must be equal.

Triangle A: third angle = 180 – (58 + 78)

= 44 *Angles are: 58° 78° 44°.*

Triangle B: third angle = 180 – (78 + 44)

= 58 *Angles are: 58° 78° 44°.*

Triangles A and B are similar.

EXAMPLE

Are these triangles similar?

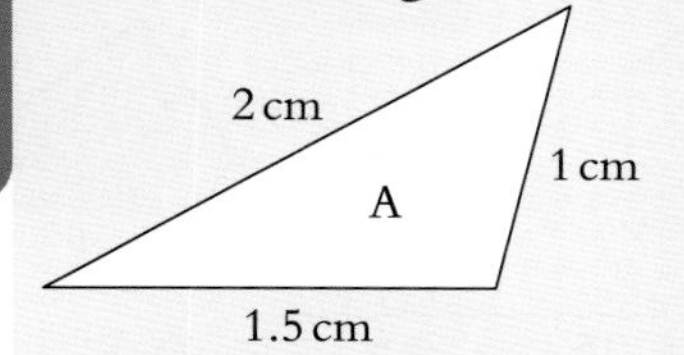

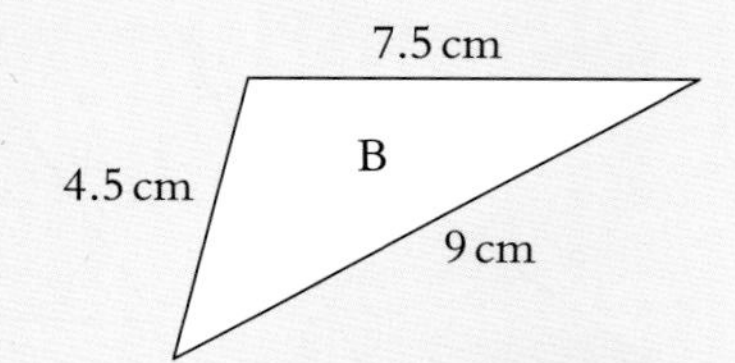

For the triangles to be similar, corresponding sides must be in the same ratio.

Ratio of longest sides = $\frac{9}{2} = 4.5$

Ratio of shortest sides = $\frac{4.5}{1} = 4.5$

Ratio of remaining sides = $\frac{7.5}{1.5} = 5$

Triangles A and B are not similar.

EXAMPLE

Find x and y.

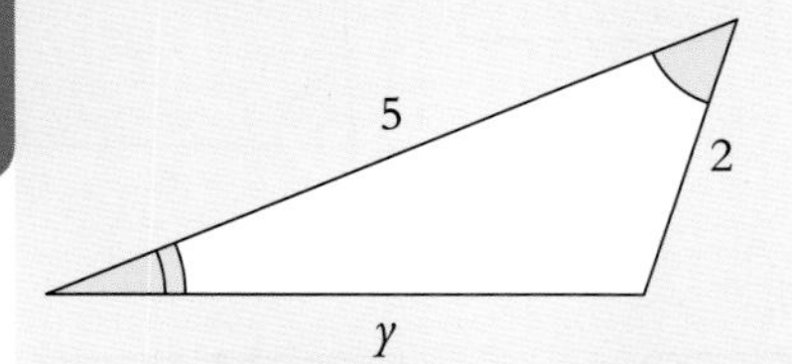

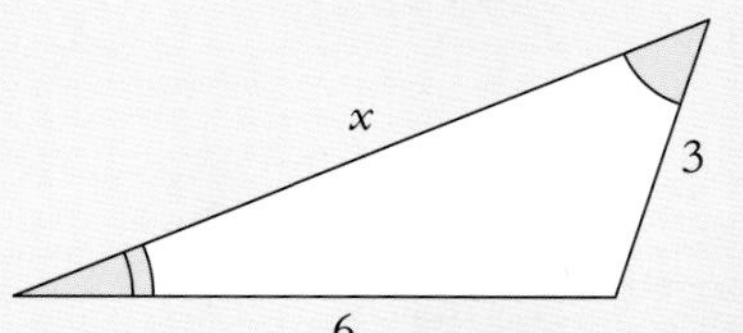

Using ratios of corresponding sides $\frac{3}{2} = \frac{x}{5} = \frac{6}{y}$

separating gives $\frac{3}{2} = \frac{x}{5}$ and $\frac{3}{2} = \frac{6}{y}$

'cross multiply' $15 = 2x$ $3y = 12$

$x = 7.5$ $y = 4$

Alternative method:

using scale factors

$3 = 2 \times 1.5$

So $x = 5 \times 1.5 = 7.5$

and $y = 6 \div 1.5 = 4$

EXERCISE 3.10

1 Which of these triangles are similar in shape?

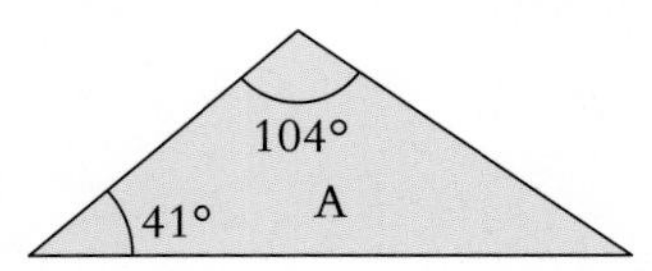

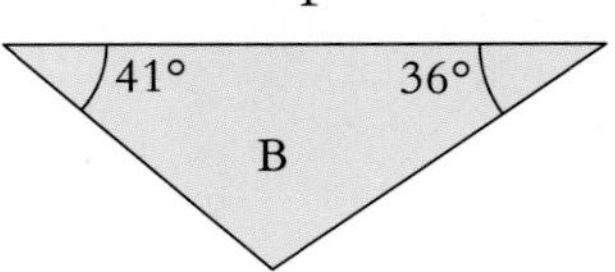

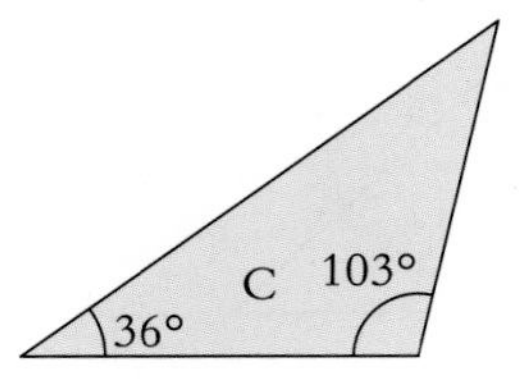

2 Which of these triangles are similar in shape?

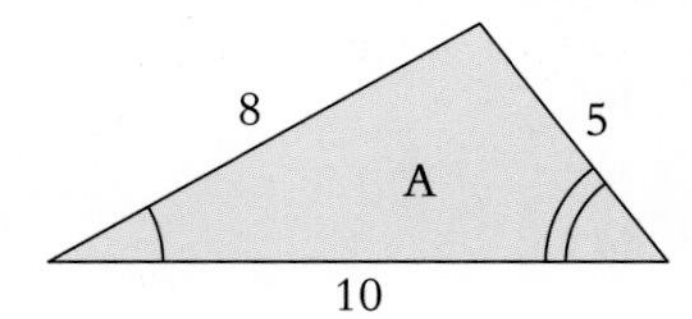

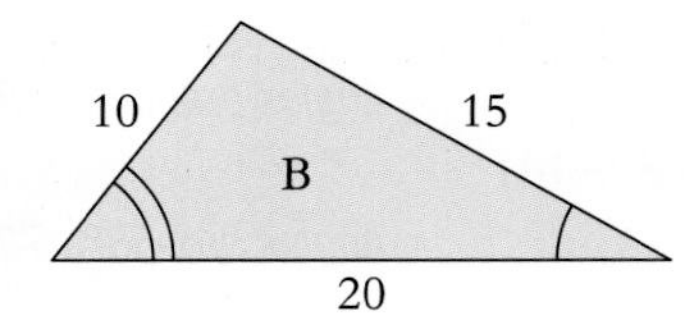

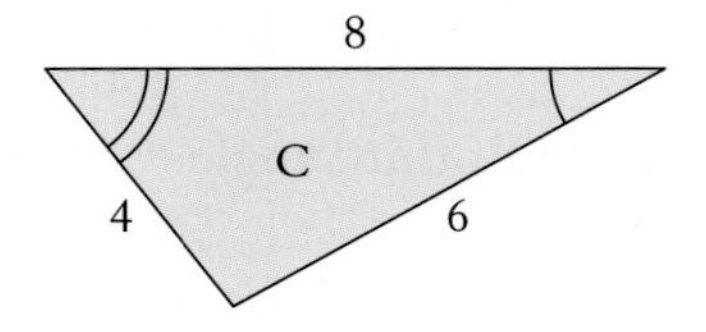

3 The following pairs of shapes are similar. Find the unknown lengths.

a

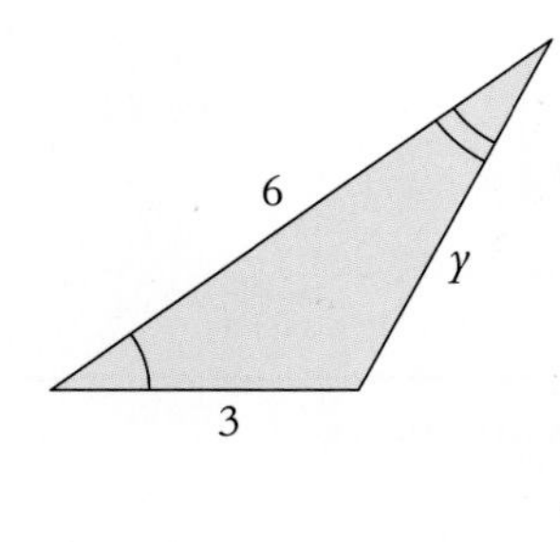

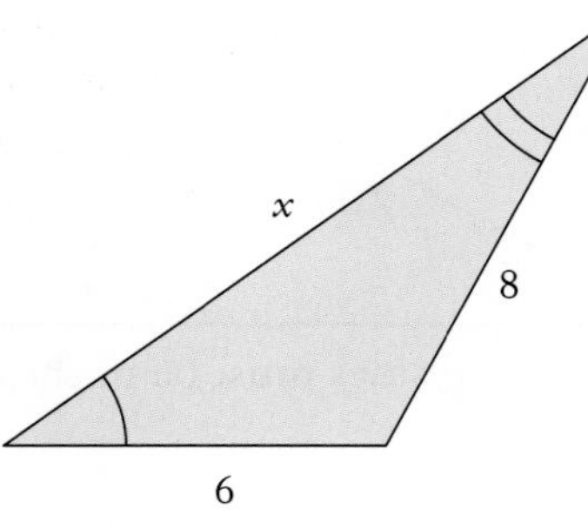

b

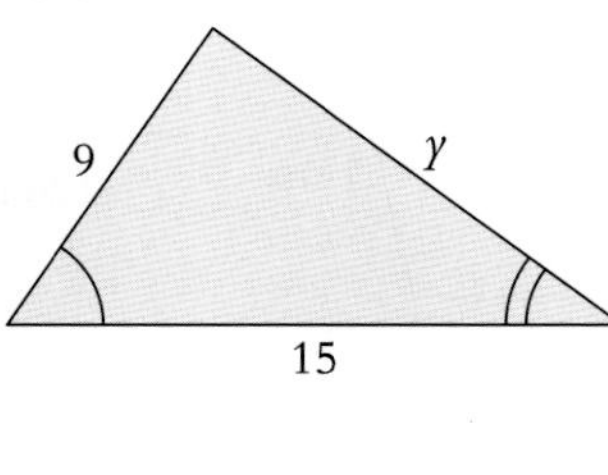

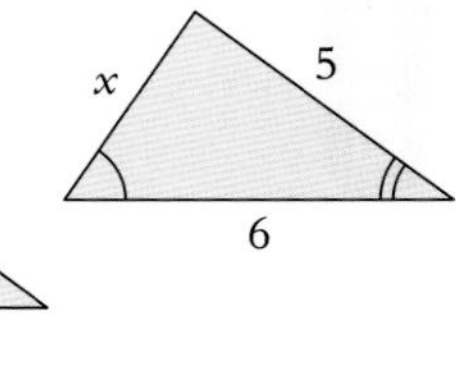

c

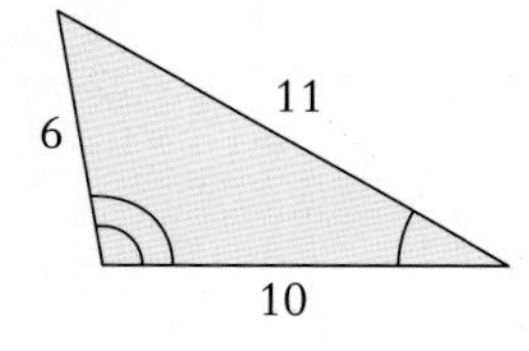

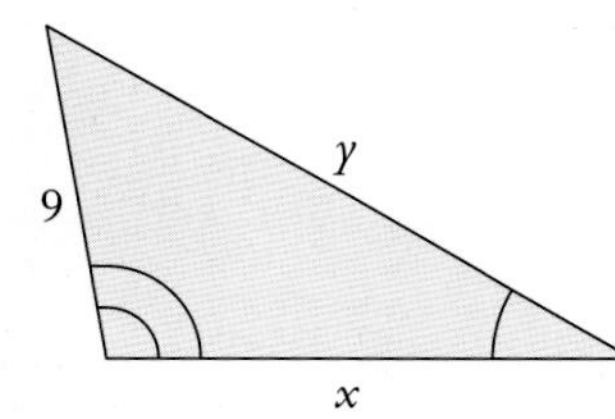

d

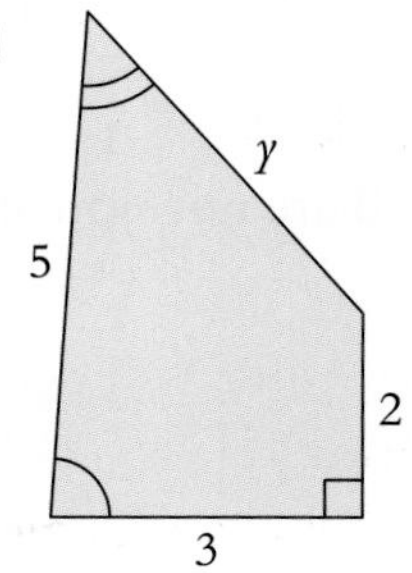

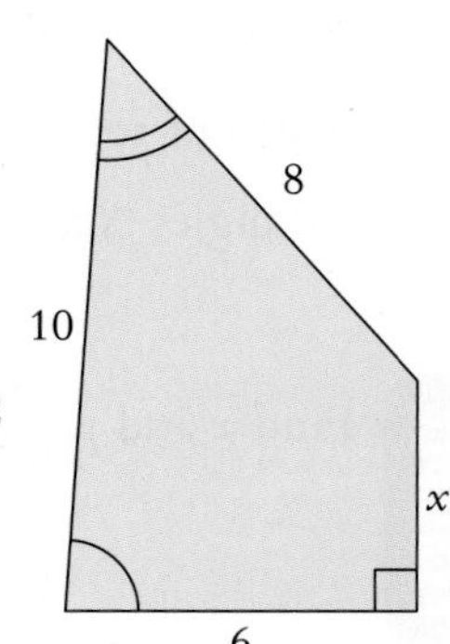

e

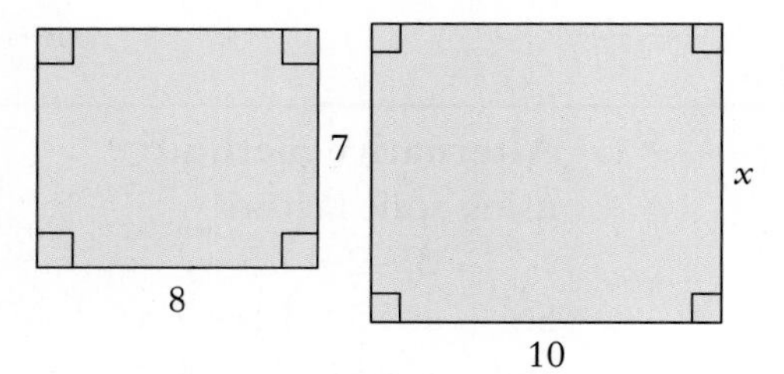

f

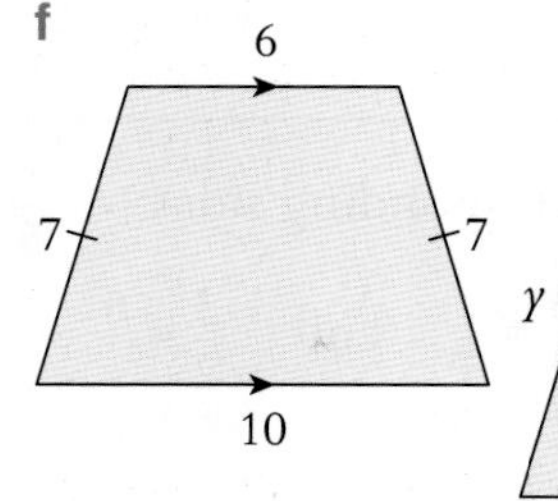

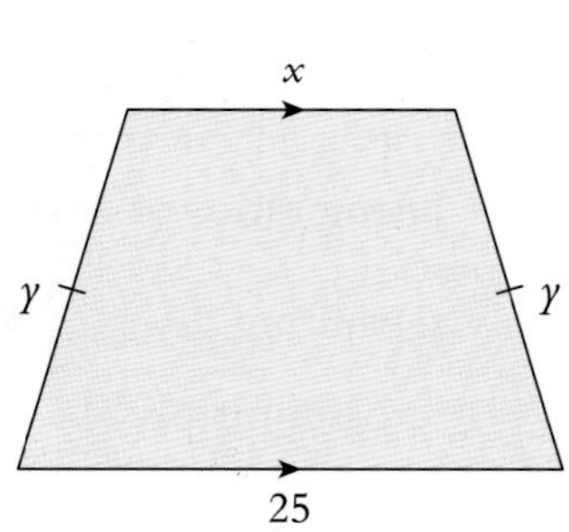

EXAMPLE

Triangles ABC and PQR are similar. Find the values of x and y.

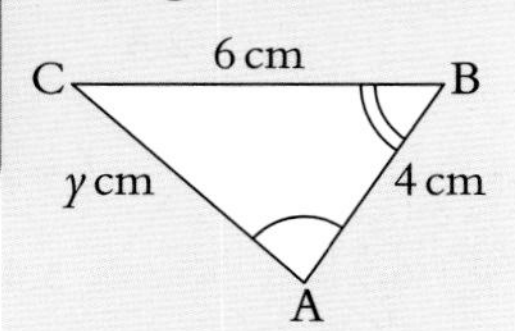

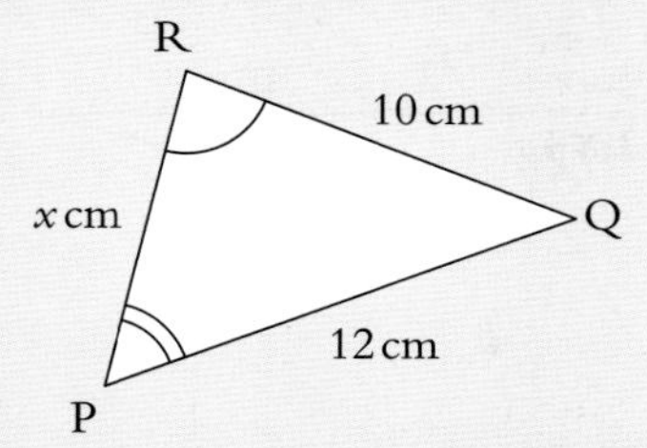

Rotating the second triangle makes it easier to see which are the corresponding sides.

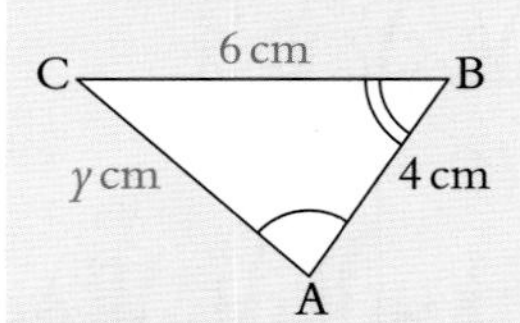

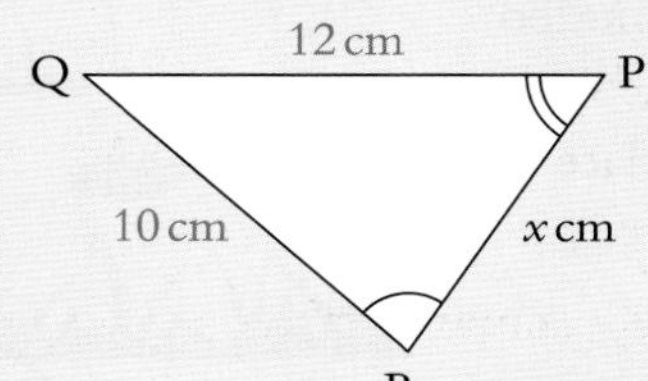

Using ratios of corresponding sides:

$$\frac{12}{6} = \frac{x}{4} = \frac{10}{y}$$

so $\frac{12}{6} = \frac{x}{4}$ and $\frac{12}{6} = \frac{10}{y}$

$2 = \frac{x}{4}$ $\quad$ $2 = \frac{10}{y}$

$x = 8$ $\quad$ $y = 5$

Alternative method:
using scale factors

$12 = 6 \times 2$

So $x = 4 \times 2 = 8$

and $y = 10 \div 2 = 5$

EXAMPLE

AB is parallel to DE.
AB =14 cm, AD = 6 cm, CD = 18 cm and CE = 12 cm.
BE = x cm and DE = y cm.
Find the values of x and y.

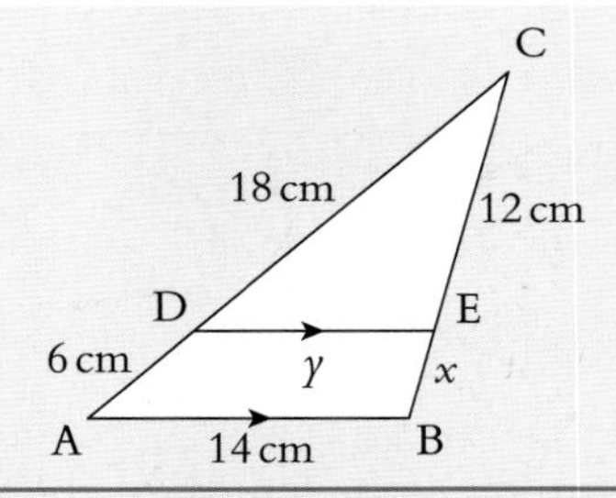

Triangles ABC and DEC are similar because:
∠DCE = ∠ACB (common angle)
∠CDE = ∠CAB (corresponding angles)
∠CED = ∠CBA (corresponding angles)
(Draw the two triangles separately to help you to identify the corresponding sides.)
Using ratios of corresponding sides:

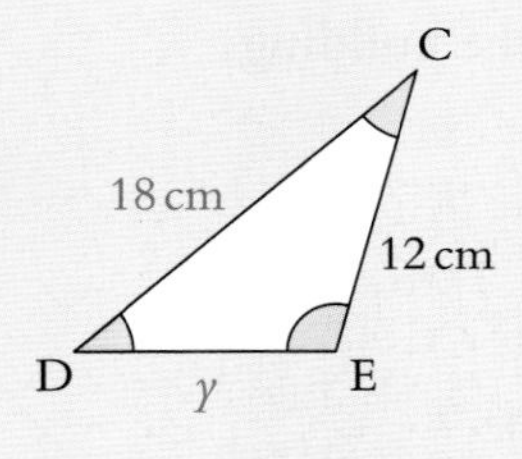

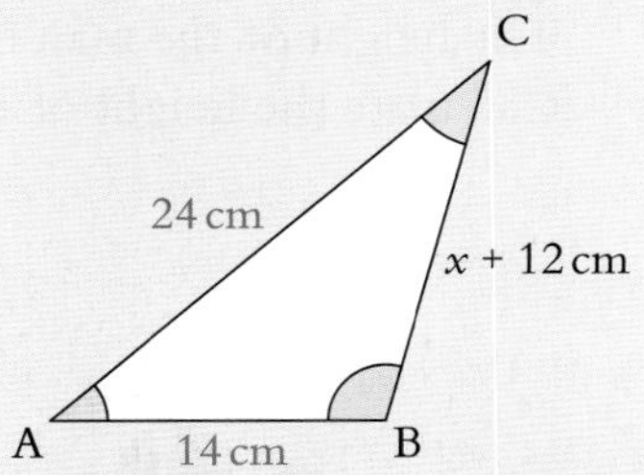

$$\frac{24}{18} = \frac{x+12}{12} = \frac{14}{y}$$

so $\frac{24}{18} = \frac{x+12}{12}$ and $\frac{24}{18} = \frac{14}{y}$

$288 = 18x + 216$ $\quad$ $24y = 252$ 'cross multiplying'

$18x = 72$ $\quad$ $y = 10.5$

$x = 4$

UNIT 3

EXERCISE 3.11

1 Find the values of x and y for each of the following diagrams.

a

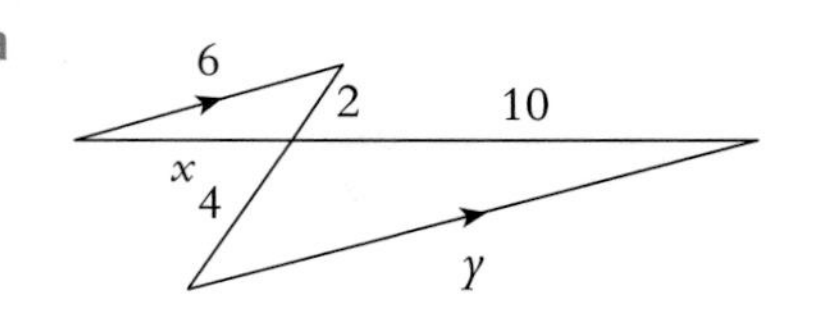

b

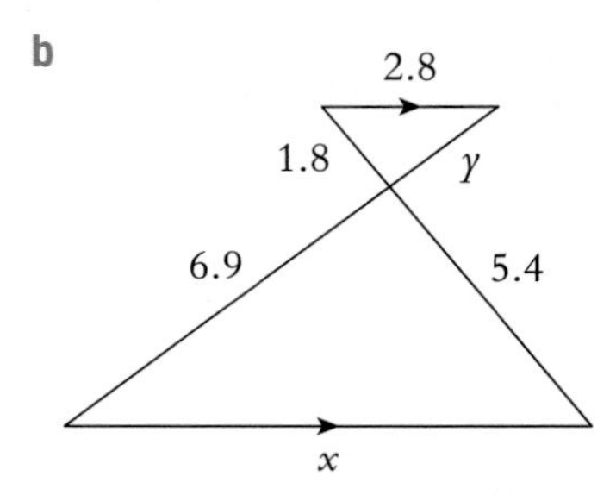

2 The width of a river can be estimated without having to cross over the river. Use the diagram below to calculate the width (w) of this river.

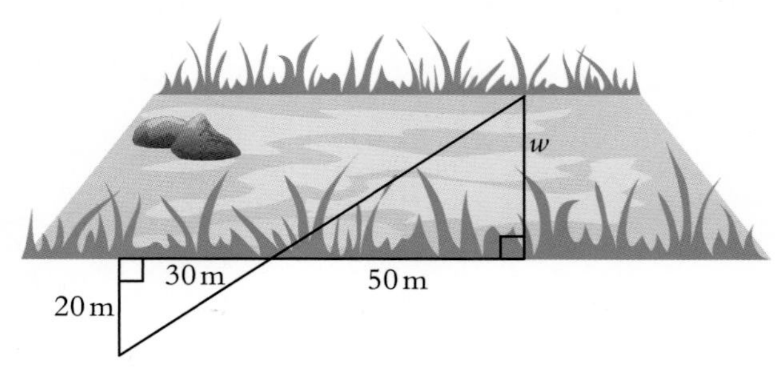

3 Find the value of x for each of the following diagrams.

a

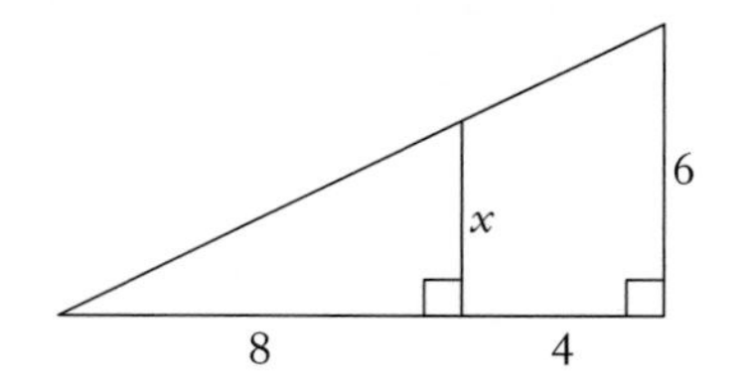

b

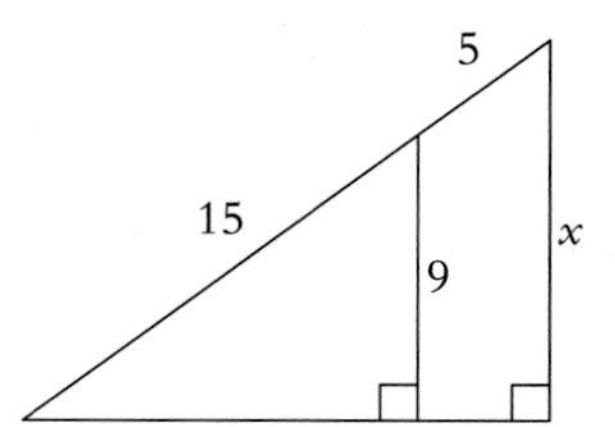

c

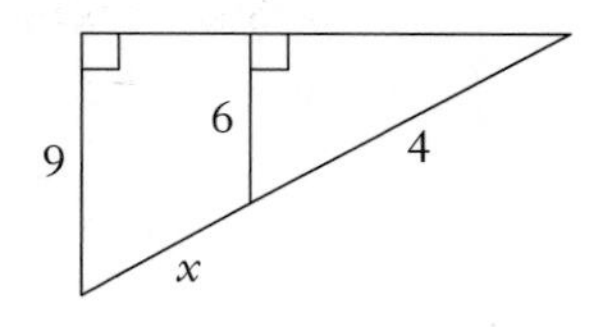

d

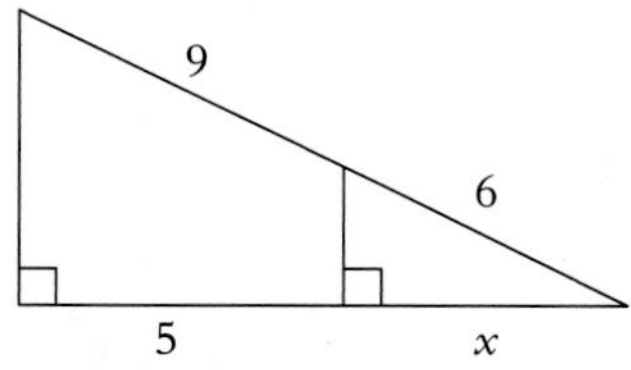

4 At midday the length of the shadow from a building is 50 m and the length of the shadow from a man is 1.65 m.
The height of the man is 1.8 m.
Calculate the height of the building.

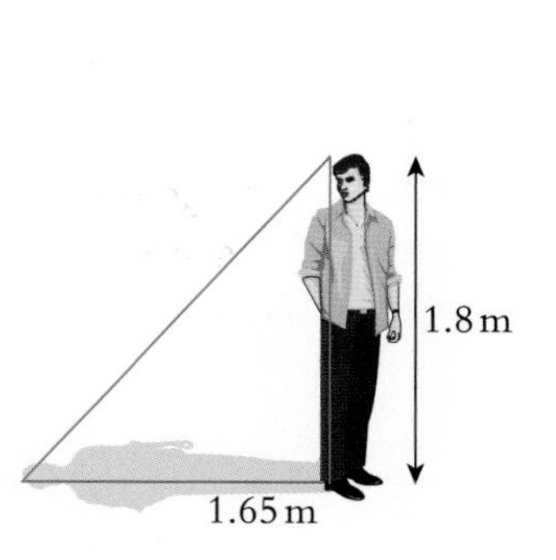

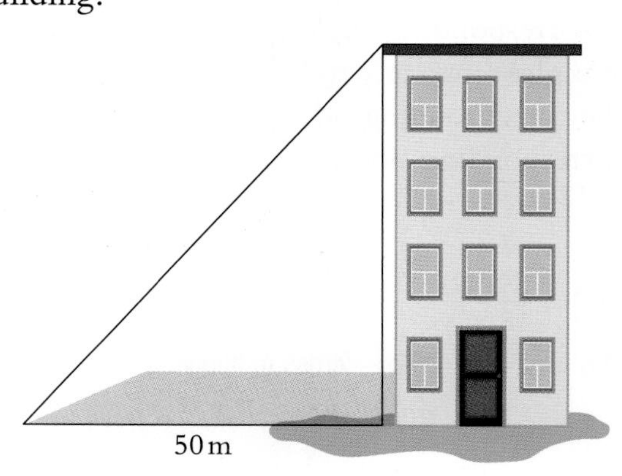

5 Find the values of x and y for each of the following diagrams.

a

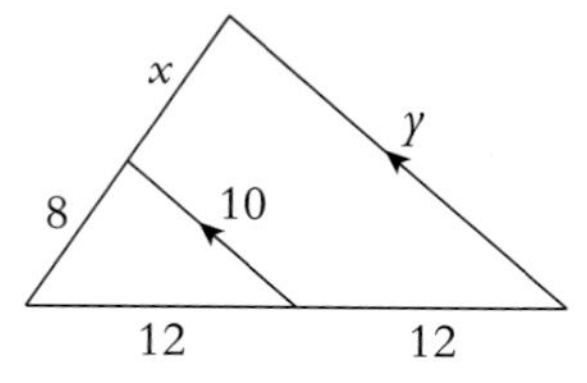

b

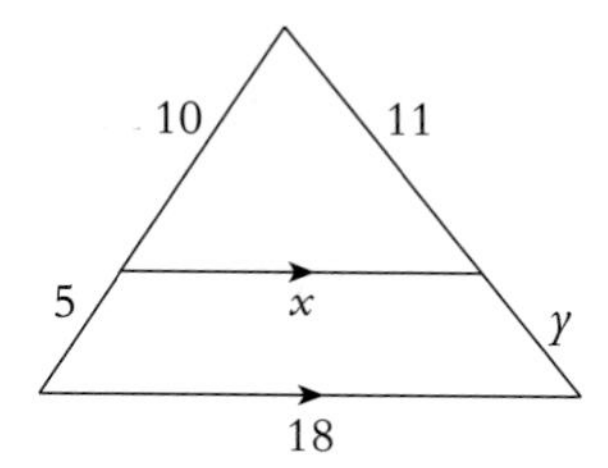

6 DE is parallel to AC.

$AD = \frac{1}{4}AB$.

a Find the coordinates of D and E.

b Find the coordinates of the midpoint of the line DE.

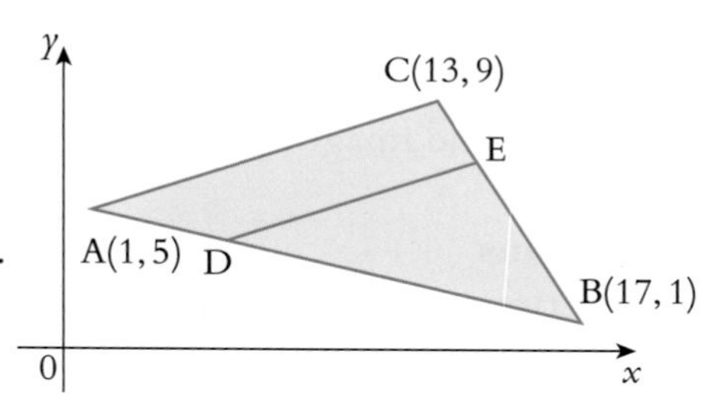

7 A is the point (3, −7) and B is the point (13, −2).
P and Q are points that lie on the line segment joining A to B.

a Find the coordinates of the midpoint of the line AB.

b AP : PB = 3 : 2. Find the coordinates of the point P.

c AQ : AB = 1 : 5. Find the coordinates of the point Q.

8 Find the values of x and y for each of the following diagrams.

a

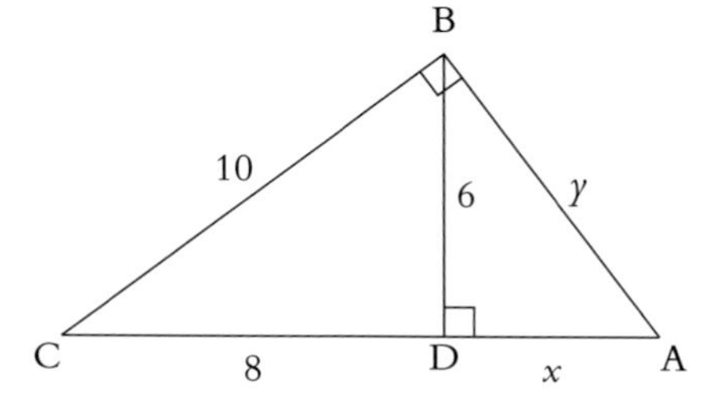

b

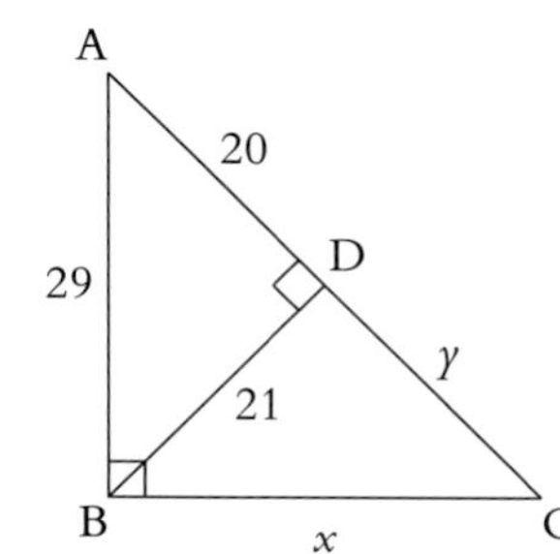

TOP TIP

∠BCD = ∠ABD
in each of these triangles.

9 Find the values of x and y for each of the following diagrams.

a ∠CAD = ∠DCB

b ∠DAB = ∠CBD

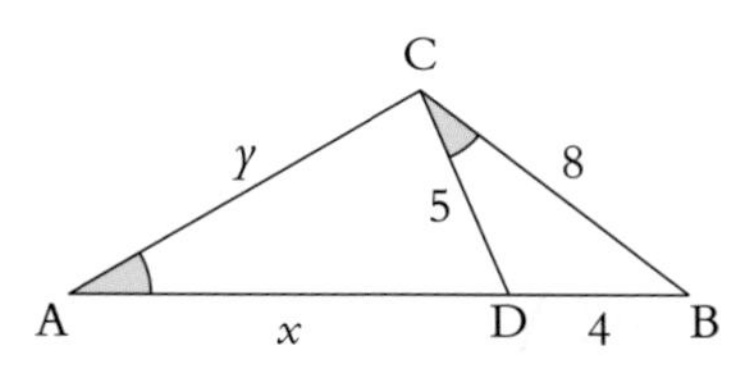

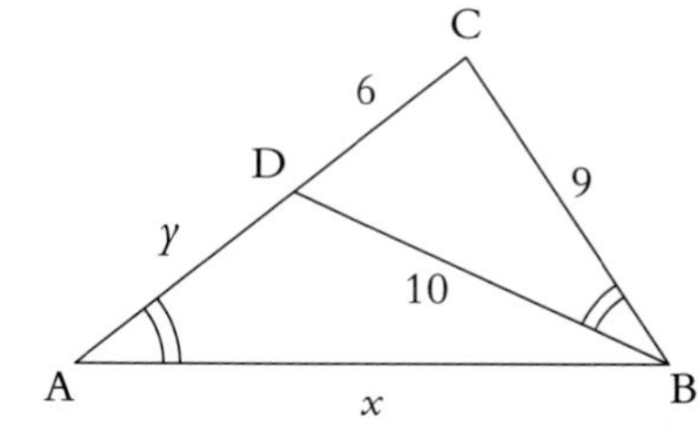

UNIT 3

Reflections, rotations and translations

THIS SECTION WILL SHOW YOU HOW TO

- Reflect, rotate and translate a shape
- Use a combination of these transformations

A **transformation** changes the position or size of a shape.
The original shape is called the **object** and the final shape is called the **image**.
Reflections, **rotations** and **translations** are transformations that change only the position of a shape.
The object and image are congruent for reflections, rotations and translations.

Reflections

A reflection is a transformation that involves a mirror line.

In the diagram, triangle A′B′C′ is the image of triangle ABC after a reflection in the line $x = 4$.

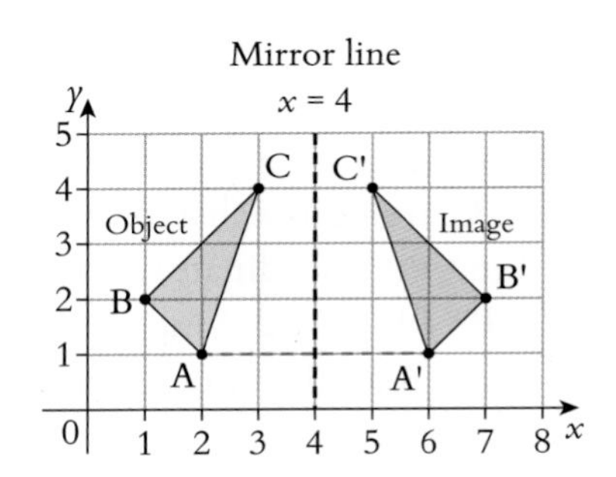

The line joining the point A to its image A′ is at right angles to the mirror line **and** A is the same distance in front of the mirror line as A′ is behind the mirror line.

EXAMPLE

On a graph plot the points P(3, 1), Q(7, 3) and R(7, 7).
Reflect triangle PQR in the line $y = x$. Label the image P′Q′R′.

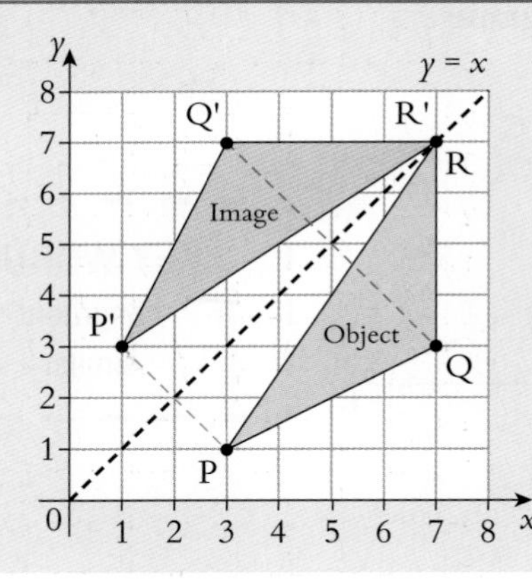

The point R lies on the line $y = x$ so it does not move when reflected in the line. It is called an **invariant** point.

The lines PP′ and QQ′ are at right angles to the mirror line $y = x$.

EXERCISE 3.12

1 Copy the diagrams and draw the image of each shape when reflected in the mirror line.

a

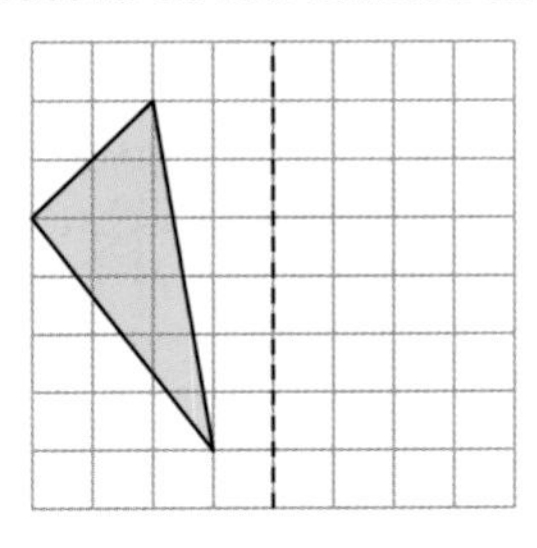

b

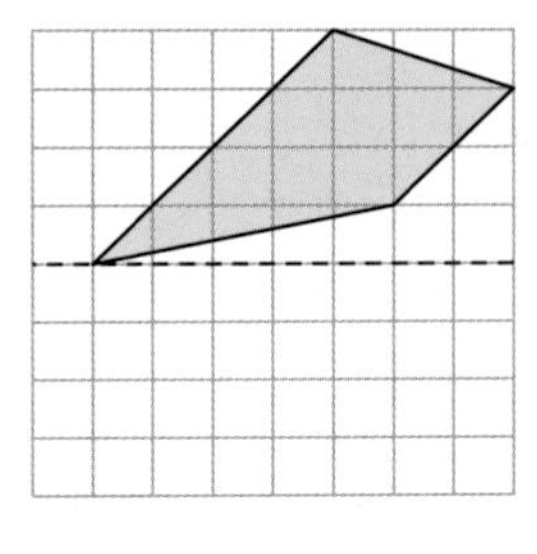

c

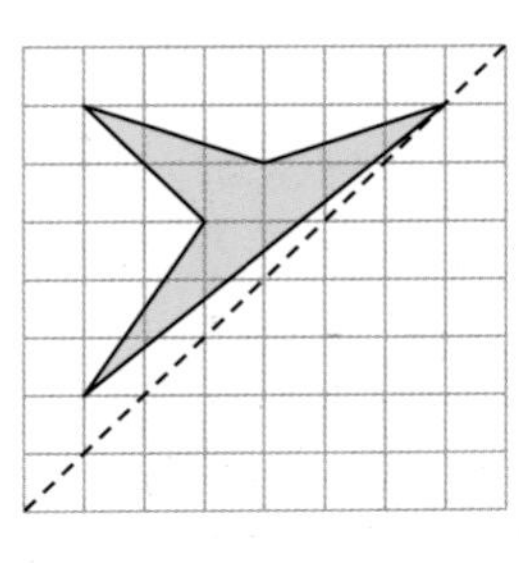

d

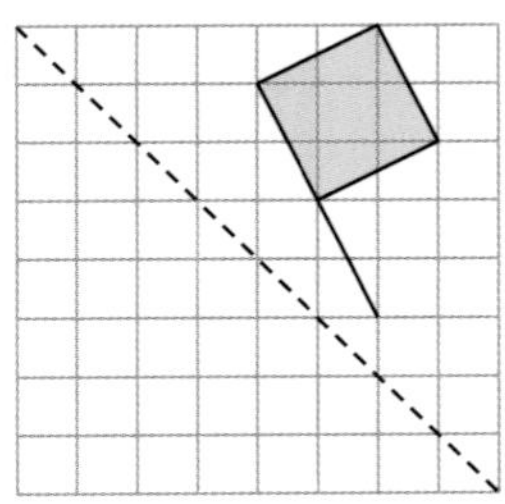

e

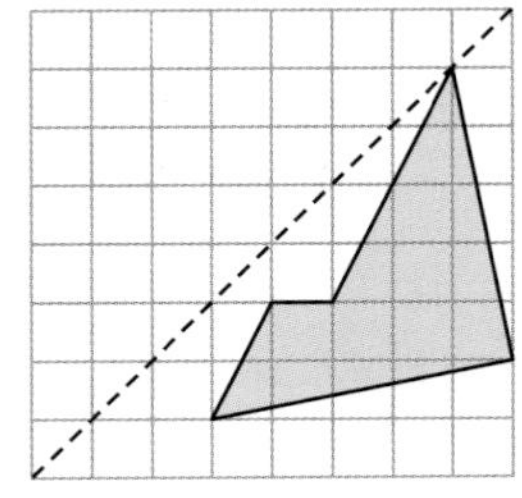

f

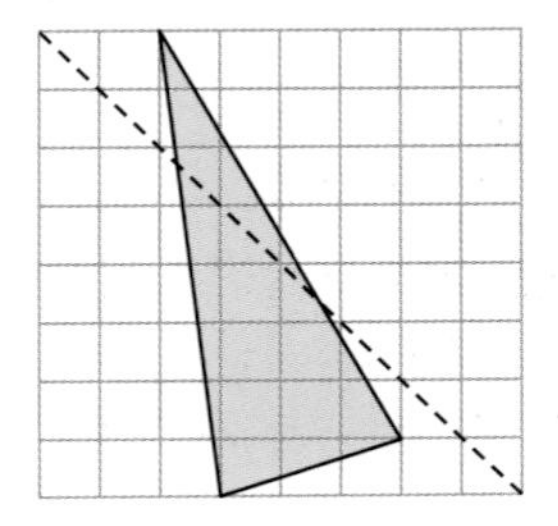

2 Write the equation of the mirror line for each of these reflections.

a A onto B
b B onto C
c B onto D
d E onto F
e E onto G
f E onto I
g F onto H
h G onto H

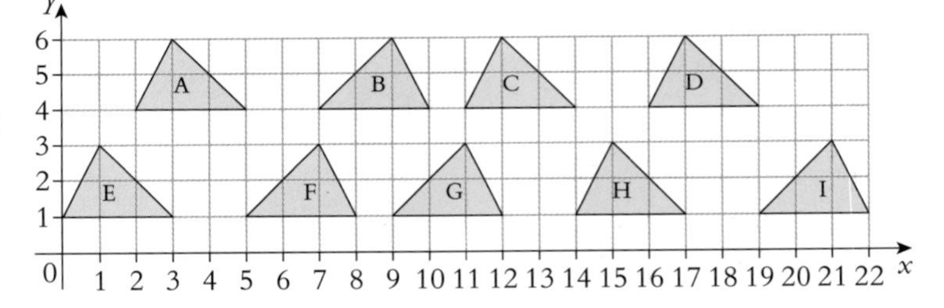

3 Name the mirror line for each of these reflections.

a A onto B
b A onto C
c D onto F
d E onto F
e G onto H
f H onto I

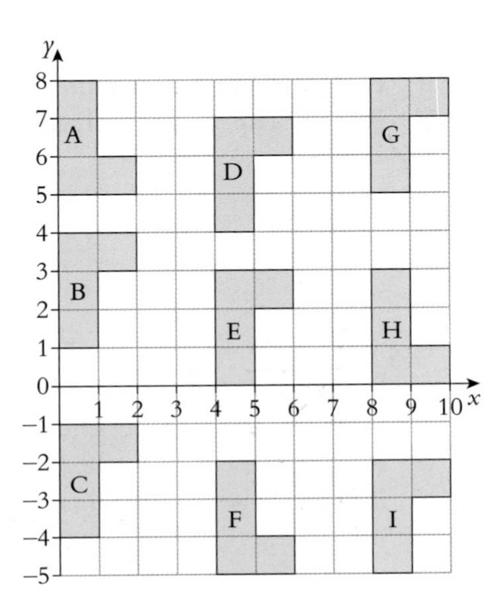

4 Name the mirror line for each of these reflections.

- **a** A onto B
- **b** A onto D
- **c** B onto C

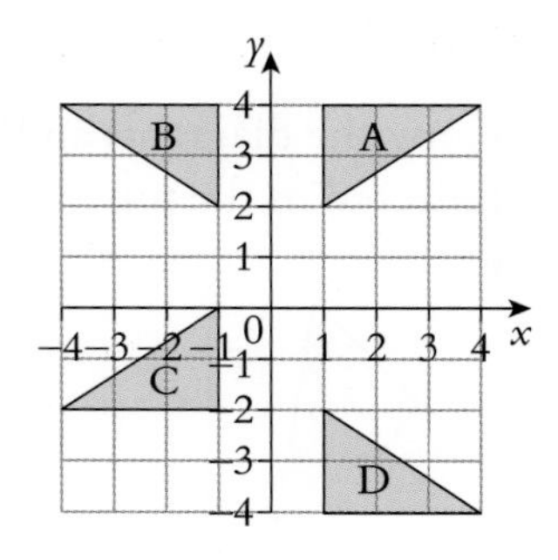

5 Name the mirror line for each of these reflections.

- **a** A onto B
- **b** A onto C
- **c** C onto E
- **d** B onto E
- **e** D onto E

6 Name the mirror line for each of these reflections.

- **a** A onto D
- **b** F onto G
- **c** C onto E
- **d** B onto H
- **e** C onto D
- **f** C onto H
- **g** D onto G
- **h** A onto E
- **i** H onto G
- **j** F onto B

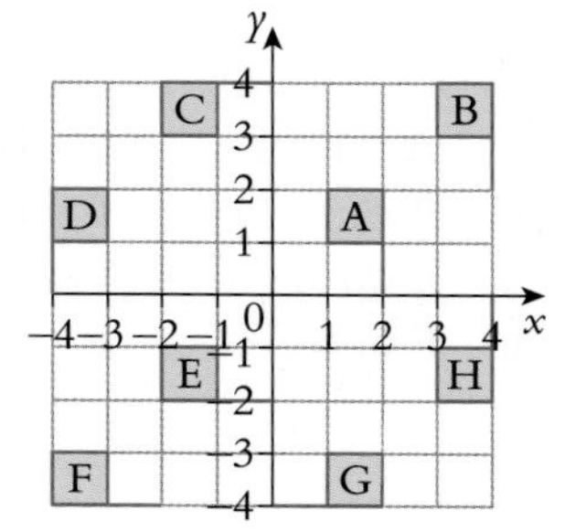

7 Copy the diagram onto graph paper.

- **a** Reflect shape A in the line $x = 0$ and label the image B.
- **b** Reflect shape A in the line $y = 1$ and label the image C.
- **c** Reflect shape A in the line $y = x$ and label the image D.

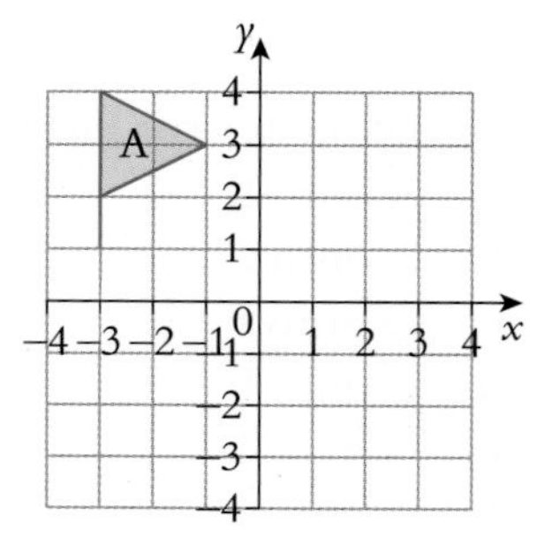

8 Copy the diagram onto graph paper.

- **a** Reflect shape A in the line $y = 2$ and label the image B.
- **b** Reflect shape A in the line $y = -1$ and label the image C.
- **c** Reflect shape A in the line $x = 0$ and label the image D.
- **d** Reflect shape A in the line $y = -x$ and label the image E.

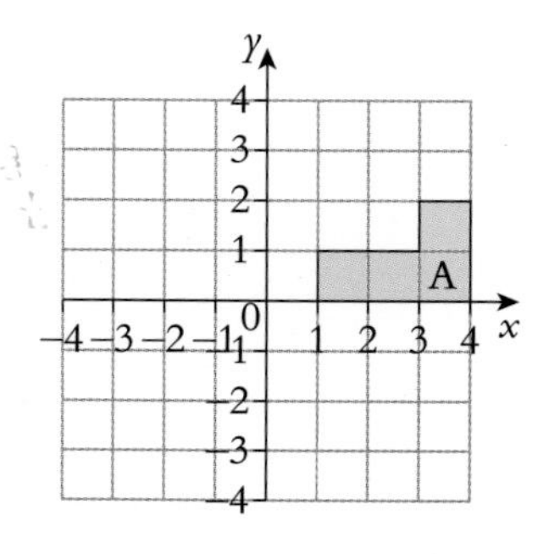

9 Copy the diagram onto graph paper.

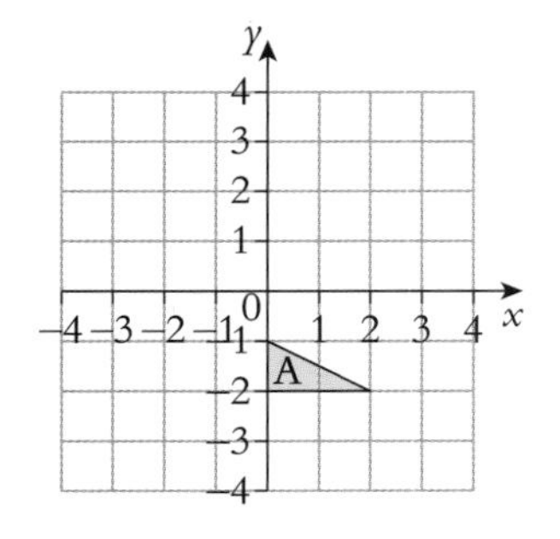

- **a** Reflect shape A in the line $y = 1$ and label the image B.
- **b** Reflect shape A in the line $x = -1$ and label the image C.
- **c** Reflect shape A in the line $x = 2$ and label the image D.
- **d** Reflect shape A in the line $y = x$ and label the image E.
- **e** Reflect shape A in the line $y = x + 2$ and label the image F.
- **f** Reflect shape A in the line $x + y = 1$ and label the image G.

10 Copy the diagram onto graph paper.

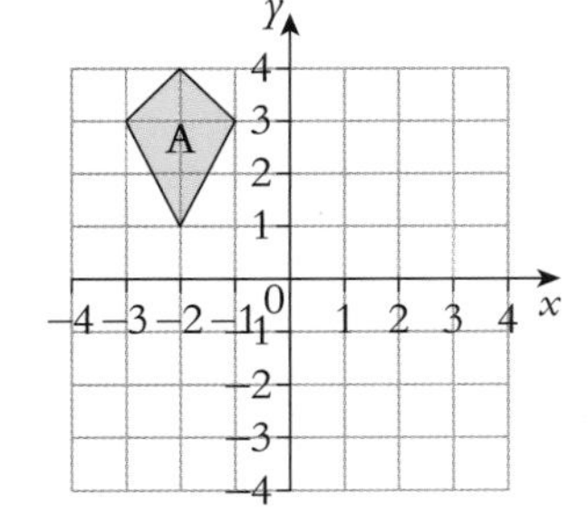

- **a** Reflect shape A in the x-axis and label the image B.
- **b** Reflect shape A in the line $x = 0.5$ and label the image C.
- **c** Reflect shape A in the line $y = x$ and label the image D.
- **d** Reflect shape A in the line $y = x + 3$ and label the image E.

11 Use axes from −4 to 4.
Draw the pentagon whose vertices are (−1, −1), (2, 3), (1, 4), (−2, 3) and (−3, 1).
Reflect the pentagon in the line $y = x$.
The object and image are congruent.
Mark the sides and angles to show which are equal to each other.

12 Copy the diagram onto graph paper.

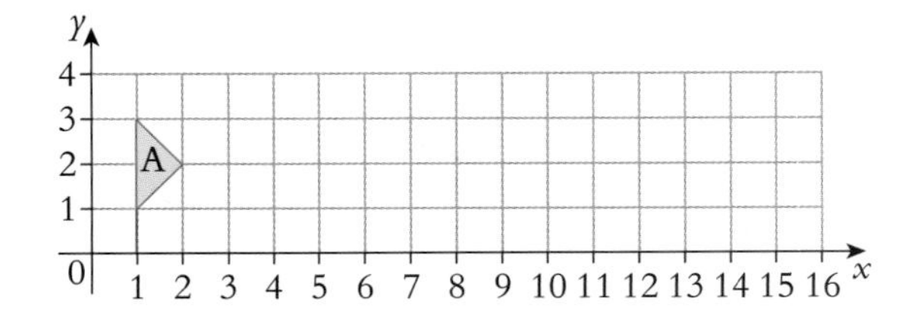

- **a** Reflect shape A in the line $x = 4$ and label the image B.
- **b** Reflect shape B in the line $x = 8$ and label the image C.
- **c** Reflect shape C in the line $x = 10.5$ and label the image D.
- **d** Shape A can be transformed onto shape D by a **single** reflection. Name the mirror line for this single reflection.

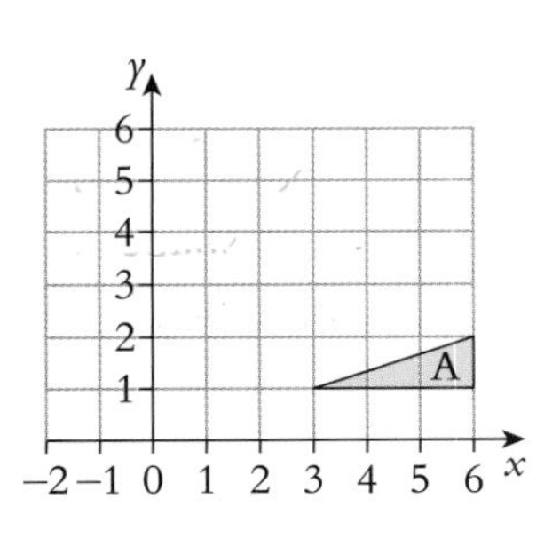

13 Copy the diagram onto graph paper.

- **a** Reflect shape A in the line $y = x$ and label the image B.
- **b** Reflect shape B in the line $x + y = 4$ and label the image C.
- **c** Reflect shape C in the line $x = 2$ and label the image D.
- **d** Shape A can be transformed onto shape D by a **single** reflection. Name the mirror line for this single reflection.

Rotations

A **rotation** is a transformation that turns a shape about a fixed point.
The fixed point is called the **centre of rotation**.
The shape can be rotated in a clockwise (↻) or anticlockwise (↺) direction.

To describe a rotation fully, you need to give
- the angle of rotation
- the direction of rotation
- the centre of rotation.

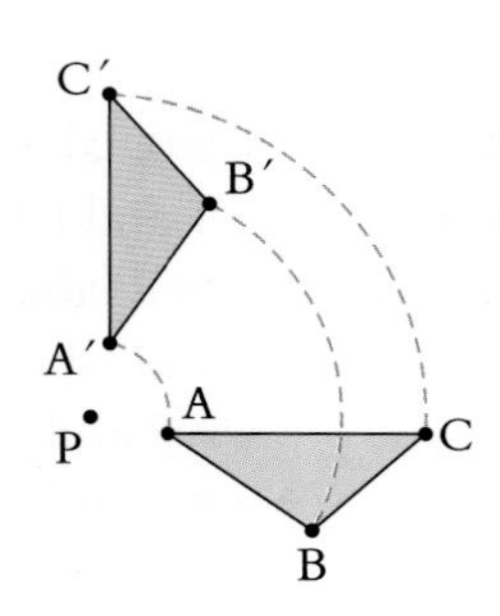

In the diagram above, triangle A′B′C′ is the image of triangle ABC after a rotation of 90° anticlockwise about the point P.

➡ **NOTE:** you can use tracing paper to help you to do rotations.

➡ **NOTE:** you can use compasses to check the paths traced out by the points.

EXAMPLE

Rotate shape A 180° about the point (8, 2).

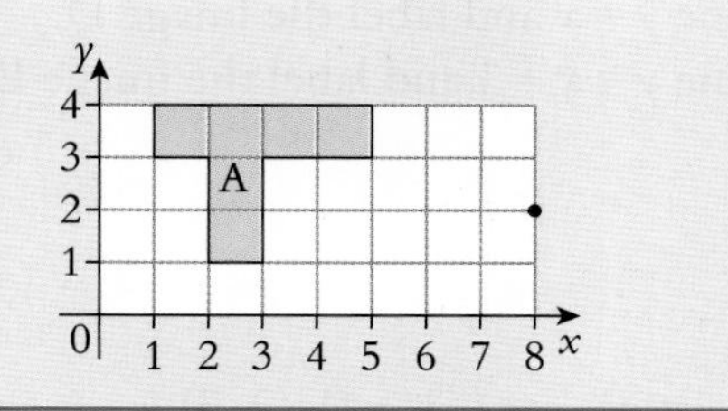

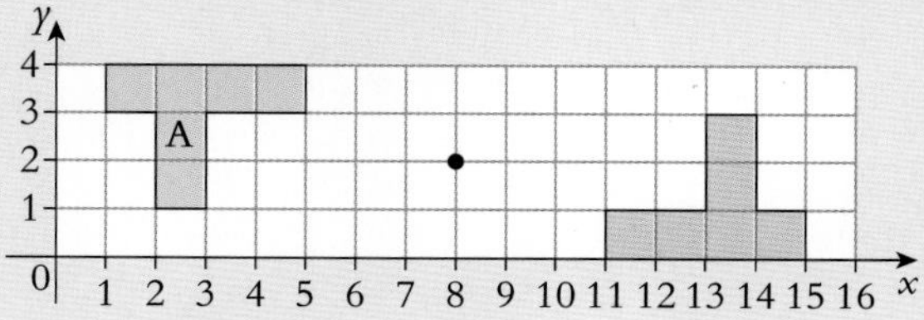

A rotation of 180° clockwise is the same as a rotation of 180° anticlockwise so the direction does not need to be given.

EXERCISE 3.13

1 Copy these shapes on graph paper.
Draw the image of each shape for the given rotations.
Use the marked point as the centre of rotation.

a

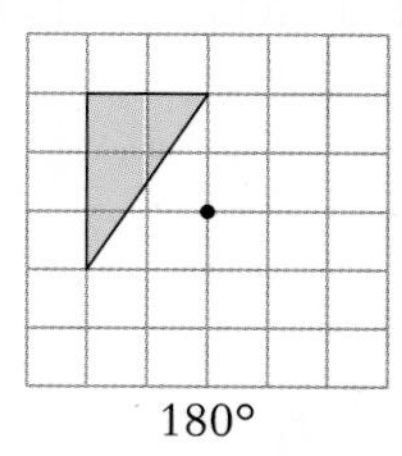

180°

b

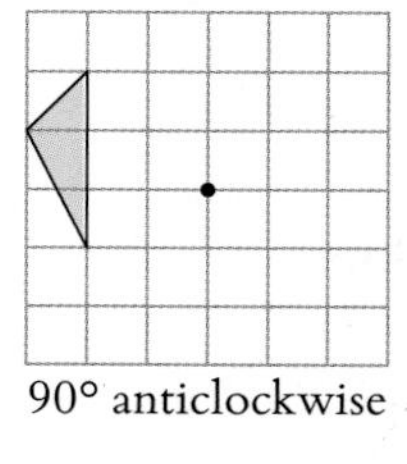

90° anticlockwise

c

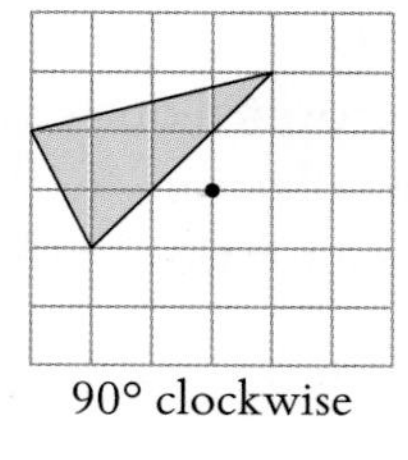

90° clockwise

d

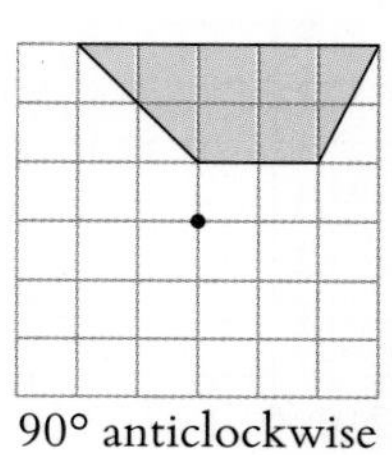

90° anticlockwise

e

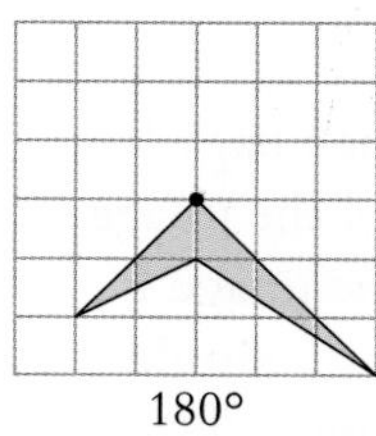

180°

f

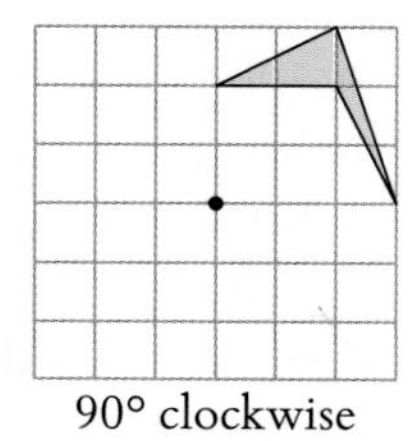

90° clockwise

2 Use axes from −4 to 4.

a Draw the triangle with vertices $(-4, 1)$, $(-4, 4)$ and $(-3, 1)$. Label the triangle A.

b Rotate A through 90° clockwise about the origin. Label the image B.

c Rotate A through 180° about the origin. Label the image C.

d Rotate A through 90° anticlockwise about the origin. Label the image D.

3 Use axes from −4 to 4.

a Draw the triangle with vertices $(3, 1)$, $(3, 4)$ and $(2, 3)$. Label the triangle A.

b Rotate A through 180° about the point $(2, 0)$. Label the image B.

c Rotate A through 180° about the point $(0, 1)$. Label the image C.

d Rotate A through 90° clockwise about the point $(1, 3)$. Label the image D.

4 Use axes from −4 to 4.

a Draw the kite with vertices $(1, 0)$, $(2, 2)$, $(1, 3)$ and $(0, 2)$. Label the kite A.

b Rotate A through 90° anticlockwise about the point $(-1, 1)$. Label the image B.

c Rotate A through 180° about the point $(-1, 1)$. Label the image C.

d Rotate A through 90° clockwise about the point $(-1, 1)$. Label the image D.

5 Use axes from −6 to 6.

a Draw the triangle with vertices $(3, -4)$, $(6, -1)$ and $(5, 1)$. Label the triangle A.

b Rotate A through 180° about the point $(2, -1)$. Label the image B.

c Rotate B through 90° clockwise about the point $(0, 3)$. Label the image C.

d Rotate C through 90° anticlockwise about the point $(-2, 1)$. Label the image D.

Describing a rotation

EXAMPLE

Describe fully the single transformation that takes triangle ABC onto triangle A′B′C′.

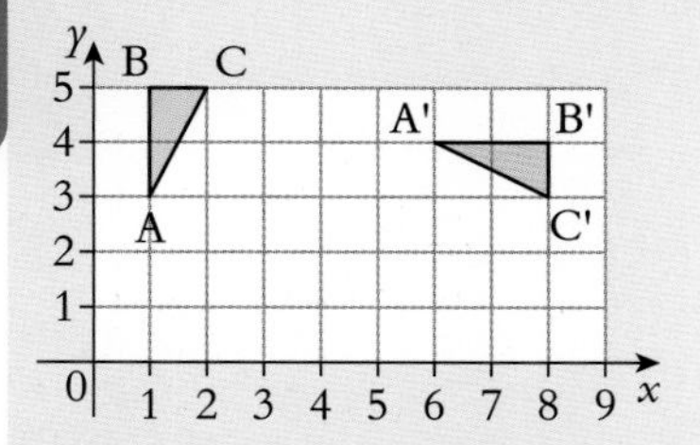

The line AB is vertical and the line A′B′ is horizontal so the angle of rotation is 90°.
The shape has been rotated clockwise.
There are two methods for finding the centre of rotation.

Method 1 (By trial and error)

Use tracing paper to try to find the centre of rotation.
Through trial and error you should find that the centre of rotation is at the point (4, 1).

Method 2 (By construction)

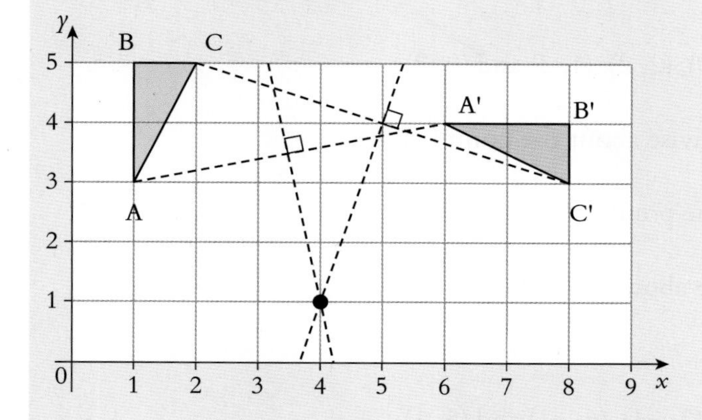

➡ **NOTE:** only two perpendicular bisectors need to be drawn to locate the centre of rotation.

Join A to A′ and draw the perpendicular bisector of the line AA′
Join C to C′ and draw the perpendicular bisector of the line CC′
The point where the two perpendicular bisectors meet is the centre of rotation.
The transformation that takes triangle ABC onto triangle A′B′C′ is a rotation 90° clockwise about the point (4, 1).

EXERCISE 3.14

1 Describe fully the single transformation that takes shape A onto shape B for each of the following diagrams.

a

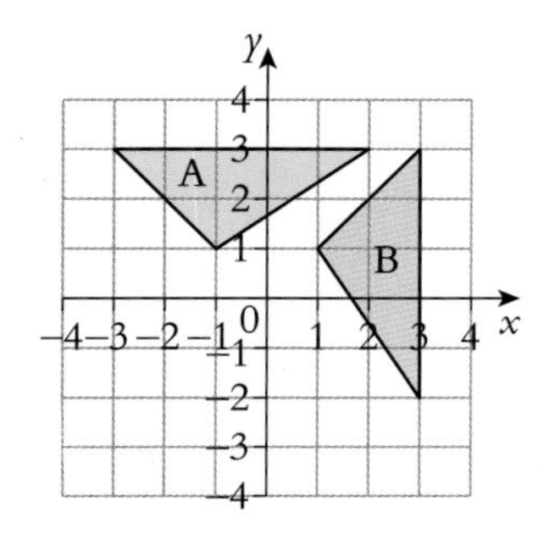

b

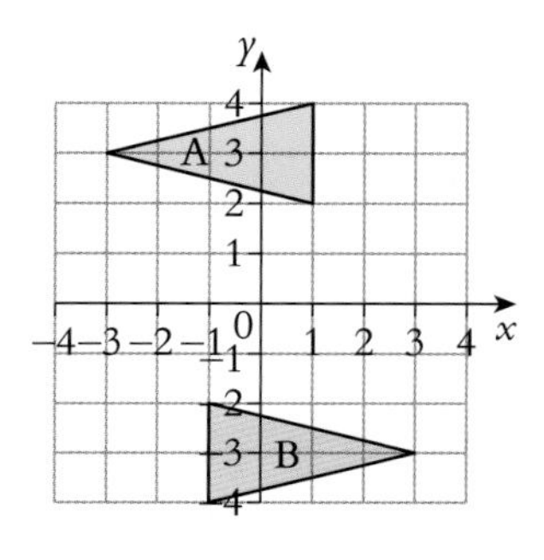

c

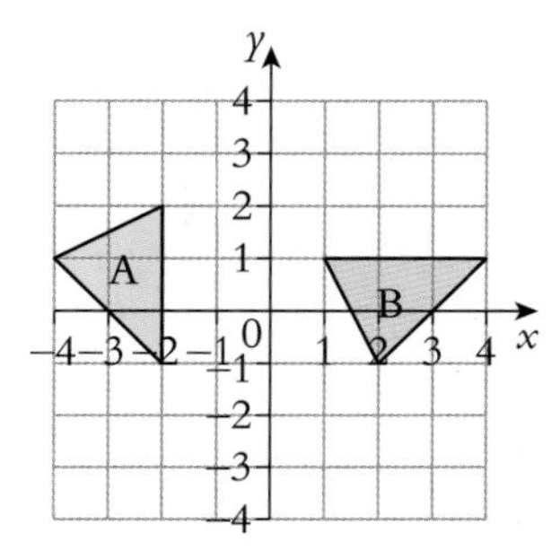

d

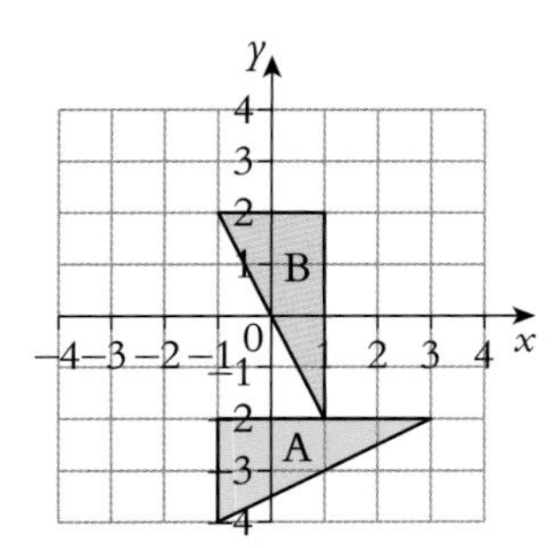

e

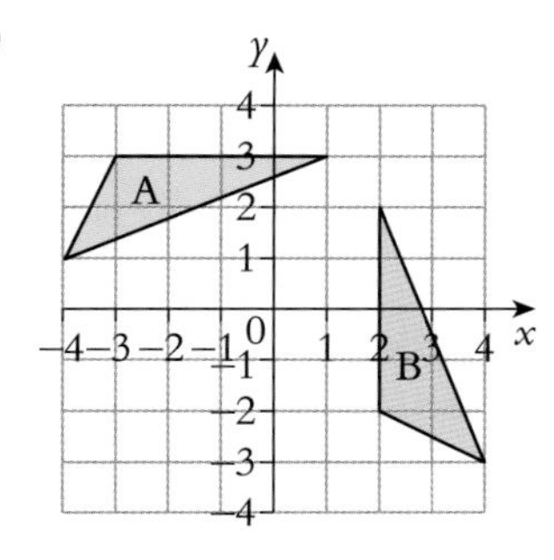

f

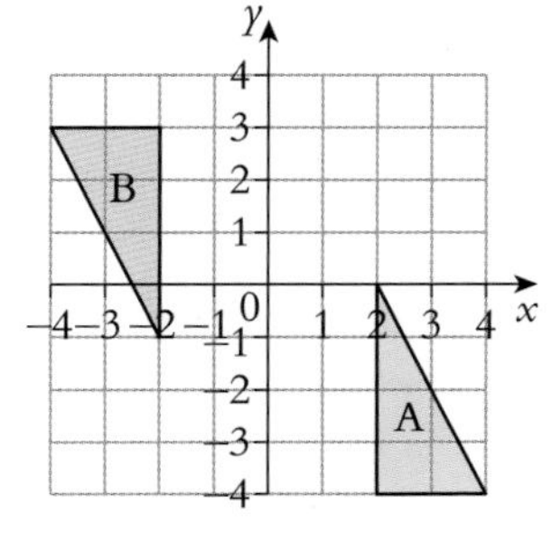

2

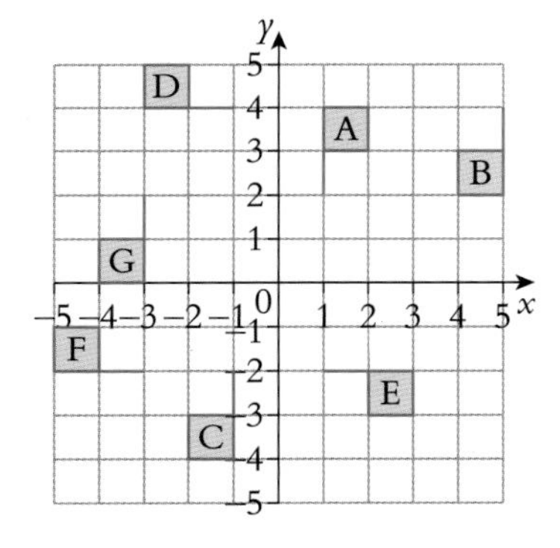

Describe fully the single transformation that moves shape:

a A onto B
b A onto C
c A onto D
d A onto E
e A onto F
f A onto G
g C onto D
h C onto E
i C onto F
j D onto E
k D onto G
l E onto G.

UNIT 3

Translations

A **translation** is a transformation that 'slides' a shape from one place to another. In a translation every point on the original shape moves the same distance and in the same direction.

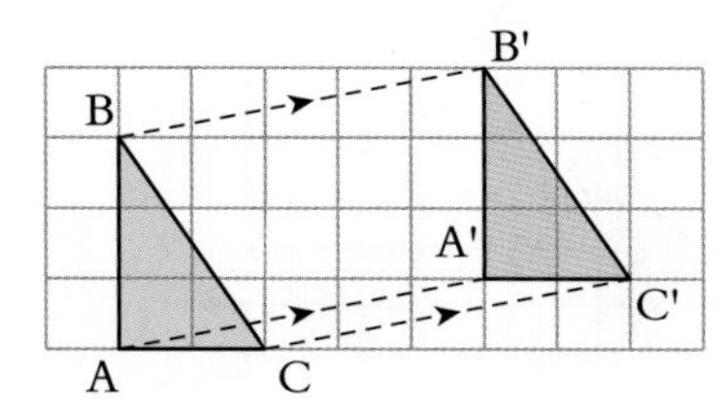

This is a translation of 5 squares to the right and 1 square up.
The translation can be described using a **column vector**:

This is a translation of $\begin{pmatrix} 5 \\ 1 \end{pmatrix}$

← The top number is the movement in the x direction.
← The bottom number is the movement in the y direction.

For the top number: a movement to the right is positive
a movement to the left is negative.

For the bottom number: a movement up is positive
a movement down is negative.

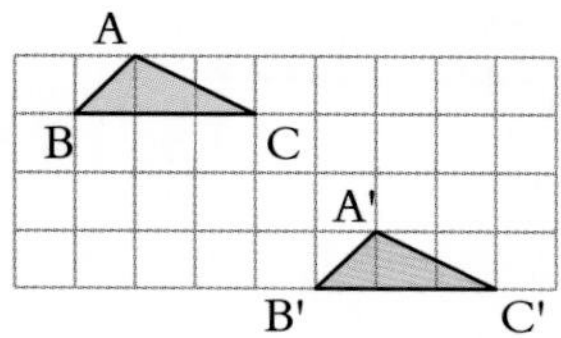

This is a translation of $\begin{pmatrix} 4 \\ -3 \end{pmatrix}$.

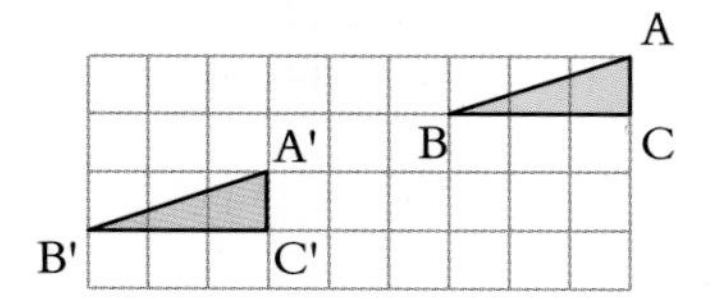

This is a translation of $\begin{pmatrix} -6 \\ -2 \end{pmatrix}$.

Translate this shape by the vector $\begin{pmatrix} -6 \\ 1 \end{pmatrix}$.

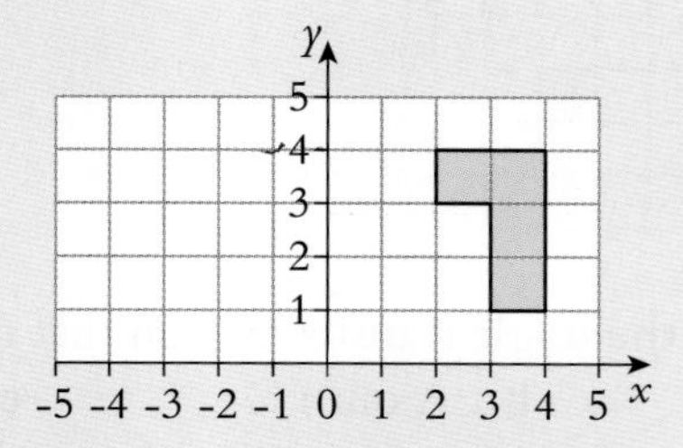

$\begin{pmatrix} -6 \\ 1 \end{pmatrix}$ means:

- move 6 squares to the left and
- move 1 square up.

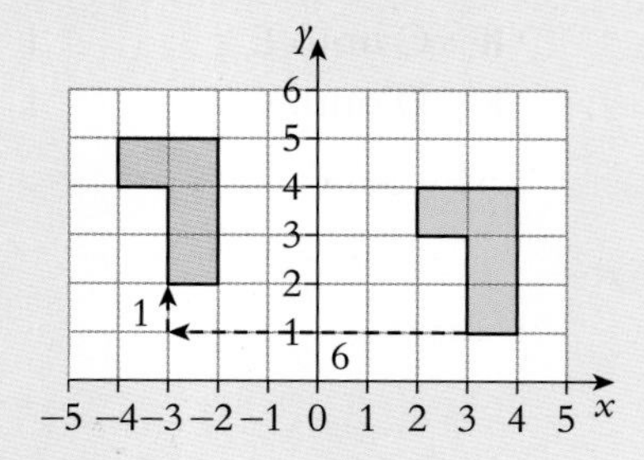

EXERCISE 3.15

1 Write down the vector for the translation that maps shape A onto shape B.

a 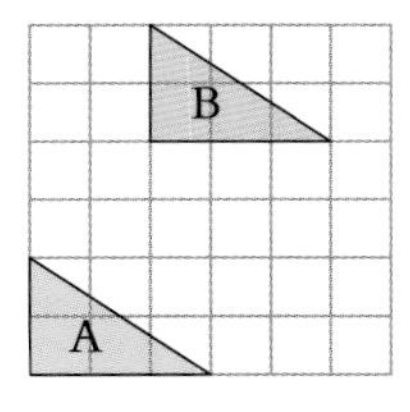

b 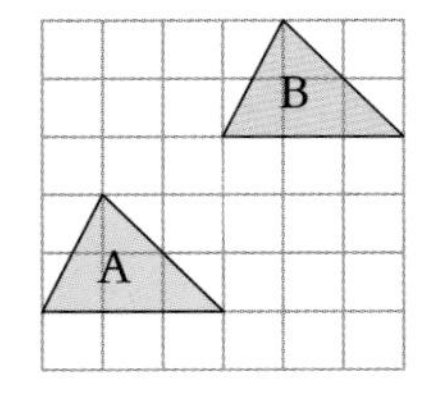

c 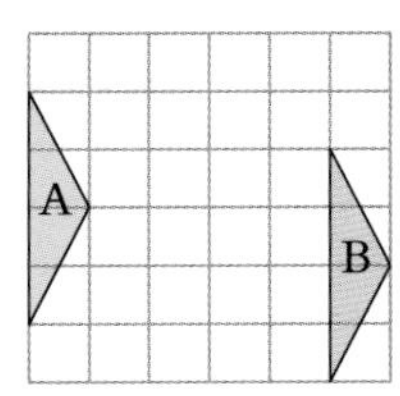

d 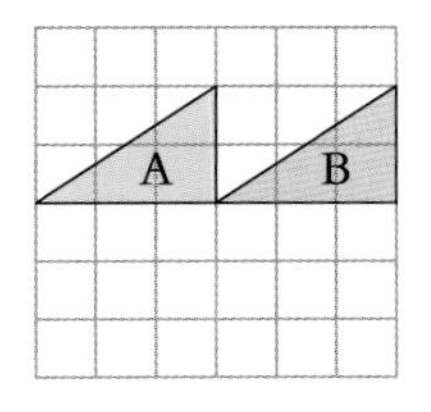

e 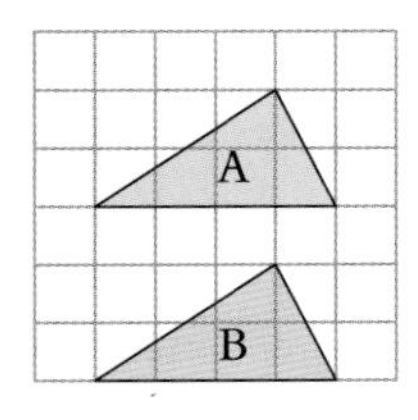

f 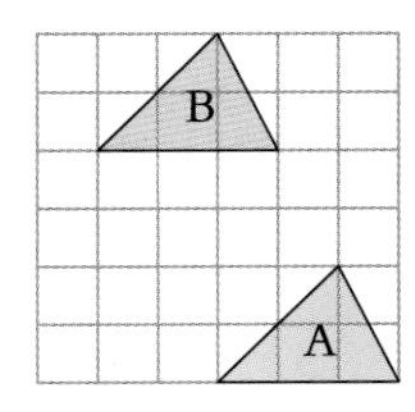

2 Copy these shapes onto squared paper. Translate each shape by the given vector.

a 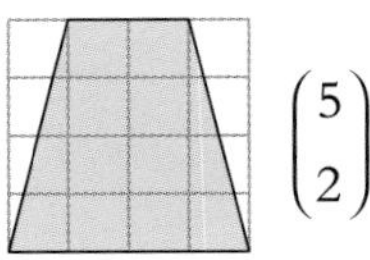 $\begin{pmatrix}5\\2\end{pmatrix}$

b 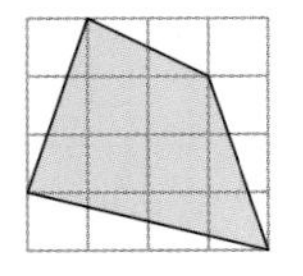 $\begin{pmatrix}-4\\3\end{pmatrix}$

c 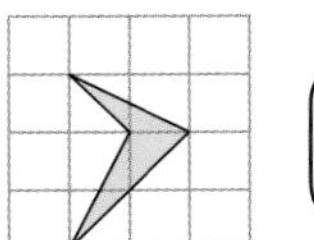 $\begin{pmatrix}0\\-3\end{pmatrix}$

d 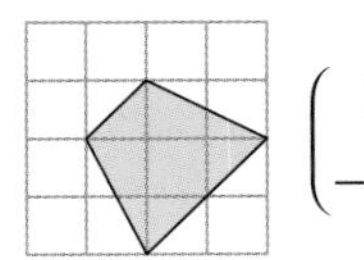 $\begin{pmatrix}3\\-4\end{pmatrix}$

e 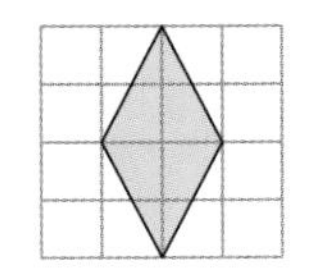 $\begin{pmatrix}-3\\-4\end{pmatrix}$

f 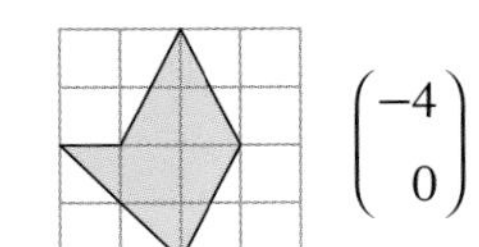 $\begin{pmatrix}-4\\0\end{pmatrix}$

3 Copy and complete the following table.

	Point	**Transformation**	**Image**
a	(3, 6)	Translation of $\begin{pmatrix}5\\2\end{pmatrix}$	
b	(1, 0)	Translation of $\begin{pmatrix}4\\-2\end{pmatrix}$	
c	(2, −5)	Translation of $\begin{pmatrix}-6\\4\end{pmatrix}$	
d	(−2, 8)	Translation of $\begin{pmatrix}-7\\-2\end{pmatrix}$	
e	(−10, −4)	Translation of $\begin{pmatrix}6\\-8\end{pmatrix}$	
f	(0, −3)	Translation of $\begin{pmatrix}7\\-5\end{pmatrix}$	

4

Describe fully the single transformation that moves triangle:

a A onto B **b** A onto C **c** A onto D **d** A onto E
e A onto F **f** A onto G **g** A onto H **h** H onto G
i F onto E **j** B onto H **k** C onto F **l** G onto D.

5 **a** Translate shape A by the vector $\begin{pmatrix} 3 \\ 3 \end{pmatrix}$
Label the image B.

b Translate shape B by the vector $\begin{pmatrix} 4 \\ -1 \end{pmatrix}$
Label the image C.

c Describe the single transformation that moves shape A onto shape C.

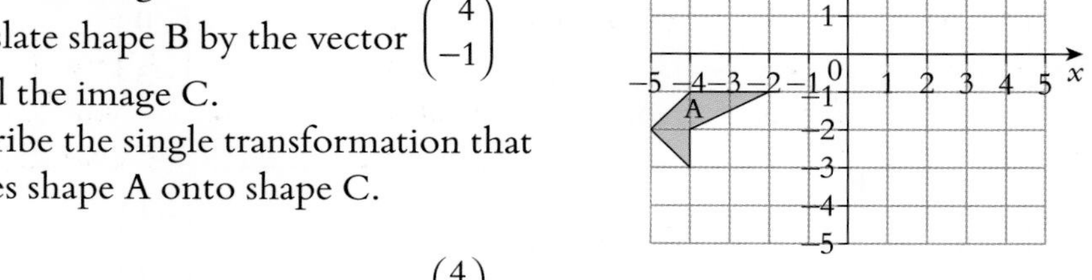

6 **a** Translate shape A by the vector $\begin{pmatrix} 4 \\ 1 \end{pmatrix}$
Label the image B.

b Translate shape B by the vector $\begin{pmatrix} -8 \\ -3 \end{pmatrix}$
Label the image C.

c Describe the single transformation that moves shape A onto shape C.

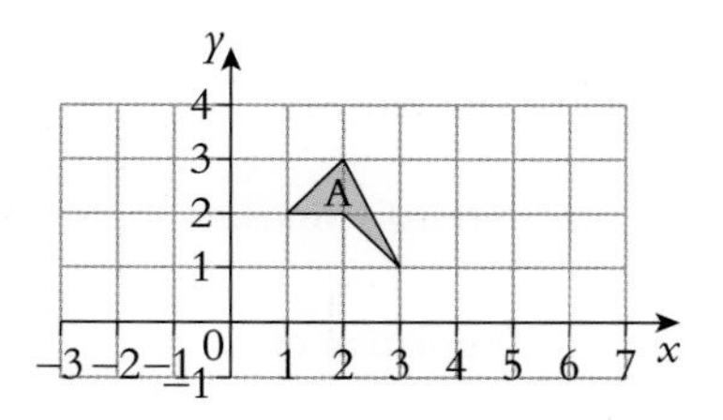

7 Triangle ABC is translated onto triangle A′B′C′.
Copy and complete the table.

A = (1, 1)	B = (3, 2)	C = (…, …)
A′ = (4, −4)	B′ = (…, …)	C′ = (4, −1)

8 Quadrilateral ABCD is translated onto quadrilateral A′B′C′D′.
Copy and complete the table.

A = (…, …)	B = (2, 2)	C = (…, …)	D = (1, 2)
A′ = (2, −3)	B′ = (0, −2)	C′ = (−3, −1)	D′ = (…, …)

EXERCISE 3.16

1

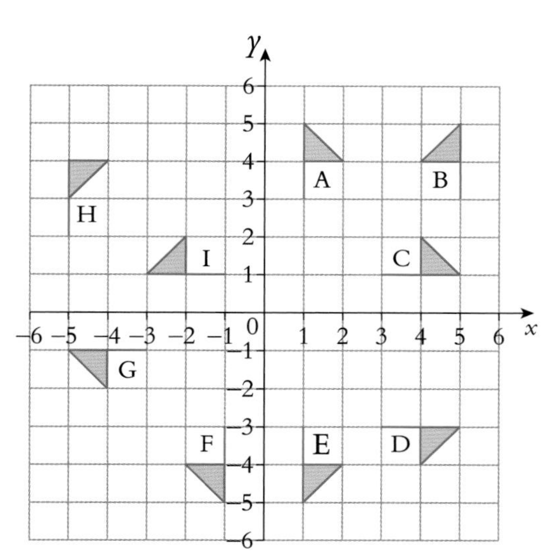

Describe fully the single transformation that moves shape:

a	A onto B	b	A onto C	c	A onto E	d	A onto F
e	A onto G	f	A onto H	g	A onto I	h	G onto F
i	E onto G	j	C onto D	k	E onto F	l	C onto I.

2

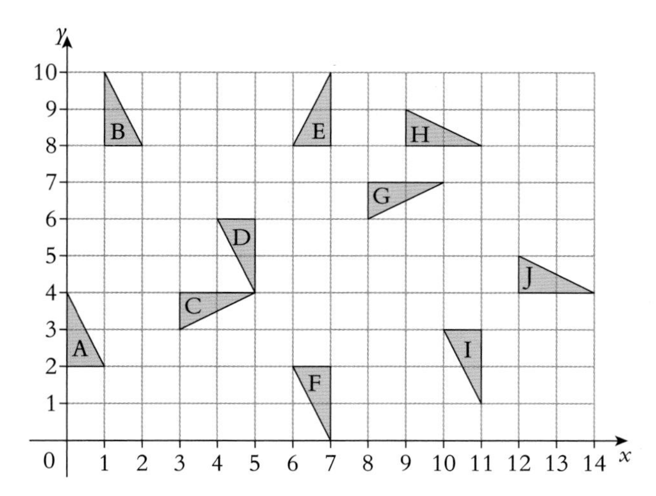

Describe fully the single transformation that moves triangle:

a	A onto B	b	A onto F	c	B onto E	d	B onto D
e	C onto D	f	C onto G	g	D onto I	h	E onto F
i	I onto J	j	J onto H	k	E onto H	l	F onto C.

Combined transformations

Transformations can be applied one after another. A combination of transformations can sometimes be described using a single transformation.

EXAMPLE

a Draw triangle A whose vertices are (2, 1), (4, 1) and (2, 2).
b Reflect triangle A in the line $y = x$. Label the image B.
c Rotate triangle B 90° anticlockwise about the point (0, 0). Label the image C.
d Describe fully the single transformation that maps triangle A onto triangle C.

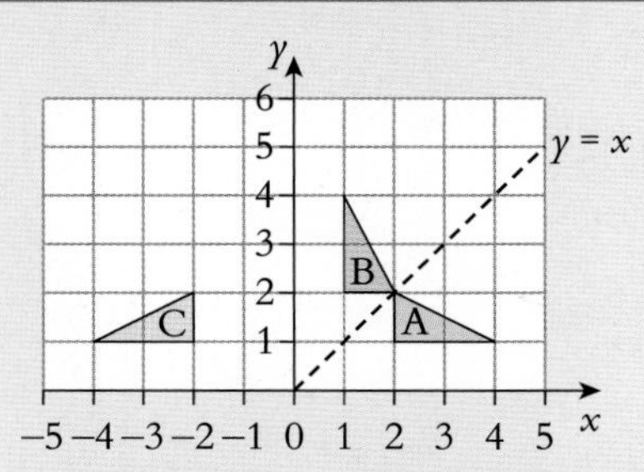

The single transformation that maps triangle A onto triangle C is a reflection in the y-axis.

EXAMPLE

a Draw triangle A whose vertices are (−6, 1), (−2, 1) and (−5, 3).
b Reflect triangle A in the line $x = -1$. Label the image B.
c Reflect triangle B in the line $y = -1$. Label the image C.
d Describe fully the single transformation that maps triangle A onto triangle C.

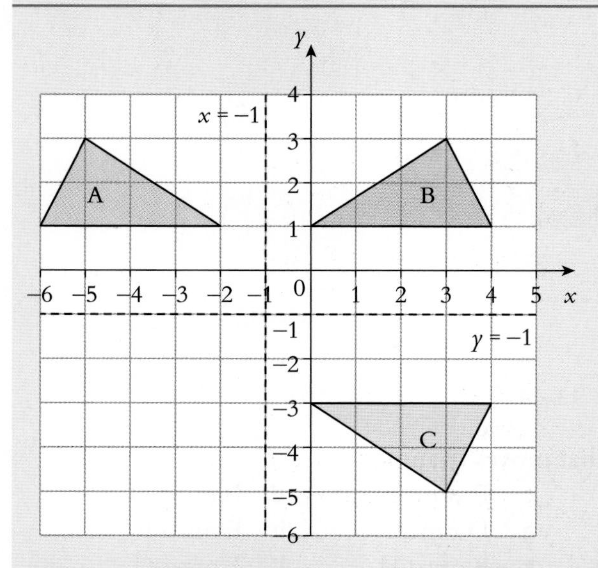

The single transformation that maps triangle A onto triangle C is a rotation of 180° about the point (−1, −1).

EXERCISE 3.19

1 In each diagram shape A has been enlarged to make shape B.
State the scale factor and the centre of enlargement for each diagram.

a

b

c

d

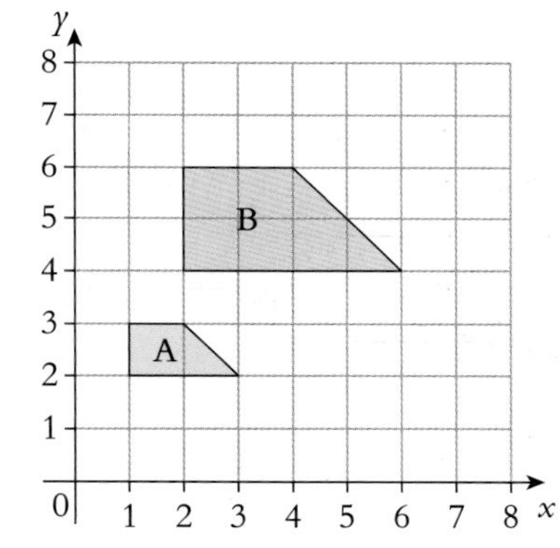

e

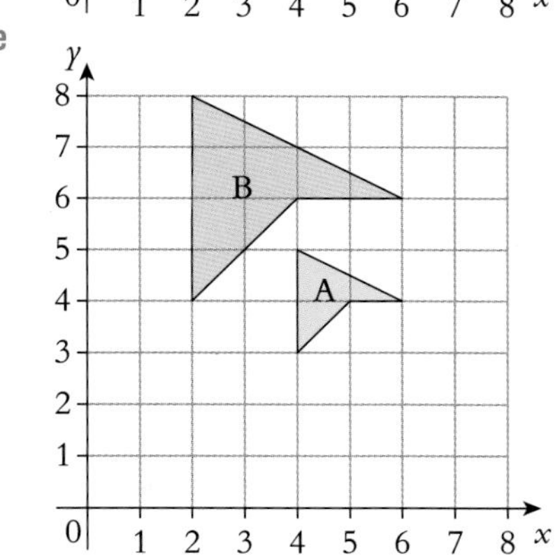

f

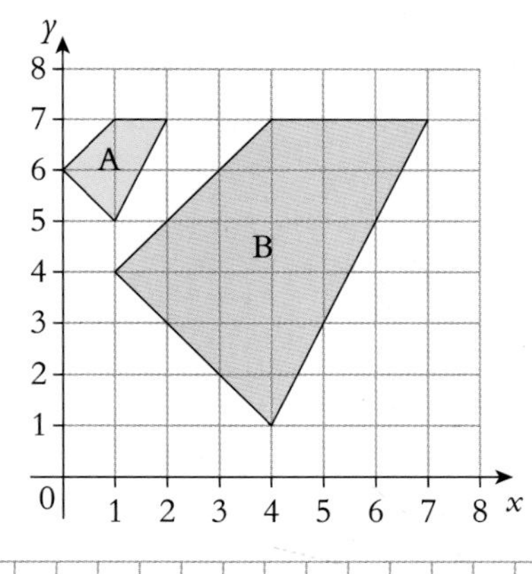

2 Describe fully the single transformation that maps shape:

a A onto C b A onto D
c D onto F d B onto E
e B onto C f B onto D

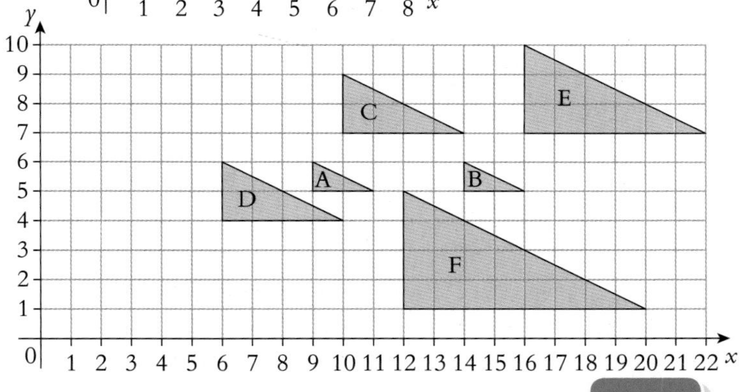

EXAMPLE

Enlarge triangle ABC using a scale factor of 2 and centre of enlargement P(1, 2).

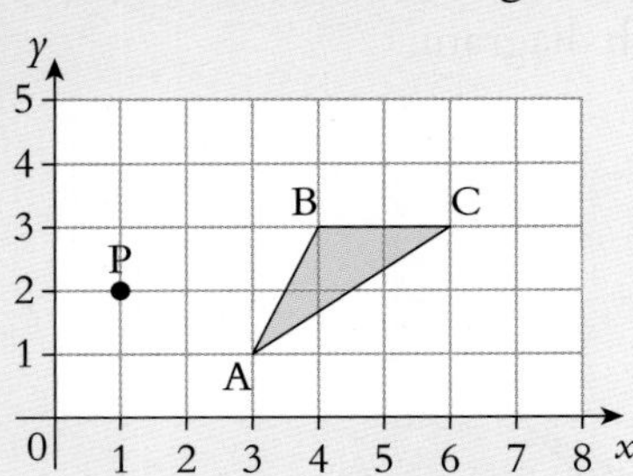

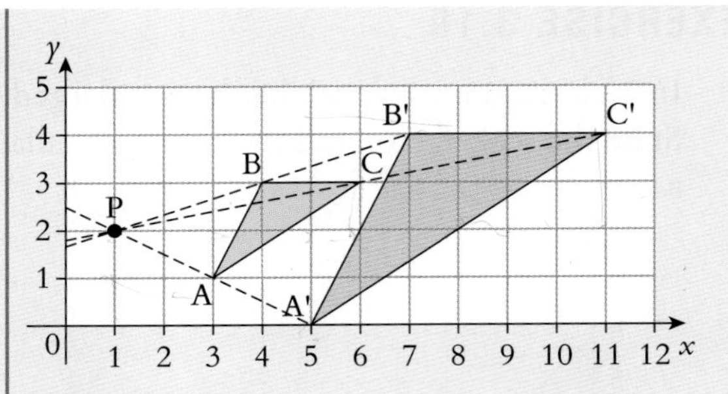

$PA' = 2 \times PA$
$PB' = 2 \times PB$
$PC' = 2 \times PC$

EXAMPLE

Enlarge quadrilateral ABCD using a scale factor of $\frac{1}{2}$ and centre of enlargement P(11, 3).

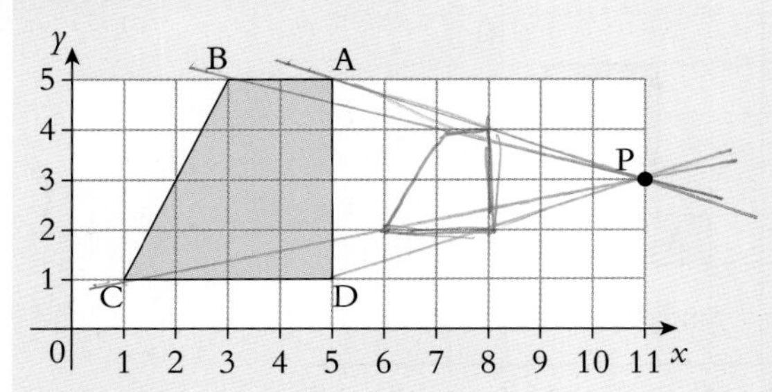

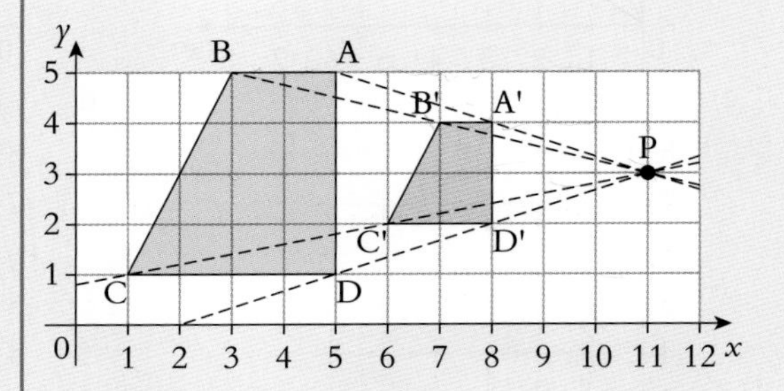

$PA' = \frac{1}{2} \times PA$

$PB' = \frac{1}{2} \times PB$

$PC' = \frac{1}{2} \times PC$

$PD' = \frac{1}{2} \times PD$

An enlargement scale factor $\frac{1}{2}$ makes the shape smaller.

EXAMPLE

Enlarge triangle ABC using a scale factor of −2 and centre of enlargement P(4, 2).

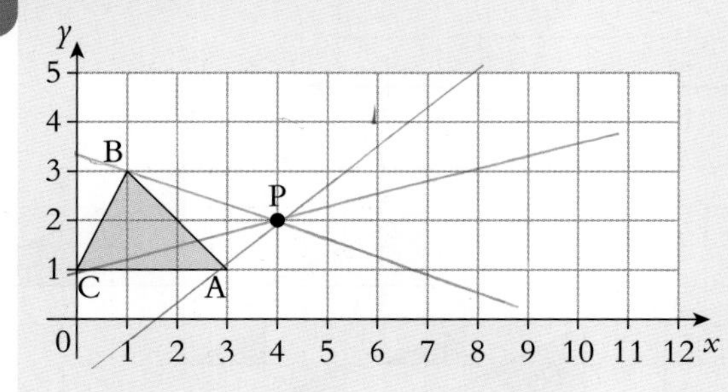

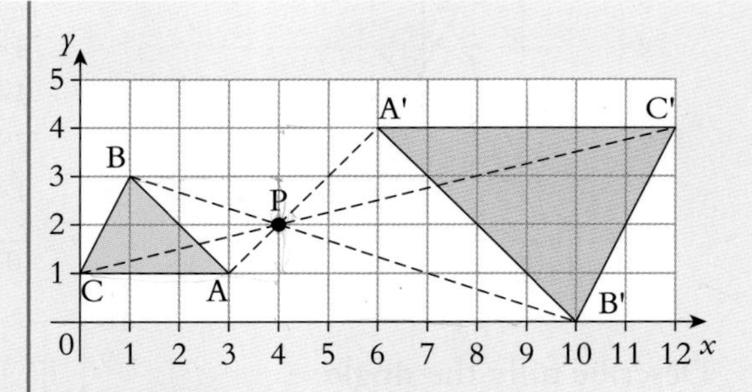

$PA' = 2 \times PA$
$PB' = 2 \times PB$
$PC' = 2 \times PC$

For a negative enlargement the object and image are on opposite sides of the centre of enlargement.

EXERCISE 3.20

1 Copy each diagram and enlarge with scale factor 2. Use point P as the centre of enlargement.

a
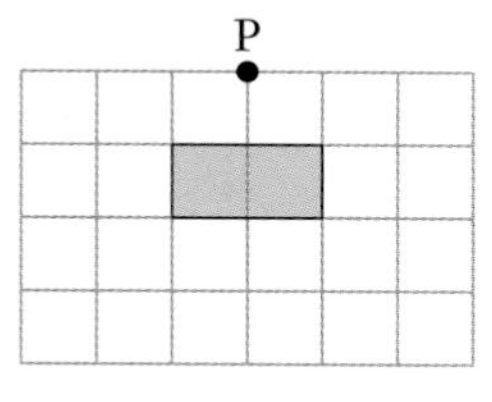

b
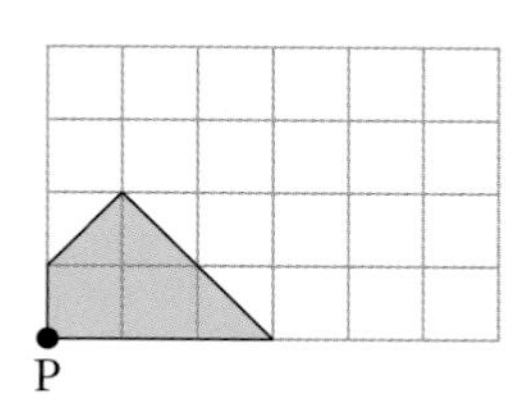

c
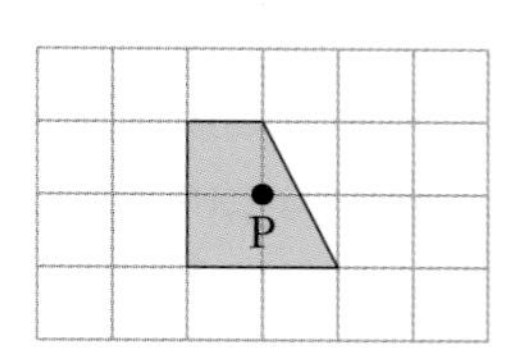

2 Copy each diagram and enlarge with scale factor $\frac{1}{2}$. Use point P as the centre of enlargement.

a
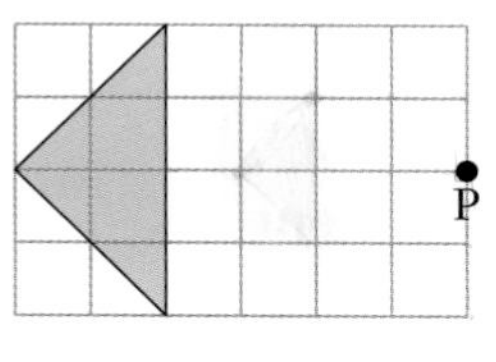

b
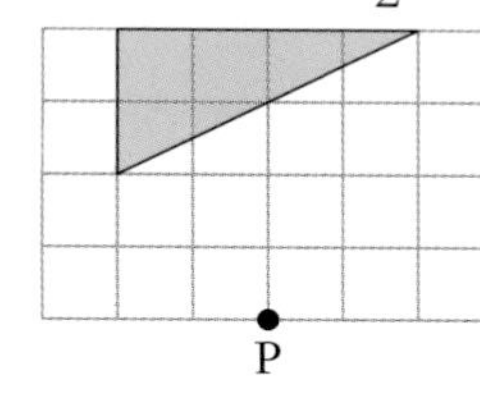

c
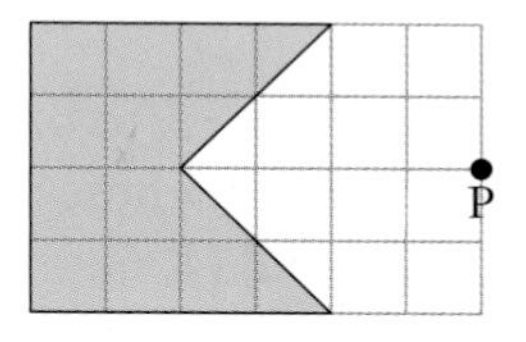

3 Copy each diagram and enlarge with scale factor −2. Use point P as the centre of enlargement.

a
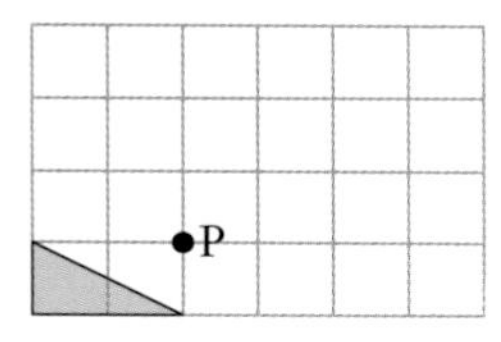

b
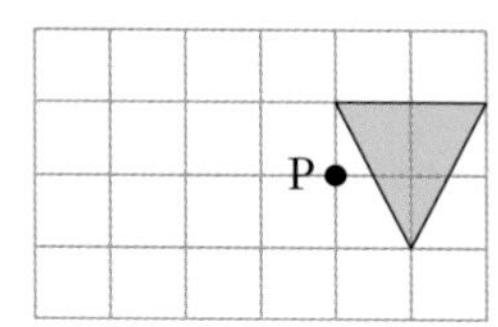

c
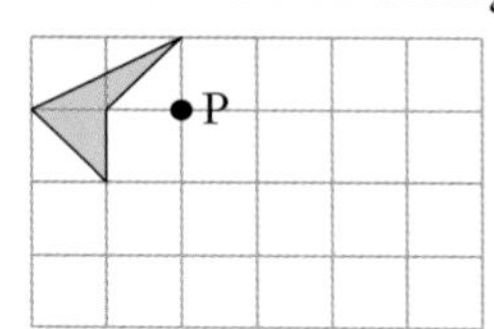

4 Copy the following diagrams. Enlarge each shape using the given scale factor and the point P as the centre of enlargement.

a
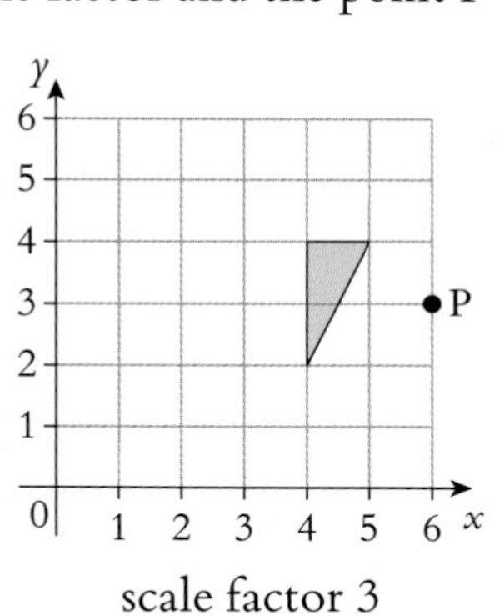

scale factor 3

b
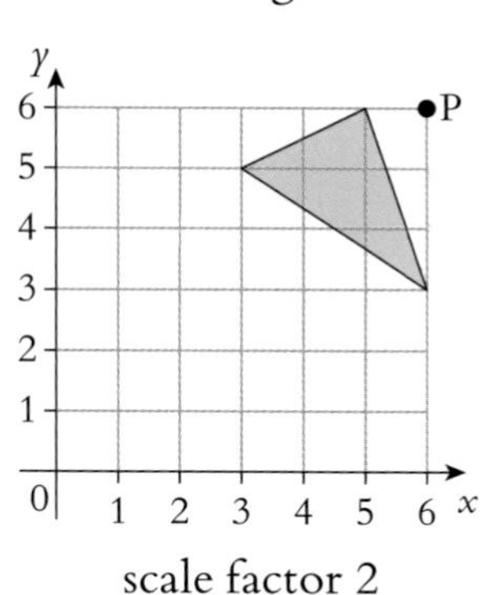

scale factor 2

c
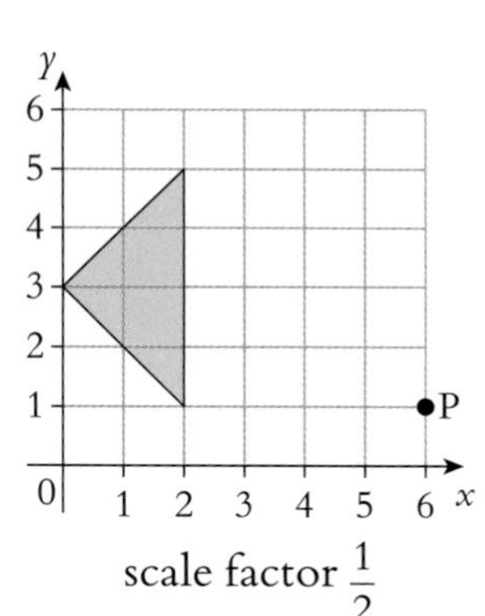

scale factor $\frac{1}{2}$

d
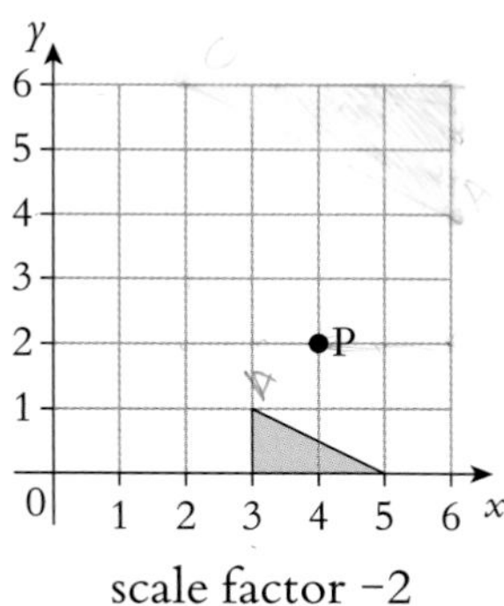

scale factor −2

e
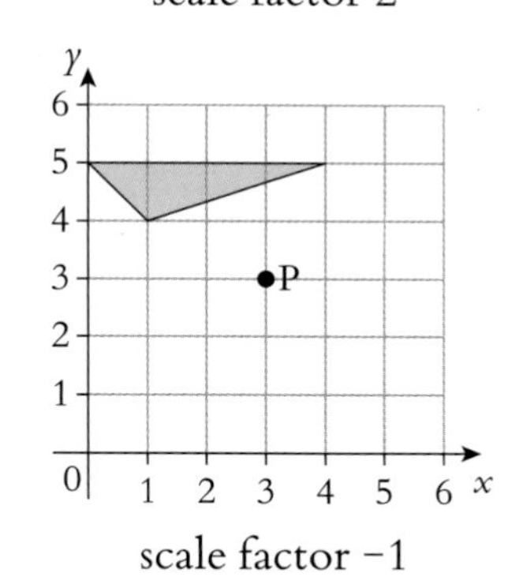

scale factor −1

f
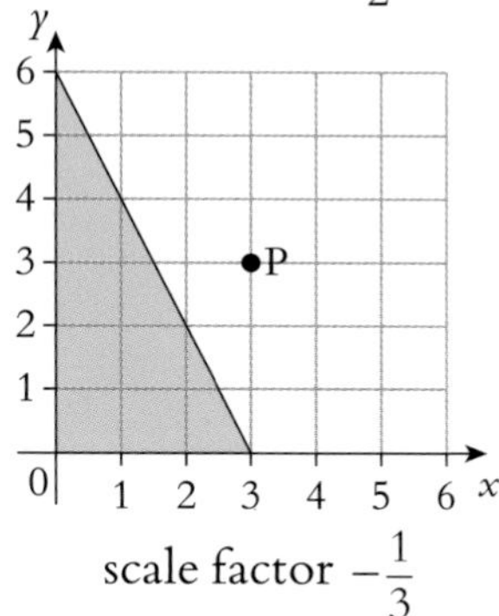

scale factor $-\frac{1}{3}$

5 Draw axes from −7 to 7.
Draw triangle T whose vertices are (2, 0), (0, 2) and (−2, 2).
Draw the image of triangle T after each of these enlargements.
Label your images A, B, C, and D.

Scale factor	Centre of enlargement	Image
2	(−2, 0)	A
$\frac{1}{2}$	(0, −4)	B
−1	(−2, 1)	C
−2	(0, −1)	D

6 Draw axes from −6 to 6.
Draw triangle T whose vertices are (4, 2), (6, 2) and (4, 6).
Draw the image of triangle T after each of these enlargements.
Label your images P, Q, R and S.

Scale factor	Centre of enlargement	Image
$\frac{1}{2}$	(0, 0)	P
$\frac{1}{2}$	(−6, 4)	Q
−1	(4, 0)	R
−2	(2, 3)	S

7 Draw axes from −6 to 6.
Draw triangle T whose vertices are (−6, −2), (−4, −4) and (−2, −4).
Enlarge triangle T with scale factor $2\frac{1}{2}$, centre (−6, −6).

8 Draw axes from −6 to 6.
Draw triangle T whose vertices are (1, 0), (−3, 4) and (0, −2).
Enlarge triangle T with scale factor $1\frac{1}{3}$, centre (6, 1).

9 Draw axes from −7 to 7.
Draw triangle T whose vertices are (5, −2), (5, 2) and (−3, 2).
Enlarge triangle T with scale factor $-1\frac{1}{2}$, centre (1, 0).

EXAMPLE

Enlarge triangle P with scale factor 3, centre (0, 0).
Label the image Q.
Enlarge triangle Q with scale factor $\frac{1}{3}$, centre (−9, 6).
Label the image R.
Describe fully the single transformation that maps triangle P onto triangle R.

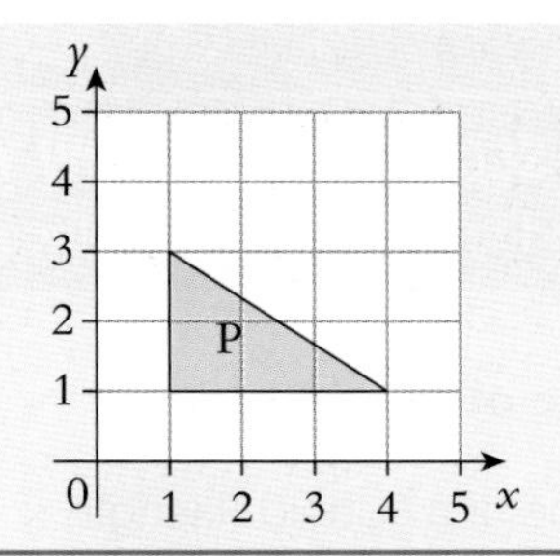

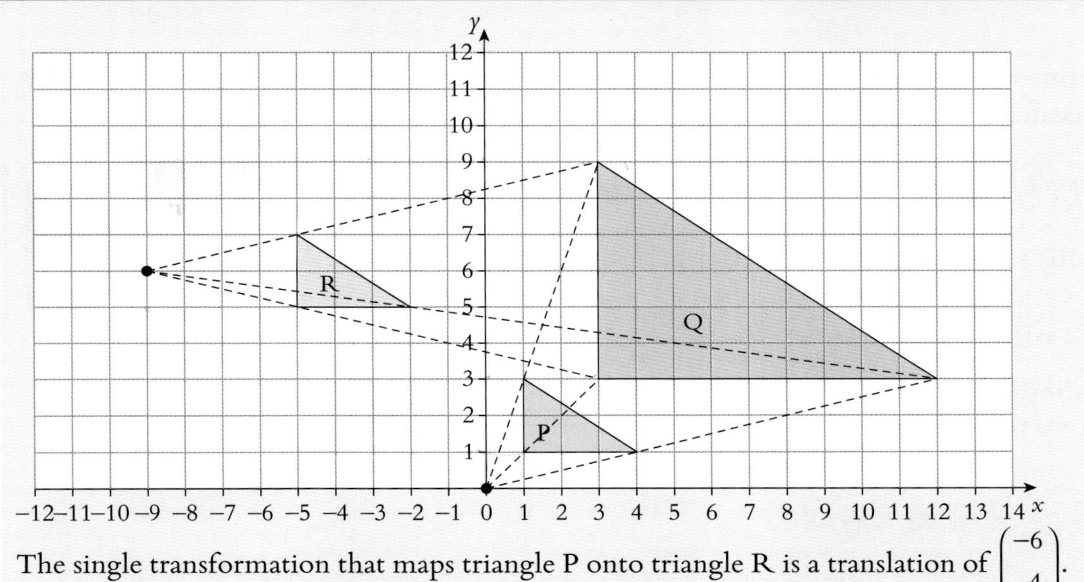

The single transformation that maps triangle P onto triangle R is a translation of $\begin{pmatrix} -6 \\ 4 \end{pmatrix}$.

EXERCISE 3.21

1 Draw axes from −6 to 6.

- a Draw triangle P with vertices (1, 0), (0, 1) and (−2, −1).
- b Enlarge triangle P with scale factor 2, centre (2, −2). Label the image Q.
- c Enlarge triangle Q with scale factor $\frac{1}{2}$, centre (−6, −6). Label the image R.
- d Describe fully the single transformation that maps triangle P onto triangle R.

2 Draw axes from −6 to 6.

- a Draw triangle P with vertices (−3, 0), (−6, 0) and (−5, −1).
- b Enlarge triangle P with scale factor −2, centre (−3, −2). Label the image Q.
- c Enlarge triangle Q with scale factor −2, centre (0, −2). Label the image R.
- d Describe fully the single transformation that maps triangle P onto triangle R.

3 Draw axes from −6 to 6.

- a Draw triangle P with vertices (−4, 1), (−4, 3) and (−3, 1).
- b Enlarge triangle P with scale factor 4, centre (−6, 2). Label the image Q.
- c Enlarge triangle Q with scale factor $\frac{1}{2}$, centre (−6, −2). Label the image R.
- d Describe fully the single transformation that maps triangle P onto triangle R.

KEY WORDS
enlargement
scale factor
centre of enlargement

Surface area and volume 1

THIS SECTION WILL SHOW YOU HOW TO

- Find the volume and surface area of a prism

Volume of a cuboid

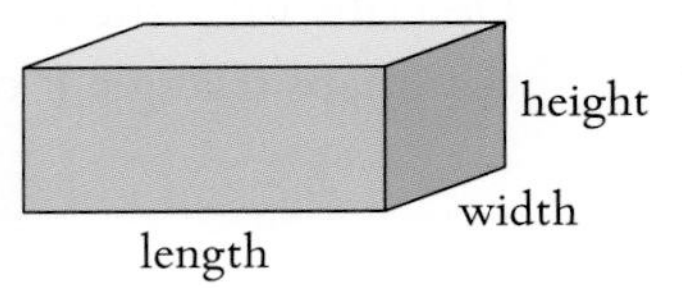

Volume of a cuboid = length × width × height

Volume of a prism

A **prism** is a three-dimensional shape with a uniform **cross-section**.

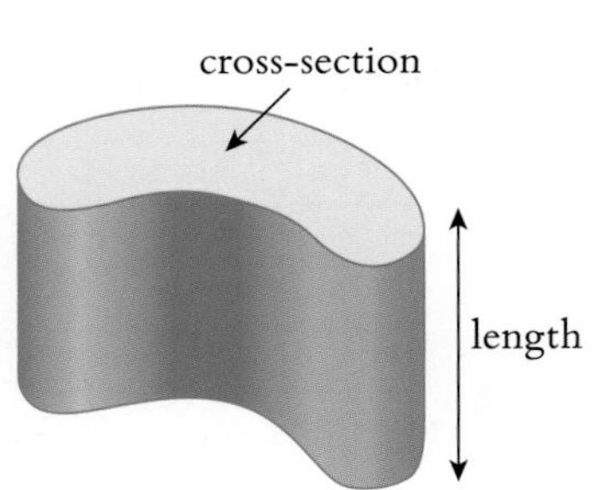

Volume of a prism = area of cross-section × length

➡ **NOTE:** a cuboid is a prism with a rectangular cross-section.

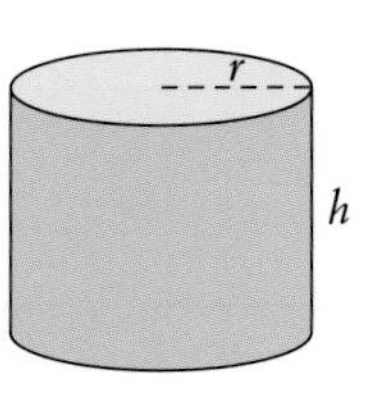

A **cylinder** is a prism with a circular cross-section.
Volume of cylinder = area of cross-section × length
$= \pi r^2 \times h$

Volume of cylinder = $\pi r^2 h$

Converting units of volume

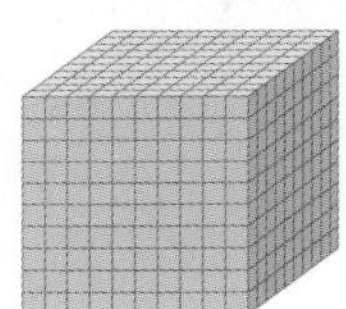

$1 \text{ cm}^3 = 10 \times 10 \times 10 \text{ mm}^3$

Similarly,
$1 \text{ m}^3 = 100 \times 100 \times 100 \text{ cm}^3$

$1 \text{ cm}^3 = 1000 \text{ mm}^3$

$1 \text{m}^3 = 1\,000\,000 \text{ cm}^3$

Capacity

The **capacity** is the amount that a container can hold.
Units for capacity are litres (ℓ) and millilitres (mℓ).

1 litre = 1000 millilitres

1 cm^3 contains 1 mℓ

'The bottle has a capacity of 200 mℓ' means that the bottle has a volume of 200 cm^3.

EXAMPLE

Find the volume of the prism.

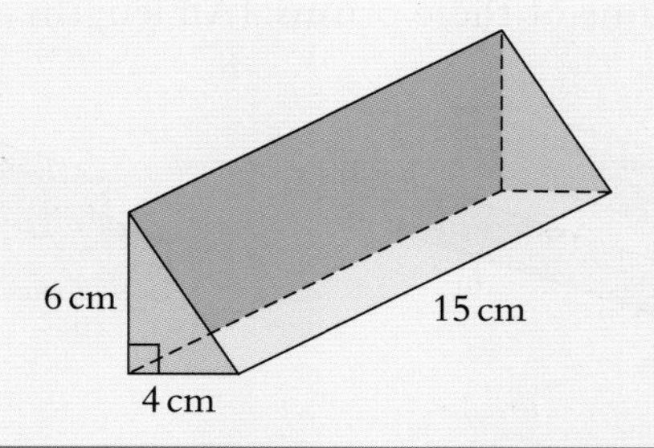

Area of cross-section = $\frac{1}{2} \times 4 \times 6 = 12$ cm^2.

Volume of prism = area of cross-section × length

$= 12 \times 15$

$= 180$ cm^3

EXAMPLE

a Find the volume of the cylinder.

b Find the capacity of the cylinder in litres.

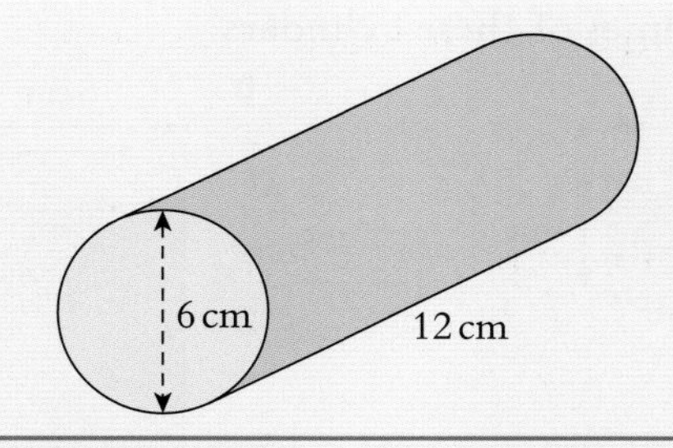

a Volume of cylinder $= \pi r^2 h$

$= \pi \times 3^2 \times 12$

$= 339$ cm^3 (to 3 s.f.)

b Capacity = 339 mℓ

$= 0.339$ litres

Remember: 1 cm^3 = 1 mℓ and 1000 mℓ = 1 litre

EXERCISE 3.22

1 Find the volume of these cuboids.

a

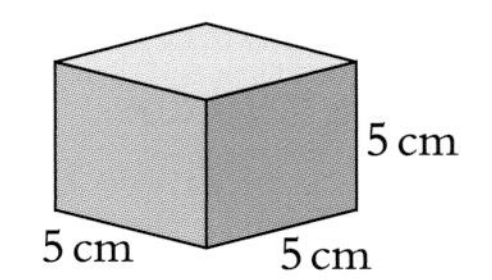

b

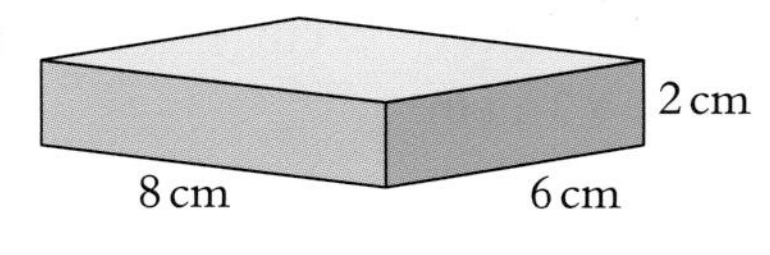

c

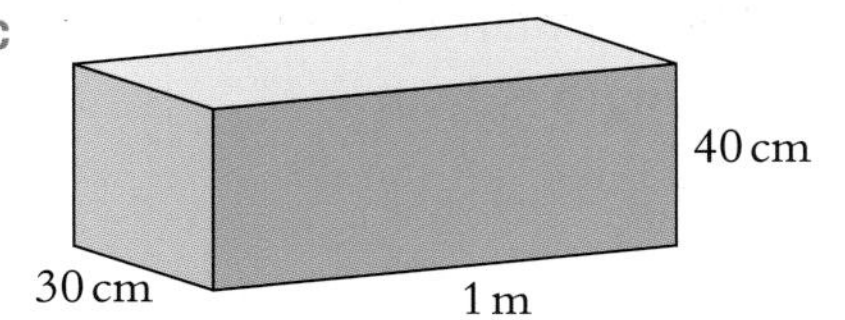

2 Find the volume of these prisms.

a

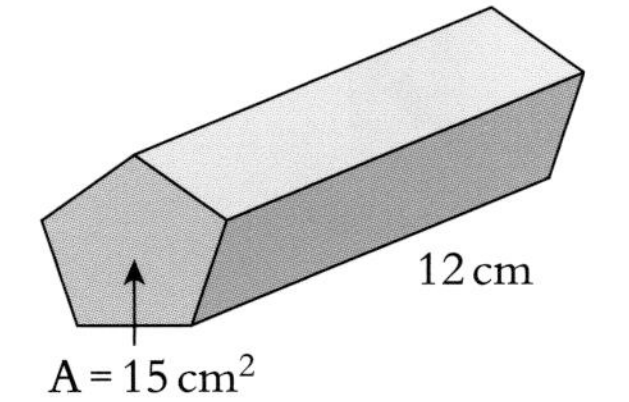

b

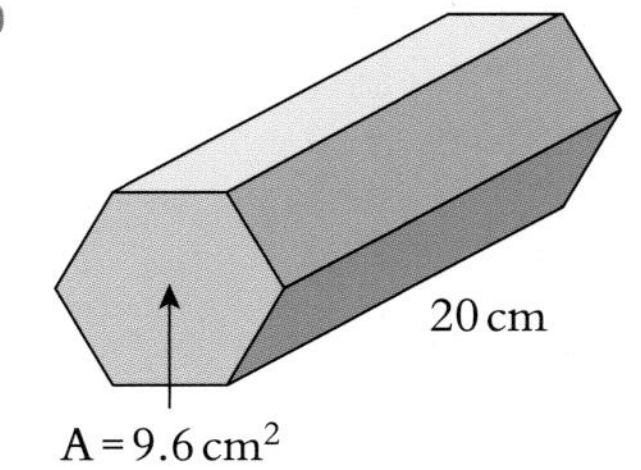

c

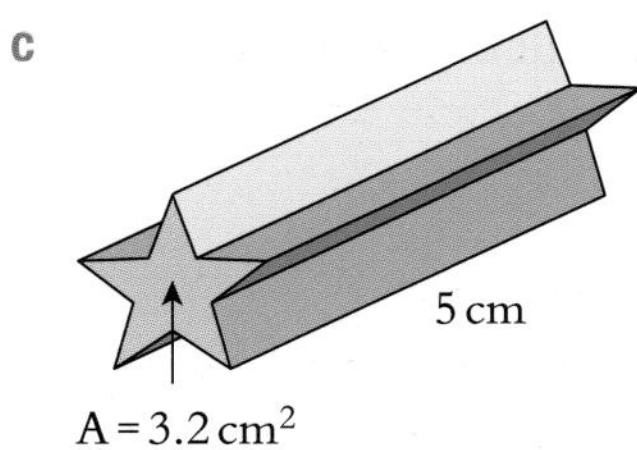

3 Find the volume of these prisms. (All lengths are in cm.)

a

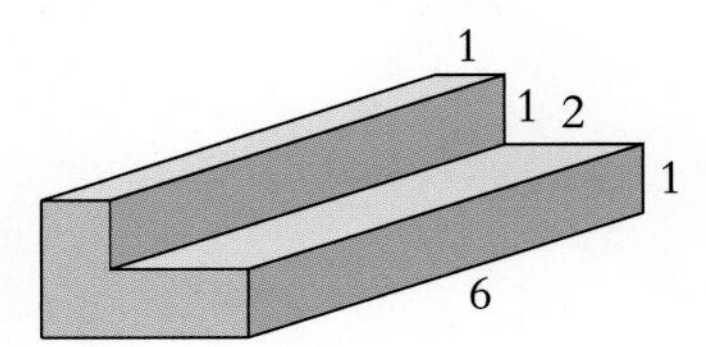

b

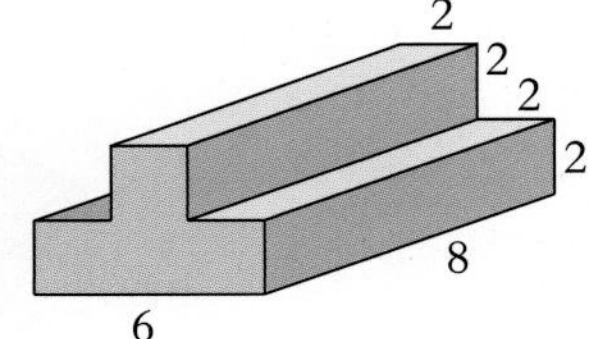

c

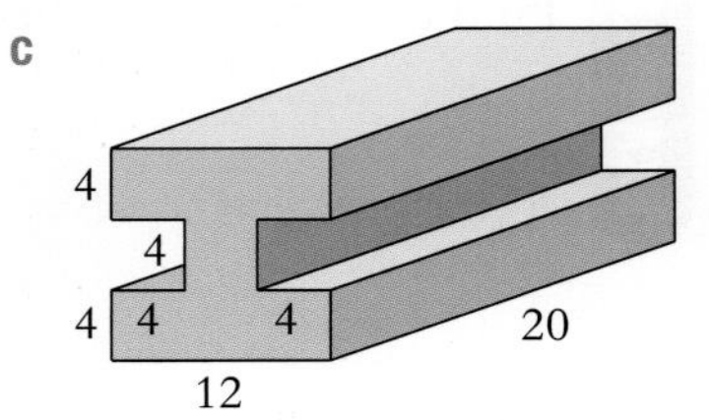

d

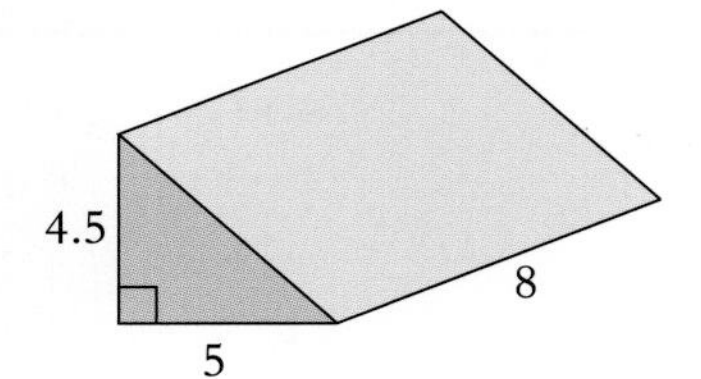

e

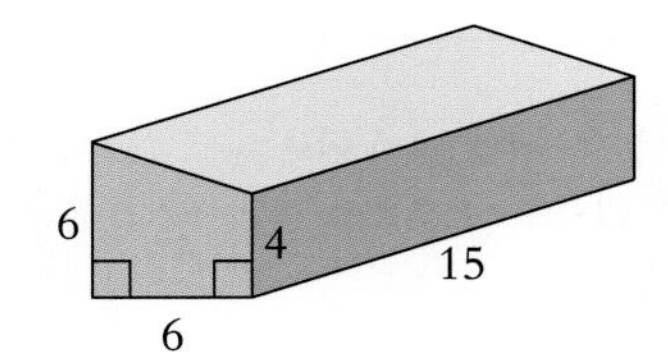

f

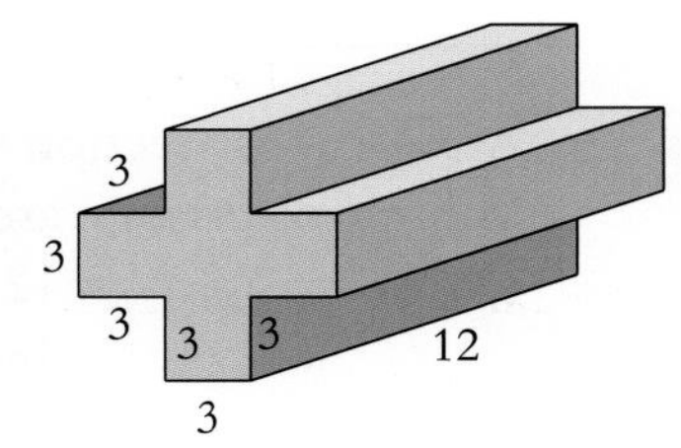

4 Find the volumes of these cylinders.

a

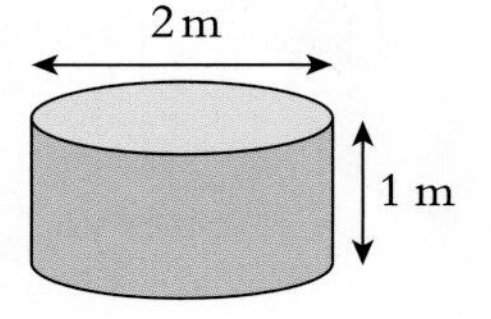

b

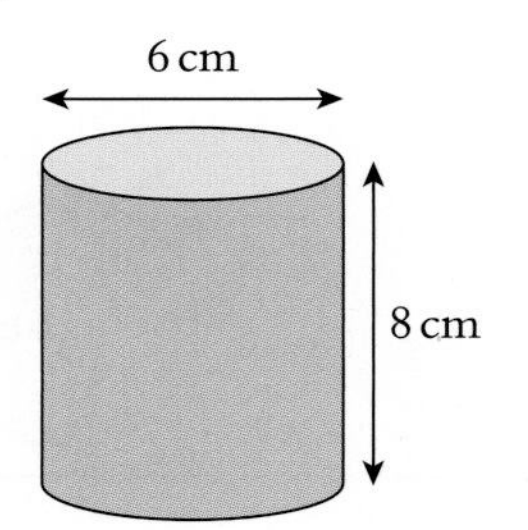

c

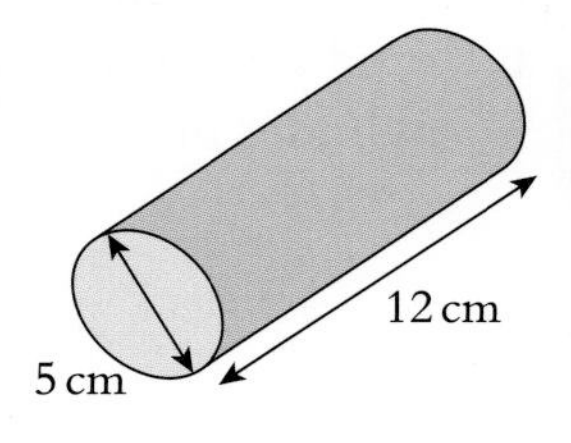

5 Find the volume of these hollow cylinders.

a

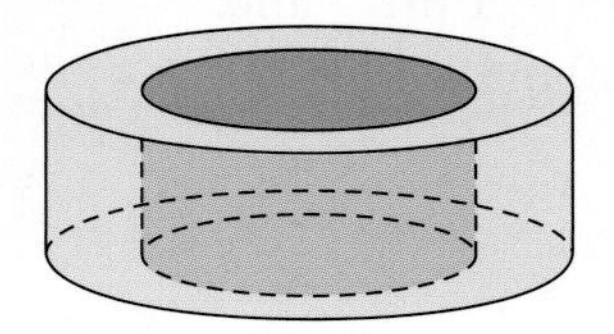

outer diameter = 10 cm
inner diameter = 6 cm
length = 3 cm

b

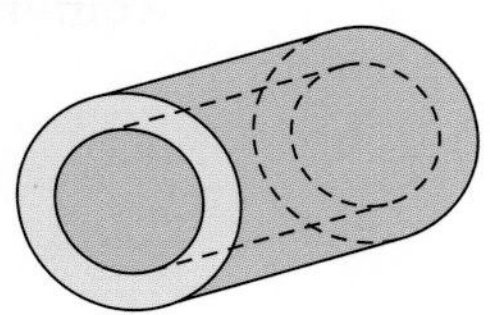

outer diameter = 8 cm
inner diameter = 5 cm
length = 16 cm

c

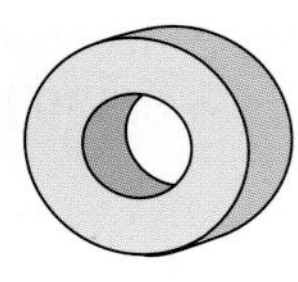

outer diameter = 50 cm
inner diameter = 30 cm
length = 20 cm

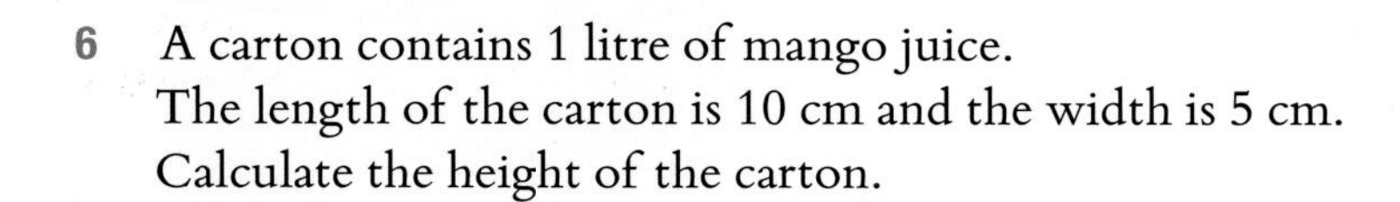

6 A carton contains 1 litre of mango juice.
The length of the carton is 10 cm and the width is 5 cm.
Calculate the height of the carton.

7 This container is a prism.
The cross-section of the prism is a trapezium.

a Find the volume of the container in m^3.

b Find the volume of the container in cm^3.

c How many litres of water does the container hold when it is full?

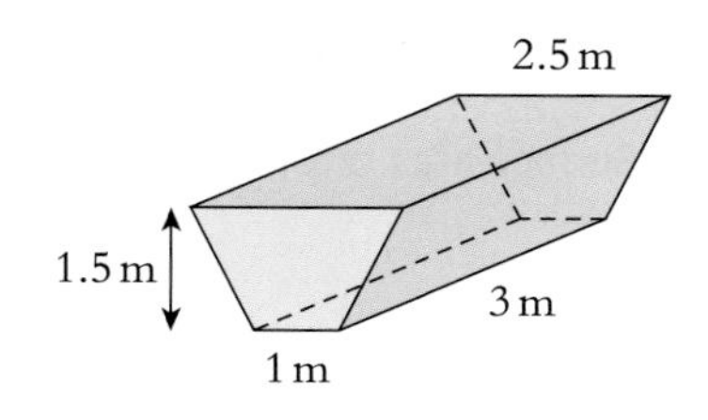

8 This piece of cheese is a prism.
The cross-section is a sector of a circle of radius 14 cm and angle 25°.
The height of the piece of cheese is 6 cm.
Calculate the volume.

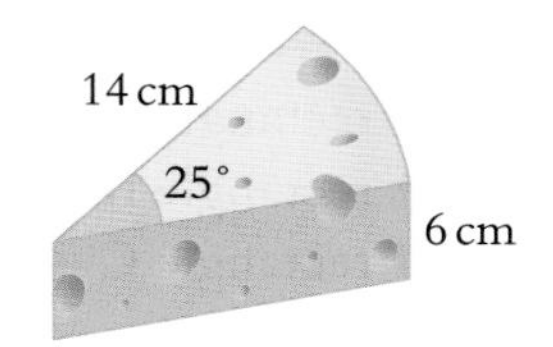

9 A cylindrical can contains 1.5 litres of coconut milk.
The height of the can is 15 cm.
Calculate the radius of the can.

10 This water trough is a prism.
The cross-section is a semi-circle of diameter 50 cm.
When full the water trough holds 300 litres of water.
Calculate the length of the water trough.

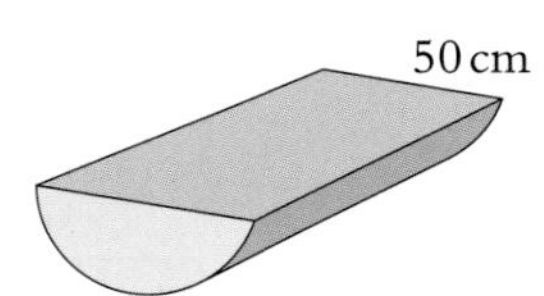

11 The diagram shows a cube with a circular hole.
The cube has sides of length 5 cm.
The circular hole has radius 1 cm.
Calculate the volume of the solid.

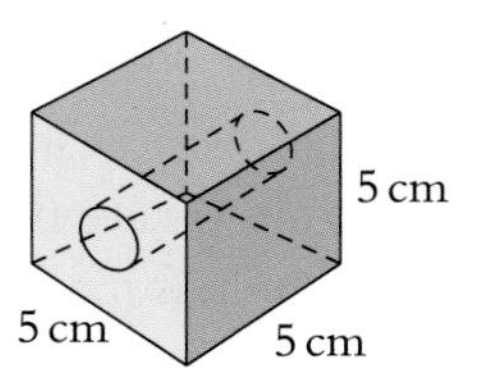

12 The cross-section of this solid prism is a regular hexagon with sides of length 8 cm.
The length of the prism is 25 cm.

a Find the area of the regular hexagon.

b Find the volume of the prism.

The prism is made of wood.
1 cm^3 of wood has a mass of 0.65 g.

c Find the mass of the hexagonal prism.

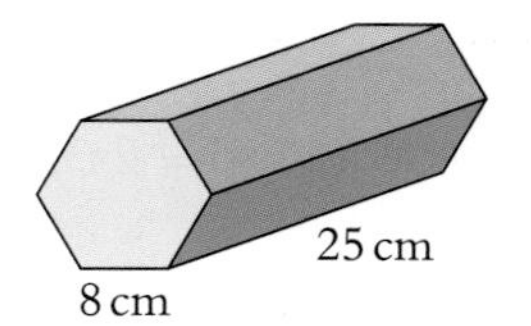

Prism

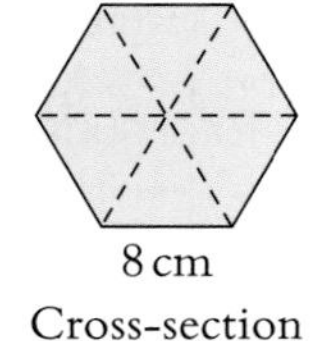

Cross-section

TOP TIP

The hexagon is made from six equilateral triangles. Find the area of one of the triangles and multiply by six.

Surface area of a prism

Surface area of prism = sum of area of all the faces

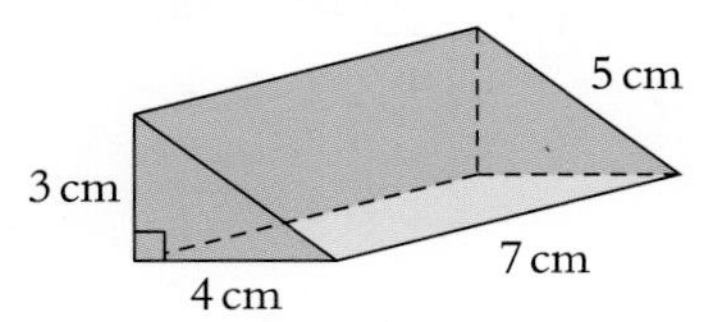

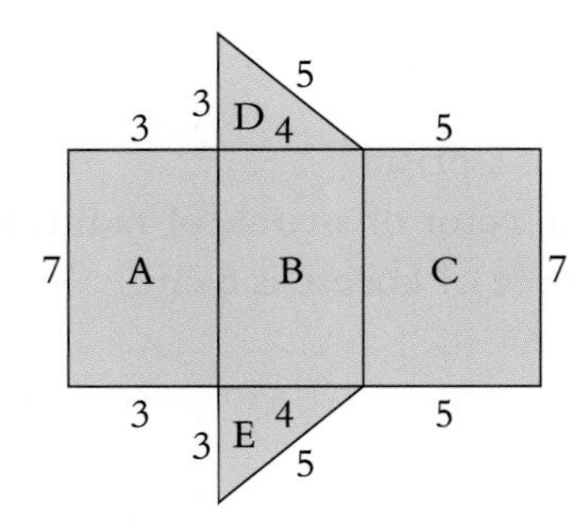

Area of rectangle A = $3 \times 7 = 21\ \text{cm}^2$
Area of rectangle B = $4 \times 7 = 28\ \text{cm}^2$
Area of rectangle C = $5 \times 7 = 35\ \text{cm}^2$
Area of triangle D = $\frac{1}{2} \times 3 \times 4 = 6\ \text{cm}^2$

Area of triangle E = $\frac{1}{2} \times 3 \times 4 = 6\ \text{cm}^2$

Total surface area = $96\ \text{cm}^2$

Surface area of a cylinder

The surface of a solid cylinder consists of two circles and a rectangle.

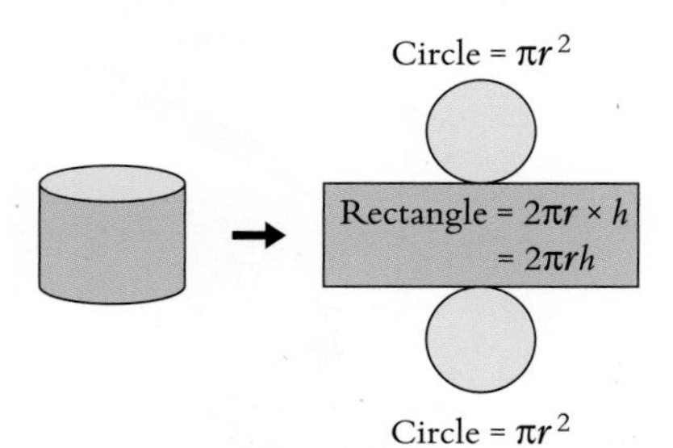

The two circles are flat surfaces.
The rectangle is made from the **curved surface** of the cylinder.

Curved surface area of cylinder = $2\pi rh$

Total surface area of cylinder = $2\pi rh + 2\pi r^2$

EXAMPLE

A solid cylinder has radius 3 cm and length 10 cm.
Calculate the total surface area of the cylinder.

Total surface area of cylinder $= 2\pi rh + 2\pi r^2$
$= 2 \times \pi \times 3 \times 10 + 2 \times \pi \times 3^2$
$= 60\pi + 18\pi$
$= 78\pi$
$= 245\ \text{cm}^2$ (to 3 s.f.)

EXERCISE 3.23

1 Calculate the surface area of these prisms.

a

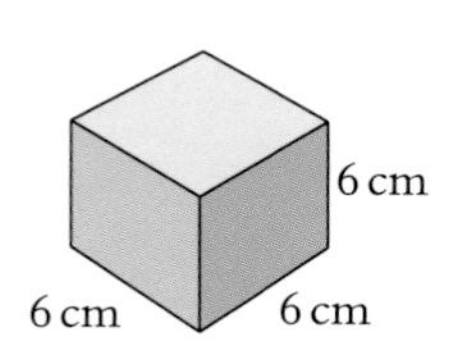

b

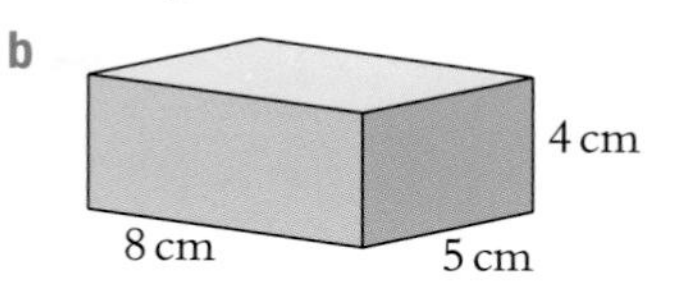

c

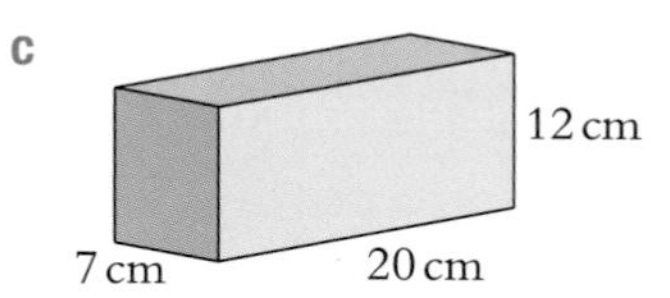

2 Find the surface area of these prisms. (All lengths are in cm.)

a

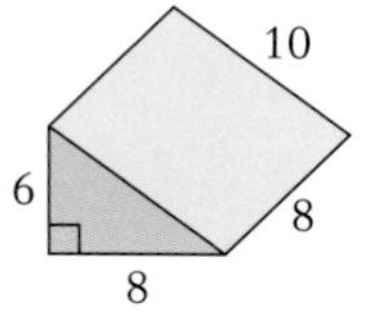

b

c

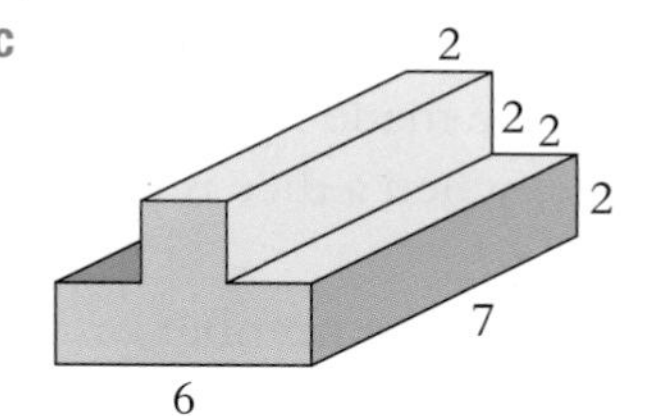

3 Find the surface area of these prisms. (All lengths are in cm.)

a

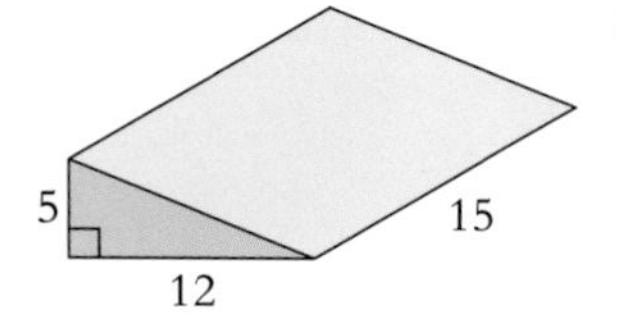

b

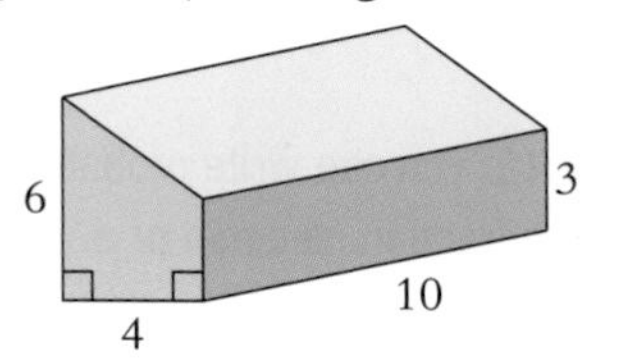

TOP TIP

You will need to use Pythagoras to find the length of the sloping edges.

4 Find the **curved** surface areas of these cylinders.

a

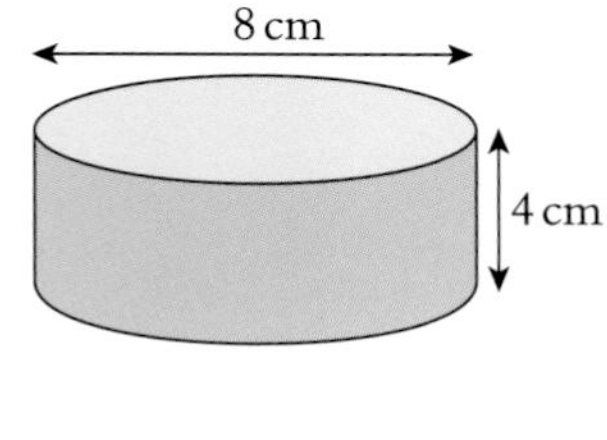

b

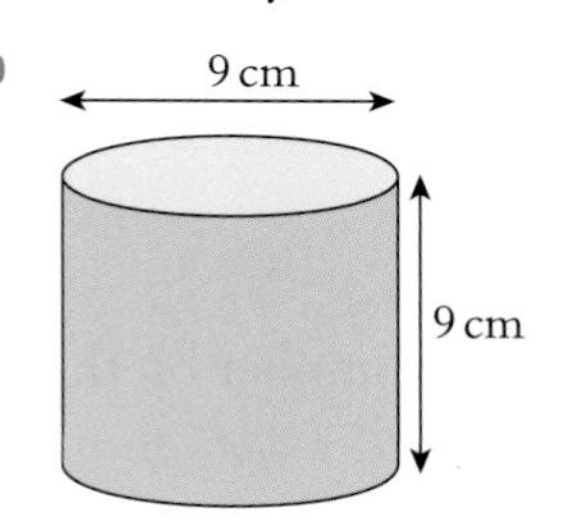

c

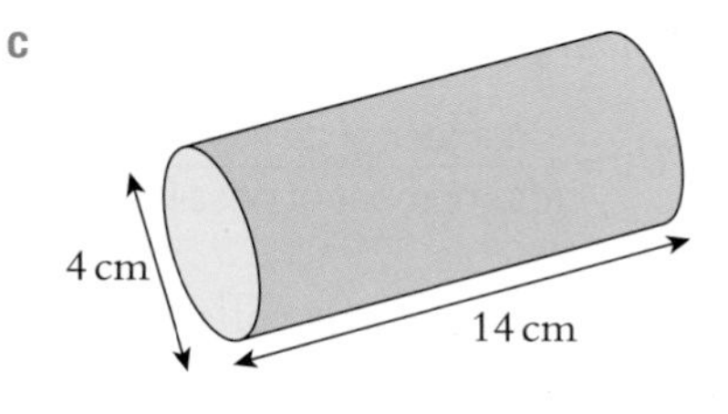

5 Find the **total** surface area of these solid cylinders.

a

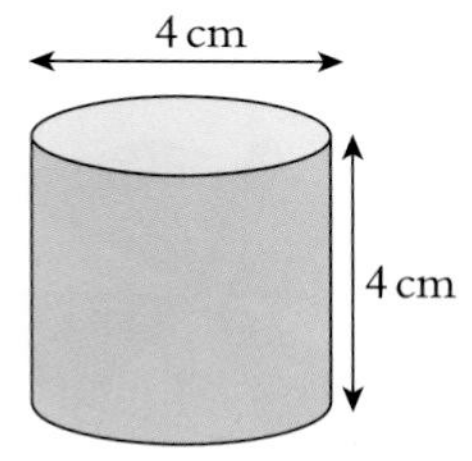

b

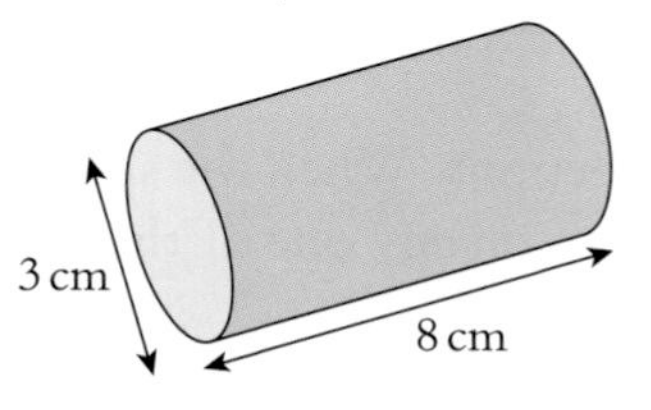

c

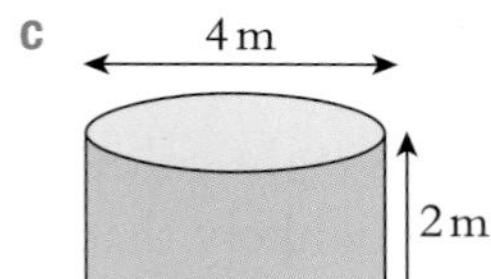

6 This piece of cheese is a prism.
The cross-section is a sector of a circle of radius 12 cm and angle 32°.
The height of the piece of cheese is 5 cm.
Calculate the total surface area.

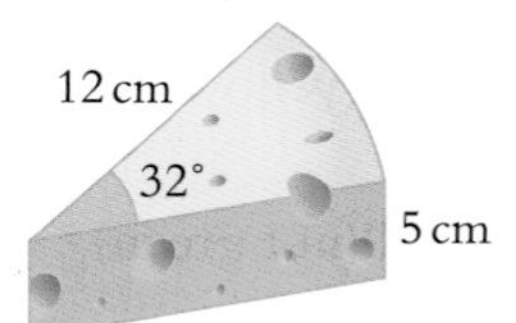

KEY WORDS

prism, cross-section
cylinder, capacity
volume, surface area
curved surface

Probability 1

THIS SECTION WILL SHOW YOU HOW TO
- Understand and use the probability scale from 0 to 1
- Calculate the probability of a single event
- Use relative frequency
- Use possibility diagrams to calculate the probability of simple combined events

Probability is a measure of how likely an **event** is to happen.
An event can be made up of one or more possible **outcomes**.
For example, when a dice is rolled then the possible outcomes are 1, 2, 3, 4, 5 and 6.
If the event A is 'the number on the dice is greater than 4' then there are two desired outcomes for event A which are 5 and 6.

Probability is measured on a scale from 0 to 1

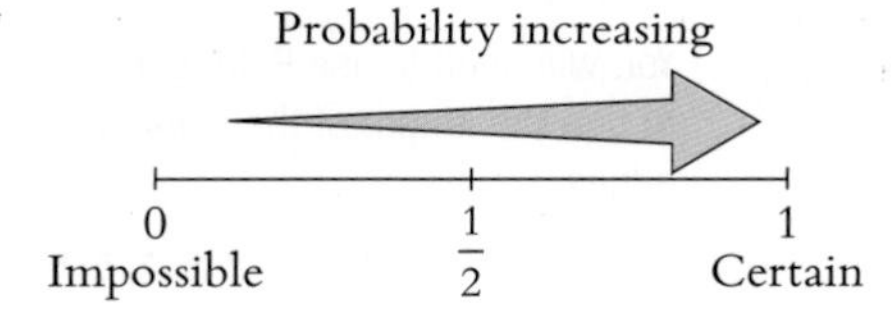

➡ **NOTE:** you can write probability as a fraction, decimal or a percentage.

If all possible outcomes are equally likely, you can calculate the theoretical probability, P(A), of the event happening.

$$P(A) = \frac{\text{number of desirable outcomes}}{\text{total number of possible outcomes}}$$

So, when a **fair** dice is rolled the probability the number on the dice is greater than $4 = \frac{2}{6} = \frac{1}{3}$

A fair (or **unbiased**) dice means the possible outcomes are equally likely.
A **biased** dice means the possible outcomes are not equally likely.

EXAMPLE

The spinner has eight coloured sections. When the spinner is spun it is equally likely that it stops on any one of the sections.
Find the probability that it stops on

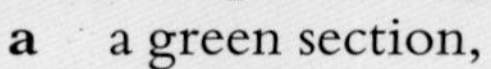

a a green section, **b** a yellow section,
c a blue section, **d** a red section,
e a section that is not green, **f** a blue or green section.

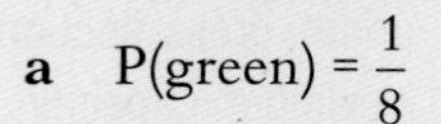

a $P(\text{green}) = \frac{1}{8}$ **b** $P(\text{yellow}) = \frac{3}{8}$ **c** $P(\text{blue}) = \frac{4}{8} = \frac{1}{2}$

d $P(\text{red}) = 0$ **e** $P(\text{not green}) = \frac{7}{8}$ **f** $P(\text{blue or green}) = \frac{4+1}{8} = \frac{5}{8}$

In the last example $P(\text{green}) = \frac{1}{8}$ and $P(\text{not green}) = \frac{7}{8}$

This is an example of the rule $P(\bar{A}) = 1 - P(A)$ where $\bar{A}$ means 'not A'.

The probability that Omar is late for school is 0.17.
Find is the probability that he is not late to school.

P(not late) = 1 – P(late) = 1 – 0.17 = 0.83

EXERCISE 3.24

1 An unbiased eight-sided dice is thrown. Find the probability that the dice

a lands on a 5
b lands on an even number
c lands on a square number
d lands on a number greater than 4
e lands on a 6 or a 7
f lands on a number smaller than 9

2 One of the five cards is chosen at random.
Find the probability of choosing a card with a number that is

3 6 6 8 9

a a six
b an odd number
c not prime
d square
e a multiple of 3
f a factor of 72

3 A box contains 5 orange balls, 4 yellow balls and 1 green ball.
A ball is picked out of the box at random. Find the probability that it is

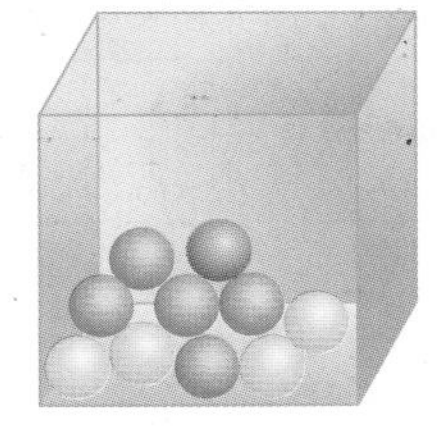

a green
b orange
c yellow
d not green
e orange or green
f green or yellow
g blue
h not blue

4

One of the nine cards is chosen at random. Find the probability that
a the letter on the card is an A
b the letter on the card is not an A
c a green card is chosen
d the card is orange with a letter B

5 The table shows the number of boys and girls in Years 8 and 9 at a school.

a A student is chosen at random from Year 8.
Find the probability that the student is a girl.
b A student is chosen at random from Year 9.
Find the probability that the student is a boy.

	Year 8	Year 9
Boys	52	48
Girls	68	52

EXAMPLE

One of these six cards is chosen at random.
The first card is **not** replaced before a second card is chosen at random.

7	8	8	7	7	9

a Find the probability that the first card chosen is a
i 7 **ii** 8 **iii** 9

b If the first card chosen is a 8, find the probability that the second card chosen is a
i 7 **ii** 8 **iii** 9

---

a i $P(7) = \frac{3}{6} = \frac{1}{2}$ *because there are* ***three*** *cards with a number 7 and* ***six*** *cards to chose from*

ii $P(8) = \frac{2}{6} = \frac{1}{3}$ *because there are* ***two*** *cards with a number 8 and* ***six*** *cards to chose from*

iii $P(9) = \frac{1}{6}$ *because there is* ***one*** *card with a number 9 and* ***six*** *cards to chose from*

b If the first card chosen was a 8 then the cards remaining are

7	8	7	7	9

i $P(7) = \frac{3}{5}$ *because there are* ***three*** *cards with a number 7 and* ***five*** *cards to chose from*

ii $P(8) = \frac{1}{5}$ *because there is now only* ***one*** *card with a number 8 and* ***five*** *cards to chose from*

iii $P(9) = \frac{1}{5}$ *because there is* ***one*** *card with a number 9 and* ***five*** *cards to chose from*

Sometimes the probabilities of all the outcomes are put into a table.
The probabilities for part **a** in the example above could be displayed as:

Number	7	8	9
Probability	$\frac{1}{2}$	$\frac{1}{3}$	$\frac{1}{6}$

➡ **NOTE:** the bottom row of the table will always add up to 1.

EXAMPLE

Score	1	2	3	4	5
Probability	0.1	x	0.2	0.3	0.15

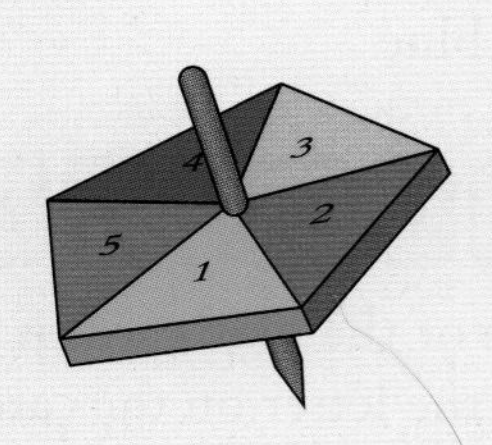

The table shows the probabilities of getting scores on a biased 5-sided spinner.

a Find the value of x.

b Find the probability of a score greater than 2.

c Find the probability of not scoring a 3.

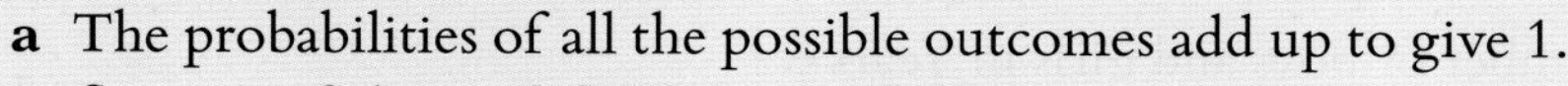

a The probabilities of all the possible outcomes add up to give 1.

So $0.1 + x + 0.2 + 0.3 + 0.15 = 1$

$x + 0.75 = 1 \Rightarrow x = 0.25$

b $P(\text{score} > 2) = P(3) + P(4) + P(5)$

$= 0.2 + 0.3 + 0.15 = 0.65$

c $P(\text{not } 3) = 1 - P(3)$

$= 1 - 0.2 = 0.8$

EXERCISE 3.25

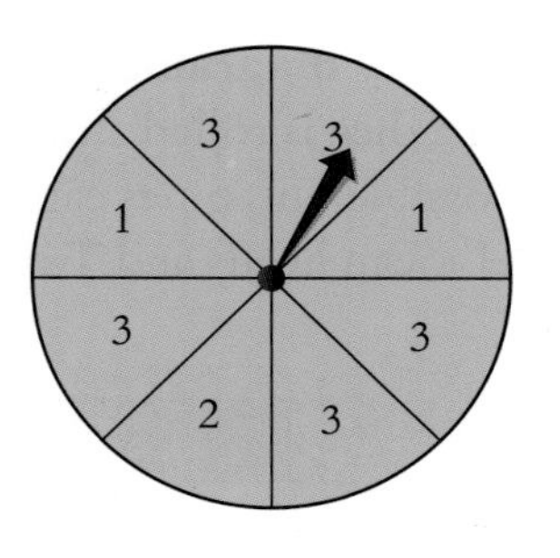

1 The circular board is divided into eight numbered sections. When the arrow is spun it is equally likely to stop in any of the eight sections.

a Copy and complete the table which shows the probability of the arrow stopping at each number.

b The arrow is spun once. Find

i the probability of a number smaller than 3

ii the probability of an odd number

iii the probability of a prime number

iv the probability the number is a 4.

Number	1	2	3
Probability		$\frac{1}{8}$	

UNIT 3

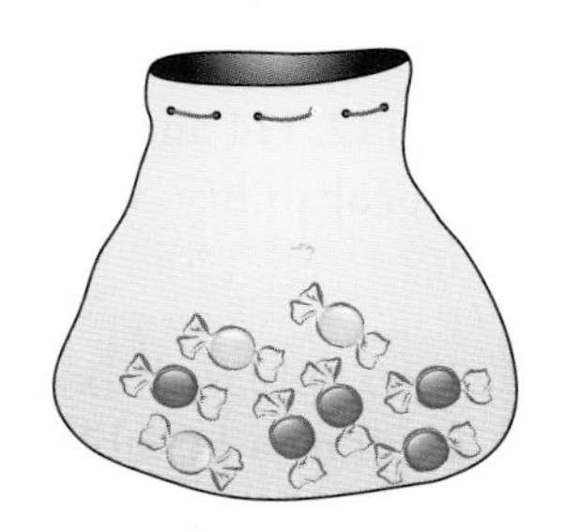

2 There are five red sweets and three yellow sweets in a bag. Rosa takes a sweet at random from the bag and eats it. Rosa then takes a second sweet at random from the bag.

a Find the probability that the first sweet that Rosa took from the bag was

i red ii yellow.

b If the first sweet that Rosa took from the bag was red, find the probability that the second sweet she takes from the bag is

i red ii yellow.

c If the first sweet that Rosa took from the bag was yellow, find the probability that the second sweet she takes from the bag is

i red ii yellow.

3 The table shows the probability of the spinner landing on each of the four colours.

Colour	Red	Yellow	Green	Blue
Probability	$\frac{1}{6}$	$\frac{1}{4}$	$\frac{1}{3}$	x

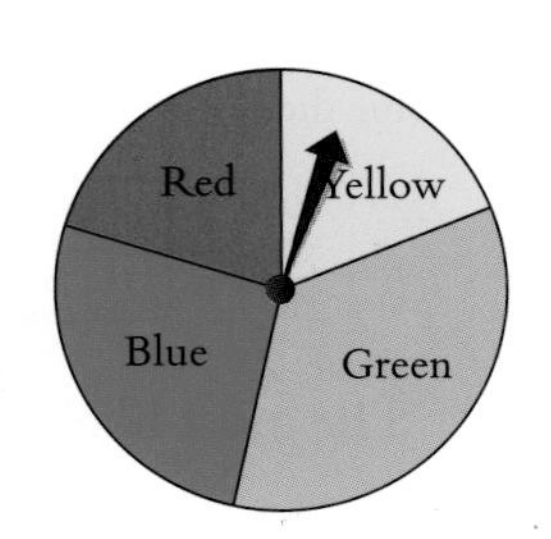

a Find the value of x.

b What is the most likely colour?

c Find the probability of not getting a red.

4

Number	1	2	3	4	5	6
Probability	0.1	0.15	0.05	0.1	$2x$	x

The table shows the probabilities of getting scores on a biased 6-sided dice.

a Find the value of x.

b What is the most likely score?

c Find the probability of a score greater than 2.

Possibility diagrams

When a dice is rolled and a one-dollar coin is tossed, the possible outcomes can be recorded in a possibility diagram. Let H mean heads and T mean tails

		Dice					
		1	2	3	4	5	6
Coin	H	H1	H2	H3	H4	H5	H6
	T	T1	T2	T3	T4	T5	T6

The possibility diagram shows that there are 12 equally likely outcomes.

The probability of each of the outcomes = $\frac{1}{12}$.

There are three places on the diagram where you can get a head and an even number; they are H2, H4 and H6.

So, the probability of getting a head and an even number = $\frac{3}{12} = \frac{1}{4}$

EXAMPLE

Two unbiased six-sided dice are rolled. The scores are added together.

a Draw a possibility diagram to show all the outcomes.

b What is the most likely total score?

c Use the diagram in part **a** to find the probability of

i a total score of 9 **ii** a total score less than 4.

a

Red dice

Blue dice	1	2	3	4	5	6
1	2	3	4	5	6	7
2	3	4	5	6	7	8
3	4	5	6	7	8	9
4	5	6	7	8	9	10
5	6	7	8	9	10	11
6	7	8	9	10	11	12

➡ **NOTE:** there are 36 possible outcomes.

The probability of each outcome = $\frac{1}{36}$

b The most likely total score is 7

there are more outcomes with a total of 7 than any other number

c **i** P(total score of 9) = $\frac{4}{36} = \frac{1}{9}$

there are 4 scores of 9 on the diagram

ii P(total score less than 4) = $\frac{3}{36} = \frac{1}{12}$

there are 3 scores less than 4 on the diagram

EXERCISE 3.26

1

	1	2	3	4	5
1					
2				6	
3					
4					
5			8		

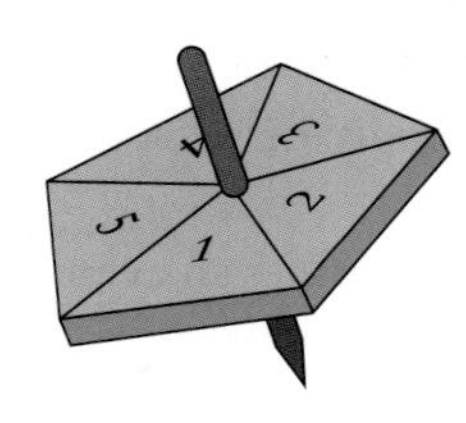

An unbiased five-sided spinner is spun twice and the scores are added.

a Copy and complete the possibility diagram.

b Use the diagram to find the probability that the total score is

i 9 ii 3 iii 1

iv an odd number v a square number, vi a prime number.

2

	1	2	3	4
1	2	3		
2				
3				
4			7	

Two unbiased four-sided dice are thrown and their scores are added together.

a Copy and complete the possibility diagram.

b Use the diagram to find the probability the total score is

i 7 ii 5 iii even iv less than 5.

3

	2	4	6	8
1	2			
2		8		
3				24

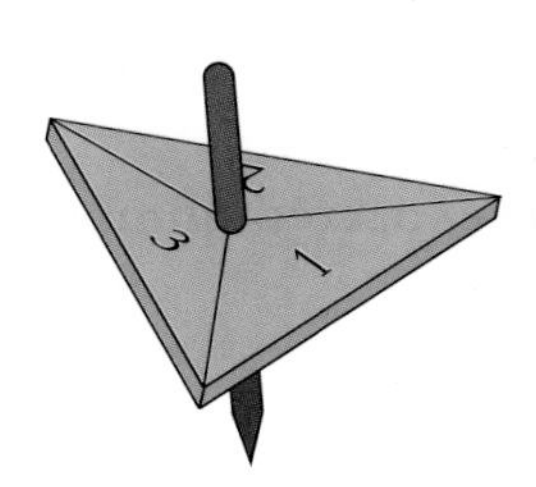
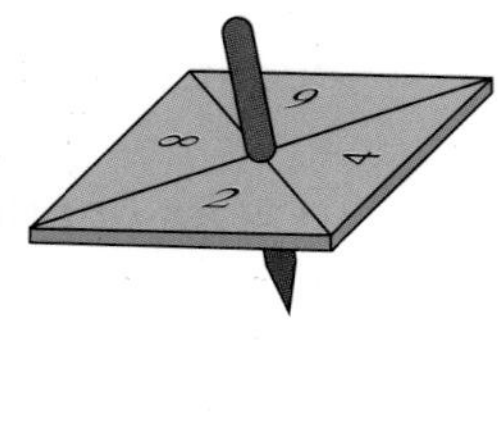

An unbiased three-sided spinner numbered 1, 2, 3 and an unbiased four-sided spinner numbered 2, 4, 6, 8 are spun.
The scores are multiplied together.

a Copy and complete the possibility diagram.

b Use the diagram to find the probability the product of the scores is

i 8 ii odd iii even iv smaller than 7.

Experimental probability

You can use data from a survey or experiment to estimate **experimental probability**.
The more trials that are done, then the more reliable the estimate will be.
Relative frequency is an estimate of experimental probability.

$$\text{Relative frequency} = \frac{\text{number of times an event occurs}}{\text{total number of trials}}$$

EXAMPLE

Sami thinks that a particular spinner is biased.
To test his theory he spins it 100 times.
The results are shown in the table.

Score	1	2	3	4	5
Frequency	13	15	16	42	14

a Calculate the relative frequency for each number on the spinner.
b Compare the relative frequencies with the theoretical probabilities to decide if the spinner is biased.
c How could the experiment be improved?

a

Score	1	2	3	4	5
Frequency	13	15	16	42	14
Relative frequency	$\frac{13}{100}$ = 0.13	$\frac{15}{100}$ = 0.15	$\frac{16}{100}$ = 0.16	$\frac{42}{100}$ = 0.42	$\frac{14}{100}$ = 0.14

b The theoretical probability for obtaining each number is $\frac{1}{5} = 0.2$.
The figures suggest that the spinner is biased towards landing on the number 4.
c The experiment could be improved by doing more trials.

You may need to work out the **expected number** of times that an event might happen.

Expected number = total number × probability

EXAMPLE

A biased coin is tossed 150 times. It lands on heads 51 times.
a Estimate the probability that the coin lands on a head on the next throw.
b Estimate the number of heads you would expect when you toss the coin 650 times.

a $\text{P(Head)} = \frac{51}{150} = \frac{17}{50}$

b Expected number of heads = total number × probability

$$= 650 \times \frac{17}{50} = 221$$

EXERCISE 3.27

1 The probability that a biased coin will land on heads is 0.72
The coin is tossed 400 times.
Estimate the number of times the coin will land on a head.

2 The probability that a biased dice will land on a five is 0.29
The dice is rolled 500 times.
Estimate the number of times the dice will land on a five.

3 A fair dice is rolled 900 times.
Estimate the number of times the dice will land on a five.

4 The probability that a light bulb is defective is 0.015
Find the number of defective light bulbs that you would expect to find in a batch of 2000 light bulbs.

5 The probability that a drawing pin lands point down when it is dropped is 0.82
The drawing pin is dropped 600 times.
Estimate the number of times it will land point down.

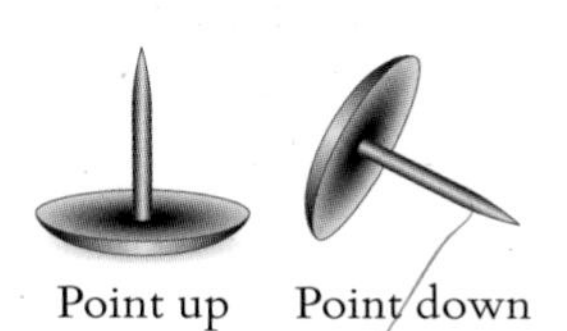

6 The table shows the results when the three-sided spinner is spun 200 times.

Score	1	2	3
Frequency	51	84	65

a Copy and complete the table.

Score	1	2	3
Relative frequency			

b Estimate the number of times the spinner will land on the number 3 if the spinner is spun 720 times.

7 The table shows the results when the circular spinner is spun 200 times.

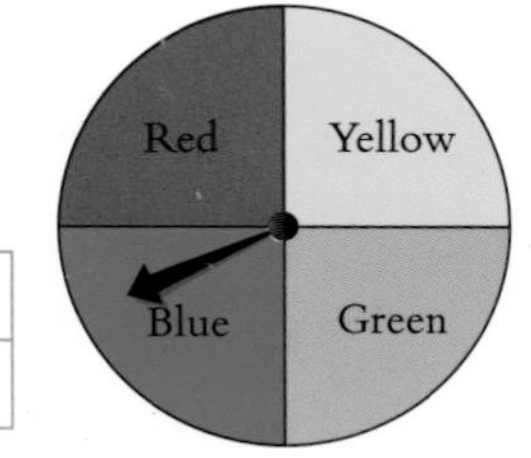

Colour	Red	Yellow	Green	Blue
Frequency	36	72	54	38

a Copy and complete the table.

Colour	Red	Yellow	Green	Blue
Relative frequency				

b Estimate the number of times the spinner will land on each of the four colours if the spinner is spun 900 times.

KEY WORDS
probability
event
outcome
fair
unbiased
biased
possibility diagram
experimental probability
relative frequency
expected number

Unit 3 Examination-style questions

1 **(a)** Write 562 000 000 in standard form. [1]

(b) Calculate $2.4 \times 10^5 + 1.3 \times 10^4$, giving your answer in standard form. [2]

2 **(a)** Write 0.000 000 78 correct to 1 significant figure. [1]

(b) Write 0.000 000 78 in standard form. [1]

3 Solve the simultaneous equations.
You must show all your working.

$$5x + 2y = 8$$
$$3x - 2y = 8$$ [2]

4 Solve the simultaneous equations.
You must show all your working.

$$7x + 3y = 16$$
$$2x + 9y = 29$$ [3]

5 Solve the simultaneous equations.
You must show all your working.

$$4x - 5y + 4 = 0$$
$$6x - 14y - 7 = 0$$ [4]

6 Factorise.

$$x^2 - 7x$$ [1]

7 Factorise.

(a) $5x^3y^2 - 10x^2y$ [2]

(b) $2xy + 3y - 8x - 12$ [2]

8 Simplify.

$$\frac{5x^2}{x - 3x^2}$$ [2]

9 Make a the subject of the formula.

$$c = \frac{a}{5} - b$$ [2]

10 Make x the subject of the formula.

$$5(x + 2y) = 7x - 4$$ [2]

11 Make h the subject of the formula.

$$A = 2\pi rh + \pi r^2$$ [2]

12 Make r the subject of the formula.

$$p = 2q - 5r^2$$ [3]

13 Make P the subject of the formula.

$$R = \frac{2}{3}\sqrt{PQ}$$ [3]

14

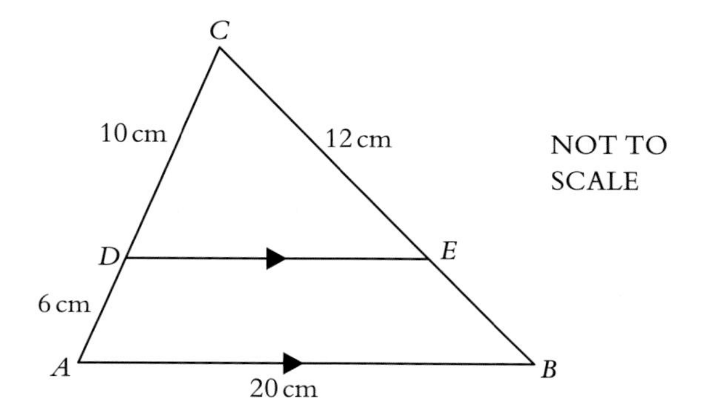

In the diagram, ABC is a triangle and AB is parallel to DE.
$AD = 6$ cm, $DC = 10$ cm, $CE = 12$ cm and $AB = 20$ cm.

(a) Find DE. [2]

(b) Find EB. [3]

15 Change 47 500 000 cm^3 to m^3. [1]

16

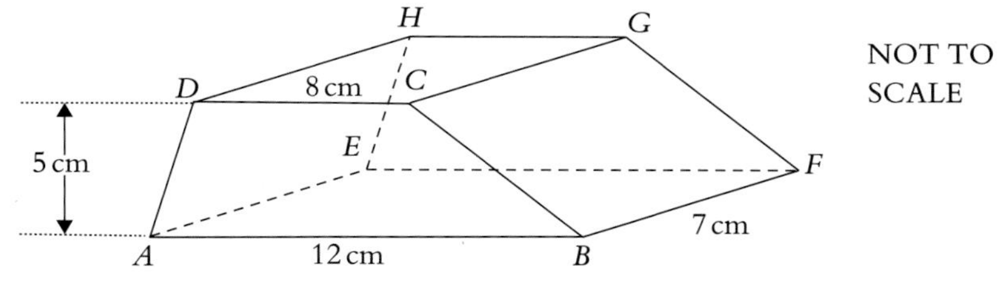

The diagram shows a prism $ABCDEFGH$ of length 7 cm.
Trapezium $ABCD$ is a cross section of the prism.
AB is parallel to DC and the distance between these two sides is 5 cm.
$AB = 12$ cm and $DC = 8$ cm.

(a) Calculate the area of trapezium $ABCD$. [2]

(b) Calculate the volume of the prism. [1]

17

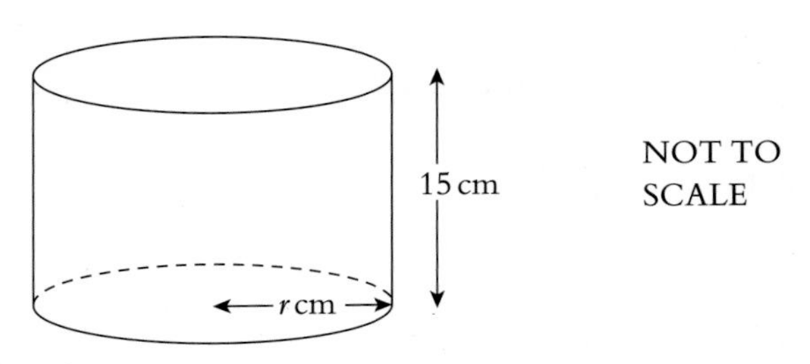

The diagram shows a solid circular cylinder.
The height is 15 cm and the radius is r cm.
The volume of the cylinder is 3600 cm^3.

(a) Find the value of r. [3]

(b) Calculate the total surface area of the cylinder. [4]

18

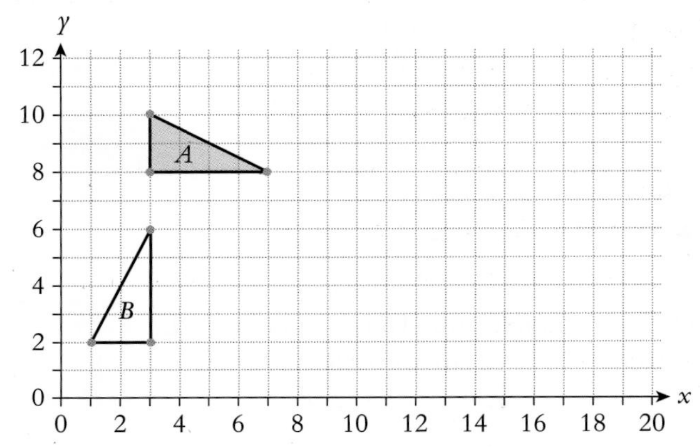

(a) Describe fully the **single** transformation that maps triangle A onto triangle B. [3]

(b) Draw the image of triangle A after translation by the vector $\begin{pmatrix} 2 \\ -5 \end{pmatrix}$. [2]

(c) Draw the image of triangle A after enlargement, scale factor −2, centre (8, 8). [3]

19

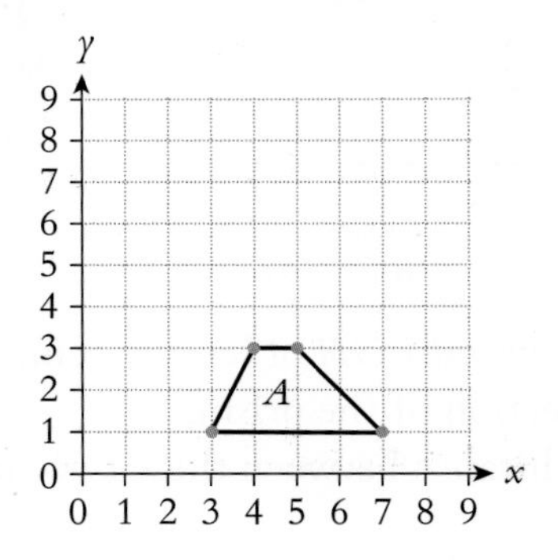

(a) Draw the image of shape A after refection in the line $y = x$.
Label the mage B. [2]

(b) Draw the image of shape A after rotation 180° centre (4.5, 4.5).
Label the image C. [3]

(c) Describe fully the single transformation that maps shape B onto shape C. [2]

20 P E P P E R

One of the six cards is chosen at random.
Find the probability that

(a) the letter on the card is P, [1]
(b) the letter on the card is not R, [1]
(c) the letter on the card is W. [1]

21

Score	1	2	3	4	5	6
Probability	0.2	x	0.15	0.05	0.3	0.2

The table shows the probabilities of getting scores on a biased 6-sided dice.

(a) Find the value of x. [2]
(b) What is the most likely score? [1]
(c) Find the probability of a score greater than 3. [1]

22 Two fair six-sided dice are rolled.
The numbers are added together to give the score.

(a) Complete the possibility diagram.

First dice

Second dice	1	2	3	4	5	6
1	1		3			
2				8		
3			9			18
4	4					
5		10				
6				24		36

[2]

(b) **(i)** Find the probability of a score of 12. [2]
(ii) Find the probability of a score less than 8. [2]

23 The probability that a biased dice lands on the number 4 is 0.18
The dice is rolled 500 times.
Find the number of times the dice is expected to land on the number 4. [1]

UNIT 3

Percentages 2

THIS SECTION WILL SHOW YOU HOW TO

- Express one quantity as a percentage of another quantity
- Calculate percentage increase and decrease
- Calculate percentage profit and loss

Expressing one quantity as a percentage of a second quantity

To write one quantity as a percentage of a second quantity

- Write the first quantity as a fraction of the second quantity
- Multiply by 100 to change the fraction into a percentage

EXAMPLE

Write 52 as a percentage of 65.

$$\text{Percentage} = \frac{\text{first quantity}}{\text{second quantity}} \times 100\% = \frac{52}{65} \times 100\% = 80\%$$

Percentage increase and percentage decrease

$$\%\ \text{increase} = \frac{\text{increase}}{\text{original value}} \times 100\% \qquad \%\ \text{decrease} = \frac{\text{decrease}}{\text{original value}} \times 100\%$$

EXAMPLE

The population of a village increases from 2400 to 2520.
Calculate the percentage increase.

Increase = 2520 – 2400 = 120

$$\%\ \text{increase} = \frac{\text{increase}}{\text{original value}} \times 100\%$$

$$= \frac{120}{2400} \times 100\% = 5\%$$

Percentage profit and loss

When a question involves money you are often asked to work out the **percentage profit** or the **percentage loss**.

$$\%\ \text{profit} = \frac{\text{profit}}{\text{original cost}} \times 100\% \qquad \%\ \text{loss} = \frac{\text{loss}}{\text{original cost}} \times 100\%$$

EXAMPLE

A car bought for \$5500 is sold for \$5280.
Calculate the percentage loss.

Loss = 5500 – 5280 = 220

$$\%\ \text{loss} = \frac{\text{loss}}{\text{original cost}} \times 100\%$$

$$= \frac{220}{5500} \times 100\% = 4\%$$

EXERCISE 4.1

1 Jarred scores 69 out of 120 in a test. Write this as a percentage.

2 Express \$44.40 as a percentage of \$120.

3 Eugenio buys a washing machine for \$845 and then pays an installation charge of \$30. Express the installation charge as a percentage of the washing machine price.

4 In a sale the price of a bag is reduced from \$50 to \$45. Find the percentage decrease.

5 A factory produces 55 000 torches one year and 88 000 torches the next year. Find the percentage increase.

6 A set of kitchen scales records a mass of 2.1 kg when the true mass is 2 kg. Calculate the percentage error.

7 The number 3.1 is used as an approximation for π. Calculate the percentage error.

8 A computer is bought for \$1358 and is then sold for \$1400. Calculate the percentage profit.

9 A clock valued at \$600 is sold for \$570. Calculate the percentage loss.

10 Cecilia's income increases from \$70 000 a year to \$85 000 a year. Calculate the percentage increase.

11 The number of students in a school decreases from 640 to 620. Calculate the percentage decrease.

12 In 2005 the population of Malaysia was 2.56×10^7. In 2009 the population of Malaysia was 2.75×10^7. Calculate the percentage increase in the population.

13 Fernando is a tax payer. He earns \$60 000 a year.
- The first \$12 000 of his earnings is tax free.
- He pays 10% tax on the next \$8 000 of his income.
- He pays 22% tax on the rest of his earnings.

Express the total amount of tax that he pays as a percentage of the \$60 000.

KEY WORDS
percentage profit
percentage loss

Matrix algebra

THIS SECTION WILL SHOW YOU HOW TO

- Describe the size of a matrix
- Find the sum and product of two matrices
- Calculate the determinant of a matrix
- Find the inverse of a matrix

These tables show the sales of TVs and radios over three days in shop A and shop B.

SHOP A	TV	Radio
DAY 1	5	3
DAY 2	7	8
DAY 3	4	5

SHOP B	TV	Radio
DAY 1	9	4
DAY 2	8	5
DAY 3	6	3

These can be combined to give the total sales of TVs and radios for shops A and B.

TOTAL	TV	Radio
DAY 1	14	7
DAY 2	15	13
DAY 3	10	8

This can be written in **matrix** form as follows.

$$\begin{pmatrix} 5 & 3 \\ 7 & 8 \\ 4 & 5 \end{pmatrix} + \begin{pmatrix} 9 & 4 \\ 8 & 5 \\ 6 & 3 \end{pmatrix} = \begin{pmatrix} 14 & 7 \\ 15 & 13 \\ 10 & 8 \end{pmatrix}$$

Two **matrices** can be added together if they are the same size.
The size of a matrix is called the **order** of a matrix.

The matrix $\begin{pmatrix} 5 & 3 \\ 7 & 8 \\ 4 & 5 \end{pmatrix}$ has order 3×2. (3 **rows** and 2 **columns**)

➡ **NOTE:** the number of rows is always written first.

EXAMPLE

Write down the order of this matrix $\begin{pmatrix} 2 & 5 & -1 & 6 & -3 \\ 3 & 0 & 1 & 5 & -1 \end{pmatrix}$

$$\begin{pmatrix} 2 & 5 & -1 & 6 & -3 \\ 3 & 0 & 1 & 5 & -1 \end{pmatrix}$$

The matrix has 2 rows and 5 columns.
The order is 2 × 5

EXAMPLE

$\mathbf{A} = \begin{pmatrix} 1 & 6 & 2 \\ 0 & -4 & 3 \end{pmatrix}$ $\mathbf{B} = \begin{pmatrix} -7 & 10 & -3 \\ 2 & 5 & 7 \end{pmatrix}$.

Find **a** $\mathbf{A} + \mathbf{B}$ **b** $\mathbf{A} - \mathbf{B}$

a $\mathbf{A} + \mathbf{B} = \begin{pmatrix} 1+-7 & 6+10 & 2+-3 \\ 0+2 & -4+5 & 3+7 \end{pmatrix} = \begin{pmatrix} -6 & 16 & -1 \\ 2 & 1 & 10 \end{pmatrix}$

b $\mathbf{A} - \mathbf{B} = \begin{pmatrix} 1--7 & 6-10 & 2--3 \\ 0-2 & -4-5 & 3-7 \end{pmatrix} = \begin{pmatrix} 8 & -4 & 5 \\ -2 & -9 & -4 \end{pmatrix}$

EXERCISE 4.2

1 Write down the order of these matrices.

a $(3\ \ 2\ \ 5\ \ 1)$ b (2) c $(4\ \ -1\ \ 0)$ d $(5\ \ 5)$

e $\begin{pmatrix} 2 & 5 \\ 1 & -3 \end{pmatrix}$ f $\begin{pmatrix} 6 & -2 & -5 \\ 0 & -1 & 3 \end{pmatrix}$ g $\begin{pmatrix} 5 \\ -4 \end{pmatrix}$ h $\begin{pmatrix} 1 & 7 & -3 & 6 \\ 2 & -5 & 4 & 7 \end{pmatrix}$

i $\begin{pmatrix} 4 \\ 3 \\ -1 \end{pmatrix}$ j $\begin{pmatrix} 2 & 3 & 4 \\ 5 & 6 & 7 \\ 8 & 9 & 10 \end{pmatrix}$ k $\begin{pmatrix} -1 & 2 \\ 4 & 0 \\ 6 & 9 \end{pmatrix}$ l $\begin{pmatrix} 5 & 3 & 5 \\ 2 & 0 & -1 \\ -5 & 7 & 2 \\ 0 & 6 & 0 \end{pmatrix}$

2 $\mathbf{A} = \begin{pmatrix} 5 & 4 \\ 3 & 8 \end{pmatrix}$ $\mathbf{B} = \begin{pmatrix} 2 & 0 \\ 1 & 5 \end{pmatrix}$ Find a $\mathbf{A} + \mathbf{B}$ b $\mathbf{A} - \mathbf{B}$

3 $\mathbf{A} = (5\ \ 3\ \ 7)$ $\mathbf{B} = (3\ \ 8\ \ -2)$ Find a $\mathbf{A} + \mathbf{B}$ b $\mathbf{A} - \mathbf{B}$

4 $\mathbf{A} = \begin{pmatrix} 5 & 1 & 7 \\ -2 & 0 & 6 \end{pmatrix}$ $\mathbf{B} = \begin{pmatrix} 7 & 3 & -1 \\ 0 & -2 & 5 \end{pmatrix}$ Find a $\mathbf{A} + \mathbf{B}$ b $\mathbf{A} - \mathbf{B}$

5 $\mathbf{A} = \begin{pmatrix} 6 & 2 \\ -8 & 0 \\ 2 & -5 \\ 8 & -4 \end{pmatrix}$ $\mathbf{B} = \begin{pmatrix} 10 & 5 \\ -8 & 0 \\ -2 & 5 \\ -6 & -3 \end{pmatrix}$ Find a $\mathbf{A} + \mathbf{B}$ b $\mathbf{A} - \mathbf{B}$

6 $\mathbf{A} = \begin{pmatrix} 5 & 1 \\ -3 & 2 \end{pmatrix}$ $\mathbf{B} = \begin{pmatrix} 2 & 9 \\ -7 & 0 \end{pmatrix}$ $\mathbf{C} = \begin{pmatrix} 6 & -2 \\ 2 & -4 \end{pmatrix}$ Find

a $\mathbf{A} + \mathbf{B}$ b $\mathbf{A} + \mathbf{C}$ c $\mathbf{A} - \mathbf{C}$ d $\mathbf{B} - \mathbf{C}$
e $\mathbf{A} + \mathbf{B} + \mathbf{C}$ f $\mathbf{A} + \mathbf{B} - \mathbf{C}$ g $\mathbf{A} - (\mathbf{B} - \mathbf{C})$ h $\mathbf{A} - (\mathbf{B} + \mathbf{C})$

7 $\mathbf{A} = \begin{pmatrix} 3 & 1 \\ -4 & 2 \\ 7 & 8 \end{pmatrix}$ $\mathbf{B} = \begin{pmatrix} 0 & -8 \\ 4 & 0 \\ 7 & 5 \end{pmatrix}$ $\mathbf{C} = \begin{pmatrix} -1 & 2 \\ 6 & 4 \\ 3 & -9 \end{pmatrix}$ Find

a $\mathbf{A} + \mathbf{B}$ b $\mathbf{A} + \mathbf{C}$ c $\mathbf{B} + \mathbf{C}$ d $\mathbf{B} - \mathbf{C}$
e $\mathbf{A} - \mathbf{B}$ f $\mathbf{A} + \mathbf{B} - \mathbf{C}$ g $\mathbf{A} - (\mathbf{B} - \mathbf{C})$ h $\mathbf{B} - (\mathbf{A} + \mathbf{C})$

8 $\begin{pmatrix} x & 6 \\ -2 & y \end{pmatrix} + \begin{pmatrix} -3 & 4 \\ 0 & 7 \end{pmatrix} = \begin{pmatrix} 4 & 10 \\ -2 & -1 \end{pmatrix}$ Find the values of x and y.

9 $\begin{pmatrix} 2 & x \\ -5 & 4 \\ 0 & -3 \end{pmatrix} + \begin{pmatrix} 7 & 1 \\ 8 & y \\ 4 & 0 \end{pmatrix} = \begin{pmatrix} 9 & -6 \\ 3 & -1 \\ 4 & -3 \end{pmatrix}$ Find the values of x and y.

UNIT 4

Multiplying a matrix by a number

If $\mathbf{A} = \begin{pmatrix} 1 & 2 \\ 4 & -3 \end{pmatrix}$ then $2\mathbf{A} = \mathbf{A} + \mathbf{A} = \begin{pmatrix} 1 & 2 \\ 4 & -3 \end{pmatrix} + \begin{pmatrix} 1 & 2 \\ 4 & -3 \end{pmatrix} = \begin{pmatrix} 2 & 4 \\ 8 & -6 \end{pmatrix}$

So 2A is the same as multiplying each term in the matrix A by the number 2.

EXAMPLE

$\mathbf{A} = \begin{pmatrix} 5 & -1 & 0 \\ -3 & 4 & 2 \end{pmatrix}$ Find 4**A**.

$$4\mathbf{A} = \begin{pmatrix} 4 \times 5 & 4 \times -1 & 4 \times 0 \\ 4 \times -3 & 4 \times 4 & 4 \times 2 \end{pmatrix} = \begin{pmatrix} 20 & -4 & 0 \\ -12 & 16 & 8 \end{pmatrix}$$

Multiplying a row matrix by a column matrix

If a TV costs \$500 and a radio costs \$40, you can work out the amount of money shop A takes for the TVs and radios on day 1 as follows

SHOP A	TV	Radio
DAY 1	5	3
DAY 2	7	8
DAY 3	4	5

$5 \times 500 + 3 \times 40 = 2500 + 120 = 2620$

You can write this in matrix form as $\begin{pmatrix} 5 & 3 \end{pmatrix}\begin{pmatrix} 500 \\ 40 \end{pmatrix} = \begin{pmatrix} 2620 \end{pmatrix}$

Similarly for day 2 $\begin{pmatrix} 7 & 8 \end{pmatrix}\begin{pmatrix} 500 \\ 40 \end{pmatrix} = \begin{pmatrix} 3820 \end{pmatrix}$

Similarly for day 3 $\begin{pmatrix} 4 & 5 \end{pmatrix}\begin{pmatrix} 500 \\ 40 \end{pmatrix} = \begin{pmatrix} 2200 \end{pmatrix}$

EXAMPLE

Work out **a** $\begin{pmatrix} 2 & 5 & 3 \end{pmatrix}\begin{pmatrix} 4 \\ 6 \\ 1 \end{pmatrix}$ **b** $\begin{pmatrix} -3 & 0 & 1 & 2 \end{pmatrix}\begin{pmatrix} 4 \\ -2 \\ 1 \\ 5 \end{pmatrix}$

a $\begin{pmatrix} 2 & 5 & 3 \end{pmatrix}\begin{pmatrix} 4 \\ 6 \\ 1 \end{pmatrix} = \begin{pmatrix} 2 \times 4 + 5 \times 6 + 3 \times 1 \end{pmatrix} = \begin{pmatrix} 41 \end{pmatrix}$

b $\begin{pmatrix} -3 & 0 & 1 & 2 \end{pmatrix}\begin{pmatrix} 4 \\ -2 \\ 1 \\ 5 \end{pmatrix} = \begin{pmatrix} -3 \times 4 + 0 \times -2 + 1 \times 1 + 2 \times 5 \end{pmatrix} = \begin{pmatrix} -1 \end{pmatrix}$

EXERCISE 4.3

1 $\mathbf{A} = \begin{pmatrix} 3 & -4 \end{pmatrix}$ Find $4\mathbf{A}$

2 $\mathbf{A} = \begin{pmatrix} 4 & -1 \\ 0 & 2 \end{pmatrix}$ Find $\frac{1}{2}\mathbf{A}$

3 $\mathbf{A} = \begin{pmatrix} 6 & -3 & 2 \\ 0 & 1 & -1 \end{pmatrix}$ Find $8\mathbf{A}$

4 $\mathbf{A} = \begin{pmatrix} -2 & 1 & 6 & -5 \\ 1 & 0 & 2 & 1 \end{pmatrix}$ Find $-3\mathbf{A}$

5 $\mathbf{A} = \begin{pmatrix} 2 & -3 \\ 1 & 1 \end{pmatrix}$ $\mathbf{B} = \begin{pmatrix} 5 & 0 \\ -2 & 1 \end{pmatrix}$ $\mathbf{C} = \begin{pmatrix} -4 & -6 \\ 0 & 2 \end{pmatrix}$ Find

a $2\mathbf{A}$ b $5\mathbf{C}$ c $3\mathbf{B}$ d $\frac{1}{2}\mathbf{C}$

e $-3\mathbf{A}$ f $2\mathbf{A} + 3\mathbf{B}$ g $5\mathbf{C} - 3\mathbf{B}$ h $2\mathbf{A} + \frac{1}{2}\mathbf{C}$

6 Calculate the following.

a $\begin{pmatrix} 1 & 5 \end{pmatrix}\begin{pmatrix} 2 \\ 3 \end{pmatrix}$ b $\begin{pmatrix} -1 & 2 \end{pmatrix}\begin{pmatrix} 0 \\ 7 \end{pmatrix}$ c $\begin{pmatrix} 6 & 6 \end{pmatrix}\begin{pmatrix} -6 \\ -6 \end{pmatrix}$

d $\begin{pmatrix} 2 & 5 & 3 \end{pmatrix}\begin{pmatrix} 1 \\ 2 \\ 1 \end{pmatrix}$ e $\begin{pmatrix} 4 & -1 & 0 \end{pmatrix}\begin{pmatrix} 2 \\ 0 \\ 1 \end{pmatrix}$ f $\begin{pmatrix} -2 & 2 & -2 \end{pmatrix}\begin{pmatrix} -1 \\ 1 \\ -1 \end{pmatrix}$

g $\begin{pmatrix} 6 & 2 & 1 & 3 \end{pmatrix}\begin{pmatrix} 5 \\ 1 \\ 3 \\ 3 \end{pmatrix}$ h $\begin{pmatrix} 5 & -2 & 0 & 4 \end{pmatrix}\begin{pmatrix} -2 \\ 0 \\ 3 \\ 1 \end{pmatrix}$ i $\begin{pmatrix} 0 & 2 & 6 & 2 \end{pmatrix}\begin{pmatrix} 8 \\ 2 \\ 1 \\ 4 \end{pmatrix}$

7 $\begin{pmatrix} x & 3 \end{pmatrix}\begin{pmatrix} 2 \\ -1 \end{pmatrix} = \begin{pmatrix} 15 \end{pmatrix}$ Find the value of x.

TOP TIP

Form an equation in x and then solve.

8 $\begin{pmatrix} 1 & x & 3 \end{pmatrix}\begin{pmatrix} 2 \\ 4 \\ 2 \end{pmatrix} = \begin{pmatrix} 36 \end{pmatrix}$ Find the value of x.

9 $\begin{pmatrix} 2 & -3 & -1 & -4 \end{pmatrix}\begin{pmatrix} 1 \\ -6 \\ x \\ 3 \end{pmatrix} = \begin{pmatrix} 10 \end{pmatrix}$ Find the value of x.

Multiplying matrices

In the last section you saw that when a TV costs $500 and a radio costs $40

SHOP A	TV	Radio
DAY 1	5	3
DAY 2	7	8
DAY 3	4	5

DAY 1: $\begin{pmatrix} 5 & 3 \end{pmatrix}\begin{pmatrix} 500 \\ 40 \end{pmatrix} = \begin{pmatrix} 2620 \end{pmatrix}$ DAY 2: $\begin{pmatrix} 7 & 8 \end{pmatrix}\begin{pmatrix} 500 \\ 40 \end{pmatrix} = \begin{pmatrix} 3820 \end{pmatrix}$

DAY 3: $\begin{pmatrix} 4 & 5 \end{pmatrix}\begin{pmatrix} 500 \\ 40 \end{pmatrix} = \begin{pmatrix} 2200 \end{pmatrix}$

These can be combined into just one matrix equation.

$$\begin{pmatrix} 5 & 3 \\ 7 & 8 \\ 4 & 5 \end{pmatrix}\begin{pmatrix} 500 \\ 40 \end{pmatrix} = \begin{pmatrix} 2620 \\ 3820 \\ 2200 \end{pmatrix}$$

order 3×2 2×1 3×1

same

It is only possible to multiply two matrices when the number of columns in the first matrix is the same as the number of rows in the second matrix.

In general First matrix Second matrix

Order $a \times b$ $c \times d$

You can multiply the matrices if $b = c$.

The product of the two matrices will be of order $a \times d$.

EXAMPLE

Work out $\begin{pmatrix} 2 & 1 \\ 3 & 4 \end{pmatrix}\begin{pmatrix} 5 \\ 8 \end{pmatrix}$

First consider the order of the matrices: 2×2 2×1

The numbers in the middle are the same so it is possible to multiply the matrices. The answer will be a 2×1 matrix.

$\begin{pmatrix} 2 & 1 \\ 3 & 4 \end{pmatrix}\begin{pmatrix} 5 \\ 8 \end{pmatrix} = \begin{pmatrix} \star \\ \star \end{pmatrix}$

You must now multiply each row in the first matrix with the column in the second matrix to find the missing numbers.

$2 \times 5 + 1 \times 8 = 18$

$3 \times 5 + 4 \times 8 = 47$

$\begin{pmatrix} 2 & 1 \\ 3 & 4 \end{pmatrix}\begin{pmatrix} 5 \\ 8 \end{pmatrix} = \begin{pmatrix} 18 \\ 47 \end{pmatrix}$

EXAMPLE

Work out $\begin{pmatrix} 2 & 1 & 4 \\ 1 & 0 & 3 \\ 5 & 3 & 2 \end{pmatrix}\begin{pmatrix} 3 & 4 \\ 1 & 2 \\ 2 & 1 \end{pmatrix}$

First consider the order of the matrices: $3 \times 3 \quad 3 \times 2$ The answer will be a 3×2 matrix.

$$\begin{pmatrix} 2 & 1 & 4 \\ 1 & 0 & 3 \\ 5 & 3 & 2 \end{pmatrix}\begin{pmatrix} 3 & 4 \\ 1 & 2 \\ 2 & 1 \end{pmatrix} = \begin{pmatrix} \star & \star \\ \star & \star \\ \star & \star \end{pmatrix}$$

You must now multiply each row in the first matrix with each column in the second matrix to find the missing numbers.

$2 \times 3 + 1 \times 1 + 4 \times 2 = 15$ $\quad$ $2 \times 4 + 1 \times 2 + 4 \times 1 = 14$

$1 \times 3 + 0 \times 1 + 3 \times 2 = 9$ $\quad$ $1 \times 4 + 0 \times 2 + 3 \times 1 = 7$

$5 \times 3 + 3 \times 1 + 2 \times 2 = 22$ $\quad$ $5 \times 4 + 3 \times 2 + 2 \times 1 = 28$

$$\begin{pmatrix} 2 & 1 & 4 \\ 1 & 0 & 3 \\ 5 & 3 & 2 \end{pmatrix}\begin{pmatrix} 3 & 4 \\ 1 & 2 \\ 2 & 1 \end{pmatrix} = \begin{pmatrix} 15 & 14 \\ 9 & 7 \\ 22 & 28 \end{pmatrix}$$

EXERCISE 4.4

1 Multiply these matrices.

a $\begin{pmatrix} 1 & 4 \\ 5 & 2 \end{pmatrix}\begin{pmatrix} 3 & 1 \\ 0 & 4 \end{pmatrix}$

b $\begin{pmatrix} 2 & 0 \\ 1 & 1 \end{pmatrix}\begin{pmatrix} 5 & 1 & 2 \\ 0 & 4 & 1 \end{pmatrix}$

c $\begin{pmatrix} 6 & 2 \\ 1 & 0 \end{pmatrix}\begin{pmatrix} 5 \\ 3 \end{pmatrix}$

d $\begin{pmatrix} 6 & 1 \\ 0 & 8 \end{pmatrix}\begin{pmatrix} 5 & 3 \\ 2 & 1 \end{pmatrix}$

e $\begin{pmatrix} 5 & 1 & 2 \\ 4 & 0 & 1 \end{pmatrix}\begin{pmatrix} 2 & 1 \\ 0 & -1 \\ 3 & 0 \end{pmatrix}$

f $\begin{pmatrix} 1 & 6 \end{pmatrix}\begin{pmatrix} 7 \\ 2 \end{pmatrix}$

g $\begin{pmatrix} 5 & 1 & 2 & 3 \\ 0 & 1 & -2 & 4 \end{pmatrix}\begin{pmatrix} 5 & -1 \\ 0 & 2 \\ 1 & 0 \\ 3 & 1 \end{pmatrix}$

h $\begin{pmatrix} 5 & 1 \\ 1 & 2 \\ -3 & 0 \\ 6 & 4 \end{pmatrix}\begin{pmatrix} 3 & 0 & 1 & 2 \\ 4 & 3 & 2 & 1 \end{pmatrix}$

2 $\mathbf{A} = \begin{pmatrix} 2 & 8 \\ -5 & 3 \\ 2 & -1 \end{pmatrix}$ and $\mathbf{B} = \begin{pmatrix} 6 & 1 & 7 & 3 \\ 2 & 0 & 5 & 4 \end{pmatrix}$

a The order of the matrix **AB** is $x \times y$. Write down the values of x and y.

b Explain why it is impossible to calculate the matrix **BA**.

3 Work out $\begin{pmatrix} 4 & 2 & -1 \\ 3 & 0 & 5 \\ 2 & -6 & 4 \end{pmatrix}\begin{pmatrix} 3 \\ -2 \\ 1 \end{pmatrix}$

2 × 2 matrices

$$\begin{pmatrix} 0 & 0 \\ 0 & 0 \end{pmatrix}\begin{pmatrix} 2 & 3 \\ -4 & 1 \end{pmatrix} = \begin{pmatrix} 0 & 0 \\ 0 & 0 \end{pmatrix} \text{ and } \begin{pmatrix} 2 & 3 \\ -4 & 1 \end{pmatrix}\begin{pmatrix} 0 & 0 \\ 0 & 0 \end{pmatrix} = \begin{pmatrix} 0 & 0 \\ 0 & 0 \end{pmatrix}$$

The matrix $\begin{pmatrix} 0 & 0 \\ 0 & 0 \end{pmatrix}$ is called the **zero** matrix.

$$\begin{pmatrix} 1 & 0 \\ 0 & 1 \end{pmatrix}\begin{pmatrix} 2 & 3 \\ -4 & 1 \end{pmatrix} = \begin{pmatrix} 2 & 3 \\ -4 & 1 \end{pmatrix} \text{ and } \begin{pmatrix} 2 & 3 \\ -4 & 1 \end{pmatrix}\begin{pmatrix} 1 & 0 \\ 0 & 1 \end{pmatrix} = \begin{pmatrix} 2 & 3 \\ -4 & 1 \end{pmatrix}$$

The matrix $\begin{pmatrix} 1 & 0 \\ 0 & 1 \end{pmatrix}$ is called the **identity** matrix for multiplication and is denoted by the letter I.

When you multiply a matrix by the identity matrix (**I**) the numbers in the matrix stay the same.

IA = **A** and **AI** = **A**

If $\mathbf{A} = \begin{pmatrix} a & b \\ c & d \end{pmatrix}$, the **determinant** of matrix **A** (denoted by $|\mathbf{A}|$) is defined as follows.

$|\mathbf{A}| = ad - bc$

The steps for finding the determinant of the matrix $\mathbf{A} = \begin{pmatrix} 6 & 7 \\ 2 & 3 \end{pmatrix}$ are:

STEP 1: $\begin{pmatrix} 6 & 7 \\ 2 & 3 \end{pmatrix}$ Multiply the numbers on the leading diagonal. 6×3

STEP 2: Multiply the numbers on the other diagonal. 2×7

STEP 3: Subtract

$|\mathbf{A}| = 6 \times 3 - 2 \times 7$

$|\mathbf{A}| = 18 - 14$

$|\mathbf{A}| = 4$

EXAMPLE

$\mathbf{A} = \begin{pmatrix} -2 & 4 \\ -2 & 3 \end{pmatrix}$ Find $|\mathbf{A}|$.

$\begin{pmatrix} -2 & 4 \\ -2 & 3 \end{pmatrix}$

$|\mathbf{A}| = -2 \times 3 - 4 \times -2$

$|\mathbf{A}| = -6 - -8$

$|\mathbf{A}| = 2$

EXERCISE 4.5

1 Work out the determinant of each of these matrices.

a $\begin{pmatrix} 1 & 1 \\ 3 & 4 \end{pmatrix}$ b $\begin{pmatrix} 5 & 3 \\ 3 & 2 \end{pmatrix}$ c $\begin{pmatrix} 9 & -2 \\ -4 & 1 \end{pmatrix}$ d $\begin{pmatrix} 3 & -2 \\ -2 & -1 \end{pmatrix}$

e $\begin{pmatrix} 4 & 11 \\ 3 & 8 \end{pmatrix}$ f $\begin{pmatrix} 0.5 & 0.5 \\ 6 & 10 \end{pmatrix}$ g $\begin{pmatrix} 2 & 5 \\ -2 & -3 \end{pmatrix}$ h $\begin{pmatrix} 2 & 4 \\ 3 & 6 \end{pmatrix}$

2 The determinant of the matrix $\begin{pmatrix} 2 & -1 \\ 3 & x \end{pmatrix}$ is 10.
Find the value of x.

3 The determinant of the matrix $\begin{pmatrix} 2 & -4 \\ x & 3 \end{pmatrix}$ is 2.
Find the value of x.

TOP TIP

AB means $\begin{pmatrix} 3 & 0 \\ 2 & -1 \end{pmatrix}\begin{pmatrix} -1 & 4 \\ 0 & 2 \end{pmatrix}$

BA means $\begin{pmatrix} -1 & 4 \\ 0 & 2 \end{pmatrix}\begin{pmatrix} 3 & 0 \\ 2 & -1 \end{pmatrix}$

4 $\mathbf{A} = \begin{pmatrix} 3 & 0 \\ 2 & -1 \end{pmatrix}$ $\mathbf{B} = \begin{pmatrix} -1 & 4 \\ 0 & 2 \end{pmatrix}$

Find a **AB** b **BA**

5 $\mathbf{A} = \begin{pmatrix} 5 & 1 \\ 1 & -2 \end{pmatrix}$ $\mathbf{B} = \begin{pmatrix} -3 & 1 \\ 2 & 3 \end{pmatrix}$

Find a **AB** b **BA** c $\mathbf{A}^2$ d $\mathbf{B}^2$

TOP TIP

$\mathbf{A}^2$ means $\begin{pmatrix} 5 & 1 \\ 1 & -2 \end{pmatrix}\begin{pmatrix} 5 & 1 \\ 1 & -2 \end{pmatrix}$

6 $\mathbf{A} = \begin{pmatrix} 2 & -1 \\ -3 & 2 \end{pmatrix}$ and $\mathbf{B} = \begin{pmatrix} 2 & 3 \\ 1 & 4 \end{pmatrix}$ Find

a $3\mathbf{A}$ b $5\mathbf{B}$ c $\mathbf{A} - \mathbf{B}$ d $3\mathbf{A} + 5\mathbf{B}$
e **AB** f **BA** g $\mathbf{A}^2$ h $\mathbf{B}^2$
i $|\mathbf{A}|$ j $|\mathbf{B}|$

7 $\mathbf{A} = \begin{pmatrix} 1 & -1 \\ 1 & 2 \end{pmatrix}$ and $\mathbf{B} = \begin{pmatrix} 5 & -3 \\ 1 & -1 \end{pmatrix}$ Find

a $2\mathbf{A}$ b $4\mathbf{B}$ c $2\mathbf{A} - 4\mathbf{B}$ d **AB**
e **BA** f $\mathbf{A}^2$ g $\mathbf{B}^2$ h $(\mathbf{A} - \mathbf{B})^2$
i $|\mathbf{A}|$ j $|\mathbf{B}|$

8 $\mathbf{A} = \begin{pmatrix} 5 & 3 \\ 2 & 1 \end{pmatrix}$. Find the 2×2 matrix **B** such that $\mathbf{A} + \mathbf{B} = \begin{pmatrix} 0 & 0 \\ 0 & 0 \end{pmatrix}$.

Inverse matrices

If **AB** = **I**, where **I** is the identity matrix then **B** is called the **inverse** of matrix **A**.
The inverse of matrix **A** is denoted by $\mathbf{A}^{-1}$.
The inverse matrix is calculated using the following formula.

If $\mathbf{A} = \begin{pmatrix} a & b \\ c & d \end{pmatrix}$ then $\mathbf{A}^{-1} = \frac{1}{|\mathbf{A}|}\begin{pmatrix} d & -b \\ -c & a \end{pmatrix}$

The steps for finding the inverse of the matrix $\mathbf{A} = \begin{pmatrix} 4 & 6 \\ 1 & 2 \end{pmatrix}$ are:

STEP 1: Calculate the determinant. $|\mathbf{A}| = 4 \times 2 - 1 \times 6 = 8 - 6 = 2$

STEP 2: Swap the numbers on the leading diagonal to give $\begin{pmatrix} 2 & 6 \\ 1 & 4 \end{pmatrix}$

STEP 3: Change the signs of the numbers on the other diagonal to give $\begin{pmatrix} 2 & -6 \\ -1 & 4 \end{pmatrix}$

STEP 4: Divide by the determinant. $\mathbf{A}^{-1} = \begin{pmatrix} 1 & -3 \\ -\frac{1}{2} & 2 \end{pmatrix}$ or $\mathbf{A}^{-1} = \frac{1}{2}\begin{pmatrix} 2 & -6 \\ -1 & 4 \end{pmatrix}$

EXAMPLE

Find the inverse of $\mathbf{A} = \begin{pmatrix} 1 & 7 \\ -1 & -4 \end{pmatrix}$

$\mathbf{A} = \begin{pmatrix} 1 & 7 \\ -1 & -4 \end{pmatrix}$

$|\mathbf{A}| = 1 \times -4 - 7 \times -1 = -4 - -7 = 3$

$\mathbf{A}^{-1} = \frac{1}{3}\begin{pmatrix} -4 & -7 \\ 1 & 1 \end{pmatrix}$

$\mathbf{A}^{-1} = \frac{1}{3}\begin{pmatrix} -4 & -7 \\ 1 & 1 \end{pmatrix}$ or $\mathbf{A}^{-1} = \begin{pmatrix} -1\frac{1}{3} & -2\frac{1}{3} \\ \frac{1}{3} & \frac{1}{3} \end{pmatrix}$

- *swap the numbers on the leading diagonal*
- *change the signs of the numbers on the other diagonal*
- *divide by the determinant*

→ **CHECK:** $\mathbf{A}^{-1}\mathbf{A} = \frac{1}{3}\begin{pmatrix} -4 & -7 \\ 1 & 1 \end{pmatrix}\begin{pmatrix} 1 & 7 \\ -1 & -4 \end{pmatrix} = \frac{1}{3}\begin{pmatrix} 3 & 0 \\ 0 & 3 \end{pmatrix} = \begin{pmatrix} 1 & 0 \\ 0 & 1 \end{pmatrix}$ ✓

A matrix has no inverse when the determinant is zero because you cannot divide by zero.
A matrix with no inverse is called a **singular** matrix.
A matrix with an inverse is called a **non-singular** matrix.

EXERCISE 4.6

1 Find the inverse matrix for each of these matrices.

a $\begin{pmatrix} 2 & 5 \\ 1 & 3 \end{pmatrix}$ b $\begin{pmatrix} 3 & 4 \\ 2 & 3 \end{pmatrix}$ c $\begin{pmatrix} 5 & 2 \\ 4 & 2 \end{pmatrix}$ d $\begin{pmatrix} -1 & 1 \\ 1 & -2 \end{pmatrix}$

e $\begin{pmatrix} 3 & 2 \\ -2 & -1 \end{pmatrix}$ f $\begin{pmatrix} 6 & -1 \\ -4 & 1 \end{pmatrix}$ g $\begin{pmatrix} 2 & 4 \\ -3 & -7 \end{pmatrix}$ h $\begin{pmatrix} 1 & 1 \\ -3 & -4 \end{pmatrix}$

i $\begin{pmatrix} -9 & -1 \\ -7 & -1 \end{pmatrix}$ j $\begin{pmatrix} 3 & -5 \\ -2 & 4 \end{pmatrix}$ k $\begin{pmatrix} 4 & -1 \\ -1 & 0.5 \end{pmatrix}$ l $\begin{pmatrix} 1 & -2 \\ 1 & -1 \end{pmatrix}$

2 Explain why $\begin{pmatrix} 4 & -8 \\ -2 & 4 \end{pmatrix}$ does not have an inverse.

3 The matrix $\begin{pmatrix} x & 5 \\ -2 & 2 \end{pmatrix}$ has no inverse. Find the value of x.

4 The matrix $\begin{pmatrix} 2 & x \\ 4 & 3 \end{pmatrix}$ has no inverse. Find the value of x.

5 $\mathbf{X} = \begin{pmatrix} 7 & 3 \\ 2 & 1 \end{pmatrix}$ and $\mathbf{XY} = \mathbf{I}$. Find the matrix $\mathbf{Y}$.

6 $\mathbf{Y} = \begin{pmatrix} -2 & 1 \\ 8 & 3 \end{pmatrix}$ and $\mathbf{XY} = \mathbf{I}$. Find the matrix $\mathbf{X}$.

7 $\mathbf{A} = \begin{pmatrix} 5 & 3 \\ 2 & 1 \end{pmatrix}$ Find the 2×2 matrix $\mathbf{C}$ such that $\mathbf{AC} = \begin{pmatrix} 1 & 0 \\ 0 & 1 \end{pmatrix}$.

8 $\mathbf{A} = \begin{pmatrix} 3 & 4 \\ 1 & 1 \end{pmatrix}$ Find

a $6\mathbf{A}$ b $\mathbf{A}^2$ c $\mathbf{A}^2 + 2\mathbf{A}$ d $(3\mathbf{A})^2$

e $\mathbf{A}^3$ f $|\mathbf{A}|$ g $\mathbf{A}^{-1}$ h $\mathbf{A}^{-1}\mathbf{A}$

9 $\mathbf{XY} = \begin{pmatrix} 7 & -1 \\ 11 & -5 \end{pmatrix}$ and $\mathbf{X} = \begin{pmatrix} 2 & -1 \\ 4 & -1 \end{pmatrix}$ Find the matrix $\mathbf{Y}$.

10 $\mathbf{XY} = \begin{pmatrix} 12 & 1 \\ 14 & -3 \end{pmatrix}$ and $\mathbf{Y} = \begin{pmatrix} 2 & 1 \\ 5 & 0 \end{pmatrix}$ Find the matrix $\mathbf{X}$.

KEY WORDS

matrix
matrices
order
rows
columns
zero matrix
identity matrix
inverse matrix
singular matrix
non-singular matrix

UNIT 4

Expanding double brackets

THIS SECTION WILL SHOW YOU HOW TO

- Expand and simplify expressions with two brackets

The area of a rectangle that has sides of length $x + 3$ and length $x + 8$ can be found by splitting the rectangle into four parts.

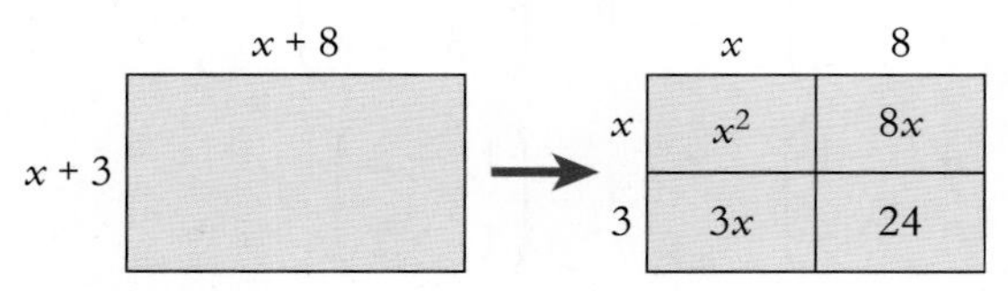

Area = $(x + 3)(x + 8)$ Area = $x^2 + 8x + 3x + 24$

So $(x + 3)(x + 8) = x^2 + 8x + 3x + 24$
$= x^2 + 11x + 24$

EXAMPLE

Use a diagram to expand $(x + 2)(x + 7)$

	x	7
x	x^2	$7x$
2	$2x$	14

$(x + 2)(x + 7) = x^2 + 7x + 2x + 14$
$= x^2 + 9x + 14$

EXERCISE 4.7

1 Copy and complete the following.

	x	10
x	....	$10x$
3	....	....

$(x + 3)(x + 10) =$......$+10x +$.......$+$......
$=$

Use diagrams to help you expand these brackets.

2 $(x + 3)(x + 4)$
3 $(x + 7)(x + 3)$
4 $(x + 4)(x + 2)$
5 $(x + 1)(x + 5)$
6 $(x + 8)(x + 7)$
7 $(x + 2)(x + 6)$
8 $(x + 6)(x + 7)$
9 $(x + 4)(x + 4)$
10 $(x + 15)(x + 4)$
11 $(x + 6)(x + 12)$
12 $(2x + 3)(x + 5)$
13 $(3x + 2)(x + 5)$

First – Outside – Inside – Last

You can expand double brackets without drawing a diagram.

> To expand double brackets you multiply each term in the first bracket by each term in the second bracket.

The mnemonic **FOIL** will help you remember what to do when expanding double brackets.

F	first
O	outside
I	inside
L	last

F L

$(x + 2)(x + 7) = x^2 + 7x + 2x + 14 = x^2 + 9x + 14$

I O

EXAMPLE

Expand and simplify **a** $(x + 8)(x + 7)$ **b** $(x + 4)(x - 5)$ **c** $(x - 3)(x - 10)$

a $(x + 8)(x + 7)$ **F** $x \times x = x^2$ **O** $x \times 7 = 7x$ **I** $8 \times x = 8x$ **L** $8 \times 7 = 56$

$= x^2 + 7x + 8x + 56$

$= x^2 + 15x + 56$

b $(x + 4)(x - 5)$ **F** $x \times x = x^2$ **O** $x \times -5 = -5x$ **I** $4 \times x = 4x$ **L** $4 \times -5 = -20$

$= x^2 - 5x + 4x - 20$

$= x^2 - x - 20$

c $(x - 3)(x - 10)$ **F** $x \times x = x^2$ **O** $x \times -10 = -10x$ **I** $-3 \times x = -3x$ **L** $-3 \times -10 = 30$

$= x^2 - 10x - 3x + 30$

$= x^2 - 13x + 30$

➡ **NOTE:** remember to look out for double negatives: $-3 \times -10 = +30$

You need to be careful when squaring a bracket.

$(x + 6)^2$ means $(x + 6)(x + 6)$

$= x^2 + 6x + 6x + 36$

$= x^2 + 12x + 36$

EXAMPLE

Expand and simplify **a** $(x - 7)^2$ **b** $(3x + 4)(2x - 5)$

a $(x - 7)^2$

$= (x - 7)(x - 7)$ **F** $x \times x = x^2$ **O** $x \times -7 = -7x$ **I** $-7 \times x = -7x$ **L** $-7 \times -7 = 49$

$= x^2 - 7x - 7x + 49$

$= x^2 - 14x + 49$

b $(3x + 4)(2x - 5)$ **F** $3x \times 2x = 6x^2$ **O** $3x \times -5 = -15x$ **I** $4 \times 2x = 8x$ **L** $4 \times -5 = -20$

$= 6x^2 - 15x + 8x - 20$

$= 6x^2 - 7x - 20$

EXERCISE 4.8

1 Rose has made mistakes in her expanding brackets homework.

Copy out each question and correct her mistakes.

a $(x + 3)(x + 4) = x^2 + 4x + 3x + 7 = x^2 + 7x + 7$

b $(x + 4)(x - 6) = x^2 - 6x + 4x - 2 = x^2 - 10x - 2$

c $(x - 5)(x - 4) = x^2 - 4x - 5x - 20 = x^2 - 9x - 20$

d $(x + 4)^2 = x^2 + 16$

Expand and simplify:

2 a $(x + 1)(x + 2)$ b $(x + 5)(x + 7)$ c $(x + 3)(x + 8)$ d $(x + 4)(x + 2)$
e $(x + 4)(x + 5)$ f $(x + 3)(x + 1)$ g $(x + 9)(x + 3)$ h $(x + 5)(x + 6)$
i $(x + 4)(x + 3)$ j $(x + 2)(x + 3)$ k $(x + 8)(x + 2)$ l $(1 + x)(4 + x)$

3 a $(x - 1)(x + 2)$ b $(x - 2)(x + 6)$ c $(x + 6)(x - 3)$ d $(x + 4)(x - 1)$
e $(x - 9)(x + 7)$ f $(x - 2)(x + 8)$ g $(x - 8)(x + 4)$ h $(x - 7)(x + 2)$
i $(x + 2)(x - 5)$ j $(x - 3)(x + 8)$ k $(x - 3)(x + 1)$ l $(9 + x)(4 - x)$

4 a $(x + 3)(x - 3)$ b $(x + 8)(x - 8)$ c $(x - 6)(x + 6)$ d $(x + 10)(x - 10)$
e $(x - 4)(x + 4)$ f $(x - 9)(x + 9)$ g $(x - 1)(x + 1)$ h $(12 - x)(12 + x)$

5 a $(x - 5)(x - 2)$ b $(x - 4)(x - 1)$ c $(x - 3)(x - 8)$ d $(x - 6)(x - 2)$
e $(x - 7)(x - 7)$ f $(x - 6)(x - 3)$ g $(x - 5)(x - 4)$ h $(x - 4)(x - 7)$
i $(x - 9)(x - 4)$ j $(x - 1)(x - 1)$ k $(x - 5)(x - 1)$ l $(2 - x)(2 - x)$

6 a $(x + 3)^2$ b $(y + 5)^2$ c $(x - 4)^2$ d $(x + 7)^2$
e $(x + 8)^2$ f $(a - 10)^2$ g $(6 - x)^2$ h $(2 + x)^2$

7 a $(2y + 1)(y - 3)$ b $(4x - 3)(x + 2)$ c $(5x + 2)(x + 4)$ d $(7y - 2)(y + 5)$
e $(4y + 3)(2y - 7)$ f $(2x - 11)(5x - 3)$ g $(5x - 6)(3x - 4)$ h $(9y - 2)(3y - 10)$
i $(8y - 5)(3y - 4)$ j $(3a - 7)(2a + 3)$ k $(1 - 3a)(3 - 4a)$ l $3(5x - 3)(2x + 1)$

8 a $(2x + 1)(2x - 1)$ b $(3x + 2)(3x - 2)$ c $(5x + 7)(5x - 7)$ d $(5x - 2)(5x + 2)$

9 a $(2x - 1)^2$ b $(5x + 2)^2$ c $(3x + 2y)^2$ d $(8 - 3y)^2$

10 a $(x + 1)(x - 9) + (x + 4)(x + 7)$ b $(2x + 3)(x - 1) + (3x - 5)(2x + 7)$
c $(2x + 1)(2x - 5) - (3x - 1)(x + 2)$ d $(4x - 3)(3x + 2) + (2x - 9)(x - 1)$
e $(3x + 1)^2 + (2x - 5)^2$ f $(2x - 3)^2 + (2x + 1)^2$
g $(6x + 5)^2 - (2x - 3)^2$ h $2(3x + 1)^2 - 3(2x + 5)^2$

EXERCISE 4.9

1 Solve these equations.

a $(x + 3)(x + 4) = (x - 2)(x + 5)$

b $(3x + 2)(x + 4) = 3x(x + 6)$

c $(x + 2)^2 = (x - 5)(x + 2)$

2 Find x for each of these right angled triangles.

a

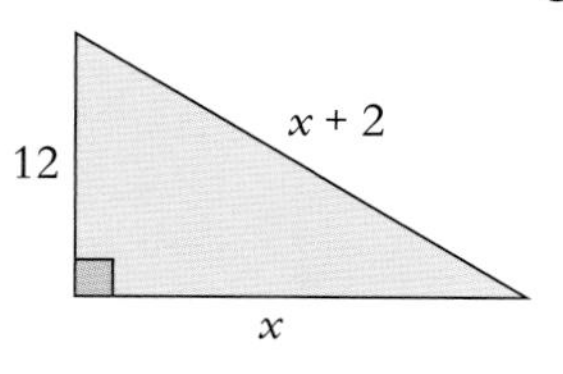

b

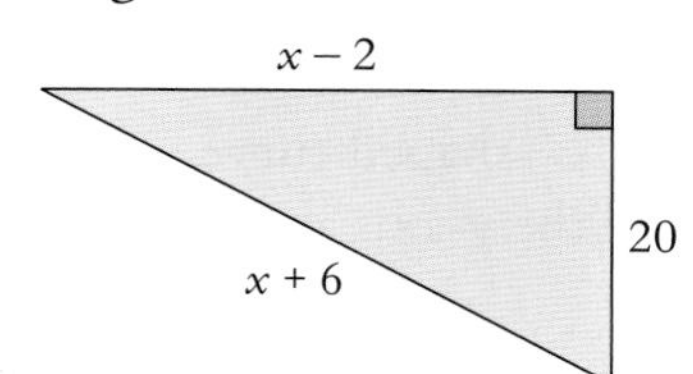

3

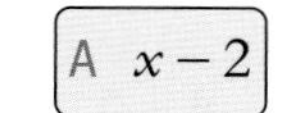

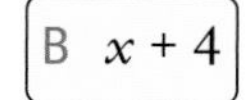

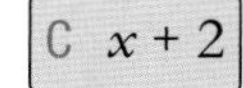

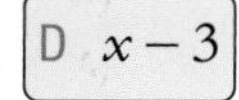

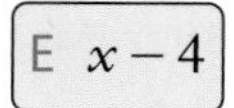

Which two cards when multiplied together give:

a $x^2 + 2x - 8$ b $x^2 - 7x + 12$ c $x^2 - 16$ d $x^2 + 6x + 8$

e $x^2 + x - 12$ f $x^2 - 4$ g $x^2 - 5x + 6$ h $x^2 - 6x + 8$

4 These rectangles have the same area. Find the value of x.

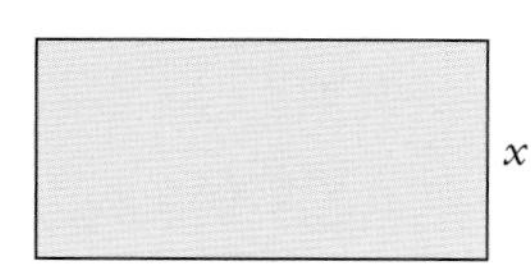

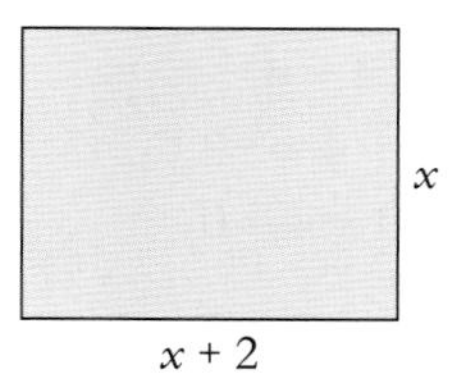

5 Write down and simplify an expression for

a the volume of the cuboid

b the surface area of the cuboid.

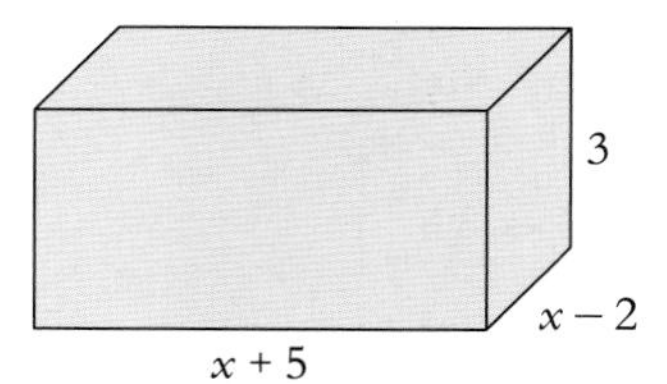

6 Show that $(n - 2)^2 + (n - 1)^2 + n^2 + (n + 1)^2 + (n + 2)^2$ simplifies to $5(n^2 + 2)$.

7 a Write down an expression for the area of the white square in terms of c.

b Write down an expression for the area of the large square in terms of a and b.

c Write down an expression for the area of all four of the shaded right-angled triangles in terms of a and b.

d Use your answers from parts **a**, **b** and **c** to prove that $a^2 + b^2 = c^2$.

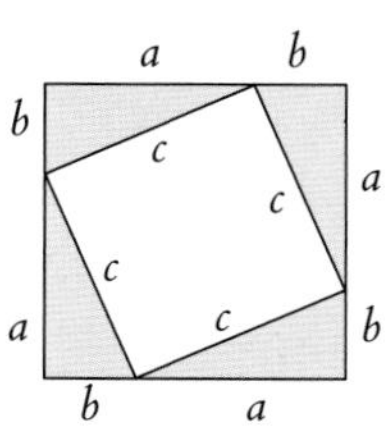

KEY WORDS

FOIL (first, outside, inside, last)

Quadratic graphs

THIS SECTION WILL SHOW YOU HOW TO

- Recognise quadratic graphs
- Draw quadratic graphs
- Use quadratic graphs in practical situations

$y = ax^2 + bx + c$ is called a **quadratic function**.
The simplest quadratic function is $y = x^2$.
The graph of $y = x^2$ is a smooth ∪-shaped curve.
The graph is symmetrical about the y-axis and it passes through the point (0, 0).

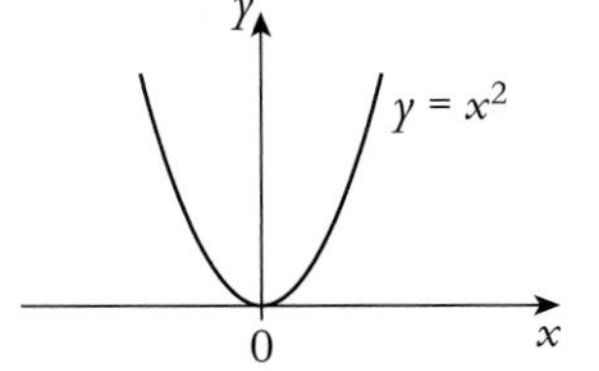

EXAMPLE

a Complete the table of values for $y = x^2 - 5x + 4$
b Draw the graph of $y = x^2 - 5x + 4$
c Name the line of symmetry of the curve.
d Use your graph to find the values of x when $y = 2$

x	0	1	2	3	4	5
y		0		−2		4

a

x	0	1	2	3	4	5	2.5
y	4	0	−2	−2	0	4	−2.25

The extra column in the table is needed to find where the curve turns

When $x = 0$, $y = (0)^2 - (5 \times 0) + 4 = 0 - 0 + 4 = 4$
When $x = 2$, $y = (2)^2 - (5 \times 2) + 4 = 4 - 10 + 4 = -2$
When $x = 4$, $y = (4)^2 - (5 \times 4) + 4 = 16 - 20 + 4 = 0$
When $x = 2.5$, $y = (2.5)^2 - (5 \times 2.5) + 4$
$= 6.25 - 12.5 + 4 = -2.25$

b

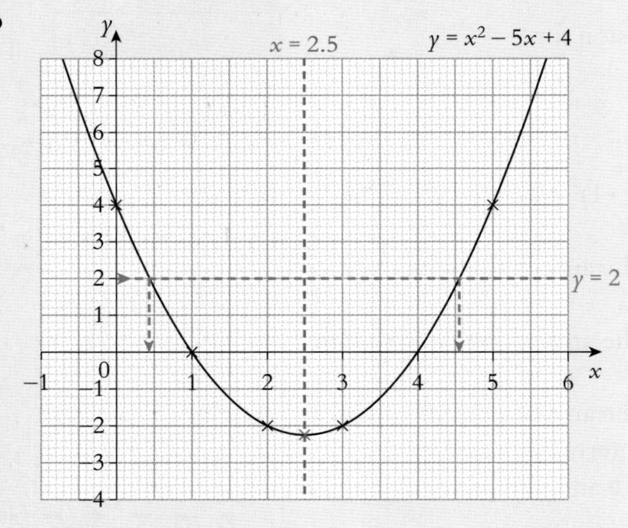

The points should be marked with a cross.

The points must be joined with a smooth curve.
They must not be joined with straight lines.

The axes and the curve should be labelled.

c The line of symmetry is $x = 2.5$
d When $y = 2$, $x \approx 0.4$ or $x \approx 4.6$

The symbol '≈' means 'is approximately equal to'.

EXAMPLE

B is 5 km due east of A. C is 5 km from A on a bearing of 060°.
Find the bearing of C from B.

First draw a diagram to show the positions of A, B and C.
$\angle CAB = 90° - 60° = 30°$

Next draw a north line at B and indicate the required angle.

Triangle ABC is isosceles.

$\angle ACB = \angle ABC = 75°$

Bearing of C from B $= 270° + 75°$
$= 345°$

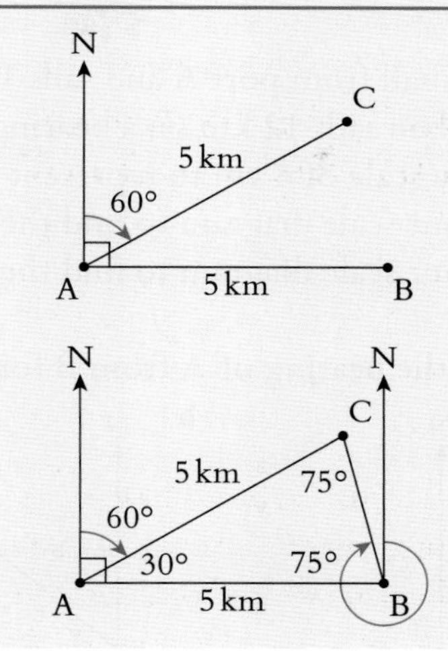

UNIT 4

EXERCISE 4.13

1 Draw diagrams to show these bearings:
   a The bearing of B from A is 056°
   b The bearing of F from G is 270°
   c The bearing of N from M is 135°
   d The bearing of P from Q is 340°

2 Write down the bearing for each of these compass directions.
   a south east
   b due south
   c due west
   d south west
   e north east
   f north west

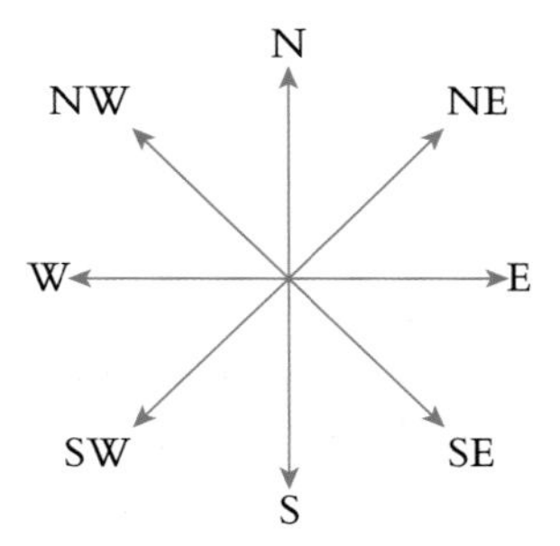

3 Write down the bearing of
   a Dubai from Doha
   b Doha from Kuwait City
   c Shiraz from Doha
   d Kuwait City from Dubai.

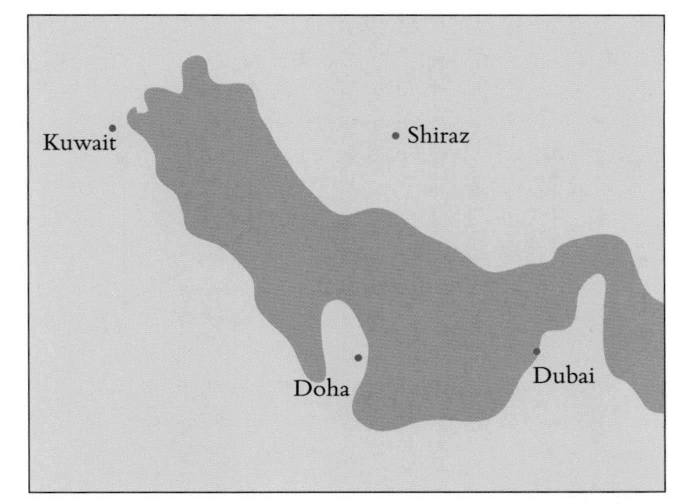

4 Bermuda is 1650 km from Orlando on a bearing of 077°.
Jamaica is 1990 km from Bermuda on a bearing of 220°.
Draw a scale diagram to show the positions of Bermuda, Orlando and Jamaica.

5 A boat sets sail from port A and sails 16 km due east to port B.
The boat then sails 12 km on a bearing of 140° to port C.

a Using a scale of 1 cm to represent 2 km draw a scale diagram to show the journey.
b Use your scale diagram to find the bearing of port C from port A.
c Use your scale diagram to find the distance between port A and port C.

6 Work out the bearing of A from B for these diagrams.

a b

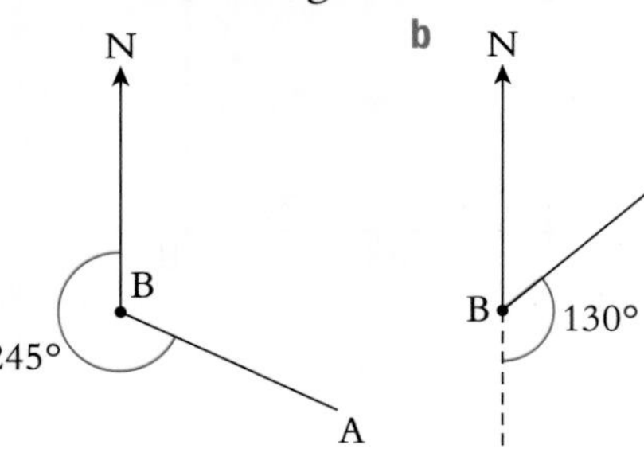

c

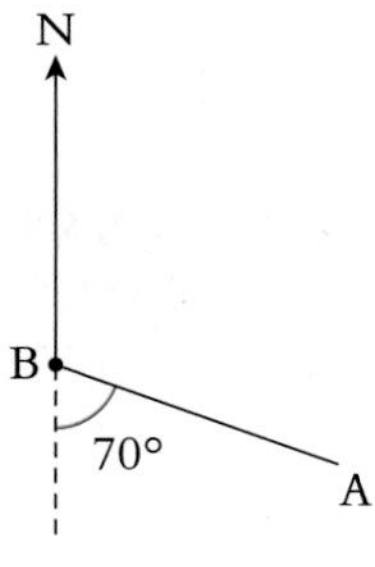

d

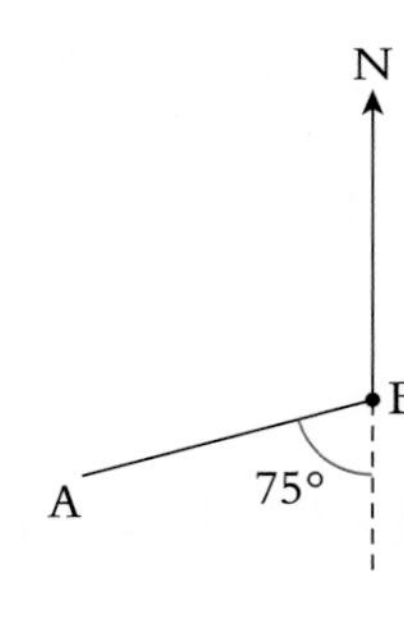

e

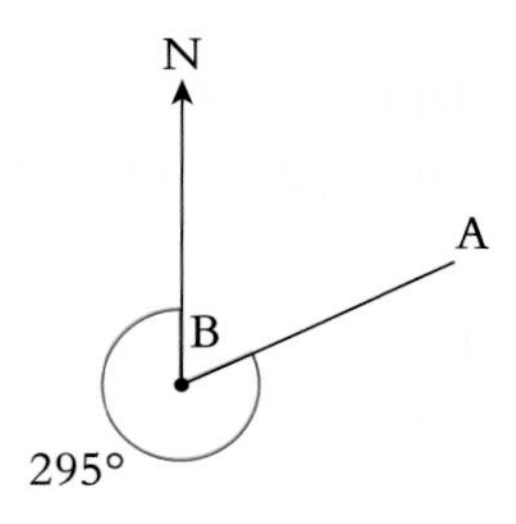

f

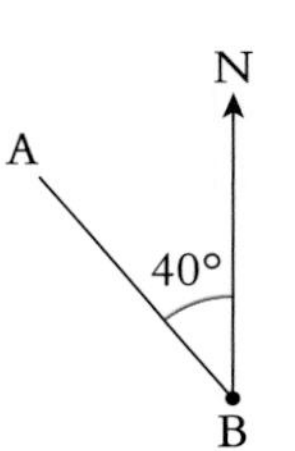

g

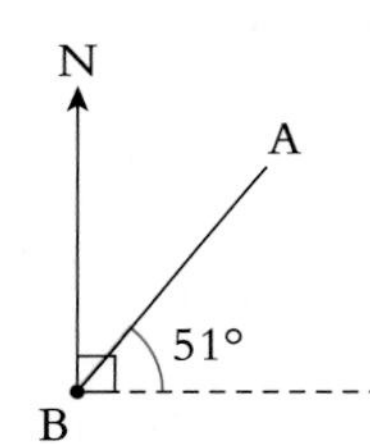

h

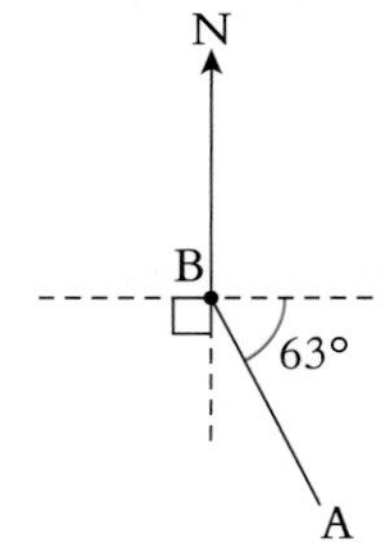

7 B is 8 km due east of A.
C is 8 km from A on a bearing of 130°.
Find the bearing of C from B.

8 Work out the bearing of P from Q for these diagrams:

a

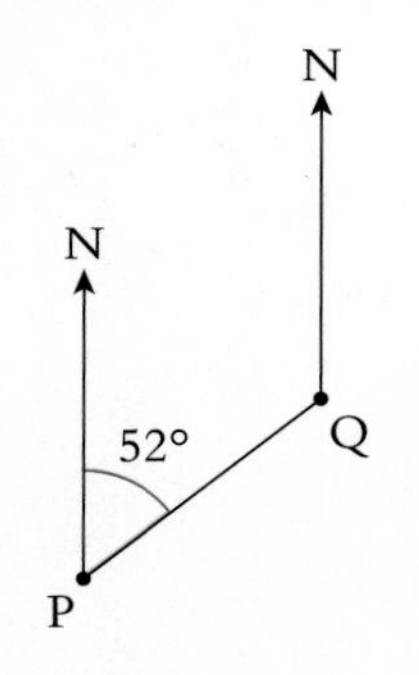

b

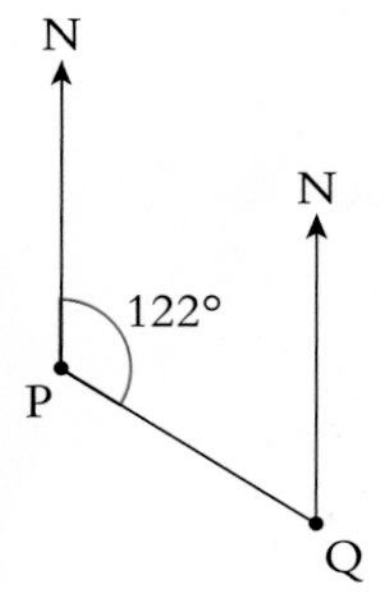

c

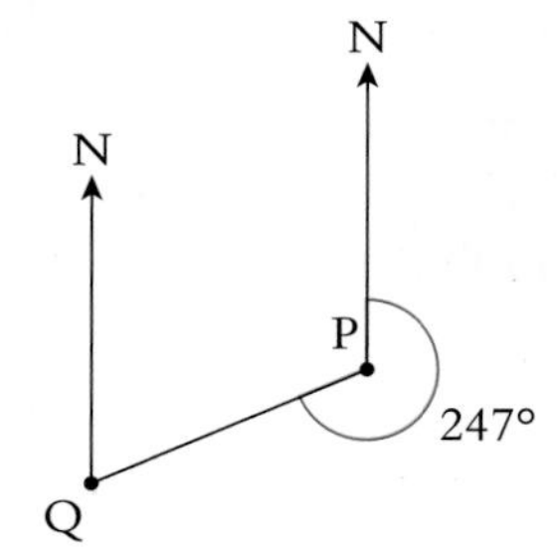

d

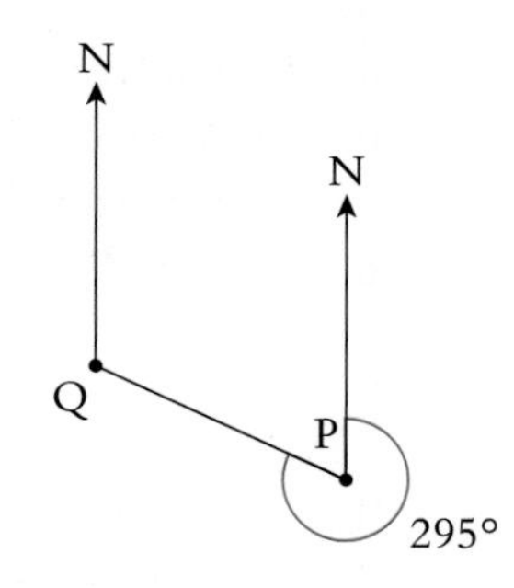

9 The diagram shows the positions of three towns A, B and C.
Find the bearing of

a B from C
b C from B
c A from B
d B from A
e A from C
f C from A.

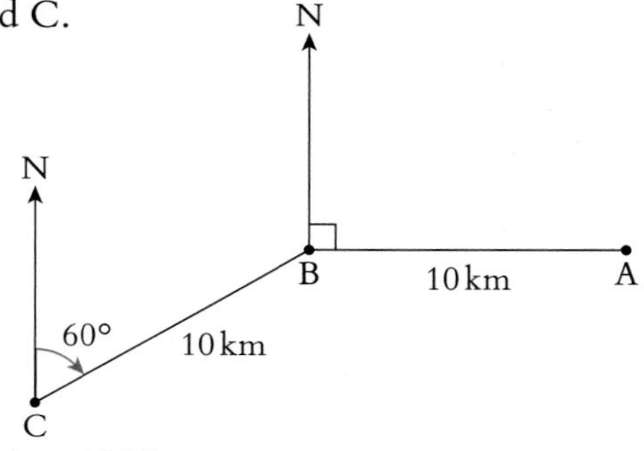

10 The bearing of B from A is 070°, the bearing of C from A is 120° and the bearing of C from B is 125°.

a Find the bearing of A from C.
b Find the bearing of B from C.

11 Two runners, Deepika and Khadeeja, set off from the point A.
Deepika runs 8 km on a bearing of 095° to the point B.
Khadeeja runs 8 km on a bearing of 155° to the point C.

a Find the distance between B and C.
b Find the bearing of B from C.
c Find the bearing of C from B.

12 B is 5 km from A on a bearing of 038°.
C is 12 km from B on a bearing of 128°.
Find the distance between A and C.

13 A ship sails on a bearing of 110° from a port P.
A lighthouse L is 8 km south east of port P.

a Draw a scale diagram to show the positions of P and L and the path of the ship.
b Mark clearly on your scale diagram the locus of points where the ship is less than 4 km from the lighthouse.

14 The bearing of Q from P is 120°.
The bearing of R from Q is 240°.
The bearing of P from R is 020°.
Find the values of x, y and z.

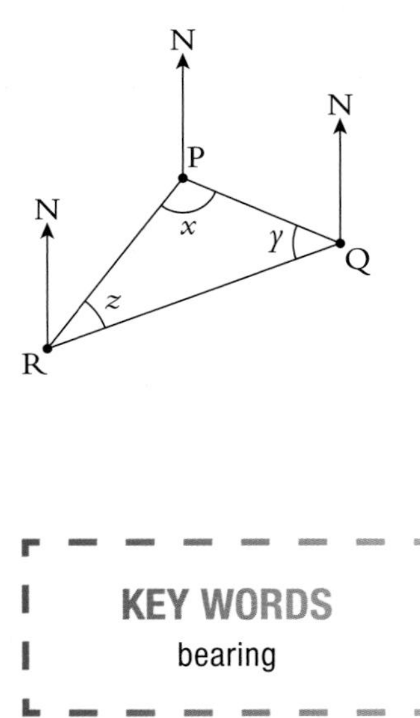

15 The bearing of B from A is 140°.
The bearing of C from B is 275°.
The bearing of A from C is 043°.
Find the values of x, y and z.

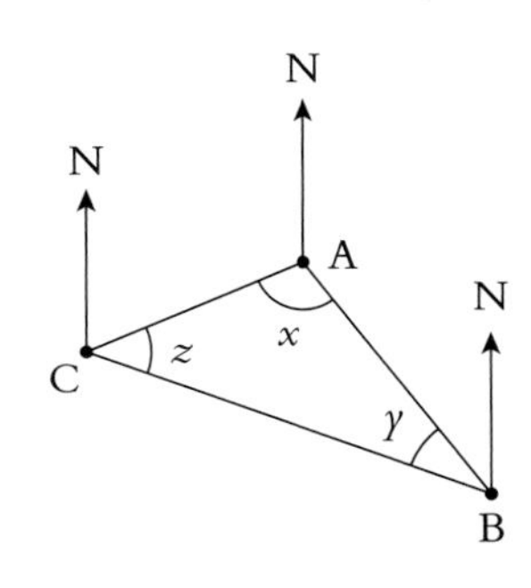

KEY WORDS
bearing

UNIT 4

Trigonometry

THIS SECTION WILL SHOW YOU HOW TO

- Use the tangent, sine and cosine ratios to find sides and angles

Trigonometry is used to find lengths and angles in a right-angled triangle without using scale drawings.
The sides of a right-angled triangle are given special names.
The longest side is called the **hypotenuse**.
The side opposite the angle $x°$ is called the **opposite** side.
The side next to the angle $x°$ is called the **adjacent** side.

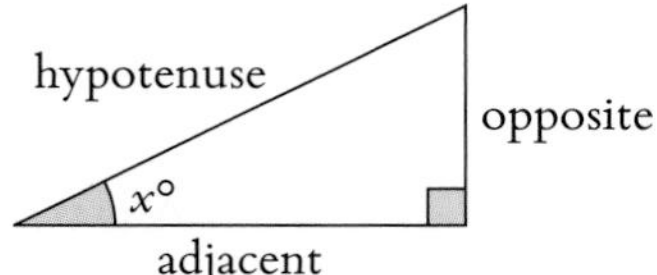

The different relationships or ratio of the sides to each other in a triangle make up trigonometric formulae. The three trigonometrical ratios that you have to know are tangent, cosine and sine.

The tangent ratio

$$\tan x = \frac{\text{opposite}}{\text{adjacent}}$$

Calculating sides

The following examples show you how to use the tangent ratio to find the length of a side.

EXAMPLE

Find the length of x.

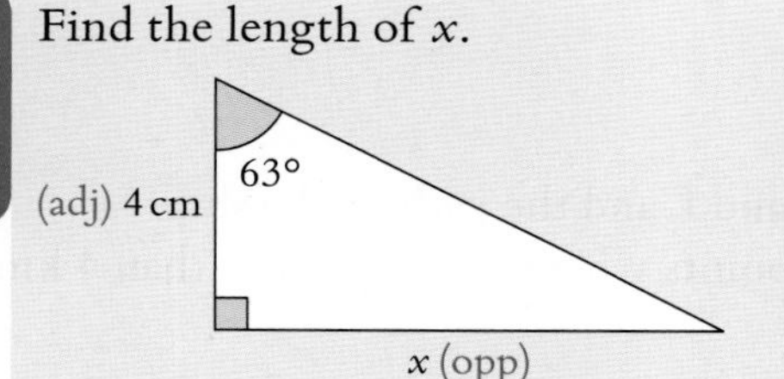

$\tan 63° = \frac{\text{opp}}{\text{adj}}$ *replace opp by x and adj by 4*

$\tan 63° = \frac{x}{4}$ *multiply both sides by 4*

$x = 4 \tan 63°$

$= 7.85$ cm (to 3 s.f.)

EXAMPLE

Find the length of x.

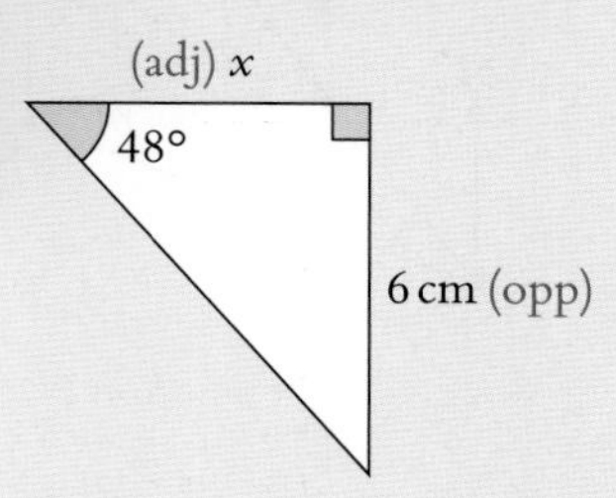

$\tan 48° = \frac{\text{opp}}{\text{adj}}$ *replace opp by 6 and adj by x*

$\tan 48° = \frac{6}{x}$ *multiply both sides by x*

$x \times \tan 48° = 6$ *divide both sides by tan 48°*

$x = \frac{6}{\tan 48°}$

$= 5.40$ cm (to 3 s.f.)

EXERCISE 4.14

In this exercise all lengths are in cm.

1 Find the length x for each triangle.

a

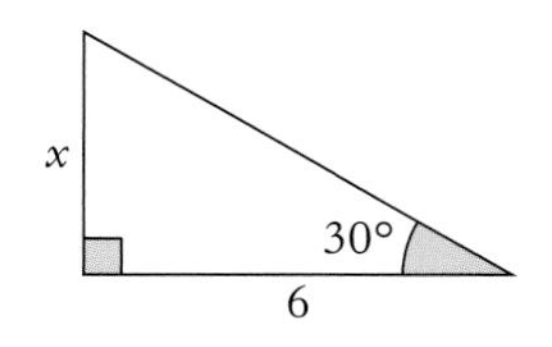

b

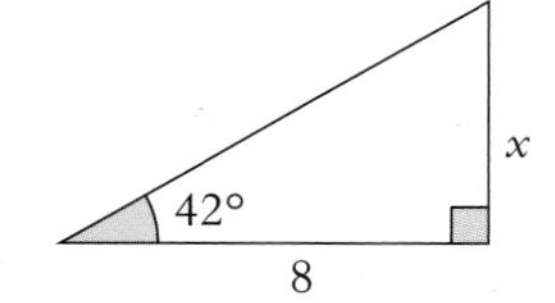

c

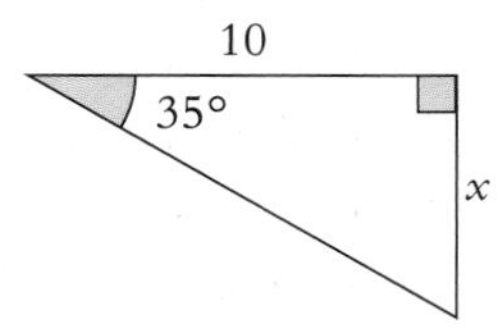

d

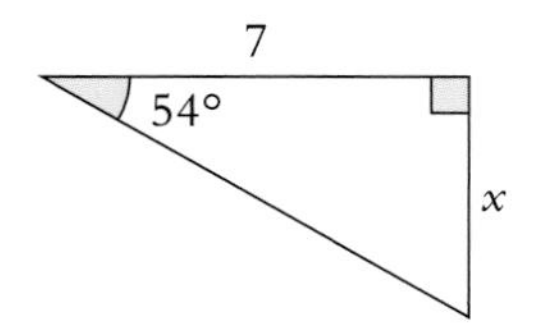

e

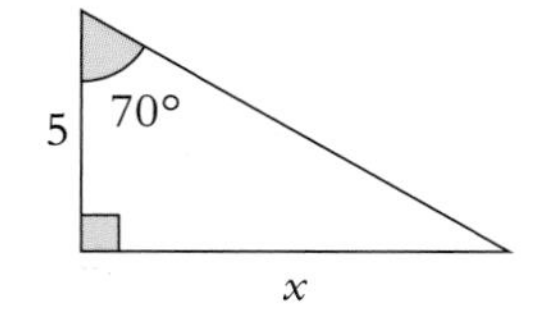

f

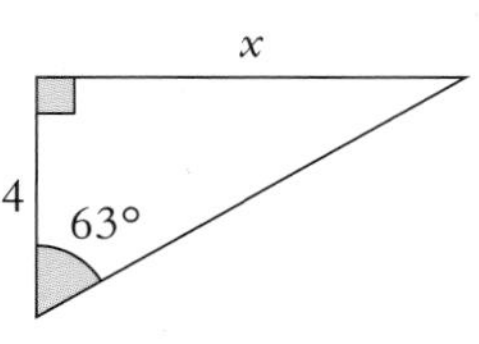

g

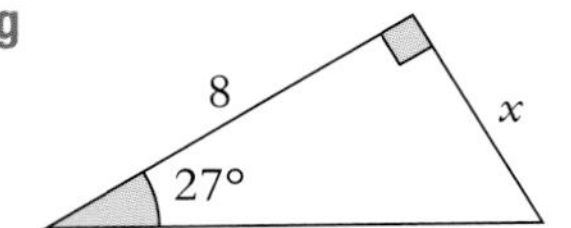

h

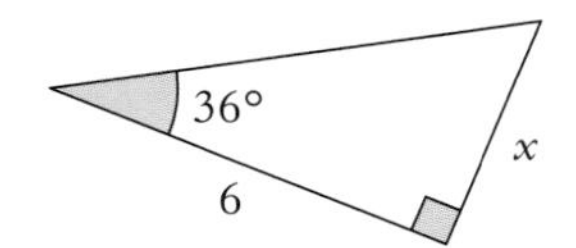

i

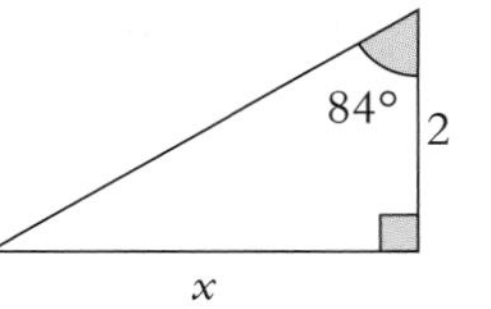

j

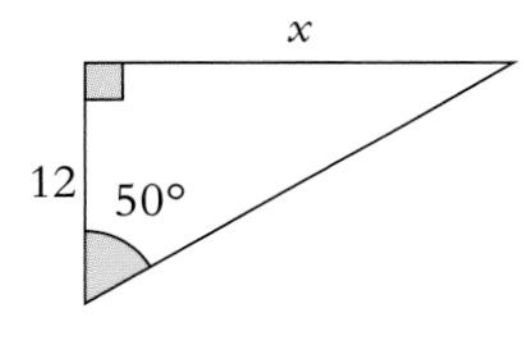

k

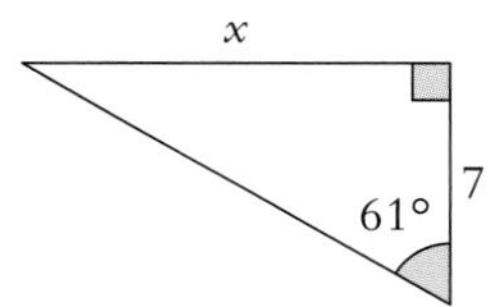

l

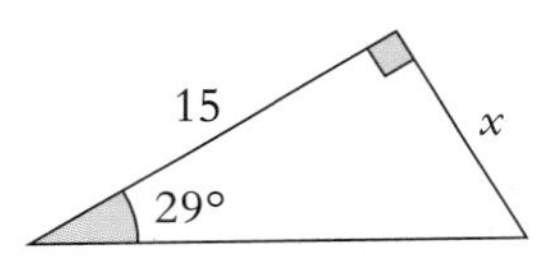

m

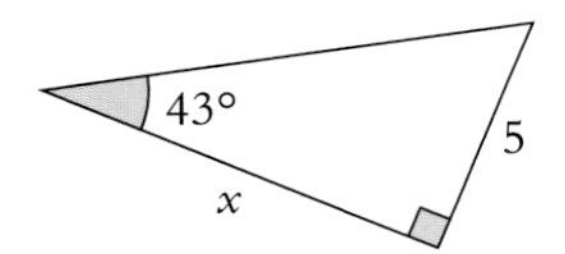

n

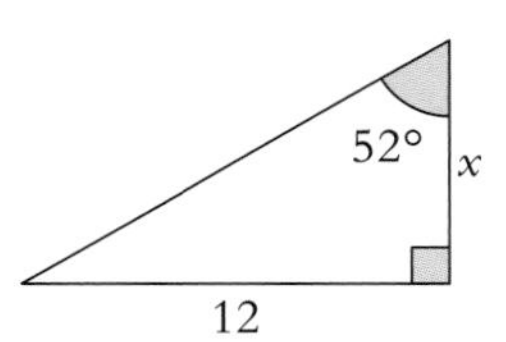

o

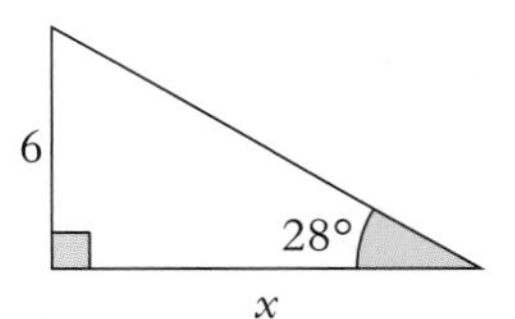

2 Find the lengths of x and y.

a

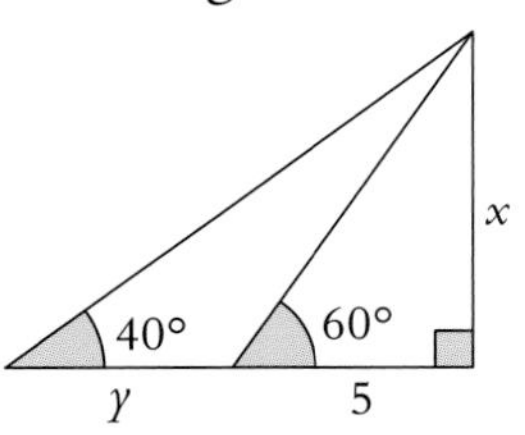

b

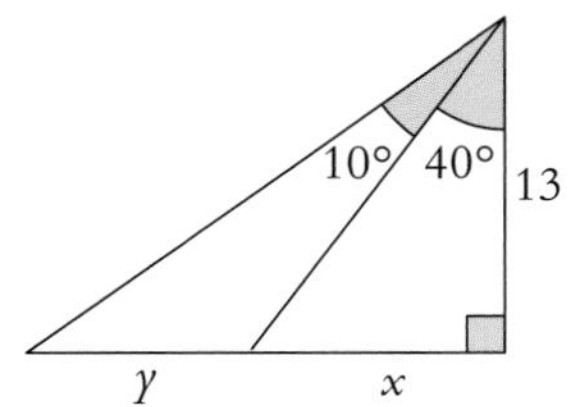

c

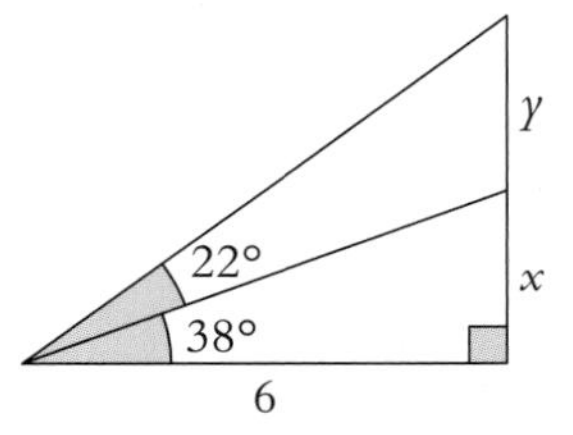

d

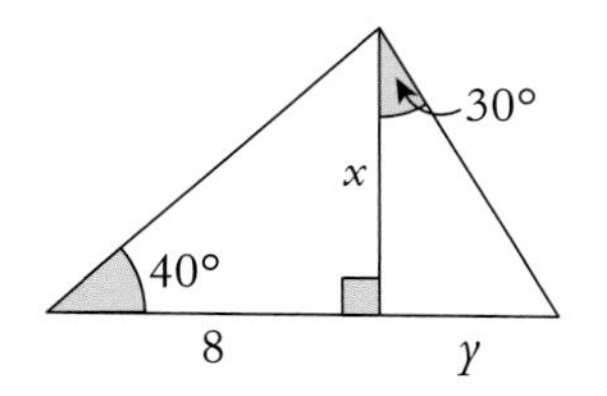

e

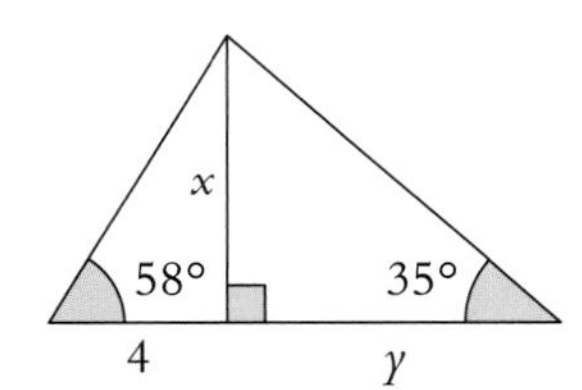

f

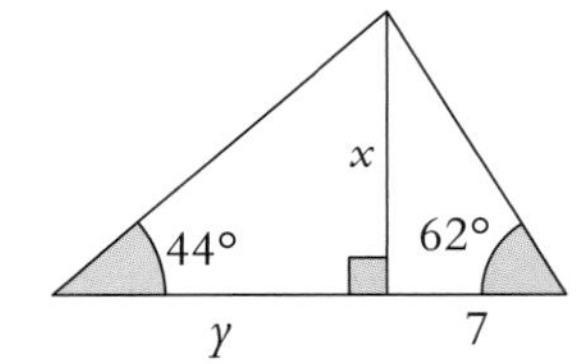

UNIT 4

Calculating angles using the tangent ratio

If you know both the opposite and adjacent sides in a right-angled triangle, you can calculate the size of an angle inside the triangle.

To do this you must use the 'inverse tan' button on your calculator. (On most calculators this is labelled $\tan^{-1}$)

EXAMPLE

Find the size of angle x.

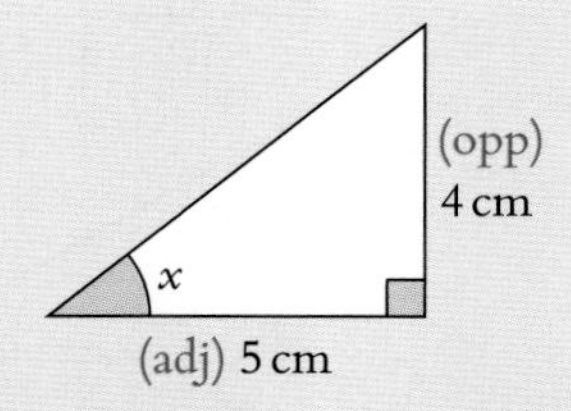

$\tan x = \dfrac{\text{opp}}{\text{adj}}$ *replace opp by 4 and adj by 5*

$\tan x = \dfrac{4}{5}$ *to find x use the $\tan^{-1}$ button on your calculator*

$x = \tan^{-1}\left(\dfrac{4}{5}\right)$

$x = 38.7°$ (to 1 d.p.) ➡ **NOTE:** you should give angle answers correct to 1 d.p.

EXAMPLE

ABCD is a rectangle.
Calculate the size of angle x.

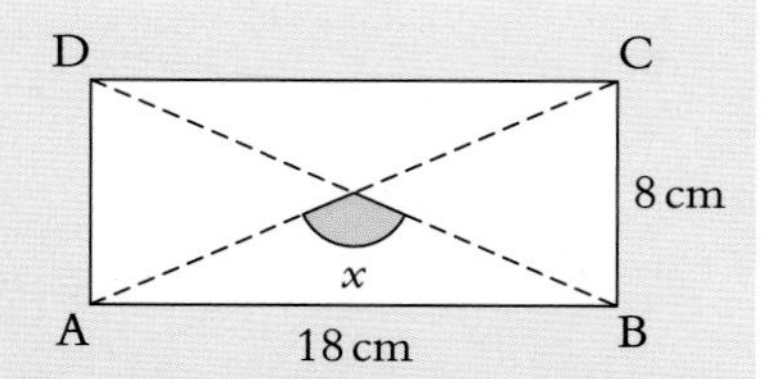

The tangent ratio can only be used for right-angled triangles, so split the bottom triangle into two identical right-angled triangles.

D C y 4 (adj) A (opp) 9 B

$\tan y = \dfrac{\text{opp}}{\text{adj}}$ *replace opp by 9 and adj by 4*

$\tan y = \dfrac{9}{4}$ *to find y use the $\tan^{-1}$ button on your calculator*

$y = \tan^{-1}\left(\dfrac{9}{4}\right)$

$y = 66.0375.....$

$x = 2 \times y$

$= 132.1°$ (to 1 d.p.)

3 Find the values of x, y and z in each of these triangles.

a

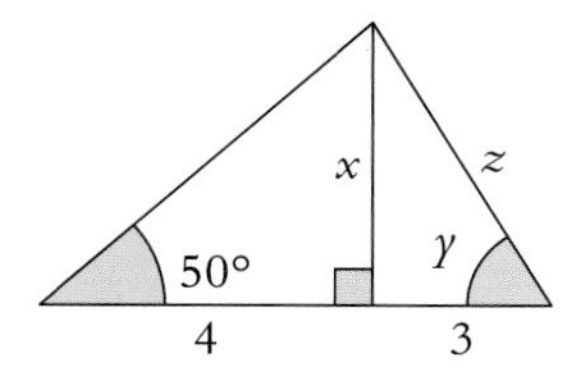

b

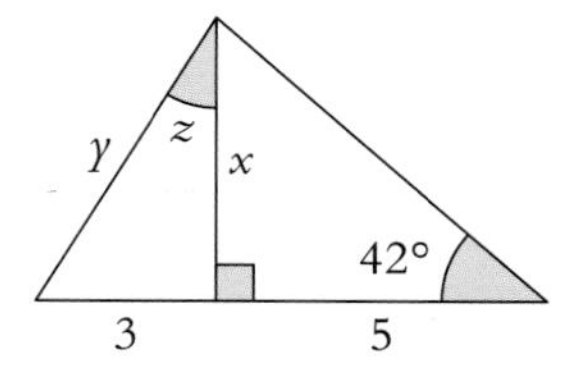

c

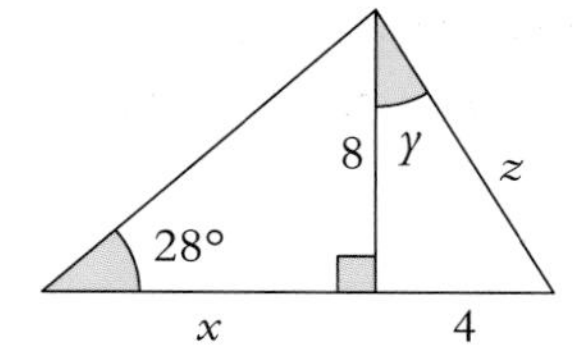

4 Calculate the size of angle x in this trapezium.

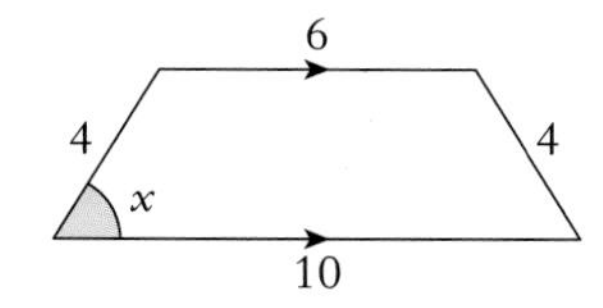

5 Find the values of x and y in these diagrams.

a

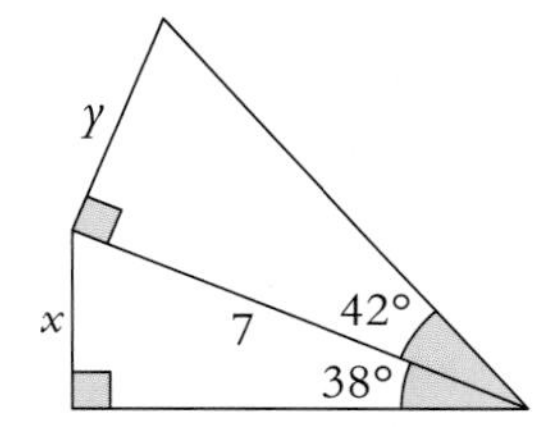

b

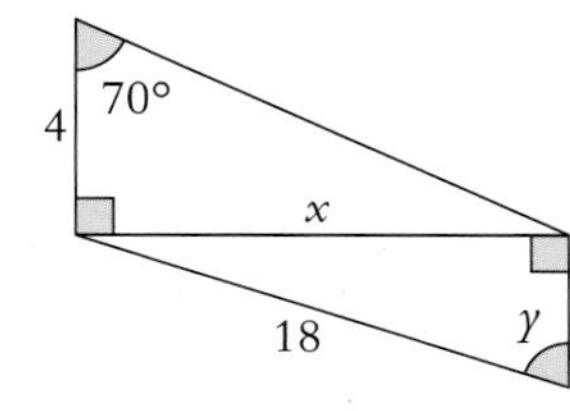

6 Find the values of x, y and z in this diagram.

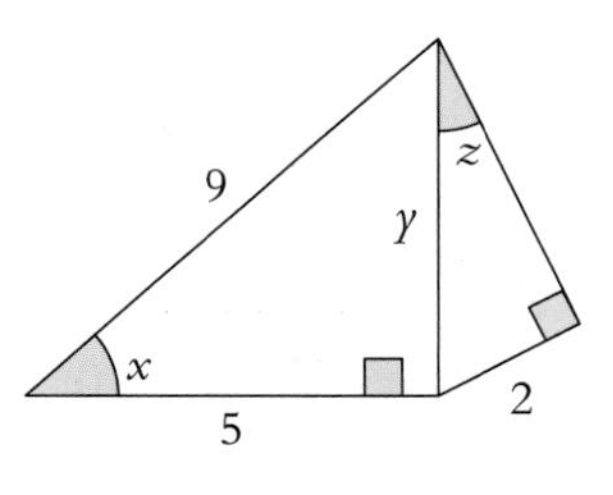

7 Find the area of each of these regular polygons.

a

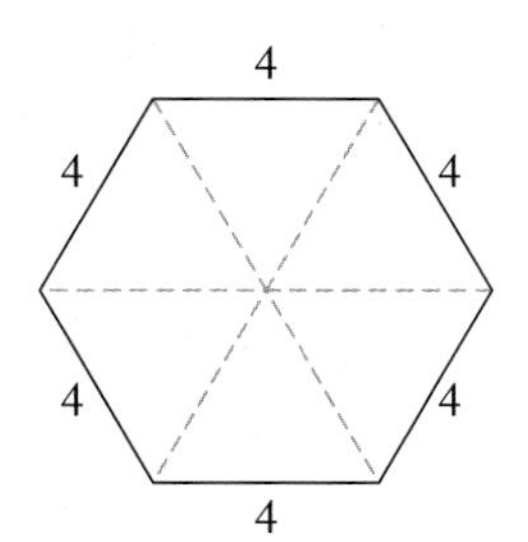

b

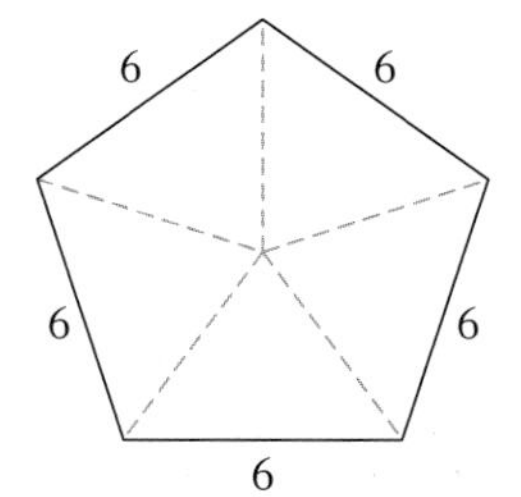

c

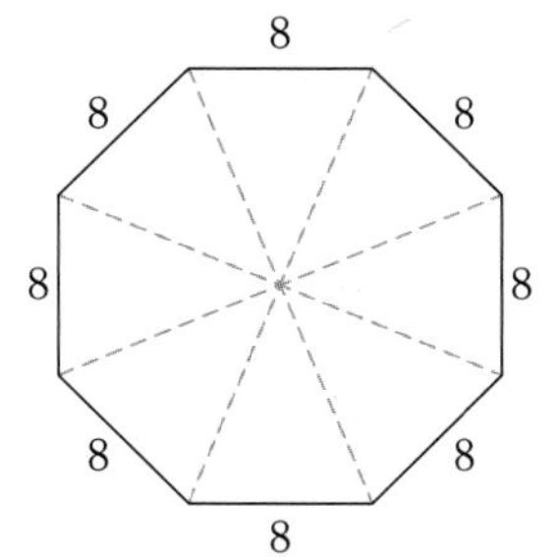

8 The cross-section of this solid prism is a regular pentagon.
The length of each side of the pentagon is 8 cm.
The length of the prism is 20 cm.
Calculate:

a the surface area of the prism,

b the volume of the prism.

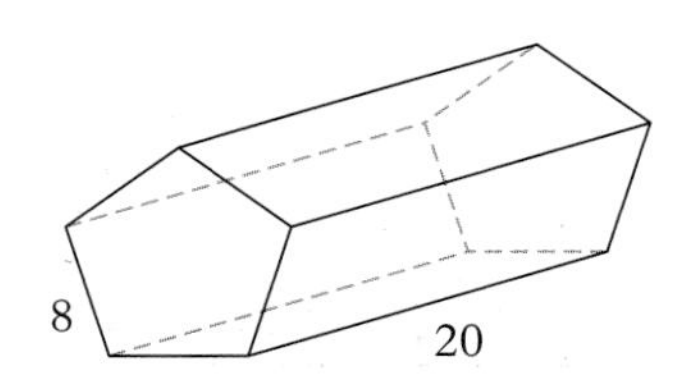

Practical applications of trigonometry and Pythagoras

EXERCISE 4.19

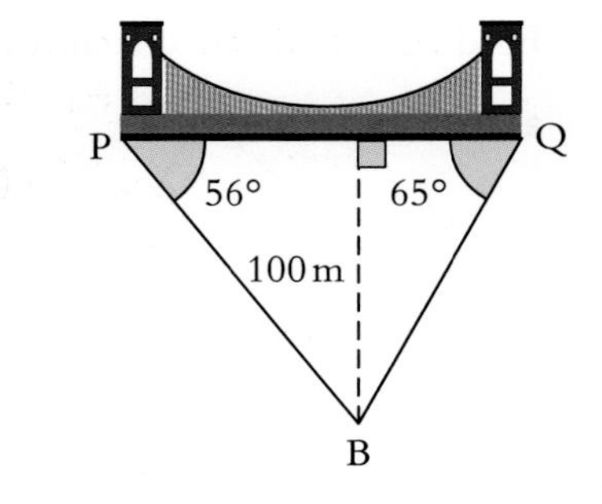

1 The diagram shows a bridge over a river.
A boat B is 100 m vertically below the bridge.
Angle QPB = 56° and angle PQB = 65°.
Calculate the length of the bridge PQ.

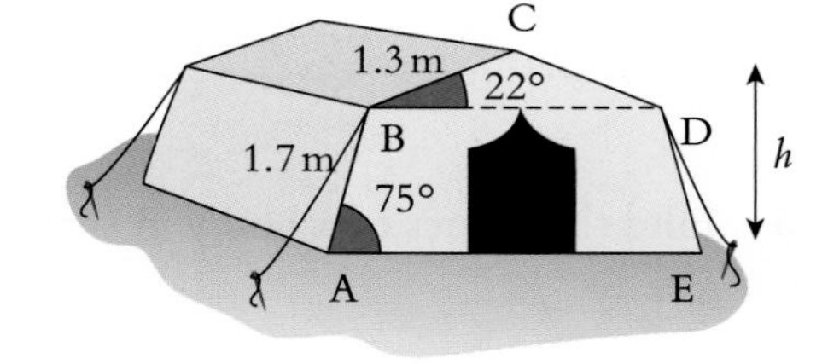

2 ABCDE is the cross-section of a tent.
AB = DE = 1.7 m BC = CD = 1.3 m
Angle EAB = angle AED = 75°
Angle CBD = angle CDB = 22°

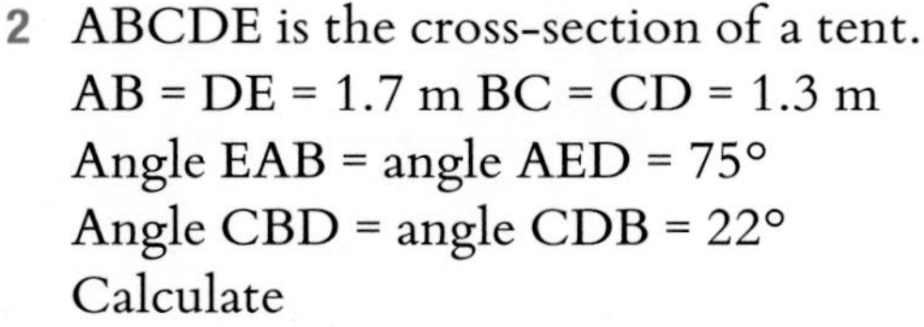

Calculate
- a the length of BD
- b the length of AE
- c the height (h) of the tent.

3 The diagram shows a 25 m high radio mast PQ.
Two wires PA and PB secure the mast.
PA = 30 m and PB = 35 m
Calculate
- a angle PAB
- b angle PBA
- c the distance between A and B.

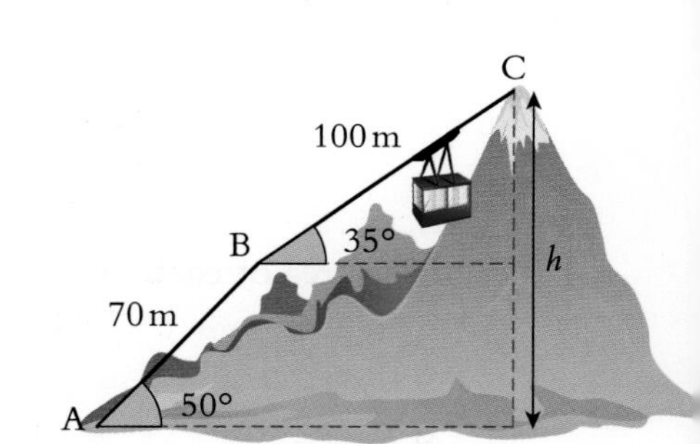

4 A cable car travels at 50° to the horizontal for 70 m and then 35° to the horizontal for 100 m. Calculate the vertical height (h) gained when travelling from A to C.

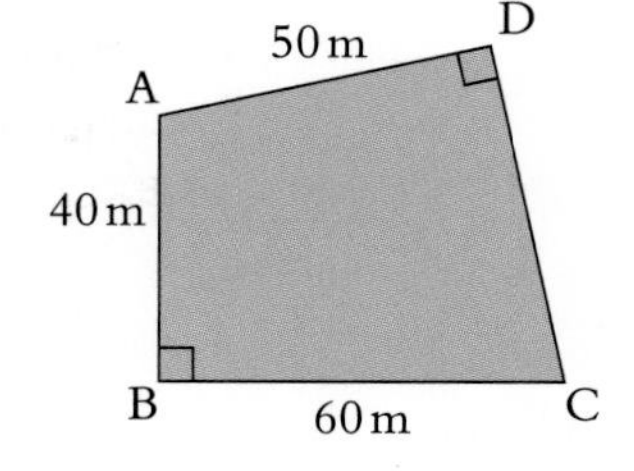

5 ABCD is a field.
Calculate
- a the length of AC
- b the length of CD
- c angle BAD
- d angle BCD
- e the area of the field.

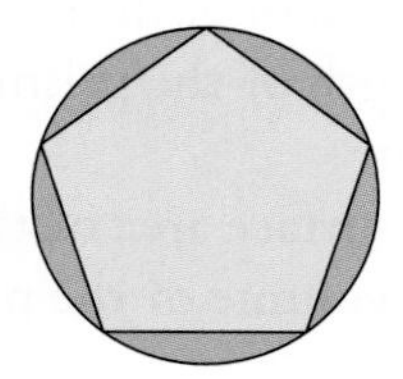

6 The vertices of a regular pentagon lie on the circumference of a circle radius 10 cm.
Calculate
- a the lengths of the sides of the pentagon
- b the difference in area between the circle and the pentagon.

7 The Tower of Pisa leans at an angle of 4° to the vertical.
The top of the tower stands 56.7 m above the ground.

a Calculate the height of the tower.

b The tower used to lean at an angle of 5.5° to the vertical.
How much lower did the top of the tower used to stand above the ground?

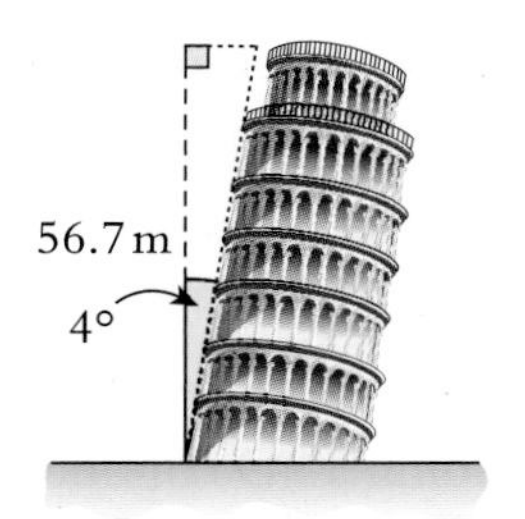

8 A boat sails 50 km on a bearing of 080°.
Calculate:

a How far north it travels,

b How far east it travels.

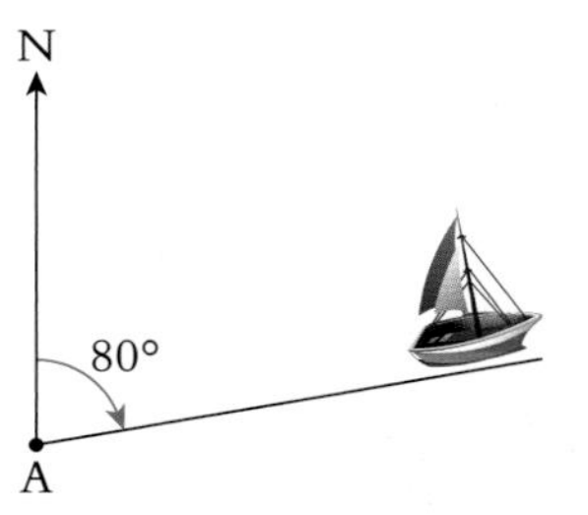

9 A hot air balloon flies 3000 km on a bearing of 245°.
Calculate:

a How far south it travels

b How far west it travels.

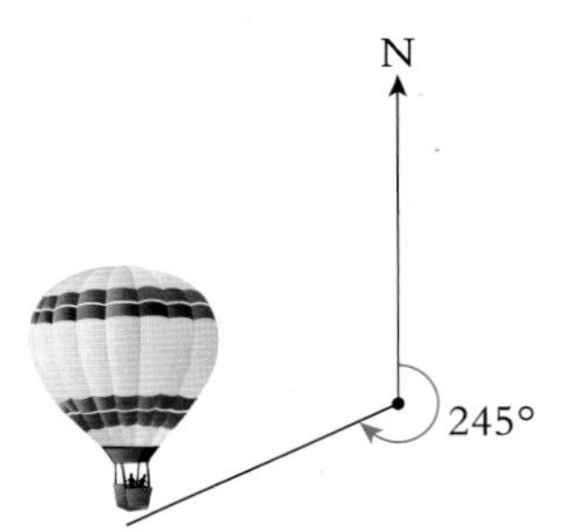

10 An aircraft flies 500 km from an airport A on a bearing of 070° to arrive at airport B.
The plane then flies 800 km from airport B on a bearing of 035° to arrive at airport C.

a Calculate how far north C is from A

b Calculate how far east C is from A

c Calculate the direct distance between A and C

d Calculate the bearing of C from A

e What is the bearing of A from C?

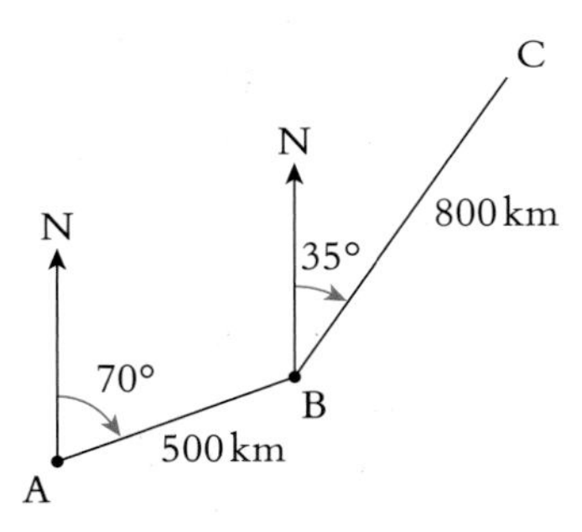

UNIT 4

Three-dimensional trigonometry

To find the angle between the line AB and the plane PQRS:

- drop a perpendicular from the point B to meet the plane at X
- draw the right-angled triangle BAX
- the angle between AB and the plane is angle BAX

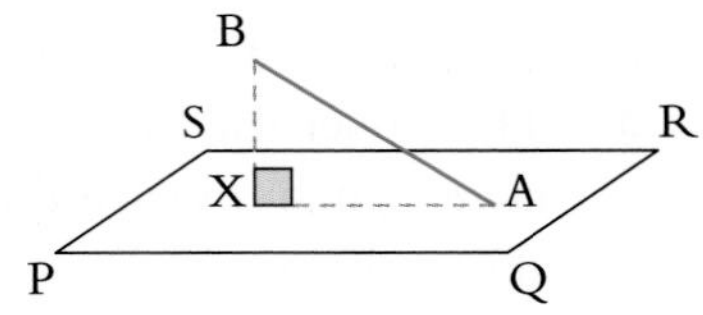

When doing a 3D trigonometry question it is important to draw clearly the triangle (usually right-angled) that you are going to use in your calculations. On your triangle show all the known lengths and angles.

EXAMPLE

ABCDEFGH is a cuboid.
AB = 7 cm, BC = 4 cm and CG = 6 cm.
Calculate

a the length of BD

b the angle between the line BH and the base ABCD.

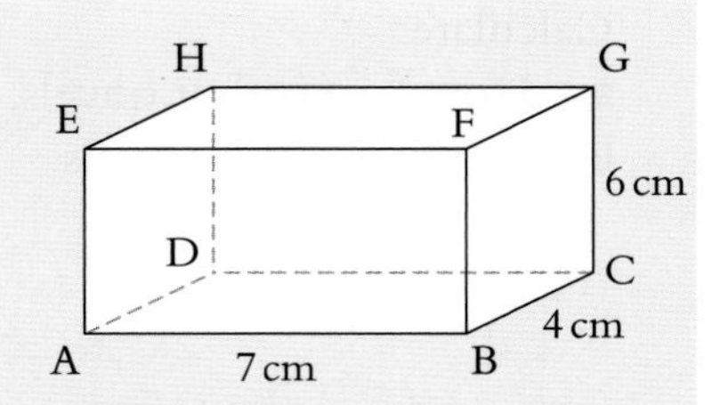

a Draw the rectangular base of the cuboid and on it show the line BD.
Use Pythagoras on the right-angled triangle DAB.

$BD^2 = 7^2 + 4^2$

$BD^2 = 65$

$BD = \sqrt{65}$

$BD = 8.062...$

$BD = 8.06$ cm (to 3 s.f.)

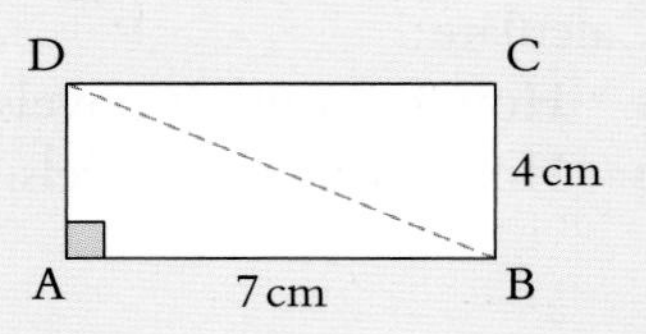

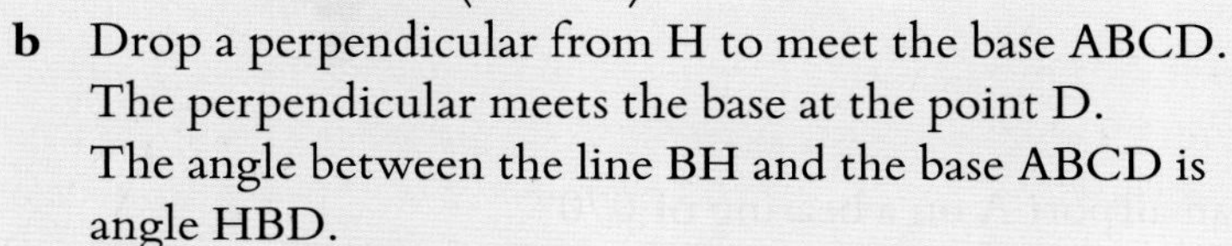

b Drop a perpendicular from H to meet the base ABCD.
The perpendicular meets the base at the point D.
The angle between the line BH and the base ABCD is angle HBD.
Draw triangle HBD. (It has a right angle at D.)

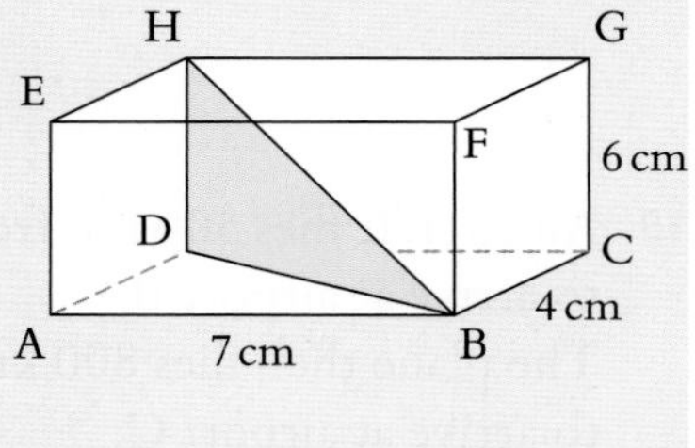

$\tan x = \frac{\text{opp}}{\text{adj}}$

$\tan x = \frac{6}{\sqrt{65}}$

$x = \tan^{-1}\left(\frac{6}{\sqrt{65}}\right)$

$x = 36.7°$ (to 1 d.p.)

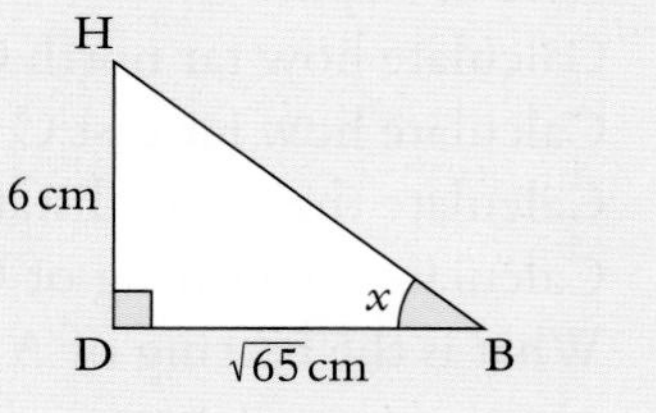

EXERCISE 4.20

1 ABCDEFGH is a cube.
Calculate:

a the length of AC

b the length of AG

c the angle between AG and the base ABCD.

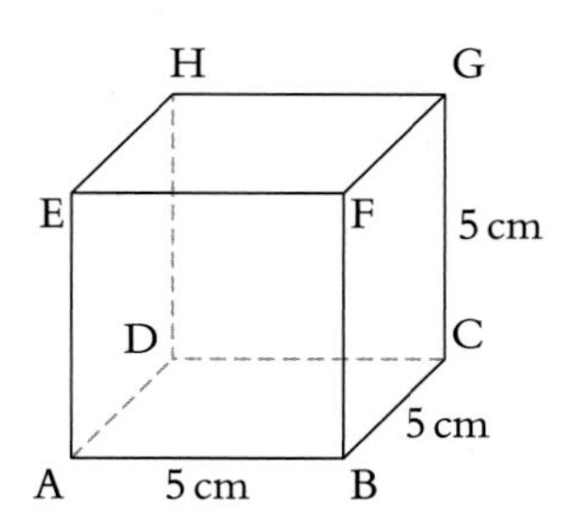

2 ABCDEFGH is a cuboid.
Calculate:

a the length of AC
b the length of AG
c the angle between AG and the base ABCD.

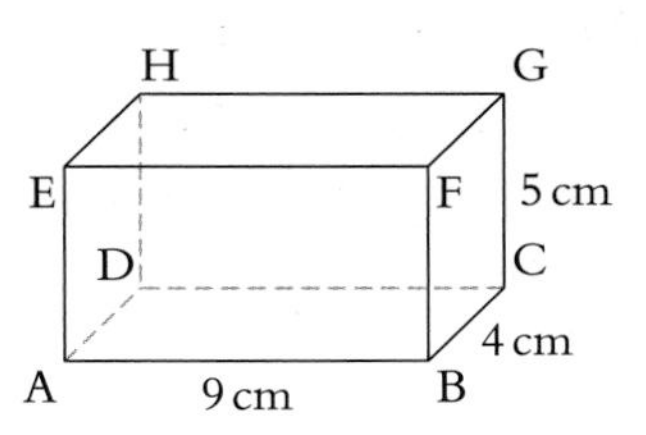

3 ABCDEF is a triangular prism.
Calculate:

a the length of BD
b the length of DE
c the angle between DE and the base ABCD.

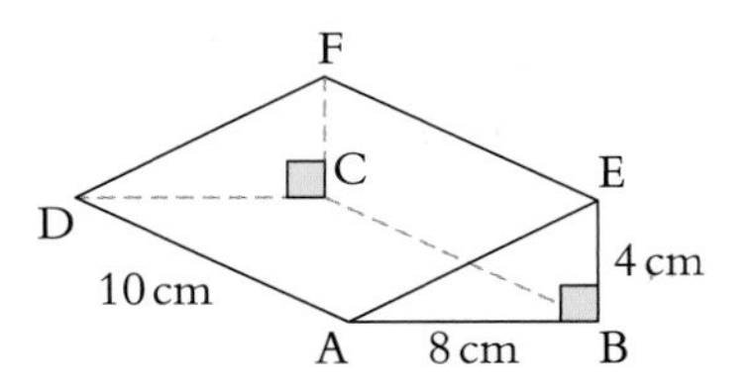

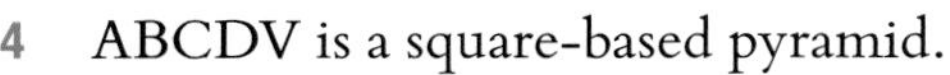

4 ABCDV is a square-based pyramid.
AB = 4 cm and VC = 5 cm.
O is the middle of the square base.
V is vertically above O.
X is the midpoint of AD.
Y is the midpoint of BC.
Calculate:

a the lengths of AC, VO and VY
b the angle between VC and the base ABCD
c angle XVY.

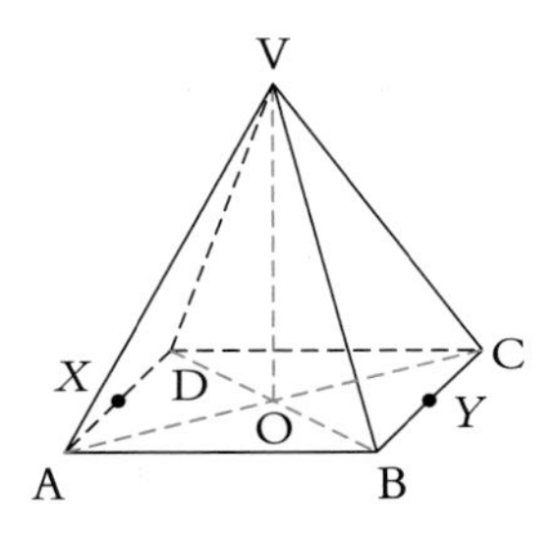

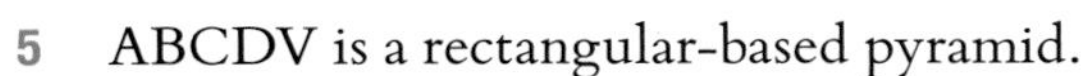

5 ABCDV is a rectangular-based pyramid.
AB = 12 cm, BC = 9 cm and VO = 6 cm.
O is the middle of the rectangular base.
V is vertically above O.
X is the midpoint of AD.
Y is the midpoint of BC.
Calculate:

a the lengths of OB, VB and VY
b the angle between VB and the base ABCD
c angle XVY.

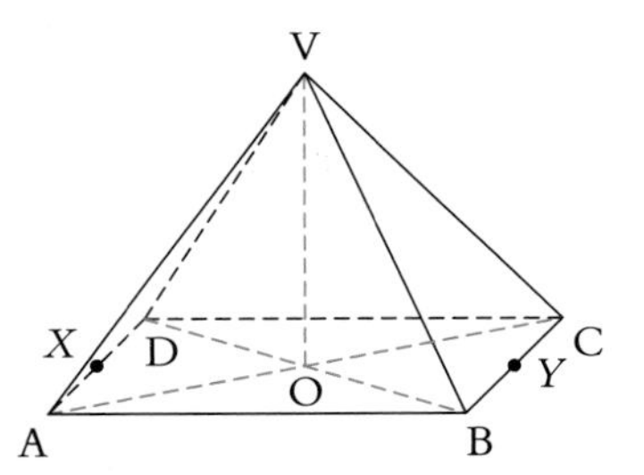

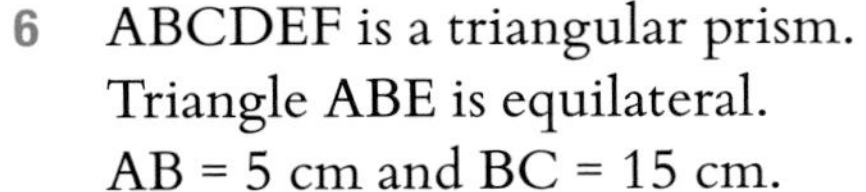

6 ABCDEF is a triangular prism.
Triangle ABE is equilateral.
AB = 5 cm and BC = 15 cm.
Calculate the angle between AF and the base ABCD.

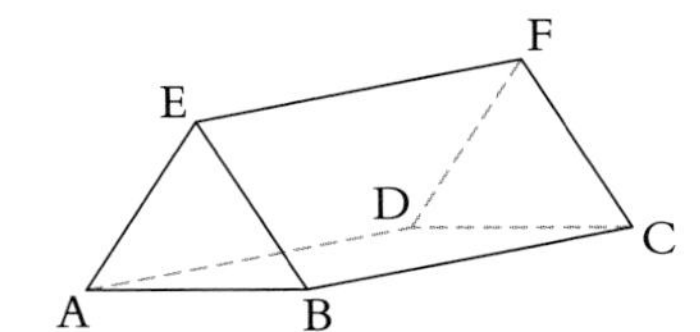

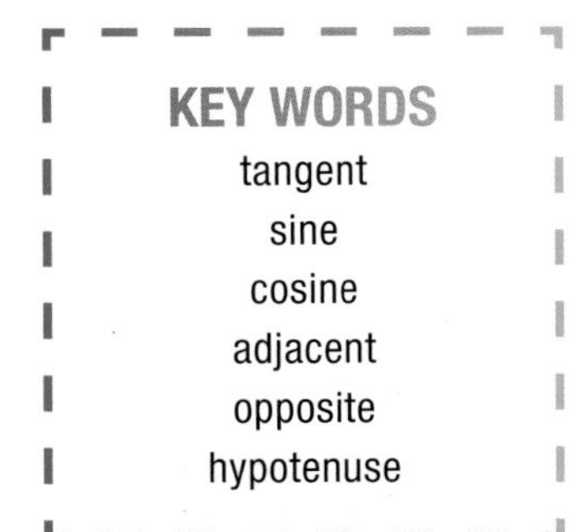

KEY WORDS

tangent
sine
cosine
adjacent
opposite
hypotenuse

Angles of elevation and depression

THIS SECTION WILL SHOW YOU HOW TO

- Use angles of elevation and depression

The **angle of elevation** is the angle measured upwards from the horizontal.

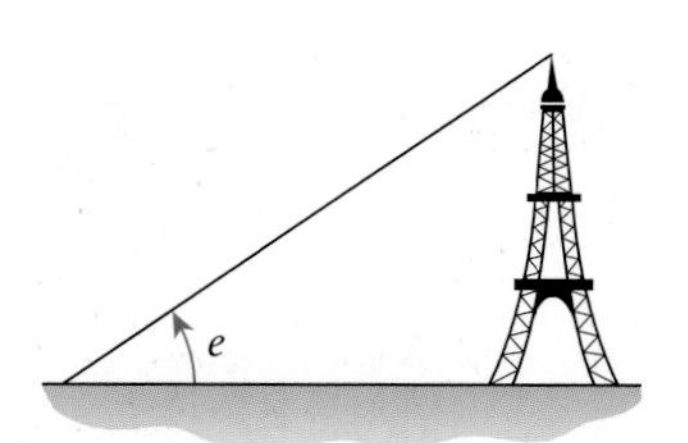

e = angle of elevation

The **angle of depression** is the angle measured downwards from the horizontal.

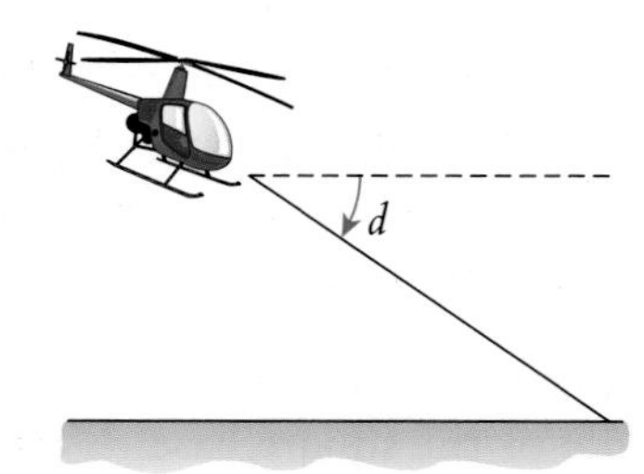

d = angle of depression

EXAMPLE

A radio mast stands on horizontal ground. The angle of elevation of the top of the mast from a point 200 m away from the base of the mast is 17°.
Calculate the height, h, of the radio mast.

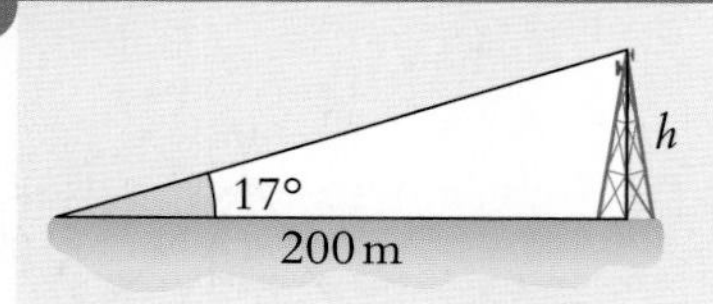

$$\tan 17° = \frac{h}{200}$$

$$h = 200 \times \tan 17°$$

$$h = 61.1 \text{ m}$$

EXERCISE 4.21

1. An electricity pylon stands on horizontal ground. The angle of elevation of the top of the pylon from a point 150 m away from the base of the pylon is 44°. Calculate the height of the pylon.

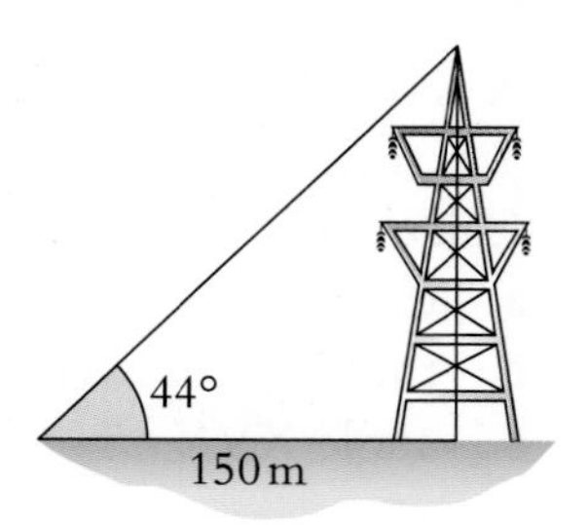

2. A tree stands on horizontal ground. The angle of elevation of the top of the tree from a point 50 m away from the base of the tree is 11.3°. Calculate the height of the tree.

3. The picture shows the CN tower in Canada. The angle of elevation of the top of the tower from a point 100 m away from the base of the tower is 79.75°. Calculate the height of the CN tower.

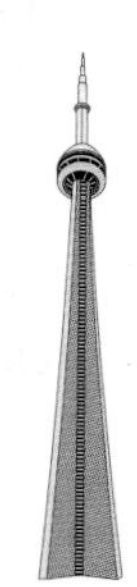

4 The picture shows the Burj Khalifa tower in Dubai.
The angle of elevation of the top of the tower from a point 100 m away from the base of the tower is 83.11°.
Calculate the height of the Burj Khalifa tower.

5 Use your answers to questions 3 and 4 to find the difference in heights between the CN tower and the Burj Khalifa tower.

6 An aeroplane is flying at a height of 1000 m.
The angle of depression of airport A from the plane is 15°.
Calculate the horizontal distance between the aeroplane and the airport.

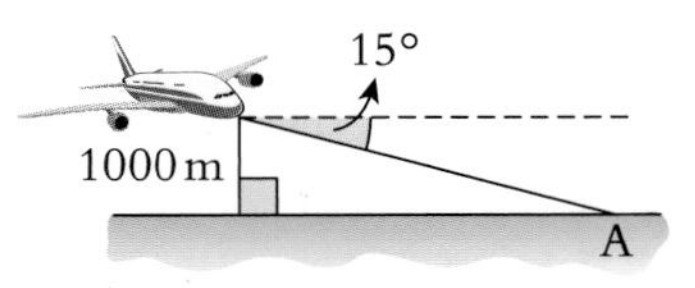

7 The lighthouse is 36 m high. Two boats A and B are due South of the lighthouse.
From the top of the lighthouse the angles of depression of boats A and B are 22° and 15°.
Calculate the distance AB.

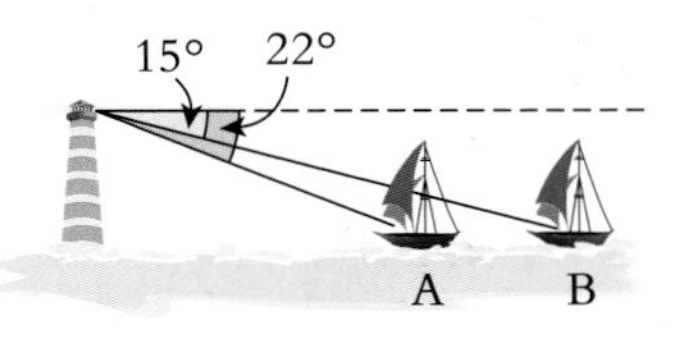

8 A hot air balloon is 70 m above the sea.
Two buoys A and B are due East of the balloon.
The angle of depression from the balloon to buoy A is 60°.
The angle of depression from the balloon to buoy B is 40°.
Calculate the distance AB.

9 The diagram shows a tower.
The point A is 30 m from the base of the tower.
On top of the tower is a flagpole PQ.
The angle of elevation of P from the point A is 53°.
The angle of elevation of Q from the point A is 50°.
Calculate the length of PQ.

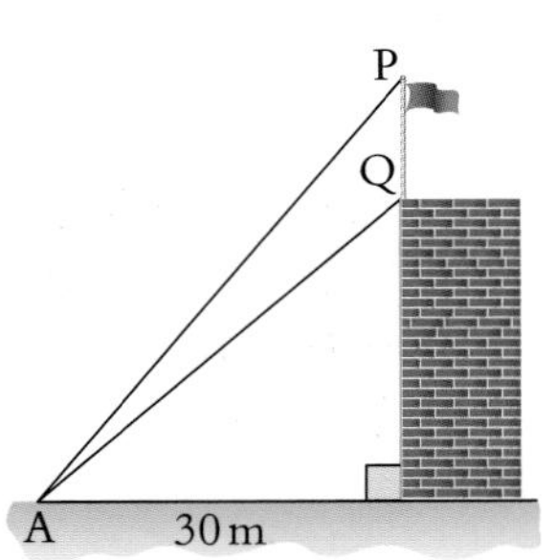

10 PQRS is a rectangular horizontal car park.
PT is a vertical lamp-post.
PS = 50 m and PQ = 40 m.
The angle of elevation of T from S is 8°.
Calculate:

- **a** PT and TR
- **b** the angle of elevation of T from Q
- **c** the angle of depression of R from T.

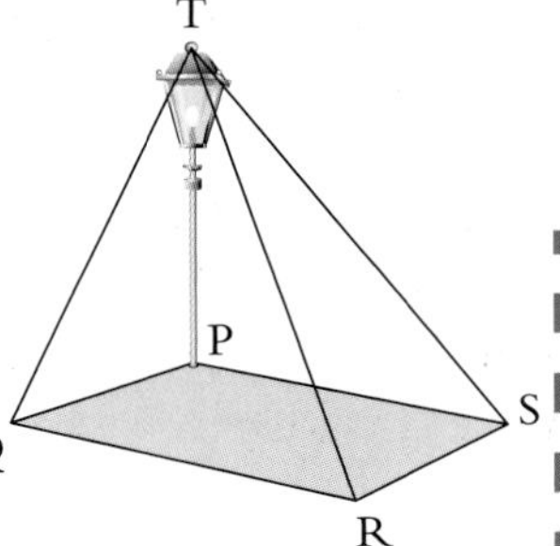

KEY WORDS
angle of elevation
angle of depression

UNIT 4

Scatter diagrams

THIS SECTION WILL SHOW YOU HOW TO
- Draw scatter diagrams and lines of best fit
- Understand positive, negative and zero correlation

You can use a **scatter diagram** to represent paired **data**.

EXAMPLE

The table shows the heights and masses of ten adult males.

Height (cm)	160	166	170	175	180	185	188	190	195	200
Mass (kg)	66	76	70	90	82	100	94	90	110	104

Draw a scatter diagram to show this information.

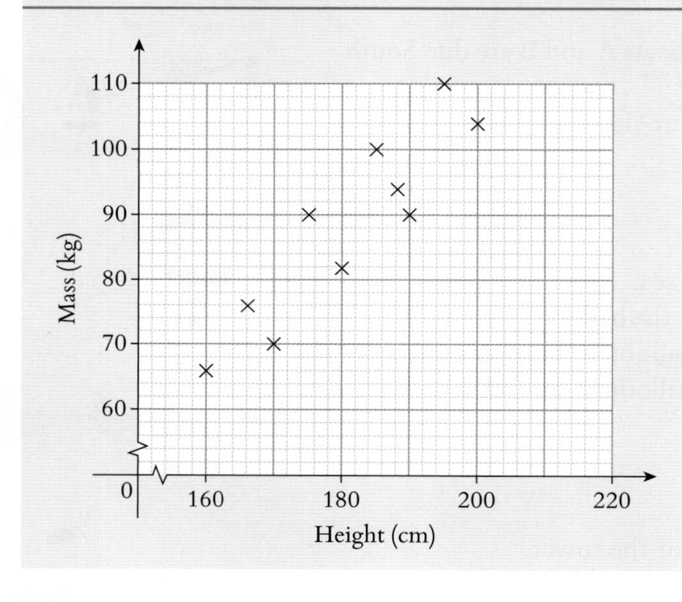

➡ **NOTE:** the information is plotted as coordinate points: (160, 66), (166, 76), (170, 70)...

➡ **NOTE:** a jagged line shows that the scale does not start at zero.

Correlation

A scatter diagram shows you if there is **correlation** (a connection) between the two variables.

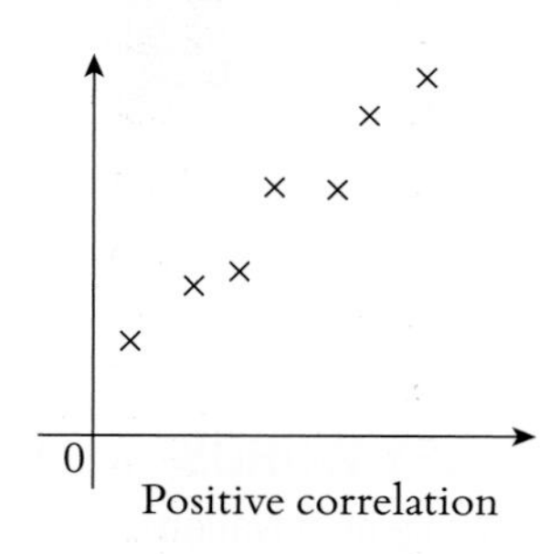

Positive correlation

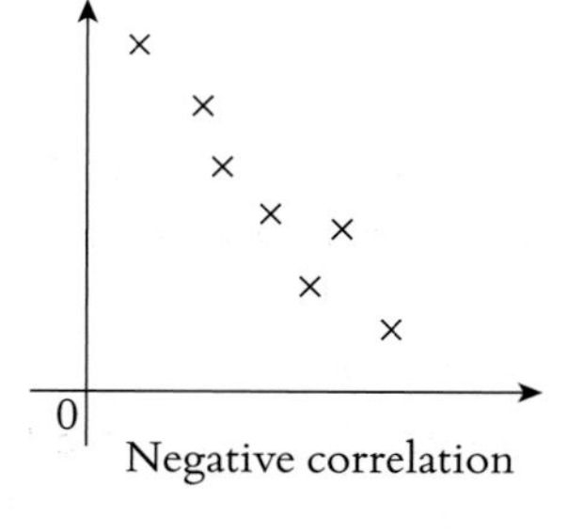

Negative correlation

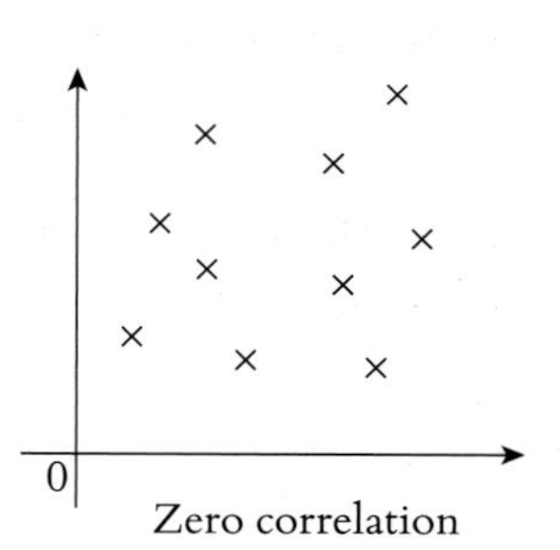

Zero correlation

In the example above, the scatter diagram shows that there is positive correlation between the two variables.

EXERCISE 4.22

1 Name the type of correlation shown by each of these scatter diagrams.

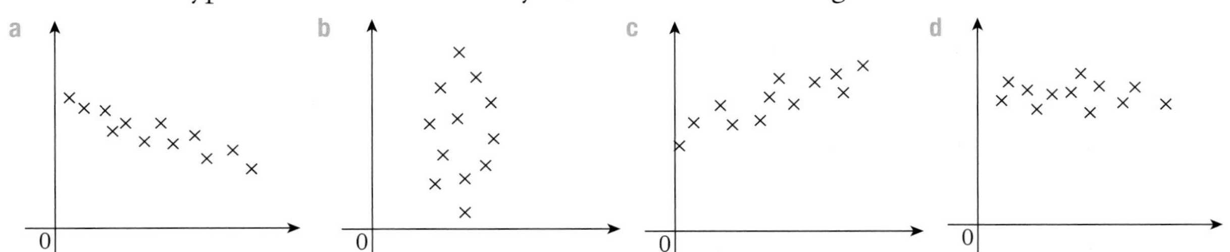

2 The table shows the test results for eight students in mathematics and art.

Mathematics	5	8	8	10	13	15	17	18
Art	10	6	18	13	13	19	15	11

a Show this information on a scatter diagram.
b Name the type of correlation shown.

3 The table shows information collected by a café owner for eight days.

Midday temperature (°C)	11	30	18	24	25	12	17	22
Number of hot drinks sold	24	4	11	8	11	20	18	15

a Show this information on a scatter diagram.
b Name the type of correlation shown.

4 The table shows the test results of ten students in chemistry and physics.

Chemistry	6	8	9	9	11	12	13	15	16	19
Physics	6	7	9	11	15	15	17	14	16	18

a Show this information on a scatter diagram.
b Name the type of correlation shown.

5 The table shows the age and value of eight cars.
a Show this information on a scatter diagram.
b Name the type of correlation shown.

Age (years)	Value ($)
0	18 000
1	15 000
2	12 000
4	8 000
4	10 000
5	7 000
5	8 000
6	5 000

6 Name the type of correlation that you would expect for each of these:
a The number of hours of sunshine in a day and the number of cold drinks sold.
b The midday temperature and the number of coats sold.
c The distance travelled by a car and the amount of money spent on petrol.
d The height of an adult and the IQ of the adult.
e The age of a child and the mass of the child.

Lines of best fit

When a scatter diagram shows positive or negative (linear) correlation you can draw a **line of best fit**.

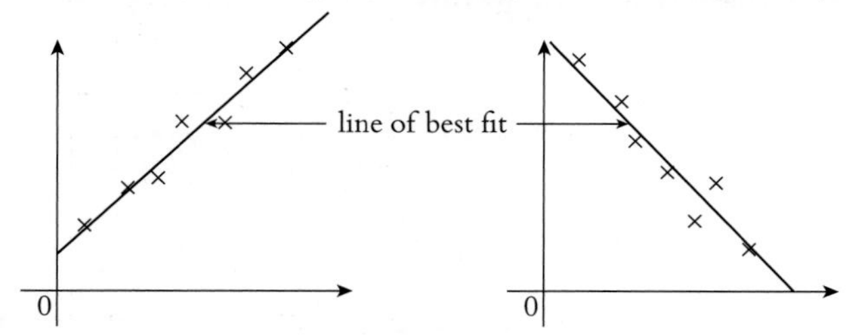

A line of best fit is usually drawn 'by eye' and should

- follow the general trend of the points
- have a similar number of points above and below the line.

The closer the points are to the line of best fit, the stronger the correlation.
The line of best fit can be used to predict data values within the range of data collected.
If you are asked to calculate the mean of the two data sets, then the mean point can also be plotted on the scatter graph and the line of best fit drawn through it.

UNIT 4

EXAMPLE

The table shows the daily amount of rainfall (mm) and the number of sun hats sold in a shop over a period of ten days.

Rainfall (mm)	0	1	3	5	2	6	8	4	2	6
Number of sunhats sold	26	23	18	11	21	9	2	15	17	5

a Calculate the mean rainfall for the ten days.
b Calculate the mean number of sunhats sold for the ten days.
c Show the information from the table on a scatter diagram and draw the line of best fit.
d Name the type of correlation shown by the diagram.
e Use your line of best fit to predict the number of sunhats sold on a day with 7 mm of rainfall.

a Mean rainfall $= \frac{0+1+3+5+2+6+8+4+2+6}{10} = \frac{37}{10} = 3.7$

b Mean number of hats sold $= \frac{26+23+18+11+21+9+2+15+17+5}{10} = \frac{147}{10} = 14.7$

c

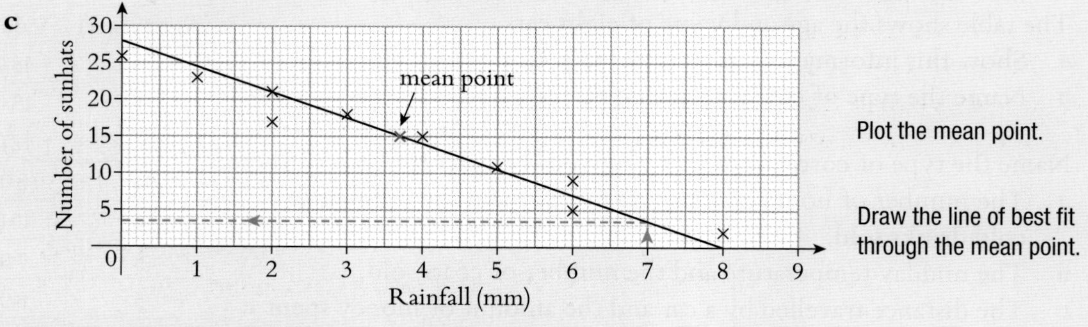

d The scatter graph shows negative correlation

e From the graph the predicted number of sunhats is approximately 4.

EXERCISE 4.23

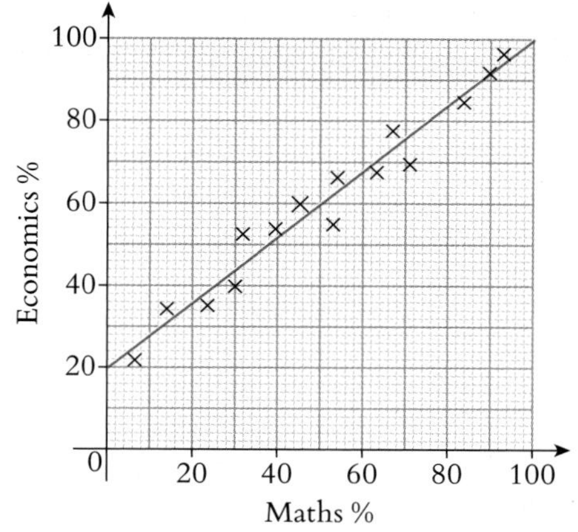

1 The scatter diagram shows information about the marks earned by a group of students in two examinations.

a Use the line of best fit to predict the economics mark for a student who scored 60% in the mathematics examination.

b Use the line of best fit to predict the mathematics mark for a student who scored 40% in the economics examination.

2 The table shows information about the number of hours spent doing homework per week and the number of hours spent watching television per week for ten students.

Homework (hours)	3	13	10	6	1	5	8	14	12	11
Television (hours)	9	3	7	8	12	9	8	2	5	5

a Calculate the mean number of hours spent doing homework per week.

b Calculate the mean number of hours spent watching television per week.

c Show the information from the table on a scatter diagram and draw the line of best fit.

d Name the type of correlation shown by the diagram.

e A student watched six hours of television in one week. Use your line of best fit to predict the number of hours this student spent doing homework that week.

3 The table shows information about the hand span and the height of ten adult females.

Hand span (cm)	20.5	18	22.4	22.7	23.5	19	19.7	21.5	22.2	21
Height (m)	170	160	186	183	190	167	171	177	181	177

a Calculate the mean hand span.

b Calculate the mean height.

c Show the information from the table on a scatter diagram and draw the line of best fit.

d Name the type of correlation shown by the diagram.

e An adult female has a hand span of 20.7 cm. Use your line of best fit to predict the height of this female.

Unit 4 Examination-style questions

1 Work out 45 cents as a percentage of \$18. [1]

2 A car is bought for \$5600 and it is then sold for \$6272.

Calculate the percentage profit. [3]

3 In 2010 the population of a country was 3.24×10^6.
In 2015 the population of the same country was 3.19×10^6.

Calculate the percentage decrease in the population. [3]

4 Work out.

$$\begin{pmatrix} 5 & 3 & 2 \\ 4 & 1 & 0 \\ -3 & 0 & 7 \end{pmatrix}\begin{pmatrix} 2 \\ -1 \\ 6 \end{pmatrix}$$ [2]

5 $\mathbf{A} = \begin{pmatrix} 2 & 1 \\ -4 & 3 \end{pmatrix}$ $\mathbf{B} = \begin{pmatrix} 5 & -1 \\ 2 & 7 \end{pmatrix}$

(a) Find $\mathbf{A} + \mathbf{B}$. [1]
(b) Find $\mathbf{BA}$. [2]
(c) Find $\mathbf{B}^2$. [2]
(d) Find $\mathbf{A}^{-1}$, the inverse of matrix $\mathbf{A}$. [2]

6 **(a)** Find $(5 \quad 1)\begin{pmatrix} 3 \\ -2 \end{pmatrix}$. [2]

(b) $\begin{pmatrix} 3 \\ -2 \end{pmatrix}(a \quad b) = \begin{pmatrix} -21 & 15 \\ 14 & -10 \end{pmatrix}$

Find the value of a and the value of b. [2]

(c) Explain why $\begin{pmatrix} 9 & -6 \\ -6 & 4 \end{pmatrix}$ has no inverse. [1]

7 Expand and simplify.

$$(x + 5)(x - 8)$$ [2]

8 Expand and simplify.

$$(2a - 3b)(3a - 5b)$$ [3]

9 Simplify.

$$100 - 3(4y - 5)^2$$ [3]

EXERCISE 5.1

1 Twelve metres of material costs \$31.56
Calculate the cost of five metres of the same material.

2 Three kilograms of fish costs \$11.25
Calculate the cost of eight kilograms of fish.

3 Jarred earns \$101.50 for working 7 hours in a café.
Calculate how much he will be paid for working 10 hours.

4 A car uses 20 litres of petrol to travel 120 kilometres.
- **a** How far will the car travel if it uses 25 litres of petrol?
- **b** How many litres of petrol are needed for a journey of 315 kilometres?

5 The recipe for pancakes will make 8 pancakes.
Anna wants to make 60 pancakes.
Calculate the amount of each ingredient needed.

RECIPE:
200g flour
480ml milk
2g salt
2 eggs

6 A machine fills 600 bottles in 4 minutes.
How long does the machine take to fill 40 bottles?
Give your answer in seconds.

7 Three people take 4 hours to paint a fence.
How long will it take five people?
Give your answer in hours and minutes.

8 Six people take 5 days to pick a crop of oranges.
How long would it take 4 people to pick the same crop?

9 A bag of sweets is shared between 8 children so that they each receive 9 sweets.
How many sweets would each child receive if there were 12 children?

10 It takes 48 hours to fill a swimming pool with water using 3 identical pumps.
- **a** How long would it take to fill the pool if there were eight of these pumps?
- **b** How many pumps would be needed to fill the pool in 12 hours?

11 A hotel has 120 rooms.
It takes 6 cleaners 4 hours to clean all the rooms.
How many cleaners would be needed to clean all the rooms in 3 hours?

12 One sack of rice will feed 100 people for 10 days.
- **a** For how many days will the sack of rice feed 250 people?
- **b** How many people will the sack of rice feed for 40 days?

Converting money from one currency to another currency

You can use a conversion graph to convert one currency to another currency.

The graph below can be used to convert between US dollars ($) and South African Rand (ZAR).

The blue dashed lines show how to convert 500 ZAR into US dollars ($).

500 ZAR = $73.50

The red dashed lines show how to convert $40 into South African Rand (ZAR).

$40 = 272 ZAR

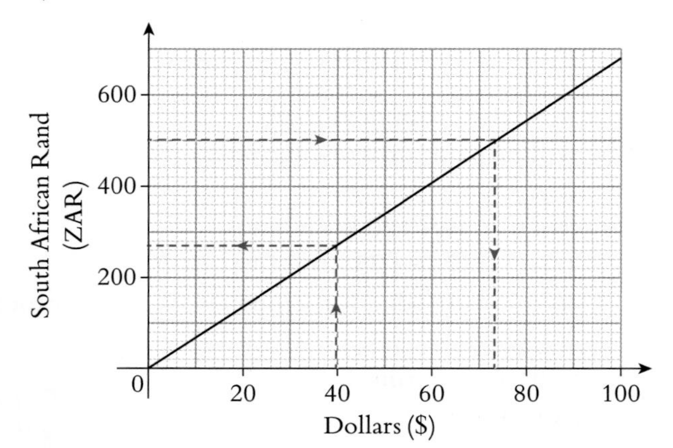

If you know the exchange rate you can use proportion to convert money from one currency to another.

EXAMPLE

Rafael changed $500 into Japanese Yen when the exchange rate was $1 = 83.7 Yen.
How many Japanese Yen did he receive?

$1 = 83.7 Yen
$500 = 500 × 83.7 Yen
$500 = 41850 Yen

EXAMPLE

Jani changed $350 into pounds (£) when the exchange rate was £1 = $1.315
How many pounds did he receive?
Give your answer correct to 2 decimal places.

$1.315 = £1

$\$1 = £\frac{1}{1.315}$

$\$350 = 350 \times £\frac{1}{1.315} = £266.159\ldots = £266.16$

EXERCISE 5.2

1

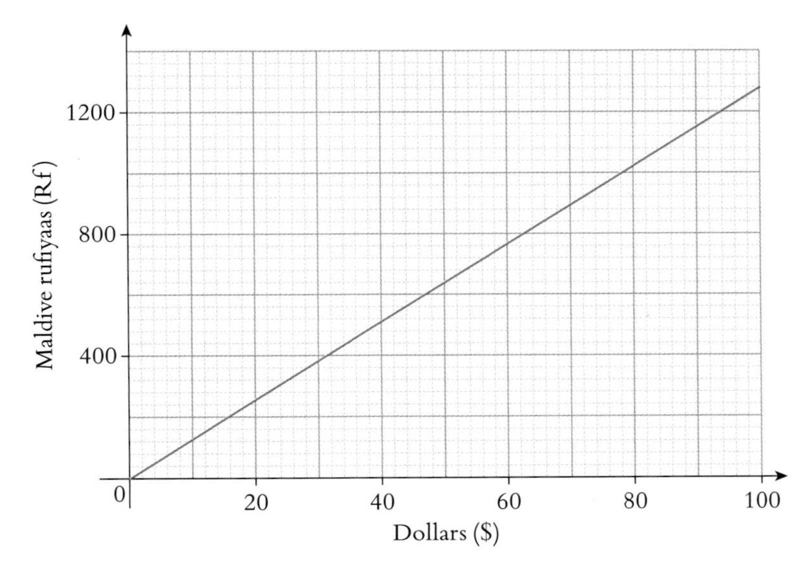

The graph can be used to convert between US dollars ($) and Maldives Rufiyaa (Rf).

a Use the graph to convert the amounts in dollars into Maldives Rufiyaa (Rf).

i $25 ii $54 iii $81 iv $200

b Use the graph to convert the amounts in Maldives Rufiyaa (Rf) into dollars.

i 600 Rf ii 1100 Rf iii 240 Rf iv 2000 Rf

2 If the exchange rate between US dollars and Argentine Peso (ARS) is $1 = 3.98 ARS,

a Change the following amounts of money into Argentine Peso (ARS).

i $50 ii $300 iii $72 iv $2000

b Change the following amounts of money into US dollars ($).

i 40 ARS ii 82 ARS iii 500 ARS iv 36.4 ARS

3 If the exchange rate between Malaysian Ringgit (MYR) and US dollars ($) is 1 MYR = $0.319,

a change the following amounts of money into Malaysian Ringgit (MYR).

i $80 ii $20 iii $74 iv $500

b change the following amounts of money into US dollars ($).

i 60 MYR ii 18 MYR iii 300 MYR iv 470 MYR

4 Zara changed 800 US dollars ($) into Thai Baht (THB) when the exchange rate was 1 THB = $0.032
One month later she changed all the Thai Baht back into US dollars when the exchange rate was 1 THB = $0.035
Calculate how much profit she made.
Give your answer in dollars.

KEY WORDS
direct proportion
inverse proportion

Increase and decrease in a given ratio

THIS SECTION WILL SHOW YOU HOW TO

- Increase or decrease a quantity in a given ratio

If you want to increase \$60 in the ratio 5 : 4 there are two possible methods.

Method 1

4 parts = \$60

1 part = \$60 ÷ 4 = \$15

5 parts = 5 × \$15 = \$75

So the answer is \$75.

Method 2

$\$60 \times \frac{5}{4} = \$75$

EXAMPLE

Decrease 540 kg in the ratio 17 : 20

Method 1

20 parts = 540 kg

1 part = 540 kg ÷ 20 = 27 kg

17 parts = 17 × 27 kg = 459 kg

So the answer is 459 kg.

Method 2

$540\,\text{kg} \times \frac{17}{20} = 459\,\text{kg}$

EXAMPLE

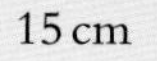

A photograph measures 15 cm by 10 cm.

The photograph is enlarged in the ratio 3 : 2

a Calculate the measurements of the enlarged photograph.

b Calculate **i** the area of the original photograph

ii the area of the enlarged photograph.

c Write down and simplify the following ratio

area of enlarged photograph : area of original photograph

---

a $15\,\text{cm} \times \frac{3}{2} = 22.5\,\text{cm}$ and $10\,\text{cm} \times \frac{3}{2} = 15\,\text{cm}$

Enlarged photograph is 22.5 cm by 15 cm.

b **i** Area of original photograph = 15 × 10 = 150 cm^2

ii Area of enlarged photograph = 22.5 × 15 = 337.5 cm^2

c area of enlarged photograph : area of original photograph

= 337.5 : 150 *divide both sides by 37.5*

= 9 : 4

➡ **NOTE:** it is important to notice that $9 : 4 = 3^2 : 2^2$

EXERCISE 5.3

1 Increase \$60 in the following ratios

a 5 : 3 b 7 : 4 c 13 : 12 d 11 : 8

2 Decrease 45 kg in the following ratios

a 4 : 9 b 11 : 15 c 2 : 3 d 4 : 5

3 One year a farmer grows 5000 kg of potatoes.
The next year he uses a fertiliser to increase the total mass of potatoes that he grows.
The mass of potatoes that he grows increases in the ratio 5 : 4
Calculate the new mass of potatoes that he grows.

4 Robert runs in an 800-metres race and he takes 2 minutes 15 seconds.
He trains hard for his next competition to try to improve the time that he takes.
He decreases the time he takes in the ratio 13 : 15
Calculate the new time that he takes to complete the race.

5 A car manufacturer produces 480 cars a week.
The demand for new cars increases so the car manufacturer increases the number of cars that are produced each week in the ratio 7 : 6
Calculate the new number of cars produced each week.

6 Daisy has a mass of 120 kg.
She is told that she must diet.
She diets and decrease her mass in the ratio 7 : 8
Calculate her new mass.

7 A shop has a sale and reduces the cost of a computer in the ratio 23 : 25
The original cost of the computer was \$1800.
Calculate the sale price of the computer.

8 A rectangle measures 6 cm by 8 cm.
The lengths of the sides of the rectangle are increased in the ratio 5 : 4

a What are the lengths of the new rectangle?

b What is the area of i the new rectangle ii the original rectangle?

c Write down and simplify the following ratio
the area of the new rectangle : the area of the original rectangle.

9 A cuboid measures 6 cm by 15 cm by 21 cm.
The lengths of the sides of the cuboid are increased in the ratio 4 : 3

a What are the measurements of the new cuboid?

b What is the volume of i the new cuboid ii the original cuboid?

c Write down and simplify the following ratio
the volume of the new cuboid : the volume of the original cuboid.

Functions

THIS SECTION WILL SHOW YOU HOW TO

- Use function notation
- Form composite and inverse functions

Function notation

A **mapping** transforms one set of numbers into a different set of numbers.
A **function** is a special mapping that maps each number of a set to only one number in another set.

$$\begin{array}{ccc} & \text{f} & \\ 1 & \longrightarrow & 4 \\ 2 & \longrightarrow & 5 \\ 3 & \longrightarrow & 6 \\ 4 & \longrightarrow & 7 \end{array}$$

The function f for the diagram above is 'add 3'.
This can be written using function notion in two ways:

$\text{f}: x \rightarrow x + 3$	which is read as	f is such that x is mapped to $x + 3$
or $\text{f}(x) = x + 3$	which is read as	f of $x = x + 3$

$\text{f}(1) = 4$ means that the function f maps the number 1 to the number 4.
$\text{f}(2) = 5$ means that the function f maps the number 2 to the number 5.

EXAMPLE

If $\text{f}(x) = x^2 + 5x$ Find **a** $\text{f}(6)$ **b** $\text{f}(-4)$

a f is the function that squares the number and then adds 5 lots of the number, so

$\text{f}(6) = 6^2 + 5 \times 6$
$= 36 + 30$
$= 66$

b $\text{f}(-4) = (-4)^2 + (5 \times -4)$
$= 16 + -20$
$= -4$

Sometimes you have to work in reverse.

EXAMPLE

If $\text{f}(x) = 3x + 4$. Find the value of x for which $\text{f}(x) = 25$.

Replace $\text{f}(x)$ in the function $\text{f}(x) = 3x + 4$ by the number 25.

$3x + 4 = 25$
$3x = 21$
$x = 7$

➡ **CHECK:**
$3 \times 7 + 4 = 25$ ✓

EXERCISE 5.4

1 $f(x) = 2x + 3$. Find the value of

a $f(4)$ b $f(-6)$ c $f\left(\frac{1}{2}\right)$ d $f(0)$

2 $f(x) = 5 - 3x$. Find the value of

a $f(8)$ b $f(-4)$ c $f\left(\frac{1}{2}\right)$ d $f\left(-\frac{1}{2}\right)$

3 $f(x) = x^2 + 3x$. Find the value of

a $f(2)$ b $f(-3)$ c $f(0)$ d $f\left(\frac{1}{2}\right)$

4 $f(x)= 2x^2 + x - 5$. Find the value of

a $f(3)$ b $f(-1)$ c $f\left(\frac{1}{2}\right)$ d $f(0)$

5 $f(x) = x^3 + 2$. Find the value of

a $f(4)$ b $f(-2)$ c $f\left(\frac{1}{2}\right)$ d $f(0)$

6 $f(x) = \frac{3x^2 + 1}{4}$. Find the value of

a $f(4)$ b $f(-6)$ c $f\left(\frac{1}{2}\right)$ d $f(0)$

7 $g(x) = 5x - 2$. Find the value of x when $g(x) = 13$

8 $f(x) = x^2 + 2$. Find the values of x when $f(x) = 38$

9 $f(x) = \frac{1}{2x+4}$. Find the value of x when $f(x) = 2$

10 $f(x) = \frac{5}{2x+3}$. Find the value of x when $f(x) = -2$

11 $f(x) = 3x - 2$ and $g(x) = 4x + 1$.
Find the value of x such that $f(x) = g(x)$.

TOP TIP

Put the two functions equal to each other and solve to find x.

12 $f(x) = x^2 + 7x$ and $g(x) = x^2 - x + 4$.
Find the value of x such that $f(x) = g(x)$.

13 $f(x) = x^2 + 3x - 8$ and $g(x) = 3x + 1$. Find the values of x such that $f(x) = g(x)$.

14 $f(x) = \frac{1}{x+2}$ and $g(x) = \frac{3}{2x-1}$. Find the value of x such that $f(x) = g(x)$.

Factorising 2

THIS SECTION WILL SHOW YOU HOW TO

- Factorise a quadratic expression

A **quadratic** expression is an expression of the form $ax^2 + bx + c$ where a, b and c are numbers and $a \neq 0$.

Examples of quadratic expressions are $3x^2 - 5x + 4$ $2x^2 - 6x$ $x^2 - 9$

Factorising quadratic expressions of the form $ax^2 + bx$

Reminder: $x^2 + 3x = x(x + 3)$

$5x - x^2 = x(5 - x)$

$6x^2 - 4x = 2x(3x - 2)$

Factorising quadratic expressions of the form $x^2 + bx + c$

Factorising is the reverse of expanding.

To factorise you need to find two numbers that multiply to give the **constant** and also add to give the **coefficient** of x.

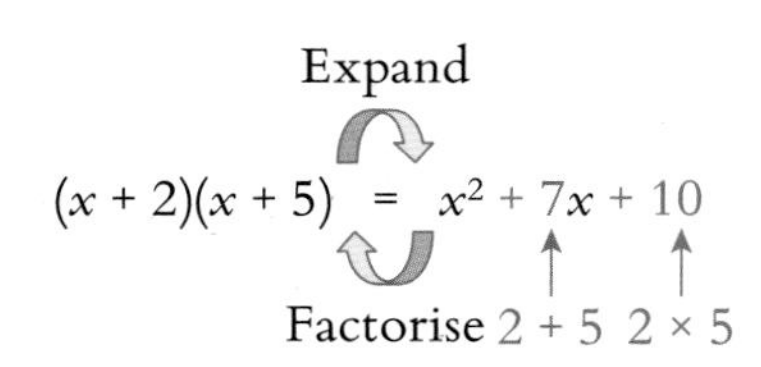

EXAMPLE

Factorise **a** $x^2 + 10x + 16$ **b** $x^2 - 2x - 15$ **c** $x^2 - 5x + 6$

a $x^2 + 10x + 16$

*List the pairs of numbers that **multiply** to give 16 and then check to see which pair **add** to give 10.*

1 and 16 (2 and 8) 4 and 4

so

$x^2 + 10x + 16 = (x + 2)(x + 8)$

CHECK:

$(x + 2)(x + 8) = x^2 + 8x + 2x + 16 = x^2 + 10x + 16$ ✓

b $x^2 - 2x - 15$

*List the pairs of numbers that **multiply** to give −15 and then check to see which pair **add** to give −2.*

1 and −15 (3 and −5) −1 and 15 −3 and 5

so

$x^2 - 2x - 15 = (x + 3)(x - 5)$

CHECK:

$(x + 3)(x - 5) = x^2 - 5x + 3x - 15 = x^2 - 2x - 15$ ✓

c $x^2 - 5x + 6$

*List the pairs of numbers that **multiply** to give 6 and then check to see which pair **add** to give −5.*

1 and 6 2 and 3 −1 and −6 (−2 and −3)

so

$x^2 - 5x + 6 = (x - 2)(x - 3)$

CHECK:

$(x - 2)(x - 3) = x^2 - 3x - 2x + 6 = x^2 - 5x + 6$ ✓

EXERCISE 5.8

1 Factorise these expressions.

a $x^2 - 81$ b $y^2 - 1$ c $x^2 - 64$ d $a^2 - 36$

e $y^2 - 100$ f $16 - a^2$ g $x^2 - y^2$ h $144 - x^2$

i $4x^2 - 9$ j $16y^2 - 1$ k $25x^2 - 36$ l $9 - 16x^2$

m $100x^2 - 81y^2$ n $25a^2 - 49b^2$ o $81x^2 - 49y^2$ p $121x^2 - 100$

2 Factorise these expressions.

a $2x^2 - 50$ b $2y^2 - 8$ c $3x^2 - 3$

d $5y^2 - 80$ e $3y^2 - 192$ f $4x^2 - 4y^2$

g $2x^2 - 242$ h $6n^2 - 600$ i $75y^2 - 108$

TOP TIP

Take out a common factor first.

3 Factorise $x^4 - y^4$

TOP TIP

You will need to factorise twice.

4 Factorise these expressions.

a $2x^2 + 11x + 5$ b $2x^2 + 5x + 3$ c $2x^2 + 15x + 7$ d $2x^2 + 13x + 11$

e $3x^2 + 5x + 2$ f $3x^2 + 16x + 5$ g $3x^2 + 10x + 3$ h $3x^2 + 10x + 7$

i $2x^2 + 5x - 3$ j $3x^2 + 2x - 5$ k $3x^2 - 22x + 7$ l $5x^2 - 2x - 3$

m $7x^2 - 9x + 2$ n $5x^2 + 3x - 2$ o $11x^2 - 21x - 2$ p $2x^2 - 15x + 13$

5 Factorise these expressions.

a $2x^2 + 11x + 12$ b $2x^2 + 9x + 10$ c $3x^2 + 14x + 8$ d $7x^2 + 38x + 15$

e $6x^2 + 7x + 2$ f $6x^2 + 11x + 5$ g $8x^2 + 14x + 3$ h $10x^2 + 71x + 7$

i $3x^2 + 7x - 6$ j $5x^2 + 22x + 8$ k $2x^2 - 11x + 12$ l $7x^2 + 26x - 8$

m $5x^2 + 21x - 20$ n $3x^2 + x - 10$ o $2x^2 - 9x + 10$ p $3x^2 + 14x + 16$

6 Factorise these expressions.

a $4x^2 + 2x - 42$ b $6x^2 + 21x + 9$ c $6x^2 + 16x + 8$

d $12x^2 + 28x + 8$ e $16x^2y - 20xy - 6y$ f $6x^3 - 10x^2 - 4x$

TOP TIP

Take out a common factor first.

7 Simplify these algebraic fractions.

a $\dfrac{2x^2 + 7x + 3}{x^2 + 3x}$ b $\dfrac{3x^2 + 5x - 2}{x^2 - 4}$ c $\dfrac{2x^2 + 3x - 20}{x^2 + 6x + 8}$

d $\dfrac{4x^2 - 9}{2x^2 - x - 3}$ e $\dfrac{x^2 - 5x}{2x^2 - 7x - 15}$ f $\dfrac{2x^2 - 50}{3x^2 + 7x - 40}$

TOP TIP

Factorise the numerator and the denominator before cancelling factors.

8 Factorise $(3x + 2)^4 - 4(3x + 2)^2$

KEY WORDS

quadratic

coefficient, constant

difference of two squares

UNIT 5

Cubic graphs

THIS SECTION WILL SHOW YOU HOW TO
- Recognise cubic graphs
- Draw cubic graphs
- Use cubic graphs in practical situations

$y = ax^3 + bx^2 + cx + d$ is called a **cubic function** where a, b, c and d are constants and $a \neq 0$.
The simplest cubic function is $y = x^3$.
The curve has rotational symmetry of order 2 about the origin.

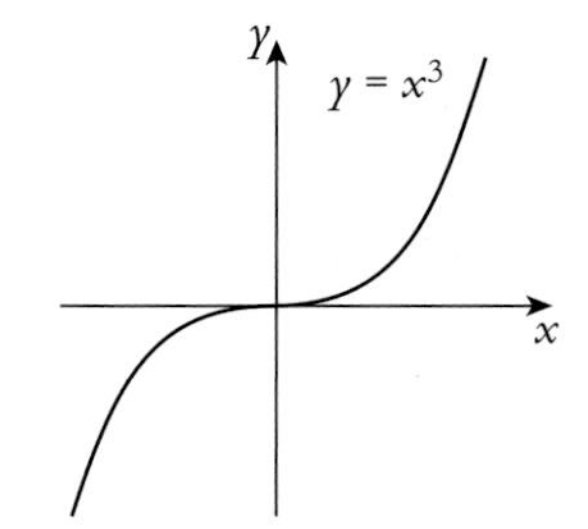

EXAMPLE

Draw the graph of $y = x^3 - 5x + 3$ for $-3 \leq x \leq 3$.
Use your graph to find the values of x when $y = 5$.

First work out a table of values for the graph.

When $x = -3$, $y = (-3)^3 - (5 \times -3) + 3$
$y = -27 - -15 + 3$
$y = -27 + 15 + 3$
$y = -9$ etc

x	−3	−2	−1	0	1	2	3
y	−9	5	7	3	−1	1	15

Use your table of values to draw the graph of the function.

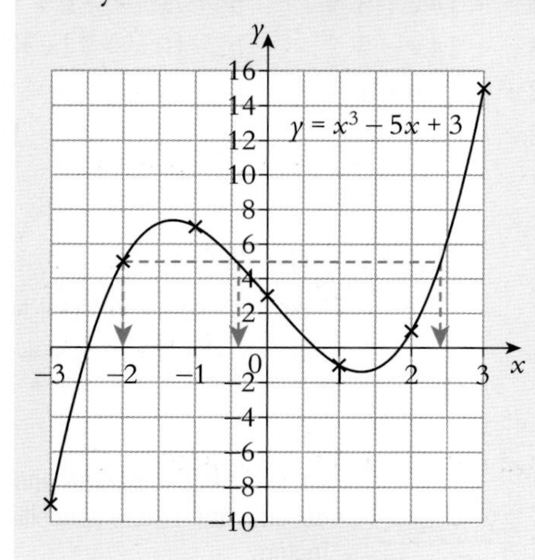

To find the values of x when $y = 5$ you need to draw the line $y = 5$ and then find the x coordinates of the three points where the line intersects the curve.

The graphs intersect in three places.
$x = -2$ $\quad x = -0.4$ $\quad x = 2.4$

The general shape of a cubic graph depends on the value of a

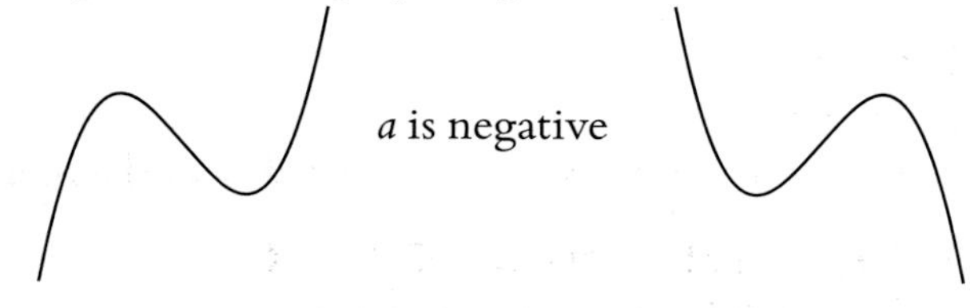

EXERCISE 5.9

1 Copy and complete the table for each of these cubic functions.
Draw the graph of each function.

a $y = x^3 + 1$

x	−3	−2	−1	0	1	2	3
y						9	

b $y = 2 - x^3$

x	−3	−2	−1	0	1	2	3
y			3				

c $y = -x^3$

x	−3	−2	−1	0	1	2	3
y						−8	

d $y = x^3 + 2x$

x	−3	−2	−1	0	1	2	3
y					3		

e $y = 3x^2 - x^3$

x	−2	−1	0	1	2	3	4
y					4		

f $y = x^3 - 2x^2 - 4x$

x	−2	−1	0	1	2	3	4
y		1					

g $y = x^3 - 4x^2 + 2$

x	−1	0	1	2	3	4
y				−6		

h $y = \frac{1}{2}x^3 - 3x^2 - 2$

x	−6	−5	−4	−3	−2	−1	0	1	2
y					10				

2 a Copy and complete the table of values for $y = x^3 + 2$

x	−3	−2	−1	0	1	2	3
y			1				

b Draw the graph of $y = x^3 + 2$

c Use your graph to find the value of x when $y = 7$

d By drawing a tangent to the graph estimate the gradient of the curve when $x = 1$.

3 a Copy and complete the table of values for $y = \frac{1}{2}x^3 - 2x^2 + 6$

x	−2	−1	0	1	2	3	4
y					2		

b Draw the graph of $y = \frac{1}{2}x^3 - 2x^2 + 6$

c Use your graph to find the values of x when $y = 4$

d By drawing a tangent to the graph estimate the gradient of the curve when $x = 2$.

4 The function $f(x) = \frac{1}{2}x^3$, draw the graph of $y = f(x)$ for $-3 \leq x \leq 3$
On the same axes draw the graph of $y = 2x$
Write down the coordinates of the points of intersection of the curve and the line.

5 Draw the graph of $y = x^3 - 4x^2 + 3$ for $-1 \leq x \leq 4$
On the same axes draw the graph of $x + y = 1$
Write down the coordinates of the points of intersection of the curve and the line.

6 The function $f(x) = x^3 - 3x^2$, draw the graph of $y = f(x)$ for $-2 \leq x \leq 4$
Write down the values of x where the graph has negative gradient.

Practical applications of cubic graphs

EXAMPLE

A square piece of card has sides of length 20 cm.
A square of length x cm is removed from each of the four corners of the card.
The remaining piece of card is then folded to make an open box as shown below.

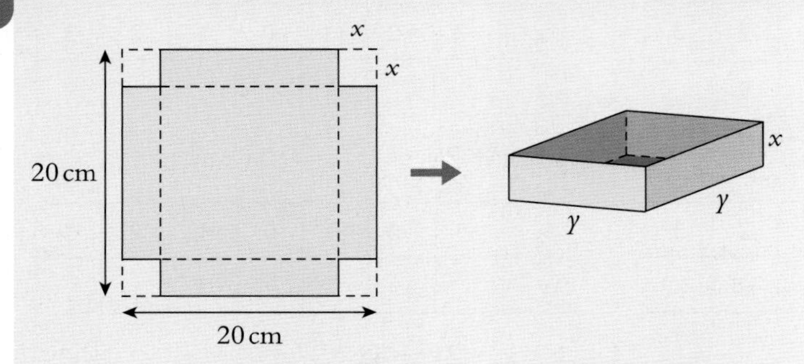

a Show that the volume, V cm^3, of the box is given by the formula

$$V = 400x - 80x^2 + 4x^3$$

b Draw the graph of $V = 400x - 80x^2 + 4x^3$ for $0 \le x \le 10$.

c Use your graph to estimate the maximum possible volume of the box and the value of x that gives the maximum volume.

---

a $y = 20 - 2x$

Volume $= y \times y \times x$
$= (20 - 2x)(20 - 2x)x$ *expand brackets*
$= (400 - 40x - 40x + 4x^2)x$
$= 400x - 80x^2 + 4x^3$

b

x	0	1	2	3	4	5	6	7	8	9	10
V	0	324	512	588	576	500	384	252	128	36	0

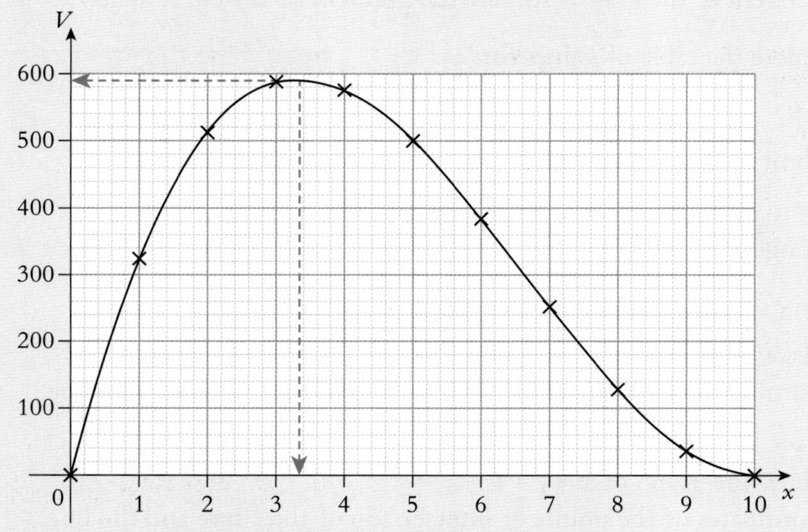

c Maximum volume ≈ 590 cm^3 when $x \approx 3.3$

EXERCISE 5.10

1

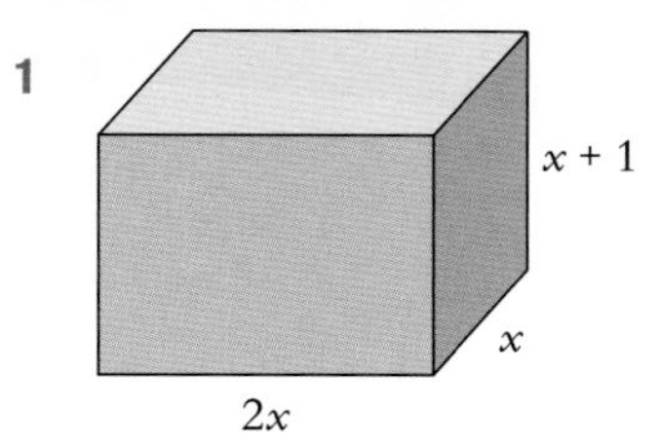

The rectangular tank has sides of length $2x$, x and $(x + 1)$ m.

a Show that the volume, V cm^3, of the tank is given by the formula
$V = 2x^3 + 2x^2$

b Copy and complete the table of values.

x	0	0.2	0.4	0.6	0.8	1
V	0					

c Draw the graph of $V = 2x^3 + 2x^2$ for $0 \le x \le 1$

d Use your graph to estimate the value of x that gives a volume of 2 m^3.

2 A rectangular piece of card has sides of length 20 cm and 16 cm.
A square of length x cm is removed from each of the four corners of the card.
The remaining piece of card is then folded to make an open box as shown below.

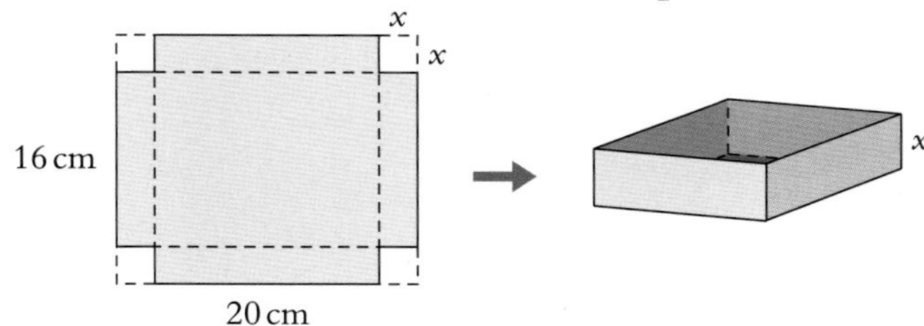

a Show that the volume, V cm^3, of the box is given by the formula
$V = 320x - 72x^2 + 4x^3$

b Copy and complete the table of values.

x	0	1	2	3	4	5	6	7	8
V	0								

c Draw the graph of $V = 320x - 72x^2 + 4x^3$ for $0 \le x \le 8$

d Use your graph to estimate the maximum possible volume of the box and the value of x that gives the maximum volume.

Surface area and volume 2

THIS SECTION WILL SHOW YOU HOW TO

- Find the surface area and volume of pyramids, cones and spheres

Pyramids

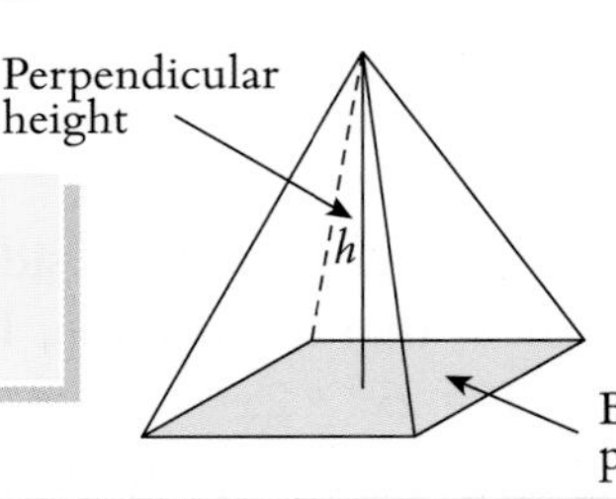

Volume of a pyramid = $\frac{1}{3}$ × base area × perpendicular height
Surface area = sum of areas of all the faces of the pyramid

EXAMPLE

ABCDE is a square-based pyramid.
AB = BC = 6 cm.
The perpendicular height of the pyramid is 4 cm.
P is the mid-point of BC and O is the middle of the base.

a Calculate the volume of the pyramid.
b Calculate the length of EP.
c Calculate the area of △ BCE.
d Calculate the surface area of the pyramid.

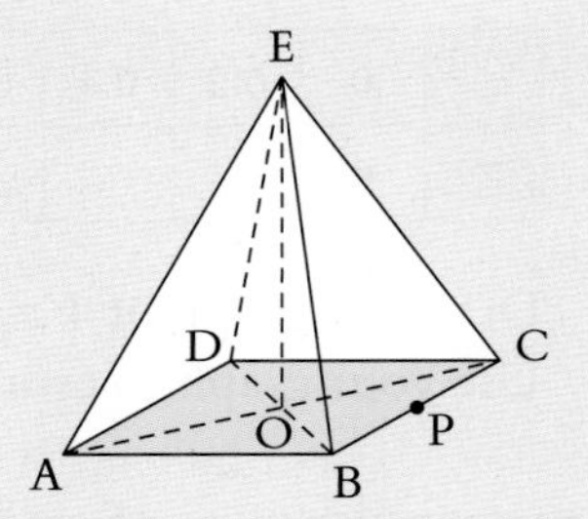

a Volume of pyramid $= \frac{1}{3} \times \text{base area} \times \text{perpendicular height}$

$= \frac{1}{3} \times (6 \times 6) \times 4$

$= 48\,\text{cm}^3$

b Using Pythagoras on △ EOP: $EP^2 = 4^2 + 3^2$

$EP^2 = 25$

$EP = 5\,\text{cm}$

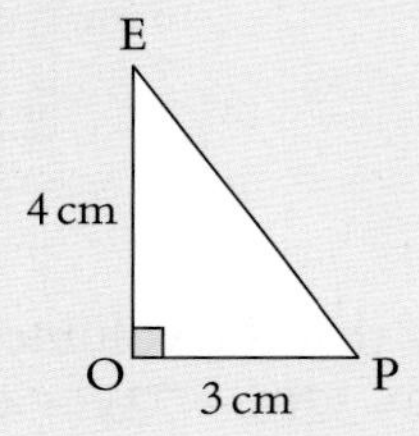

c Area of △ BCE $= \frac{1}{2} \times \text{base} \times \text{height}$

$= \frac{1}{2} \times 6 \times 5 = 15\,\text{cm}^2$

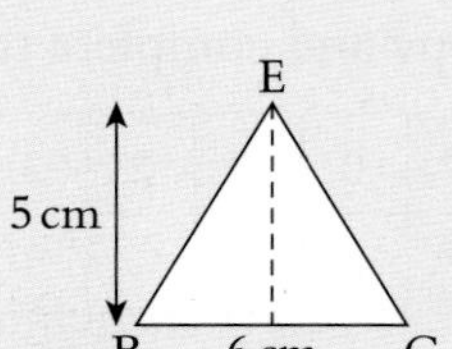

d Surface area = sum of all the faces of the pyramid
= area of base + area of four triangular faces
= 36 + (4 × 15)
= $96\,\text{cm}^2$

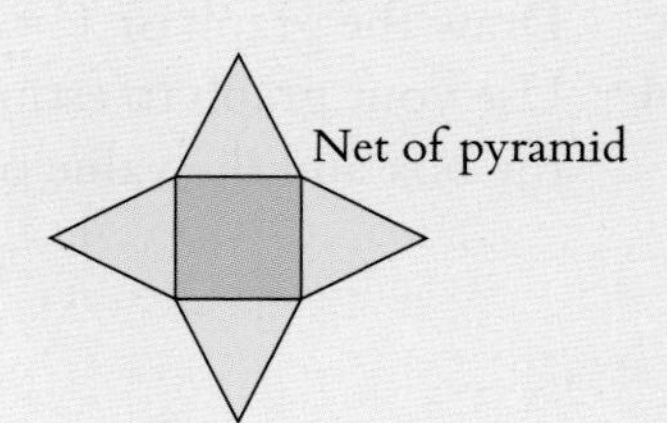

EXERCISE 5.11

1 Find the volume of each of these pyramids.
The base area (A cm^2) and the perpendicular height (h cm) are given for each pyramid.

a
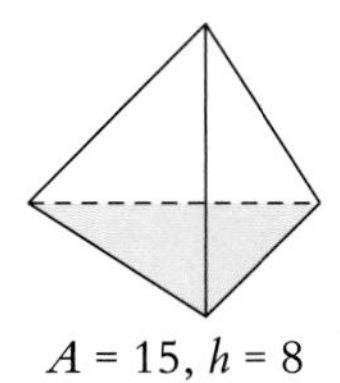
$A = 15$, $h = 8$

b
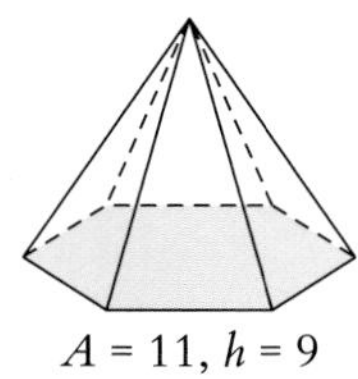
$A = 11$, $h = 9$

c
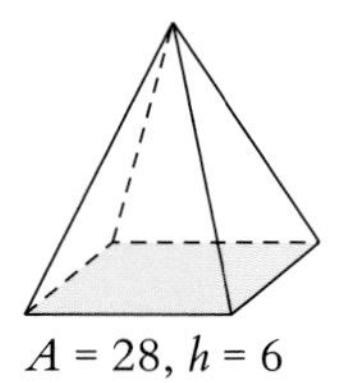
$A = 28$, $h = 6$

d
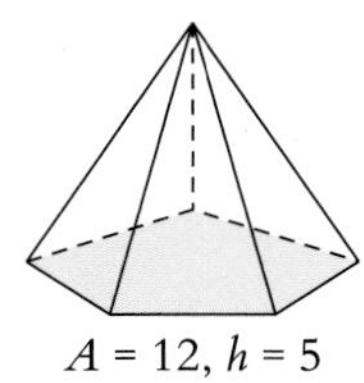
$A = 12$, $h = 5$

2 A pyramid has a square base of side 8 cm.
The perpendicular height is 6 cm.
Calculate the volume of the pyramid.

3 A pyramid has a rectangular base of side 6 cm and 10 cm.
The perpendicular height is 7 cm.
Calculate the volume of the pyramid.

4 A pyramid has an equilateral triangular base of side 5 cm.
The perpendicular height is 4 cm.
Calculate the volume of the pyramid.

5 A pyramid has a square base of side x cm.
The perpendicular height is x cm.
The volume of the pyramid is 72 cm^3. Find the value of x.

6 The Great Pyramid of Khutu in Egypt is a square-based pyramid.
The square base has sides of length 241 m.
The perpendicular height of the pyramid is 153 m.
Calculate the volume of the pyramid.
Give your answer in standard form correct to 3 s.f.

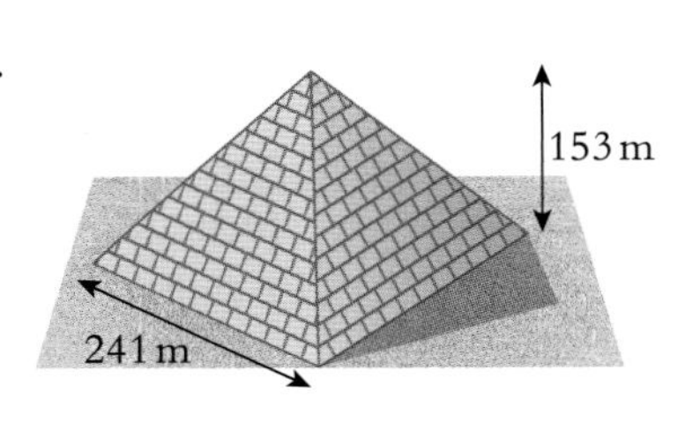

7 The Louvre Pyramid in France is a large glass and metal square-based pyramid.
The square base has sides of length 35.42 m.
The perpendicular height of the pyramid is 21.64 m.
Calculate the volume of the pyramid.
Give your answer in standard form correct to 3 s.f.

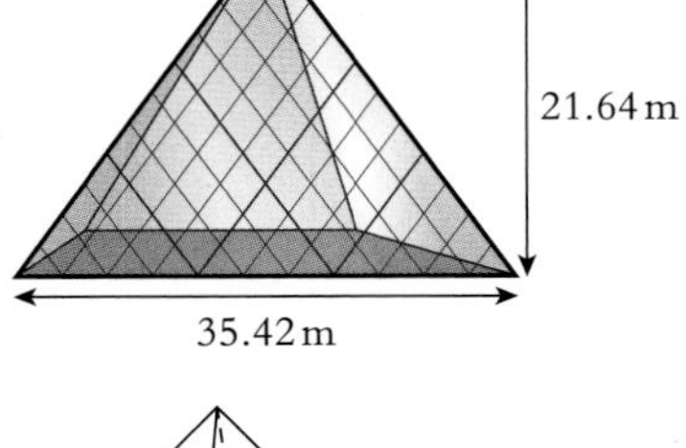

8 A crystal is in the form of a **regular octahedron**.
The edges of the octahedron are all 4 mm long.
Calculate the volume of the crystal.

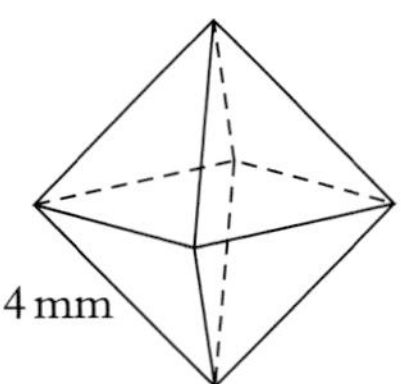

9 A square-based pyramid has a base of side 8 cm and a height of 12 cm.
A square-based pyramid of height 6 cm is removed by cutting parallel to the base, leaving the solid shown.
Calculate the volume of the **truncated pyramid**.

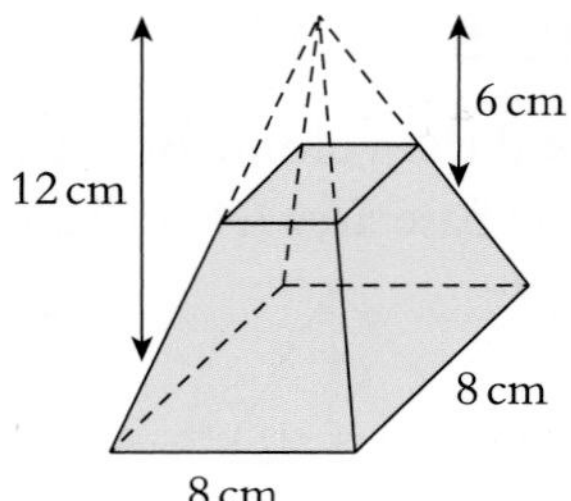

10 A rectangular-based pyramid has a base of sides 12 cm and 9 cm and a height of 15 cm.
A rectangular-based pyramid of height 5 cm is removed by cutting parallel to the base, leaving the solid shown.
Calculate the volume of the truncated pyramid.

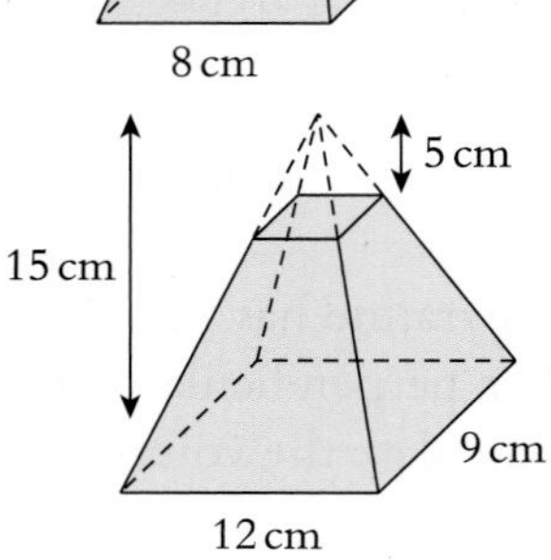

11 Find the surface area of each of these pyramids.

a

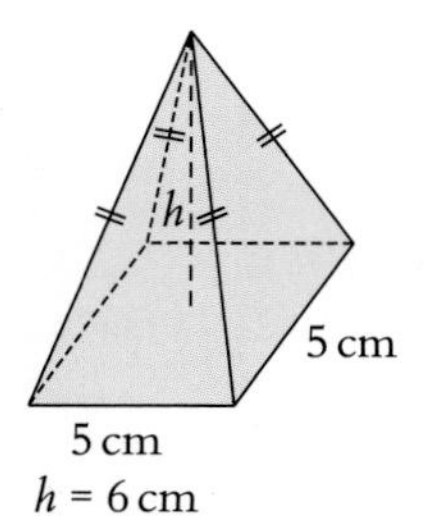

b

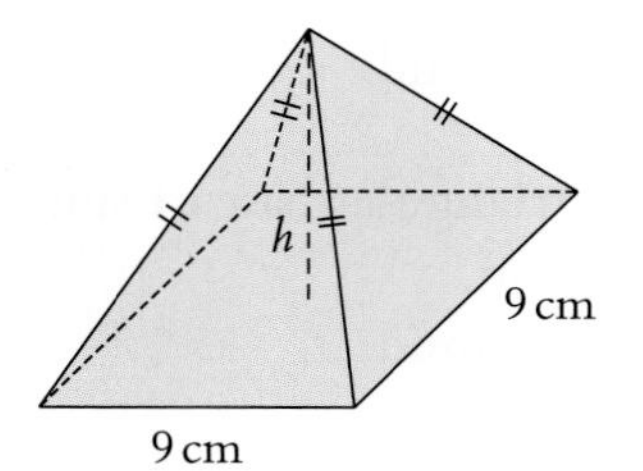

c

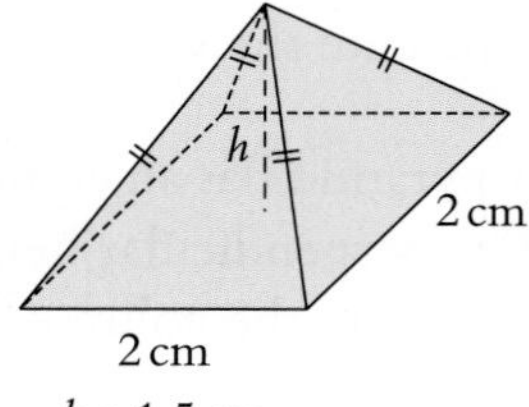

d

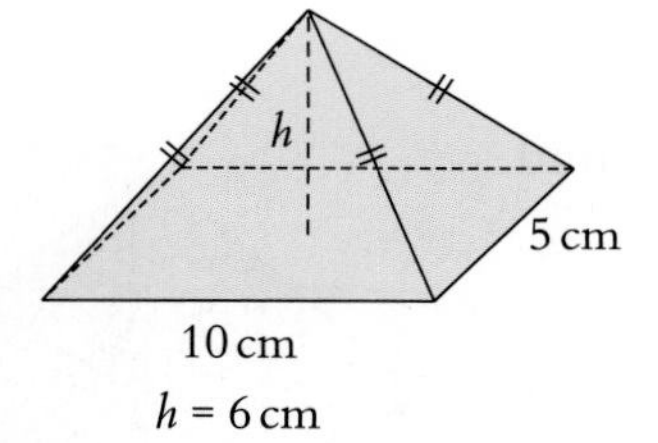

e

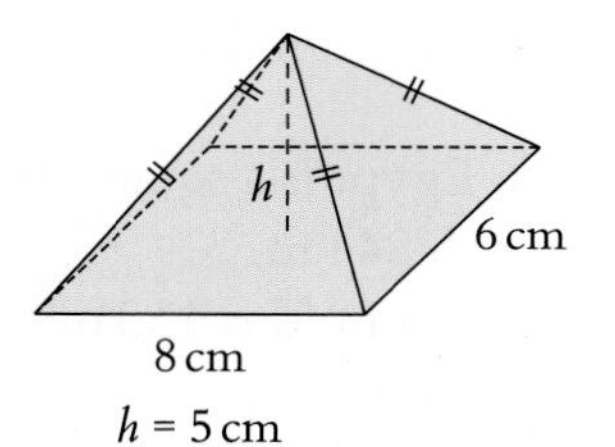

f

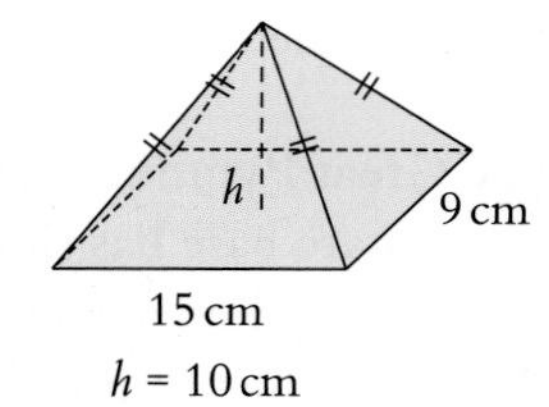

12 Find the surface area of these **regular tetrahedrons**.

a

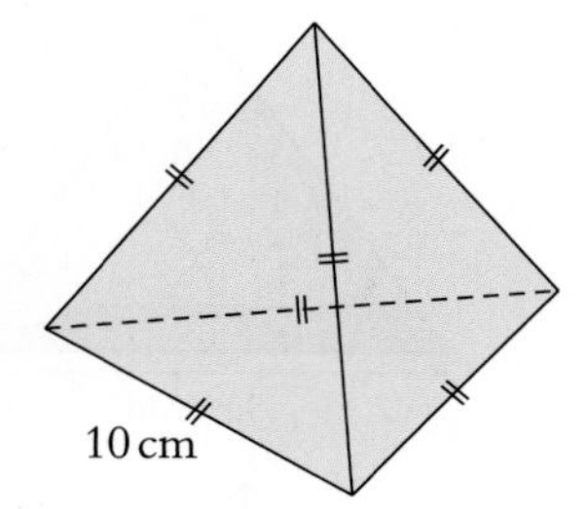

b

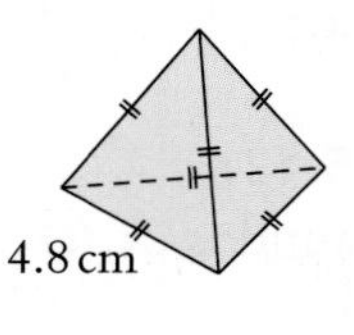

TOP TIP

Tetrahedron is the name for a triangular-based pyramid (tetra- for 4 faces).
Regular means that all the sides are the same length so each face is an equilateral triangle.

13 Find the volume of each regular tetrahedron in question 12.

Cones

A **cone** is a pyramid with a circular base.

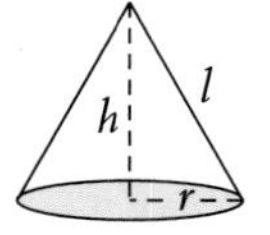

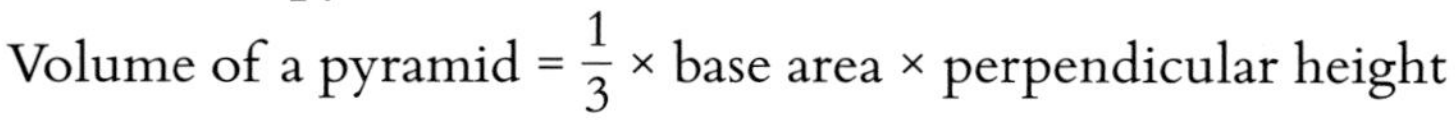

Volume of a pyramid = $\frac{1}{3}$ × base area × perpendicular height

So, Volume of a cone = $\frac{1}{3} \times \pi r^2 \times h$

The surface of a cone is made from a flat circular base and a curved surface.
The curved surface is made from a sector of a circle.

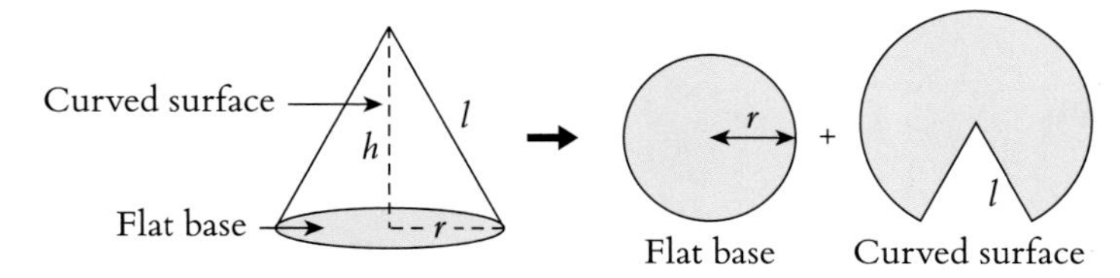

> Volume of a cone = $\frac{1}{3}\pi r^2 h$
> Curved surface area of a cone = πrl, where l is the slant height
> Total surface area of cone = $\pi r^2 + \pi rl$

When you make a cut parallel to the base of a cone and remove the top part, the part that is left is called a **frustum**.

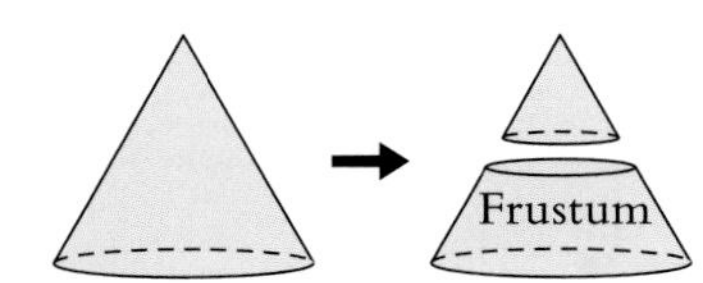

> Volume of frustum = volume of whole cone – volume of smaller cone

The cone has base radius 3 cm and perpendicular height 4 cm.

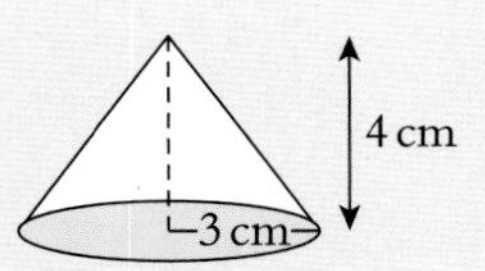

a Calculate the volume of the cone.

b Calculate the curved surface area of the cone.

c Calculate the total surface area of the cone

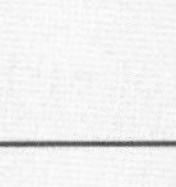

a Volume = $\frac{1}{3} \times \pi r^2 h = \frac{1}{3} \times \pi \times 3^2 \times 4 = 37.7\text{ cm}^3$

b Curved surface area = πrl

(so you must first calculate the slant height, l)

Using Pythagoras: $l^2 = 3^2 + 4^2$

$l^2 = 25$

$l = 5$

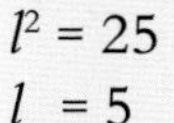

Curved surface area = $\pi rl = \pi \times 3 \times 5 = 47.1\text{ cm}^2$

c Total surface area = $\pi r^2 + \pi rl = \pi \times 3^2 + 47.1 = 75.4\text{ cm}^2$

4 cm
l
3 cm

EXERCISE 5.12

1 Calculate the volume of these cones.

a
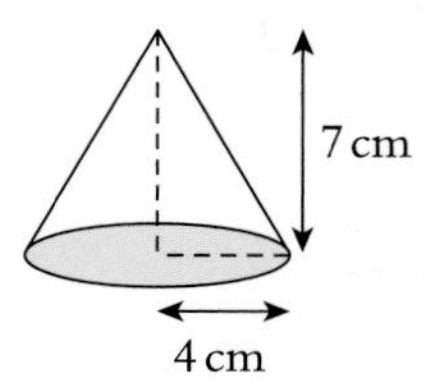

b
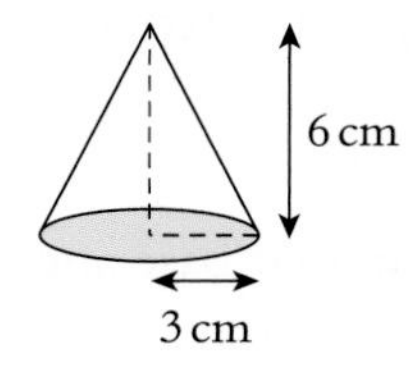

c
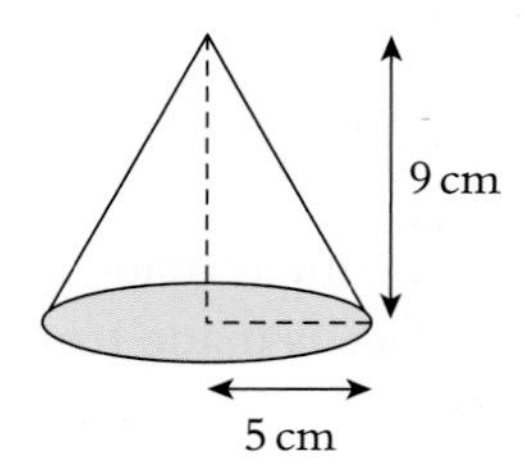

2 A solid cone has a base radius of 4 cm.
The volume of the cone is 100 cm^3.
Calculate the height of the cone.

3 A solid cone has a height of 12 cm.
The volume of the cone is 250 cm^3.
Calculate the radius of the base of the cone.

4 The diagram shows a toy rocket made from a cylinder and a cone.
Calculate the volume of the toy rocket.

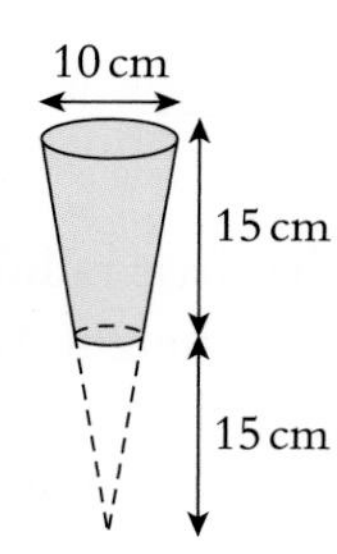

5 A drinking glass is a truncated cone.
Calculate the volume of the drinking glass.

10 cm
15 cm
15 cm

6 Calculate the volume of each of these frustums.

a
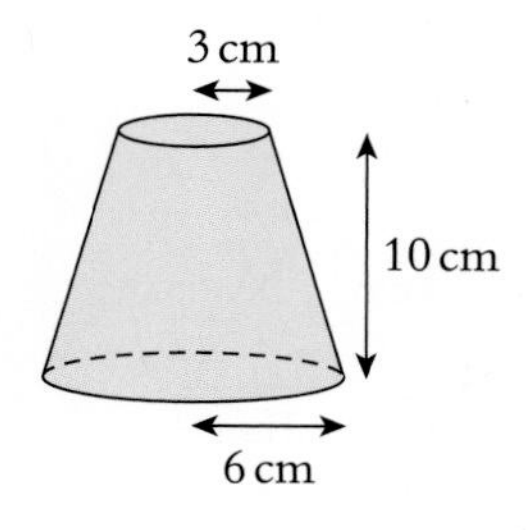

b
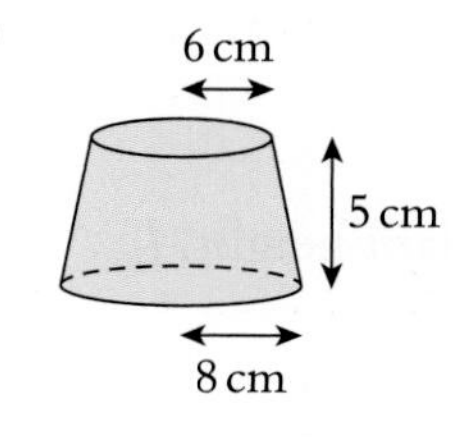

c
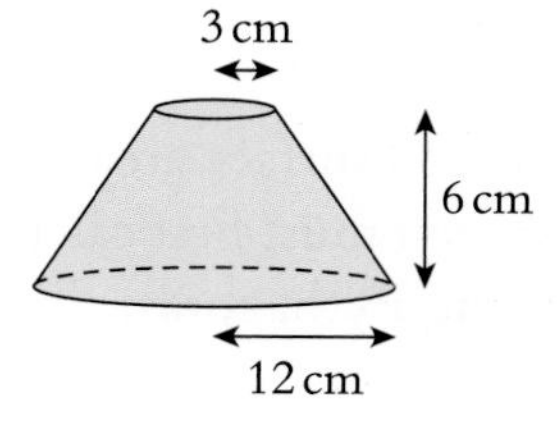

7 The diagram shows a frustum of a cone.
Find an expression in terms of r and h for the volume of the frustum.

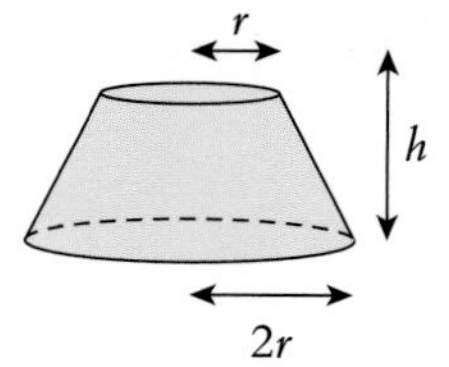

8 Calculate the **curved** surface area of each of these cones.

a

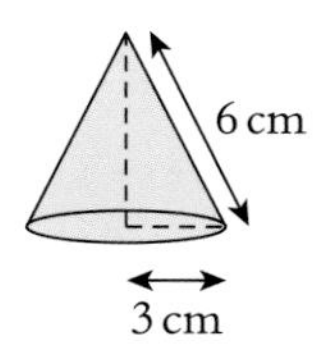

b

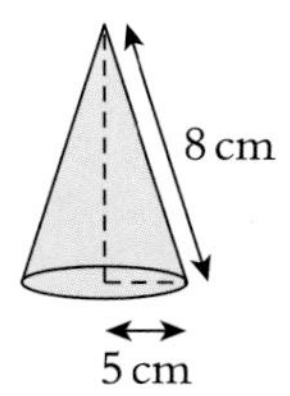

c

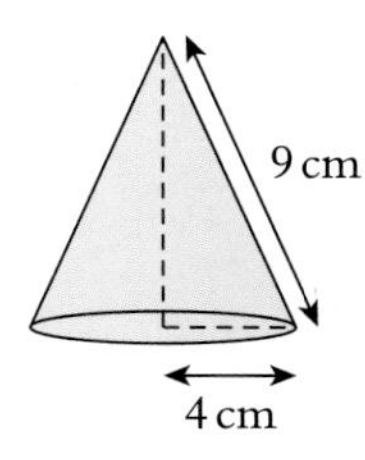

9 Calculate the **total** surface area of each of these cones.

a

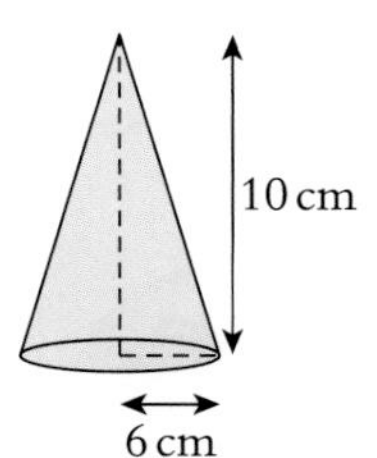

b

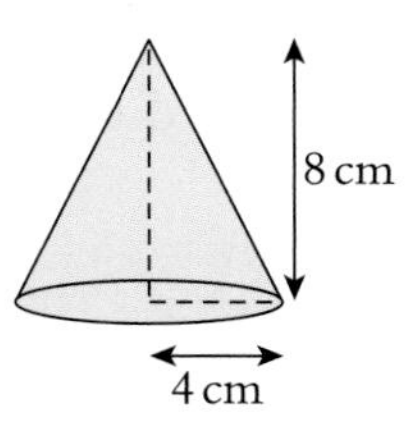

c

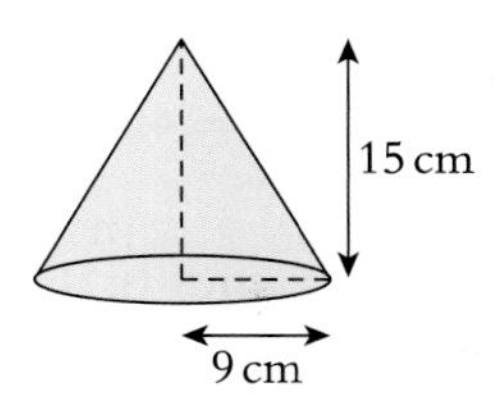

10 Each of these sectors forms the curved surface of a cone.
Calculate the curved surface area of each cone.

a

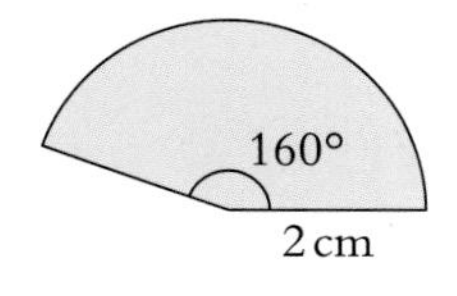

b

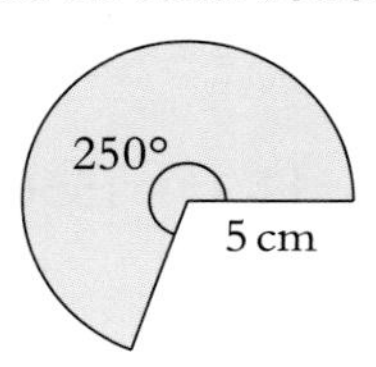

c

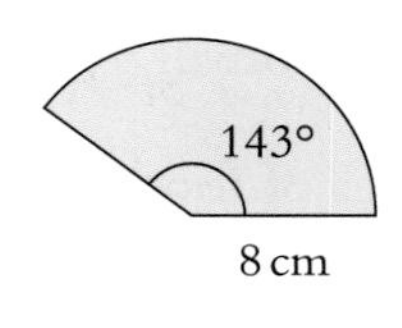

11 The straight edges of each of these sectors are joined together to make a cone.
Calculate: **i** the curved surface area of each cone,
ii the radius of the base of each cone,
iii the height of each cone.

a

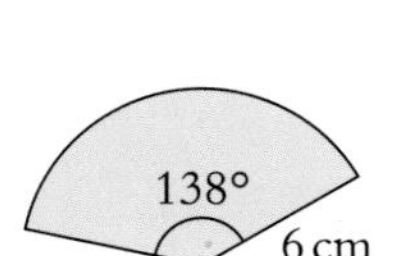

b

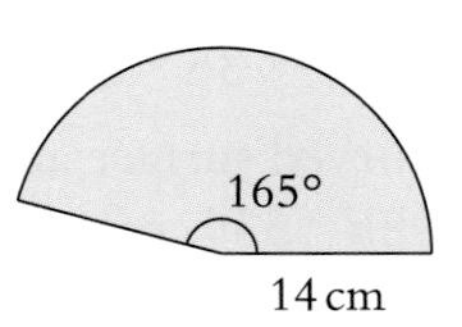

c

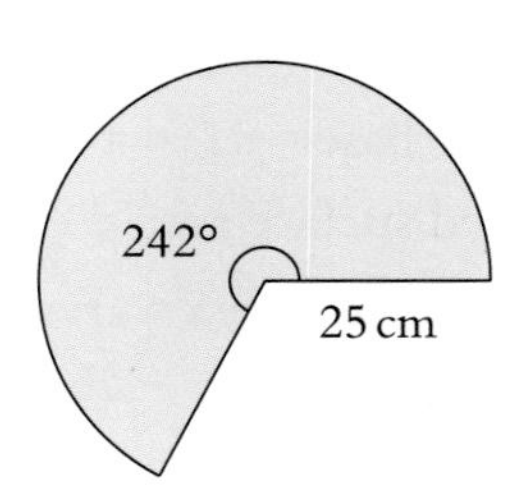

12 Calculate the **total** surface area of each of these solid frustums.

a

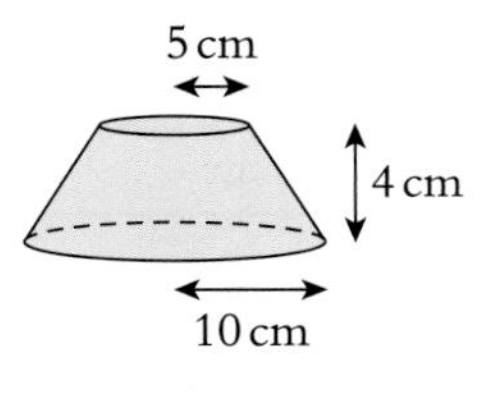

b

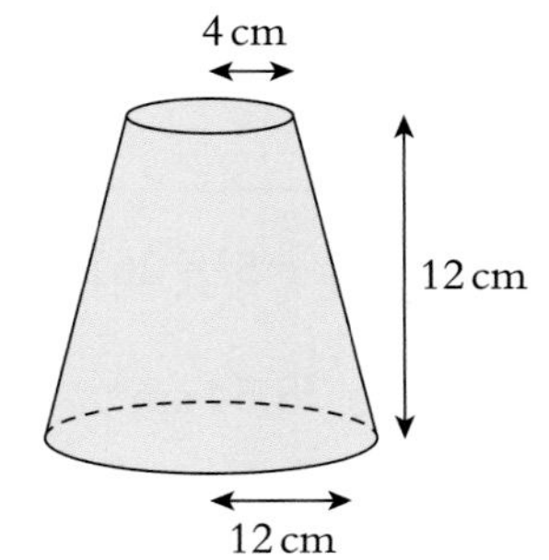

c

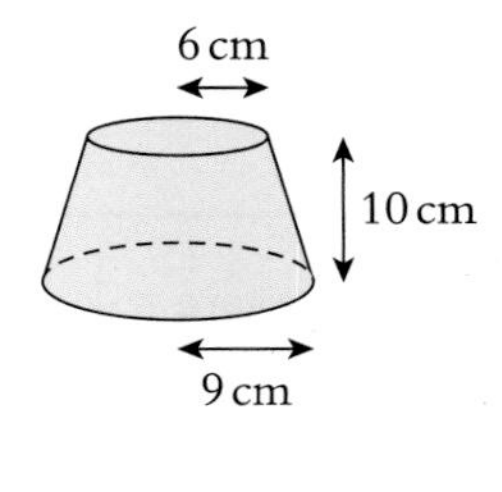

Spheres

r

northern hemisphere

southern hemisphere

Volume of a **sphere** $= \frac{4}{3}\pi r^3$
Surface area of a sphere $= 4\pi r^2$

A **hemisphere** is half a sphere.
The earth is divided into two hemispheres.

EXAMPLE

A football has a radius of 11.1 cm.
Calculate:

a the volume

b the surface area of the football.

a Volume of football $= \frac{4}{3}\pi r^3 = \frac{4}{3} \times \pi \times 11.1^3 = 5730\,\text{cm}^3$ (to 3 s.f.)

b Surface area of football $= 4\pi r^2 = 4 \times \pi \times 11.1^2 = 1550\,\text{cm}^2$ (to 3 s.f.)

EXAMPLE

A solid hemisphere has radius 6 cm.
Calculate:

a the volume

b the total surface area of the hemisphere.

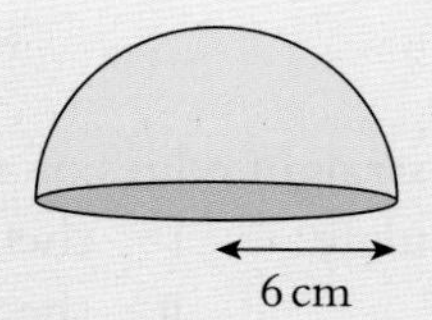

a Volume $= \frac{1}{2} \times \frac{4}{3}\pi r^3 = \frac{1}{2} \times \frac{4}{3} \times \pi \times 6^3 = 452\,\text{cm}^3$

b Area of base $= \pi r^2 = \pi \times 6^2 = 36\pi$
Curved surface area $= \frac{1}{2} \times 4\pi r^2 = \frac{1}{2} \times 4\pi \times 36 = 72\pi$
Total surface area = area of base + curved surface area
$= 36\pi + 72\pi = 108\pi = 339\,\text{cm}^2$

EXAMPLE

The diagram shows a toy that is made from a cone on top of a hemisphere.
Calculate the volume of the toy.

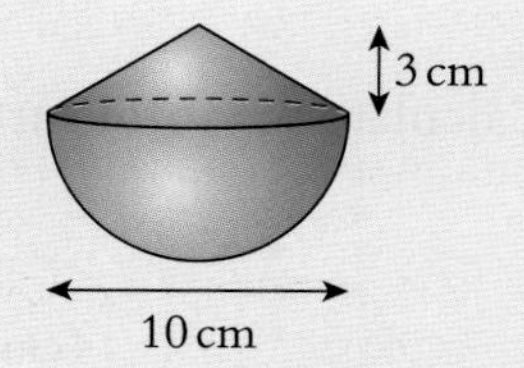

Volume of hemisphere $= \frac{1}{2} \times \frac{4}{3}\pi r^3 = \frac{1}{2} \times \frac{4}{3}\pi \times 5^3 = 261.8\,\text{cm}^3$

Volume of cone $= \frac{1}{3}\pi r^2 h = \frac{1}{3} \times \pi \times 5^2 \times 3 = 78.54\,\text{cm}^3$

Volume of toy $= 261.8 + 78.54 = 340\,\text{cm}^3$ (to 3 s.f.)

EXERCISE 5.13

1 Find the volume of each of these spheres.

a

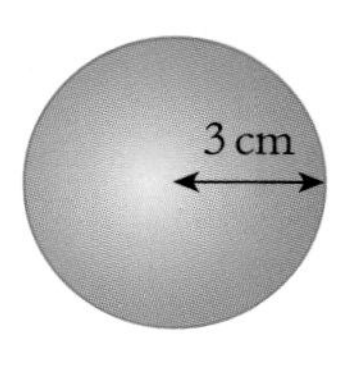

b

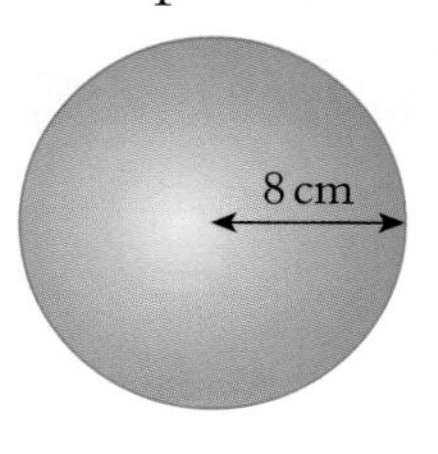

c

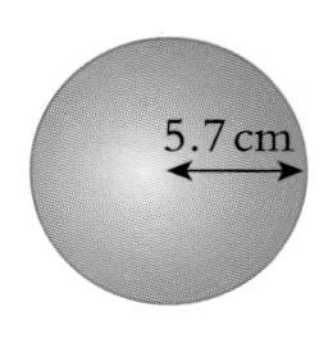

2 Find the surface area of each sphere in question 1.

3 Find the volume of each of these solid hemispheres.

a

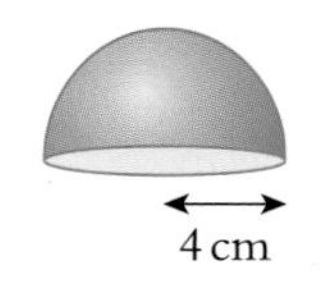

b

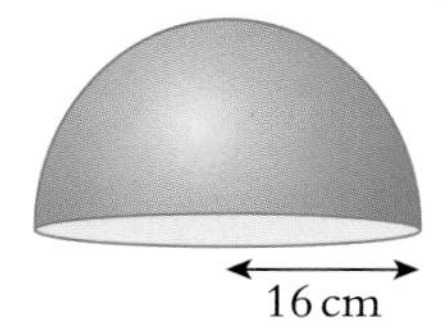

c

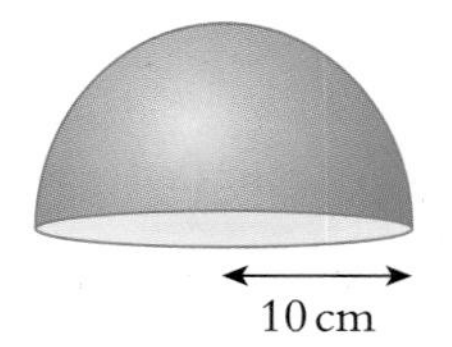

4 Find the total surface area of each of the solid hemispheres in question 3.

5 The volume of a sphere is 432 cm^3. Calculate the radius of the sphere.

6 The surface area of a sphere is 500 cm^2. Calculate the radius of the sphere.

7 A cylindrical tube contains two tennis balls.
The radius of each tennis ball is 3.75 cm.
The tennis balls touch the sides, top and bottom of the tube.
Calculate the volume of empty space inside the tube.

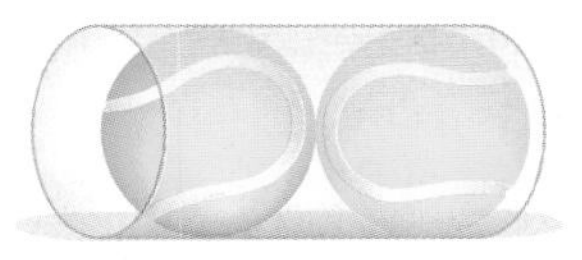

8 A cricket ball is dropped into a cylindrical tank of water.
The ball sinks to the bottom of the tank.
The diameter of the cricket ball is 7.2 cm
The diameter of the cylindrical tank is 15 cm.
Calculate the increase in height of the water in the tank.

9 A solid metal cube of side 8 cm is melted and then recast to make a solid sphere.
Calculate the radius of the sphere.

10 Calculate the volume of these solid shapes.

a

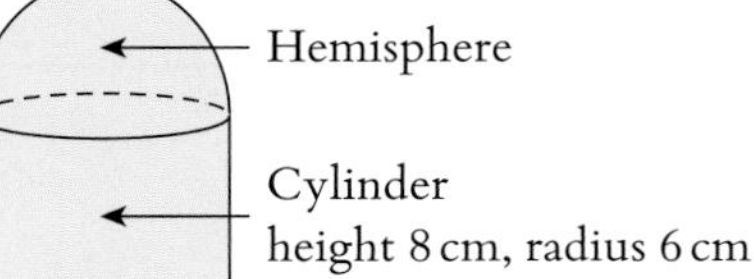

b

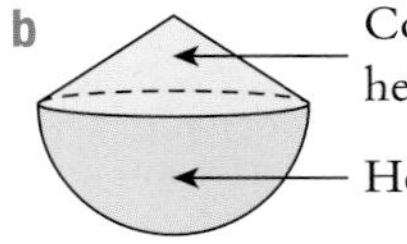

11 Calculate the total surface area of the solid shapes in question 10.

> **KEY WORDS**
> pyramid
> truncated pyramid
> regular octahedron
> regular tetrahedron
> cone, frustum, sphere
> hemisphere

UNIT 5

Areas of similar shapes

THIS SECTION WILL SHOW YOU HOW TO:

- Find lengths and areas of similar shapes using scale factors

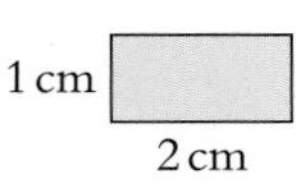

Area = $2\,\text{cm}^2$

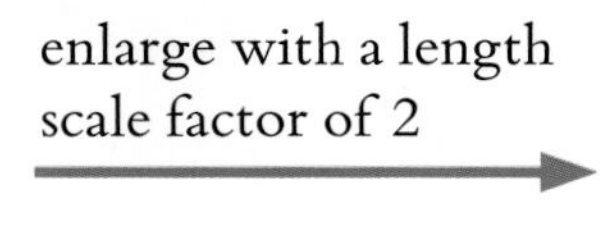

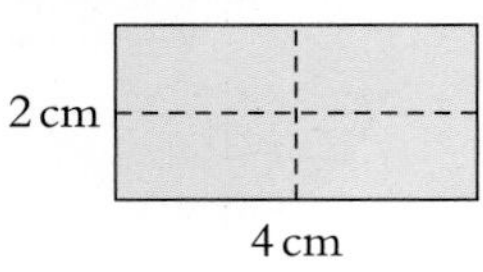

Area = $8\,\text{cm}^2$

The diagram shows that if the **length scale factor** is 2, then the **area scale factor** is 4.

1 cm

2 cm

Area = $2\,\text{cm}^2$

enlarge with a length scale factor of 3

3 cm

6 cm

Area = $18\,\text{cm}^2$

The diagram shows that if the length scale factor is 3, then the area scale factor is 9.

General rule: If the length scale factor is k, then the area scale factor is k^2.

EXAMPLE

The two quadrilaterals are similar.
The area of the smaller quadrilateral is $5\,\text{cm}^2$.
Find the area of the larger quadrilateral.

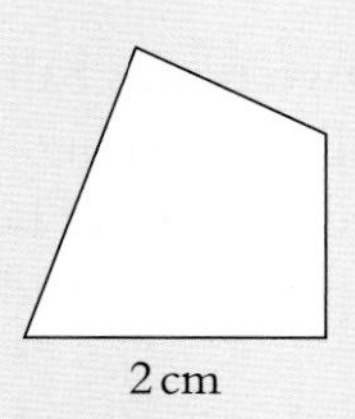

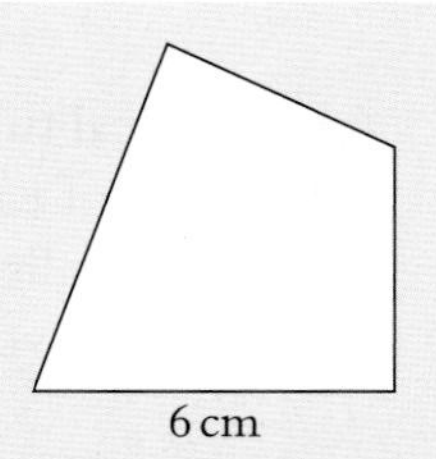

The length scale factor $k = \frac{6}{2} = 3$.

The area scale factor $k^2 = 3^2$.

So the area of the larger quadrilateral $= 3^2 \times 5 = 45\,\text{cm}^2$.

EXAMPLE

The two triangles are similar.
The area of the larger triangle is $10.8\,\text{cm}^2$.
Find the area of the smaller triangle.

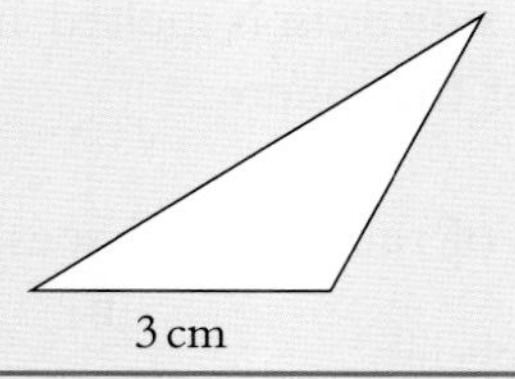

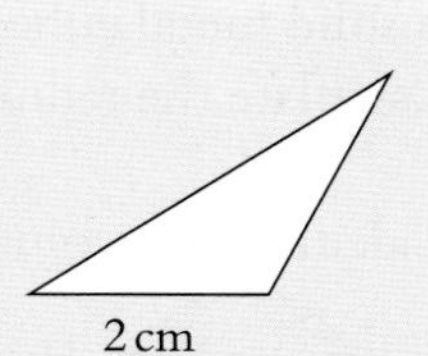

The length scale factor $k = \frac{2}{3}$

The area scale factor $k^2 = \left(\frac{2}{3}\right)^2$

So the area of the smaller triangle $= \left(\frac{2}{3}\right)^2 \times 10.8 = 4.8\,\text{cm}^2$

EXERCISE 5.16

In questions 1 to 10 each pair of objects are similar.
The volume of each object is written inside the object.
Find the value of x for each question.

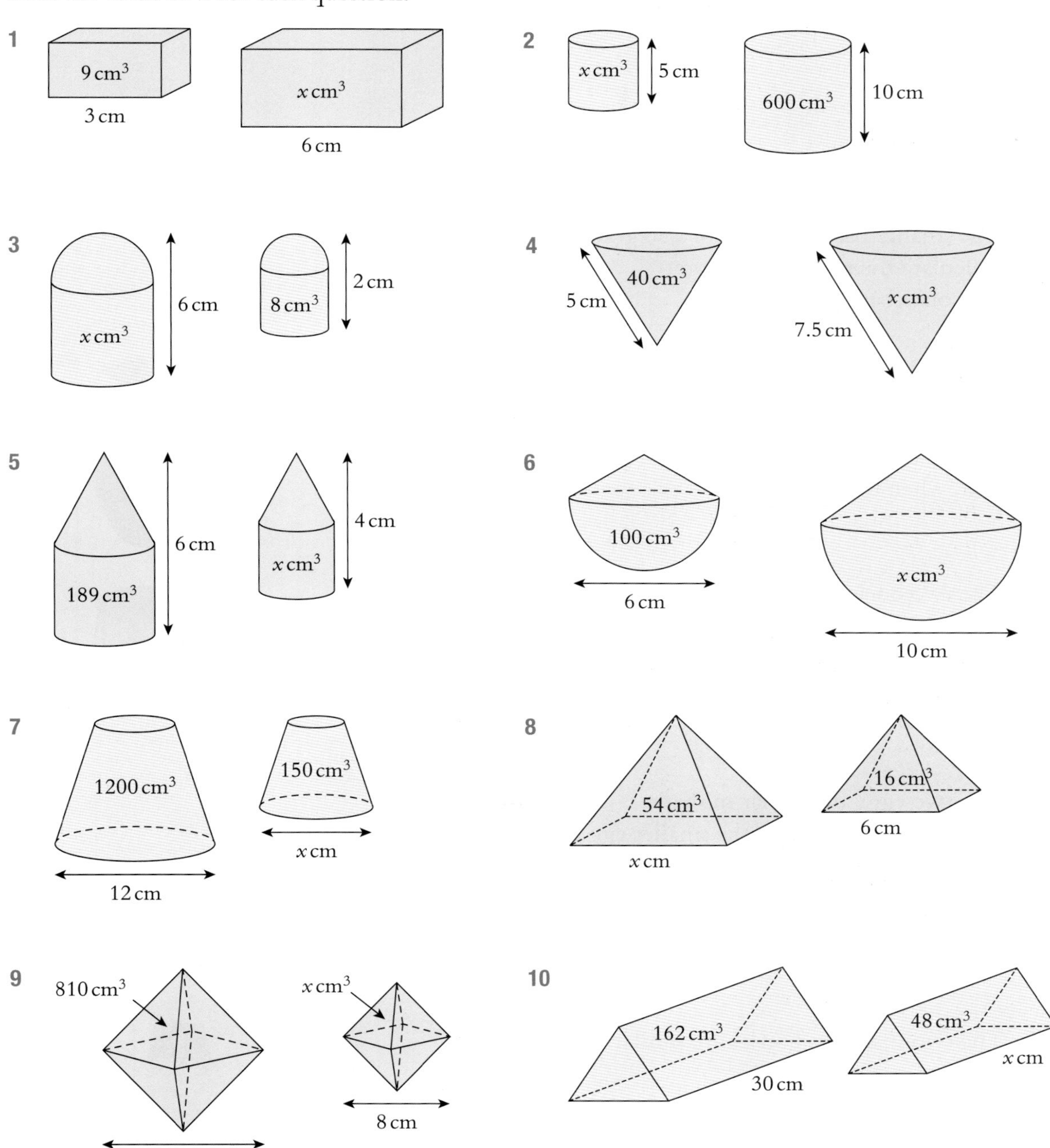

11 The two bottles are similar in shape.
The larger bottle holds 100 mℓ of perfume.
Calculate how many millilitres of perfume the smaller bottle holds.

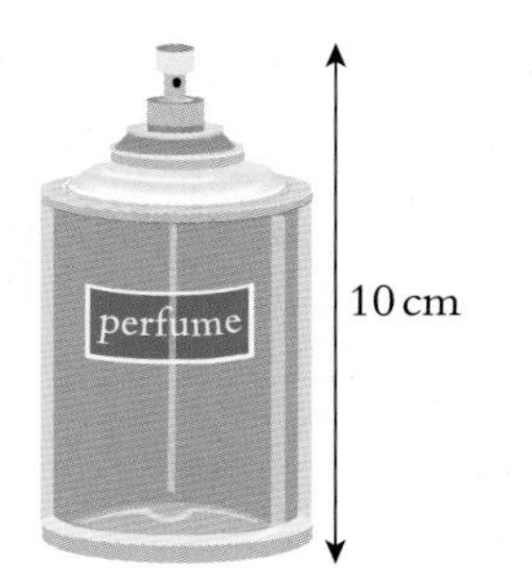

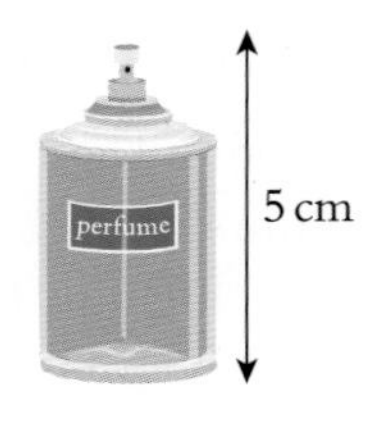

12 The two bars of chocolate are similar in shape.
The smaller bar has a mass of 150 grams.
Calculate the mass of the larger chocolate bar.

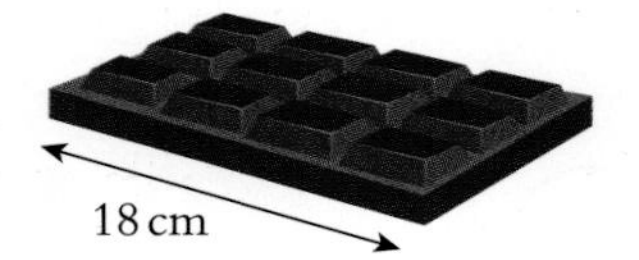

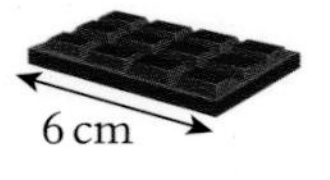

13 The two vases are similar in shape.
The largest vase holds 1.5 litres of water when full.
How much water does the smaller vase hold when full?

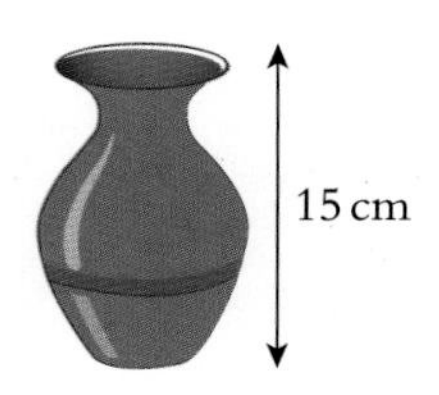

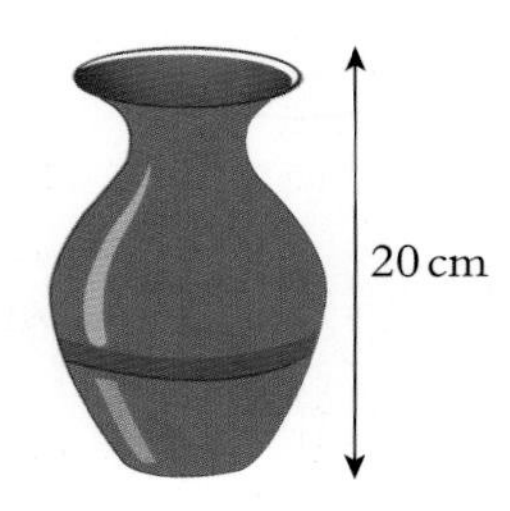

14 The two buckets are similar in shape.
The larger bucket holds 10 litres of water when full.
How much water does the smaller bucket hold when full?

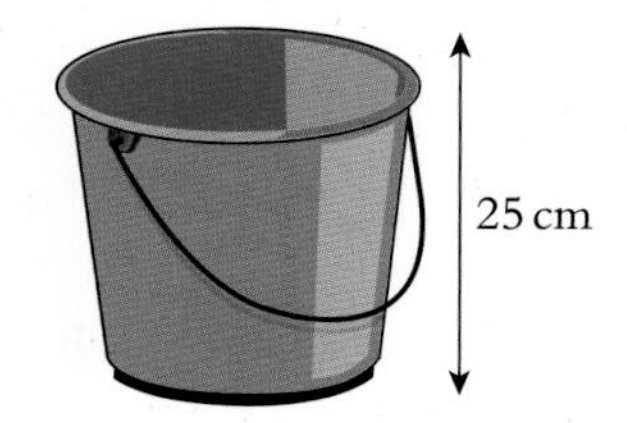

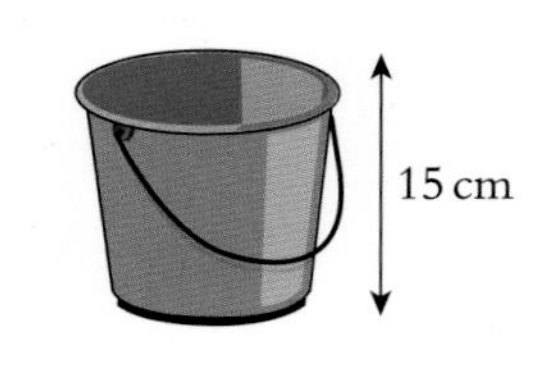

15 The two eggs are similar in shape.
Calculate the length of the smaller egg.

mass = 50 g

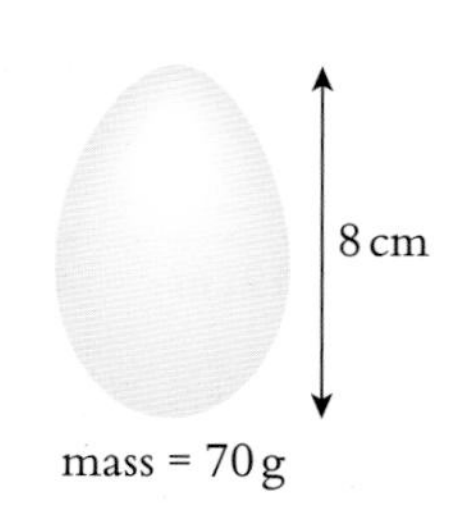

mass = 70 g

16 The two shampoo bottles are similar in shape.
Calculate the height of the larger bottle.

500 ml

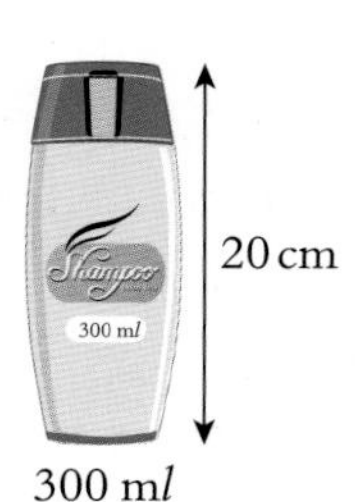

300 ml

17 A cylindrical cup has a height of 6 cm and a base radius of 4 cm.
A large cylindrical jug is a similar shape to the cup.
The jug is full of mango juice.
The cup can be filled with mango juice from the jug exactly 125 times.
Calculate the height and radius of the jug.

18 Two vases are similar in shape.
The ratio of their heights is 3:2

a The volume of the larger vase is 810 cm^3.
Calculate the volume of the smaller vase.

b The surface area of the smaller vase is 240 cm^2.
Calculate the surface area of the larger vase.

19 Two similar objects have a surface area of 4 cm^2 and 9 cm^2.
The mass of the smaller object is 200 grams.
Find the mass of the larger object.

20 Two similar cones have surface areas in the ratio 9:16

a Find the ratio of their lengths.

b Find the ratio of their volumes.

21 The two Russian dolls are similar.

a The surface area of the larger doll is 150 cm^2.
Calculate the surface area of the smaller doll.

b The volume of the larger doll is 120 cm^3.
Calculate the volume of the smaller doll.

22 A model plane is made on a scale of 1:50
Copy and complete the table.

	Real plane	Model plane
Wingspan	11.23 m	22.46 cm
Length of plane	9.12 m	.... cm
Wing area	22.48 m^2	.... cm^2

23 A model car is made on a scale of 1:40
Copy and complete the table.

	Real car	Model car
Length of car	3.723 m	.... cm
Area of windscreen	0.4835 m^2	.... cm^2
Storage space in boot	160 litres	.... m*l*

KEY WORDS
volume scale factor

Grouped frequency tables

THIS SECTION WILL SHOW YOU HOW TO

- Find the modal group from a grouped frequency table
- Estimate the mean from a grouped frequency table

Large sets of data are often shown in a **grouped frequency table.**
The **modal class** is the group with the highest frequency.
In a grouped frequency table you do not know the individual data values so you can only estimate the value of the mean.
The **mid-value** of the groups is used to estimate the sum of the figures in each group.
For example, if you have 5 people whose heights, h cm, are in the group $164 < h \leq 168$ you can estimate the sum of their heights by calculating $5 \times 166 = 830$ cm.

EXAMPLE

The table shows the time, t minutes, taken by 100 people to complete a Sudoku puzzle.

a Write down the modal class.

b Calculate an estimate of the mean time taken.

Time (t minutes)	Frequency
$0 < t \leq 10$	20
$10 < t \leq 20$	32
$20 < t \leq 30$	41
$30 < t \leq 40$	5
$40 < t \leq 50$	2

a The highest frequency is 41, so the modal class is $20 < t \leq 30$.

b The mid-value of the group $0 < t \leq 10$ is 5, the mid-value of the group $10 < t \leq 20$ is 15 etc.
Two extra columns need to be added to the table for the calculations.

➡ **NOTE:** to find the mid-value of the group $10 < t \leq 20$ you calculate $\frac{10+20}{2} = \frac{30}{2} = 15$

Time (t minutes)	Frequency (f)	mid-value (x)	($f \times x$)
$0 < t \leq 10$	20	5	$20 \times 5 = 100$
$10 < t \leq 20$	32	15	$32 \times 15 = 480$
$20 < t \leq 30$	41	25	$41 \times 25 = 1025$
$30 < t \leq 40$	5	35	$5 \times 35 = 175$
$40 < t \leq 50$	2	45	$2 \times 45 = 90$
	Total = 100		**Total = 1870**

$$\text{The estimated mean time} = \frac{\text{total time}}{\text{total number of people}}$$

$$= \frac{1870}{100} = 18.7 \text{ minutes}$$

EXERCISE 6.1

1 In 2006 the population of Singapore was 4.5×10^6.
This was 50% greater than it was in 1990.
Calculate the population of Singapore in 1990.

2 Carla makes a profit of 15% when she sells a mobile phone for \$92.
Calculate how much Carla paid for the mobile phone.

3 In 2015 a train ticket cost \$22.
This is 10% more than the cost of the ticket in 2010.
Calculate the price of the ticket in 2010.

4 Sunita pays \$240 for a bike in a sale.
The original price had been reduced by 20%.
Calculate the original price of the bike.

5 Gregor travels to work by train.
One day, the train is slow and the journey takes 42 minutes.
This is 10% more than the normal time taken.
Calculate how long the journey normally takes.

6 The rent for a house is increased by 20% to \$19 800
Calculate the original rent.

7 The price of bananas is reduced by 25% to \$1.74 per kilogram.
Calculate the original price.

8 The cost of a holiday is reduced by 15% to \$3009.
Calculate the original cost of the holiday.

9 The cost of a cinema ticket is increased by 12% to \$9.52
Calculate the original cost of the cinema ticket.

10 The cost of renting a car is reduced by 4% to \$124.80
Calculate the original cost of renting the car.

11 The price of gold increased by 2.5% to \$45.92 per gram.
Calculate the original price.

12 The temperature decreased by 25% to 16.2 °C.
Calculate the original temperature.

13 Zoe paid \$5628 tax in 2015.
This was 5% more than the amount of tax that she paid in 2014.
Calculate the amount of tax that Zoe paid in 2014.

Simple interest

With **simple interest**, the interest earned is not reinvested.
This means that the amount of interest earned each year is unchanged.

EXAMPLE

Richard invests \$400 at a rate of 5% per year simple interest.
Calculate the amount Richard has after 3 years.

Interest earned in first year = 5% of \$400 = 0.05×400 = \$20
Interest earned in 3 years = 3×20 = \$60
Value of investment after 3 years = \$400 + \$60 = \$460

Compound interest

With **compound interest**, the interest earned each year is reinvested.
This means that the amount of interest earned each year increases.
The following example shows three methods for finding the final value of an investment.

EXAMPLE

Richard invests \$400 at a rate of 5% per year compound interest.
Calculate the amount Richard has after 3 years.

Method 1

Interest in first year = 0.05×400 = \$20
Value of investment after 1 year = 400 + 20 = \$420

Interest in second year = 0.05×420 = \$21
Value of investment after 2 years = 420 + 21 = \$441

Interest in third year = 0.05×441 = \$22.05
Value of investment after 3 years = 441 + 22.05 = \$463.05

Method 2 (Quicker method)

Value of investment after 1 year = 400×1.05 = \$420 *multiplying factor is 1.05 (= 1 + 0.05)*
Value of investment after 2 years = 420×1.05 = \$441
Value of investment after 3 years = 441×1.05 = \$463.05

Method 3 (Quickest method!)

Value of investment after 3 years = 400×1.05^3 = \$463.05

EXAMPLE

Simone invests \$200 at a rate of 4% per year compound interest.
Calculate the amount Simone has after 5 years.

Value of investment after 5 years
= 200×1.04^5
= \$243.33 (to the nearest cent)

EXERCISE 6.2

1 Amelia invests \$80 at a rate of 15% per year, simple interest.
Calculate the amount of interest Amelia earns in 5 years.

2 Karl invests \$700 at a rate of 2% per year, simple interest.
Calculate the total amount Karl has after 4 years.

3 Barbara invests \$250 at a rate of 6% per year, simple interest.
Calculate the total amount Barbara has after 3 years.

4 Sebastian invests \$450 at a rate of 3.5% per year, simple interest.
Calculate the total amount Sebastian has after 2 years.

5 Jack invests \$500 at a rate of 3% per year, compound interest.
Calculate the total amount Jack has after 2 years.

6 Omar invests \$2000 at a rate of 2.5% per year, compound interest.
Calculate the amount Omar has after 3 years.

7 Eduard invests \$700 at a rate of 2% per year, compound interest.
Calculate the amount Eduard has after 4 years.

8 Anelie invests \$400 at a rate of 5% per year, compound interest.
Calculate the amount of interest Anelie earns in 3 years.

9 Ferdinand invests \$50 at a rate of r % per year, simple interest.
After 4 years he has a total amount of \$56.
Calculate the value of r.

10 Hanna invests \$400 at a rate of r % per year, simple interest.
After 5 years she has a total amount of \$700.
Calculate the value of r.

11 Ferdinand invests \$720 at a rate of 5% per year, compound interest.
How many years will it be before his investment is worth more than \$1000?

12 Robert bought his car 4 years ago.
Each year the value of the car has depreciated by 15%.
The car is now worth \$6264.
Calculate how much the car was worth when Robert bought it.

13 In 2009 the manatee population in Florida was estimated at 3800.
The population is declining at a rate of approximately 1.1% per year.
Estimate what the manatee population will be in the year in 2020.

KEY WORDS
reverse percentage
simple interest
compound interest

Speed, distance and time

THIS SECTION WILL SHOW YOU HOW TO

- Calculate speed, distance and time

Time

You need to be able to convert a time in minutes to a fraction (or decimal) of an hour.
To do this you must divide by 60.
You also need to be able to change a fraction (or decimal) of an hour into minutes.
To do this you must multiply by 60.

EXAMPLE

a Write 21 minutes as a fraction of an hour.
b Write 36 minutes as a decimal of an hour.

a $21 \text{ minutes} = \frac{21}{60} \text{ hour}$

$= \frac{7}{20} \text{ hour}$

b $36 \text{ minutes} = \frac{36}{60} \text{ hour}$

$= 0.6 \text{ hours}$

EXAMPLE

a Change $\frac{5}{12}$ hour into minutes.
b Change 0.9 hours into minutes.

a $\frac{5}{12} \text{ hour} = \frac{5}{12} \times 60 \text{ minutes}$

$= 25 \text{ minutes}$

b $0.9 \text{ hours} = 0.9 \times 60 \text{ minutes}$

$= 54 \text{ minutes}$

Speed

Speed is a measure of how fast an object is travelling.
When the speed is constant, the formula connecting speed, distance and time is

$$\text{speed} = \frac{\text{distance}}{\text{time}}$$

If the distance is measured in metres and the time is measured in seconds then the speed is measured in metres per second (m/s or ms^{-1}).
If the distance is measured in kilometres and the time is measured in hours then the speed is measured in kilometres per hour (km/h or kmh^{-1}).

EXAMPLE

A train takes 2 hours 55 minutes to travels 350 km at a constant speed.
Find the speed of the train in km/h.

$\text{speed} = \frac{\text{distance}}{\text{time}} = \frac{350}{2\frac{11}{12}}$

$= 120 \text{ km/h}$

➡ **NOTE:** you must change the 55 minutes into hours.

$55 \text{ minutes} = \frac{55}{60} \text{ hour} = \frac{11}{12} \text{ hour}$

So, $2 \text{ hours } 55 \text{ minutes} = 2\frac{11}{12} \text{ hour}$

EXERCISE 6.5

1 List these sets.

a {square numbers less than 50}

b {prime numbers less than 25}

c {months of the year}

d {factors of 12}

e {prime factors of 25}

f {even numbers less than 15}

g {odd numbers between 20 and 30}

h {prime numbers between 150 and 200}

2 List these sets.

a {factors of 49}

b {cube numbers between 100 and 200}

c {prime factors of 100}

d {factors of 20}

e {prime factors of 36}

f {months of the year beginning with J}

3 Describe these sets by a rule.

a {1, 4, 9, 16}

b {1, 3, 5, 7, ...}

c {March, May}

d {4, 8, 12, 16, 20, 24, ...}

e {1, 8, 27, 64, ...}

f {red, blue, green, yellow, ...}

4 Describe these sets by a rule.

a {2, 3, 5, 7, 11, ...}

b {1, 2, 4, 5, 10, 20}

c {2, 4, 8, 16, 32}

d {(3, 4, 5), (6, 8, 10), (5, 12, 13), ...}

5 Which of these statements are true?

a $y = 2x + 1 \notin$ {linear graphs}

b $y = x^3 + 1 \in$ {non-linear graphs}

c $1 \in$ {prime numbers}

d parallelogram $\in$ {triangles}

e kite $\notin$ {quadrilaterals}

f trapezium $\notin$ polygons

6 Which of these statements are true?

a $5 \in$ {factors of 56}

b $3 \in$ {factors of 4350}

c $2 \notin$ {factors of $2x^2 - 4x$}

d $x \in$ {factors $x^2 - 3x$}

7 Which of these statements are true?

a $x - 2 \in$ {factors of $x^2 - 2x$}

b $x + 2 \in$ {factors of $x^2 + 2x - 8$}

c $x + 5 \notin$ {factors of $x^2 - 25$}

d $3x + 2 \in$ {factors of $6x^2 + x - 2$}

8 Which of these statements are true?

a {odd numbers divisible by 2} = $\varnothing$

b {even numbers divisible by 3} = $\varnothing$

c {prime numbers between 48 and 52} = $\varnothing$

d {months of the year beginning with the letter B} = $\varnothing$

e {common factors of 23 and 29} = $\varnothing$

f {answers to the equation $x^2 = -4$} = $\varnothing$

Venn diagrams

The **complement** of set A is the set of all elements not in A.
The complement of set A is denoted by A'.

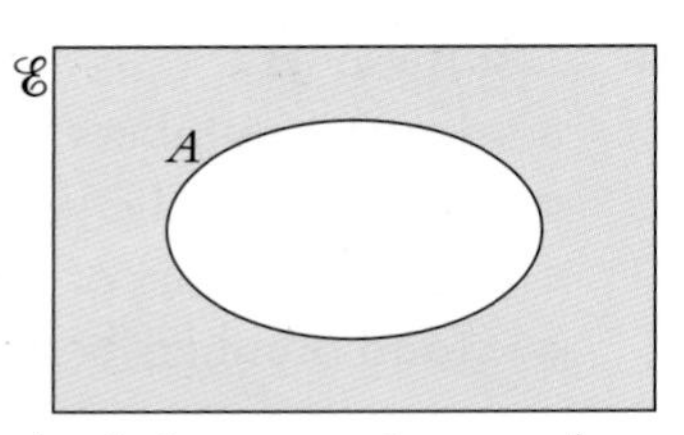

Shaded region shows A'

The **intersection** of A and B is the set of elements that are in both A and B.
The intersection of A and B is denoted by $A \cap B$.

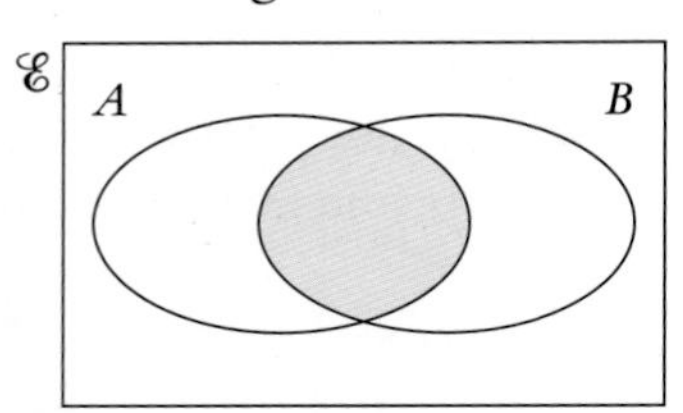

Shaded region shows $A \cap B$

The **union** of A and B is the set of elements that are in A or B or both.
The union of A and B is denoted by $A \cup B$.

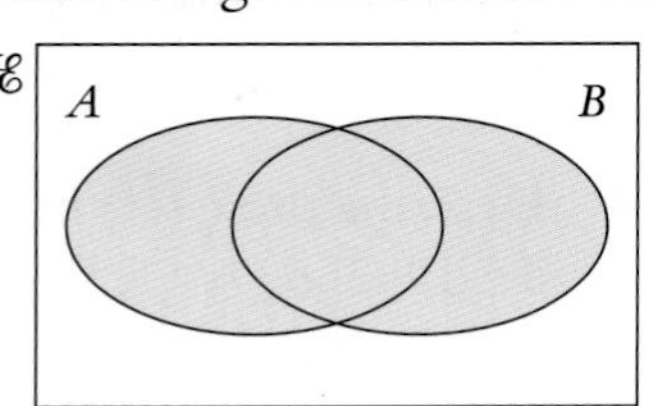

Shaded region shows $A \cup B$

Representing sets on a Venn diagram
When the expressions are more complicated you many need to use some diagrams for your working before deciding on your answer.

EXAMPLE

On a Venn diagram shade the regions **a** $A \cap B'$ **b** $A \cup B'$

Working

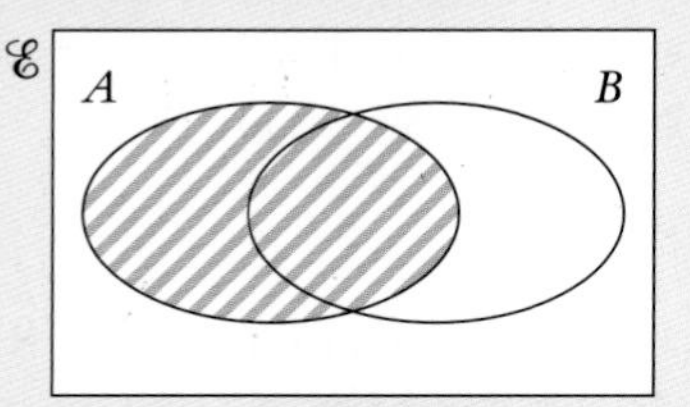

shade set A in one direction

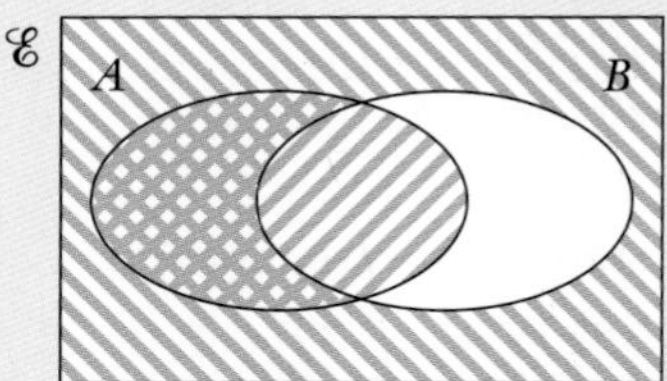

shade set B' in the opposite direction

a $A \cap B'$ is the region that is in both A and B' so you need the region with double shading.

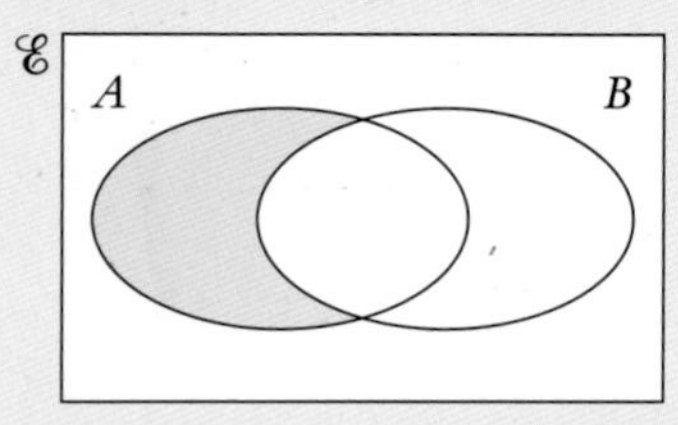

shaded region shows $A \cap B'$

b $A \cup B'$ is the region that is in A or B' or both so you need all of the shaded regions.

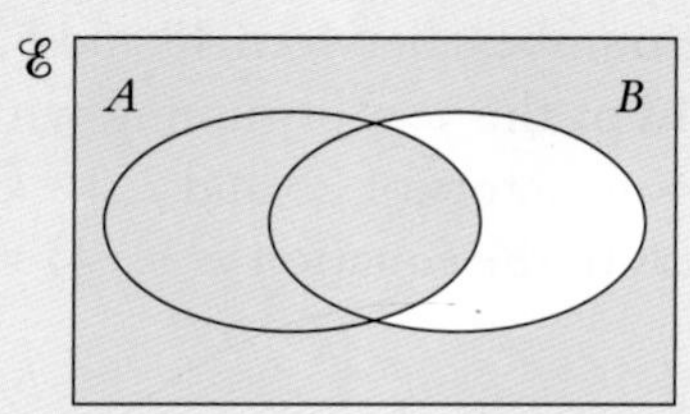

shaded region shows $A \cup B'$

EXERCISE 6.6

1 On copies of this diagram shade the following regions.

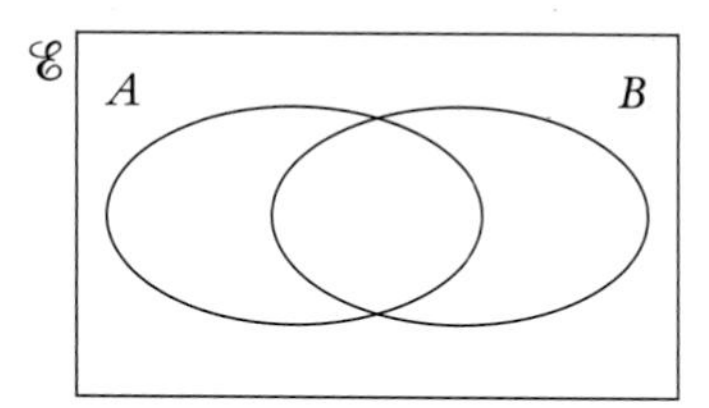

a B' b $A \cup B$
c $A \cap B$ d $A' \cap B$
e $A' \cap B'$ f $(A' \cup B)'$
g $(A \cap B') \cup (A' \cap B)$ h $(A \cap B) \cup (A \cup B)'$

If $\mathscr{E} = \{1, 2, 3, 4, 5\}$, $A = \{1, 3, 5\}$ and $B = \{3, 5\}$ then B is called a **proper subset** of A.
This is written as $B \subset A$.
On a Venn diagram the set B is drawn inside the set A.

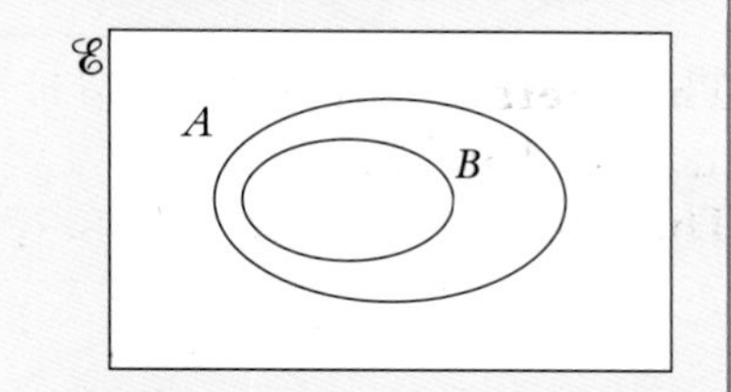

2 On copies of this diagram shade the following regions.

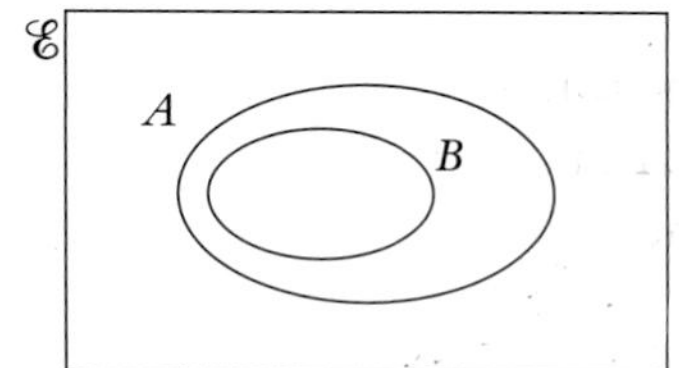

a A' b $A \cup B$ c $A \cap B$
d $A' \cap B$ e $A' \cap B'$ f $(A \cup B)'$
g $A \cup B'$ h $(A' \cup B)'$ i $(A \cap B) \cup (A \cup B)'$

If $\mathscr{E} = \{1, 2, 3, 4, 5\}$, $A = \{1, 2\}$ and $B = \{3, 5\}$ then A and B do not intersect. $A \cap B = \varnothing$.
On a Venn diagram the sets A and B are drawn so that they do not overlap.

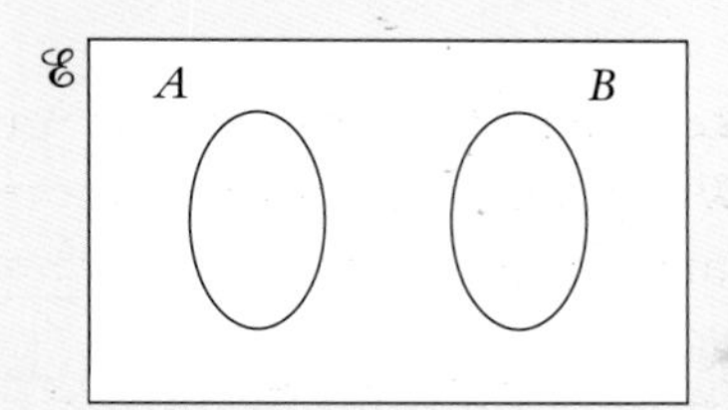

3 On copies of this diagram shade the following regions.

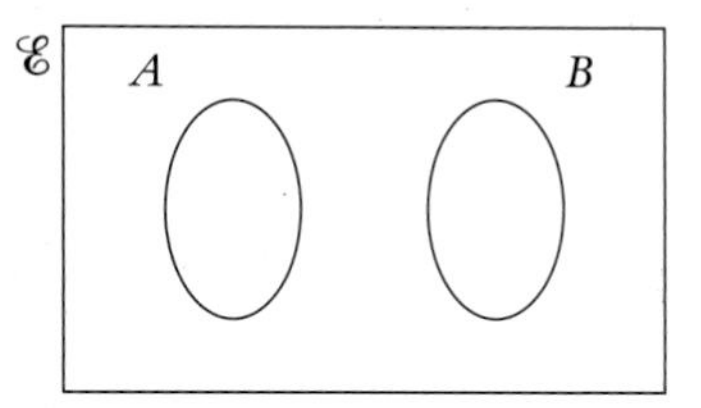

a B' b $A \cup B$ c $(A \cap B)'$
d $A \cap B'$ e $A' \cap B'$ f $(A \cup B)'$

4 On copies of this diagram shade the following regions.

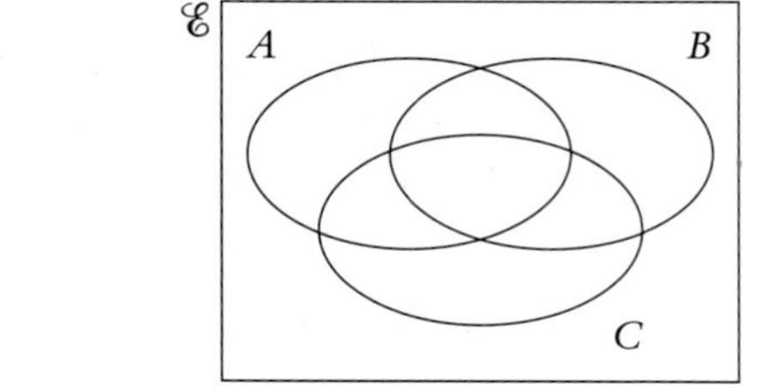

a $A \cap B \cap C$ b $(A \cap B \cap C)'$
c $A \cup B \cup C$ d $(A \cup B \cup C)'$
e $A \cup (B \cap C)$ f $A \cap (B \cup C)$
g $A' \cap B' \cap C'$ h $(A \cap B)' \cap C$
i $(A \cup C)' \cap (A \cup B)'$

You might be given a Venn diagram and then be asked to use set notation to describe the shaded region.

$A \cap B'$

EXAMPLE

Describe the shaded regions using set notation.

a

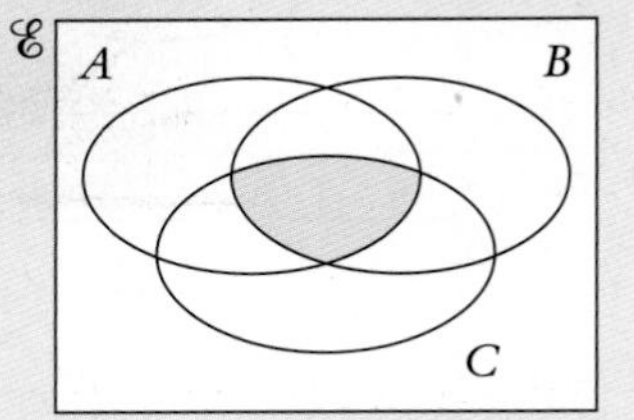

b

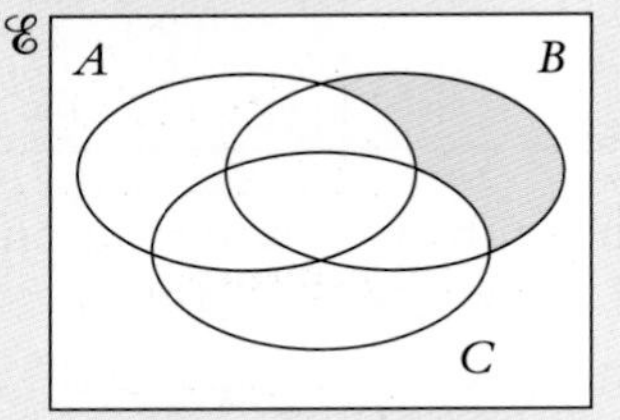

c

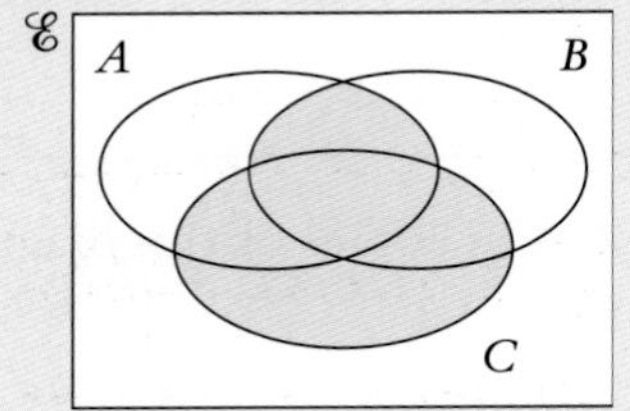

a

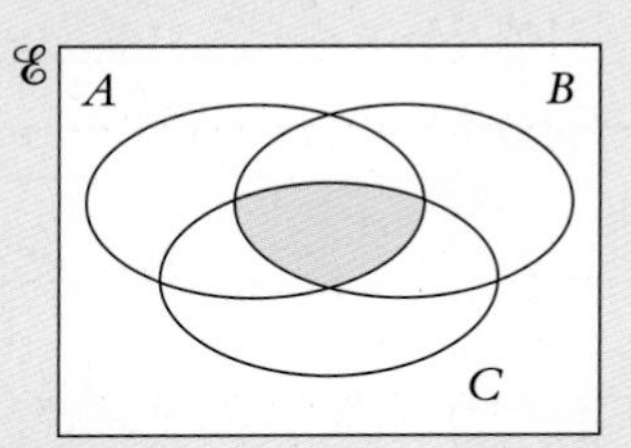

The shaded region is in A, B and C.
The region is $A \cap B \cap C$.

b

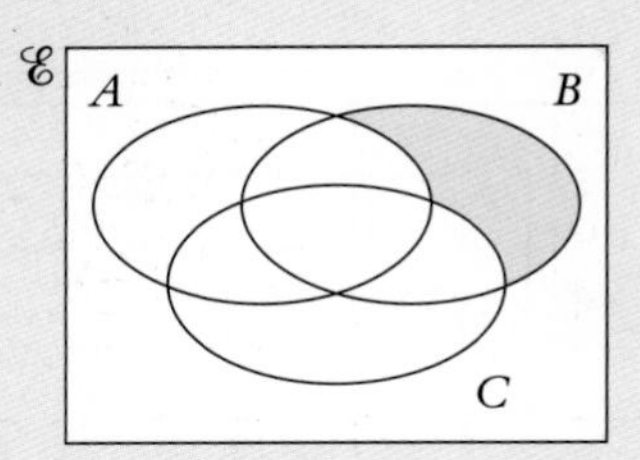

The shaded region is in B and outside $A \cup C$.
The region is $B \cap (A \cup C)'$.

c

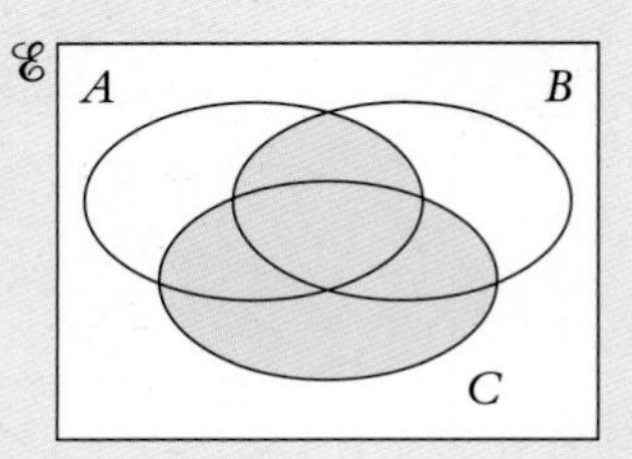

The shaded region contains C and $A \cap B$.
The region is $C \cup (A \cap B)$.

EXERCISE 6.9

1 The Venn diagram shows information about the number of elements in the sets.

For example $n(A) = 5 + 8 = 13$.

Find

a $n(A')$ **b** $n(B)$ **c** $n(B')$

d $n(A \cap B)$ **e** $n(A \cup B)$ **f** $n(A \cup B)'$

g $n(A' \cap B)$ **h** $n(A \cap B')$.

2 $n(A) = 18$, $n(B) = 13$ and $n(A \cap B) = 3$. Draw a Venn diagram to find $n(A \cup B)$.

3 $n(A) = 12$, $n(B) = 11$ and $n(A \cap B) = 4$. Draw a Venn diagram to find $n(A \cup B)$.

4 $n(A) = 35$, $n(B) = 33$ and $n(A \cap B) = 13$. Draw a Venn diagram to find $n(A \cup B)$.

5 $n(A) = 9$, $n(B) = 11$ and $n(A \cup B) = 13$. Draw a Venn diagram to find $n(A \cap B)$.

6 $n(A) = 65$, $n(B) = 57$ and $n(A \cup B) = 114$. Draw a Venn diagram to find $n(A \cap B)$.

7 $n(\mathscr{E}) = 27$, $n(A) = 12$, $n(B) = 9$ and $n(A \cap B) = 4$.

Draw a Venn diagram to find **a** $n(A \cup B)$ **b** $n(A \cup B)'$ **c** $n(A \cap B')$.

8 $n(\mathscr{E}) = 40$, $n(A) = 25$, $n(B) = 22$ and $n(A \cup B)' = 3$.

Draw a Venn diagram to find **a** $n(A \cap B)$ **b** $n(A \cap B')$.

9 $n(\mathscr{E}) = 50$, $n(A) = 26$, $n(B) = 22$ and $n(A \cap B) = 12$.

Draw a Venn diagram to find **a** $n(A \cup B)$ **b** $n(A \cup B)'$ **c** $n(A' \cap B)$.

10 $n(A) = 15$, $n(A \cap B) = 10$ and $n(A' \cap B) = 7$.

Draw a Venn diagram to find **a** $n(B)$ **b** $n(A \cup B)$.

11 $n(A) = 11$, $n(B) = 15$, $n(C) = 10$, $n(A \cap B) = 5$, $n(A \cap C) = 4$, $n(B \cap C) = 7$ and $n(A \cap B \cap C) = 3$.

Draw a Venn diagram to find $n(A \cup B \cup C)$.

12 $n(A) = 16$, $n(B) = 12$, $n(C) = 9$, $n(B \cap C) = 3$, $n(A \cap C) = 5$, $n(A \cup B) = 22$ and $n(A \cap B \cap C) = 2$.

Draw a Venn diagram to find $n(A \cup B \cup C)$.

13 $n(A) = 17$, $n(B) = 13$, $n(C) = 16$, $n(A \cap B) = 10$, $n(A \cap C) = 6$, $n(B \cap C) = 7$, $n(A \cap B \cap C) = 4$ and $n(A \cup B \cup C)' = 8$.

Draw a Venn diagram to find $n(\mathscr{E})$.

EXAMPLE

There are 29 students in a class.
22 study biology (B), 19 study chemistry (C) and 3 study neither biology or chemistry.
How many students study both biology and chemistry?

Method 1
3 students study neither biology or chemistry so the number 3 goes outside the sets B and C.

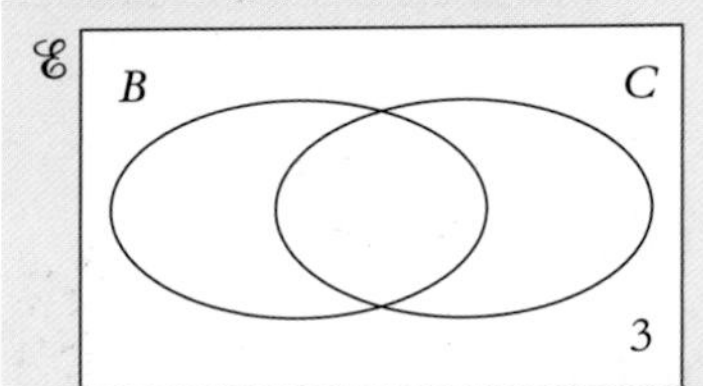

$29 - 3 = 26$ so there are now 26 students left to put on the Venn diagram.
$22 + 19 = 41$ and $41 - 26 = 15$
The number that goes in the intersection is 15 and the remaining numbers can now be filled in

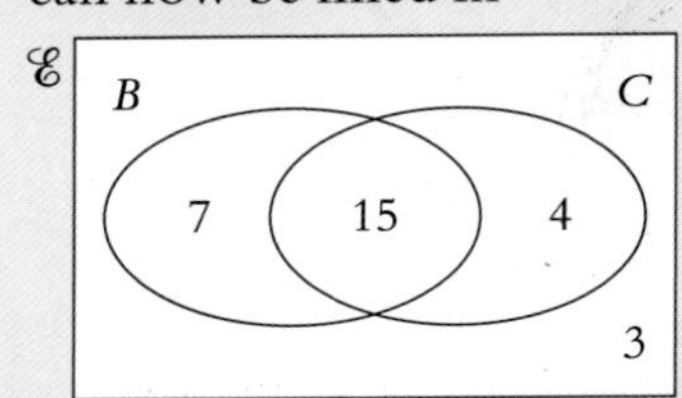

The number of students who study both biology and chemistry is 15.

Method 2 (Using algebra)
The number 3 goes outside the sets B and C on the Venn diagram.
Let x = the number of students that study both biology and chemistry.

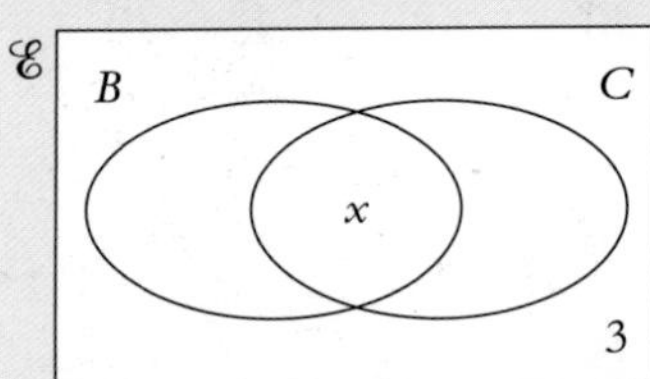

The number of students who study biology and not chemistry is $22 - x$.
The number of students who study chemistry and not biology is $19 - x$.
So the Venn diagram becomes

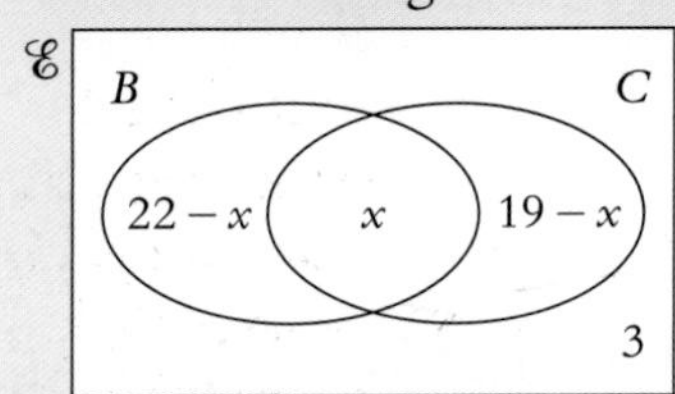

There are a total of 29 students.
So $(22 - x) + x + (19 - x) + 3 = 29$ $\quad 44 - x = 29$ $\quad x = 15$
The number of students who study both biology and chemistry is 15.

UNIT 6

EXERCISE 6.10

1 There are 30 students in a class.
17 study geography (G), 16 study history (H) and 6 study neither geography or history.
Copy and complete the Venn diagram to show this information.
How many students study both geography and history?

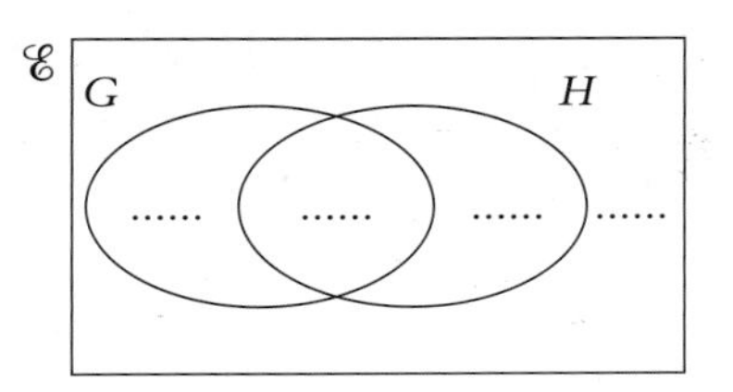

2 There are 34 students in a class.
20 like apple juice, 28 like orange juice and 2 students don't like either drink.
How many students like both apple and orange juice?

3 50 people were asked if they had ever been to Mauritius or the Seychelles on holiday.
11 had been to Mauritius, 7 had been to the Seychelles and 35 had been to neither.
How many had been to both?

4 26 students go on an activity holiday.
15 have sailed before, 8 have rock climbed before and 7 have done neither.
How many have done both before?

5 There are 27 students in a class.
22 have scientific calculators, 7 have graphical calculators and 2 students have neither.
How many students have both a scientific calculator and a graphical calculator?

6 100 students do a mathematics test.
89 had a pair of compasses, 87 had a protractor and 86 had a ruler.
78 students had a ruler and protractor.
81 had a pair of compasses and a protractor.
79 had a pair of compasses and a ruler.
74 students had all three pieces of equipment.
How many students had

a none of the three pieces of equipment,

b only a protractor.

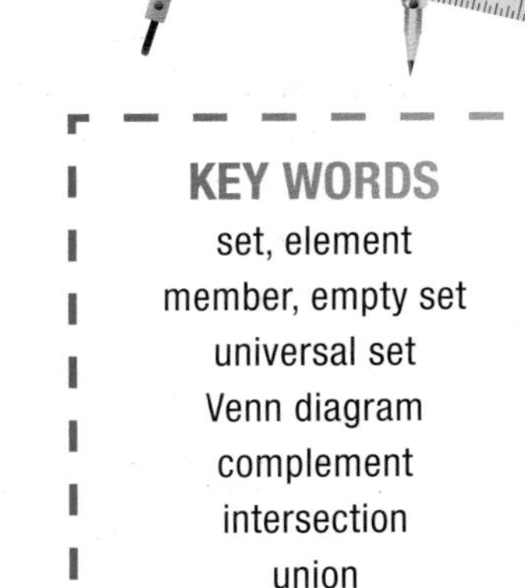

7 A group of students are given three problems to solve.
(Problem A, B and C.)
13 students solve problem A, 17 solve problem B and 7 solve problem C.
9 solve problems A and B.
6 solve problems B and C.
5 solve problems A and C.
4 solve all three problems.
2 solve none of the three problems.
How many students are there altogether.

KEY WORDS
set, element
member, empty set
universal set
Venn diagram
complement
intersection
union
subset

Indices 2

THIS SECTION WILL SHOW YOU HOW TO

- Use fraction indices

Reminder: $a^m \times a^n = a^{m+n}$ $a^m \div a^n = a^{m-n}$ $(a^m)^n = a^{mn}$

$a^0 = 1$ $a^{-n} = \frac{1}{a^n}$

Fractional indices

Consider $\sqrt{16} \times \sqrt{16}$ and $16^{\frac{1}{2}} \times 16^{\frac{1}{2}}$

$= 4 \times 4$ $= 16^{\frac{1}{2}+\frac{1}{2}} = 16^1$

$= 16$ $= 16$

so $16^{\frac{1}{2}} = \sqrt{16}$

The index $\frac{1}{2}$ means 'square root' $a^{\frac{1}{2}} = \sqrt{a}$

The index $\frac{1}{3}$ means 'cube root' $a^{\frac{1}{3}} = \sqrt[3]{a}$

This can be written more generally as:

The index $\frac{1}{n}$ means '*n*th root' $a^{\frac{1}{n}} = \sqrt[n]{a}$

EXAMPLE

Find the value of **a** $16^{\frac{1}{2}}$ **b** $27^{\frac{1}{3}}$ **c** $32^{-\frac{1}{5}}$

a $16^{\frac{1}{2}} = \sqrt{16} = 4$ **b** $27^{\frac{1}{3}} = \sqrt[3]{27} = 3$ **c** $32^{-\frac{1}{5}} = \frac{1}{32^{\frac{1}{5}}} = \frac{1}{\sqrt[5]{32}} = \frac{1}{2}$

$8^{\frac{2}{3}}$ can be written as $(8^{\frac{1}{3}})^2 = (\sqrt[3]{8})^2 = 2^2 = 4$ similarly $8^{-\frac{2}{3}} = \frac{1}{8^{\frac{2}{3}}} = \frac{1}{(\sqrt[3]{8})^2} = \frac{1}{2^2} = \frac{1}{4}$

In general: $a^{\frac{m}{n}} = (\sqrt[n]{a})^m$

EXAMPLE

Find the value of **a** $125^{\frac{4}{3}}$ **b** $16^{-\frac{3}{4}}$

a $125^{\frac{4}{3}} = (125^{\frac{1}{3}})^4 = (\sqrt[3]{125})^4 = 5^4 = 625$ **b** $16^{-\frac{3}{4}} = \frac{1}{16^{\frac{3}{4}}} = \frac{1}{(\sqrt[4]{16})^3} = \frac{1}{2^3} = \frac{1}{8}$

EXERCISE 6.11

1 Find the value of

a $9^{\frac{1}{2}}$ b $64^{\frac{1}{2}}$ c $27^{\frac{1}{3}}$ d $81^{\frac{1}{2}}$ e $64^{\frac{1}{3}}$ f $8^{\frac{1}{3}}$

g $100^{\frac{1}{2}}$ h $81^{\frac{1}{4}}$ i $144^{\frac{1}{2}}$ j $169^{\frac{1}{2}}$ k $400^{\frac{1}{2}}$ l $8000^{\frac{1}{3}}$

2 Find the value of

a $9^{-\frac{1}{2}}$ b $4^{-\frac{1}{2}}$ c $27^{-\frac{1}{3}}$ d $25^{-\frac{1}{2}}$ e $81^{-\frac{1}{4}}$ f $125^{-\frac{1}{3}}$

g $8^{-\frac{1}{3}}$ h $64^{-\frac{1}{3}}$ i $100^{-\frac{1}{2}}$ j $36^{-\frac{1}{2}}$ k $121^{-\frac{1}{2}}$ l $16^{-\frac{1}{4}}$

3 Find the value of

a $25^{\frac{3}{2}}$ b $8^{\frac{4}{3}}$ c $4^{\frac{3}{2}}$ d $9^{\frac{3}{2}}$ e $100^{\frac{5}{2}}$ f $1^{\frac{2}{3}}$

g $8^{\frac{2}{3}}$ h $9^{\frac{5}{2}}$ i $1000^{\frac{2}{3}}$ j $169^{\frac{3}{2}}$ k $81^{\frac{3}{4}}$ l $8^{\frac{4}{3}}$

4 Find the value of

a $25^{-\frac{3}{2}}$ b $8^{-\frac{4}{3}}$ c $16^{-\frac{3}{2}}$ d $4^{-\frac{3}{2}}$ e $100^{-\frac{3}{2}}$ f $8^{-\frac{2}{3}}$

g $9^{-\frac{5}{2}}$ h $64^{-\frac{5}{6}}$ i $1000^{-\frac{5}{3}}$ j $64^{-\frac{2}{3}}$ k $81^{-\frac{3}{4}}$ l $1^{-\frac{4}{3}}$

5 Simplify the following. Write your answers in the form x^n.

a $(x^3)^{\frac{2}{3}}$ b $x^{\frac{1}{2}} \times x^{1\frac{1}{2}}$ c $x^{\frac{1}{4}} \times x^{\frac{1}{4}}$ d $(x^{-4})^{\frac{1}{2}}$

e $\left(x^{\frac{1}{3}}\right)^3$ f $x^{2\frac{1}{2}} \times x^{-\frac{1}{2}}$ g $x^{-\frac{1}{2}} \times x^{-\frac{1}{2}}$ h $x^{3\frac{1}{2}} \times x^{\frac{1}{2}}$

i $\left(x^{\frac{3}{4}}\right)^4$ j $x^{\frac{2}{3}} \div x^{-\frac{1}{3}}$ k $(x^4)^{-\frac{1}{2}}$ l $\left(x^{-\frac{1}{4}}\right)^{-8}$

m $x^{1\frac{2}{5}} \div x^{-\frac{3}{5}}$ n $x^{\frac{1}{4}} \times x^{-\frac{1}{4}}$ o $x^{\frac{1}{3}} \times x^{\frac{1}{2}}$ p $x^{\frac{1}{3}} \div x^{\frac{1}{2}}$

6 Simplify

a $3x^{0.5} \times 5x^{0.5}$ b $8x^{0.5} \times 2x^{-0.5}$ c $12x^{1.5} \div 2x^{-2.5}$ d $15x^{\frac{2}{3}} \div 5x^{\frac{1}{3}}$

e $\dfrac{8x^{0.5} \times 3x^{1.5}}{4x^{0.5}}$ f $\dfrac{5x^{0.25} \times 6x^{1.25}}{2x^{-0.5} \times 5x^3}$ g $\dfrac{8x^{-3}}{2x \times 2x^{0.5}}$ h $\dfrac{5x^{0.5} + 3x^{0.5}}{2x^{-5}}$

7 Simplify

a $\dfrac{15x^6 y^{0.5}}{3x^9 y^{2.5}}$ b $\left(2x^3 y^{\frac{3}{2}}\right)^6$ c $\sqrt{25x^6 y^{\frac{1}{2}}}$ d $\sqrt{4x^8 y^6} \times \sqrt[3]{8x^3 y^{-6}}$

UNIT 6

EXAMPLE

Find the value of n in these equations. **a** $2^n = 64$ **b** $3^n = \frac{1}{81}$

a $2^n = 64$ *first write 64 as a power of 2* $64 = 2 \times 2 \times 2 \times 2 \times 2 \times 2 = 2^6$

$2^n = 2^6$

$n = 6$

b $3^n = \frac{1}{81}$ *first write $\frac{1}{81}$ as a power of 3* $\frac{1}{81} = \frac{1}{3 \times 3 \times 3 \times 3} = \frac{1}{3^4} = 3^{-4}$

$3^n = 3^{-4}$

$n = -4$

EXAMPLE

Find the value of n in these equations. **a** $5^{n+1} = 125$ **b** $2^{2n-1} = 32$.

a $5^{n+1} = 125$ *first write 125 as a power of 5* $125 = 5 \times 5 \times 5 = 5^3$

$5^{n+1} = 5^3$

$n + 1 = 3$

$n = 2$

b $2^{2n-1} = 32$ *first write 32 as a power of 2* $32 = 2 \times 2 \times 2 \times 2 \times 2 = 2^5$

$2^{2n-1} = 2^5$

$2n - 1 = 5$

$2n = 6$

$n = 3$

EXAMPLE

Find the value of n in these equations. **a** $9^{n+1} = 3^5$ **b** $8^{n-1} = 2^6$.

a $9^{n+1} = 3^5$ *the base numbers are different so replace 9 with 3^2*

$(3^2)^{n+1} = 3^5$ *use the rule $(a^m)^n = a^{mn}$*

$3^{2n+2} = 3^5$

$2n + 2 = 5$

$2n = 3$

$n = 1.5$

b $8^{n-1} = 2^6$ *the base numbers are different so replace 8 with 2^3*

$(2^3)^{n-1} = 2^6$ *use the rule $(a^m)^n = a^{mn}$*

$2^{3n-3} = 2^6$

$3n - 3 = 6$

$3n = 9$

$n = 3$

UNIT 6

EXERCISE 6.12

1 Find the value of n in these equations.

a $2^n = 32$ b $3^n = 9$ c $4^n = 4$ d $2^n = 1$

e $10^n = 100$ f $5^n = 125$ g $2^n = 256$ h $3^n = 27$

i $2^n = 2$ j $4^n = 64$ k $3^n = 81$ l $2^n = 16$

2 Find the value of n in these equations.

a $2^n = \frac{1}{16}$ b $3^n = \frac{1}{9}$ c $2^n = \frac{1}{128}$ d $4^n = \frac{1}{16}$

e $6^n = \frac{1}{36}$ f $4^n = \frac{1}{64}$ g $2^n = \frac{1}{8}$ h $3^n = \frac{1}{81}$

i $3^n = \frac{1}{3}$ j $10^n = \frac{1}{1000}$ k $5^n = \frac{1}{625}$ l $2^n = \frac{1}{2}$

3 Find the value of n in these equations.

a $2^{2n} = 64$ b $2^{n+1} = 16$ c $2^{n-1} = 8$ d $3^{n+2} = 81$

e $4^{2n} = 64$ f $10^{3n} = 1\,000\,000$ g $2^{n-2} = 128$ h $2^{2n+1} = 8$

i $2^{2n-1} = 32$ j $3^{2n+1} = 27$ k $5^{n-1} = 125$ l $7^{2n+1} = 49$

4 Find the value of n in these equations.

a $2^{n+1} = \frac{1}{2}$ b $2^{n-1} = \frac{1}{4}$ c $3^{2n} = \frac{1}{27}$ d $4^{2n} = \frac{1}{16}$

e $10^{n+1} = \frac{1}{1000}$ f $5^{2n+1} = \frac{1}{625}$ g $2^{3n+2} = \frac{1}{128}$ h $10^{2n} = \frac{1}{1\,000\,000}$

5 Find the value of x in these equations.

a $32^x = 2$ b $8^x = 2$ c $9^x = 3$ d $16^x = 4$

e $10000^x = 10$ f $27^x = 3$ g $125^x = 5$ h $169^x = 13$

i $\left(\frac{1}{2}\right)^x = 128$ j $\left(\frac{1}{3}\right)^x = 27$ k $\left(\frac{1}{5}\right)^{2x} = 25$ l $\left(\frac{1}{2}\right)^{2x} = 2$

6 Find the value of n in these equations.

a $8^{2n+1} = 2^4$ b $9^{n+1} = 3^6$ c $4^{2n+3} = 2^6$ d $25^{2n-1} = 5^4$

e $3^{n+1} = 9^{n+1}$ f $8^{2n-3} = 2^{n+6}$ g $25^{n+2} = 5^{n+1}$ h $1000^{2n} = 10^{n+10}$

KEY WORDS

fractional indices

Solving quadratic equations by factorisation

THIS SECTION WILL SHOW YOU HOW TO

- Use factorisation to solve quadratic equations

A **quadratic equation** is an equation that can be written in the form $ax^2 + bx + c = 0$ where a, b and c are numbers and $a \neq 0$.

Examples of quadratic equations are: $3x^2 - 5x + 4 = 0$ $\quad 2x^2 - 6x = 3$ $\quad x^2 - 9 = 0$

A very important concept that is used when solving quadratic equations by factorisation is:

> If $a \times b = 0$ then $a = 0$ or $b = 0$

EXAMPLE

Solve these equations **a** $x^2 - 3x = 0$ **b** $x^2 = 10x$

a $x^2 - 3x = 0$ — *factorise*

$x(x - 3) = 0$ — *the product of x and $(x - 3)$ is 0 so one of them must be 0*

$x = 0$ or $x - 3 = 0$

$x = 0$ or $x = 3$

➡ **CHECK:** $0^2 - 3 \times 0 = 0$ ✓ $\quad 3^2 - 3 \times 3 = 0$ ✓

b $x^2 = 10x$ — *rearrange so that all the terms are on one side*

$x^2 - 10x = 0$ — *factorise*

$x(x - 10) = 0$

$x = 0$ or $x - 10 = 0$

$x = 0$ or $x = 10$

➡ **CHECK:** $0^2 = 10 \times 0$ ✓ $\quad 10^2 = 10 \times 10$ ✓

EXAMPLE

Solve these equations **a** $x^2 = 25$ **b** $x^2 - 5x + 6 = 0$ **c** $2x^2 + x = 21$

a $x^2 = 25$ — *rearrange so that all the terms are on one side*

$x^2 - 25 = 0$ — *factorise*

$(x + 5)(x - 5) = 0$

$x + 5 = 0$ or $x - 5 = 0$

$x = -5$ or $x = 5$

➡ **CHECK:** $(-5)^2 = 25$ ✓ $\quad 5^2 = 25$ ✓

➡ **NOTE:** part a could also be done by simply saying that if $x^2 = 25$ then $x = \pm 5$.

b $x^2 - 5x + 6 = 0$

$(x - 2)(x - 3) = 0$ — *factorise*

$x - 2 = 0$ or $x - 3 = 0$

$x = 2$ or $x = 3$

➡ **CHECK:** $2^2 - (5 \times 2) + 6 = 0$ ✓ $\quad 3^2 - (5 \times 3) + 6 = 0$ ✓

c $2x^2 + x = 21$ — *rearrange so that all the terms are on one side*

$2x^2 + x - 21 = 0$ — *factorise*

$(2x + 7)(x - 3) = 0$

$2x + 7 = 0$ or $x - 3 = 0$

$2x = -7$ or $x = 3$

$x = -3\frac{1}{2}$ or $x = 3$

➡ **CHECK:** $2 \times \left(-3\frac{1}{2}\right)^2 + \left(-3\frac{1}{2}\right) = 21$ ✓

$2 \times 3^2 + 3 = 21$ ✓

EXERCISE 6.15

For each of these functions make a table of values and draw the graph of the function between the stated x values.

1 $y = \frac{5}{x}$ $\quad -5 \le x \le 5$

2 $y = -\frac{5}{x}$ $\quad -5 \le x \le 5$

3 $y = \frac{4}{x} + 2$ $\quad -5 \le x \le 5$

4 $y = \frac{5}{x-3}$ $\quad -2 \le x \le 8$

5 $y = \frac{6}{x} + x + 1$ $\quad -6 \le x \le 6$

6 $y = x^2 + \frac{6}{x}$ $\quad -4 \le x \le 4$

7 $y = \frac{12}{x^2}$ $\quad -6 \le x \le 6$

8 $y = \frac{4}{x^2} + x - 2$ $\quad -6 \le x \le 6$

9 Amir wants to make a rectangular enclosure next to a farm building.
He wants the enclosure to have an area of 500 m^2.
He will need fencing for three of the sides of the enclosure.

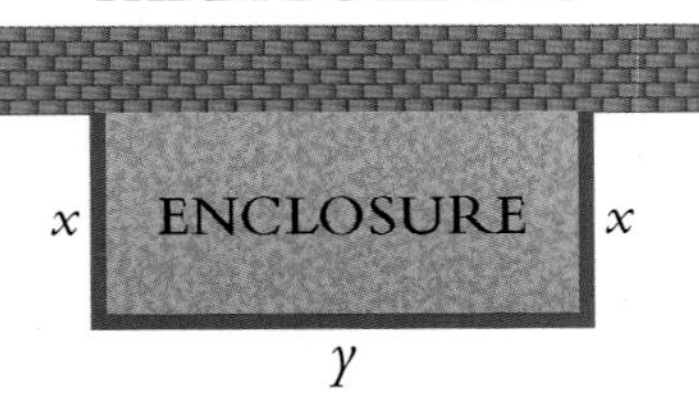

a By considering the area of the enclosure explain why $y = \frac{500}{x}$

b Write down an expression in terms of x and y for the total length, T, of fencing that is needed.

c Show that $T = 2x + \frac{500}{x}$

d Copy and complete this table of values for x and T.

x	5	10	15	20	30	40	50
T	110						110

e Draw the graph of $T = 2x + \frac{500}{x}$

f Use your graph to estimate the minimum length of fencing that is needed and the value of x that gives the minimum length of fencing.

KEY WORDS
reciprocal
asymptote

UNIT 6

The gradients of perpendicular lines

THIS SECTION WILL SHOW YOU HOW TO

- Recognise perpendicular lines
- Solve problems involving perpendicular lines

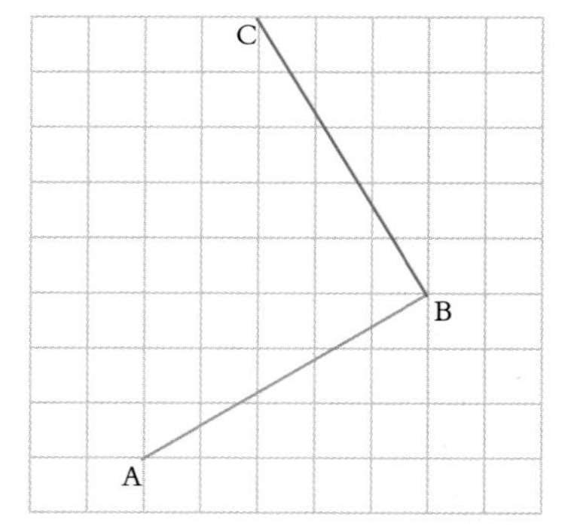

The lines AB and BC are perpendicular.

The gradient of the line AB = $\frac{3}{5}$

The gradient of the line BC = $-\frac{5}{3}$

Notice that: $\frac{3}{5} \times -\frac{5}{3} = -1$

The general rule for perpendicular lines is:

> If a line has gradient m, a line perpendicular to it has gradient $-\frac{1}{m}$.

This rule can also be written as:

> If the gradients of two perpendicular lines are m_1 and m_2, then $m_1 \times m_2 = -1$.

EXAMPLE

Find the gradient of a line which is perpendicular to the line joining A(7, −2) and B(5, 4).

Gradient of AB = $\frac{4-(-2)}{5-7} = \frac{6}{-2} = -3$ *using* $\frac{y_2 - y_1}{x_2 - x_1}$

Let m be the gradient of the perpendicular line.
The product of the gradients of perpendicular lines is always −1.

$-3 \times m = -1$ *divide both sides by −3*

$m = \frac{-1}{-3}$ *simplify*

$m = \frac{1}{3}$

EXAMPLE

Find the equation of the perpendicular bisector of the line joining P(3, 2) and Q(7, 10).

Gradient of PQ = $\frac{10-2}{7-3} = \frac{8}{4} = 2$ *using* $\frac{y_2 - y_1}{x_2 - x_1}$

Gradient of the perpendicular is $-\frac{1}{2}$ *using* $m_1 \times m_2 = -1$

Midpoint of PQ = $\left(\frac{3+7}{2}, \frac{2+10}{2}\right) = (5, 6)$ *using* $\left(\frac{x_1+x_2}{2}, \frac{y_1+y_2}{2}\right)$

So the perpendicular bisector is the line passing through the point (5, 6) with gradient $-\frac{1}{2}$.

Substitute $x = 5$, $y = 6$ and $m = -\frac{1}{2}$ into the equation of a straight line, $y = mx + c$.

$$6 = -\frac{1}{2} \times 5 + c$$
$$6 = -2.5 + c$$
$$c = 8.5$$

Substitute for m and c in $y = mx + c$ gives

$$y = -\frac{1}{2}x + 8.5$$
$$x + 2y = 17$$

multiply both sides by 2 and rearrange

EXERCISE 6.16

1 Copy each diagram and mark a point C so that angle ABC = 90° and AB = BC.

a
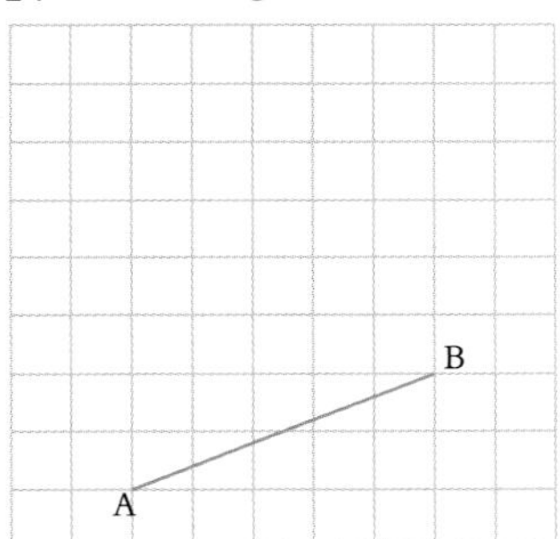

b
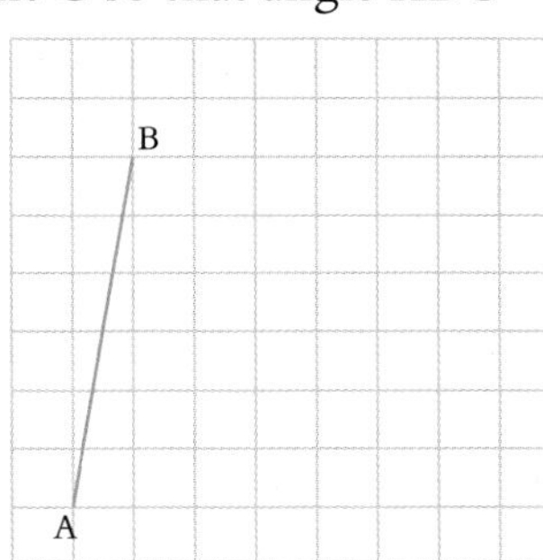

c
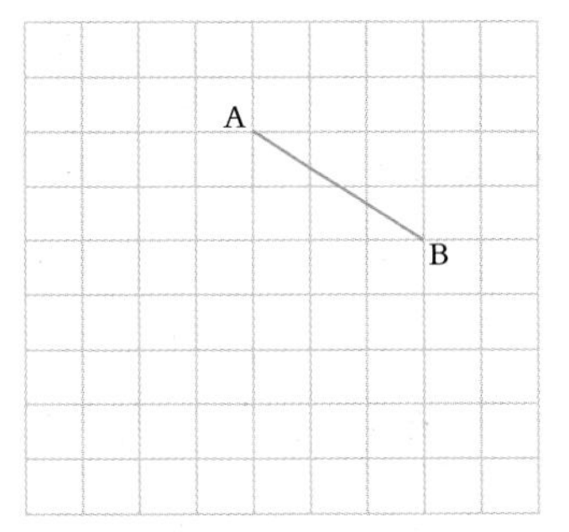

2 Copy each diagram and mark the midpoint of the line segment AB. Use the squares on the grid to help you draw the perpendicular bisector of the line segment AB.

a
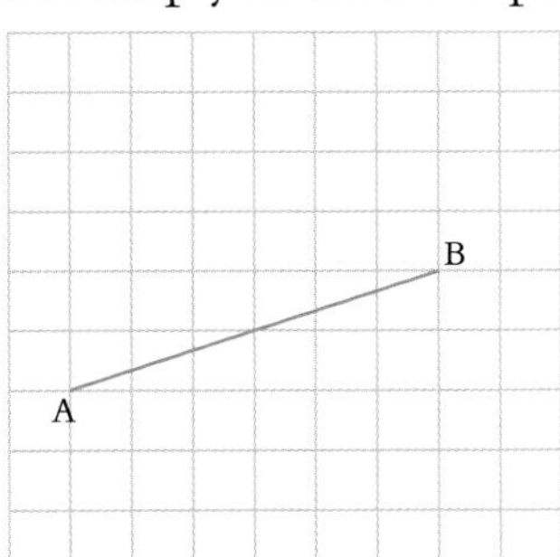

b
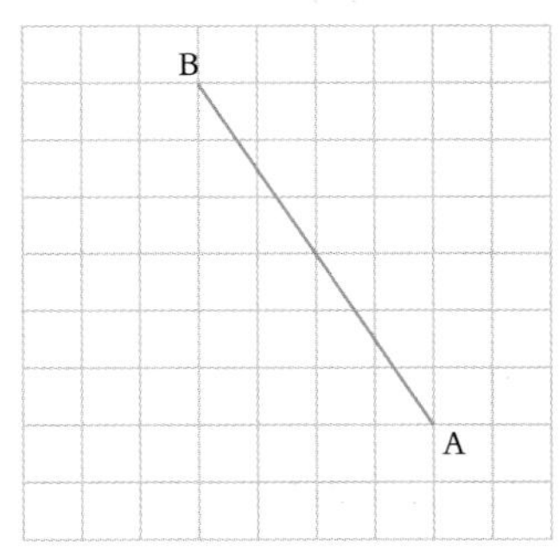

c
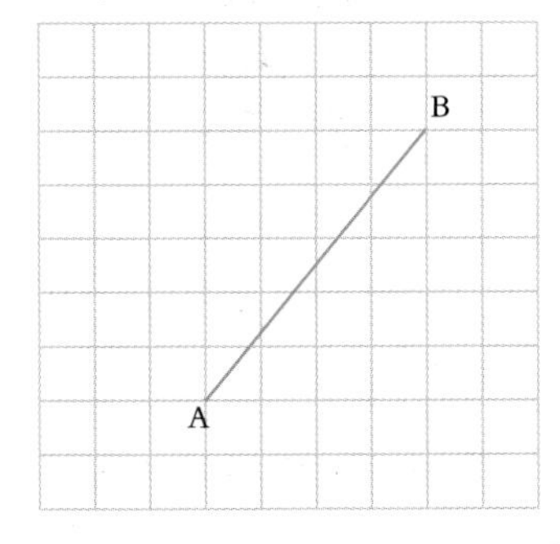

3 Find the gradient of a line perpendicular to a line with gradient:

a 5

b $\frac{1}{3}$

c $\frac{2}{5}$

d $-\frac{2}{3}$

e -4

f $\frac{5}{7}$

4 Work out if these pairs of lines are parallel, perpendicular or neither:

a $y = x + 2$
$y = 2x - 5$

b $y = 3x - 6$
$y + 3x = 1$

c $y = 4x - 1$
$y = \frac{1}{4}x + 1$

d $y = \frac{1}{2}x + 3$
$y = 5 - 2x$

e $y = 4x - 2$
$4y + x = 5$

f $y = -\frac{1}{2}x - 1$
$y = 2x + 4$

g $y = 5x + 1$
$5y + x = 3$

h $x + 2y = 3$
$x - 2y = 4$

i $2x + 3y = 7$
$3y = 8 - 2x$

5 A $3x + y = 2$ | B $2y = 3x + 4$ | C $y = 3x - 1$ | D $y = 4 - 3x$ | E $3y + x = 5$

a Find two lines from those shown here that are:
i parallel
ii perpendicular

b For the one remaining line, write down the equation of a line that is perpendicular to it.

6 Find the equation of the line that passes through the point (5, 0) and is perpendicular to the line $y = 3x - 4$.

7 Find the equation of the line:

a perpendicular to the line $y = 2x + 4$, passing through the point (6, 2).
b perpendicular to the line $y = 5x - 1$, passing through the point (0, 7).
c perpendicular to the line $2x + 3y = 12$, passing through the point (6, 1).
d perpendicular to the line $4x - y = 6$, passing through the point (4, −1).

8 Two vertices of a rectangle ABCD are: A(−4, 1) and B(−2, 2).

a Find the gradient of CD.
b Find the gradient of BC.

9 P(−3, −3), Q(3, −1) and R(2, 2) are the three vertices of a triangle.
Show that triangle PQR is a right-angled triangle.

10 Line l_1 passes through the points (−2, 3) and (6, −1).
Line l_2 passes through the points (0, −2) and (3, 4).
Show that lines l_1 and l_2 are perpendicular.

11 P is the point (1, 1) and Q is the point (6, 0).
A line l is drawn through P and perpendicular to PQ, to meet the y-axis at the point R.

a Find the equation of line l.
b Find the coordinates of point R.

12 Find the equation for the perpendicular bisector of the line joining Q(1, 3) and R(−3, 1).

13 The perpendicular bisector of the line joining A(−1, 4) and B(2, 2) intersects the x-axis at P and the y-axis at Q. Find the coordinates of P and Q.

14 Line l_1 has equation $3x + 2y = 12$.
Line l_2 has equation $y = 2x - 1$.
Lines l_1 and l_2 intersect at point A.

a Find the coordinates of A.

b Find the equation of the line through A which is perpendicular to line l_1.

15 A(−1, −2), B(5,1) and C(1, 4)
ABCD is a trapezium.
AB is parallel to DC and angle BAD is 90°.
Find the coordinates of D.

16 E is the point (−6, 0) and F is the point (−2, 6).
Find the point G on the x-axis such that angle EFG is 90°.

17 Angle ABC is 90° and M is the midpoint of the line AB.
The point C lies on the y-axis.
Find the coordinates of B and C.

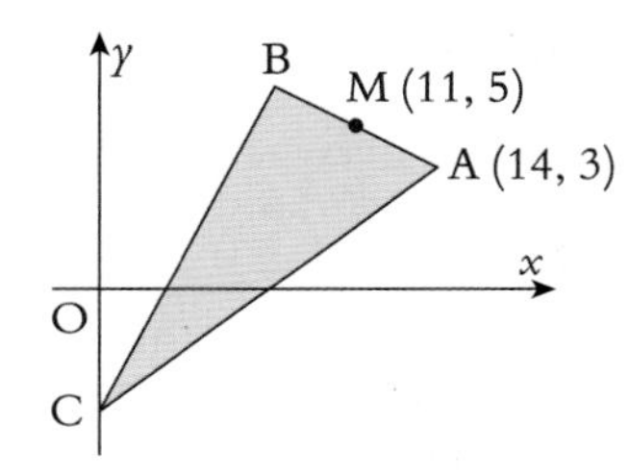

18 The coordinates of three points are A(−2, 6), B(4, 4) and C(k, −7).
M is the midpoint of AB and MC is perpendicular to AB.

a Find the coordinates of M.

b Find the value of k.

19 The coordinates of triangle ABC are A(−2, 2), B(5, 6) and C(12, −5).
P is the foot of the perpendicular from B to AC.

a Find the equation of BP.

b Find the coordinates of P.

20 The vertices of triangle ABC are A(−3, 2), B(6, −5) and C(k, k + 1).
Find the possible values of k if angle ACB is 90°.

21 The vertices of triangle PQR are P(−4, −k), Q(−1, k) and R(3k, −2).
Find the possible values of k if angle PQR is 90°.

22 The coordinates of triangle EFG are E(−2, 1), F(6, 5) and G(7, −2).

a Find the equation of the perpendicular bisectors of
 i EF **ii** FG.

b Find the coordinates of the point which is equidistant from E, F and G.

TOP TIP

The point is where the perpendicular bisectors of the sides intersect.

UNIT 6

Circle theorems

THIS SECTION WILL SHOW YOU HOW TO

- Use the properties of tangents and chords to solve problems
- Use the circle theorems to solve problems

Tangents

A straight line can intersect a circle in three possible ways. It can be:

A DIAMETER  A CHORD A TANGENT

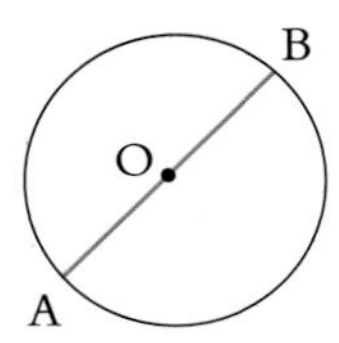

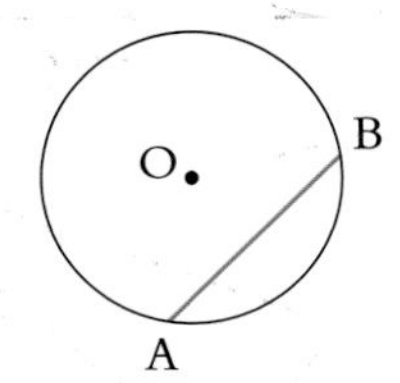

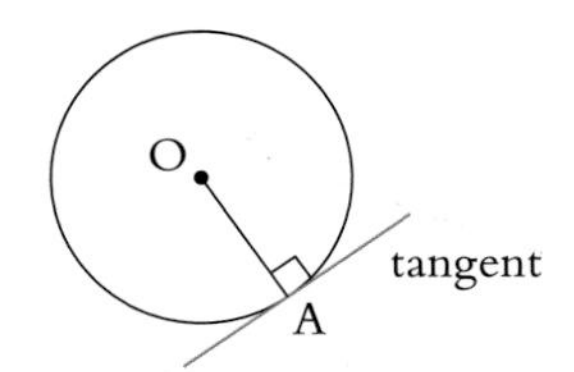

2 points of intersection 2 point of intersection 1 point of intersection

PROPERTY 1

A **tangent** to a circle 'touches' the circle at one point.
The angle between a tangent and a radius is a right angle.

$O\hat{A}P = 90°$.

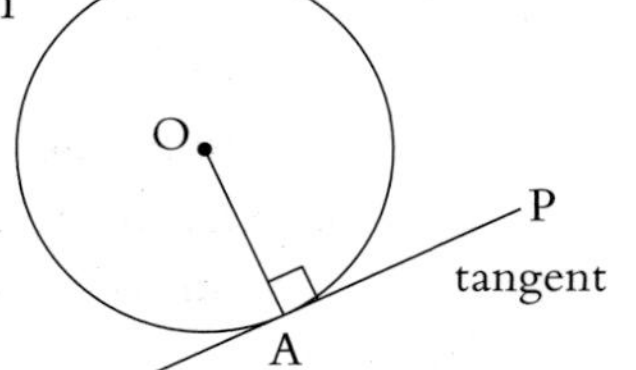

PROPERTY 2

The two tangents drawn from a point P outside a circle are equal in length.

AP = BP.

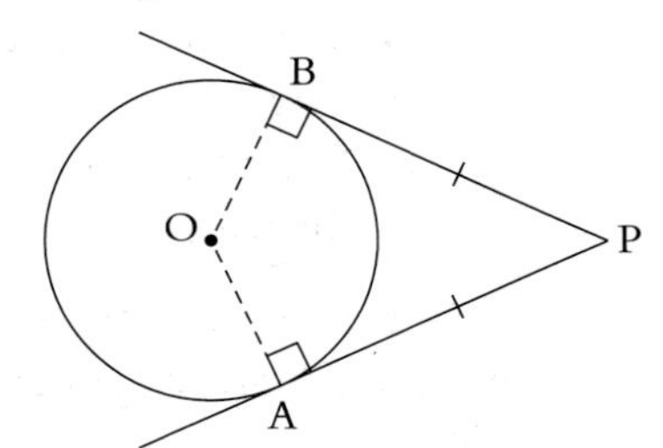

EXAMPLE

In the diagram, O is the centre of the circle radius 5 cm.
AP is a tangent to the circle at A and AP = 15 cm.

a Calculate the length of OP and the size of angle AOP.

b Calculate the shaded area.

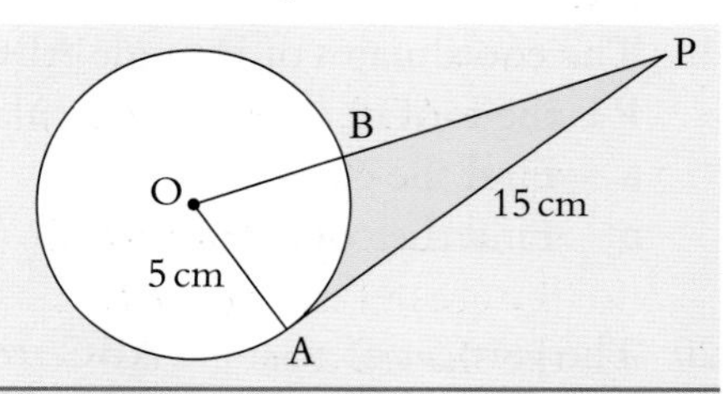

a $O\hat{A}P = 90°$ *tangent and radius at right angles*

$15^2 + 5^2 = OP^2$ $\tan A\hat{O}P = \frac{15}{5}$

$OP^2 = 250$ $A\hat{O}P = \tan^{-1}\left(\frac{15}{5}\right)$

$OP = 15.8$ cm $A\hat{O}P = 71.6°$

b Area of triangle AOP $= \frac{1}{2} \times \text{base} \times \text{height} = \frac{1}{2} \times 15 \times 5 = 37.5\text{ cm}^2$

Area of sector OAB $= \frac{71.6}{360} \times \pi \times 5^2 = 15.6\text{ cm}^2$

Shaded area = area of triangle AOP – area of sector OAB = $37.5 - 15.6 = 21.9\text{ cm}^2$

EXERCISE 6.17

1 O is the centre of the circle and AP is a tangent to the circle. Calculate the value of x for each diagram.

a

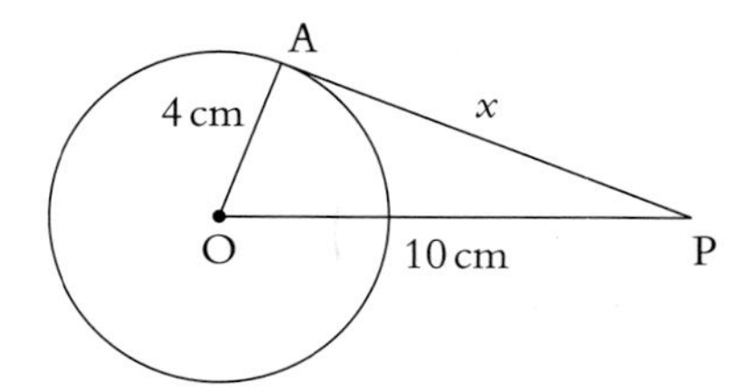

b

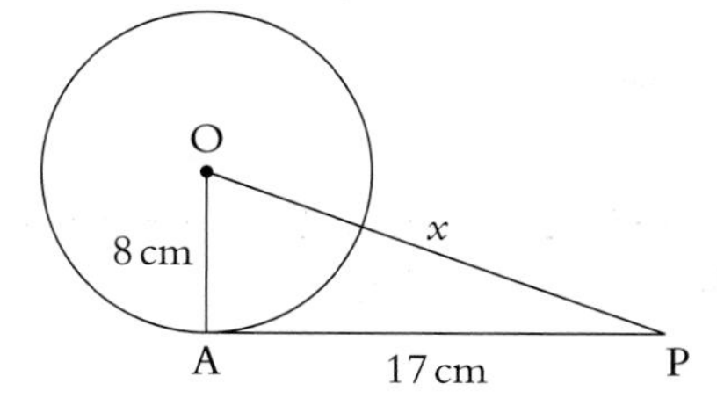

c

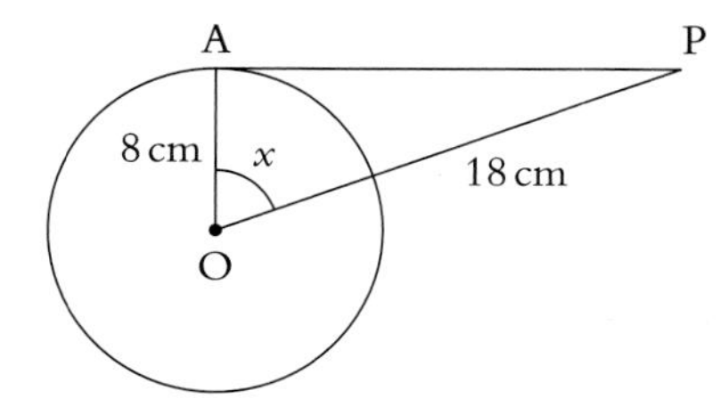

d

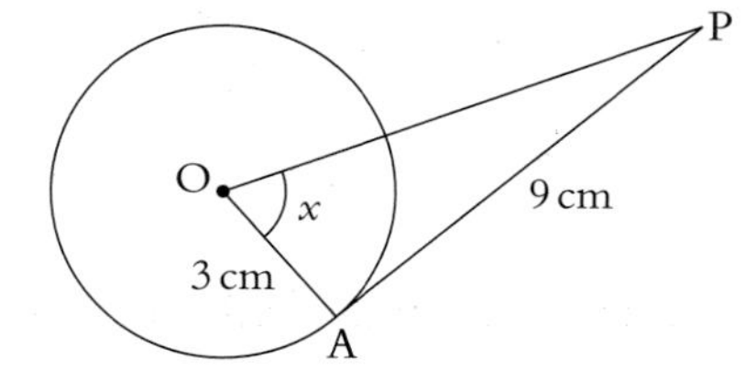

2 O is the centre of the circle.
AP is a tangent to the circle at A.
OP = 12 cm and AP = 10 cm.
Calculate:

a the radius of the circle,
b angle AOP,
c the area of triangle OAP,
d the shaded area.

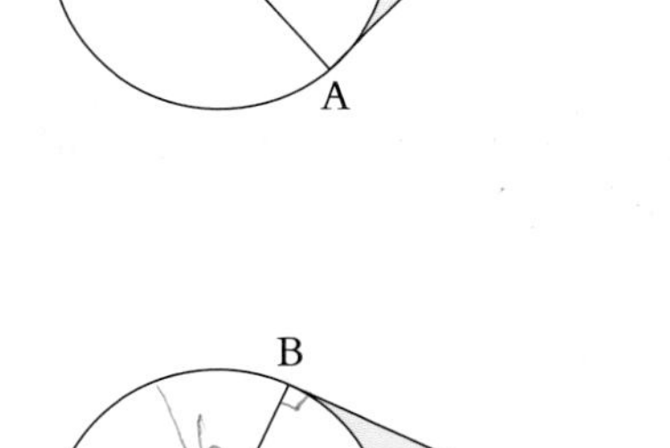

3 O is the centre of the circle, radius 6 cm.
AP and BP are tangents to the circle.
AP = 8 cm.
Calculate:

a angle AOP,
b angle AOB,
c the area of quadrilateral OAPB,
d the area of sector OAB,
e the shaded area.

4 ABC is an equilateral triangle.
AB, BC and AC are tangents to the circle at the points P, Q and R.
O is the centre of the circle, radius 8 cm.
Calculate:

a the length PO,
b the length PC,
c the length PA,
d the area of triangle ABC,
e the shaded area.

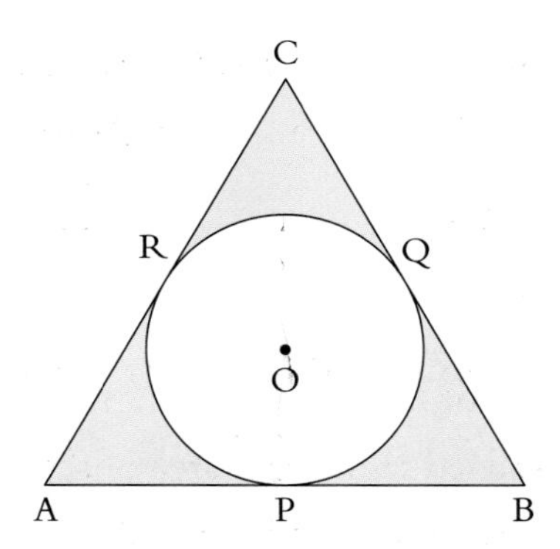

UNIT 6

Chords and segments

A straight line joining two points on the circumference of a circle is called a **chord**.

A chord divides a circle into two **segments**.

major segment

minor segment

Symmetry properties of chords

PROPERTY 1

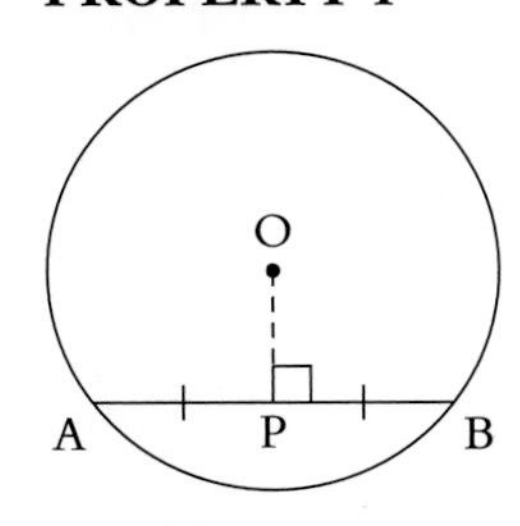

The perpendicular line from the centre of a circle to a **chord** bisects the chord.

AP = PB.

➡ **NOTE:** triangle OAB is isosceles.

PROPERTY 2

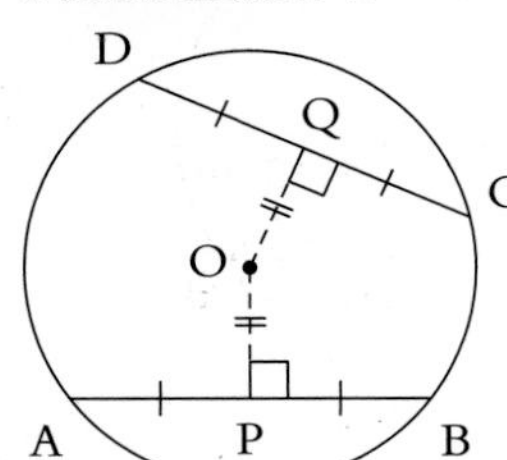

If two chords AB and CD are the same length then they will be the same perpendicular distance from the centre of the circle.

If AB = CD then OP = OQ.

EXAMPLE

A chord AB lies on a circle of radius 12 cm.
The chord is 18 cm long. M is the midpoint of AB.

a Calculate the length of OM and angle AOB.

b Calculate the shaded area.

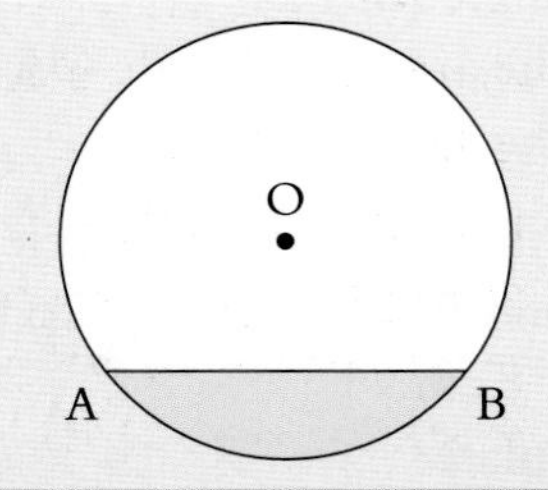

a

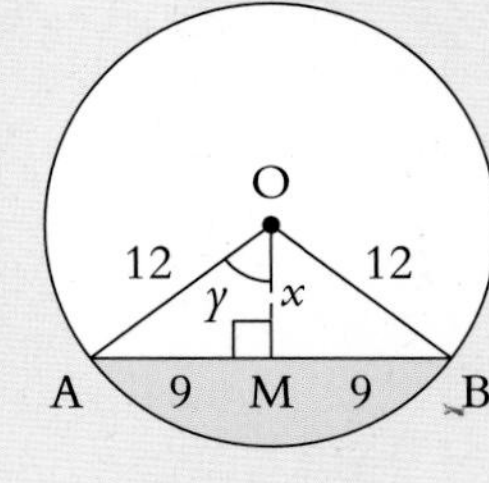

Using Pythagoras

$x^2 + 9^2 = 12^2$

$x^2 + 81 = 144$

$x^2 = 63$

$x = 7.937...$

OM is 7.94 cm (to 3 s.f.)

Let $A\hat{O}M = y$

$\sin y = \dfrac{9}{12}$

$y = \sin^{-1}\left(\dfrac{9}{12}\right)$

$y = 48.59...$

$A\hat{O}B = 2 \times 48.59...$

$A\hat{O}B = 97.2°$ (to 1 d.p.)

b Area of triangle AOB $= \dfrac{1}{2} \times \text{base} \times \text{height} = \dfrac{1}{2} \times 18 \times 7.937 = 71.44\text{ cm}^2$

Area of sector AOB $= \dfrac{97.2}{360} \times \pi \times 12^2 = 122.12\text{ cm}^2$

Area of shaded segment = area of sector – area of triangle

Area of shaded segment $= 122.12 - 71.44 = 50.7\text{ cm}^2$

EXERCISE 6.18

1 A chord AB lies on a circle of radius 5 cm.
The chord is 2 cm from the centre of the circle.
Calculate the length of the chord.

2 A chord AB lies on a circle of radius 8 cm.
The chord is 12 cm long.
Calculate the distance of the chord from the centre of the circle.

3 A chord AB is 24 cm long.
The distance of the chord from the centre of the circle is 5 cm.
Calculate the radius of the circle.

4 O is the centre of the circle, radius 15 cm.
The chord AB is 24 cm long.
  a Calculate the distance of the chord from O.
  b Calculate the area of triangle OAB.

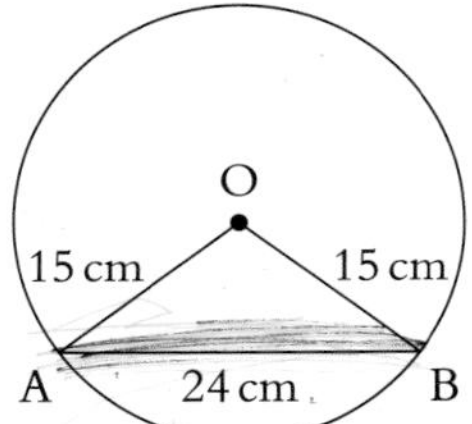

5 O is the centre of the circle, radius 5 cm.
$A\hat{O}B = 140°$. M is the midpoint of AB.
Calculate:
  a the length of OM,
  b the length of AM,
  c the length of the chord AB,
  d the area of triangle OAB,
  e the area of sector OAB,
  f the area of the shaded segment.

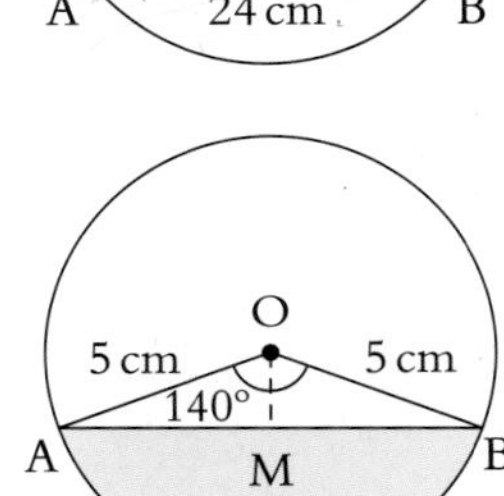

6 O is the centre of the circle, radius 12 cm.
The chords AB and CD are parallel.
AB = 20 cm and CD = 15 cm.
Calculate the distance between the two chords.

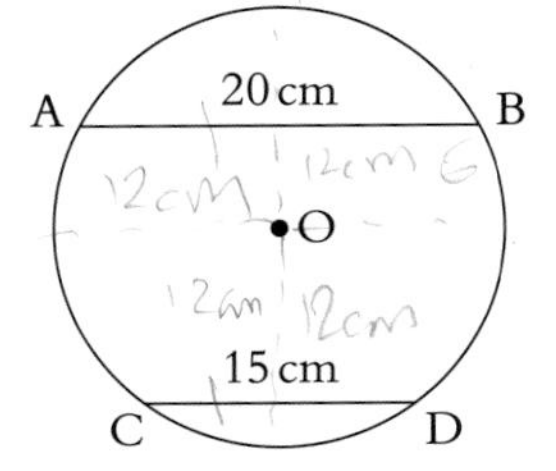

7 O is the centre of the circle, radius 10 cm.
AB = 16 cm and AC = BC.
M is the midpoint of AB.
Calculate:
  a the length of OM,
  b the length of CM,
  c the area of triangle ABC.

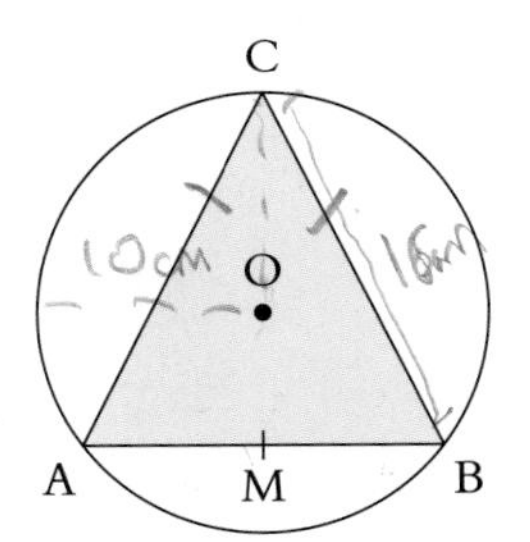

UNIT 6

EXAMPLE

Find the values of x and y.
Give reasons for your answers.

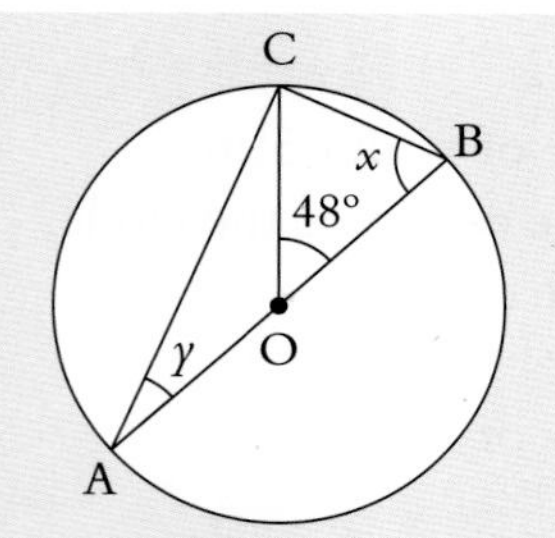

$x = \frac{180 - 48}{2} = 66°$ $\triangle$BOC *is isosceles*

$\hat{AOC} = 180 - 48 = 132°$ *angles on a straight line*

$y = \frac{180 - 132}{2} = 24°$ $\triangle$AOC *is isosceles*

EXERCISE 6.19

Find the size of each lettered angle.

1 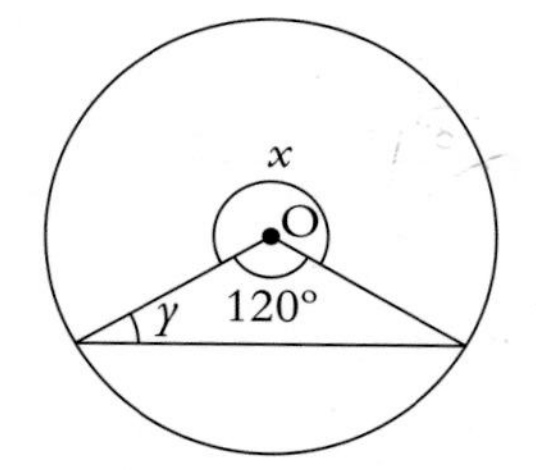

2 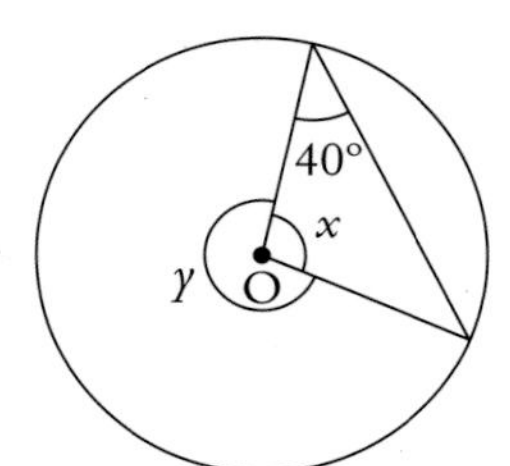

3 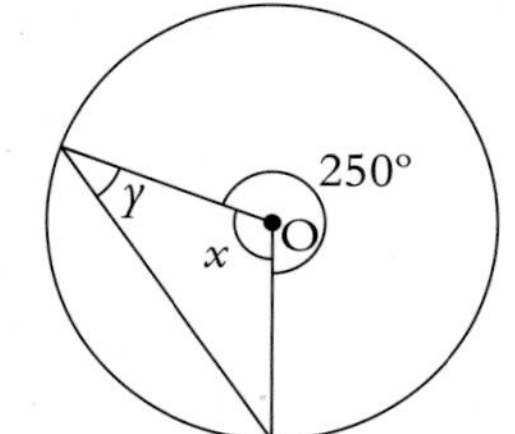

4 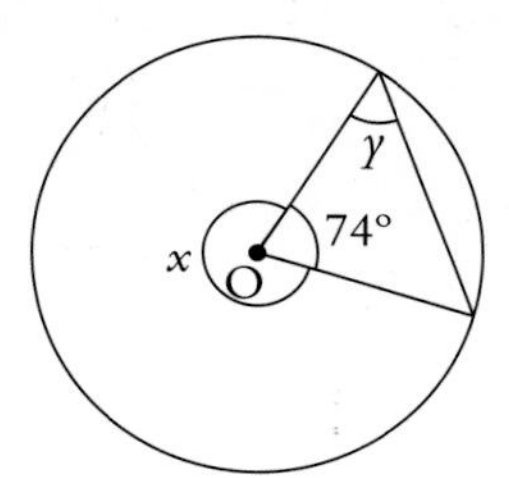

5 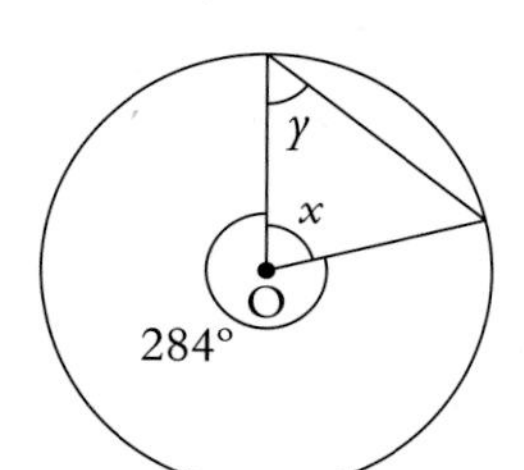

6 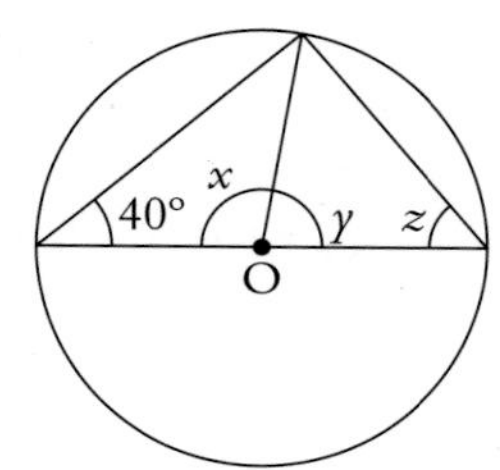

7 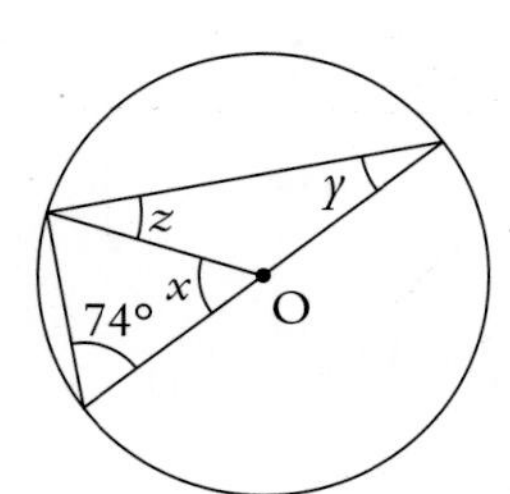

8 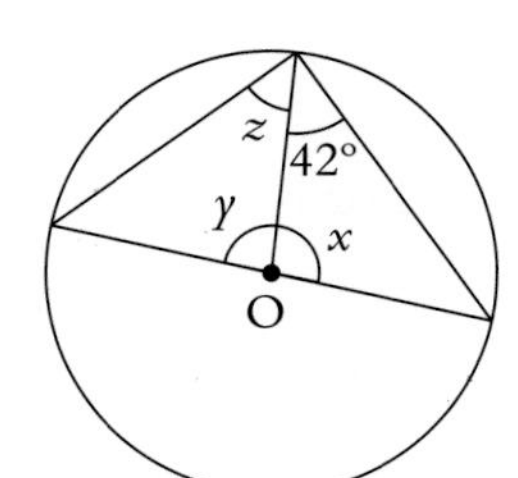

9 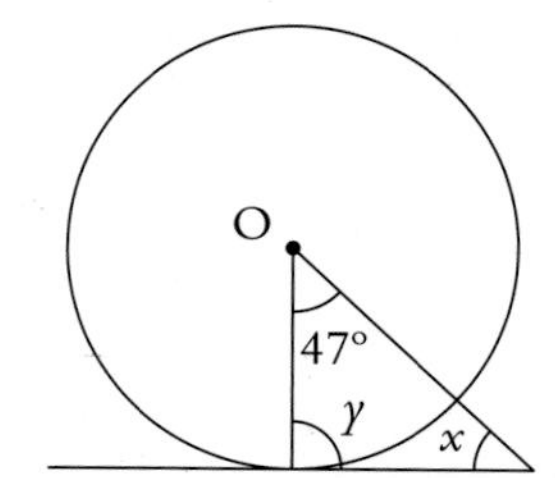

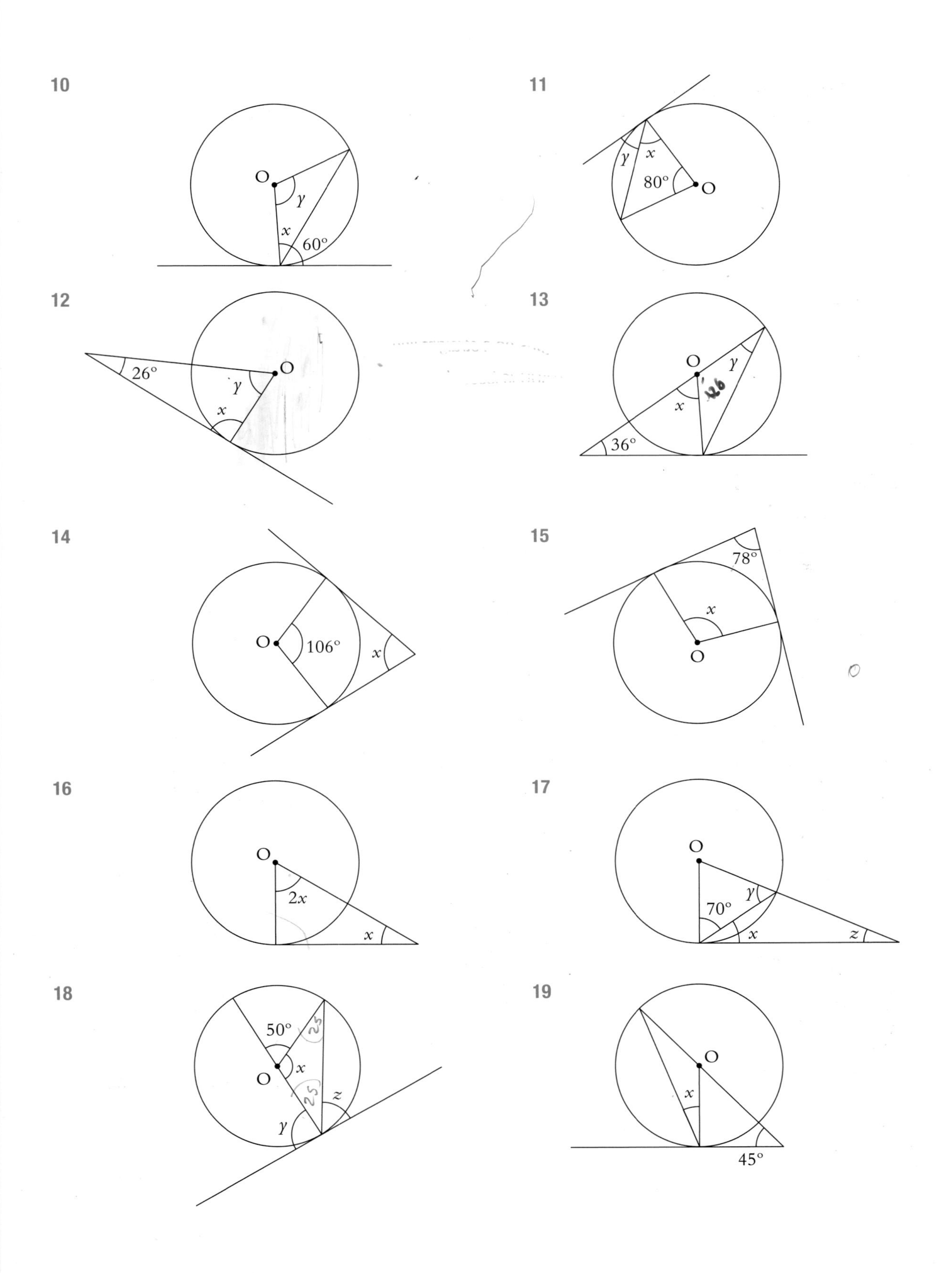
10
O
y
x
60°
11
y
x
80°
O
12
26°
O
y
x
13
O
y
x
36°
14
O
106°
x
15
78°
x
O
16
O
2x
x
17
O
y
70°
x
z
18
50°
x
O
z
y
19
O
x
45°

UNIT 6

THEOREM 1

The angle at the centre is twice the angle at the circumference.

THEOREM 2

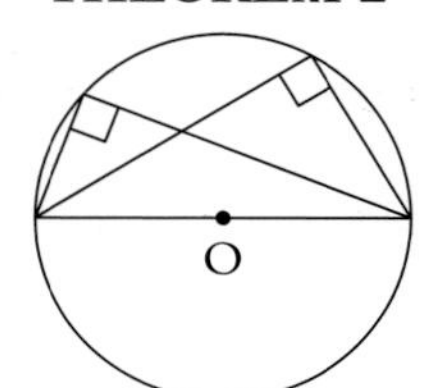

An angle in a semi-circle is always a right angle.

Proof of theorem 1

Let $O\hat{C}A = a$ and $O\hat{C}B = b$

$C\hat{A}O = a$ and $C\hat{B}O = b$ *isosceles triangles*

$A\hat{O}D = 2a$ and $B\hat{O}D = 2b$ *exterior angles of triangles*

$A\hat{C}B = a + b$ and $A\hat{O}B = 2a + 2b$

So $A\hat{O}B = 2 \times A\hat{C}B$

Proof of theorem 2

Angle at centre = 180°

Angle at centre = 2 × angle at circumference

$180° = 2 \times a$

$a = 90°$

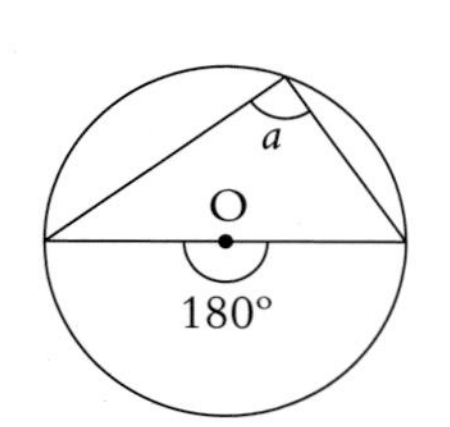

EXAMPLE

Find the value of x for each of these diagrams.

a

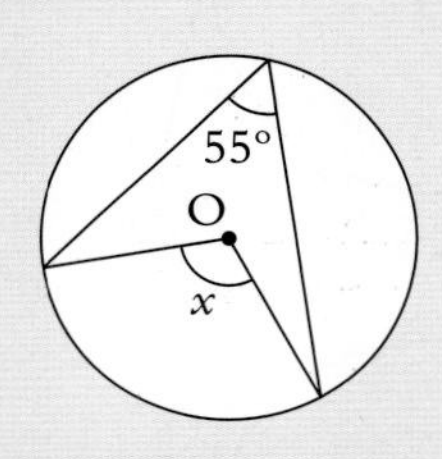

b

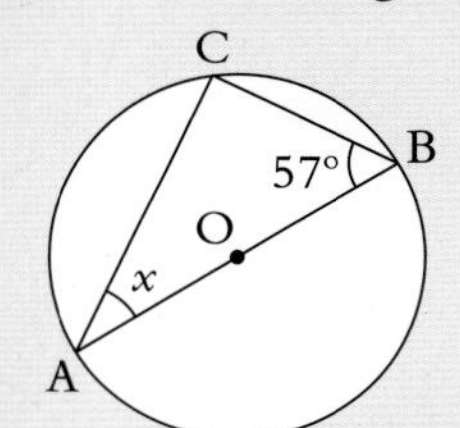

c

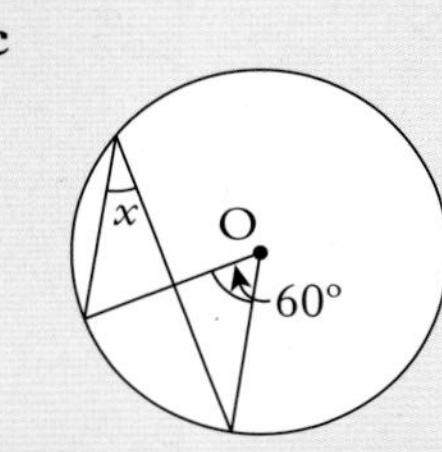

d

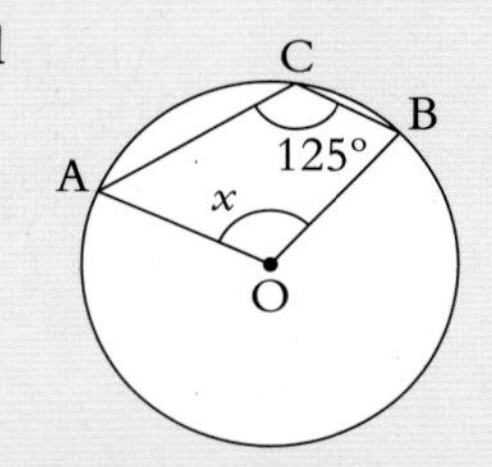

a $x = 2 \times 55°$

$x = 110°$

b $x = 180° - (90° + 57°)$

$x = 33°$

c $60° = 2 \times x$

$x = 30°$

d (reflex) $A\hat{O}B = 2 \times 125 = 250°$

$x = 360 - 250 = 110°$

EXAMPLE

Find the values of x and y.

Give reasons for your answers.

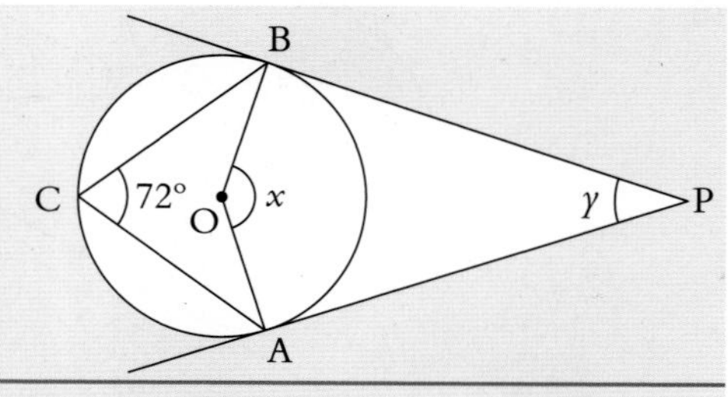

$x = 2 \times 72 = 144°$ *angle at centre = 2 × angle at circumference*

$O\hat{B}P = O\hat{A}P = 90°$ *tangent and radius at right angles*

$y = 360 - (90 + 90 + 144) = 36°$ *angles in a quadrilateral add up to 360°*

EXERCISE 6.20

Find the value of the letters in each of the following diagrams.

1

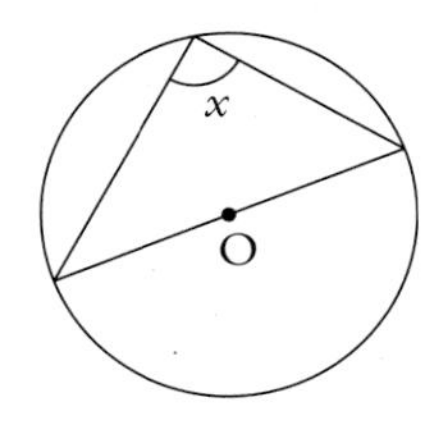

2

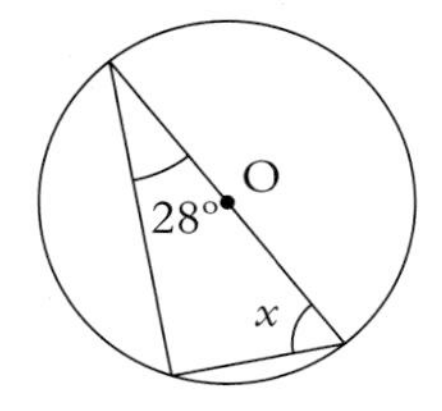

3

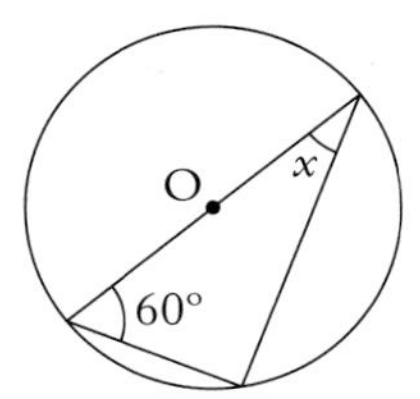

4

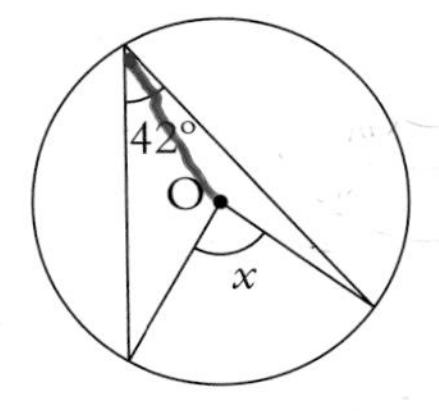

5

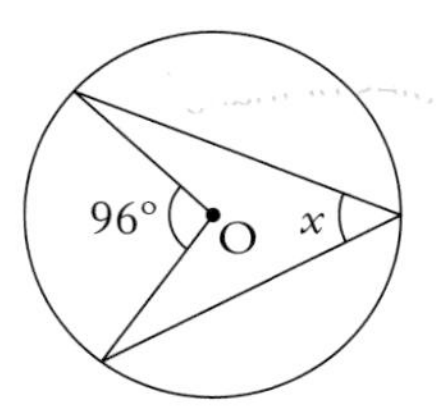

6

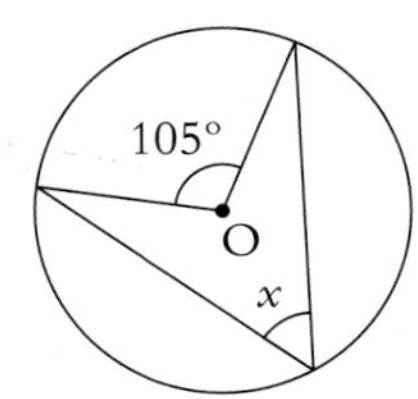

7

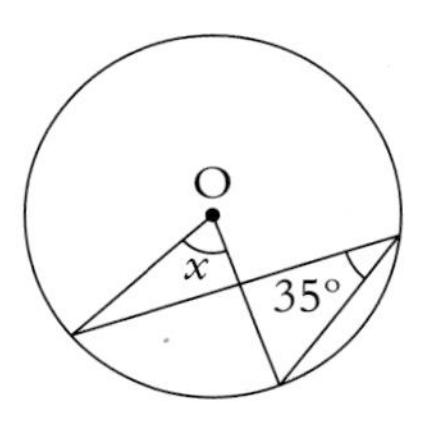

8

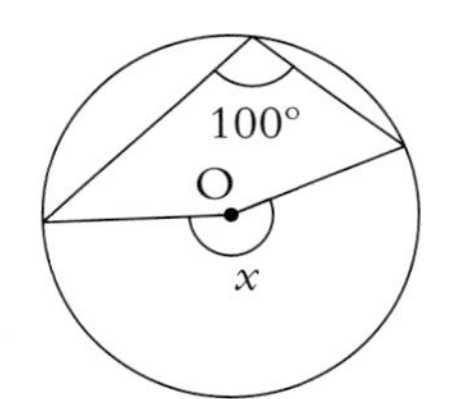

9

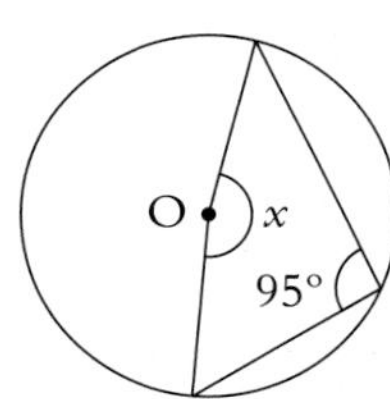

10

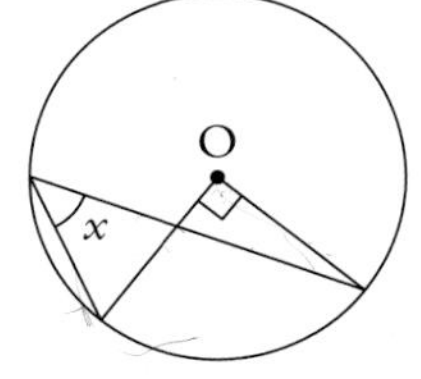

11

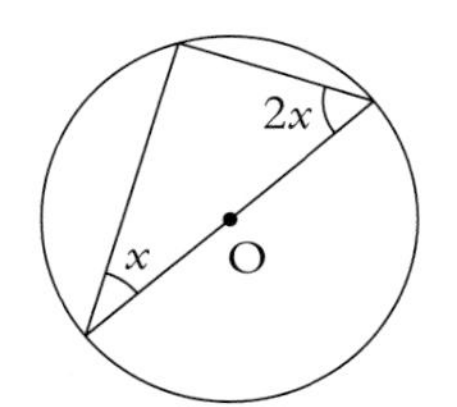

12

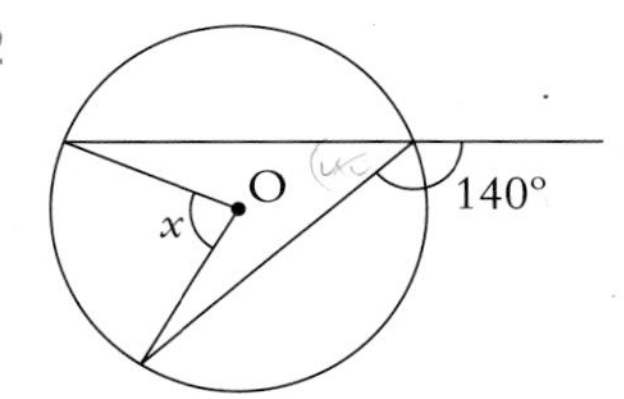

13

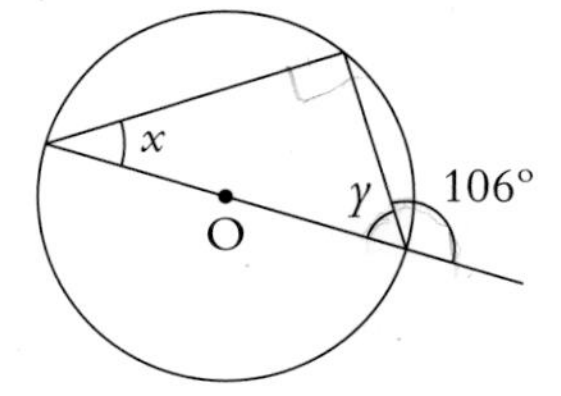

14

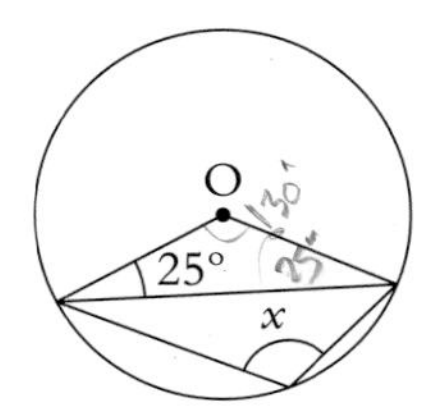

15

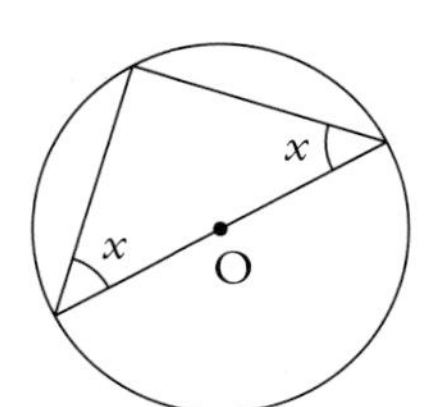

16

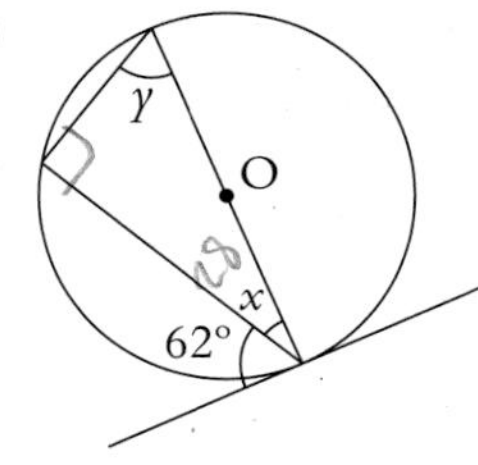

17

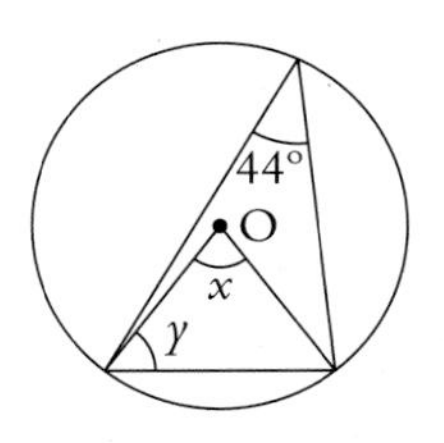

18

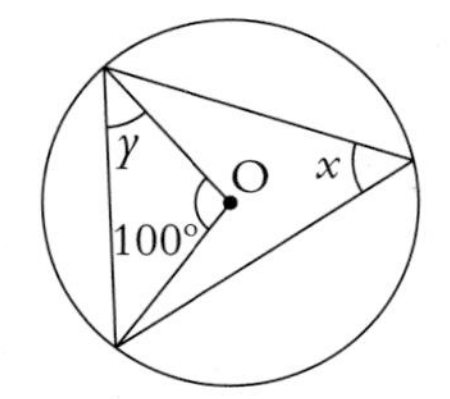

THEOREM 3

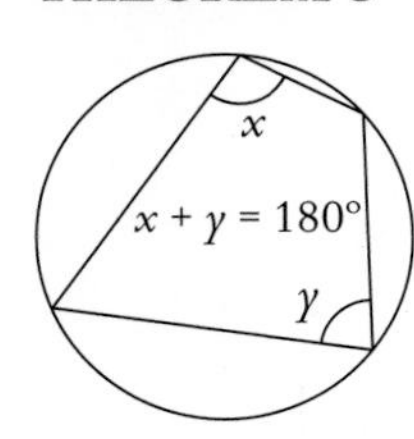

Opposite angles of a **cyclic quadrilateral** add up to 180°.

THEOREM 4

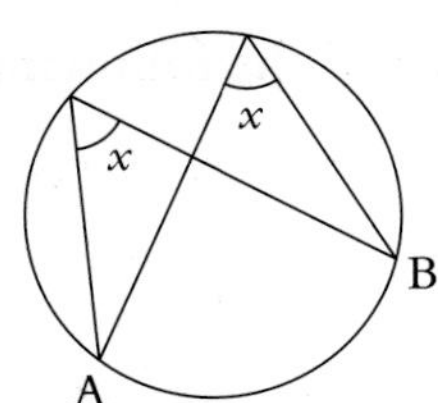

Angles from the same arc in the same segment are equal.

➡ **NOTE:** all four vertices of a cyclic quadrilateral lie on the circumference of a circle.

Proof of theorem 3

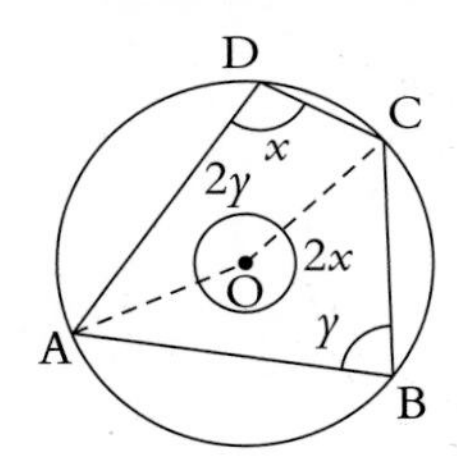

Let	$A\hat{D}C = x$ and $A\hat{B}C = y$	
(reflex)	$A\hat{O}C = 2x$	*angle at centre = 2 × angle at circumference*
(obtuse)	$A\hat{O}C = 2y$	*angle at centre = 2 × angle at circumference*
	$2x + 2y = 360°$	*angles at a point*
So	$x + y = 180°$	

Proof of theorem 4

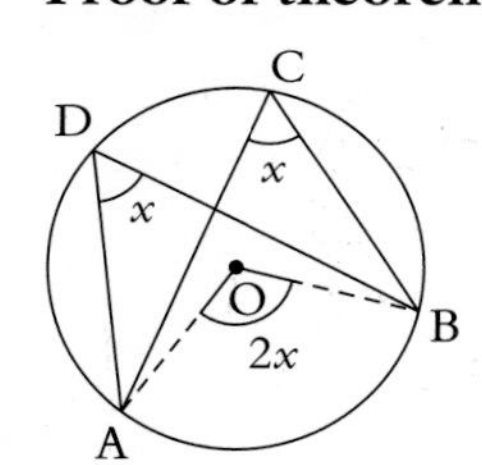

Let	$A\hat{D}B = x$	
Then	$A\hat{O}B = 2x$	*angle at centre = 2 × angle at circumference*
So	$A\hat{C}B = x$	*angle at centre = 2 × angle at circumference*

EXAMPLE

Find the value of x and y for each of these diagrams.

a

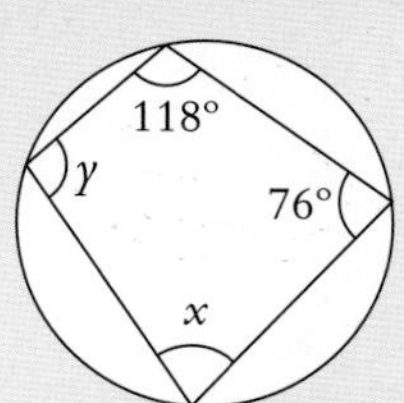

b

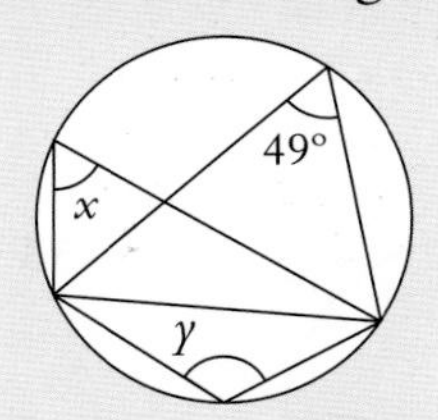

a $x + 118° = 180°$ and $y + 76° = 180°$ *opposite angles in a cyclic quadrilateral*

$x = 62°$ $\quad y = 104°$

b $x = 49°$ *angles from the same arc in the same segment*

$y + 49° = 180°$ *opposite angles in a cyclic quadrilateral*

$y = 131°$

EXERCISE 6.21

Find the value of the letters in each of the following diagrams.

1

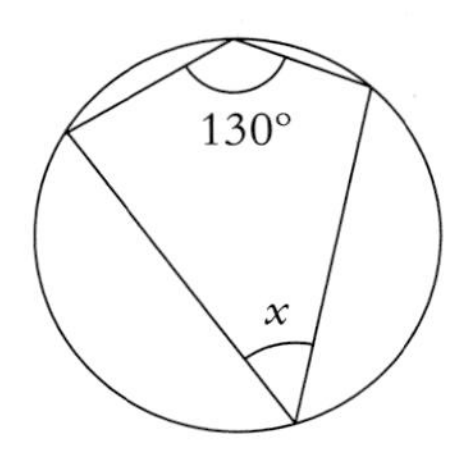

2

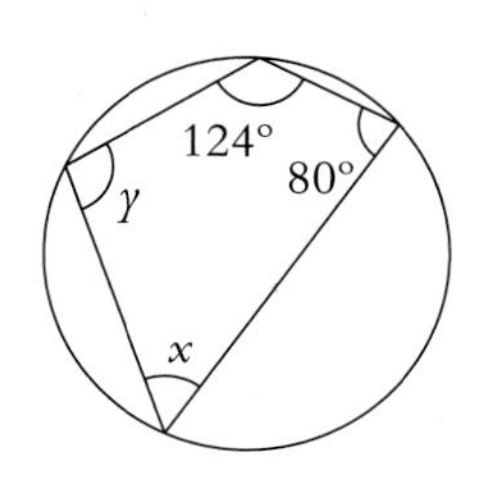

3

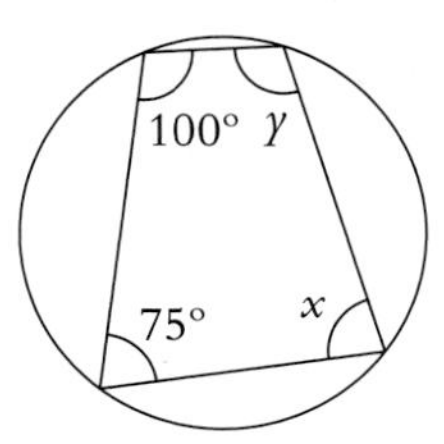

4

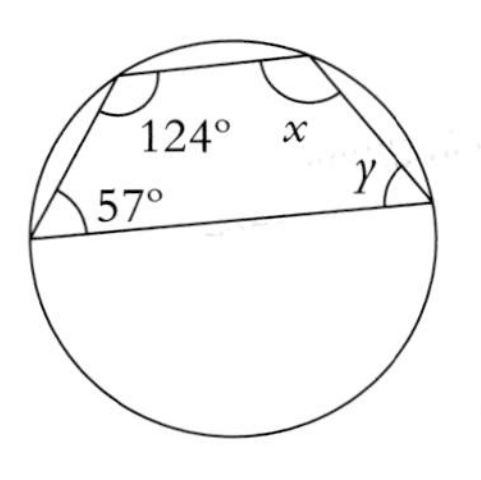

5

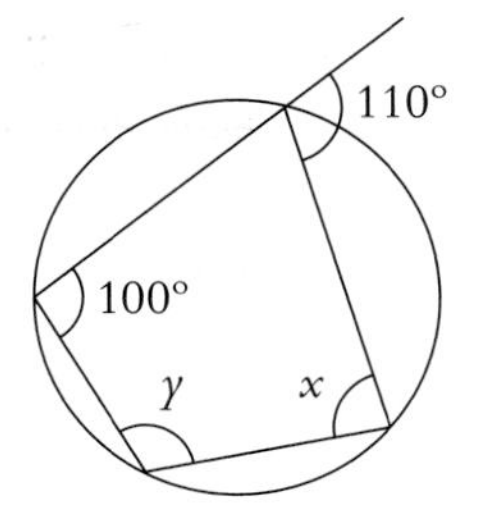

6

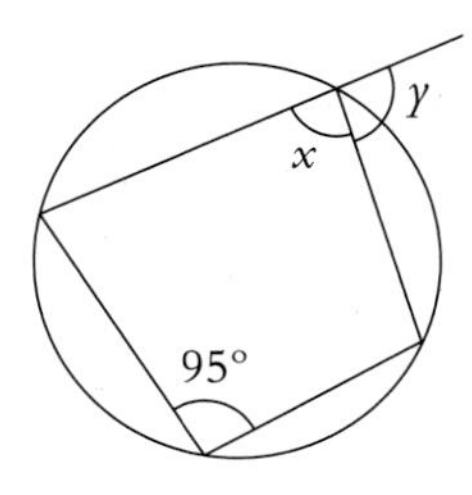

7

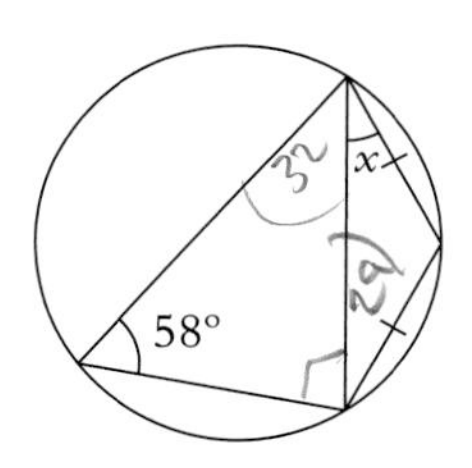

8

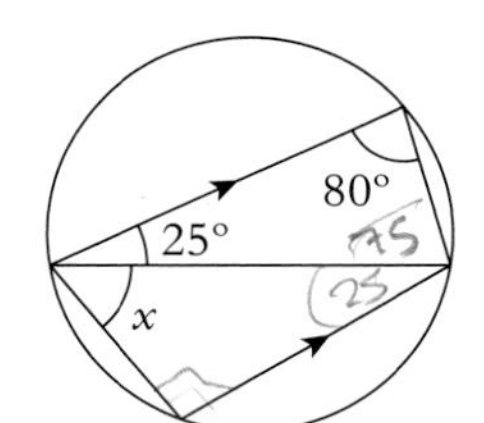

9

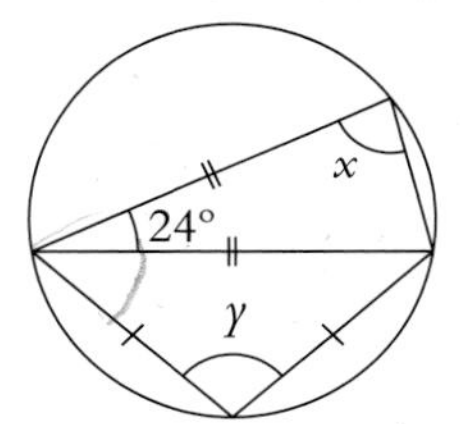

10

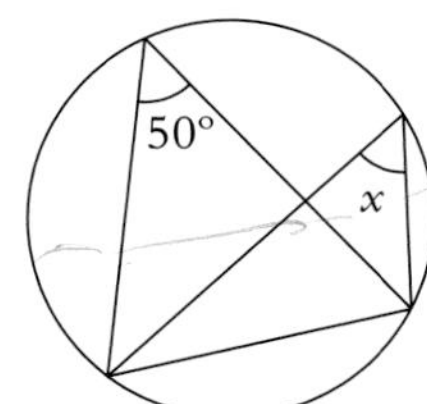

11

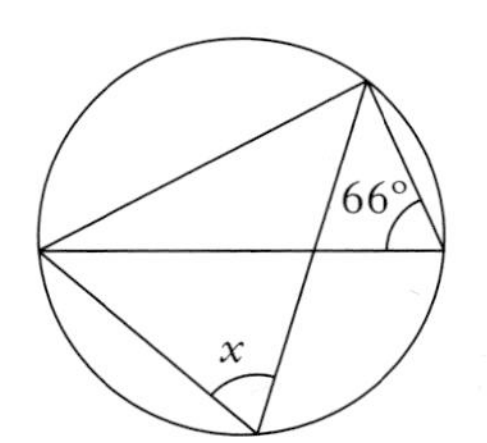

12

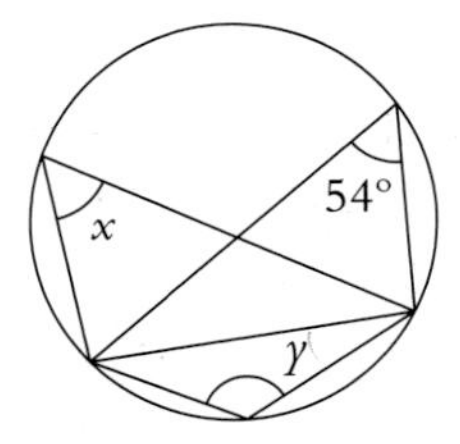

13

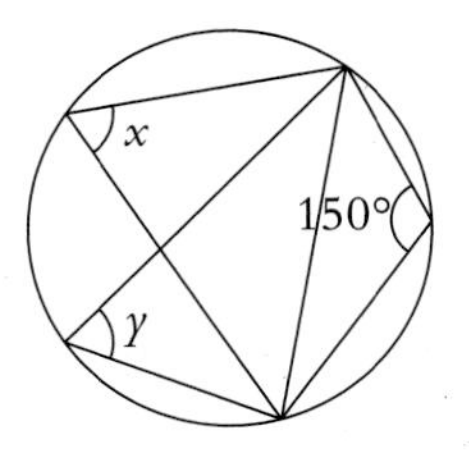

14

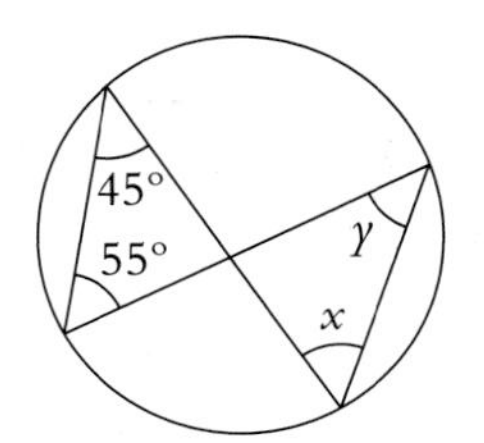

15

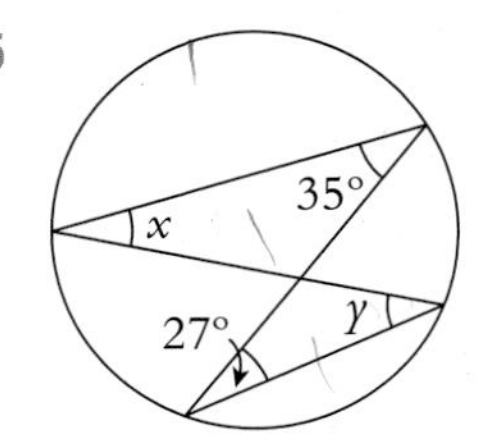

UNIT 6

Summary of circle theorems:

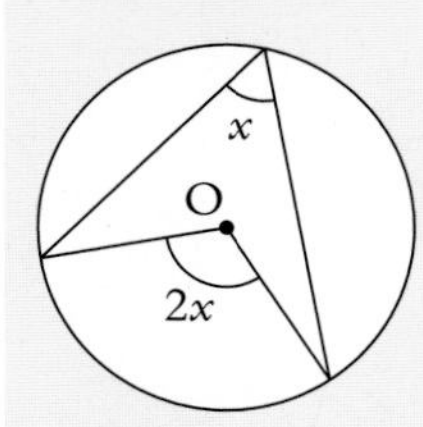

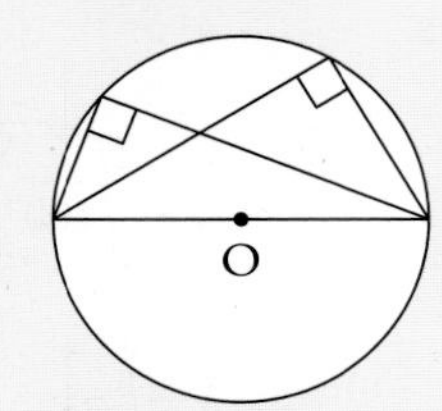

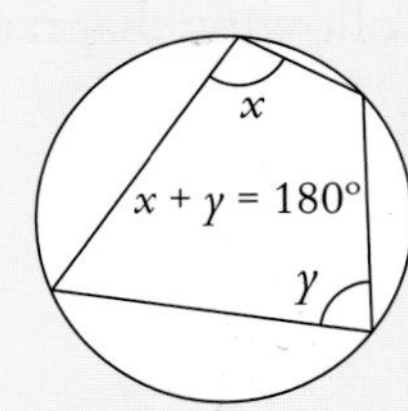

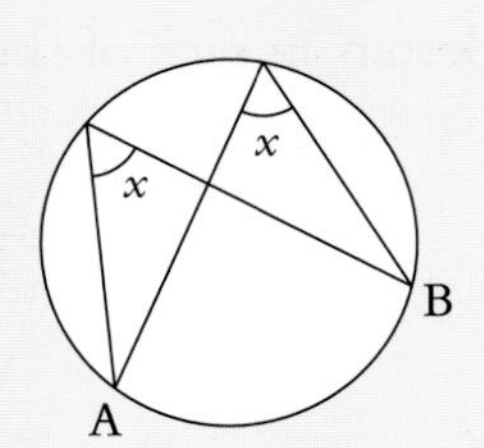

You need to learn all these rules together with the properties of tangents and chords.

EXAMPLE

Find the values of x and y.
Give reasons for your answers.

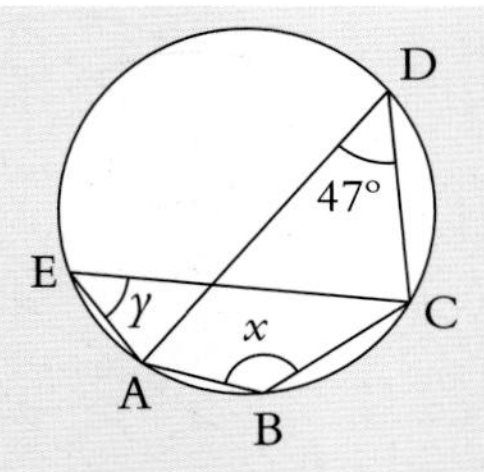

$x = 133°$ *opposite angles of a cyclic quadrilateral*
$y = 47°$ *angles from the same arc in the same segment*

EXAMPLE

Find the values of x, y and z.
Give reasons for your answers.

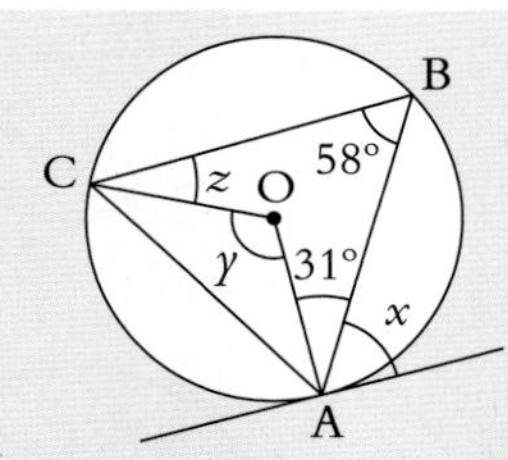

$x = 59°$ *tangent and radius at right angles*
$y = 116°$ *angle at centre = 2 × angle at circumference*

To find z you need to use quadrilateral OABC.
First find the angle at O.

reflex $A\hat{O}C = 244°$ *angles at a point add up to 360°*
$z = 360 - (58 + 31 + 244)$ *angles in a quadrilateral add up to 360°*
$z = 27°$

EXERCISE 6.22

Find the value of the letters in each of the following diagrams.

1

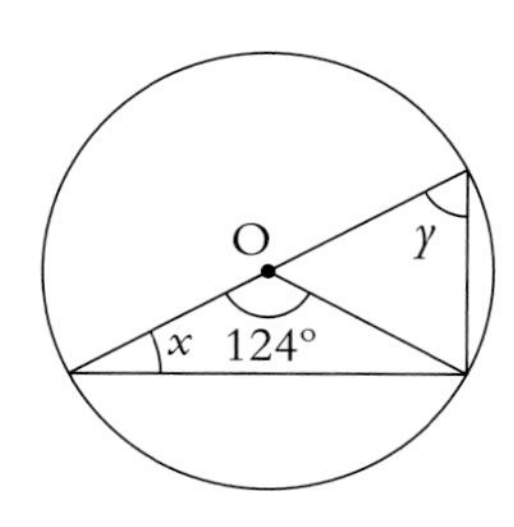

2

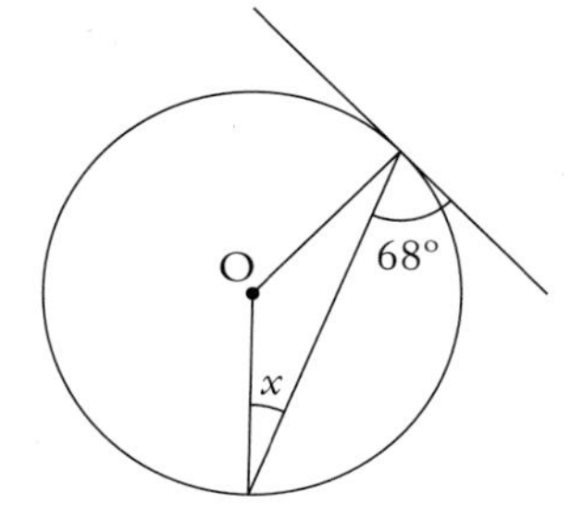

3

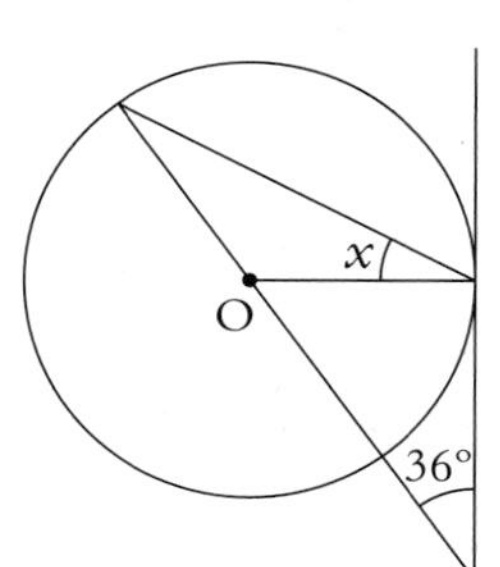

4

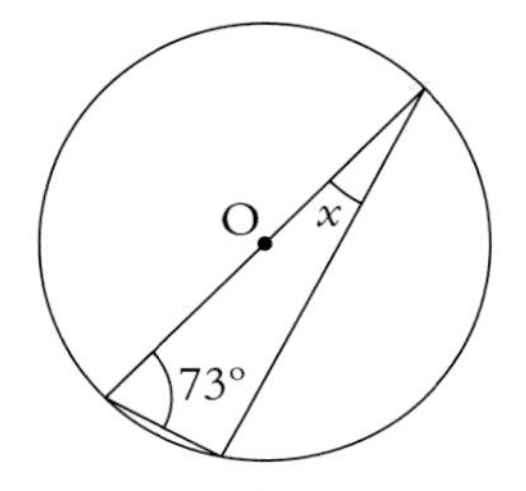

5

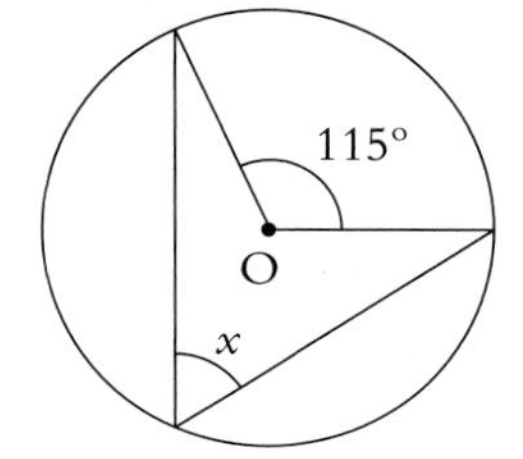

6

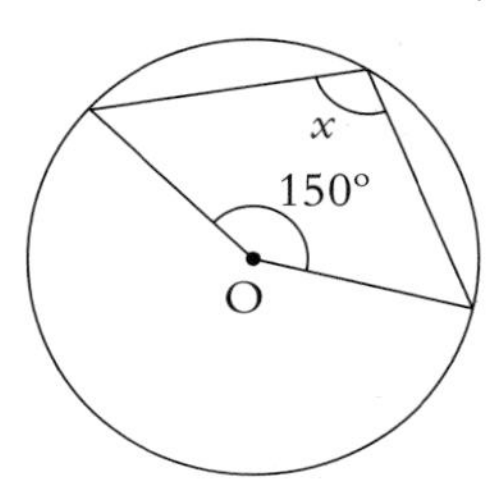

7

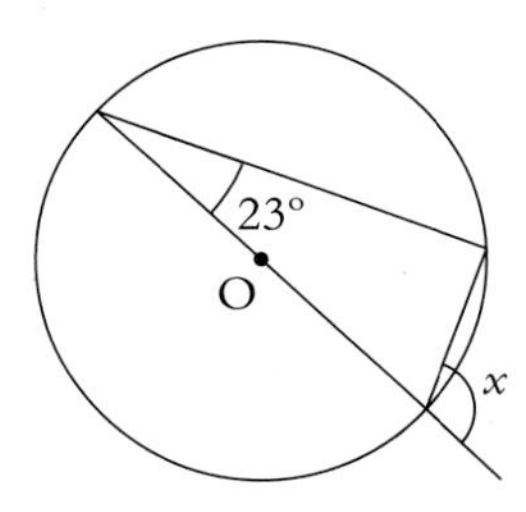

8

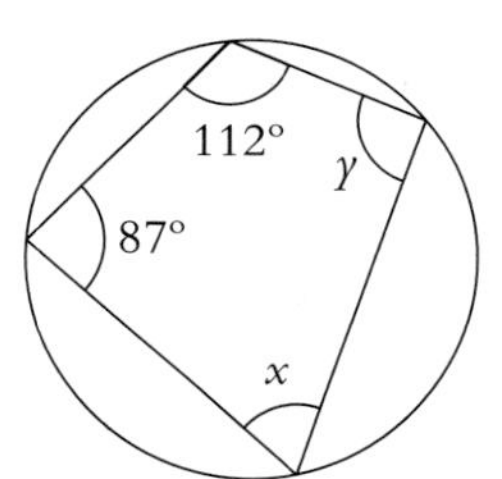

9

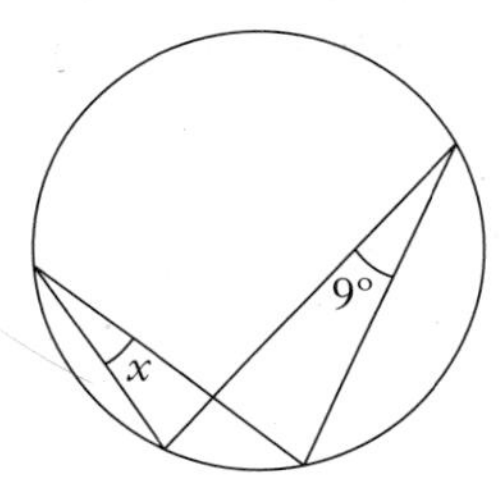

10

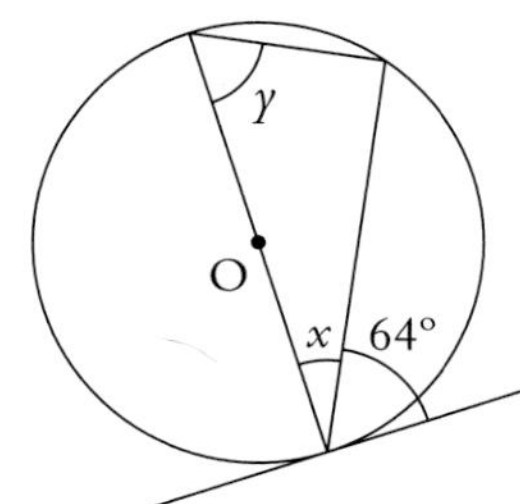

11

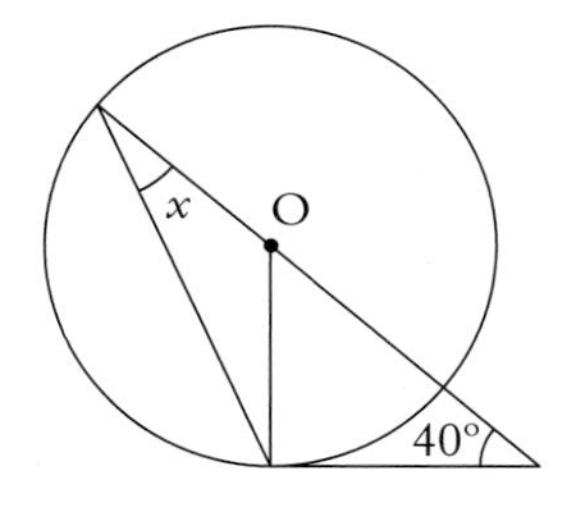

12

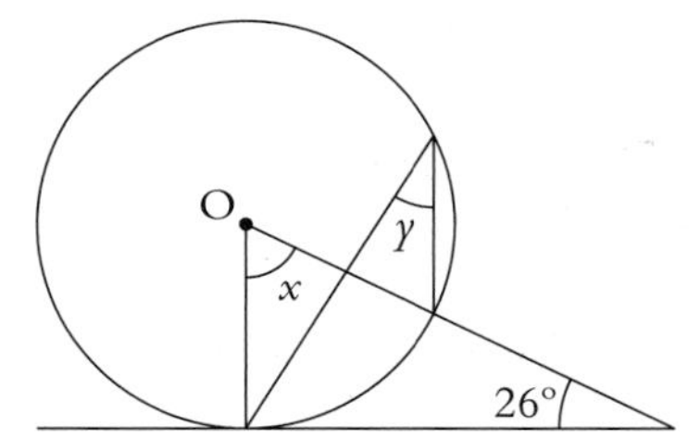

13

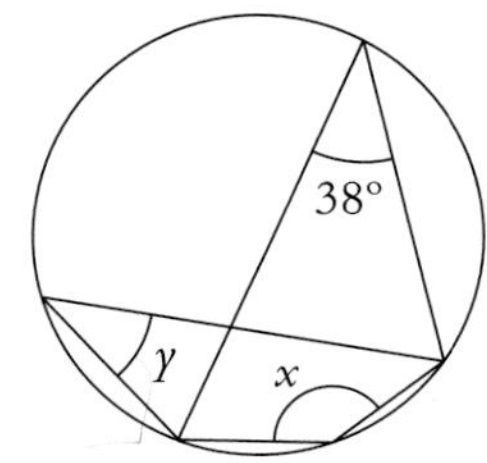

14

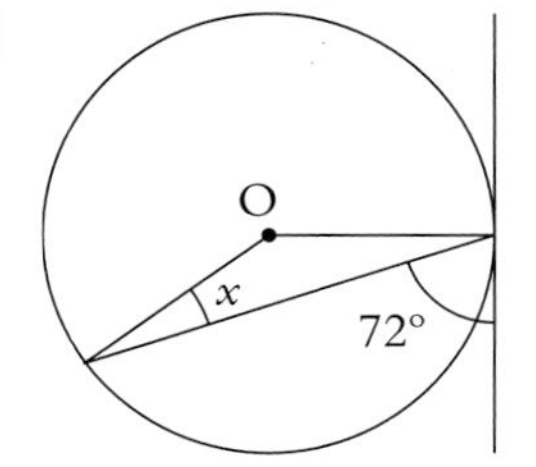

15

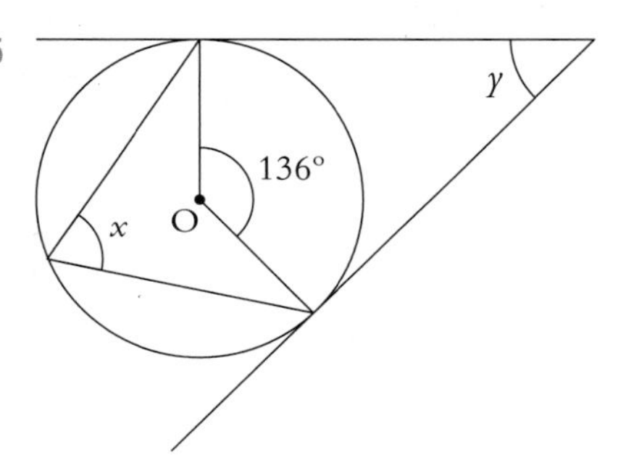

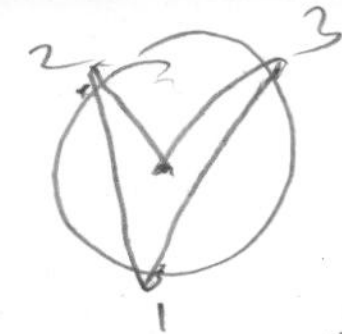

16

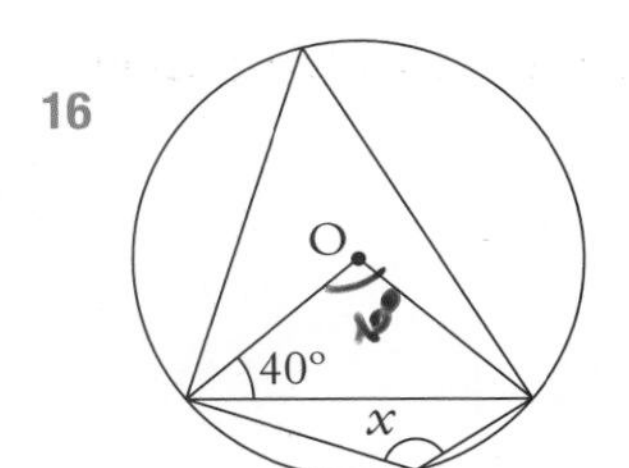

17

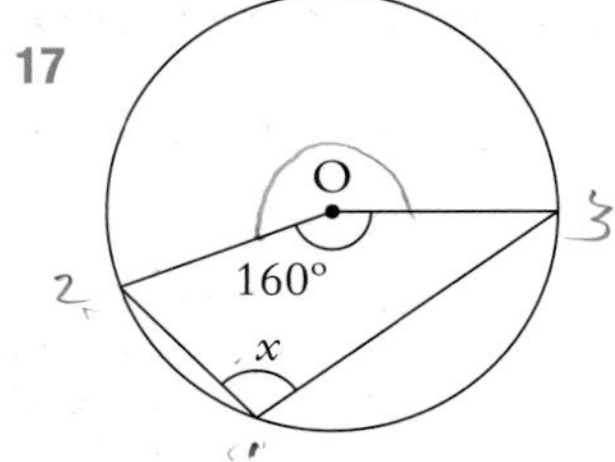

18

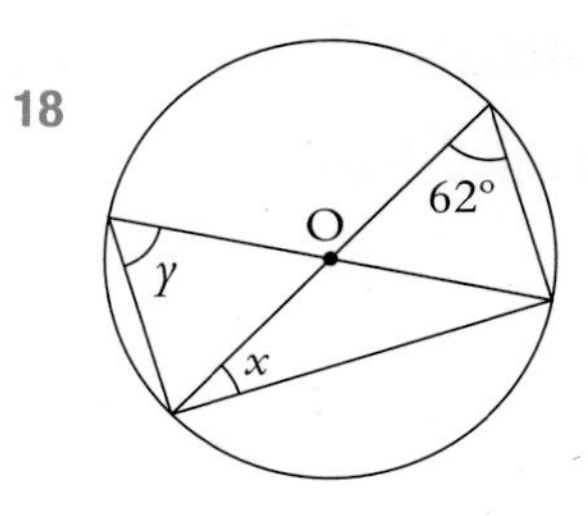

19

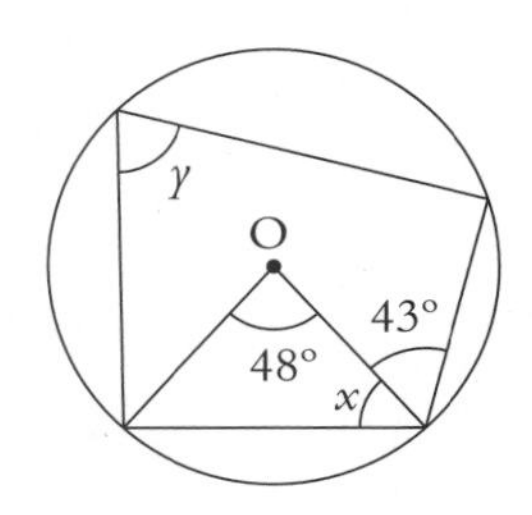

20

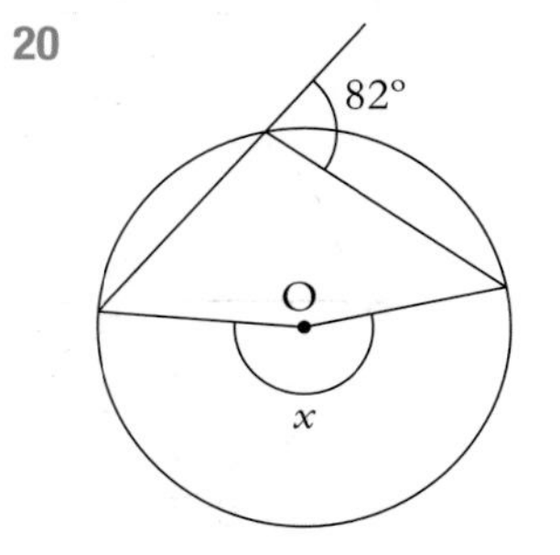

21

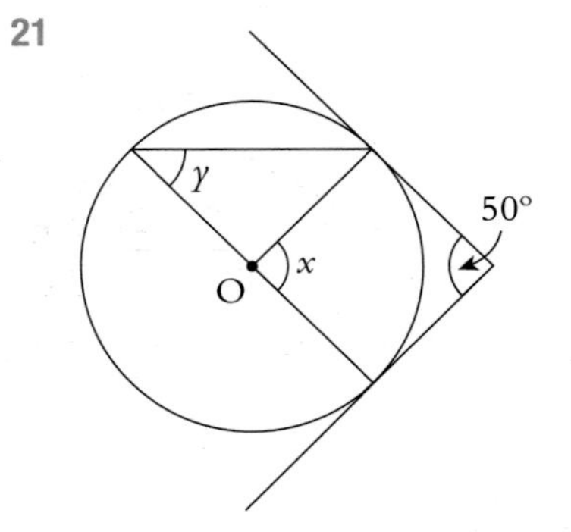

22

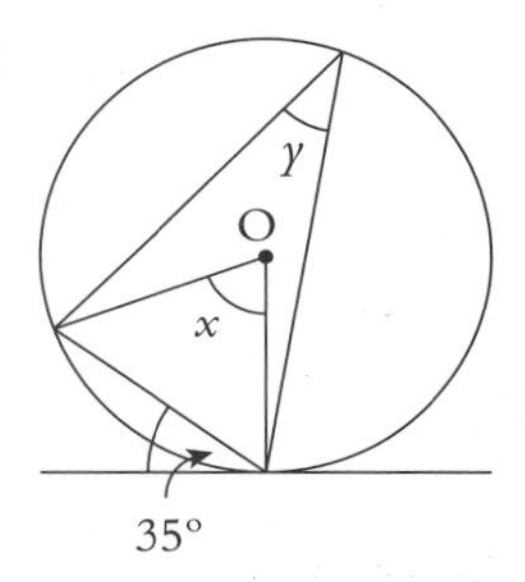

23

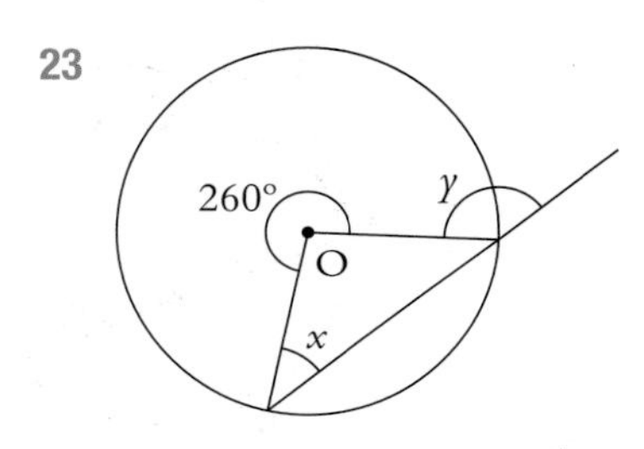

24

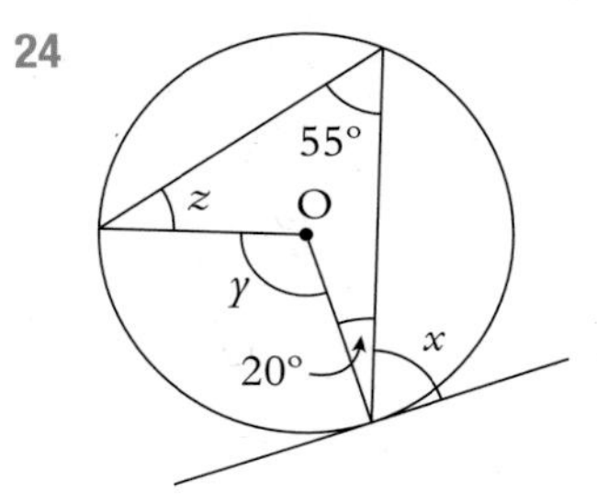

25

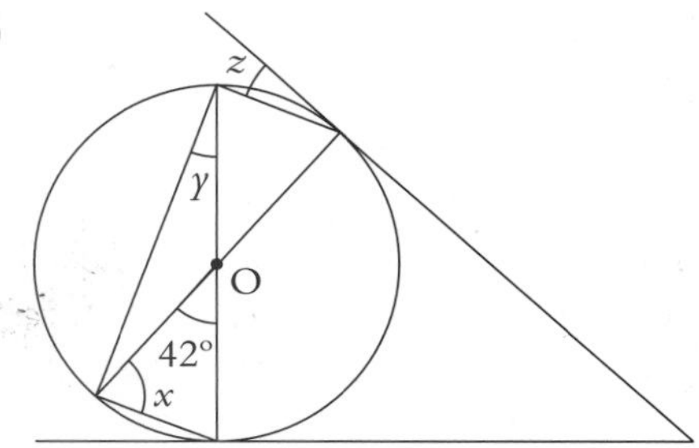

26

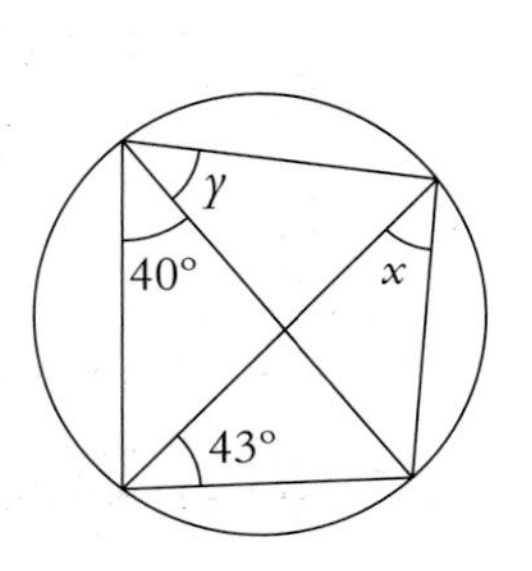

27

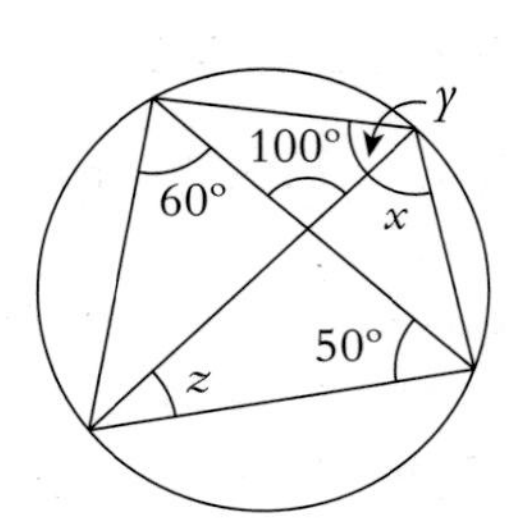

28

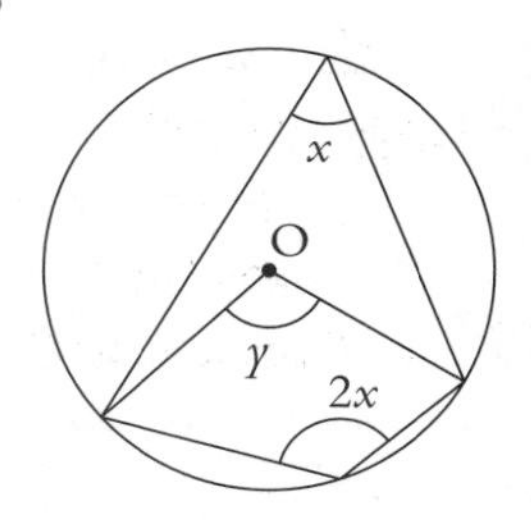

29

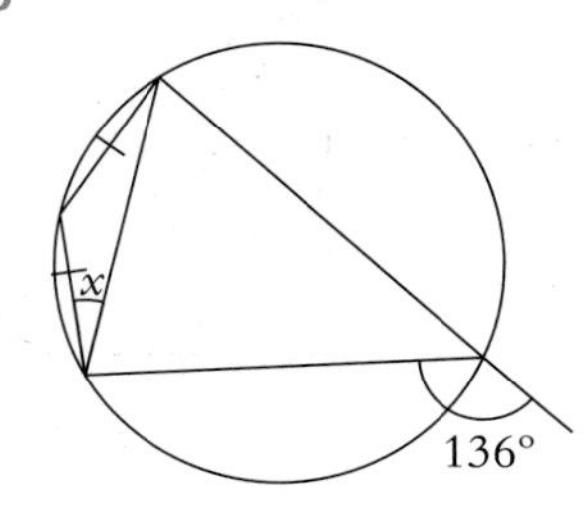

30

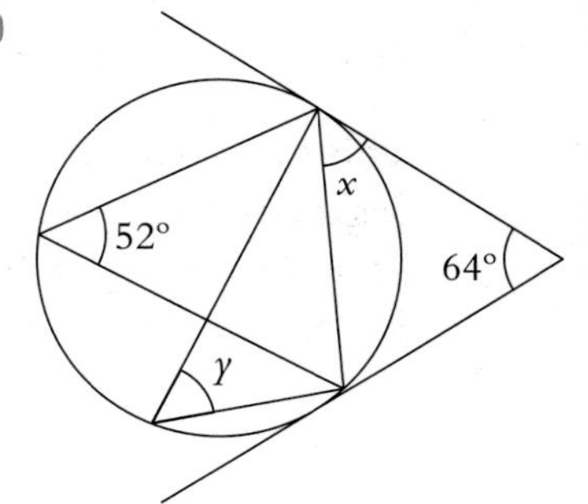

For questions 31–36 explain how you calculate each variable using the various circle theorems etc.

31

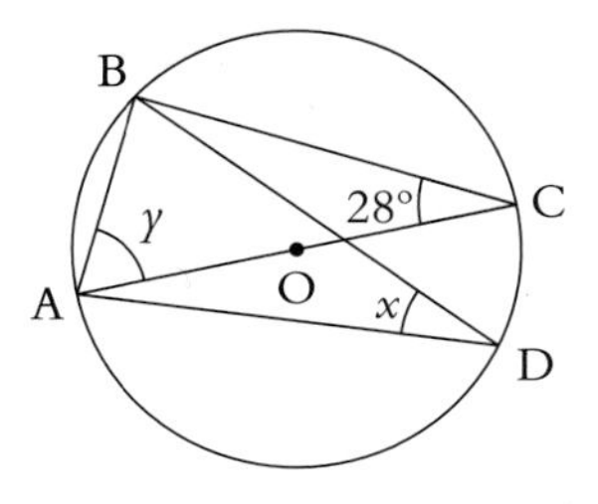

32

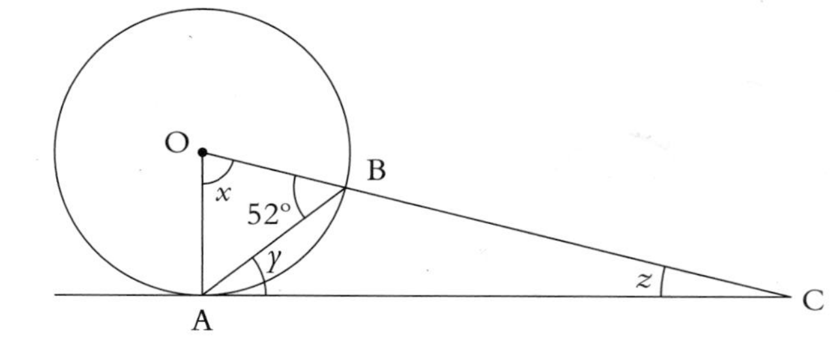

33

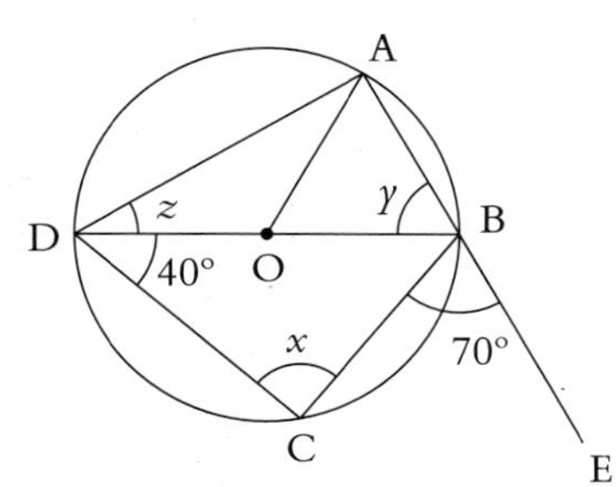

34

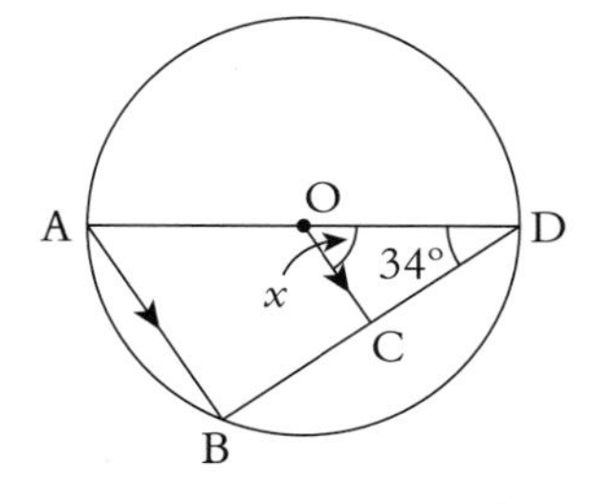

35

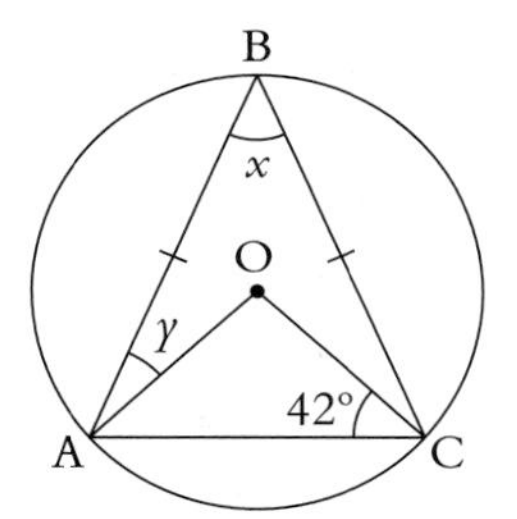

36

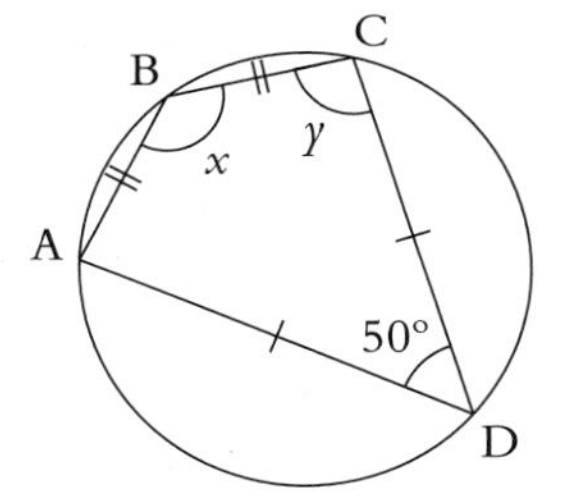

37 Use your answers to question 36 to make a general statement about cyclic kites.

CHALLENGE 1

Prove that $B\widehat{C}D = x$.

This means that the exterior angle of a cyclic quadrilateral is equal to the opposite interior angle.
(You do not have to learn this rule for the examination.)

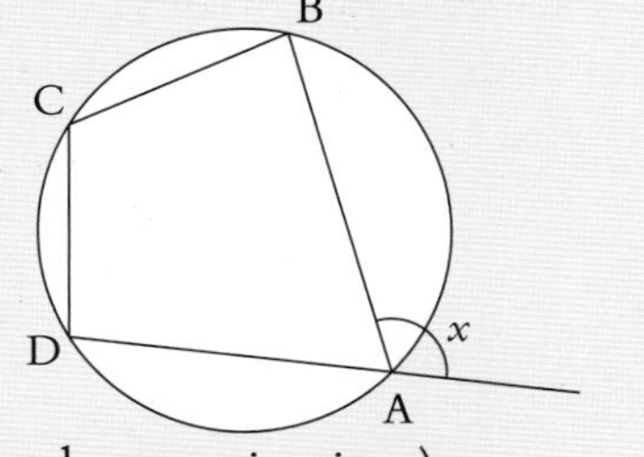

CHALLENGE 2

Prove that $A\widehat{C}B = x$.

This rule is called The Alternate Segment Theorem.
(You do not have to learn this rule for the examination.)

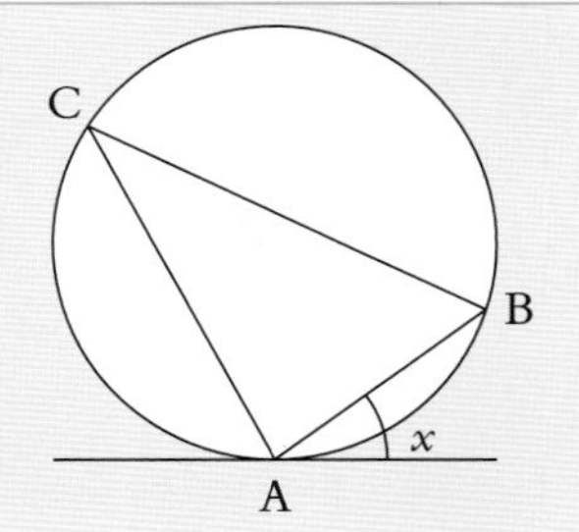

KEY WORDS

tangent
chord
segment
cyclic quadrilateral

Probability 2

THIS SECTION WILL SHOW YOU HOW TO

- Use tree diagrams to represent outcomes of compound events
- Use tree diagrams to calculate probabilities of compound events

Combined independent events

Two events A and B are **independent** if one event happening has no effect on whether or not the other event happens. If A and B are independent events, then

P(A happening and B happening) = P(A happening) × P(B happening)

P(A and B) = P(A) × P(B)

EXAMPLE

A fair coin is tossed and a fair three-sided spinner is spun.

Find the probability of getting a head on the coin and a 2 on the spinner.

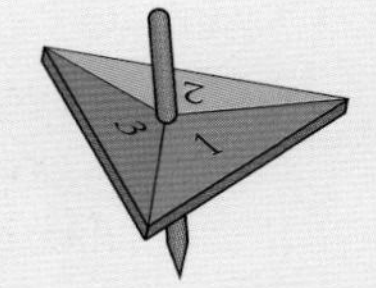

P(head and 2) = P(H and 2) = P(H) × P(2) = $\frac{1}{2} \times \frac{1}{3} = \frac{1}{6}$

Tree diagrams for independent events

A **tree diagram** is a clear way of showing all the possible outcomes of combined events. They can also be used to calculate probabilities of combined events.

You must draw your tree diagram clearly and write the probabilities on each branch.

EXAMPLE

A bag contains 3 red balls and 2 yellow balls.

A ball is chosen at random from the bag, its colour noted and it is then replaced in the bag.

A second ball is then chosen at random from the bag.

a Draw a tree diagram to show all the possible outcomes.

b Find the probability that **i** both balls are red **ii** both balls are yellow **iii** different coloured balls are chosen.

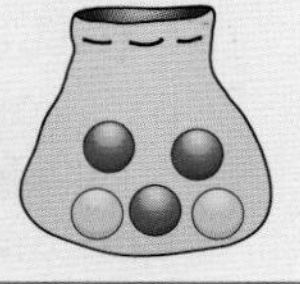

a

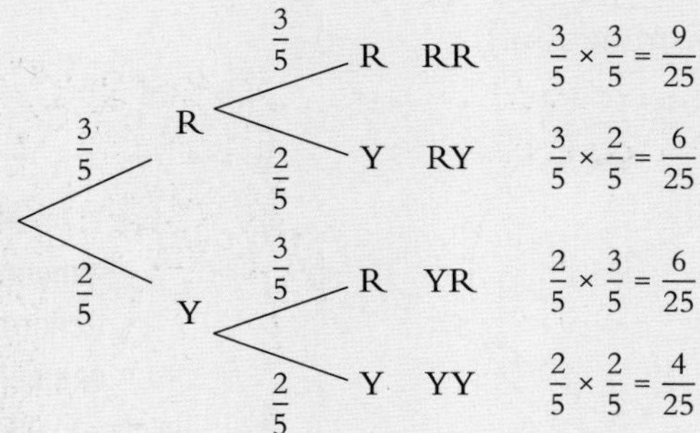

$\frac{3}{5} \times \frac{3}{5} = \frac{9}{25}$

$\frac{3}{5} \times \frac{2}{5} = \frac{6}{25}$

$\frac{2}{5} \times \frac{3}{5} = \frac{6}{25}$

$\frac{2}{5} \times \frac{2}{5} = \frac{4}{25}$

➡ **NOTE:** multiply along the branches.

➡ **CHECK:** the probabilities of all the routes through the diagram should add to 1.

➡ **NOTE:** if you need more than one of the outcomes add the probabilities together.

b

i P(RR) = $\frac{9}{25}$ **ii** P(YY) = $\frac{4}{25}$ **iii** P(different colours) = P(RY) + P(YR) = $\frac{6}{25} + \frac{6}{25} = \frac{12}{25}$

EXERCISE 6.24

1 A bag contains 5 orange balls and 2 blue balls.
Raju chooses a ball at random from the bag.
It is **not** replaced in the bag.
He then chooses a second ball at random from the bag.
Find the probability that he chooses

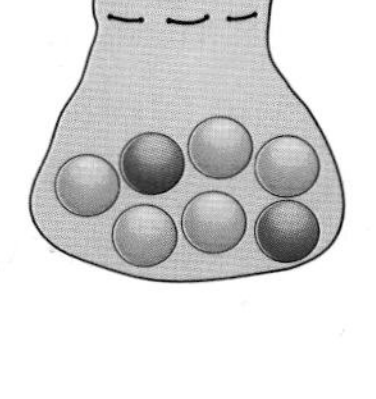

a two orange balls,
b two blue balls,
c two balls the same colour,
d two different coloured balls.

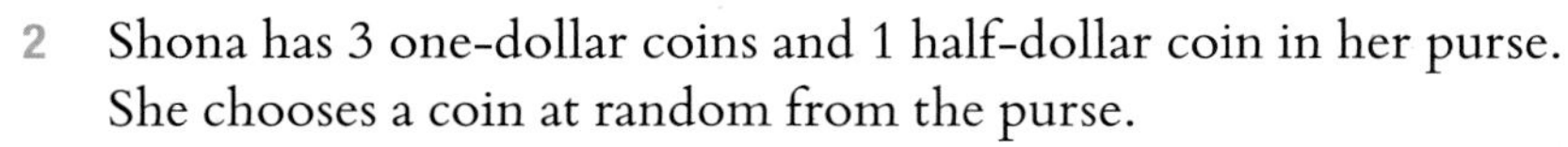

2 Shona has 3 one-dollar coins and 1 half-dollar coin in her purse.
She chooses a coin at random from the purse.
It is **not** replaced in the purse.
She then chooses a second coin at random from the purse.
Find the probability that she chooses
a 2 one-dollar coins,
b 2 half-dollar coins,
c at least 1 one-dollar coin.

3 A box contains 4 blue crayons and 2 yellow crayons.
Kurt chooses a crayon at random from the box.
It is **not** replaced in the box.
He then chooses a second crayon at random from the box.
Find the probability that he chooses

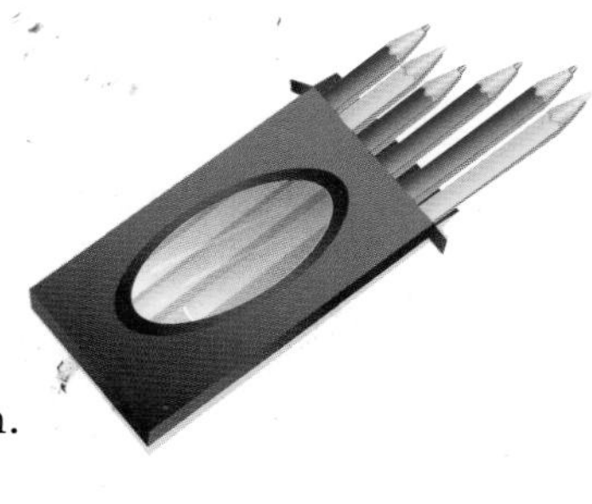

a two blue crayons,
b two yellow crayons,
c two different colour crayons,
d at least one yellow crayon.

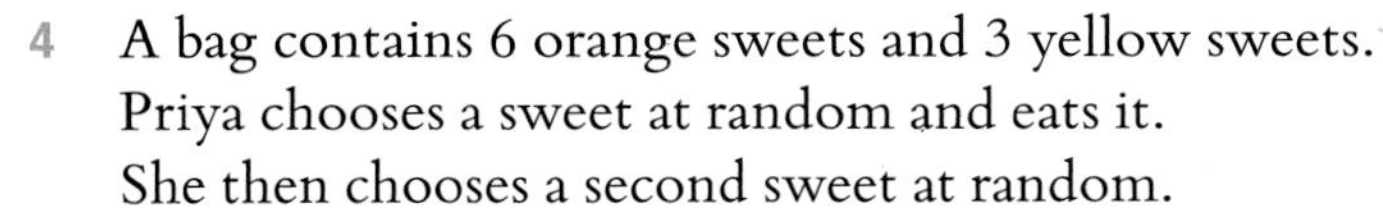

4 A bag contains 6 orange sweets and 3 yellow sweets.
Priya chooses a sweet at random and eats it.
She then chooses a second sweet at random.
Find the probability that she chooses

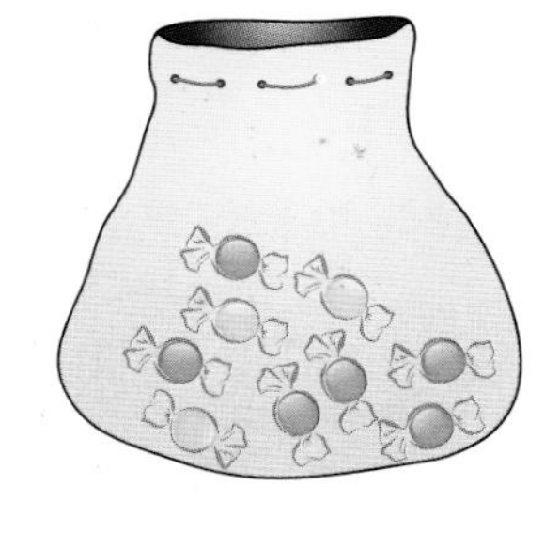

a two orange sweets,
b two yellow sweets,
c two different coloured sweets,
d at least one orange sweet.

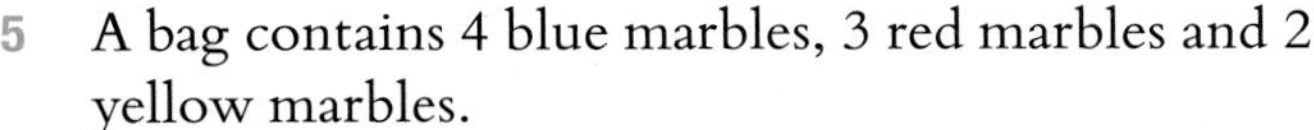

5 A bag contains 4 blue marbles, 3 red marbles and 2 yellow marbles.
Ali chooses a marble at random from the bag.
It is **not** replaced in the bag.
He then chooses a second marble at random from the bag.
Find the probability that he chooses

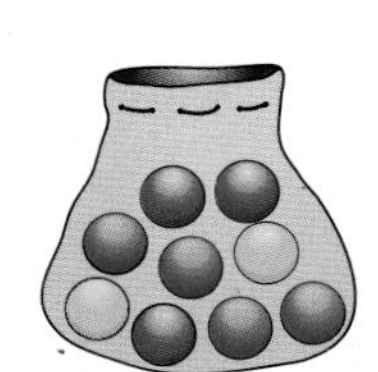

a two blues,
b two reds,
c two yellows,
d two of the same colour,
e two different colours.

6 A ball is chosen at random from bag A, the colour is noted and then the ball is placed in bag B.
A ball is then chosen at random from bag B.
Find the probability of choosing

a two blacks
b two whites
c two different colours
d at least one black.

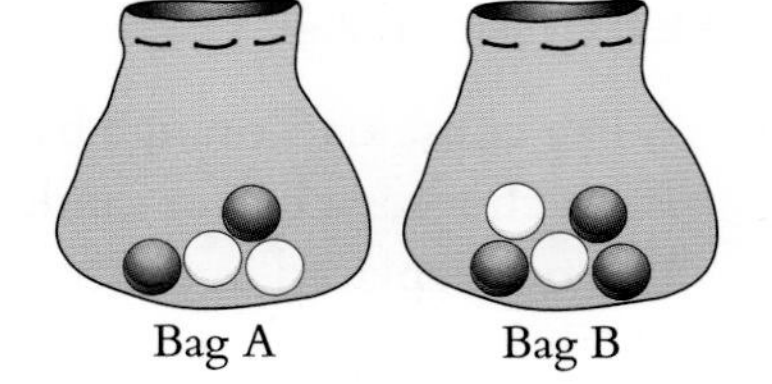

7 Omar either walks or cycles to school.
The probability that he cycles to school is 0.2.
If Omar cycles, the probability that he is late is 0.1.
If Omar walks, the probability he is late is 0.3.

a Copy and complete the tree diagram.
(C = cycle W = walk L = late and $\bar{L}$ = not late)

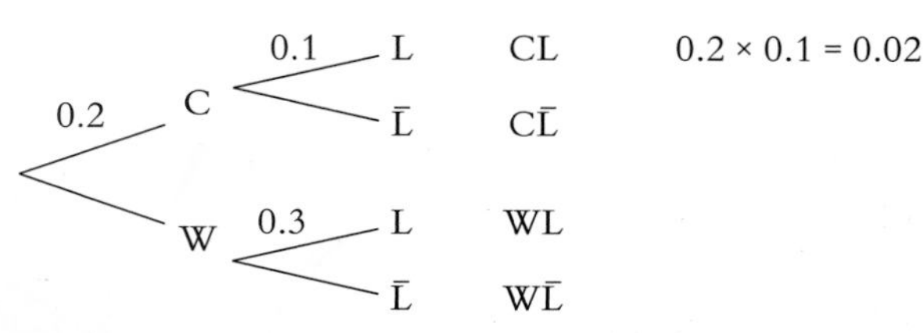

b Calculate the probability that Omar will be
i late to school,
ii not late to school.

8 a Jacob is asked to select one shape at random from either bag A or bag B.
The probability that he chooses from bag A is $\frac{2}{5}$.

i Copy and complete the tree diagram.

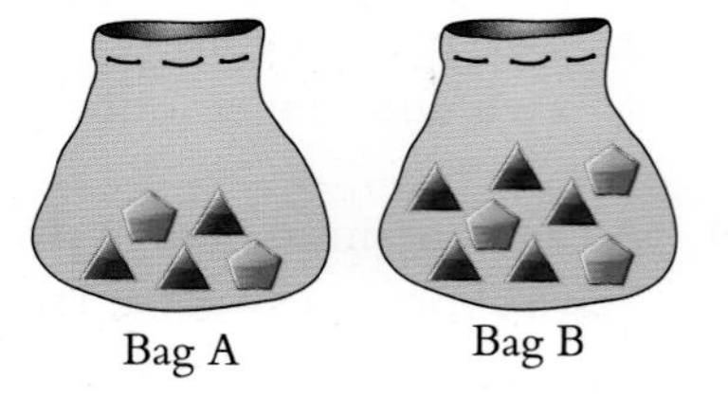

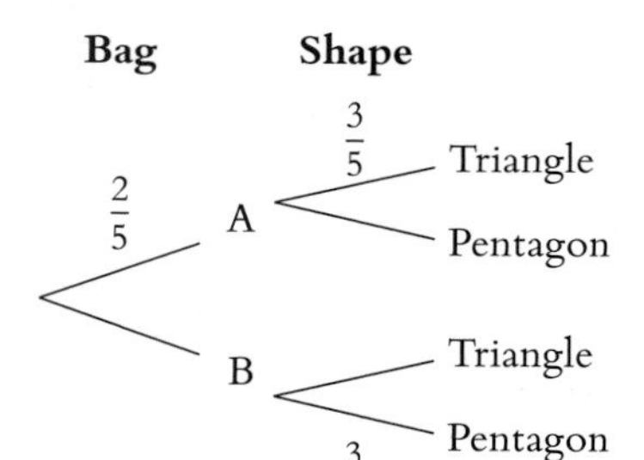

ii Calculate the probability that Jacob chooses a triangle.
iii Calculate the probability that Jacob chooses a pentagon.

b Eduardo is asked to select 2 shapes at random from bag A or two shapes at random from bag B. (He does not replace the first shape before selecting the second shape.)
The probability that he chooses from bag A is $\frac{1}{3}$.

i Calculate the probability that Eduardo chooses two triangles.
ii Calculate the probability that Eduardo chooses two different shapes.

Further probability

The probability that Anna hits the bullseye on a dartboard is 0.1.
Find the probability that she hits the bullseye for the first time on her fourth attempt.

She must miss on her first three attempts and hit on her fourth attempt.

Let M = miss and H = hit.

P(M) = 0.9 and P(H) = 0.1.

$$P(M\ M\ M\ H) = P(M) \times P(M) \times P(M) \times P(H)$$
$$= 0.9 \times 0.9 \times 0.9 \times 0.1 = 0.0729$$

EXERCISE 6.25

1 The circular board is divided into eight numbered sections. When the arrow is spun it is equally likely to stop in any of the eight sections. The spinner is spun several times until it lands on the number 2. Find the probability that this happens on the fourth spin.

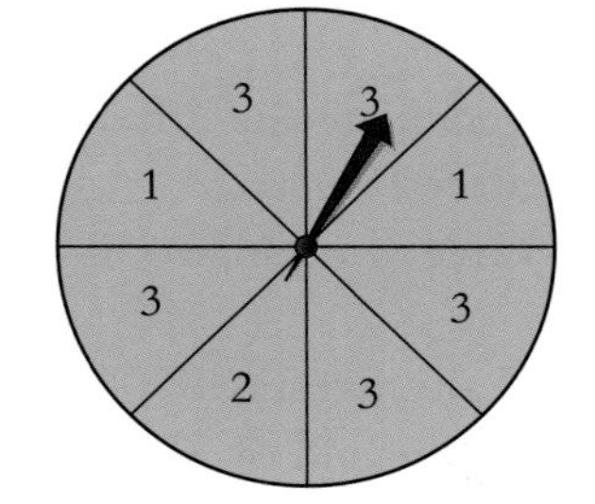

2 A fair coin is tossed repeatedly until it lands on a head. Find the probability that this happens on the sixth toss.

3 A six-sided dice is unbiased. It is numbered 1, 2, 3, 4, 5 and 6 on its faces. The dice is rolled several times until it lands on the number 5. Find the probability that this happens on the fifth roll.

4 The probability that Robert will pass his driving test on his first attempt is $\frac{2}{3}$.
If he fails, the probability that he passes on any future attempt is $\frac{3}{4}$.
Find the probability that he passes on his fifth attempt.

5 The bag contains 5 black marbles and 4 white marbles. Marbles are selected at random from the bag until a white marble is selected. (The marbles are **not** replaced.) Find the probability that this happens on the third selection.

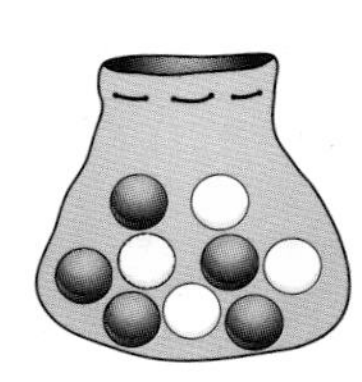

KEY WORDS
tree diagram
independent events
dependent events

UNIT 6

Unit 6 Examination-style questions

1 Tom pays \$528.90 for a computer in a sale.
The original price had been reduced by 18%.

Calculate the original price of the computer. [3]

2 The cost of a holiday increases by 6% to \$882.

Calculate the original cost of the holiday. [3]

3 Colin invests \$600 at a rate of 1.5% per year compound interest.

Calculate the amount he has after 8 years. [2]

4 Carol invests \$5000 at a rate of 2% per year compound interest.
Phillip invests \$5000 at a rate of 2.5% per year simple interest.

Calculate the difference between these two investments after 10 years.
Give your answer in dollars correct to the nearest cent. [4]

5 Raju invests \$8500.
At the end of 7 years, he has an amount of \$10 454.

Find the yearly compound interest rate. [3]

6 Amanda and Paul both invest \$20 000.
Amanda invests her \$20 000 at a rate of 2% per year compound interest.
Paul invests his \$20 000 in a bank that x% pays per year simple interest.
After 5 years their investments are worth the same amount.

Calculate the value of x correct to 3 significant figures. [5]

7 Change 54 kilometres per hour to metres per second. [2]

(d) On the grid, draw the graph of $L = 2x + \frac{48}{x}$.

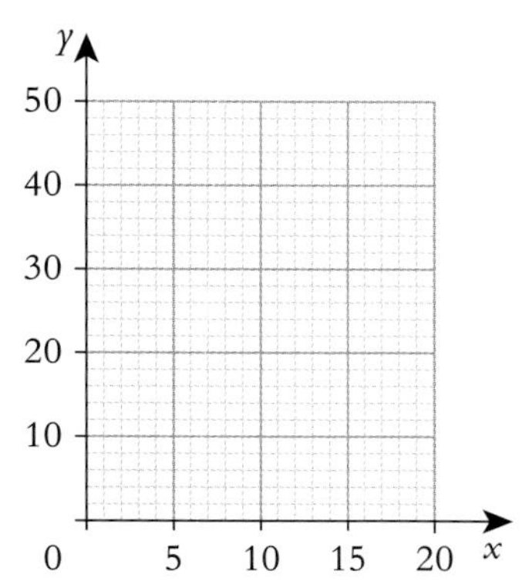

[3]

(e) What is the smallest length of fencing needed to make this rectangular enclosure? [1]

24 A is the point $(-3, 5)$ and B is the point $(5, 7)$.

(a) Find the equation of the straight line through A and B. [3]

(b) Find the equation of the perpendicular bisector of the line AB. [3]

25 Find the equation of the straight line that

- passes through the point $(0, -2)$ and
- is perpendicular to the line $4x + 3y = 12$.

Write your answer in the form $y = mx + c$. [3]

26 The coordinates of triangle ABC are $A(-5, 1)$, $B(3, -3)$ and $C(4, 4)$.
P is the foot of the perpendicular from C to the line AB.

(a) Find the equation of the line AB. [3]

(b) Find the equation of the line CP. [3]

(c) Find the coordinates of P. [2]

27

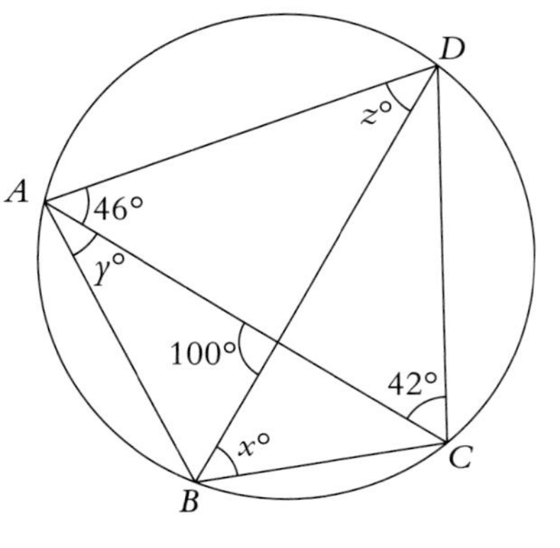

NOT TO SCALE

A, B, C and D lie on the circle.

Find the values of x, y and z. [4]

UNIT 6

28

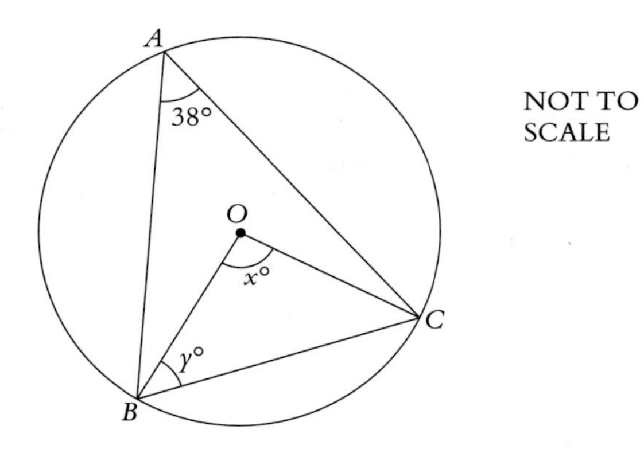

A, B and C lie on the circle, centre O.

Find the value of x and the value of y. [2]

29

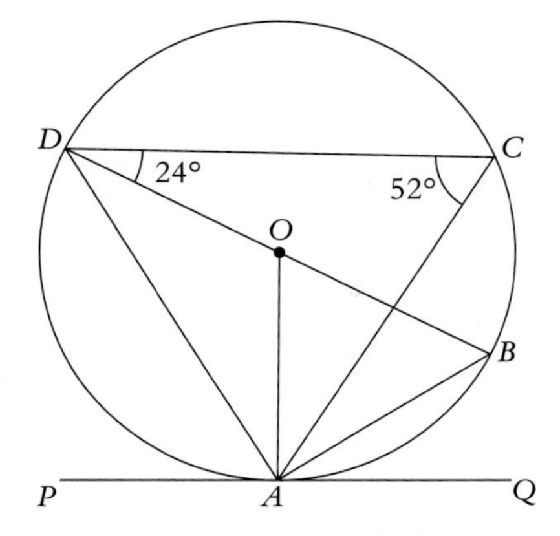

A, B, C and D lie on the circle, centre O.
BD is a diameter and PQ is a tangent at the point A.

(a) Find angle ABO. [1]
(b) Find angle BAQ. [1]
(c) Find angle AOB. [1]
(d) Find angle ODA. [1]
(e) Find angle OAC. [1]

30 The probability that Robert is late to school is 0.7

Calculate the probability that Robert is late to school for each of the next two days. [2]

31 Janet spins a 6-sided spinner numbered 1 to 6.
The probability of it landing on each number is shown in the table.

Number	1	2	3	4	5	6
Probability	0.21	0.07	x	0.15	0.22	0.21

(a) Find the value of x. [1]

(b) Janet spins the spinner twice.

(i) Find the probability that it lands on the number 4 both times. [2]

(ii) Find the probability that the sum of the two numbers is 10. [3]

32 The probability that a biased dice lands on the number six is $\frac{1}{4}$.

(a) The dice is rolled two times.

(i) Complete the tree diagram.

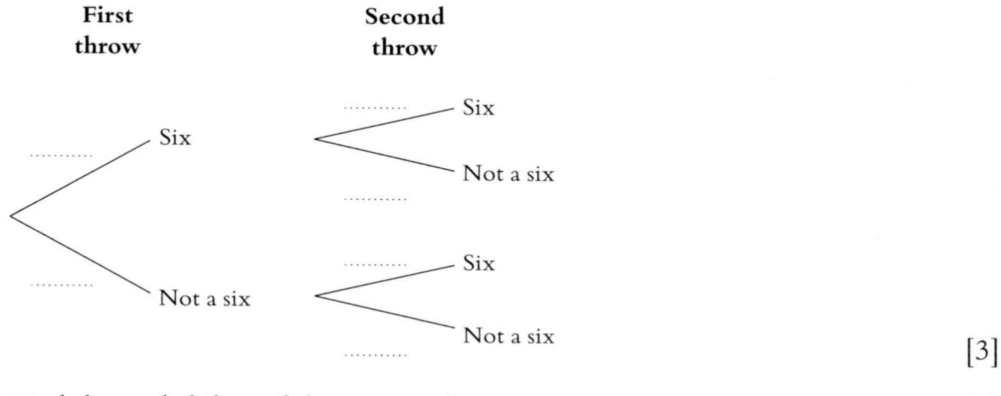

[3]

(ii) Find the probability of throwing at least one six. [3]

(b) The dice is rolled three times.
Find the probability of throwing exactly two sixes. [4]

33 A bag contains 5 red sweets and 4 yellow sweets.
Ali chooses a sweet at random from the bag and eats it.
He then chooses a second sweet at random.

Find the probability that he chooses two different coloured sweets. [3]

34

The circular board is divided into six numbered sections.

When the arrow is spun it is equally likely to stop in any of the six sections.
The spinner is spun several times until it lands on the number 2.

Find the probability that this happens on the fifth spin. [3]

35 A bag contains 5 red balls and 3 green balls.
Jane selects balls at random from the bag, without replacement, until a green ball is selected.

Find the probability that this happens on the fourth selection. [3]

36 In this question all lengths are in centimetres.

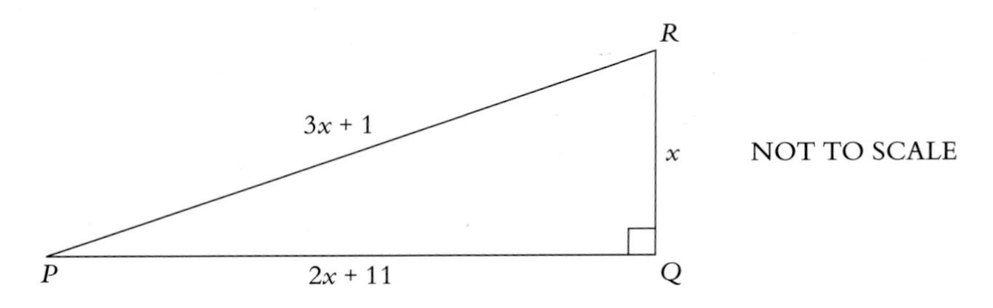

(a) Show that $2x^2 - 19x - 60 = 0$. [6]
(b) Solve the equation $2x^2 - 19x - 60 = 0$. [3]
(c) Write down the length of PQ. [1]

37 In this question all lengths are in centimetres.

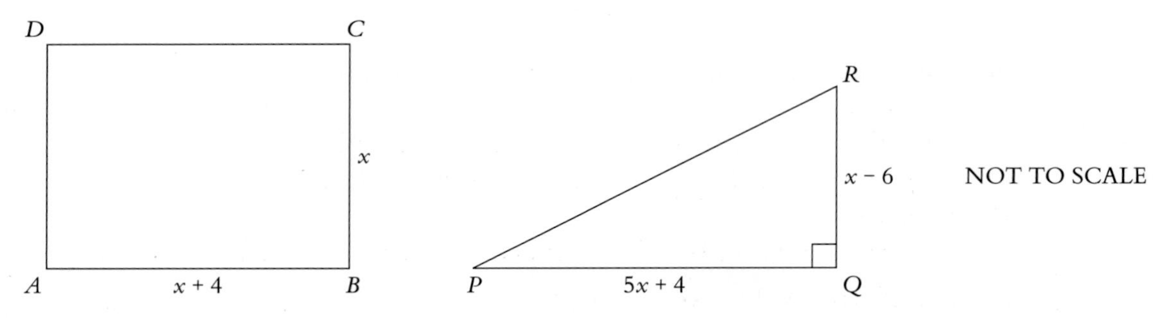

The area of the rectangle is equal to the area of the triangle.
(a) Show that $3x^2 - 34x - 24 = 0$. [3]
(b) Solve the equation $3x^2 - 34x - 24 = 0$. [3]
(c) Write down the length of AB. [1]

38

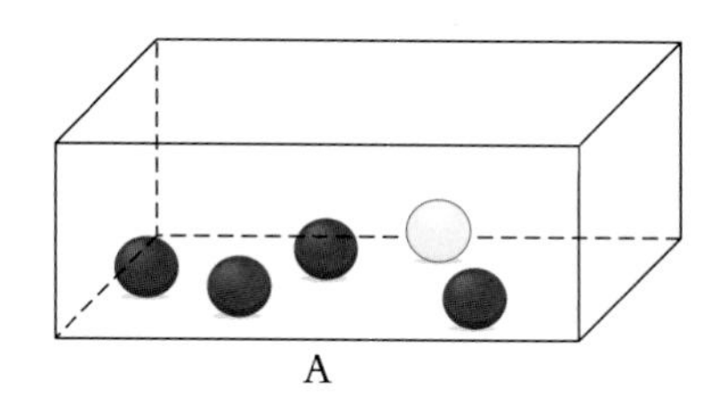

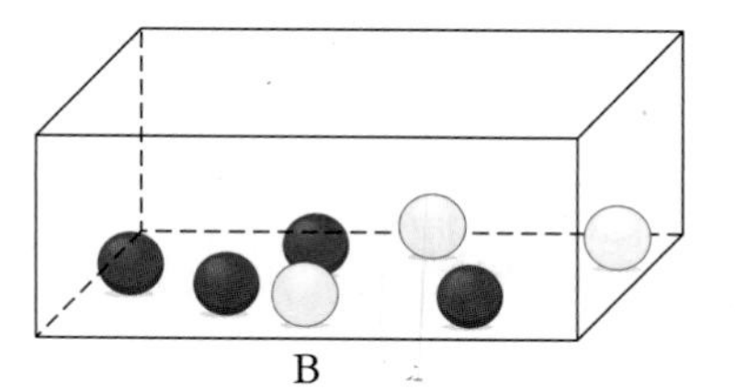

Box A contains 4 black balls and 1 white ball.
Box B contains 4 black balls and 3 white balls.

(a) A ball is chosen at random from box A.
Find the probability that this ball is white. [1]

(b) Anna chooses a box at random and then chooses a ball at random from this box.
The probability that she chooses box A is $\frac{3}{5}$.

(i) Copy and complete the tree diagram.

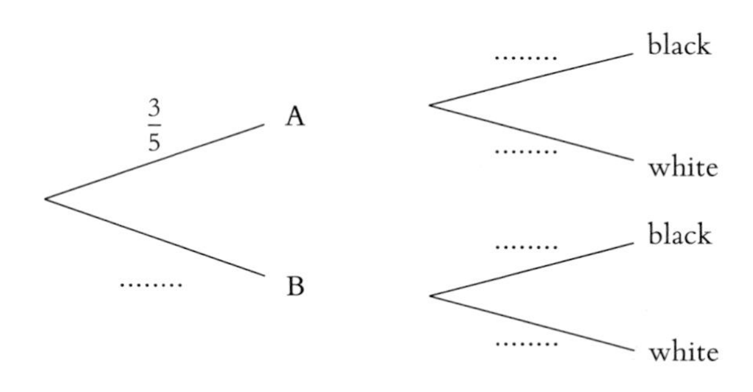

[2]

(ii) Find the probability that Anna chooses box B and a white ball. [2]

(iii) Find the probability that Anna chooses a black ball. [3]

(c) Robert is asked to choose 2 balls at random from bag A or 2 balls at random from box B.
He does not replace the first ball before choosing the second ball.
The probability that he chooses from bag A is $\frac{5}{8}$.

(i) Find the probability that Robert chooses two black balls. [2]

(ii) Find the probability that Robert chooses two different coloured balls. [3]

UNIT 6

Distance–time graphs

THIS SECTION WILL SHOW YOU HOW TO

- Read information from a distance–time graph
- Find the speed from a distance–time graph

You can use travel graphs of distance against time to find out how the distance is changing. Speed is calculate using the formula:

$$\text{speed} = \frac{\text{distance travelled}}{\text{time taken}}$$

So the steepness of a **distance–time graph** represents the speed.

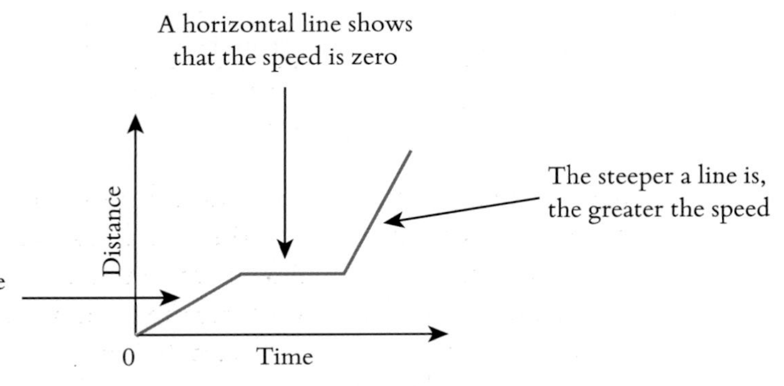

A straight line shows that the speed is constant (steady)

EXAMPLE

Roxy leaves home at 8:00 am to go on a bike ride.
She stops once for a rest.
The travel graph shows information about her journey.

a How long does she stop for a rest?
b At what time does she arrive back home?
c How far does she travel in total?
d Calculate her average speed, in km/h, for the whole journey.
e Calculate her fastest speed, in km/h, for the whole journey.

Distance from home (km)
Time (am)

---

a Four large squares on the horizontal axis represents one hour.
So, each large square represents 15 minutes. She stops for a rest for 15 minutes.

b She arrives back home at 10:30 am.

c Total distance travelled = 20 + 20 = 40 km.

d Average speed $= \frac{\text{total distance travelled}}{\text{total time taken}} = \frac{40}{2.5} = 16\text{ km/h}$

e Her fastest speed is when the graph is steepest. (This occurs between 8:00 and 8:30 am.)

Speed $= \frac{\text{distance travelled}}{\text{time taken}} = \frac{12}{0.5} = 24\text{ km/h}$

EXERCISE 7.1

1

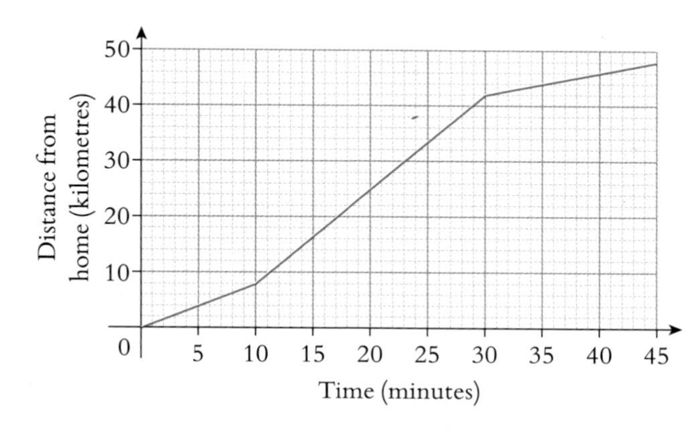

The distance–time graph shows information about Paulo's car journey to work.

a Calculate the speed of the car during the first stage of the journey.
Give your answer in **i** kilometres/minute, **ii** kilometres/hour.

b Calculate the speed of the car during the second stage of the journey.
Give your answer in **i** kilometres/minute, **ii** kilometres/hour.

c Calculate the speed of the car during the third stage of the journey.
Give your answer in **i** kilometres/minute, **ii** kilometres/hour.

d Calculate the average speed for the whole journey.
Give your answer in kilometres/hour.

2

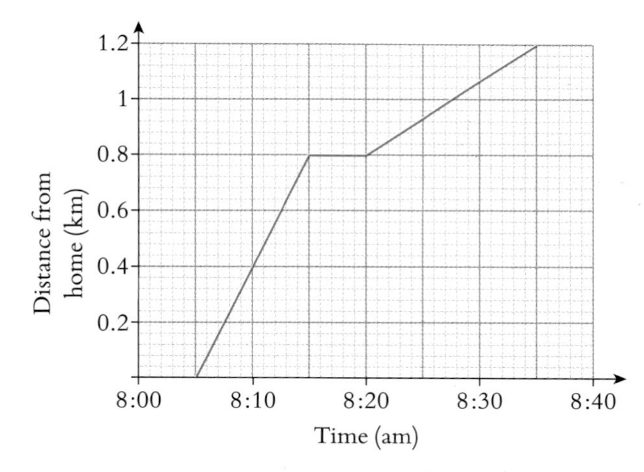

The travel graph shows information about Cara's walk to school each day.
She stops at a friend's house on the way.

a How long does she stop at her friend's house?

b Calculate her speed, in km/h, between 08:05 am and 08:15 am.
Give your answer in **i** kilometres/minute, **ii** kilometres/hour.

c Calculate her speed, in km/h, between 08:20 am and 08:35 am.
Give your answer in **i** kilometres/minute, **ii** kilometres/hour.

d Calculate her average speed, in km/h, for the whole journey.

UNIT 7

3

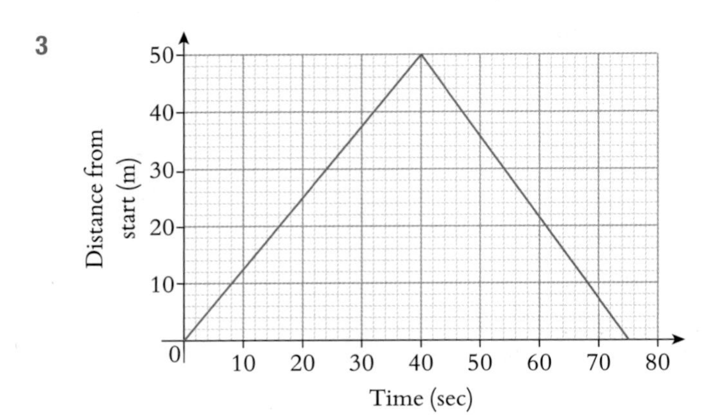

Henri swims in a 100-metre race. The swimming pool is 50 metres long. He swims the first length in 40 seconds and the second length in 35 seconds.

- **a** Calculate his speed, in m/s, for the first length.
- **b** Calculate his speed, in m/s, for the second length.
- **c** Calculate his average speed, in m/s, for the total 100 metres.

4

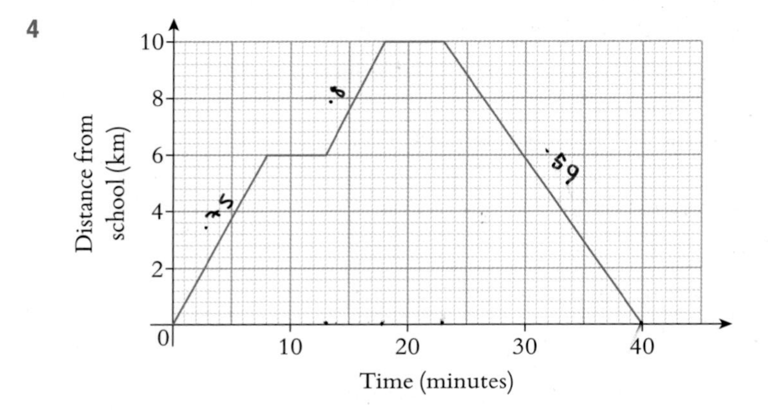

A school bus takes students home from school.
The bus stops at two villages. The bus then returns to school.
The distance–time graph shows information about the journey.

- **a** How far are the two villages from the school?
- **b** How long does the bus stop at each of the two villages?
- **c** Calculate the greatest speed of the bus.
  Give your answer in **i** kilometres/minute, **ii** kilometres/hour.
- **d** Calculate the average speed for the whole bus journey.
  Give your answer in **i** kilometres/minute, **ii** kilometres/hour.

UNIT 7

5

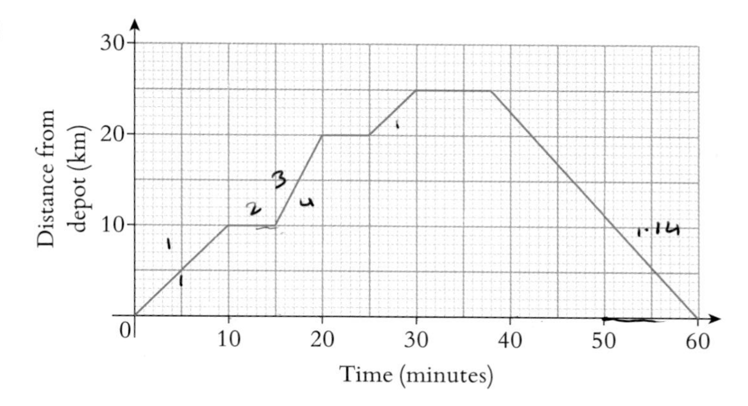

The distance–time graph shows information about the journey of a delivery van.

It starts at the depot and returns to the depot after 60 minutes.

- **a** How many times does the delivery van stop?
- **b** How many minutes did the delivery van stop altogether?
- **c** For which part of the journey did the delivery van travel fastest?
- **d** What is the total distance travelled by the delivery van?
- **e** Calculate the delivery van's average speed for the whole journey.

6 A train (P) leaves station A at 8 am and travels at a constant speed of 240 km/h towards station B.

The distance between the two stations is 180 km.

- **a** Show that the train arrives at station B at 8:45 am.

At 8:15 am a second train (Q) leaves station B and travels towards station A at a constant speed of 200 km/h.

- **b** Calculate the time that the train arrives at station A.
- **c** Copy and complete this travel graph to show the journey of the second train.

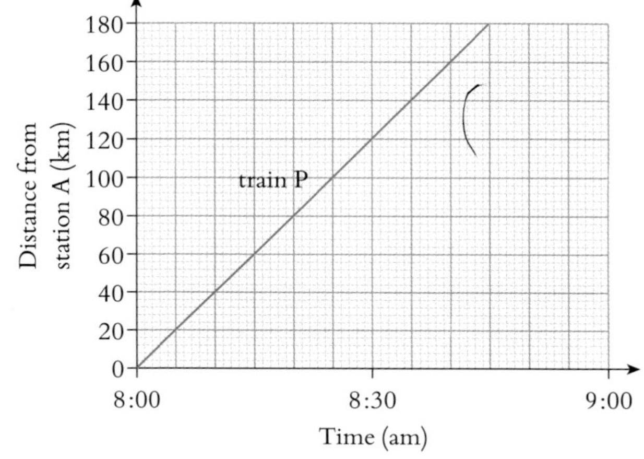

- **d** Use your graph to estimate the time at which the two trains pass each other.

Speed–time graphs

THIS SECTION WILL SHOW YOU HOW TO

- Find the acceleration and distance travelled from a speed–time graph

You can use travel graphs of speed against time to find out how the speed is changing. The rate of change of speed is called the **acceleration**.

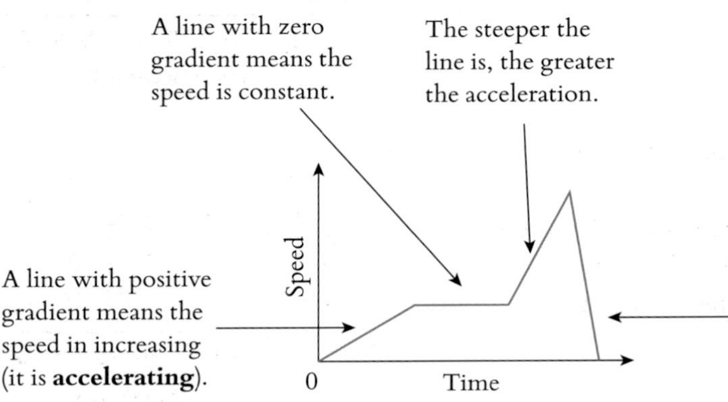

A line with negative gradient means the speed in decreasing (it is **decelerating**).

You must learn the following two rules:

In a speed–time graph,
acceleration = gradient of line = $\frac{\text{change in speed}}{\text{time}}$ distance travelled = area under the graph

The units for acceleration are metres per second or m/s^2 (or ms^{-2}).

EXAMPLE

The speed–time graph shows the speed of a car over 100 seconds.

a Calculate the acceleration of the car over the first 20 seconds.

b Calculate the acceleration of the car over the last 40 seconds.

c Calculate the distance travelled by the car in the 100 seconds.

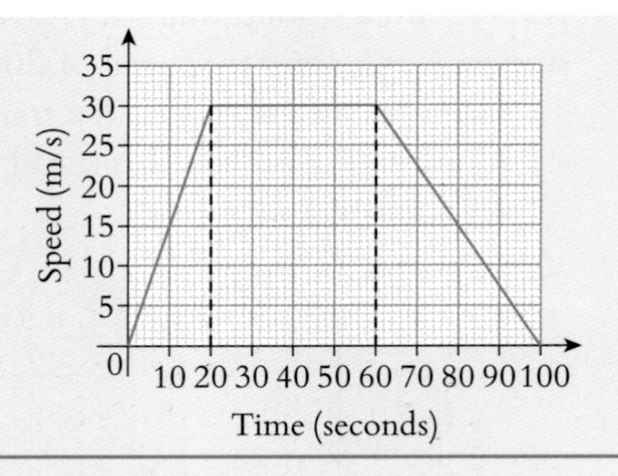

a Acceleration = gradient of line = $\frac{30}{20}$ = 1.5 m/s^2

b Acceleration = gradient of line = $-\frac{30}{40}$ = −0.75 m/s^2

➡ **NOTE:** this can also be called a deceleration of 0.75 m/s^2.

c Distance travelled = area under the graph

$= \left(\frac{1}{2} \times 20 \times 30\right) + (40 \times 30) + \left(\frac{1}{2} \times 40 \times 30\right)$

= 2100 m

➡ **NOTE:** the quickest method for finding the area under this graph is to use Area of trapezium = $\frac{1}{2}(a + b)h$

$= \frac{1}{2}(100 + 40) \times 30 = 2100$

EXERCISE 7.3

Make x the subject of these formulae.

1 $px - qx = r$

2 $ax = b + cx$

3 $ax = bx + c$

4 $x = a + bx$

5 $x + a = bx + c$

6 $ax - b = cx + 5$

7 $3 - 2x = ax + b$

8 $a^2 + ax + bx = 0$

9 $a(x - 3) = b(x + 4)$

10 $7(ax + b) = 3(cx - d)$

11 $a = \frac{x-2}{x+3}$

12 $x = \frac{ax+b}{c}$

13 $e = \frac{dx+1}{2-3x}$

14 $\frac{ax+b}{2} = \frac{cx+d}{4}$

15 $\frac{x-4}{a} = \frac{x+3}{b}$

16 $a = \frac{\sqrt{x+bx}}{c}$

17 $a = \sqrt{\frac{2+x}{x}}$

18 $2\sqrt{\frac{x-p}{x}} = q$

19 $x^2 = a + bx^2$

20 $ax^2 + b = c - dx^2$

21 $f = \frac{xy}{x+y}$

22 $2a = \frac{a^2x}{x+a}$

23

To make x the subject of the formula $a = \frac{b}{x} - c$ there are two possible methods:

Method 1

$a = \frac{b}{x} - c$	*multiply by x*
$ax = b - cx$	*collect x's*
$ax + cx = b$	*factorise*
$x(a + c) = b$	*divide by $(a + c)$*
$x = \frac{b}{a+c}$	

Method 2

$a = \frac{b}{x} - c$	*add c to both sides*
$a + c = \frac{b}{x}$	*multiply by x*
$x(a + c) = b$	*divide by $(a + c)$*
$x = \frac{b}{a+c}$	

a $\frac{5}{x} + a = b$

b $a - \frac{b}{x} = c$

c $\frac{p}{x} + 5q = 2r$

d $\frac{6f}{x} + g = 7h$

e $\frac{a}{x} + 2 = \frac{b}{x} + 3c$

f $\frac{a}{x} - b = c - \frac{d}{x}$

UNIT 7

Sequences

THIS SECTION WILL SHOW YOU HOW TO

- Write down and use the formula for the nth term of a sequence

Linear sequences

A **sequence** is a list of numbers or diagrams that are connected by a rule.

4, 7, 10, 13, … and 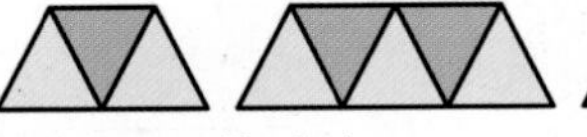 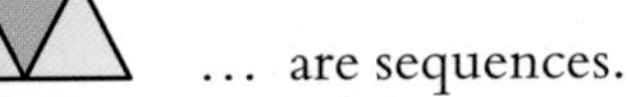 … are sequences.

The numbers in a sequence are called the **terms** of the sequence.

The ***n*th term** is used to describe a general term in a sequence.

If the nth term = $5n + 3$

1st term $= 5 \times 1 + 6 = 11$

2nd term $= 5 \times 2 + 6 = 16$

3rd term $= 5 \times 3 + 6 = 21$

4th term $= 5 \times 4 + 6 = 26$

The sequence is 11, 16, 21, 26, …

It is called a linear sequence because the differences between terms are all the same.
The rule to find the next term (the **term-to-term rule**) is +5 or add 5.

EXAMPLE

Find the nth term of these sequences

a 1, 5, 9, 13, … **b** 45, 38, 31, 24, …

a 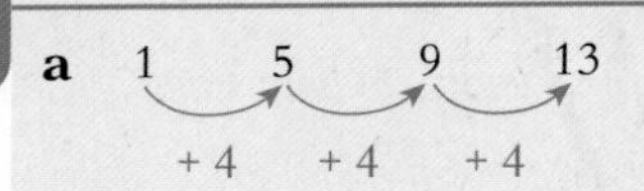

The difference between the terms is 4.
The nth term formula will contain the expression $4n$.

Term position	1	2	3	4
	↓	↓	↓	↓
Term	1	5	9	13

The rule is: multiply the term position by 4 and then subtract 3.

nth term $= 4n - 3$

b 45 38 31 24
−7 −7 −7

The difference between the terms is −7.
The nth term formula will contain the expression $-7n$

Term position	1	2	3	4
	↓	↓	↓	↓
Term	45	38	31	24

The rule is: multiply the term position by −7 and then add 52.

nth term $= -7n + 52$ or nth term $= 52 - 7n$

EXERCISE 7.4

1 Write down the next two terms for each of the following sequences.

a 7, 11, 15, 19, ... **b** 10, 15, 20, 25, ... **c** 1, 2, 4, 8, 16, ...

d 8, 9.5, 11, 12.5, ... **e** 20, 17, 14, 11, ... **f** 2, 5, 12.5, 31.25, ...

g 108, 36, 12, 4, ... **h** 100, 48, 22, 9, ... **i** 1, 1, 2, 3, 5, 8, ...

j 1, 3, 7, 15, 31, ... **k** −1, 4, 3, 7, 10, ... **l** 0.01, 0.1, 1, 10, ...

2 Write down the first 5 terms of the sequence whose nth term is

a $n + 6$ **b** $5n - 3$ **c** $4n + 2$ **d** $2n - 1$

e $3n + 5$ **f** $2n - 4$ **g** $50 - 2n$ **h** $16 - n$

i $25 - 3n$ **j** $0.5n + 2$ **k** $30 - 0.5n$ **l** $-4n + 7$

3

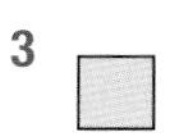

Diagram 1

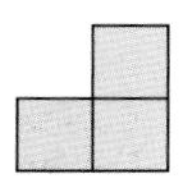

Diagram 2

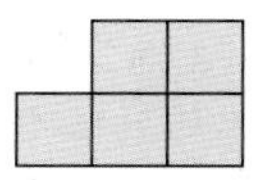

Diagram 3

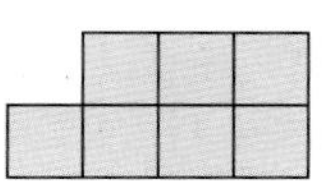

Diagram 4

Omar says the nth diagram has $n + 2$ squares.
Is he correct? Explain your answer.

4

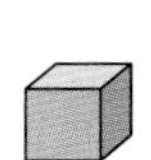

Diagram 1
1 cube

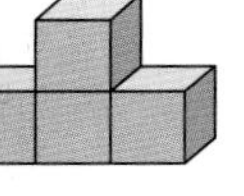

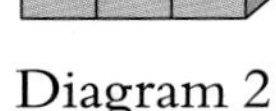

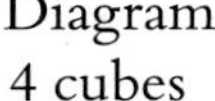

Diagram 2
4 cubes

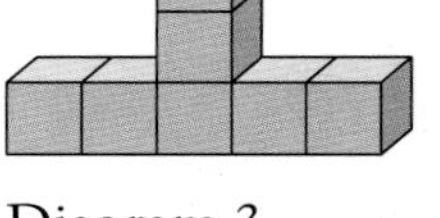

Diagram 3
7 cubes

Diagram 4
10 cubes

How many cubes are in **a** diagram 5, **b** diagram 20, **c** diagram n?

5 Find the nth term for these linear sequences.

a 7, 11, 15, 19, ... **b** 0, 3, 6, 9, ... **c** 8, 13, 18, 23, ...

d −4, −1, 2, 5, ... **e** 2, 4.5, 7, 9.5, ... **f** −8, −1, 6, 13, ...

g 1, 4, 7, 10, ... **h** −1, 3, 7, 11, ... **i** −2, 4, 10, 16, ...

j −5.5, −4, −2.5, −1, ... **k** 8, 10.5, 13, 15.5, ... **l** 6.02, 6.04, 6.06, 6.08, ...

6 Find the nth term for these linear sequences.

a 21, 19, 17, 15, ... **b** 100, 96, 92, 88, ... **c** 15, 8, 1, −6, ...

d 17, 14, 11, 8, ... **e** 5, 3, 1, −1, ... **f** 1.7, 1.3, 0.9, 0.5, ...

g −2, −5, −8, −11, ... **h** 13, 11, 9, 7, ... **i** 2, 1.5, 1, 0.5, ...

7 nth term = $8n - 5$. Which term in the sequence has a value of 155?

8 nth term = $3n - 6$. Which term in the sequence has a value of 108?

9 nth term = $54 - 4n$. Which term in the sequence has a value of −46?

10 nth term = $\frac{7n-3}{2}$. Which term in the sequence has a value of 219?

Non-linear sequences

The sequence 5, 7, 10, 14, … is called a non-linear sequence

+ 2 + 3 + 4

because the differences between the terms are not the same.

You should be able to recognise and use the following non-linear sequences:

Square numbers 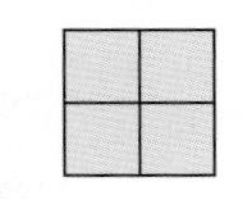 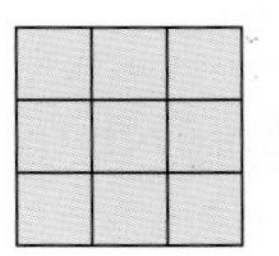 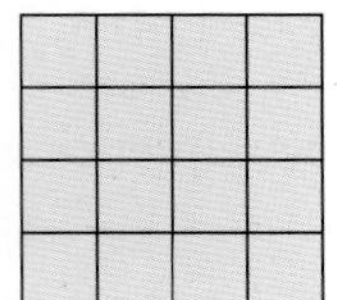 1, 4, 9, 16, … nth term = n^2

Cube numbers 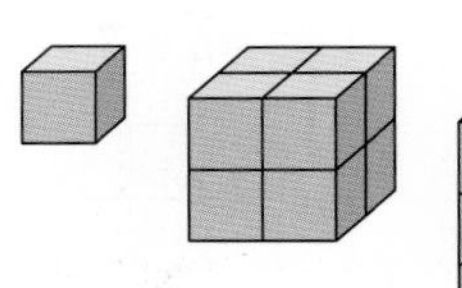 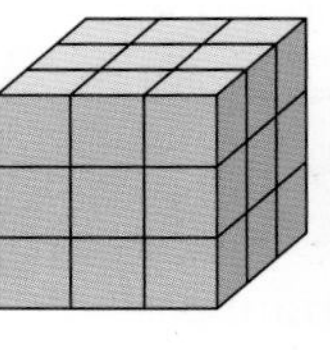 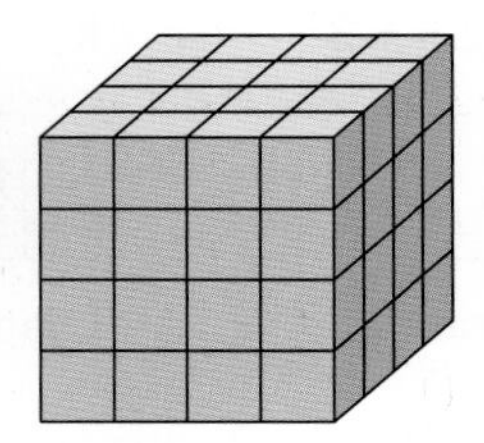 1, 8, 27, 64, … nth term = n^3

Powers For example 2, 4, 8, 16, … nth term = 2^n

EXAMPLE

Find the nth term of these sequences

a 1, 2, 4, 8, … **b** 5, 8, 13, 20, … **c** 2, 9, 28, 65, …

a Term position 1 2 3 4

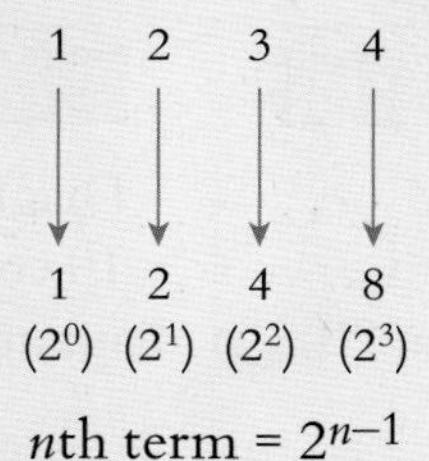

Term 1 2 4 8

(2^0) (2^1) (2^2) (2^3)

nth term = 2^{n-1}

The rule is: subtract 1 from the term position then do $2^{\text{(term position} - 1)}$

b Term position 1 2 3 4

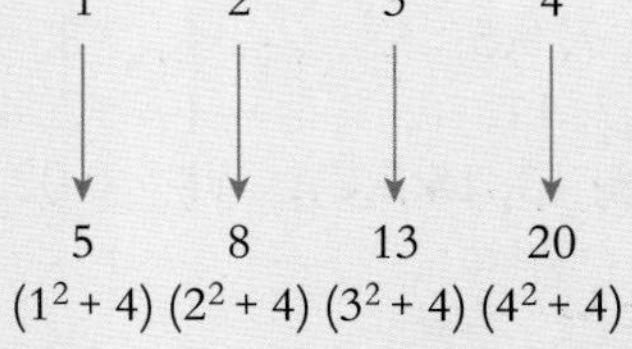

Term 5 8 13 20

$(1^2 + 4)$ $(2^2 + 4)$ $(3^2 + 4)$ $(4^2 + 4)$

nth term = $n^2 + 4$

The rule is: square the term position then add 4.

c Term position 1 2 3 4

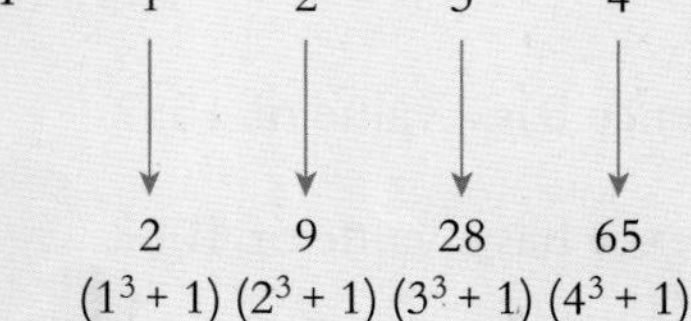

Term 2 9 28 65

$(1^3 + 1)$ $(2^3 + 1)$ $(3^3 + 1)$ $(4^3 + 1)$

nth term = $n^3 + 1$

The rule is: cube the term position and then add 1.

EXERCISE 7.5

1 Write down the first 5 terms of the non-linear sequence whose nth term is

a $n^2 + 2$ b $n^2 - 3$ c $3n^2$ d $n(n - 2)$

e $n^3 + 1$ f $n^3 - 4$ g $\frac{n+2}{n}$ h $\frac{n(n+1)(n+2)}{6}$

2 Find the nth term for these non-linear sequences.

a 1, 4, 9, 16, … b 0, 3, 8, 15, …
c 2, 8, 18, 32, … d 3, 6, 11, 18, …
e 11, 14, 19, 26, … f 3, 12, 27, 48, …

TOP TIP

These sequences are connected with square numbers.

3 Find the nth term for these non-linear sequences.

a 1, 8, 27, 64, 125, … b 0, 7, 26, 63, 124, …
c −1, 6, 25, 62, 123, … d 2, 16, 54, 128, 250, …
e 3, 10, 29, 66, 127, …

TOP TIP

These sequences are connected with cube numbers.

4 Find the nth term for these non linear sequences.

a 10, 100, 1000, 10 000, … b 2, 4, 8, 16, …
c 3, 9, 27, 81, … d 4, 8, 16, 32, …
e 1, 3, 9, 27, … f 0.1, 1, 10, 100, …

5 Find the nth term for the following sequences.

a $1 \times 3, 2 \times 4, 3 \times 5, 4 \times 6, \ldots$ b $1 \times 2 \times 3, 2 \times 3 \times 4, 3 \times 4 \times 5, 4 \times 5 \times 6, \ldots$

c $\frac{1}{6}, \frac{2}{7}, \frac{3}{8}, \frac{4}{9}, \ldots$ d $\frac{2}{5}, \frac{3}{10}, \frac{4}{17}, \frac{5}{26}, \ldots$

6 Write down the **next** term in each of the following sequences:

a $128x^2, 64x^3, 32x^4, 16x^5, \ldots$ b $6x^3, 9x^5, 12x^7, 15x^9, \ldots$
c $3x^2, 4x^3, 5x^4, 6x^5, \ldots$ d $x^6y^2, x^5y^3, x^4y^4, x^3y^5, \ldots$

7 The table shows the first four terms in three sequences A, B and C.

	Term 1	Term 2	Term 3	Term 4
Sequence A	1	8	27	64
Sequence B	4	7	10	13
Sequence C	−3	1	17	51

a Find the nth term of sequence A.
b Find the nth term of sequence B.
c Use your answers to parts **a** and **b** to write down the nth term of sequence C.

8 The diagram below shows how the fourth square pyramid number is made.

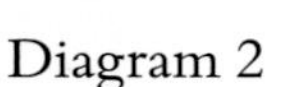

$1 + 4 + 9 + 16 = 30$

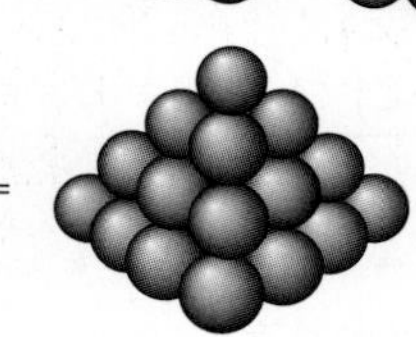

The fourth square pyramid number is 30.

Calculate the fifth and sixth square pyramid numbers.

9

Diagram 1

Diagram 2

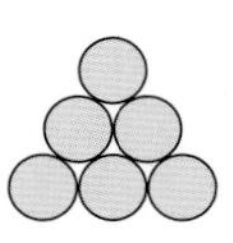
Diagram 3

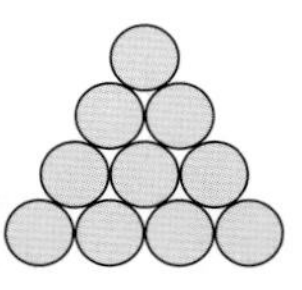
Diagram 4

a How many circles are in diagram 4?

b Draw the next diagram in the sequence.

c Copy and complete the following table.

Diagram	1	2	3	4	5
Number of circles	$1 = \frac{1 \times 2}{2}$	$3 = \frac{2 \times 3}{2}$	$6 = \frac{3 \times 4}{2}$		

d How many circles are in diagram n?

e How many circles are in diagram 15?

10 The diagram shows four polygons. The diagonals for each polygon are shown.

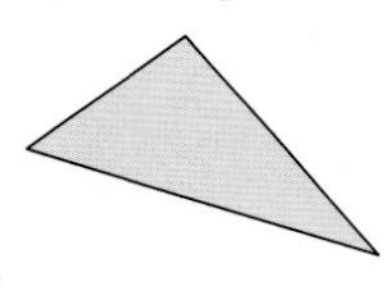
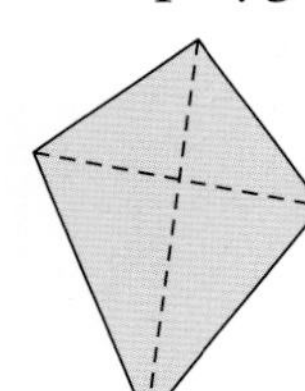
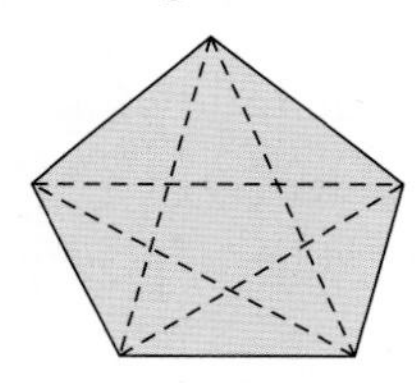
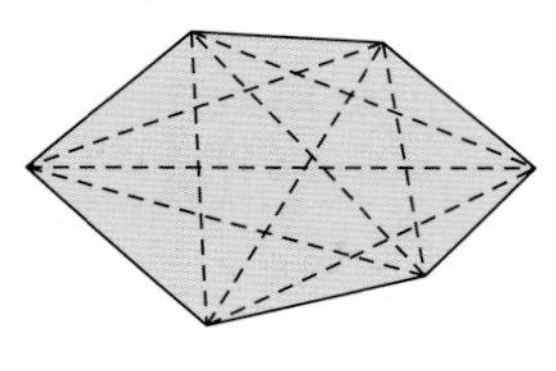

a Draw a polygon with seven sides. Show all the diagonals.

b Copy and complete the table.

Number of sides	3	4	5	6	7	8
Number of diagonals	0	2				

c How many diagonals are there in a 9-sided polygon?

d How many diagonals are there in a 20-sided polygon?

e Write down an expression for the number of diagonals in a polygon with n sides.

11 nth term = $3n^2 + 7$. Which term in the sequence has a value of 3682?

12 nth term = $n^2 + 2n - 99$. Which term in the sequence has a value of 2501?

13 **a** Write down the next row for the number pattern.

b The sum of the numbers in row 6 is $1 + 5 + 10 + 10 + 5 + 1 = 32 = 2^5$.
Copy and complete the table.

Row 1
Row 2
Row 3
Row 4
Row 5
Row 6

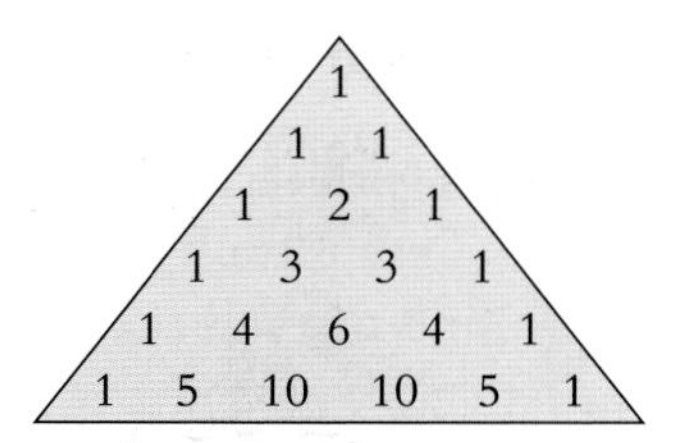

Row	1	2	3	4	5	6	7
Sum of numbers in row	$1 = 2^0$	$2 = 2^1$				$32 = 2^5$	

c Write down an expression for the sum of the numbers in row n.

d Which row has a sum of 32 768?

14

Year 1

Year 2

Year 3

The diagram shows the growth of a plant over three years.
Each year a flower is replaced by two stems and two flowers.

a Draw a diagram to show the plant in year 4.

b Copy and complete the table.

Year	1	2	3	4	5
Number of flowers	1	2	4		
Number of stems	1	3	7		

c How many flowers are there in year 6?

d How many flowers are there in year n?

e How many stems are there in year 6?

f How many stems are there in year n?

15 97, 75, 47, 40, …

The rule to find the next term in the sequence is: multiply together the digits of the last number and then add 12.
(To find the second term you work out $9 \times 7 + 12 = 63 + 12 = 75$.)
What is the 100th number in the sequence?

UNIT 7

Further non-linear sequences

EXAMPLE

The nth term of the sequence 6, 11, 18, 27, ... is given by the formula

$n\text{th term} = n^2 + bn + c$

Find the values of b and c.

The 1st term is 6, so substitute $n = 1$ into the formula.

$$1^2 + b \times 1 + c = 6$$
$$b + c = 5$$

The 2nd term is 11, so substitute $n = 2$ into the formula.

$$2^2 + b \times 2 + c = 11$$
$$2b + c = 7$$

You now have a pair of simultaneous equations to solve.

$$2b + c = 7$$
$$b + c = 5$$

Subtracting gives $b = 2$ so $2 + c = 5$ so $c = 3$

$b = 2$ and $c = 3$

If you are asked to find the values of a, b and c when the nth term $= an^2 + bn + c$ for a given sequence then the problem is more complicated.

One method is to look at the differences in the terms.

For example	6		13		24		39		58
First difference		7		11		15		19	
Second difference			4		4		4		

The second differences are all the same.

This means that it is a quadratic sequence.

A second difference of 2 $\Rightarrow$ the nth term will contain n^2

A second difference of 4 $\Rightarrow$ the nth term will contain $2n^2$

A second difference of 6 $\Rightarrow$ the nth term will contain $3n^2$

The sequence 6, 13, 24, 39, 58, ... has a second difference of 4 so the nth term will contain the term $2n^2$

So $n\text{th term} = 2n^2 + bn + c$

The 1st term is 6 $2 \times 1^2 + b \times 1 + c = 6$

$$b + c = 4$$

The 2nd term is 13 $2 \times 2^2 + b \times 2 + c = 13$

$$2b + c = 5$$

Simultaneous equations are $2b + c = 5$

and $b + c = 4$

Subtracting gives $b = 1$ so $1 + c = 4$ so $c = 3$

$a = 2$, $b = 1$ and $c = 3$

EXERCISE 7.6

1 8, 14, 22, 32, ...
The nth term of the sequence is given by the formula nth term $= n^2 + bn + c$.
Find the values of b and c.

2 8, 16, 26, 38, 52, ...
The nth term of the sequence is given by the formula nth term $= n^2 + bn + c$.
Find the values of b and c.

3 3, 4, 7, 12, 19, ...
The nth term of the sequence is given by the formula nth term $= n^2 + bn + c$.
Find the values of b and c.

4 6, 15, 28, 45, 66, ...
The nth term of the sequence is given by the formula nth term $= 2n^2 + bn + c$.
Find the values of b and c.

5 6, 16, 32, 54, 82, ...
The nth term of the sequence is given by the formula nth term $= 3n^2 + bn + c$.
Find the values of b and c.

6 9, 20, 35, 54, 77, ...
The nth term of the sequence is given by the formula nth term $= an^2 + bn + c$.
Find the values of a, b and c.

7 4, 7, 14, 25, 40, ...
The nth term of the sequence is given by the formula nth term $= an^2 + bn + c$.
Find the values of a, b and c.

8 1, 11, 27, 49, 77, ...
The nth term of the sequence is given by the formula nth term $= an^2 + bn + c$.
Find the values of a, b and c.

9 4.5, 12, 20.5, 30, 40.5, ...
The nth term of the sequence is given by the formula nth term $= an^2 + bn + c$.
Find the values of a, b and c.

10 A sequence is made from sticks.

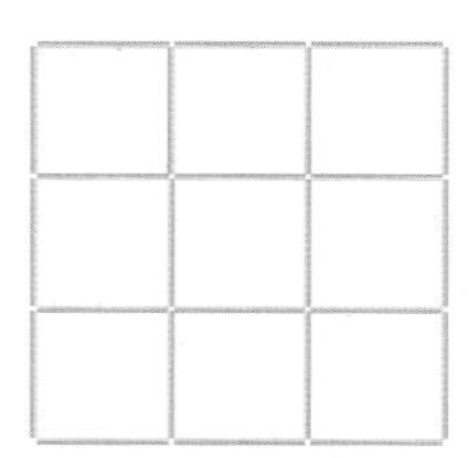

Diagram 1 4 sticks | Diagram 2 12 sticks | Diagram 3 24 sticks

How many sticks are needed to make:

a diagram 4 **b** diagram 5 **c** diagram n?

11 A sequence is made from small squares.

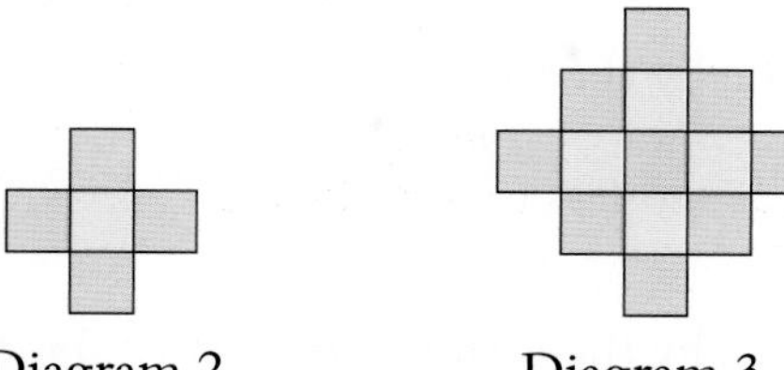

Diagram 1 1 square | Diagram 2 5 squares | Diagram 3 13 squares

How many small squares are needed to make:

a diagram 4 **b** diagram 5 **c** diagram n?

12 The nth term of the sequence 2, 8, 20, 40, 70, …. is $\frac{n(n+1)(n+2)}{3}$.

Write down the nth term of the sequence 4, 16, 40, 80, 140, …

13 a Find the value of $1^2 + 2^2 + 3^2 + 4^2$.

b The sum of the first n square numbers can be found using the formula

$$1^2 + 2^2 + 3^2 + 4^2 + 5^2 + 6^2 + \ldots\ldots + n^2 = \frac{n(n+1)(2n+1)}{k}$$

Use your answer to part **a** to find the value of k.

14 2, 21, 70, 161, 306, …

The nth term of the sequence is given by the formula nth term $= an^3 + 3n^2 + bn + 1$.
Find the values of a and b.

Further sequence notation

There are different ways of writing the terms of sequences, such as $u_1, u_2, u_3, u_4, \ldots$

EXAMPLE

A sequence is defined by $u_n = 3n + 5$.
Write down the first four terms of the sequence.

$u_1 = (3 \times 1) + 5 = 8$
$u_2 = (3 \times 2) + 5 = 11$
$u_3 = (3 \times 3) + 5 = 14$
$u_4 = (3 \times 4) + 5 = 17$
The first four terms of the sequence are 8, 11, 14 and 17.

EXAMPLE

$T_1 = 2$ $T_2 = 5$ $T_3 = 10$ $T_4 = 17$
Write down an expression for T_n

$T_1 = 1^2 + 1$
$T_2 = 2^2 + 1$
$T_3 = 3^2 + 1$
$T_4 = 4^2 + 1$
So $T_n = n^2 + 1$.

➡ **NOTE:** You should be able to recognise that the numbers 2, 5, 10 and 17 are connected to the square numbers 1, 4, 9 and 16.

EXERCISE 7.7

1 Write down the first four terms of these sequences.

a $u_n = 2n - 3$
b $u_n = 5n + 1$
c $u_n = 3n + 2$
d $u_n = n^2$
e $u_n = n^2 + 2$
f $u_n = (n - 1)^2$
g $u_n = n^3$
h $u_n = n^3 - n^2$
i $u_n = (n + 1)^3$
j $u_n = 2^n$
k $u_n = 3^{n-2}$
l $u_n = 10^{n+3}$
m $u_n = \frac{n}{n+2}$
n $u_n = \frac{n+4}{n+5}$
o $u_n = \frac{n^2}{2n+3}$

2 Find the formula for the nth term (u_n) for each of these sequences.

a $u_1 = 4$ $u_2 = 9$ $u_3 = 14$ $u_4 = 19$
b $u_1 = 8$ $u_2 = 11$ $u_3 = 14$ $u_4 = 17$
c $u_1 = 20$ $u_2 = 17$ $u_3 = 14$ $u_4 = 11$
d $u_1 = 0$ $u_2 = 3$ $u_3 = 8$ $u_4 = 15$
e $u_1 = 1$ $u_2 = 3$ $u_3 = 9$ $u_4 = 27$
f $u_1 = \frac{4}{7}$ $u_2 = \frac{5}{9}$ $u_3 = \frac{6}{11}$ $u_4 = \frac{7}{13}$

KEY WORDS
sequence
term
term-to-term rule
*n*th term

Exponential graphs

THIS SECTION WILL SHOW YOU HOW TO

- Recognise exponential graphs
- Draw exponential graphs
- Use exponential graphs in practical situations

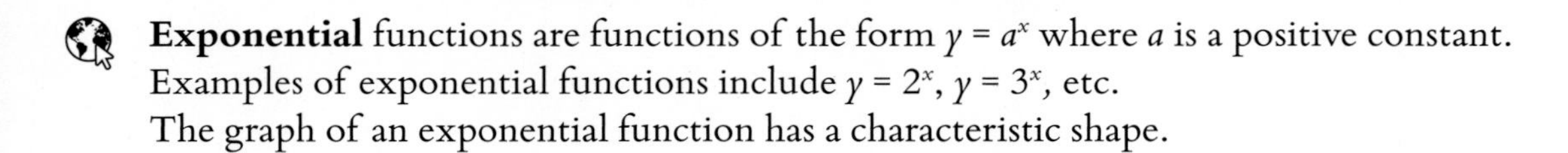

Exponential functions are functions of the form $y = a^x$ where a is a positive constant.
Examples of exponential functions include $y = 2^x$, $y = 3^x$, etc.
The graph of an exponential function has a characteristic shape.

EXAMPLE

a Draw the graph of $y = 2^x$ for $-3 \le x \le 3$.
b Use your graph to estimate the value of x when $2^x = 6$.
c By drawing a tangent to your graph estimate the gradient of the graph when $x = 1$.

a First make a table of values for x and y

when $x = -3$, $y = 2^{-3} = \frac{1}{2^3} = \frac{1}{8} = 0.125$ when $x = -2$, $y = 2^{-2} = \frac{1}{2^2} = \frac{1}{4} = 0.25$

when $x = -1$, $y = 2^{-1} = \frac{1}{2^1} = \frac{1}{2} = 0.5$ when $x = 0$, $y = 2^0 = 1$

when $x = 1$, $y = 2^1 = 2$ when $x = 2$, $y = 2^2 = 4$

when $x = 3$, $y = 2^3 = 8$

x	−3	−2	−1	0	1	2	3
y	0.125	0.25	0.5	1	2	4	8

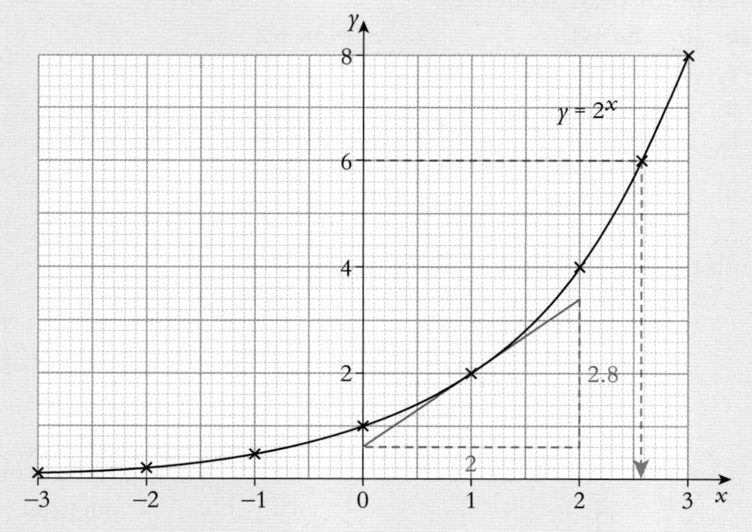

b When $2^x = 6$, $x \approx 2.6$

c When $x = 1$, gradient $\approx \frac{2.8}{2} = 1.4$

Reminder

- To describe a **reflection** you must state the mirror line.
- To describe a **rotation** you must state the direction of rotation, the angle of rotation and
  the centre of rotation.
- To describe an **enlargement** you must state the scale factor of the enlargement and the
  centre of enlargement.

EXAMPLE

A(1, 1), B(4, 1), C(3, 3) and D(1, 2).

Transform quadrilateral ABCD using the matrix $\begin{pmatrix} 0 & -1 \\ 1 & 0 \end{pmatrix}$.

Describe the transformation fully.

$$\begin{pmatrix} 0 & -1 \\ 1 & 0 \end{pmatrix}\overset{\begin{matrix}A & B & C & D\end{matrix}}{\begin{pmatrix} 1 & 4 & 3 & 1 \\ 1 & 1 & 3 & 2 \end{pmatrix}} = \overset{\begin{matrix}A' & B' & C' & D'\end{matrix}}{\begin{pmatrix} -1 & -1 & -3 & -2 \\ 1 & 4 & 3 & 1 \end{pmatrix}}$$

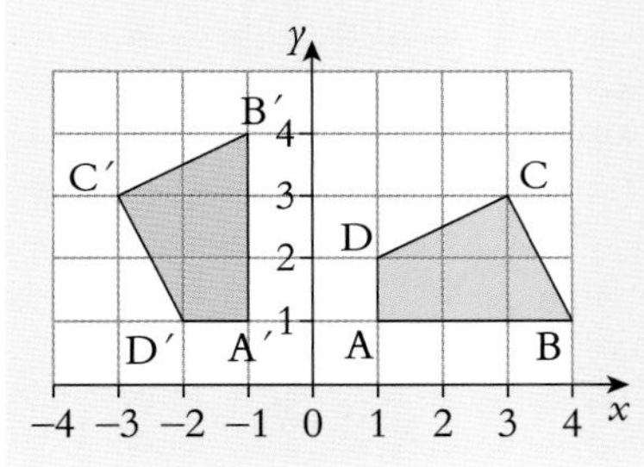

The transformation is a rotation, 90° anticlockwise about the point (0, 0).

EXERCISE 7.10

1 Transform triangle ABC using each of the matrices given below.
Label your image A′B′C′.
Describe fully the single transformation that moves ABC onto A′B′C′.
Draw a new diagram for each part of the question.

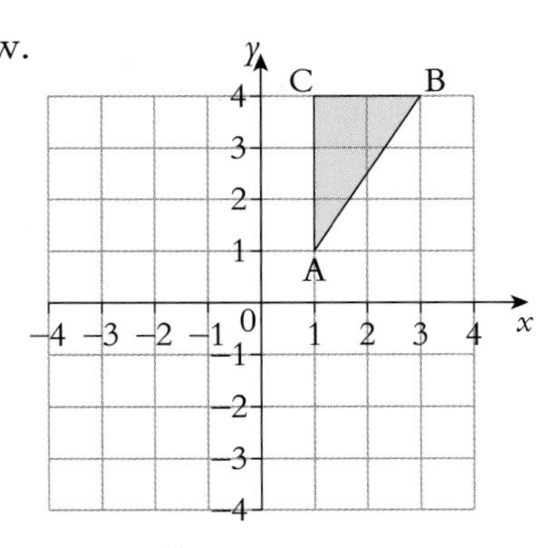

a $\begin{pmatrix} 0 & 1 \\ 1 & 0 \end{pmatrix}$ **b** $\begin{pmatrix} 1 & 0 \\ 0 & -1 \end{pmatrix}$ **c** $\begin{pmatrix} 0 & -1 \\ -1 & 0 \end{pmatrix}$

UNIT 7

2 Transform rectangle ABCD using each of the matrices given below.
Label your image A′B′C′D′.
Describe fully the single transformation that moves ABCD onto A′B′C′D′.
Draw a new diagram for each part of the question.

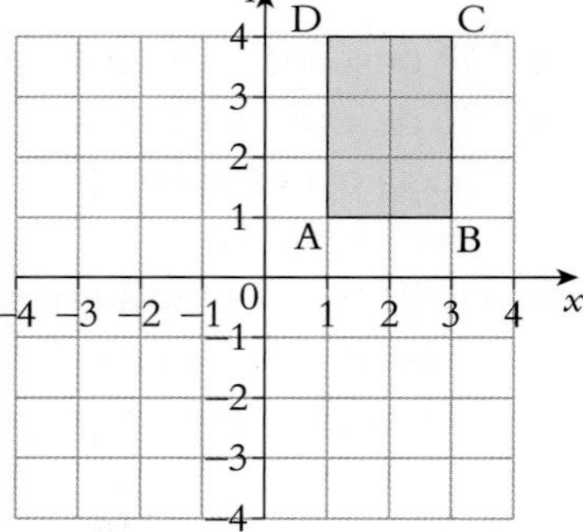

a 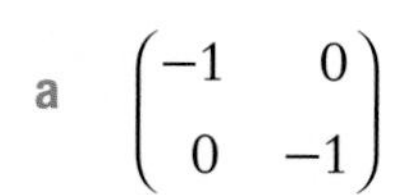 $\begin{pmatrix} -1 & 0 \\ 0 & -1 \end{pmatrix}$ **b** $\begin{pmatrix} 0 & 1 \\ -1 & 0 \end{pmatrix}$ **c** $\begin{pmatrix} 0.8 & 0.6 \\ -0.6 & 0.8 \end{pmatrix}$

3 Plot the points A (2, 1), B(3, 4), C(1, 4) and D(2, 3) on a diagram.

Transform quadrilateral ABCD using the matrix $\begin{pmatrix} 2 & 0 \\ 0 & 2 \end{pmatrix}$.
Label your image A′B′C′D′.
Describe fully the single transformation that moves ABCD onto A′B′C′D′.

4 Plot the points A(4, 1), B(3, 4), C(1, 4) and D(0, 2) on a diagram.

Transform quadrilateral ABCD using the matrix $\begin{pmatrix} 3 & 0 \\ 0 & 3 \end{pmatrix}$.
Label your image A′B′C′D′.
Describe fully the single transformation that moves ABCD onto A′B′C′D′.

5 Plot the points A(8, 2), B(4, 6) and C(2, 4) on a diagram.

Transform triangle ABC using the matrix $\begin{pmatrix} \frac{1}{2} & 0 \\ 0 & \frac{1}{2} \end{pmatrix}$.
Label your image A′B′C′.
Describe fully the single transformation that moves ABC onto A′B′C′.

6 Plot the points A(8, 2), B(4, 6), C(2, 6) and D(2, 4) on a diagram.

Transform quadrilateral ABCD using the matrix $\begin{pmatrix} -\frac{1}{2} & 0 \\ 0 & -\frac{1}{2} \end{pmatrix}$.
Label your image A′B′C′D′.
Describe fully the single transformation that moves ABCD onto A′B′C′D′.

7 Plot the points A(4, 0), B(3, 4), C(0, 2) and D(3, 2) on a diagram.

Transform quadrilateral ABCD using the matrix $\begin{pmatrix} -2 & 0 \\ 0 & -2 \end{pmatrix}$.
Label your image A′B′C′D′.
Describe fully the single transformation that moves ABCD onto A′B′C′D′.

18 Write down the matrix that maps

- **a** triangle A onto triangle B,
- **b** triangle A onto triangle C,
- **c** triangle C onto triangle D,
- **d** triangle D onto triangle C,
- **e** triangle C onto triangle A.

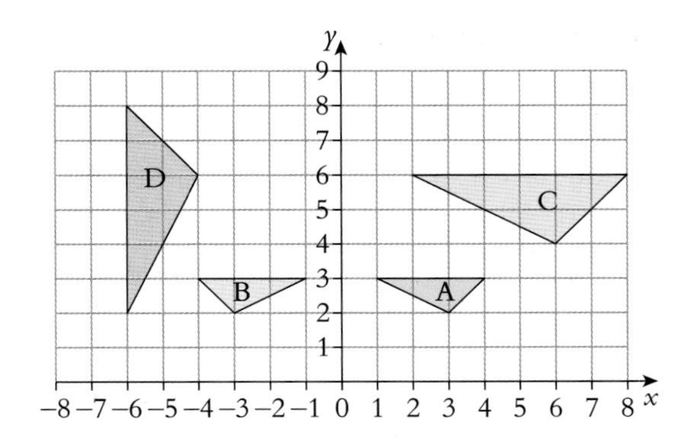

Combined transformations

EXAMPLE

A(2, 1), B(5, 1) and C(5, 3).

a Transform $\triangle ABC$ using the matrix $\begin{pmatrix} 0 & 1 \\ 1 & 0 \end{pmatrix}$. Label this $A_1B_1C_1$.

b Transform $\triangle A_1B_1C_1$ using the matrix $\begin{pmatrix} 0 & -1 \\ 1 & 0 \end{pmatrix}$. Label this $A_2B_2C_2$.

c **i** Describe fully the single transformation which maps $\triangle ABC$ onto $\triangle A_2B_2C_2$.

ii Find the 2 by 2 matrix which represents the transformation which maps $\triangle ABC$ onto $\triangle A_2B_2C_2$.

---

a

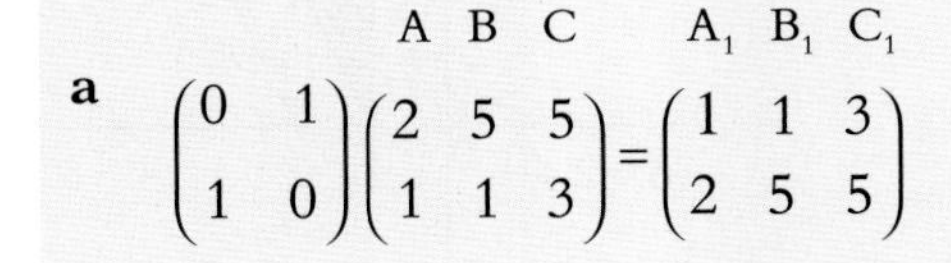

$$\begin{pmatrix} 0 & 1 \\ 1 & 0 \end{pmatrix}\begin{pmatrix} 2 & 5 & 5 \\ 1 & 1 & 3 \end{pmatrix} = \begin{pmatrix} 1 & 1 & 3 \\ 2 & 5 & 5 \end{pmatrix}$$

(columns A B C → A_1 B_1 C_1)

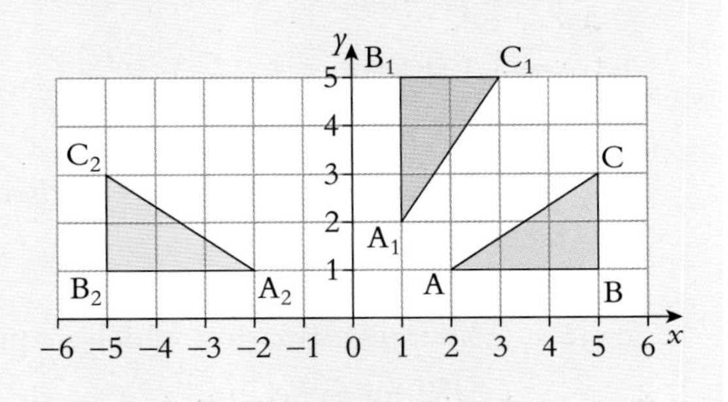

b

$$\begin{pmatrix} 0 & -1 \\ 1 & 0 \end{pmatrix}\begin{pmatrix} 1 & 1 & 3 \\ 2 & 5 & 5 \end{pmatrix} = \begin{pmatrix} -2 & -5 & -5 \\ 1 & 1 & 3 \end{pmatrix}$$

(columns A_1 B_1 C_1 → A_2 B_2 C_2)

c **i** The transformation which maps Δ ABC onto $\Delta A_2B_2C_2$ is a reflection in the y-axis.

ii

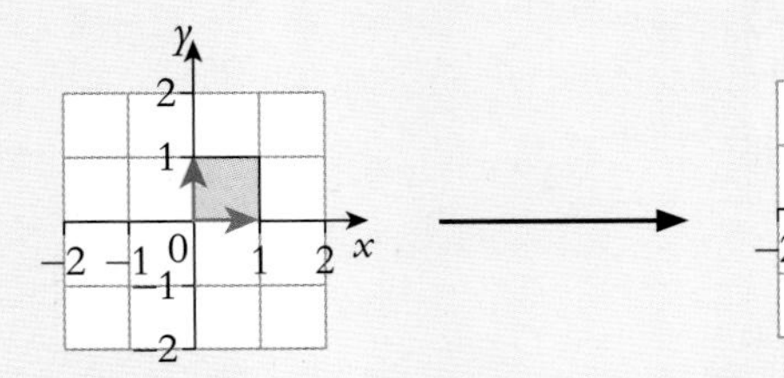

$$\begin{pmatrix} 1 \\ 0 \end{pmatrix} \rightarrow \begin{pmatrix} -1 \\ 0 \end{pmatrix} \text{ and } \begin{pmatrix} 0 \\ 1 \end{pmatrix} \rightarrow \begin{pmatrix} 0 \\ 1 \end{pmatrix}$$

$$\text{Matrix} = \begin{pmatrix} -1 & 0 \\ 0 & 1 \end{pmatrix}$$

UNIT 7

Notation

If you have a shape A and a transformation matrix **R**:

R(A) means transform shape A using the matrix **R**.

If you have a shape A and two transformation matrices **R** and **M**:

MR(A) means transform shape A using the matrix **R**
then transform the image using the matrix **M**.

EXAMPLE

$\mathbf{R} = \begin{pmatrix} -1 & 0 \\ 0 & -1 \end{pmatrix}$ and $\mathbf{E} = \begin{pmatrix} -2 & 0 \\ 0 & -2 \end{pmatrix}$

a Transform triangle A using the matrix **R**. Label this **R**(A).
Describe the transformation represented by the matrix **R**.

b Transform triangle **R**(A) using the matrix **E**. Label this **ER**(A).
Describe the transformation represented by the matrix **E**.

c **i** Describe fully the single transformation which maps triangle A onto **ER**(A).

ii Find the 2 by 2 matrix which represents the transformation which maps triangle A onto **ER**(A).

---

a 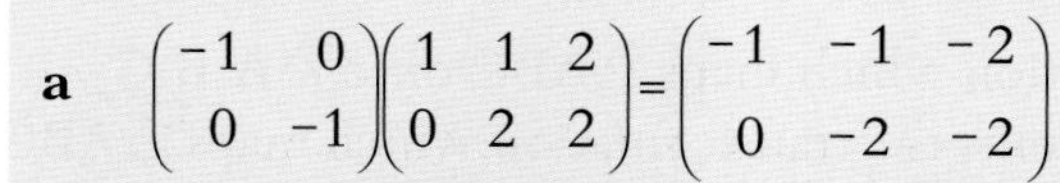

$$\begin{pmatrix} -1 & 0 \\ 0 & -1 \end{pmatrix}\begin{pmatrix} 1 & 1 & 2 \\ 0 & 2 & 2 \end{pmatrix} = \begin{pmatrix} -1 & -1 & -2 \\ 0 & -2 & -2 \end{pmatrix}$$

The matrix **R** represents a rotation of 180° about the point (0, 0).

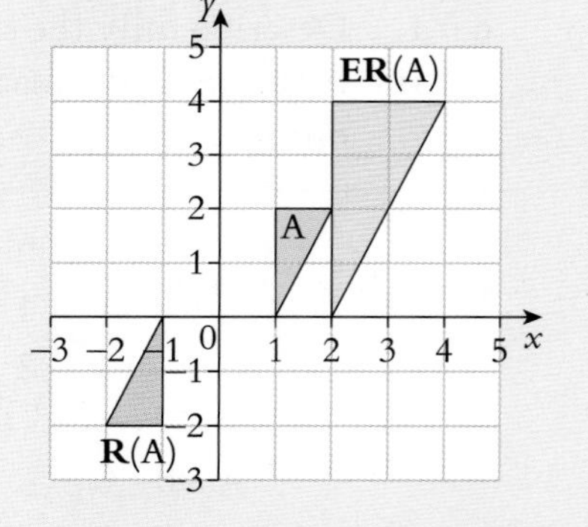

b 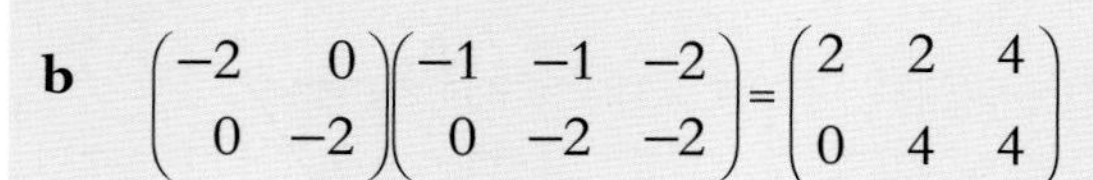

$$\begin{pmatrix} -2 & 0 \\ 0 & -2 \end{pmatrix}\begin{pmatrix} -1 & -1 & -2 \\ 0 & -2 & -2 \end{pmatrix} = \begin{pmatrix} 2 & 2 & 4 \\ 0 & 4 & 4 \end{pmatrix}$$

The matrix **E** represents an enlargement, scale factor −2, centre (0, 0).

c **i** The transformation which maps triangle A onto **ER**(A) is an enlargement, scale factor 2, centre (0, 0).

ii

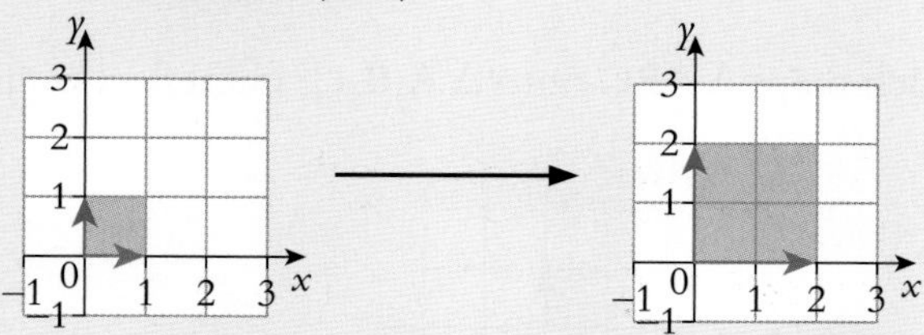

$$\begin{pmatrix} 1 \\ 0 \end{pmatrix} \rightarrow \begin{pmatrix} 2 \\ 0 \end{pmatrix} \text{ and } \begin{pmatrix} 0 \\ 1 \end{pmatrix} \rightarrow \begin{pmatrix} 0 \\ 2 \end{pmatrix}$$

$$\text{Matrix} = \begin{pmatrix} 2 & 0 \\ 0 & 2 \end{pmatrix}$$

EXERCISE 7.12

1 A(1, 1), B(4, 1) and C(2, 3).

a Transform $\triangle ABC$ using the matrix $\begin{pmatrix} 1 & 0 \\ 0 & -1 \end{pmatrix}$.
Label this $A_1B_1C_1$.

b Transform $\triangle A_1B_1C_1$ using the matrix $\begin{pmatrix} -1 & 0 \\ 0 & 1 \end{pmatrix}$.
Label this $A_2B_2C_2$.

c i Describe fully the single transformation which maps $\triangle ABC$ onto $\triangle A_2B_2C_2$.
ii Find the 2 by 2 matrix which represents the transformation which maps $\triangle ABC$ onto $\triangle A_2B_2C_2$.

2 A(0, 2), B(1, 1) and C(3, 2).

a Transform $\triangle ABC$ using the matrix $\begin{pmatrix} 0 & 1 \\ -1 & 0 \end{pmatrix}$.
Label this $A_1B_1C_1$.

b Transform $\triangle A_1B_1C_1$ using the matrix $\begin{pmatrix} -1 & 0 \\ 0 & -1 \end{pmatrix}$.
Label this $A_2B_2C_2$.

c i Describe fully the single transformation which maps $\triangle ABC$ onto $\triangle A_2B_2C_2$.
ii Find the 2 by 2 matrix which represents the transformation which maps $\triangle ABC$ onto $\triangle A_2B_2C_2$.

3 Rectangle X has vertices (1, 1), (2, 1), (2, 3) and (1, 3).

a Transform rectangle X using the matrix $\mathbf{R} = \begin{pmatrix} 0 & -1 \\ 1 & 0 \end{pmatrix}$.
Label this **R**(X).

b Describe fully the single transformation that maps
i X onto **R**(X) ii **R**(X) onto X.

c Find $\mathbf{R}^{-1}$, the inverse of the matrix **R**.

d Transform **R**(X) by the matrix $\mathbf{R}^{-1}$. Comment on your results.

4 Triangle A has vertices (−3, 1), (−9, 1) and (−7, 4).

a Transform triangle A using the matrix $\mathbf{R} = \begin{pmatrix} 0 & 1 \\ -1 & 0 \end{pmatrix}$.
Label this **R**(A).

b Transform triangle **R**(A) using the matrix $\mathbf{M} = \begin{pmatrix} 0 & 1 \\ 1 & 0 \end{pmatrix}$.
Label this **MR**(A).

c Describe fully the single transformation which maps triangle A onto **MR**(A).

KEY WORDS
transformation matrix

Cumulative frequency

THIS SECTION WILL SHOW YOU HOW TO
- Construct and use cumulative frequency diagrams
- Estimate and interpret the median, percentiles, quartiles and inter-quartile range

Cumulative frequency tables and diagrams

You can construct a **cumulative frequency** table from a grouped frequency table.
The grouped frequency table below shows information about the times taken, t seconds, for 25 students to solve a puzzle.
The cumulative frequency (cf) of the data is found by calculating the running total of the frequencies (f).
For example, to find the number of people taking twenty seconds or less ($t \le 20$) you add 3 and 6.
To find the number of people taking thirty seconds or less ($t \le 30$) you add 3, 6 and 10.

Grouped frequency table

Time (t)	f
$0 < t \le 10$	3
$10 < t \le 20$	6
$20 < t \le 30$	10
$30 < t \le 40$	4
$40 < t \le 50$	2

Time (t)	Cumulative frequency
$t \le 10$	**3**
$t \le 20$	3 + 6 = **9**
$t \le 30$	3 + 6 + 10 = **19**
$t \le 40$	3 + 6 + 10 + 4 = **23**
$t \le 50$	3 + 6 + 10 + 4 + 2 = **25**

Cumulative frequency table

Time (t)	cf
$t \le 10$	3
$t \le 20$	9
$t \le 30$	19
$t \le 40$	23
$t \le 50$	25

To draw a cumulative frequency diagram (or c.f. graph), you must use the values from the cumulative frequency table.
So for the above example, you must plot (10, 3), (20, 9), (30, 19), (40, 23) and (50, 25).
It is important to note that the points are plotted at the upper end of the class.

Cumulative frequency diagram

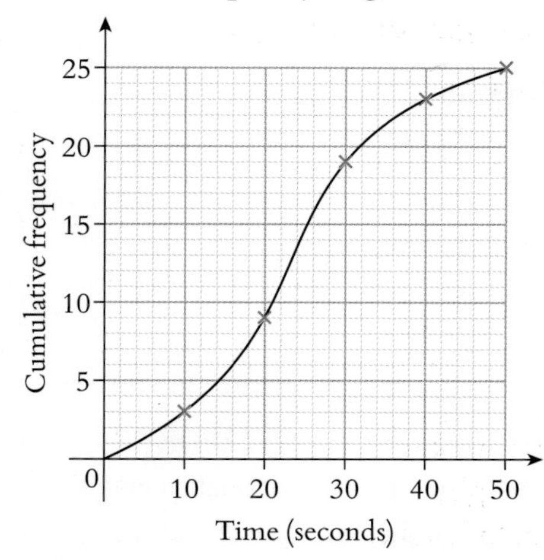

There are no students who took 0 seconds or less so the point (0, 0) can also be plotted. Draw a smooth curve through the points.

➡ **NOTE:** it is important to remember that the cumulative frequency diagram shows the number of students taking less than or equal to a particular time.

Comparing data sets

You can compare data sets using information obtained from a cumulative frequency diagram. To do this you compare:

- The medians (this is a measure of average),
- The inter-quartile ranges (this is a measure of spread).

EXAMPLE

The table shows information about the examination results of a group of boys and girls.
Compare the boys' results with the girls' results.

	Median mark	IQR
Boys	56	24
Girls	52	35

The boys' median is greater than the girls' median, so on average the boys achieved higher marks.
The IQR for the girls is greater than the IQR for the boys, so the marks for the girls are more varied.

EXERCISE 7.14

1 For each of the cumulative frequency diagrams find estimates for
 i the median, ii the LQ, iii the UQ,
 iv the IQR, v the 40th percentile.

a

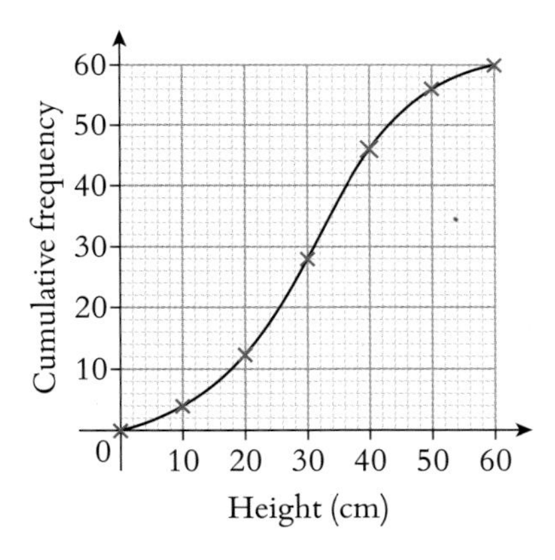

b

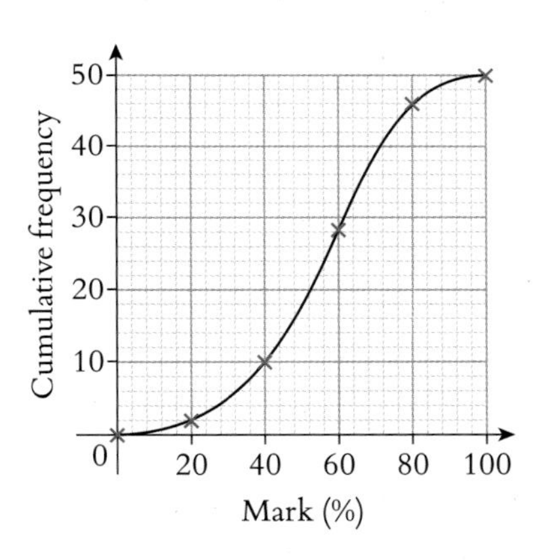

c

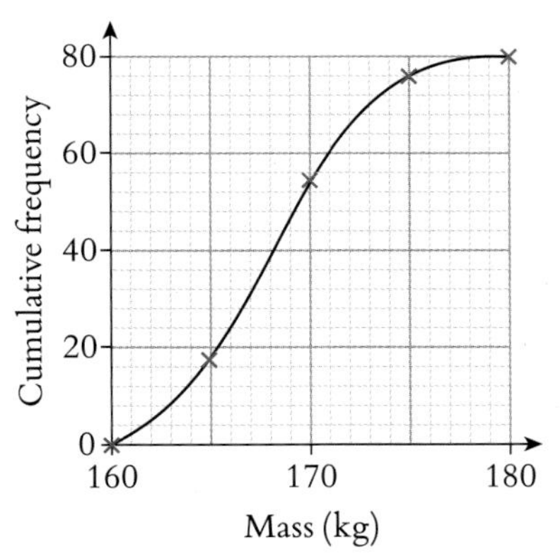

d

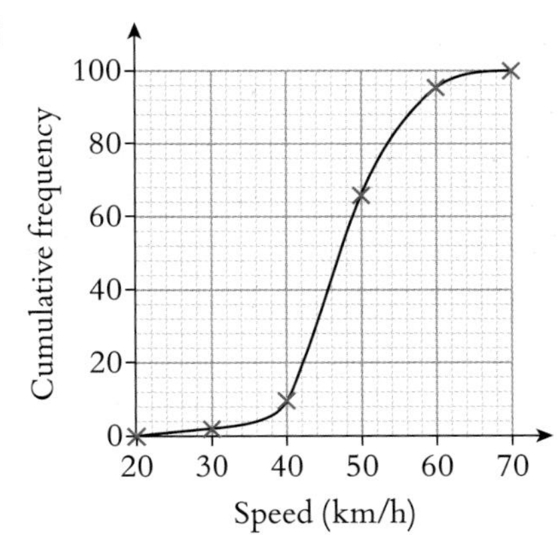

2 200 apples are weighed.
The table shows the results.

a Draw a cumulative frequency diagram to show the data.

b Use your cumulative frequency diagram to find
 i the median,
 ii the lower quartile,
 iii the upper quartile,
 iv the inter-quartile range,
 v the 68th percentile.

Mass (m kg)	Frequency
$100 < m \le 110$	2
$110 < m \le 120$	2
$120 < m \le 130$	9
$130 < m \le 140$	19
$140 < m \le 150$	40
$150 < m \le 160$	48
$160 < m \le 170$	40
$170 < m \le 180$	24
$180 < m \le 190$	12
$190 < m \le 200$	4

3 150 passengers on an aircraft had their baggage weighed.
The table shows the results.

a Draw a cumulative frequency diagram to show the data.

b Use your cumulative frequency diagram to find
 i the median,
 ii the lower quartile,
 iii the upper quartile,
 iv the inter-quartile range,
 v the 84th percentile.

Mass (m kg)	Frequency
$0 < m \le 5$	10
$5 < m \le 10$	20
$10 < m \le 15$	38
$15 < m \le 20$	35
$20 < m \le 25$	25
$25 < m \le 30$	12
$30 < m \le 35$	10

4 60 potatoes are weighed.
The table shows the results.

a Draw a cumulative frequency diagram to show the data.

b Use your cumulative frequency diagram to find
 i the median,
 ii the lower quartile,
 iii the upper quartile,
 iv the inter-quartile range,
 v the 36th percentile.

Mass (m g)	Frequency
$200 < m \le 250$	4
$250 < m \le 300$	7
$300 < m \le 350$	3
$350 < m \le 400$	16
$400 < m \le 450$	21
$450 < m \le 500$	9

5

	Median time	IQR
10 to 20 year olds	150	20
60 to 70 year olds	300	28

The table shows information about the reaction times (milliseconds) of a group of 10 to 20 year olds and the reaction times of a group of 60 to 70 year olds.
Compare the two sets of data.

6

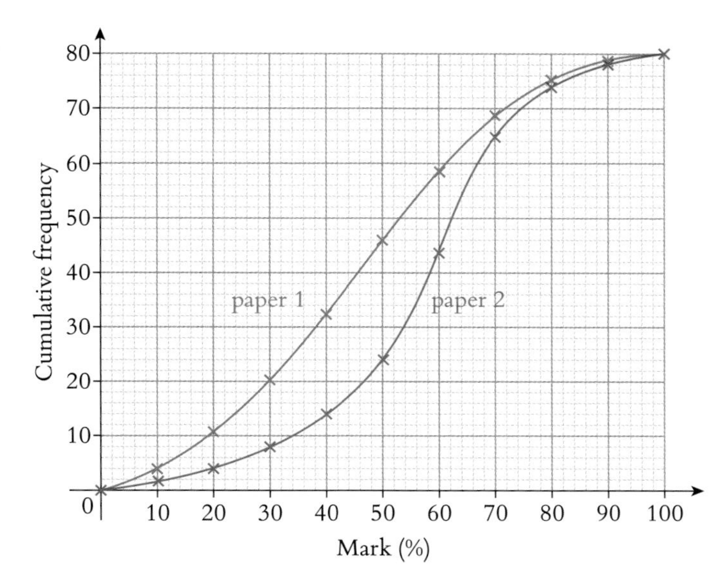

The cumulative frequency diagram shows information about the marks scored by a group of eighty students on Paper 1 and Paper 2 of a mathematics examination.

a Use the cumulative frequency diagram to copy and complete the following table.

	Median	LQ	UQ	IQR
Paper 1				
Paper 2				

b Compare the marks for Paper 1 and Paper 2.

7 In a survey, 100 shoppers leaving two different supermarkets were asked how much they had just spent. The results are shown in the two tables.

Supermarket A

Amount ($\$ x$)	Frequency
$0 < x \le 30$	2
$30 < x \le 60$	25
$60 < x \le 90$	47
$90 < x \le 120$	16
$120 < x \le 150$	10

Supermarket B

Amount ($\$ x$)	Frequency
$0 < x \le 30$	14
$30 < x \le 60$	24
$60 < x \le 90$	34
$90 < x \le 120$	26
$120 < x \le 150$	2

a Draw a cumulative frequency diagram to show the information about the amounts spent in supermarket A and in supermarket B.

b Use the cumulative frequency diagram to copy and complete the following table.

	Median	LQ	UQ	IQR
Supermarket A				
Supermarket B				

c Compare the amounts of money spent in supermarkets A and B.

KEY WORDS

cumulative frequency
lower quartile
upper quartile
inter-quartile range
percentile

UNIT 7

Unit 7 Examination-style questions

1 The diagram shows a speed–time graph for the journey of a car.

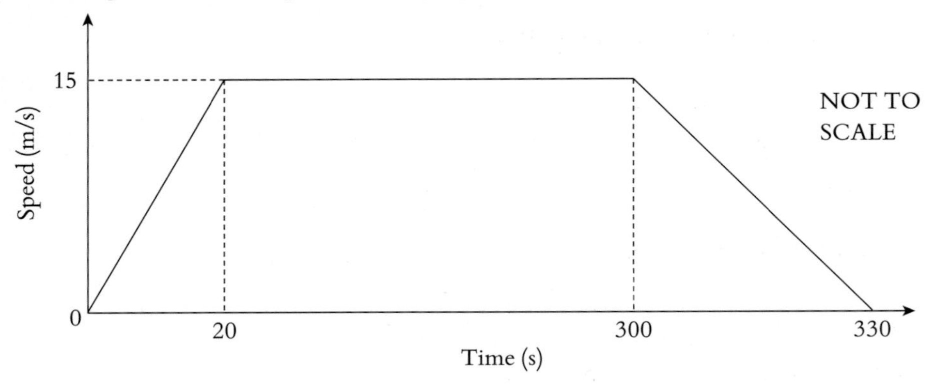

(a) Calculate the acceleration during the first 20 seconds. [1]
(b) Calculate the total distance travelled. [3]

2 The diagram shows information about the first 200 seconds of a train journey.

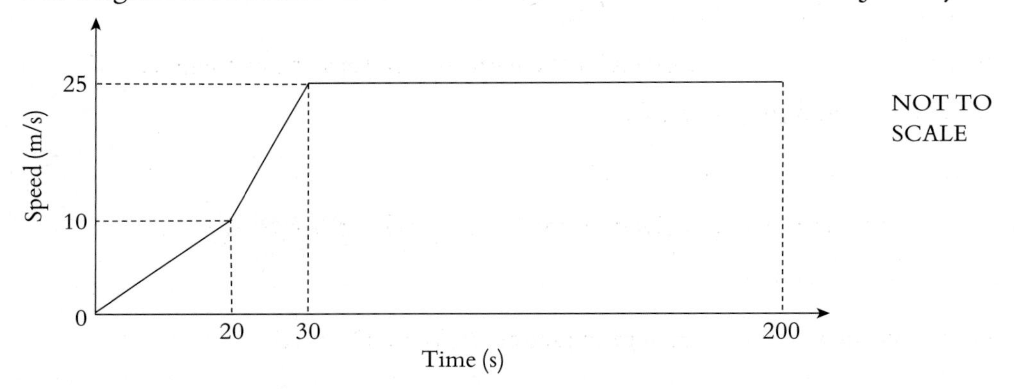

(a) Calculate the acceleration during the first 20 seconds. [1]
(b) Calculate the distance travelled by the car in the first 200 seconds. [3]
(c) Calculate the average speed of the car for the first 200 seconds. [2]

3 Rearrange the formula to make c the subject.

$$a = \frac{2bc}{b + c}$$ [4]

4 Rearrange the formula to make p the subject.

$$\frac{3}{p} - 4q = 5r$$ [4]

5 Rearrange the formula to make x the subject.

$$ax^2 - c = bx^2 - d$$ [4]

$$\sqrt[12]{1.3706} = 1 + \frac{r}{100}$$

$$1.02599\ldots = 1 + \frac{r}{100}$$ *subtract 1 from both sides*

$$0.02599\ldots = \frac{r}{100}$$ *multiply both sides by 100*

$r = 2.60\%$ correct to 3 sf

EXERCISE 8.2

1 Flynn invests \$800 in a bank account paying 1.8% compound interest per year.
Calculate the amount he has after 6 years.

2 Bianca invests \$480 in a bank account paying 2% compound interest per year.
Show that at the end of 6 years, she has an amount of \$541, correct to the nearest dollar.

3 Wei invests \$4000 in a bank account paying 2.5% compound interest per year.
Calculate the interest he receives after 15 years.
Give your answer correct to the nearest dollar.

4 Velimir invests \$1800 in an account paying 1.7% compound interest per year.
Calculate how much interest he will receive after 10 years.
Give your answer correct to the nearest dollar.

5 Jane invests \$4000 in an account paying 2.5% compound interest per year.
Carol invests \$4000 in an account paying 3% simple interest per year.
Calculate the difference between these two investments after 20 years.
Give your answer in dollars correct to the nearest cent.

6 Chris invested some money at a rate of 4% compound interest per year.
After 5 years the value of the investment totalled \$669.16.
Calculate how much Chris had invested.

7 Ali invests \$$x$ at a rate of 3% compound interest per year.
After 20 years the value of the investment is \$7495.36.
Calculate the value of x.

8 Haroon invests \$2000. At the end of 4 years, his account is worth \$2295.
Find the yearly compound interest rate.

9 Fiona invests \$1200 for 10 years at a rate of x% compound interest per year.
At the end of the 10 years she has a total amount of \$1652.27 correct to the nearest cent.
Find the value of x.

10 Sarah and Azhar both invest \$10 000.
Sarah invests her \$10 000 at a rate of x% compound interest per year.
Azhar invests his \$10 000 in a bank that pays 2% simple interest per year.
After 7 years their investments are worth the same amount.
Calculate the value of x.

UNIT 8

Exponential growth and decay

An example of an exponential function is $y = 2^x$. You learned about exponential graphs and their applications in Unit 7.

When the growth rate of an exponential function is positive, it is called **exponential growth**. Some examples of exponential growth are:

- the value of an investment
- population growth
- bacteria growth

When the growth rate of an exponential function is negative, it is called **exponential decay**. Some examples of exponential decay are:

- population decrease
- the value of a car
- radioactive decay
- temperature cooling

The formula that is used for solving exponential growth and decay problems is:

$$\text{Final value} = x_0\left(1+\frac{r}{100}\right)^n$$

where x_0 is the initial value, r is the rate of growth or decay, and n is the time interval.

Note For exponential growth: $r > 0$
For exponential decay: $r < 0$

EXAMPLE

At the start of an experiment there are 5000 bacteria.
The number of bacteria increases at a rate of 25% per hour.
Work out the number of bacteria after 10 hours.

$$\text{Number of bacteria} = x_0\left(1+\frac{r}{100}\right)^n \qquad \textit{use } x_0 = 5000,\ r = 25 \textit{ and } n = 10$$

$$= 5000\left(1+\frac{25}{100}\right)^{10}$$

$$= 500(1.25)^{10}$$

$$= 46\,566.128\ldots$$

$$= 46\,600 \text{ correct to 3 s.f.}$$

EXAMPLE

Farhat buys a boat for \$2500.
Each year the value of the boat decreases by 10% of its value at the beginning of that year.
Find the value of the boat after 4 years.

$$\text{Value after 4 years} = P\left(1+\frac{r}{100}\right)^n \qquad \textit{use } P = 2500,\ r = -10 \textit{ and } n = 4$$

$$= 2500\left(1+\frac{-10}{100}\right)^4$$

$$= 2500(0.9)^4$$

$$= \$1640.25$$

EXERCISE 8.3

1 Francesca buys a moped for \$4000.
Each year the value of the moped decreases by 15% of its value at the beginning of that year.
Find the value of the moped after 5 years.

2 Each year the value of a car decreases by 10% of its value at the beginning of that year.
Paul buys a car for \$30 000.
Calculate the value of Paul's car after 4 years.

3 Priya buys a jet ski for \$12 000.
The value of the jet ski decreases each year by 12% of its value at the beginning of that year.
Calculate the value of the jet ski after 5 years.
Give your answer correct to the nearest dollar.

4 It is estimated that the population of an endangered species of bird is decreasing at a rate of 3.5% per year.
On January 1st, 2010 the population was 5600.
Find the expected population on January 1st, 2060.

5 It is estimated that the population of a country is growing at a rate of 0.8% per year.
On January 1st, 2010 the population was 4.44 million.
Find the expected population on January 1st, 2040.

6 At the start of an experiment there are 20 000 radioactive particles.
The number of radioactive particles decreases at a rate of 20% per day.
Work out the number of radioactive particles after 5 days.

7 At the start of an experiment there are 5000 bacteria.
The number of bacteria increases at a rate of 35% per hour.
After how many hours will the number of bacteria be greater than five million?

8 The population of a city increases exponentially at a rate of x% per year.
In 2005 the population was 40 000.
In 2015 the population was 74 000.
Find the value of x.

9 Two months ago Ricardo had 4 mice.
He now has 8 mice.
Assuming that the number of mice increases exponentially at a rate of x% per month, find

a the value of x

b the number of mice that he will have in 2 months

c the number of mice that he will have in 1 year.

Solving quadratic equations using the formula

THIS SECTION WILL SHOW YOU HOW TO

- Solve quadratic equations using the quadratic formula

Some quadratic equations do not factorise.
For example: $x^2 - 5x + 3 = 0$ does not factorise because you cannot find two integers that multiply to give 3 and add to give −5.

Quadratic equations of the form $ax^2 + bx + c = 0$ can be solved using the formula

$$x = \frac{-b \pm \sqrt{b^2 - 4ac}}{2a}$$

In the formula:

a is the coefficient of the x^2 term
b is the coefficient of the x term
c is the constant.

The ± symbol means 'plus or minus'.

➡ **NOTE:** the quadratic formula is not given to you in the examination. You must learn the formula carefully.

Factorising gives exact answers. If a question asks you to give your answer to 2 d.p., it is a clue that you need to use the formula.

EXAMPLE

Solve $2x^2 - 9x + 8 = 0$, giving your answers to two decimal places.

$2x^2 - 9x + 8 = 0$ — *first write down the values of a, b and c.*

$a = 2 \quad b = -9 \quad c = 8$

$$x = \frac{-b \pm \sqrt{b^2 - 4ac}}{2a}$$ *substitute the values into the formula*

$$x = \frac{-(-9) \pm \sqrt{(-9)^2 - 4 \times 2 \times 8}}{2 \times 2}$$ *$-(-9) = 9$ and $(-9)^2 = 81$*

$$x = \frac{9 \pm \sqrt{81 - 64}}{4}$$

$$x = \frac{9 \pm \sqrt{17}}{4}$$

$$x = \frac{9 + \sqrt{17}}{4} \quad \text{or} \quad x = \frac{9 - \sqrt{17}}{4}$$

$x = 3.28$ or $x = 1.22$ (2 d.p.)

EXAMPLE

Solve $\frac{x}{x+3} - \frac{2}{x-4} = 1$

Method 1

$$\frac{x}{x+3} - \frac{2}{x-4} = 1$$

$$\frac{x(x-4)}{(x+3)(x-4)} - \frac{2(x+3)}{(x+3)(x-4)} = 1$$

$$\frac{x(x-4)-2(x+3)}{(x+3)(x-4)} = 1$$

$$\frac{x^2-4x-2x-6}{(x+3)(x-4)} = 1$$

$$x^2 - 6x - 6 = (x+3)(x-4)$$

$$x^2 - 6x - 6 = x^2 - x - 12$$

$$6 = 5x$$

$$x = 1\frac{1}{5}$$

Method 2

$$\frac{x}{x+3} - \frac{2}{x-4} = 1$$

multiply all 3 terms by $(x + 3)(x - 4)$

$$\cancel{(x+3)}(x-4) \times \frac{x}{\cancel{x+3}} - (x+3)\cancel{(x-4)} \times \frac{2}{\cancel{x-4}} = (x+3)(x-4) \times 1$$

$$x(x-4) - 2(x+3) = (x+3)(x-4)$$

$$x^2 - 4x - 2x - 6 = x^2 - x - 12$$

$$x^2 - 6x - 6 = x^2 - x - 12$$

$$6 = 5x$$

$$x = 1\frac{1}{5}$$

EXERCISE 8.7

1 Write as a single fraction.

a $\frac{3}{x} + \frac{2}{x-5}$ b $\frac{6}{x+3} + \frac{5}{x+2}$ c $\frac{8}{x-2} - \frac{3}{x+7}$

d $\frac{8}{2x} - \frac{1}{3x+1}$ e $\frac{7}{2x+5} - \frac{2}{3x-2}$ f $\frac{6}{1-2x} + \frac{5}{x-4}$

g $\frac{x+3}{x+2} - \frac{x+2}{x+3}$ h $\frac{x+5}{x+1} - \frac{x-2}{x+3}$ i $\frac{2}{x(x+1)} + \frac{3}{x+1}$

j $\frac{1}{x^2+6x+8} + \frac{1}{x+4}$ k $\frac{2}{x^2-2x-3} - \frac{3}{x-3}$ l $\frac{x(x-y)}{x+y} + y$

2 Solve these equations.

a $\frac{x+3}{x-3} = \frac{x+2}{x+4}$ b $\frac{x-6}{x+7} = \frac{x+3}{x-2}$ c $\frac{x+2}{5x+2} = \frac{2x-1}{4x+1}$

3 Solve these equations.

a $\frac{6}{x-1} + \frac{5}{x+2} = 4$ b $\frac{7}{2x-3} - \frac{5}{2x+1} = 6$ c $\frac{8}{x+1} + \frac{20}{x+4} = 2$

d $\frac{3x}{x-2} - \frac{2}{x-3} = 4$ e $\frac{6}{x+3} - \frac{5}{2x+1} = 8$ f $\frac{4}{x-3} + \frac{9}{2x-5} = 7$

g $\frac{2x}{4-x} + \frac{8}{x} = 6$ h $\frac{2x}{6+2x} - \frac{8}{2x+1} = 3$ i $\frac{x}{x-4} - \frac{x}{x-3} = 1$

Variation

THIS SECTION WILL SHOW YOU HOW TO
- Use direct and inverse variation

Direct variation (linear relationships)

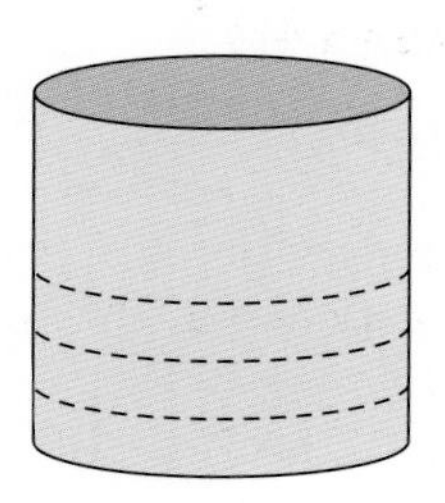

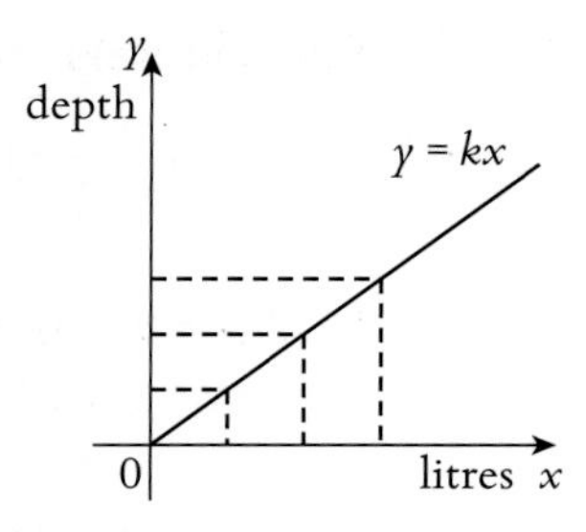

When a cylinder is filled with water, each litre of water increases the depth of water in the cylinder by the same amount.
The graph of the depth of the water against the number of litres in the cylinder is a straight line passing through the origin.
This type of relationship is known as **direct (linear) proportion** or **direct variation.**
(This means that if x is doubled then y also doubles.)

There are other ways of describing the relationship.
You need to be familiar with all of them.

- y is directly proportional to x
- y is proportional to x
- y varies directly as x
- y varies as x

The symbol for proportionality is $\propto$

> y is proportional to x
> $y \propto x$
> $y = kx$

k is called the **constant of proportionality**.
(It has the same value as the gradient of the graph.)

EXAMPLE

y is directly proportional to x and $y = 48$ when $x = 8$.

a Find the formula for y in terms of x.
b Find the value of y when $x = 2.5$
c Find the value of x when $y = 120$

a $y \propto x$ so $y = kx$
substituting $y = 48$ and $x = 8$ into $y = kx$ gives $48 = k \times 8$ $k = 6$
The formula is $y = 6x$
b substituting $x = 2.5$ into $y = 6x$ gives $y = 6 \times 2.5 \Rightarrow y = 15$
c substituting $y = 120$ into $y = 6x$ gives $120 = 6 \times x \Rightarrow x = 20$

EXERCISE 8.8

1 y is proportional to x and $y = 30$ when $x = 6$
   a Find the formula for y in terms of x.
   b Find the value of y when $x = 8$
   c Find the value of x when $y = 7$

2 y is proportional to x and $y = 56$ when $x = 7$
   a Find the formula for y in terms of x.
   b Find the value of y when $x = 3$
   c Find the value of x when $y = 6$

3 y is proportional to x and $y = 4.5$ when $x = 9$
   a Find the formula for y in terms of x.
   b Find the value of y when $x = 2.4$
   c Find the value of x when $y = 5$

4 y is proportional to x and $y = 6$ when $x = 18$
   a Find the formula for y in terms of x.
   b Find the value of y when $x = 4$
   c Find the value of x when $y = 3.4$

5 y is proportional to x and $y = 0.6$ when $x = 6$
   a Find the formula for y in terms of x.
   b Find the value of y when $x = 8$
   c Find the value of x when $y = 7$

6 If y is directly proportional to x, copy and complete this table.

x	2	5	
y		15	39

7 A stone is dropped from the top of a cliff. The speed of the stone, v m/s, varies directly with the time, t seconds, after it was dropped. After 2 seconds its speed is 19.6 m/s.
   a Find the formula for v in terms of t.
   b Find the speed when $t = 3$
   c Find the time when $v = 24.5$

8 The extension of a spring, e cm, is proportional to the mass, m kg, that hangs from it.
When a mass of 0.5 kg hangs from the spring its extension is 2.2 cm.
   a Find the formula for e in terms of m.
   b Find the extension when $m = 0.8$
   c Find the mass when $e = 1.76$

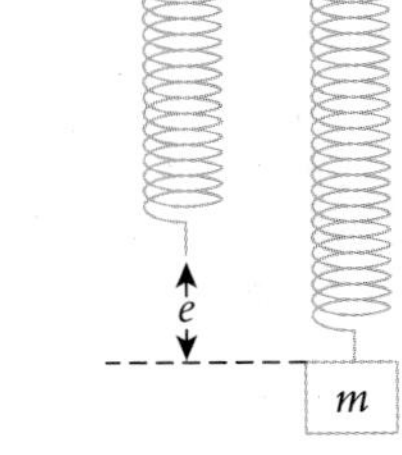

9 The depth of oil, d m, in a tank varies directly with the volume, V m^3, of oil that is in the tank. When $d = 0.9$, $V = 0.6$
   a Find the formula for d in terms of V.
   b Find the depth when $V = 1.2$
   c Find the volume when $d = 0.57$

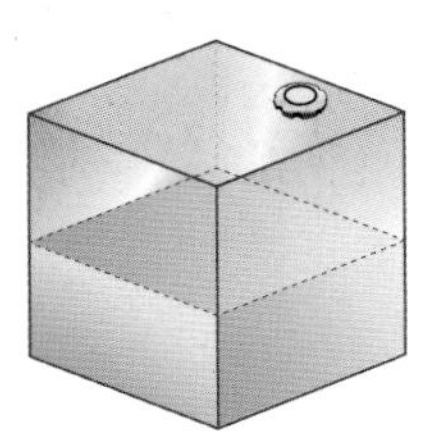

Direct variation (nonlinear relationships)

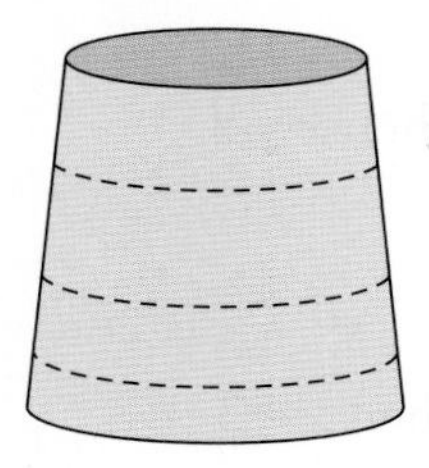

When the frustum of the cone is filled with water, each litre of water increases the depth of the water by different amounts.
A graph of the depth of the water against the number of litres in the frustum is a curve passing through the origin.

This relationship is known as a direct non-linear relationship.

EXAMPLE

y is proportional to the square of x and $y = 12$ when $x = 2$

a Find the formula for y in terms of x.

b Find the value of y when $x = 4$ **c** Find the value of x when $y = 3$

a $y \propto x^2$ so $y = kx^2$

substituting $y = 12$ and $x = 2$ into $y = kx^2$ gives $12 = k \times 2^2$

$12 = 4 \times k$

$k = 3$

The formula is $y = 3x^2$

b substituting $x = 4$ into $y = 3x^2$ gives $y = 3 \times 4^2$

$y = 3 \times 16$ $y = 48$

c substituting $y = 3$ into $y = 3x^2$ gives $3 = 3x^2$

$x^2 = 1$

$x = \pm 1$

➡ **NOTE:** there are two numbers that square to give the number 1.

EXAMPLE

The time for a pendulum to swing, t seconds, is proportional to the square root of the length, l m, of the pendulum.
$l = 0.81$ when $t = 1.8$. Find the value of t when $l = 0.49$

You must first find the formula connecting t and l.

$t \propto \sqrt{l}$ so $t = k\sqrt{l}$

substituting $t = 1.8$ and $l = 0.81$ into $t = k\sqrt{l}$ gives $1.8 = k \times \sqrt{0.81}$

$1.8 = k \times 0.9$

$k = 2$

The formula is $t = 2\sqrt{l}$

substituting $l = 0.49$ into $t = 2\sqrt{l}$ gives $t = 2 \times \sqrt{0.49}$

$t = 2 \times 0.7$

$t = 1.4$

Solving trigonometric equations

You are expected to be able to solve **trigonometric equations**, giving your answers in the range $0° \le x \le 180°$.

EXAMPLE

Solve the equation $\sin x = 0.4$ for $0° \le x \le 180°$.

You can find one answer to the equation $\sin x = 0.4$ using your calculator as follows

$x = \sin^{-1}(0.4)$

$x = 23.6°$

You must now check the sine curve to see if there are any more answers.
The graph shows that there is another answer.
Using the symmetry of the curve,
the second answer is $x = 180° - 23.6° = 156.4°$
So the solution is $x = 23.6°$ or $x = 156.4°$

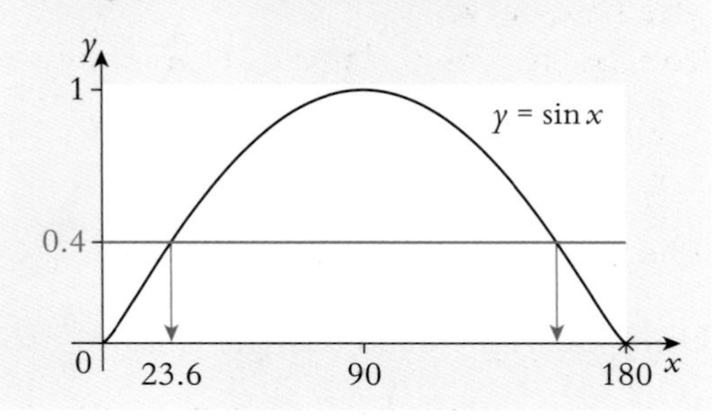

EXAMPLE

Solve the equation $\cos x = 0.624$ for $0° \le x \le 180°$.

Using a calculator

$x = \cos^{-1}(0.624)$

$x = 51.4°$

Now check the cosine curve to see if there are any more answers.
There are no more answers.
So the only solution is $x = 51.4°$

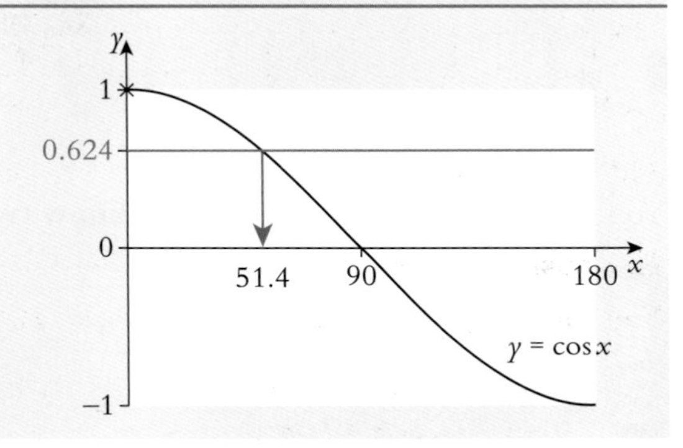

EXERCISE 8.12

Solve these equations for angles in the range $0° \le x \le 180°$.

1 $\cos x = 0.2$ **2** $\cos x = 0.7$ **3** $\sin x = 0.88$

4 $\cos x = -0.35$ **5** $\sin x = 0.2$ **6** $\sin x = 1$

7 $\cos x = -1$ **8** $\sin x = 0$ **9** $\cos x = -2$

10 $\sin x = -0.75$ **11** $\cos x = -0.76$ **12** $\sin x = 0.81$

13 $\cos x = \frac{2}{7}$ **14** $\sin x = \frac{3}{5}$ **15** $\cos x = -\frac{3}{4}$

16 $5\sin x = 4$ **17** $3\sin x = -1$ **18** $8\cos x = -9$

19 $2 + \cos x = 2.5$ **20** $1 - \sin x = 0.66$ **21** $3 + 2\sin x = 4$

22 $(\cos x)^2 = \frac{1}{4}$ **23** $(\sin x)^2 = \frac{4}{9}$ **24** $2 + (\cos x)^2 = 2.9$

KEY WORDS

sine curve

cosine curve

trigonometric equation

Area of a triangle

THIS SECTION WILL SHOW YOU HOW TO

- Calculate the area of a triangle using $\frac{1}{2}ab\sin C$

You can use trigonometry to work out the area of a triangle.

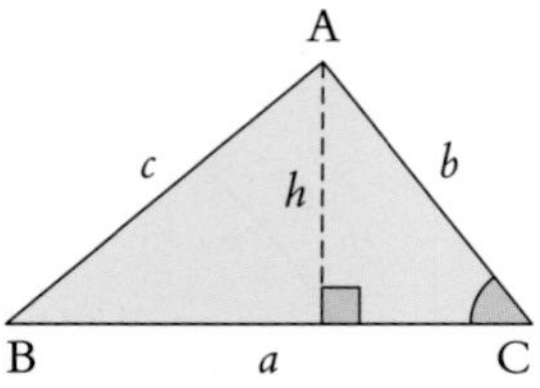

➡ **NOTE:**
angle A is opposite side a
angle B is opposite side b
angle C is opposite side c

Using the right-hand triangle $\sin C = \frac{h}{b}$ so $h = b\sin C$

Area of triangle $= \frac{1}{2} \times$ base $\times$ perpendicular height $= \frac{1}{2} \times a \times h$

replacing h by $b\sin C$ gives area $= \frac{1}{2}ab\sin C$

Area of triangle $= \frac{1}{2}ab\sin C$

You can use this formula when you know two sides and the angle between them.

EXAMPLE

Find the area of these triangles.

a

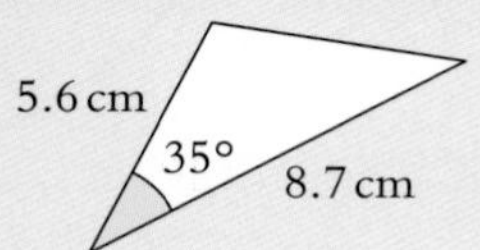

b

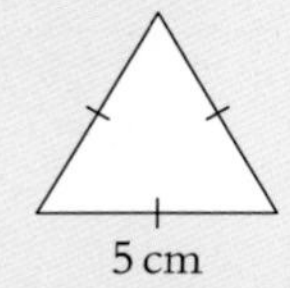

a Area $= \frac{1}{2} \times 8.7 \times 5.6 \times \sin 35°$

$= 14.0\text{ cm}^2$ (to 3 s.f.)

b Equilateral triangle so each angle is 60°

Area $= \frac{1}{2} \times 5 \times 5 \times \sin 60°$

$= 10.8\text{ cm}^2$ (to 3 s.f.)

EXAMPLE

Find the area of the parallelogram.

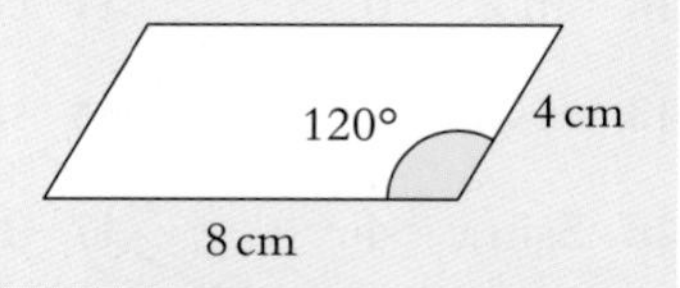

Divide the parallelogram into two congruent triangles.

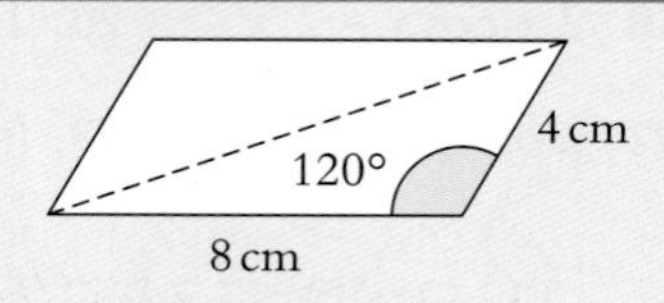

Area of triangle $= \frac{1}{2} \times 8 \times 4 \times \sin 120° = 13.856...$

Area of parallelogram $= 2 \times 13.856 = 27.7\text{ cm}^2$ (to 3 s.f.)

EXERCISE 8.15

1 Find the value of x for each of these triangles.

a

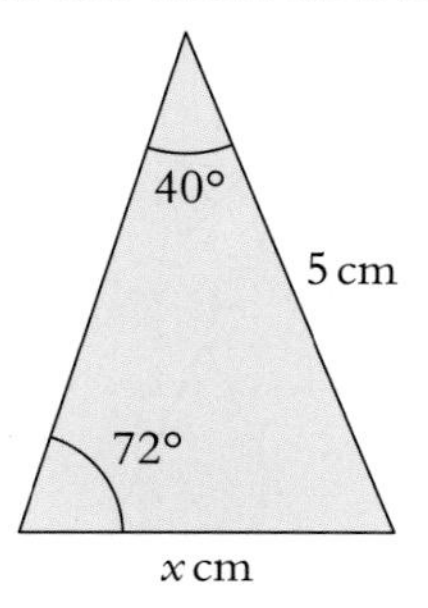

b

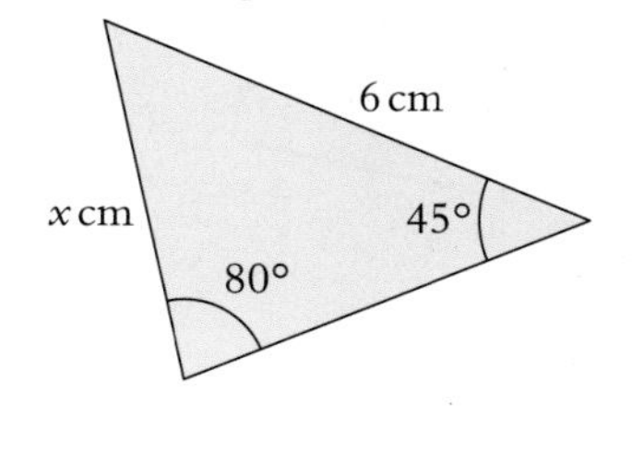

c

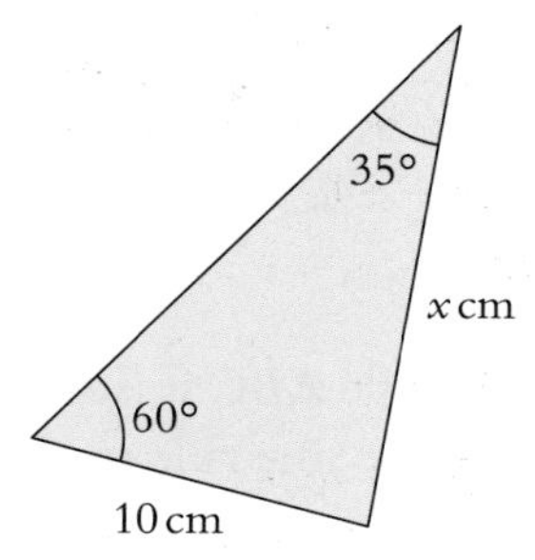

d

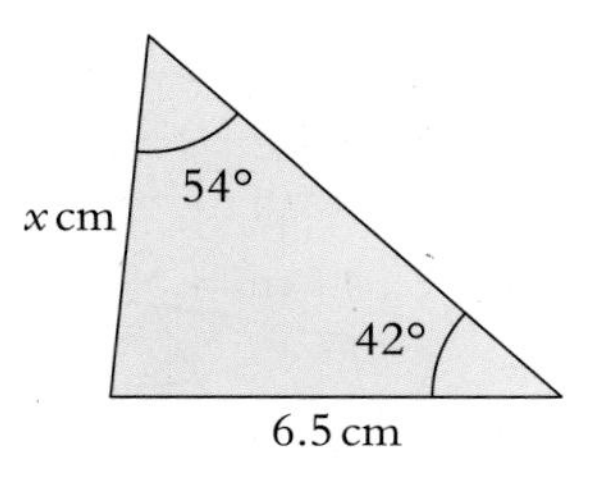

e

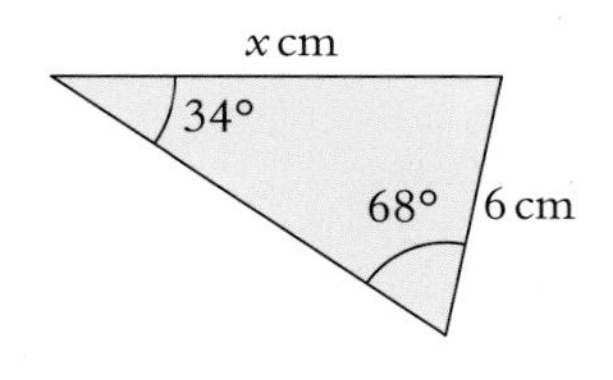

f

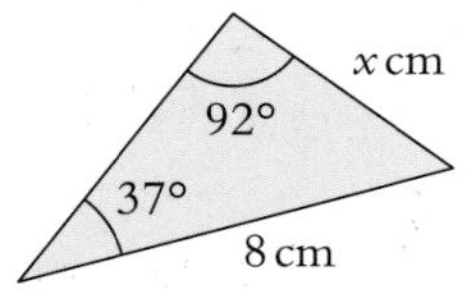

g

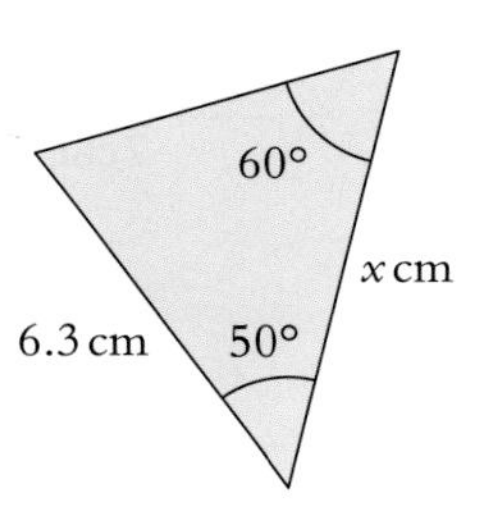

h

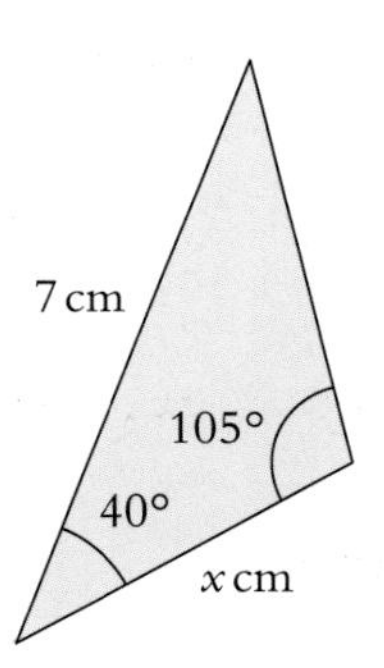

i

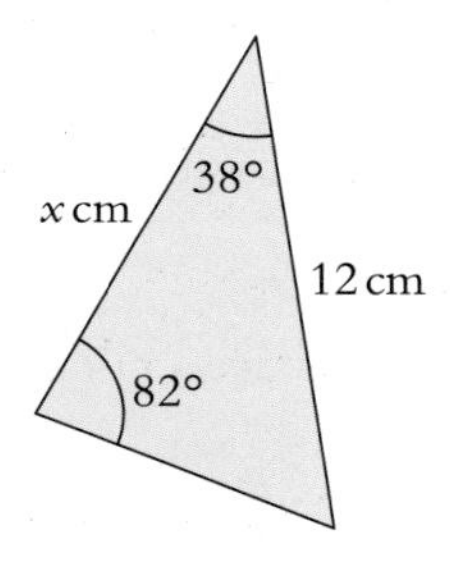

2 Find the value of x for each of these triangles.

a

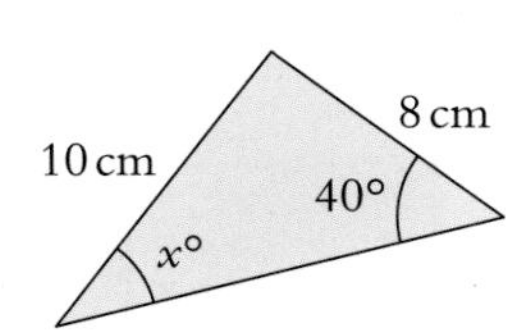

b

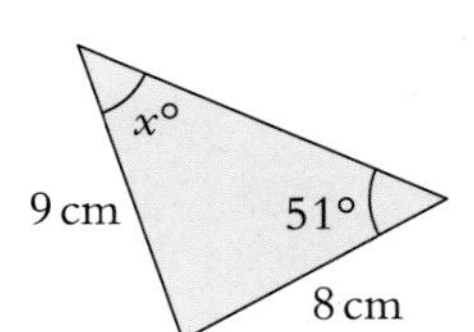

c

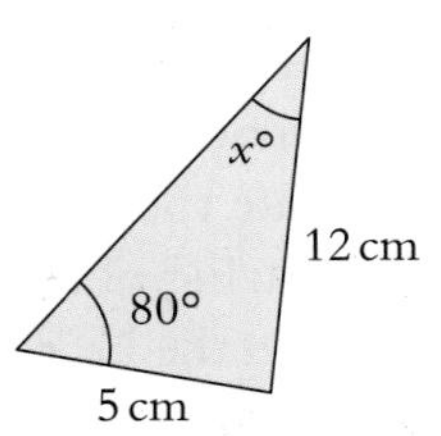

d

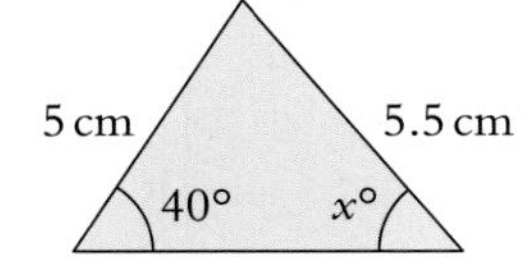

e

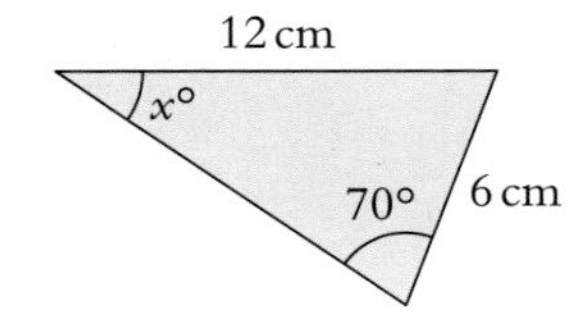

f

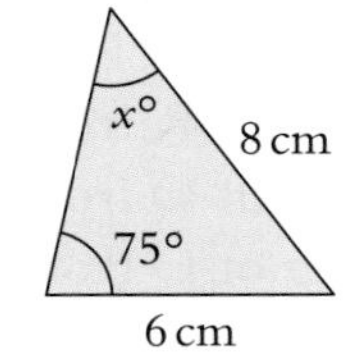

g

h

i

3 Triangle ABC is isosceles.
Use the sine rule to calculate the value of x.

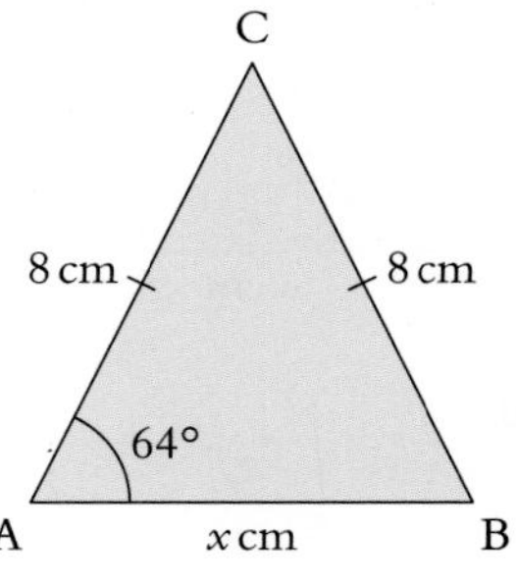

4 ABCD is a quadrilateral.
  a Calculate the length of AC.
  b Calculate the size of angle CAD.

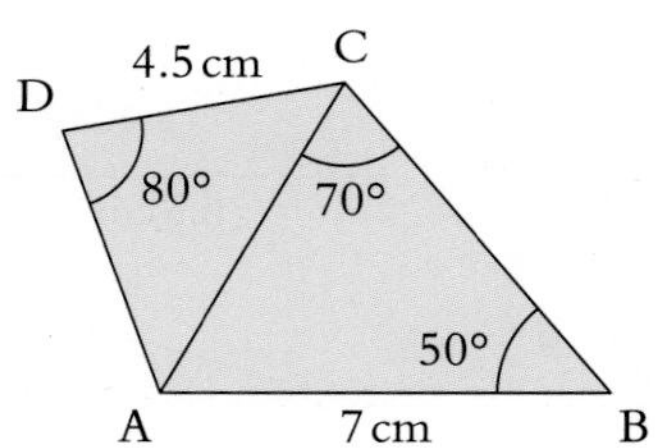

5 Calculate the size of angle ACB.

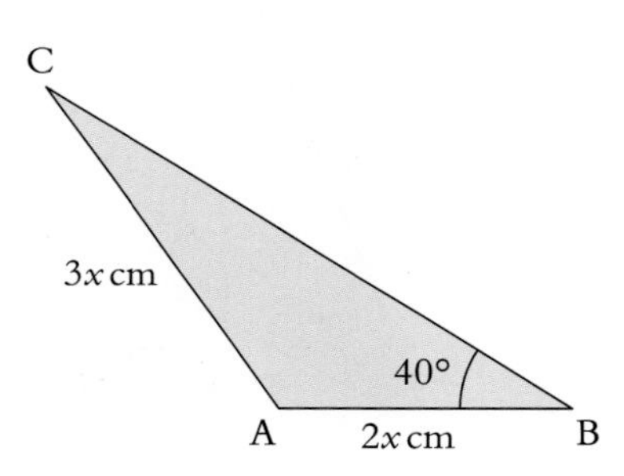

6 The diagram shows three straight roads connecting points A, B and C.
The length of AB is 15 km.
  a Calculate the length of the road BC.
  b The bearing of A from B is 290°.
    i Find the bearing of B from A.
    ii Find the bearing of B from C.

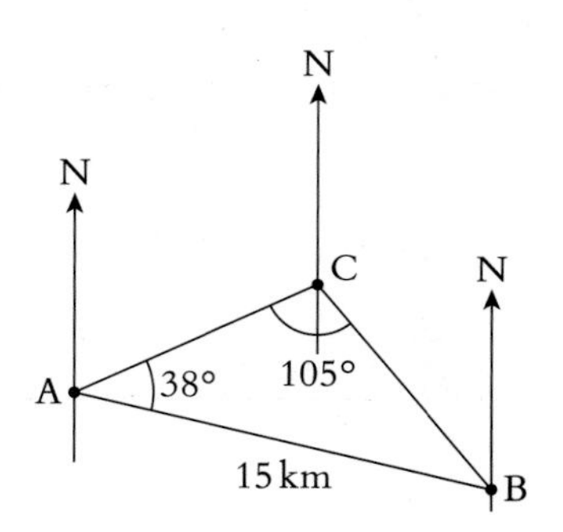

7 The diagram shows three straight roads connecting points P, Q and R.
The length of PR is 8 km.
The bearing of R from P is 055°.
The bearing of Q from P is 080°.
The bearing of Q from R is 110°.
Calculate the length of PQ.

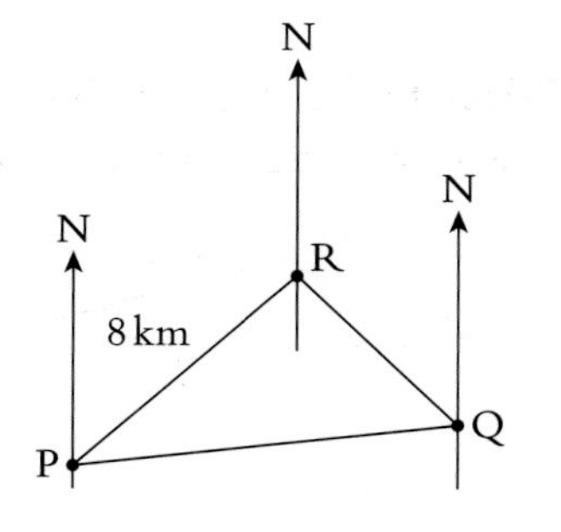

Sine rule – the ambiguous case

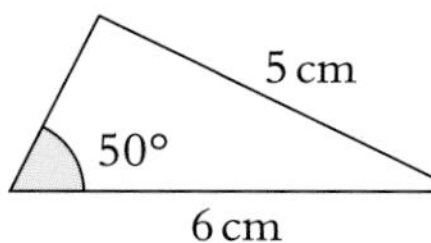

If you construct this triangle accurately, you will find there are two possible triangles that can be drawn.

ACCURATE CONSTRUCTIONS:

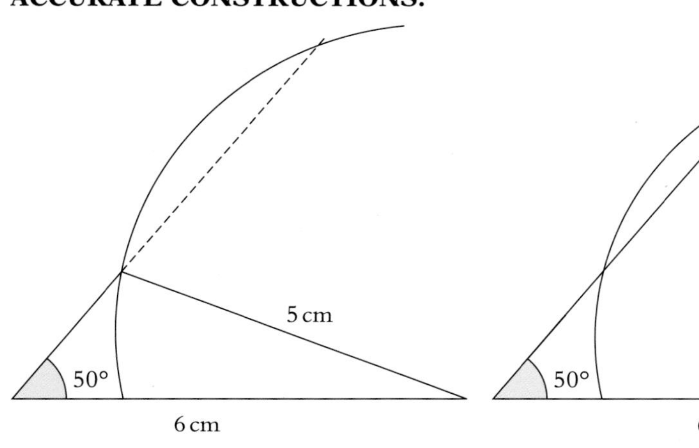

SOLVING USING SINE RULE

$$\frac{\sin x}{6} = \frac{\sin 50°}{5}$$

$$\sin x = 6 \times \frac{\sin 50°}{5}$$

$\sin x = 0.919...$

$x = 66.8°$ or $x = 180° - 66.8°$

$x = 66.8°$ or $x = 113.2°$

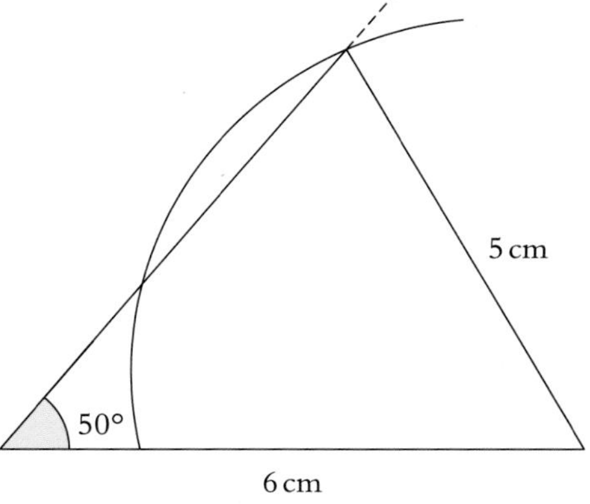

So the two possible triangles are

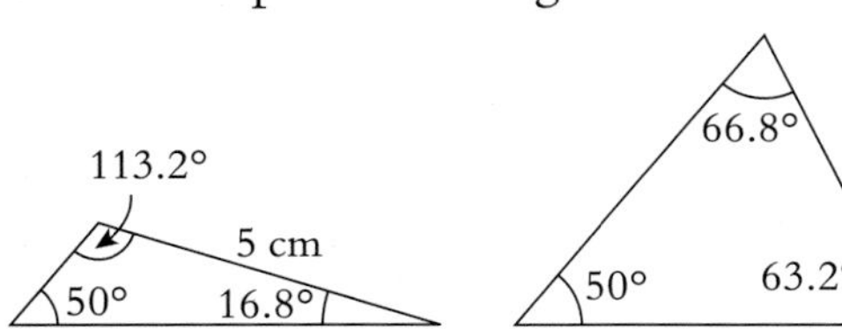

EXERCISE 8.16

Calculate the two possible values of x for each of these triangles. Check your calculations by constructing the triangles accurately.

1

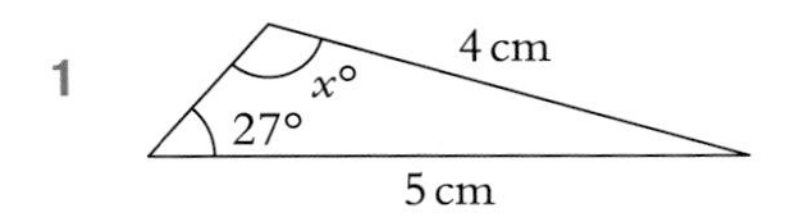

2

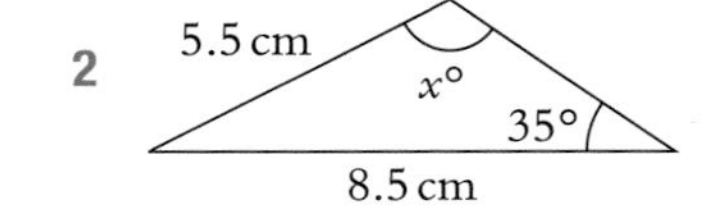

The cosine rule

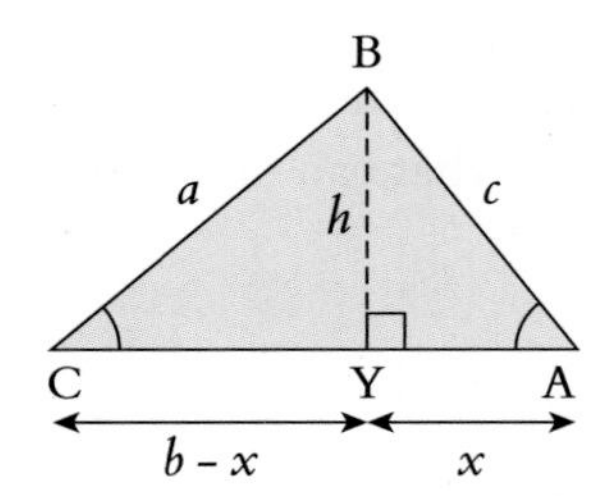

Using Pythagoras on triangle BYC:
$h^2 = a^2 - (b - x)^2$

Using Pythagoras on triangle AYB:
$h^2 = c^2 - x^2$

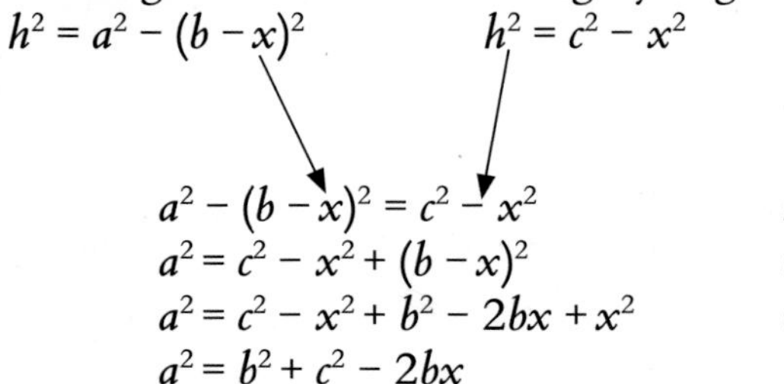

$a^2 - (b - x)^2 = c^2 - x^2$ *rearrange*
$a^2 = c^2 - x^2 + (b - x)^2$ *expand brackets*
$a^2 = c^2 - x^2 + b^2 - 2bx + x^2$ *collect like terms*
$a^2 = b^2 + c^2 - 2bx$

but, using $\triangle$AYB $\cos A = \dfrac{x}{c}$ so $x = c \cos A$
so $a^2 = b^2 + c^2 - 2bc \cos A$

The cosine rule: $a^2 = b^2 + c^2 - 2bc \cos A$

or $\cos A = \dfrac{b^2 + c^2 - a^2}{2bc}$

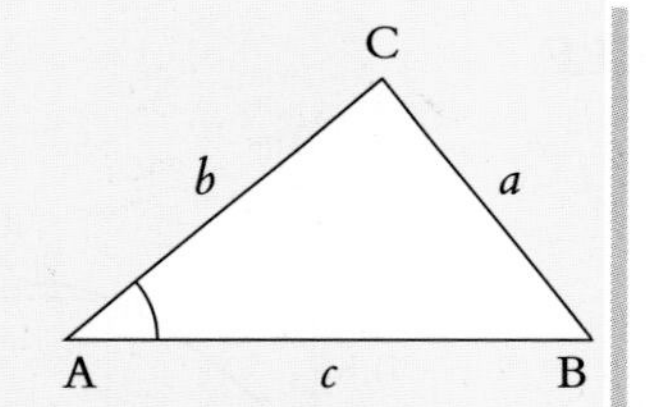

The cosine rule is used when you are working with three sides of a triangle and one angle.

Use $a^2 = b^2 + c^2 - 2bc \cos A$ to find the length of a side.

Use $\cos A = \dfrac{b^2 + c^2 - a^2}{2bc}$ to find the size of an angle.

EXAMPLE

Find the value of x in each of these triangles.

a

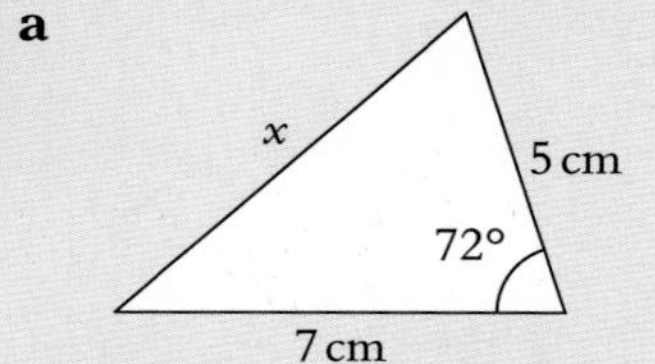

b

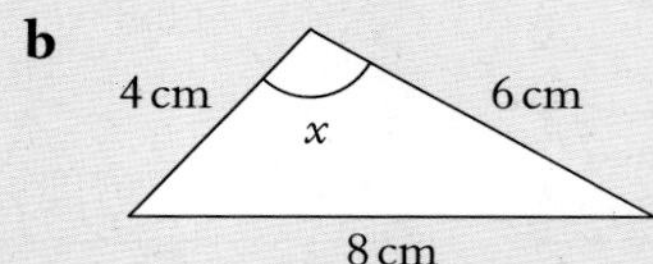

a Using the cosine rule:

$x^2 = 5^2 + 7^2 - 2 \times 5 \times 7 \times \cos 72°$
$x^2 = 52.368...$
$x = 7.24$ cm (to 3 s.f.)

b Using the cosine rule:

$\cos x = \left(\dfrac{4^2 + 6^2 - 8^2}{2 \times 4 \times 6}\right)$

$x = \cos^{-1}\left(\dfrac{4^2 + 6^2 - 8^2}{2 \times 4 \times 6}\right)$

$x = 104.5°$ (to 1 d.p.)

EXERCISE 8.19

1 The masses of 100 sweets are recorded.
The results are shown in the frequency table.
  a Calculate the frequency density for each group.
  b Draw a histogram to represent this information.

Mass (m grams)	Frequency
$5 < m \leq 10$	5
$10 < m \leq 20$	35
$20 < m \leq 30$	45
$30 < m \leq 35$	15

2 The table shows the time, t hours, spent on homework each week by 100 students.
  a Calculate the frequency density for each group.
  b Draw a histogram to represent this information.

Time (t hours)	Frequency
$0 < t \leq 1$	7
$1 < t \leq 4$	39
$4 < t \leq 7$	42
$7 < t \leq 13$	12

3 The lengths of 85 phone calls are recorded.
The results are shown in the table.
  a Calculate the frequency density for each group.
  b Draw a histogram to represent this information.

Time (t minutes)	Frequency
$0 < t \leq 1$	5
$1 < t \leq 3$	16
$3 < t \leq 7$	28
$7 < t \leq 9$	12
$9 < t \leq 15$	24

4 The heights of 200 students are measured.
The results are shown in the table.
  a Calculate the frequency density for each group.
  b Draw a histogram to represent this information.

Height (h cm)	Frequency
$150 < h \leq 160$	24
$160 < h \leq 170$	58
$170 < h \leq 175$	63
$175 < h \leq 180$	45
$180 < h \leq 200$	10

5 The speeds of 130 cars passing a 70 km/h speed limit sign are recorded.
The results are shown in the table.
Use this information to draw a histogram.

Speed (v km/h)	Frequency
$50 < v \leq 60$	15
$60 < v \leq 65$	30
$65 < v \leq 70$	40
$70 < v \leq 75$	35
$75 < v \leq 90$	10

6 200 students are given a puzzle to solve.
The results are shown in the table.
Use this information to draw a histogram.

Time (t minutes)	Frequency
$0 < t \leq 1$	4
$1 < t \leq 3$	45
$3 < t \leq 6$	90
$6 < t \leq 7$	26
$7 < t \leq 12$	35

7 The table shows information about the time spent on the internet one evening by a group of 160 students. Draw a histogram to represent this information.

Time (t minutes)	Frequency
$0 < t \leq 40$	36
$40 < t \leq 70$	90
$70 < t \leq 150$	34

8 The histogram shows information about the lengths of 35 phone calls.

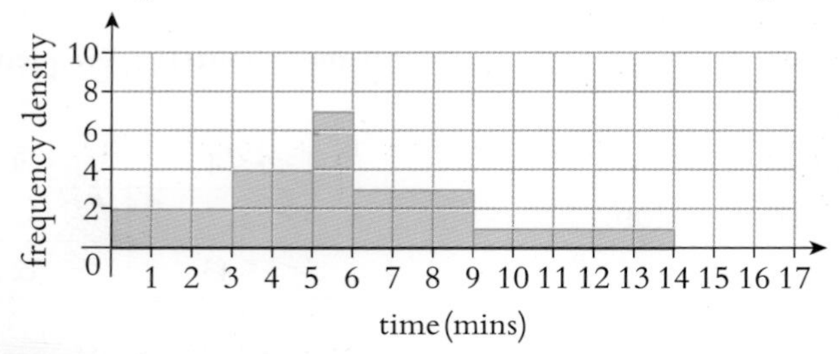

TOP TIP

frequency = frequency density × class width

a Use the information in the histogram to copy and complete the following table.

Time (t mins)	$0 < t \leq 3$	$3 < t \leq 5$	$5 < t \leq 6$	$6 < t \leq 9$	$9 < t \leq 14$
Frequency					5

b Use the information in the table to calculate an estimate of the mean length of a phone call.

9 The histogram shows information about the masses of 25 newborn babies.

a Use the information in the histogram to copy and complete the following table.

Mass (m kg)	$1 < m \leq 2$	$2 < m \leq 3$	$3 < m \leq 3.5$	$3.5 < m \leq 4$	$4 < m \leq 5$
Frequency			6		

b Use the information in the table to calculate an estimate of the mean mass.

10 75 students do a sponsored swim to raise money for charity. The table shows information about the amounts of money ($\$x$) raised.

Amount ($\$x$)	$0 < x \leq 10$	$10 < x \leq 20$	$20 < x \leq 35$	$35 < x \leq 50$
Frequency	15	40	15	5

When a histogram was drawn to show this information, the height of the column for the interval $20 < x \leq 35$ was 6 cm. Calculate the height of each of the other columns.

11 45 batteries are tested to see how long they last.
The table shows the results

Number of hours (t)	$10 < t \le 15$	$15 < t \le 17$	$17 < t \le 20$	$20 < t \le 30$
Frequency	10	8	15	12

When a histogram was drawn to show this information, the height of the column for the interval $10 < t \le 15$ was 4 cm.
Calculate the height of each of the other columns.

12 A group of students are asked how long they spent watching television one evening.
The information is shown in the histogram.

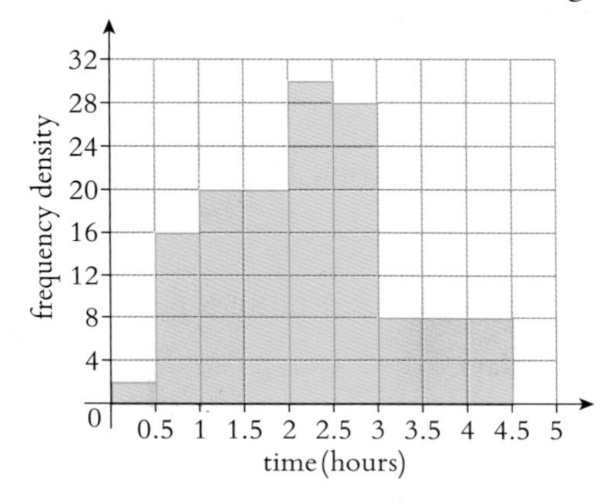

Time (t hours)	Frequency
$0 < t \le 0.5$	1
$0.5 < t \le 1$	
$1 < t \le 2$	
$2 < t \le 2.5$	
$2.5 < t \le 3$	
$3 < t \le 4.5$	

a Use the information in the histogram to copy and complete the table.
b Estimate how many students spent less than 1.5 hours watching television.
c Use the information in the table to calculate an estimate of the mean number of hours.
d A student is chosen at random.
Find the probability that the student spent more than 2 hours watching television.

13 The frequency table and histogram are incomplete.
They show information about the distances travelled to work by a group of workers.
No worker travelled more than 28 km.

Distance (d km)	Frequency
$0 < d \le 4$	8
$4 < d \le 10$	
$10 < d \le 12$	
$12 < d \le 14$	20
$14 < d \le 22$	40
$22 < d \le 28$	18

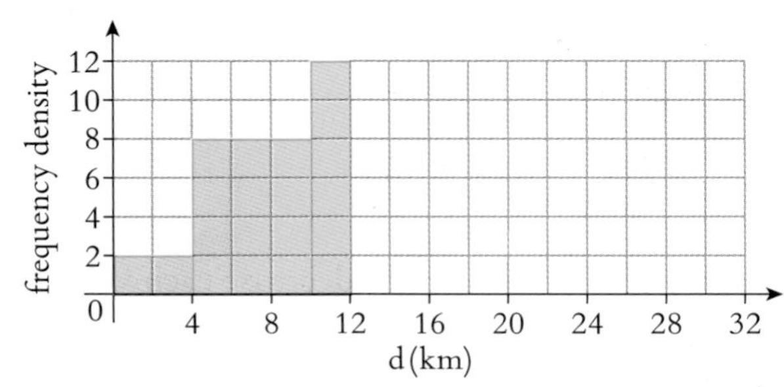

a Use the information given in the histogram to find the missing frequencies in the table.
b Use the information given in the table to draw the missing bars on the histogram.
c Calculate an estimate of the mean distance travelled.

KEY WORDS
histogram
class width
frequency density

Unit 8 Examination-style questions

1 Write the recurring decimal $0.\dot{5}$ as a fraction.
You must show all your working. [2]

2 Write the recurring decimal $0.3\dot{5}$ as a fraction in its lowest terms.
You must show all your working [3]

3 Write the recurring decimal $0.\dot{8}0\dot{4}$ as a fraction in its lowest terms.
You must show all your working. [3]

4 $2\frac{1}{4}$ π $\sqrt{\frac{4}{9}}$ $0.0\dot{4}$ -8 $\sqrt{7}$

From this list of numbers, write down two numbers that are irrational. [2]

5 At the start of an experiment there are 500 radioactive particles.
The number of radioactive particles decreases exponentially at a rate of 8.5% per day.

Work out the number of radioactive particles after 12 days. [2]

6 At the start of an experiment there are 3000 bacteria.
The number of bacteria increases exponentially at a rate of 27% per hour.

Find the numbers of complete hours it takes for the number of bacteria to be greater than one million. [2]

7 The population of a town increases exponentially at a rate of x % per year.
On 1st January 2000 the population was 50 000.
On 1st January 2015 the population was 63 000.

Find the value of x.
Give your answer correct to 3 significant figures.

8 $P = 45\,000 \times 0985^t$
A scientist uses this formula to predict the population, P, of an endangered species t years after 2010.

(a) Write down the population in 2020. [1]
(b) Write down the percentage that the population is predicted to decrease each year. [1]
(c) Find the value of P when $t = 30$.
Write your answer correct to the nearest hundred. [2]

9 Solve $2x^2 + 5x - 8 = 0$.
You must show all your working and give your answers correct to 2 decimal places. [4]

10 Solve $3x - \frac{2}{x+1} = 7$.
You must show all your working and give your answers correct to 2 decimal places. [6]

11 A solid circular cylinder has a base radius of r cm and a height of 5 cm.
The total surface are of this cylinder is $150\pi\, \text{cm}^2$.

(a) Show that $r^2 + 5r - 75 = 0$. [2]

(b) Solve the equation in **part (a)** to find the value of r.
Give your answer correct to 2 decimal places. [4]

12 Write as a single fraction in its simplest form.

$$\frac{7}{x-3} - \frac{5}{x+2}$$ [3]

13 Write as a single fraction in its simplest form.

$$\frac{3x+7}{4} + \frac{2}{x-3}$$ [3]

14 Solve.

$$\frac{5}{x-1} - \frac{2}{x+3} = 1$$ [7]

15 y is directly proportional to $\sqrt{x+2}$.
When $x = 7$, $y = 12$.

Find y when $x = 0.25$ [3]

16 y is inversely proportional to x^2.
When $x = 5$, $y = 10$.

Find y in terms of x. [2]

17 Solve the equation $\sin x = 0.5$ for $0° \leq x \leq 360°$. [2]

18

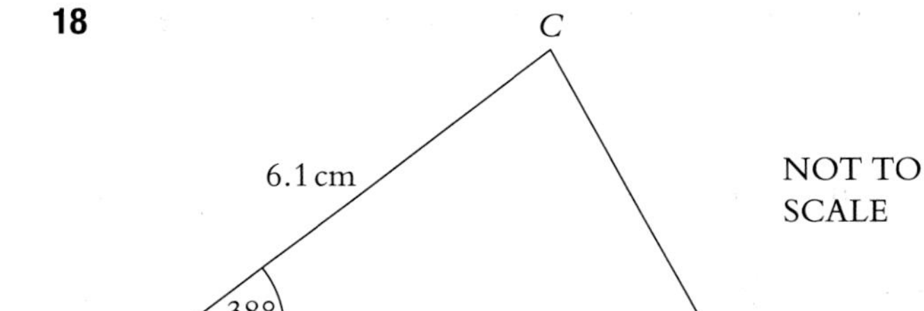

Calculate the area of triangle ABC. [2]

19

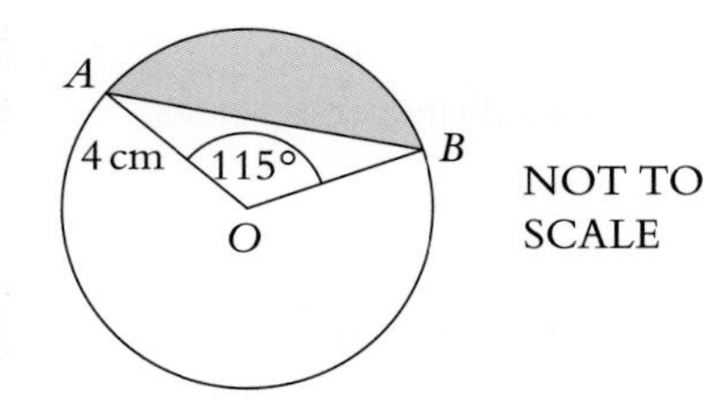

The diagram shows a circle, centre O, with radius 4 cm.
AB is a chord of the circle and angle AOB is 115°.

Calculate the shaded area. [3]

20

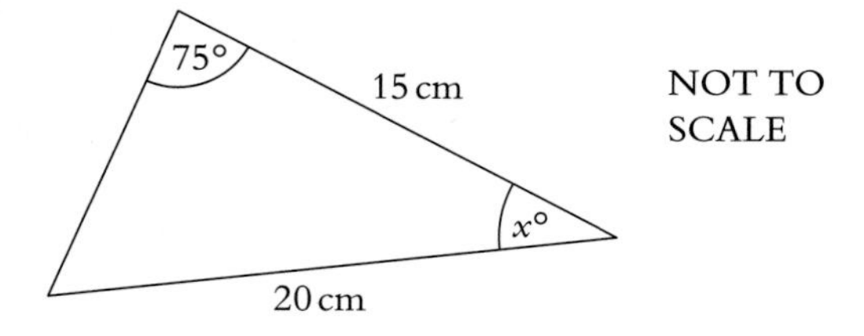

Find the value of x. [3]

21

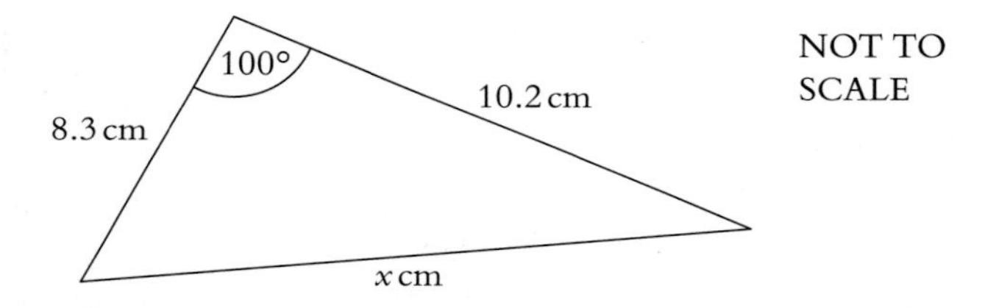

Find the value of x. [4]

22

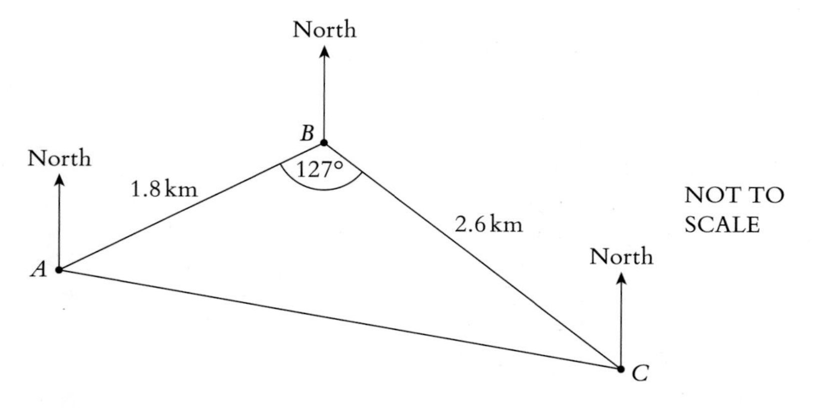

Robert sails from A to B, then from B to C and then returns to A.
$AB = 1.8$ km, $BC = 2.6$ km and angle $ABC = 127°$.

(a) **(i)** Find AC. [4]

(ii) Find angle BCA. [3]

(b) The bearing of B from A is 062°.

(i) Find the bearing of A from B. [2]

(ii) Find the bearing of A from C. [2]

23

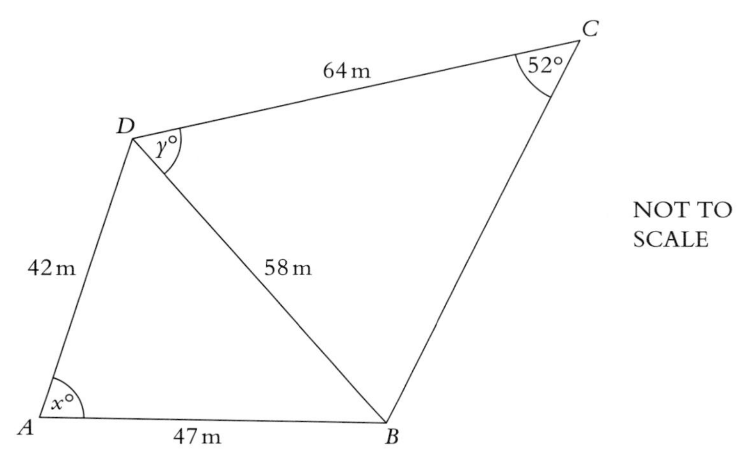

NOT TO SCALE

(a) Calculate the value of x. [4]

(b) Calculate the value of y. [4]

(c) Calculate the area of quadrilateral $ABCD$. [3]

24 The table shows information about the mass, m grams, of 100 eggs.

Mass (m grams)	$35 < m \le 40$	$40 < m \le 50$	$50 < m \le 60$	$60 < m \le 80$
Frequency	8	36	46	10

On the grid, complete the histogram to show this information.

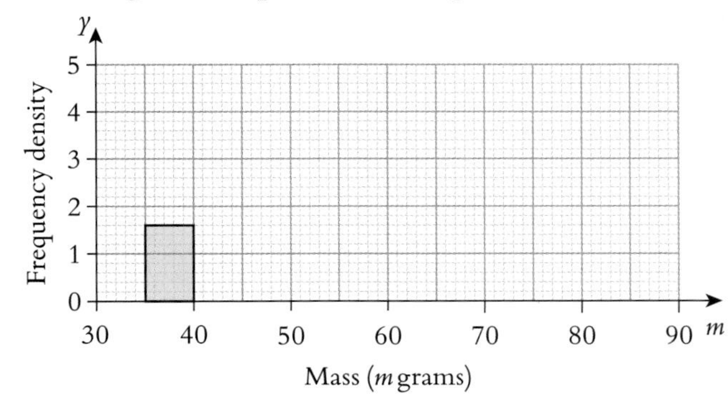

[4]

25 The histogram shows information about the mass, x kilograms, of 40 babies in a hospital.

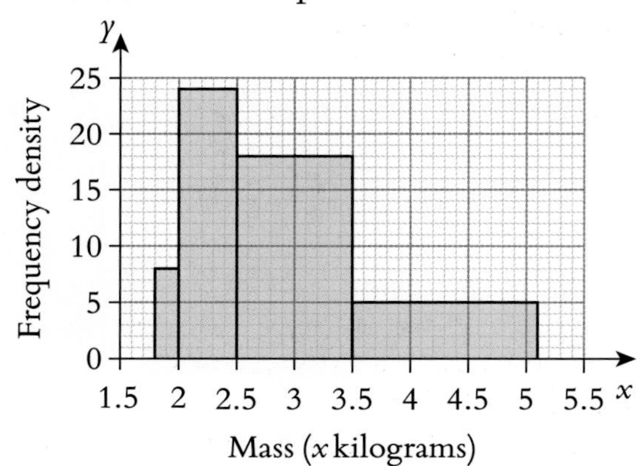

(a) Use the histogram to complete the grouped frequency table below.

Mass (x kilograms)	$1.8 < x \le 2$	$2 < x \le 2.5$	$2.5 < x \le 3.5$	$3.5 < x \le 5.1$
Frequency		12		

[3]

(b) Estimate the mean mass of the babies. [4]

26 In this question all lengths are in centimetres.

(a)

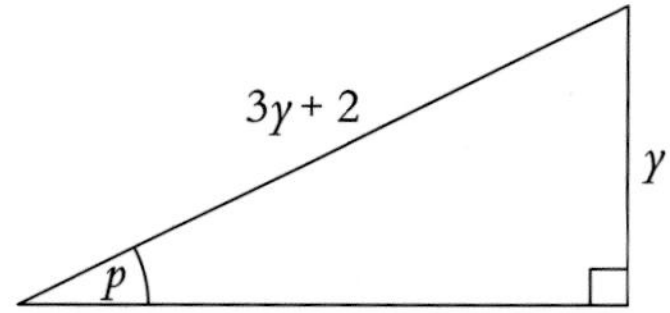

NOT TO SCALE

Given that $\sin p = 0.3$, find the value of y. [2]

(b)

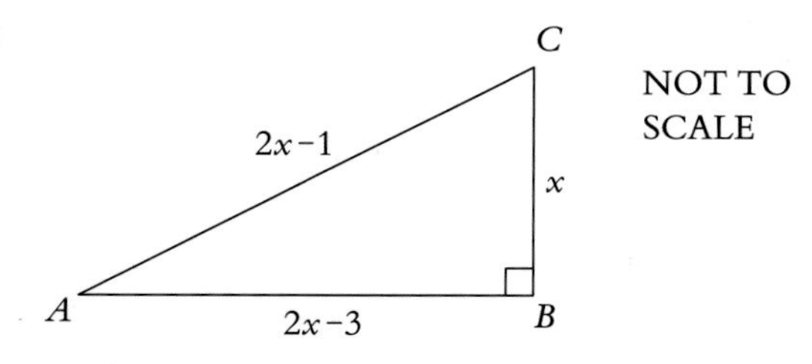

NOT TO SCALE

(i) Show that $x^2 - 8x + 8 = 0$. [6]

(ii) Solve the equation $x^2 - 8x + 8 = 0$.
Show all your working and give your answers correct to 2 decimal places. [4]

(iii) Find the two possible values for the perimeter of triangle ABC. [2]

UNIT 8

27 Haroon and Farhat each cycle a distance of 50 km.
Haroon's average speed is x km/h.
Farhat's average speed is 2 km/h faster than Haroon's average speed.
Farhat takes 6 minutes less than Haroon to cycle the 50 km.

(a) Write down an expression, in terms of x, for the time, in hours, it takes for Haroon to cycle the 50 km. [1]

(b) Write down an expression, in terms of x, for the time, in hours, it takes for Farhat to cycle the 50 km. [1]

(c) Show that $x^2 + 2x - 1000 = 0$. [4]

(d) Show the equation $x^2 + 2x - 1000 = 0$.
Show all your working and give your answers correct to 2 decimal places. [4]

(e) Calculate the time it takes for Farhat to cycle the 50 km.
Give your answer in hours and minutes, correct to the nearest minute. [3]

28

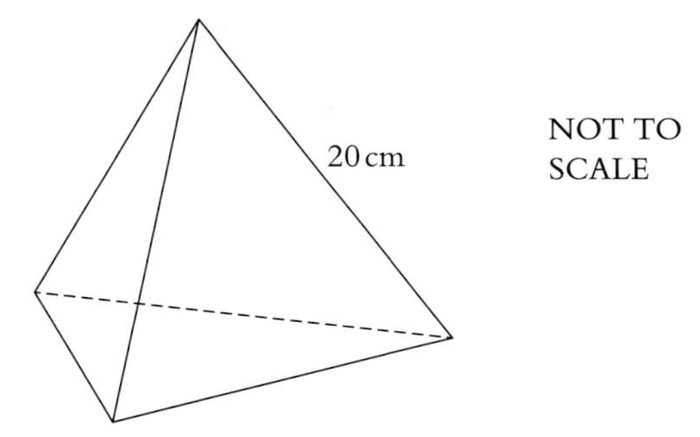

NOT TO SCALE

The diagram shows a solid triangular-based pyramid.
The length of each edge of the pyramid is 20 cm.

Calculate the total surface area of this pyramid. [3]

UNIT 8

Upper and lower bounds

THIS SECTION WILL SHOW YOU HOW TO

- Find the upper and lower bounds of calculations

If the length of a nail is given as 6 cm correct to the nearest cm, you can find the **lower bound** and the **upper bound** for the length of the nail.

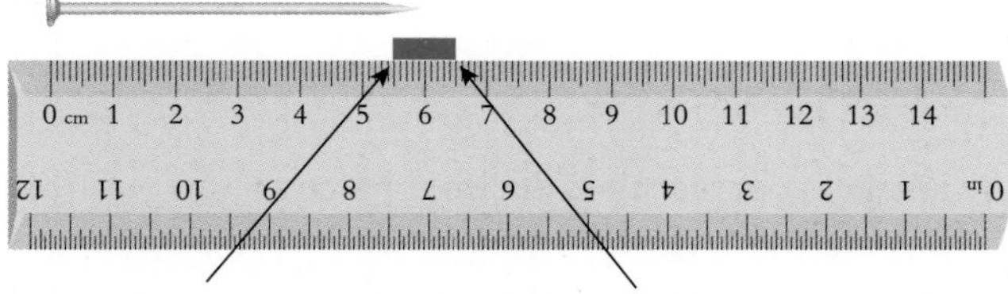

The lower bound is halfway between 5 cm and 6 cm. Lower bound = 5.5 cm.
The upper bound is halfway between 6 cm and 7 cm. Upper bound = 6.5 cm.
The range of possible values for the length of the nail can be written as:

$$5.5 \text{ cm} \le \text{length} < 6.5 \text{ cm}$$

EXAMPLE

The mass of a parcel is 3.2 kg, correct to one decimal place.
Write down the upper and lower bounds for the mass of the parcel.

You need to think of the numbers (to 1 d.p.) directly below 3.2 and directly above 3.2.
These numbers are 3.1 and 3.3.
The lower bound is halfway between 3.1 and 3.2.
The upper bound is halfway between 3.2 and 3.3.
Lower bound = 3.15 cm Upper bound = 3.25 cm.

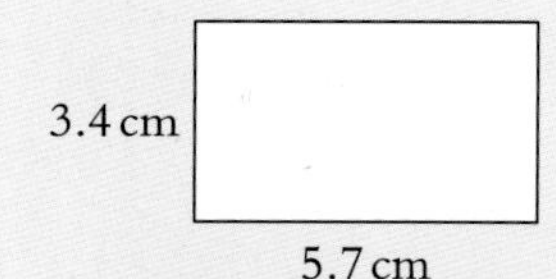

EXAMPLE

A rectangle has sides of length 5.7 cm and 3.4 cm, each correct to the nearest millimetre.
Calculate the upper and lower bounds for the area of the rectangle.

3.4 cm

5.7 cm

First write down the lower and upper bounds for the length and width of the rectangle.

Lower bound length = 5.65 Upper bound length = 5.75
Lower bound width = 3.35 Upper bound width = 3.45

So, the lower bound for the area = lower bound length × lower bound width
$= 5.65 \times 3.35$
$= 18.9275 \text{ cm}^2$

and the upper bound for the area = upper bound length × upper bound width
$= 5.75 \times 3.45$
$= 19.8375 \text{ cm}^2$

➡ **NOTE:** you must not round your answers when you are asked for a bound. You must write the full calculator display.

EXAMPLE

$p = 8$ cm $q = 2$ cm

The lengths of p and q have been given to the nearest centimetre.

Find the lower bound for each of the following calculations

a $p + q$ **b** $p - q$ **c** $p \times q$ **d** $p \div q$

---

a First write down the lower and upper bounds for the lengths of p and q.

Lower bound of $p = 7.5$ Upper bound of $p = 8.5$
Lower bound of $q = 1.5$ Upper bound of $q = 2.5$

Lower bound of $(p + q)$ = lower bound of p + lower bound of q
$= 7.5 + 1.5$
$= 9$ cm

b Lower bound of $(p - q)$ = lower bound of p – upper bound of q
$= 7.5 - 2.5$
$= 5$ cm

➡ **NOTE:** you must choose carefully which bounds to use in the calculation.

c Lower bound of $(p \times q)$ = lower bound of p × lower bound of q
$= 7.5 \times 1.5$
$= 11.25$

d Lower bound of $(p \div q)$ = lower bound of p ÷ upper bound of q
$= 7.5 \div 2.5$
$= 3$

EXERCISE 9.1

1 Write down the upper and lower bounds for each of these measurements.

a 6 cm (to nearest cm)   b 32 min (to nearest minute)
c 32 kg (to nearest kg)   d 9.2 cm (to nearest millimetre)
e 7.7 cm (to 1 d.p.)   f 2.63 kg (to 3 s.f.)
g 62.9 g (to 1 d.p.)   h 476 s (to nearest second)
i 4.94 m (to 3 s.f.)   j 0.245 km (to nearest metre)

2 A bottle contains 250 ml of oil, correct to the nearest millilitre.
Write down the range of possible values for the amount of oil, x millilitres, in the bottle.

3 A carton has a mass of 1.2 kg correct to 2 significant figures.
Calculate the minimum possible total mass of 20 cartons.

4 The cost of making a necklace is $124 correct to the nearest dollar.
Calculate the upper and lower bounds for the cost of making 150 necklaces.

5 When a bicycle wheel turns once, the bicycle travels 205 cm, correct to the nearest cm. Calculate the upper and lower bounds for the distance travelled by the bicycle when the wheel has turned 150 times.

6 The length of each side of an equilateral triangle is 46 mm, correct to the nearest millimetre.
Calculate the upper and lower bounds for the perimeter of the triangle.

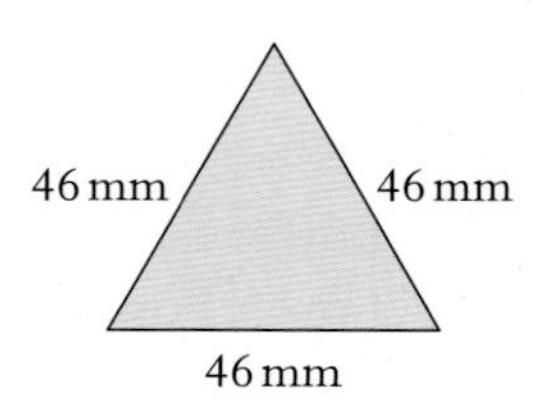

7 The length of each side of a regular octagon is 8.2 cm, correct to one decimal place.
Calculate the upper and lower bounds for the perimeter of the octagon.

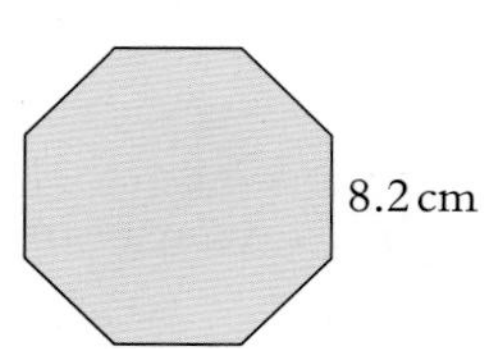

8 A rectangle has sides of length 5.3 cm and 7.4 cm, each correct to one decimal place.
Calculate the upper and lower bounds for

a the perimeter of the rectangle,

b the area of the rectangle.

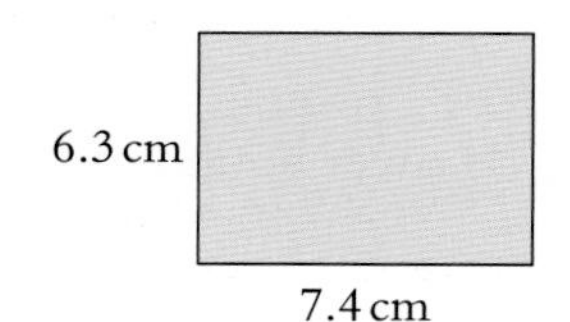

9 A rectangular photograph measures 22.4 cm by 18.6 cm, each correct to one decimal place.
Calculate the upper and lower bounds for

a the perimeter of the photograph,

b the area of the photograph.

10 The diagram shows a quadrilateral.
The lengths of the sides are given to the nearest cm.
Calculate the upper and lower bounds for the perimeter of the quadrilateral.

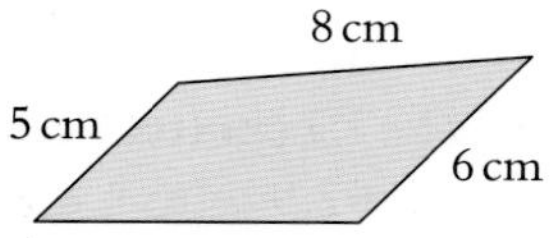

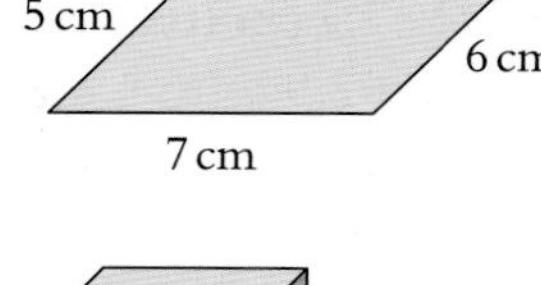

11 A cube has sides of length 4.7 cm correct to the nearest millimetre.
Calculate the upper and lower bounds for

a the volume of the cube,

b the surface area of the cube.

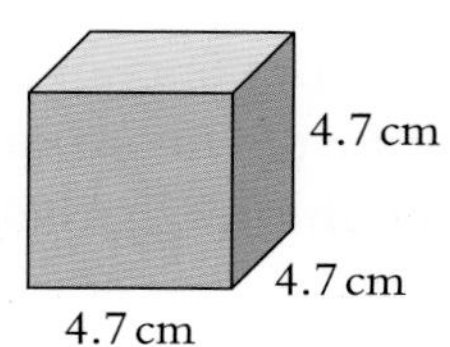

12 A cuboid has sides of length 8.1 cm, 5.2 cm and 3.5 cm correct to the nearest millimetre.
Calculate the upper and lower bounds for

a the volume of the cuboid,

b the surface area of the cuboid.

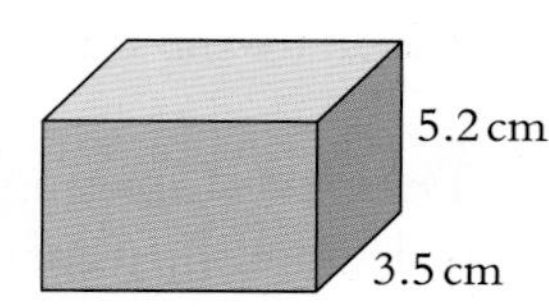

13 The lengths of the edges of the triangular prism are correct to the nearest centimetre.
Calculate the upper and lower bounds for the volume of the prism.

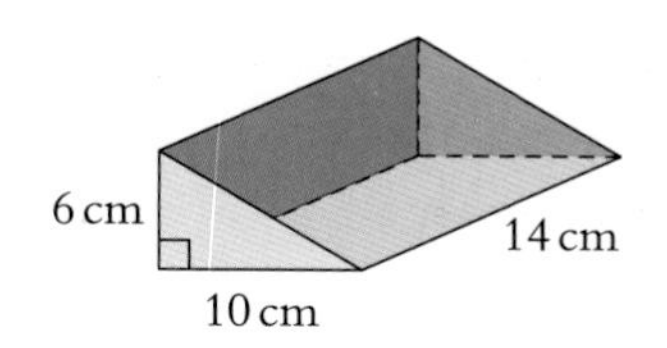

14 A truck can carry a maximum load of 8000 kg.
The truck is loaded with packages each with a mass of 35 kg correct to two significant figures.
Calculate the maximum number of packages that the truck can carry.

15 The area of a circle is 85 cm², correct to the nearest square centimetre.
Calculate the lower bound for the radius, r, of the circle.
(Write down all the numbers on your calculator display.)

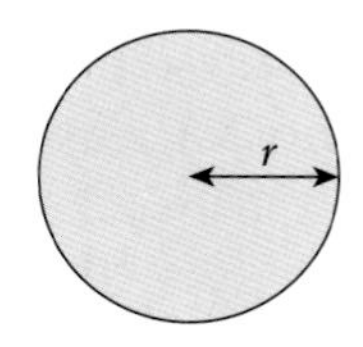

16 An equilateral triangle has sides of length 18 cm correct to the nearest centimetre.
Calculate the lower bound for the area of the triangle.
(Write down all the numbers on your calculator display.)

17 The lengths of AB and AC are correct to the nearest millimetre.
Angle BAC is correct to the nearest degree.

a Calculate the lower bound for the length of BC.

b Calculate the upper bound for the area of the triangle.

(In each case write down all the numbers on your calculator display.)

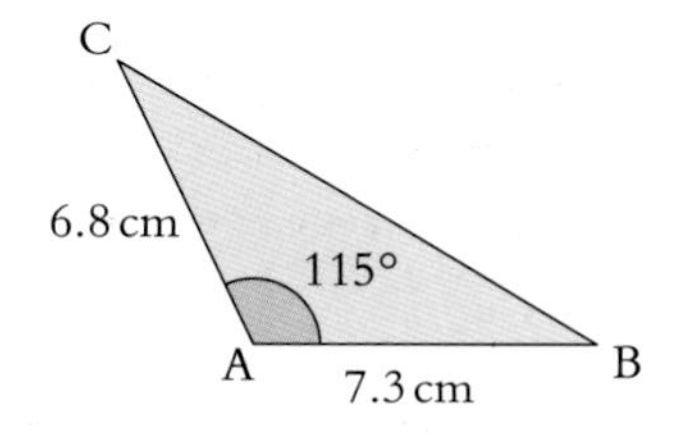

18 The lengths of the two planks of wood, correct to the nearest centimetre, are 168 cm and 259 cm.
Calculate the upper and lower bounds for

a the total length of the two planks of wood.

b the difference in length of the two planks of wood.

19 The radius and height of the cylinder have been measured correct to the nearest centimetre.
Calculate the upper bound for the volume of the cylinder.
(Write down all the numbers on your calculator display.)

20 The formula $a = \frac{v - u}{t}$ is used to find the acceleration for an object that increases its speed from u to v in a time t.
If u = 4.5 m/s (to 1 d.p.), v = 10.2 m/s (to 1 d.p.) and t = 3.6 s (to 1 d.p.), calculate the upper bound for the acceleration.

KEY WORDS
upper bound
lower bound

UNIT 9

Simultaneous equations 2

THIS SECTION WILL SHOW YOU HOW TO

- Solve simultaneous equations using the substitution method

Substitution method

In the substitution method you make x or y the subject of one of the equations and then substitute into the other equation.

EXAMPLE

Solve these simultaneous equations. $y = 3x - 10$
$x + y = -2$

$y = 3x - 10 \quad (1)$
$x + y = -2 \quad (2)$

Substitute for y from equation (1) into equation (2).

$x + (3x - 10) = -2$ *remove brackets and collect like terms*
$4x - 10 = -2$ *add 10 to both sides*
$4x = 8$ *divide both sides by 4*
$x = 2$

Substitute $x = 2$ in equation (1) and then solve to find y.

$y = 3 \times 2 - 10 = -4$

The solution is $x = 2$, $y = -4$

➡ **CHECK:** $-4 = 3 \times 2 - 10$ ✓
$2 + -4 = -2$ ✓

EXAMPLE

Solve these simultaneous equations. $2x + 3y = 0$
$x + 6y = 3$

$2x + 3y = 0 \quad (1)$
$x + 6y = 3 \quad (2)$

There is only one x in equation (2) so make x the subject of equation (2).

$x = 3 - 6y$

Substitute for x into equation (1).

$2(3 - 6y) + 3y = 0$ *remove brackets*
$6 - 12y + 3y = 0$ *collect like terms*
$6 - 9y = 0$ *add 9y to both sides*
$6 = 9y$ *divide both sides by 9*
$y = \frac{6}{9} = \frac{2}{3}$

Substitute $y = \frac{2}{3}$ in the equation $x = 3 - 6y$ to find x.

$x = 3 - \left(6 \times \frac{2}{3}\right) = 3 - 4 = -1$

The solution is $x = -1$, $y = \frac{2}{3}$

➡ **CHECK:** $(2 \times -1) + \left(3 \times \frac{2}{3}\right) = -2 + 2 = 0$ ✓
$-1 + \left(6 \times \frac{2}{3}\right) = -1 + 4 = 3$ ✓

EXERCISE 9.2

Solve these simultaneous equations

1 a $x + 2y = 4$
$x = 3y - 11$

b $x + 3y = 21$
$y = 2x - 14$

c $5x + 2y = 1$
$x = 3y + 7$

d $x + 2y = 11$
$x = 2y + 5$

e $y = 6 + 2x$
$4x + 2y = 16$

f $x = 3 - y$
$2x - 3y = -19$

g $x = 2y + 10$
$5x + 4y = 8$

h $y = 4x - 6$
$2x - 3y = -7$

i $4x + 2y - 14 = 0$
$x = y - 8$

j $5x - 3y = -17$
$x = 3y - 25$

k $x = 18 + 7y$
$x - 4y = 9$

l $y = 10x + 47$
$2y - 5x = 34$

2 a $y = 2x - 7$
$y = x - 3$

b $x = 5y + 1$
$x = 25 - 3y$

c $y = 3x - 9$
$y = 4x + 14$

d $x = 3y + 5$
$x = 2y - 4$

3 a $2x + 3y = 0$
$x - 2y = 14$

b $3x + y = 15$
$5x - 3y = -3$

c $2x + y = -3$
$3x - 2y = -8$

d $x - 5y = 18$
$x + 3y = 2$

e $2x + 3y + 10 = 0$
$x + 4y + 15 = 0$

f $x + 2y = 1$
$3x - 2y = 19$

g $5x + 3y = -18$
$2x + y = -7$

h $x + 2y = 11$
$4y - x = -2$

i $3x + y = -1$
$6x - 2y = -2$

j $x + 2y = 8$
$3x + 4y - 6 = 0$

k $3x + 2y = 18$
$x - 2y = -6$

l $3x + y = 0$
$6x - 5y = 14$

4 Find the coordinates of the point where the lines $y = x + 2$ and $3x + 2y = 14$ intersect.

5 Find the coordinates of the point where the lines $y = 3x - 1$ and $2x - 5y = 18$ intersect.

6 A bag contains one-dollar coins and 50-cent coins. There are twice as many one-dollar coins as 50-cent coins and the total value of the coins is \$42.50. How many one-dollar coins are there?

7 A magazine costs \$2.50 more than a newspaper. The total cost of the magazine and the newspaper is \$4.80. Find the cost of the magazine.

8 Adam is three times the age of Billy. The sum of their ages is 56 years. How old is Adam?

9 Sheena has some \$10 and \$5 notes. She has four times as many \$10 notes as \$5 notes and the total value of all the notes is \$585. How many \$5 notes does she have?

10 Theatre tickets cost either \$30 or \$42. One evening twice as many \$30 tickets are sold as \$42 tickets and the total money taken is \$18 564. Find the total number of tickets that are sold.

KEY WORDS
substitution method

Linear programming

THIS SECTION WILL SHOW YOU HOW TO

- Solve problems using linear programming

EXAMPLE

A company plans to make two types of chair.
The machine time and craftsman's time needed for each type of chair are shown in the table.

	Chair A	Chair B
Machine time (hours)	2	3
Craftsman's time (hours)	4	2

The company has 30 hours of machine time and 32 hours of craftsman's time available.
The company wants to make at least 2 of chair A and at least 6 of chair B.
Let x be the number of chair A and y be the number of chair B.
Write down four inequalities in x and y.
Represent these four inequalities on a graph.
What is the largest number of type A chair that can be made?

Machine time gives $2x + 3y \leq 30$
Craftsman's time gives $4x + 2y \leq 32$ (this simplifies to $2x + y \leq 16$)
Number of chair A gives $x \geq 2$
Number of chair B gives $y \geq 6$
The boundary lines $2x + 3y = 30$, $2x + y = 16$, $x = 2$ and $y = 6$ need to be drawn.

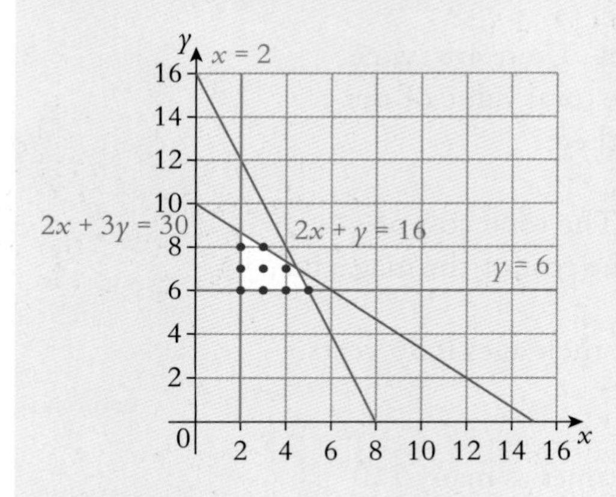

- The boundary lines are all **solid** lines in this example.
- Remember to shade the **unwanted** region.
- The unshaded region represents the required region.
- The red dots represent all the possible combinations of chairs in the required region (these are also shown in the table below).

Chair A (x)	2	2	2	3	3	3	4	4	5
Chair B (y)	6	7	8	6	7	8	6	7	6

➡ **NOTE:** the maximum or minimum numbers always occur at/near one of the corners of the required region.

The largest number of type A chair that can be made is 5.

EXERCISE 9.3

1 A car park has space for x cars and y motorcycles.
The car park has an area of $900\,m^2$.
Each car needs $15\,m^2$ of space and each motorcycle needs $3\,m^2$ of space.
There must be space for at least 40 cars.
There must be space for at least 20 motorcycles.

a Explain why $5x + y \leq 300$

b Write down two more inequalities in x and y.

c Represent these three inequalities on a graph.

d What is the largest possible number of cars in the car park?

2 Petra works in a hotel during the school holidays.
In one week she spends x hours cleaning rooms and y hours waitressing.
She spends no more than 15 hours working.
She spends at least 6 hours waitressing.
She spends at least twice as much time waitressing as she does cleaning.

a Write down three inequalities in x and y.

b Represent these three inequalities on a graph.

c What is the largest possible number of hours she can spend cleaning rooms?

3 A builder plans to build some executive homes and some standard homes.
He has $9000\,m^2$ of land available.
He builds x executive homes and y standard homes.
Each executive home requires $400\,m^2$ of land and each standard home requires $200\,m^2$ of land.
The number of standard homes is to be less than the number of executive homes.
There must be at least 10 standard homes.

a Explain why $2x + y \leq 45$

b Write down two more inequalities in x and y.

c Represent these three inequalities on a graph.

d What is the largest possible number of executive homes that can be built?

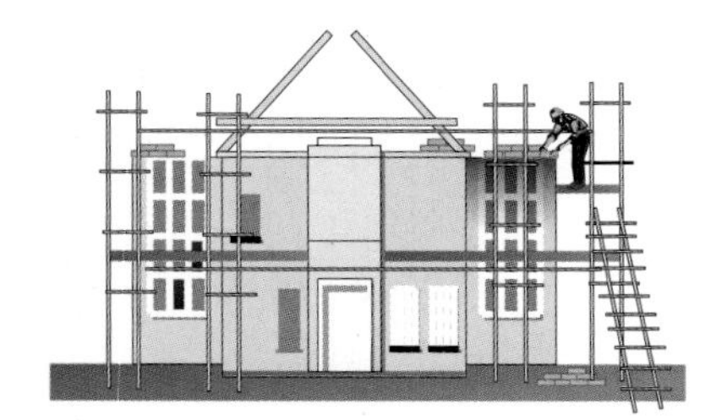

4 A school trip is planned for 100 students.
The school uses x type A minibuses and y type B minibuses to transport the students.
The type A minibus can carry 16 students and the type B minibus can carry 10 students.
The school wishes to use no more than eight minibuses.
The school wishes to use at least one type A minibus.
The school wishes to use at least three type B minibuses.

a Explain why $8x + 5y \geq 50$

b Write down three more inequalities in x and y.

c Represent these four inequalities on a graph.

d What is the largest possible number of type A minibus that can be used?

EXAMPLE

An aircraft has 30 first class seats available and 220 economy class seats available.
A first class seat costs \$240 and an economy class seat costs \$120.
A flight will be cancelled if the money taken for the seats is less than \$28 800.
Let x = the number of first class seats sold and y = the number of economy class seats sold.

a Write down three inequalities in x and y.

b Represent these three inequalities on a graph.

c What is the minimum number of seats to be sold for the flight not to be cancelled?

---

a First class seats $x \leq 30$

Economy class seats $y \leq 220$

Money taken $240x + 120y \geq 28\,800$ divide by 120 to simplify the inequality

$2x + y \geq 240$

The three inequalities are $x \leq 30$, $y \leq 220$ and $2x + y \geq 240$.

b The boundary lines $x = 30$, $y = 220$ and $2x + y = 240$ need to be drawn.
To draw the line $2x + y = 240$ substitute $x = 0$ and $y = 0$ into the equation to find the axis crossing points.

x	0	120
y	240	0

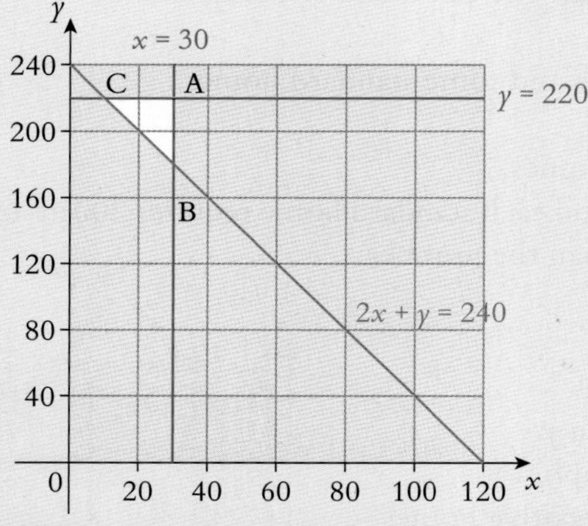

- The boundary lines are all **solid** lines in this example.
- Remember to shade the **unwanted** region.
- The unshaded region represents the required region.

c The maximum or minimum occur at/near one of the corners of the required region.
The three corners of the required region are A(30, 220), B(30, 180) and C(10, 220).

At A number of seats = 30 + 220 = 250
At B number of seats = 30 + 180 = 210
At C number of seats = 10 + 220 = 230

The minimum number of seats sold for the flight not to be cancelled is 210.

EXERCISE 9.4

1 A company makes model animals out of wood. They make giraffes and elephants. The carving time and sanding time needed for each model animal are shown below.

	Giraffe	Elephant
Carving time (hours)	4	3
Sanding time (hours)	1	1

The company has 48 hours of carving time and 14 hours of sanding time available.
The company wants to make at least 5 giraffes and at least 5 elephants.
Let x be the number of giraffes and y be the number of elephants.

a Write down four inequalities in x and y.

b Represent these four inequalities on a graph.

c The company makes a profit of \$10 for each giraffe sold and a profit of \$20 for each elephant sold.
Calculate the greatest possible profit.

2 4500 DVDs need to be packed into storage units.
There are two sizes of storage units available.
A large unit can hold 500 DVDs and a small unit can hold 300 DVDs.
No more than 11 storage units are to be used.
At least 2 of the storage units must be small.
Let x be the number of large units used and y be the number of small units used.

a Write down three inequalities in x and y.

b Represent these three inequalities on a graph.

c A large storage unit costs \$10 and a small storage unit costs \$8.
Calculate the least possible cost of the storage units.

3 Lara is making some small cakes and biscuits to raise money for charity.
She only has 600 g of flour and 580 g of sugar available.
She has plenty of all the other ingredients.
The amount of flour and sugar needed for each cake and biscuit are shown below.

	Small cake	Biscuit
Flour (g)	6	3
Sugar (g)	4	5

Lara wants to make at least 50 small cakes and at least 40 biscuits.
Let x be the number of small cakes and y be the number of biscuits made.

a Write down four inequalities in x and y.

b Represent these four inequalities on a graph.

c Lara sells each cake for \$1.50 and each biscuit for \$1.
Calculate the largest possible amount of money that Lara can make if she sells all her cakes and biscuits.

Solving quadratic equations by completing the square

THIS SECTION WILL SHOW YOU HOW TO

- Complete the square
- Solve quadratic equations by completing the square

Completing the square

If you expand the expressions $(x + b)^2$ and $(x - b)^2$ you will obtain the results:

$$(x + b)^2 = x^2 + 2bx + b^2 \quad \text{and} \quad (x - b)^2 = x^2 - 2bx + b^2$$

Rearranging these give you the following important results:

$$x^2 + 2bx = (x + b)^2 - b^2$$
$$x^2 - 2bx = (x - b)^2 - b^2$$

This is known as **completing the square**.

To complete the square for $x^2 + 6x$:

$$6 \div 2 = 3$$

$$x^2 + 6x = (x + 3)^2 - 3^2$$
$$= (x + 3)^2 - 9$$

➡ **CHECK:** $(x + 3)(x + 3) - 9 = x^2 + 3x + 3x + 9 - 9 = x^2 + 6x$ ✓

EXAMPLE

Complete the square for the expressions

a $x^2 - 12x$ **b** $x^2 - 7x$

a $x^2 - 12x$:

$$12 \div 2 = 6$$

$$x^2 - 12x = (x - 6)^2 - 6^2$$
$$= (x - 6)^2 - 36$$

➡ **CHECK:** $(x - 6)(x - 6) - 36$
$= x^2 - 6x - 6x + 36 - 36$
$= x^2 - 12x$ ✓

b $x^2 - 7x$:

$$7 \div 2 = 3.5$$

$$x^2 - 7x = (x - 3.5)^2 - 3.5^2$$
$$= (x - 3.5)^2 - 12.25$$

➡ **CHECK:** $(x - 3.5)(x - 3.5) - 12.25$
$= x^2 - 3.5x - 3.5x + 12.25 - 12.25$
$= x^2 - 7x$ ✓

EXERCISE 9.5

1 Complete the square for these expressions.

a $x^2 - 2x$ b $x^2 + 4x$ c $x^2 - 10x$ d $x^2 + 20x$
e $x^2 - 16x$ f $x^2 - 12x$ g $x^2 + 18x$ h $x^2 + 36x$
i $x^2 + 22x$ j $x^2 + 24x$ k $x^2 - 3x$ l $x^2 + 5x$
m $x^2 + 7x$ n $x^2 + 9x$ o $x^2 - 20x$ p $x^2 + x$

You may need to take out a numerical factor before completing the square.

$5x^2 - 20x \longrightarrow 5[x^2 - 4x] \longrightarrow 5[(x-2)^2 - 4] \longrightarrow 5(x-2)^2 - 20$

take 5 outside; complete the square; multiply both terms by 5

2 Complete the square for these expressions.

a $5x^2 - 20x$ b $3x^2 + 12x$ c $4x^2 + 20x$ d $3x^2 - 21x$
e $2x^2 + 12x$ f $2x^2 - 5x$ g $3x^2 - 4x$ h $2x^2 + 9x$

You may also have a constant term in your quadratic expression.

$x^2 + 12x - 3 \longrightarrow (x+6)^2 - 6^2 - 3 \longrightarrow (x+6)^2 - 39$

complete the square for $x^2 + 12x$
$x^2 + 12x = (x+6)^2 - 6^2$

combine the numbers
$-6^2 - 3 = -39$

3 Write these expressions in the form $(x + p)^2 + q$.

a $x^2 - 8x + 15$ b $x^2 - 4x - 12$ c $x^2 - 6x - 4$ d $x^2 + 4x - 3$
e $x^2 + 12x - 5$ f $x^2 - 8x + 1$ g $x^2 + 10x - 3$ h $x^2 + 9x - 22$

4 $x^2 - 10x + 4 = (x - a)^2 + b$. Find the values of a and b.

5 $x^2 + 8x - 3 = (x + p)^2 + q$. Find the values of p and q.

6 $2x^2 - 8x + 3 = 2(x - p)^2 + q$. Find the values of p and q.

7 You can make x the subject of the equation $x^2 + 8x = a$ by completing the square.

$x^2 + 8x = a \longrightarrow (x+4)^2 - 4^2 = a \longrightarrow (x+4)^2 = a + 16 \longrightarrow x + 4 = \pm\sqrt{a+16} \longrightarrow x = -4 \pm \sqrt{a+16}$

Make x the subject of these equations. (You will need to complete the square first.)

a $x^2 - 6x = b$ b $x^2 + 10x = a + b$

You can solve quadratic equations by completing the square.

EXAMPLE

Solve the equation $x^2 + 4x - 3 = 0$, giving your answers to two decimal places.

$x^2 + 4x - 3 = 0$ — *add 3 to both sides*

$x^2 + 4x = 3$ — *complete the square for $x^2 + 4x$*

$(x + 2)^2 - 2^2 = 3$

$(x + 2)^2 - 4 = 3$ — *add 4 to both sides*

$(x + 2)^2 = 7$ — *square root both sides*

$x + 2 = \pm\sqrt{7}$ — *subtract 2 from both sides*

$x = -2 \pm \sqrt{7}$

$x = -2 + \sqrt{7}$ or $x = -2 - \sqrt{7}$

$x = 0.65$ or $x = -4.65$ (to 2 d.p.)

The next example shows you what to do if the coefficient of x^2 is not 1.

EXAMPLE

Solve the equation $2x^2 - 8x + 5 = 0$, giving your answers to two decimal places.

$2x^2 - 8x + 5 = 0$ — *divide both sides by 2 so that the coefficient of x^2 is 1*

$x^2 - 4x + 2\frac{1}{2} = 0$ — *subtract $2\frac{1}{2}$ from both sides*

$x^2 - 4x = -2\frac{1}{2}$ — *complete the square for $x^2 - 4x$*

$(x - 2)^2 - 2^2 = -2\frac{1}{2}$

$(x - 2)^2 - 4 = -2\frac{1}{2}$ — *add 4 to both sides*

$(x - 2)^2 = 1\frac{1}{2}$ — *square root both sides*

$x - 2 = \pm\sqrt{1\frac{1}{2}}$ — *add 2 to both sides*

$x = 2 \pm \sqrt{1\frac{1}{2}}$

$x = 2 + \sqrt{1\frac{1}{2}}$ or $x = 2 - \sqrt{1\frac{1}{2}}$

$x = 3.22$ or $x = 0.78$ (to 2 d.p.)

You can use completing the square to find the **minimum** value of a quadratic expression.

$x^2 + 4x + 5 = (x + 2)^2 + 1$

➡ **NOTE:** this part of the expression is a square so it will always be ≥ 0 (positive). The smallest value it can be is 0.

The minimum value of $x^2 + 4x + 5$ is $0 + 1 = 1$.

EXERCISE 9.6

In this exercise give your answers where appropriate correct to 2 decimal places.

1 Use completing the square to solve these equations.

a $x^2 + 6x + 8 = 0$ b $x^2 + 4x + 4 = 0$ c $x^2 + 2x - 8 = 0$

d $x^2 - 12x + 34 = 0$ e $x^2 - 6x + 7 = 0$ f $x^2 + 14x - 3 = 0$

g $x^2 - 20x - 33 = 0$ h $x^2 - 10x + 15 = 0$ i $x^2 - 2x - 6 = 0$

2 Use completing the square to solve these equations.

a $x^2 + 3x + 2 = 0$ b $x^2 - 7x + 12 = 0$ c $x^2 + x - 6 = 0$

d $x^2 - 5x + 5 = 0$ e $x^2 + x - 10 = 0$ f $x^2 - 11x + 3 = 0$

3 Use completing the square to solve these equations.

a $3x^2 - 5x - 2 = 0$ b $2x^2 + 7x + 3 = 0$ c $2x^2 + 16x + 30 = 0$

d $2x^2 - 16x + 4 = 0$ e $3x^2 + 6x + 2 = 0$ f $2x^2 + 8x - 6 = 0$

4 Anna is trying to solve the equations $12 = \frac{3}{x} - x$ and $x(x - 2) = 12$.

Copy the equations and correct her working.

a

$12 = \frac{3}{x} - x$

$12x = 3 - x$

$13x = 3$

$x = \frac{3}{13}$

b

$x(x - 2) = 12$

$x = 12$ or $x - 2 = 12$

$x = 12$ or $x = 14$

5 Solve these equations.

a $x(x - 4) = 20$ b $x^2 - 7 = 2x$ c $10 + 3x = 2x^2$

d $x = 3 + \frac{1}{x}$ e $8 = \frac{2}{x} + 5x$ f $x = \frac{6}{x+2}$

6 Use completing the square to find the minimum value of $x^2 - 4x + 11$.

7 Use completing the square to show that $x^2 + 6x + 14$ is never less than 5.

8 Use completing the square to find the minimum value of $2x^2 - 4x + 5$.

CHALLENGE:

Try solving the equation $ax^2 + bx + c = 0$ by completing the square.

If you do it correctly you should obtain the answer

$$x = \frac{-b \pm \sqrt{b^2 - 4ac}}{2a}.$$

KEY WORDS

completing the square

minimum

Using graphs to solve equations

THIS SECTION WILL SHOW YOU HOW TO

- Use graphs to solve equations

You can use graphs to find approximate solutions to equations that may be difficult to solve by other methods.

EXAMPLE

Draw the graph of $y = x^2 - 5x + 4$ for $0 \leq x \leq 5$.
Use your graph to solve the equations **a** $x^2 - 5x + 4 = 2$ **b** $x^2 - 6x + 6 = 0$

a $x^2 - 5x + 4 = 2$

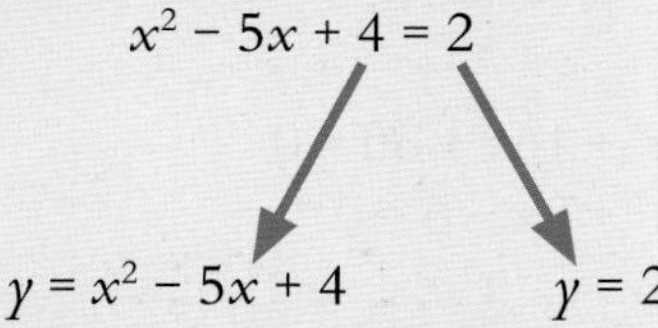

$y = x^2 - 5x + 4$ $y = 2$

To solve the equation $x^2 - 5x + 4 = 2$ draw the curve $y = x^2 - 5x + 4$ and the line $y = 2$ and find the x coordinates of the points where $y = x^2 - 5x + 4$ and $y = 2$ intersect.

The answers to $x^2 - 5x + 4 = 2$ are

$x \approx 0.4$ and $x \approx 4.6$

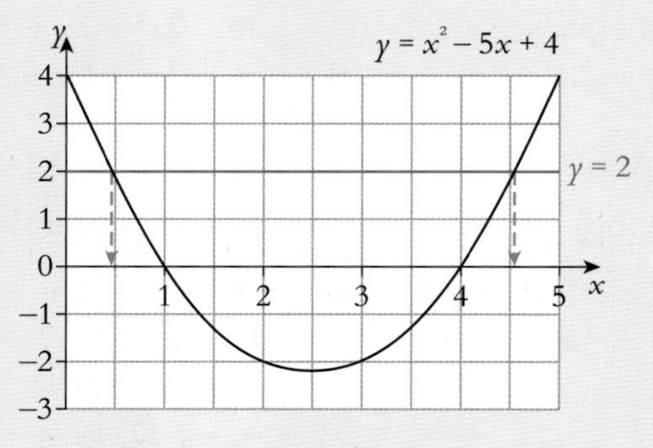

b To use the graph of $y = x^2 - 5x + 4$ to solve the equation $x^2 - 6x + 6 = 0$ you must first rearrange the equation into the form '$x^2 - 5x + 4 = ...$'

$x^2 - 6x + 6 = 0$ *add x to both sides*

$x^2 - 5x + 6 = x$ *take 2 from both sides*

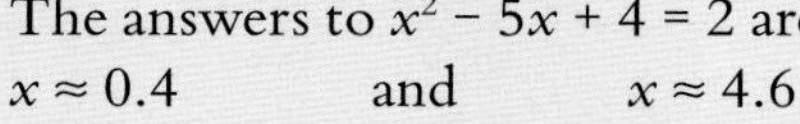

$x^2 - 5x + 4 = x - 2$

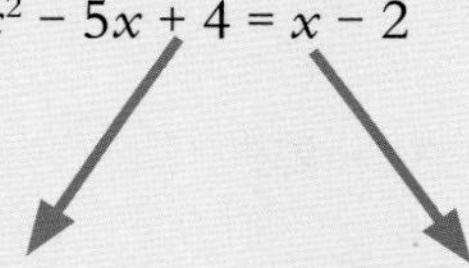

$y = x^2 - 5x + 4$ $y = x - 2$

To solve the equation $x^2 - 6x + 6 = 0$ draw the curve $y = x^2 - 5x + 4$ and the line $y = x - 2$ and find the x coordinates of the points where $y = x^2 - 5x + 4$ and $y = x - 2$ intersect.

The answers to $x^2 - 6x + 6 = 0$ are

$x \approx 1.3$ and $x \approx 4.7$

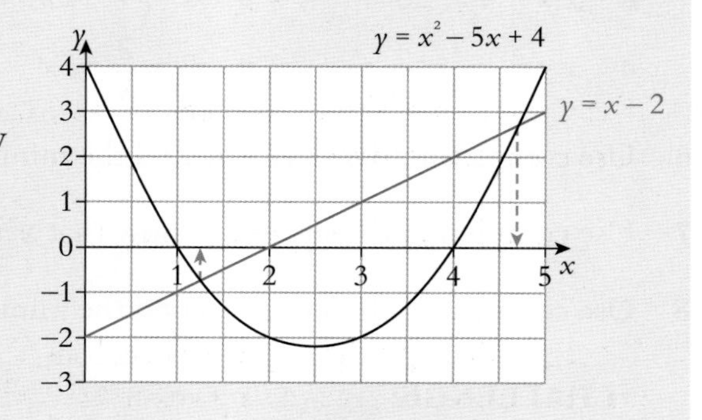

EXERCISE 9.7

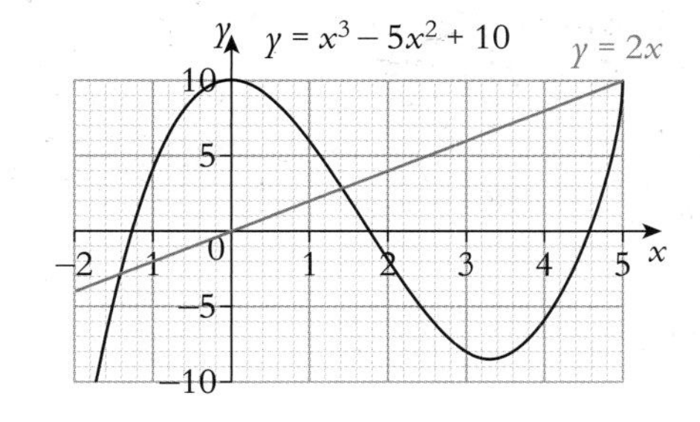

1 $y = x^3 - 5x^2 + 10$ and $y = 2x$ are shown on the graph.
Use the graph to solve the equation
$x^3 - 5x^2 + 10x = 2x$

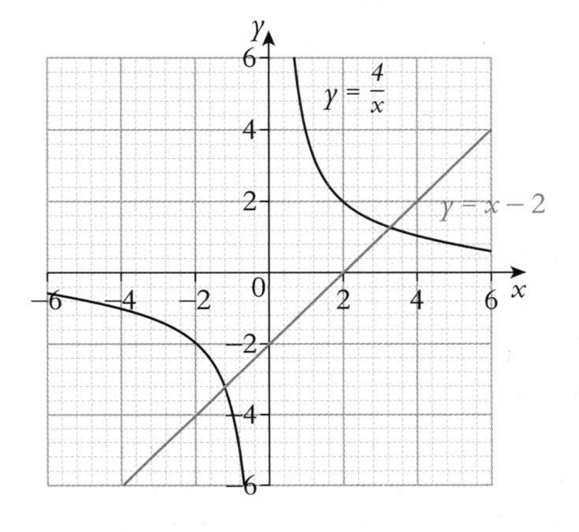

2 $y = \frac{4}{x}$ and $y = x - 2$ are shown on the graph.
Use the graph to solve the equation
$\frac{4}{x} = x - 2$

3 Draw the graph of $y = x^2$ for $-3 \leq x \leq 3$.
Use your graph to solve these equations.
a $x^2 = 6$ **b** $x^2 = x + 3$ **c** $x^2 - 2 = 2x$

4 Draw the graph of $f(x) = x^2 - 3x$ for $-2 \leq x \leq 5$
Use your graph to solve these equations.
a $x^2 - 3x = 0$ **b** $x^2 - 3x = x + 2$
c $x^2 - x - 4 = 0$

5 Draw the graph of $y = x^3 + 1$ for $-3 \leq x \leq 3$
Use your graph to solve these equations.
a $x^3 + 1 = 20$ **b** $x^3 + 1 = 5x$ **c** $x^3 + 10x - 9 = 0$

6 If the graph of $f(x) = 3^x$ has been drawn, what graph must also be drawn to solve these equations.
a $3^x = 4$ **b** $3^x = 0.5$ **c** $3^x = 4x$ **d** $3^x - 2 = x$
e $3^x - x^2 + 2 = 0$ **f** $x3^x = 1 + 3x$ **g** $1 - 3^x = 2x$ **h** $x^2 3^x = 1$

7 If the graph of $y = 3x^2 - \frac{1}{x}$ has been drawn, what graph must also be drawn to solve the equation $3x^3 - x^2 - 5x - 1 = 0$?

8 If the graph of $f(x) = \frac{2}{x^2}$ has been drawn, what graph must also be drawn to solve the equation $5x^3 - 5x^2 - 2 = 0$?

TOP TIP

You must rearrange $3x^3 - x^2 - 5x - 1 = 0$ to the form '$3x^2 - \frac{1}{x} = ...$'

UNIT 9

The roots of the equation $f(x) = k$

The graph of $f(x) = \frac{x^2}{2} - \frac{8}{x}$ is shown below.

The equation $\frac{x^2}{2} - \frac{8}{x} = k$ will have 3, 2 or 1 roots.

The number of roots depends on the value of k.

To solve the equation $\frac{x^2}{2} - \frac{8}{x} = 12$ the line $y = 12$ needs to be drawn.

To solve the equation $\frac{x^2}{2} - \frac{8}{x} = 6$ the line $y = 6$ needs to be drawn.

To solve the equation $\frac{x^2}{2} - \frac{8}{x} = -9$ the line $y = -9$ needs to be drawn.

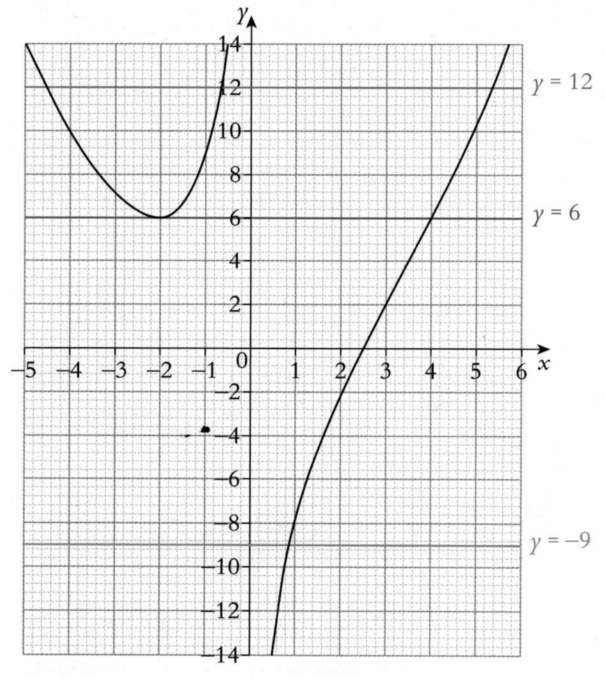

There are **three** roots to the equation $\frac{x^2}{2} - \frac{8}{x} = 12$

They are $x \approx -4.5$, $x \approx -0.7$ and $x \approx 5.2$

There are **two** roots to the equation $\frac{x^2}{2} - \frac{8}{x} = 6$

They are $x = -2$ and $x = 4$

There is **one** root to the equation $\frac{x^2}{2} - \frac{8}{x} = -9$

The root is $x \approx 0.9$

The results can be summarised as follows.

$\frac{x^2}{2} - \frac{8}{x} = k$

- $k > 6$ means there will be **three** roots
- $k = 6$ means there will be **two** roots
- $k < 6$ means there will be **one** root

EXERCISE 9.8

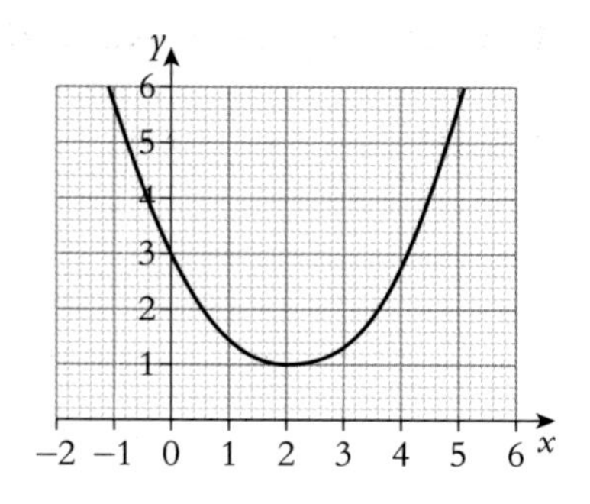

1 The graph of $f(x) = \frac{(x-2)^2}{2} + 1$ is shown.

a Use the graph to find the **two** roots of the equation $\frac{(x-2)^2}{2} + 1 = 5$

b For what value of k is there only one root of the equation $\frac{(x-2)^2}{2} + 1 = k$

c For what values of k are there no roots of the equation $\frac{(x-2)^2}{2} + 1 = k$

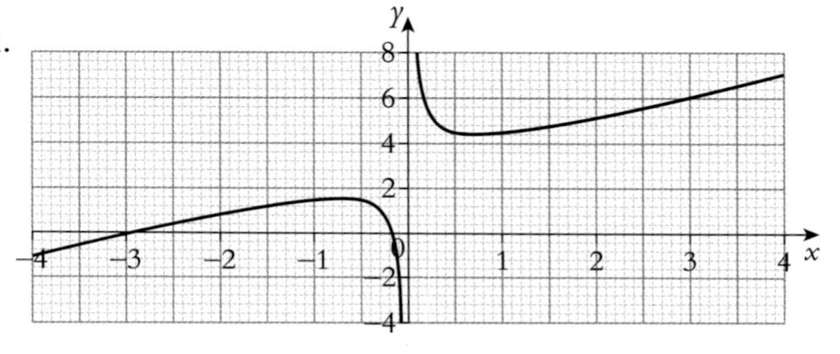

2 The graph of $f(x) = x + \frac{1}{2x} + 3$ is shown.

a Use the graph to find the **two** roots of the equation $x + \frac{1}{2x} + 3 = 6$

b There are three integer values of k for which the equation $x + \frac{1}{2x} + 3 = k$ has no roots.
Write down these three values of k.

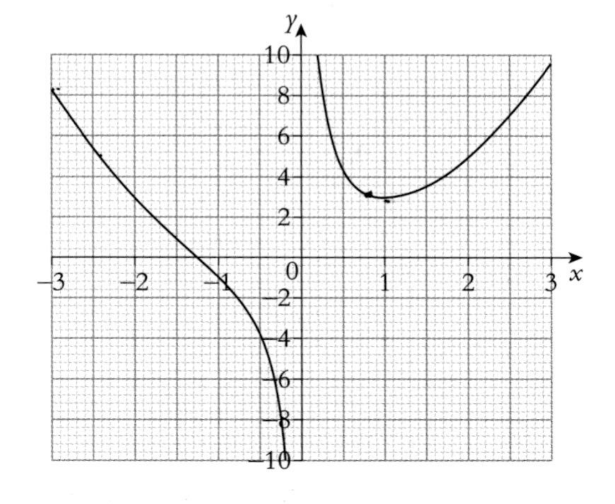

3 The graph of $f(x) = x^2 + \frac{2}{x}$ is shown.

a Use the graph to find the **three** roots of the equation $x^2 + \frac{2}{x} = 5$

b There is one value of k for which the equation $x^2 + \frac{2}{x} = k$ has two roots.
Write down this value of k.

c For what values of k is there only one root of the equation $x^2 + \frac{2}{x} = k$

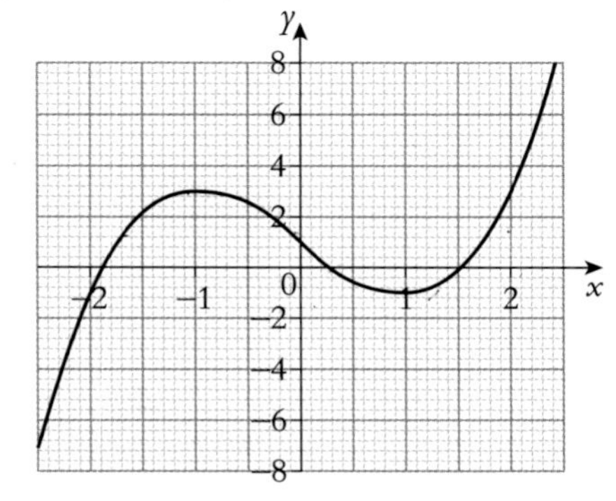

4 The graph of $y = x^3 - 3x + 1$ is shown.

a Use the graph to find the **three** roots of the equation $x^3 - 3x + 1 = 0$

b There are two values of k for which the equation $x^3 - 3x + 1 = k$ has two roots.
Write down these two values of k.

c For what values of k is there only one root of the equation $x^3 - 3x + 1 = k$

UNIT 9

Vectors and vector geometry

THIS SECTION WILL SHOW YOU HOW TO

- Use the correct notation for vectors
- Add and subtract vectors
- Solve problems in 2-D using vector geometry

A **vector** has both size and direction.
Examples of vectors are:

- displacement (5 km on a bearing of 300°)
- velocity (5 m/s due east)
- weight (20 N vertically downwards)

Vector notation

$\overrightarrow{AB}$ means the displacement from the point A to the point B.
A translation is described using a column vector.
A column vector describes the movement in both the x and y directions.

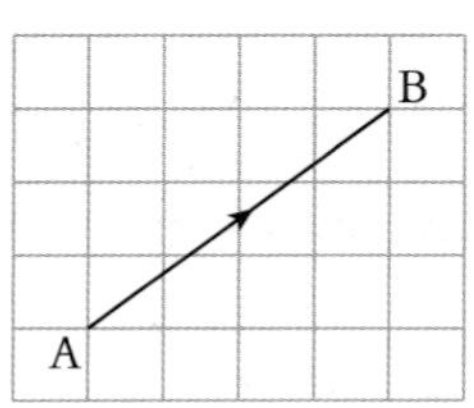

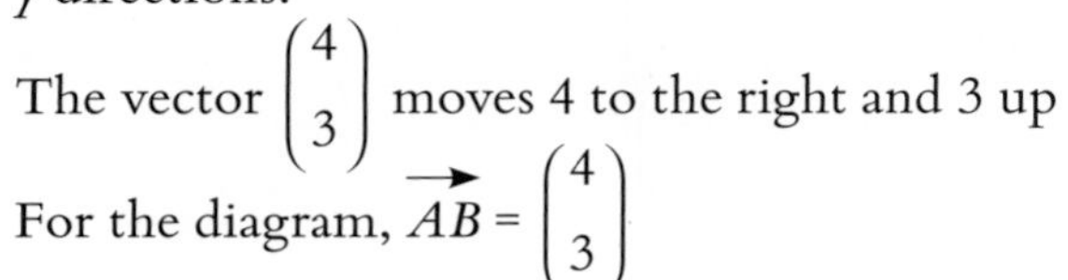

The vector $\begin{pmatrix} 4 \\ 3 \end{pmatrix}$ moves 4 to the right and 3 up

For the diagram, $\overrightarrow{AB} = \begin{pmatrix} 4 \\ 3 \end{pmatrix}$

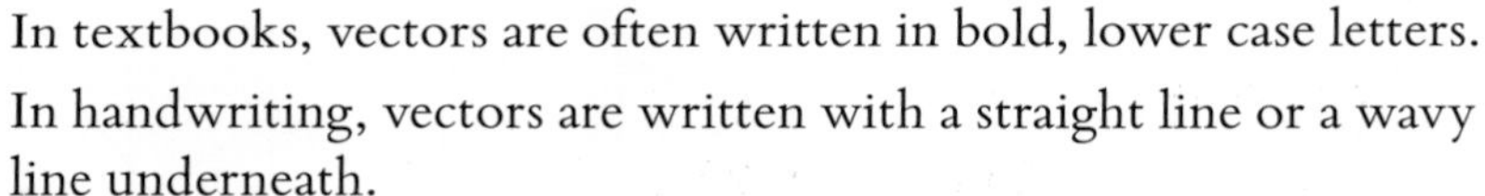

In textbooks, vectors are often written in bold, lower case letters.

In handwriting, vectors are written with a straight line or a wavy line underneath.

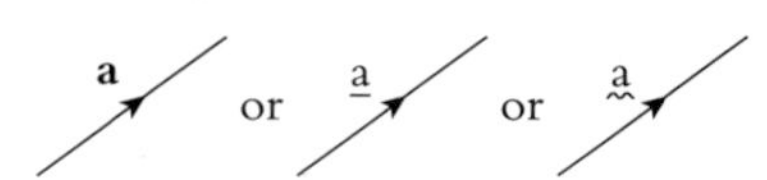

Magnitude of a vector

'The **magnitude** of the vector **a**' means 'the length of the vector **a**' and is denoted by $|\mathbf{a}|$.
$|\mathbf{a}|$ is called the **modulus** of the vector **a**.

EXAMPLE

If $\mathbf{a} = \begin{pmatrix} -4 \\ 3 \end{pmatrix}$ find the magnitude of the vector **a**.

Using Pythagoras' theorem,

$$|\mathbf{a}| = \sqrt{3^2 + (-4)^2}$$
$$= \sqrt{25}$$
$$= 5$$

The magnitude of vector **a** is 5.

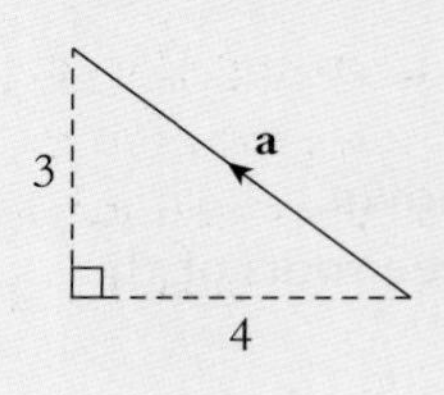

Two vectors are said to be equal if they are the same length and are in the same direction.

The vector $-\mathbf{a}$ is the same length as the vector **a** but is in the opposite direction.

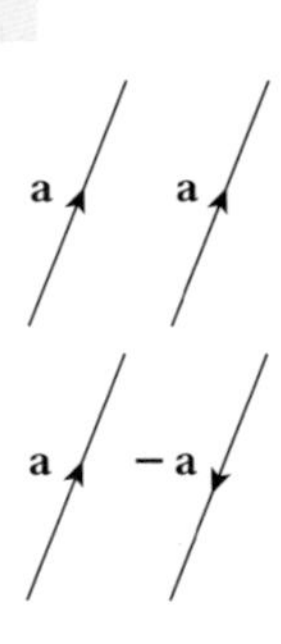

EXAMPLE

OPQ is a triangle.
X is the midpoint of OP.
Y is the midpoint of OQ.
$\overrightarrow{OP} = 2\mathbf{p}$ and $\overrightarrow{OQ} = 2\mathbf{q}$.
Find in terms of **p** and **q**

a $\overrightarrow{PQ}$ **b** $\overrightarrow{XY}$

What can you say about the lines PQ and XY?

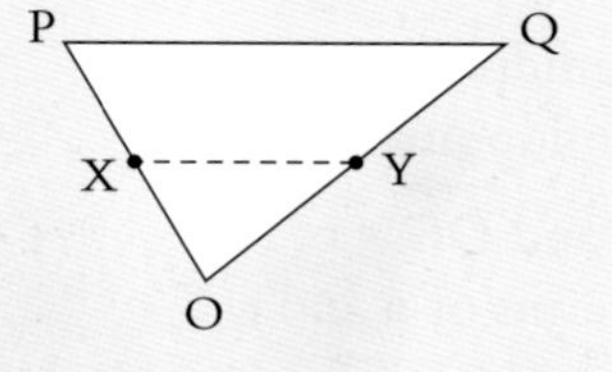

a $\overrightarrow{PQ} = \overrightarrow{PO} + \overrightarrow{OQ} = -2\mathbf{p} + 2\mathbf{q}$

b $\overrightarrow{XY} = \overrightarrow{XO} + \overrightarrow{OY} = -\mathbf{p} + \mathbf{q}$

So $\overrightarrow{PQ} = 2\overrightarrow{XY}$

This means that PQ and XY are parallel and that PQ is twice the length of XY.

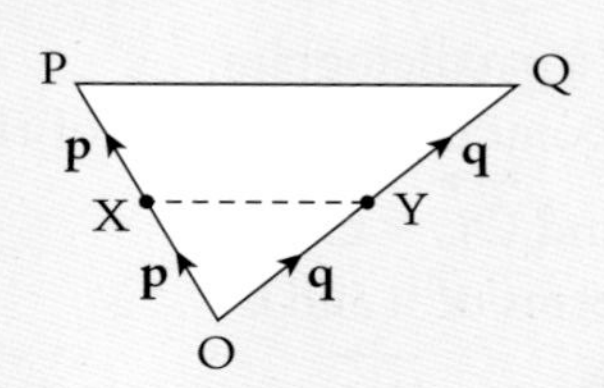

EXERCISE 9.11

1 ABC is a triangle. D is the midpoint of AC.
$\overrightarrow{AB} = \mathbf{p}$ and $\overrightarrow{AD} = \overrightarrow{DC} = \mathbf{q}$.
Find in terms of **p** and **q**

a $\overrightarrow{DA}$ b $\overrightarrow{AC}$ c $\overrightarrow{BD}$ d $\overrightarrow{CB}$

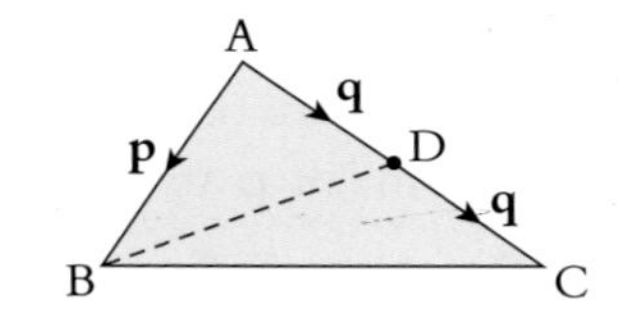

2 ABC is a triangle. M is the midpoint of AB.
$\overrightarrow{CA} = \mathbf{p}$ and $\overrightarrow{CB} = \mathbf{q}$.
Find in terms of **p** and **q**

a $\overrightarrow{AB}$ b $\overrightarrow{AM}$ c $\overrightarrow{BM}$ d $\overrightarrow{CM}$

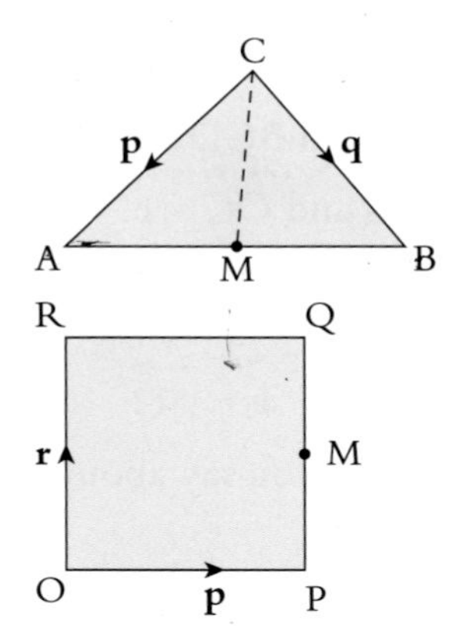

3 OPQR is a square. M is the midpoint of PQ.
$\overrightarrow{OP} = \mathbf{p}$ and $\overrightarrow{OR} = \mathbf{r}$.
Find in terms of **p** and **r**

a $\overrightarrow{OQ}$ b $\overrightarrow{MP}$ c $\overrightarrow{OM}$ d $\overrightarrow{MR}$

4 ABCD is a trapezium.
$\overrightarrow{AB} = 2\mathbf{p}$, $\overrightarrow{DC} = \mathbf{p}$ and $\overrightarrow{DA} = \mathbf{q}$.
Find in terms of **p** and **q**

a $\overrightarrow{BA}$ b $\overrightarrow{DB}$ c $\overrightarrow{CA}$ d $\overrightarrow{CB}$

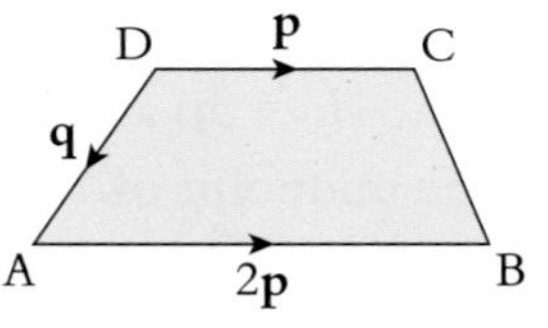

5 ABCDEF is a regular hexagon.
O is the centre of the hexagon.
$\overrightarrow{AF} = \mathbf{p}$ and $\overrightarrow{AB} = \mathbf{q}$.
Find the following in terms of **p** and **q**.

a $\overrightarrow{CD}$ b $\overrightarrow{DE}$ c $\overrightarrow{AO}$
d $\overrightarrow{CE}$ e $\overrightarrow{FD}$ f $\overrightarrow{DB}$

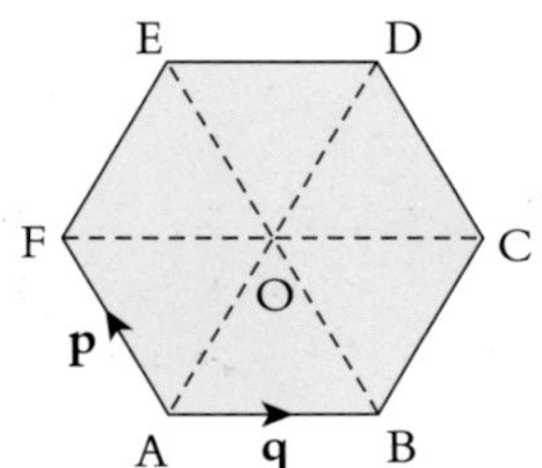

6 OABC is a parallelogram.

M is the midpoint of AB.
N is the midpoint of CB.
$\overrightarrow{OA} = \mathbf{a}$ and $\overrightarrow{OC} = \mathbf{c}$.
Find in terms of **a** and **c**

a $\overrightarrow{NB}$ b $\overrightarrow{BM}$ c $\overrightarrow{AC}$ d $\overrightarrow{MN}$

What can you say about the lines AC and MN?

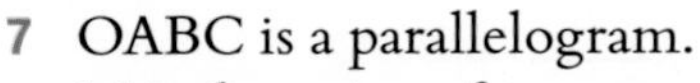

7 OABC is a parallelogram.
M is the point of intersection of the two diagonals.
$\overrightarrow{OA} = \mathbf{a}$ and $\overrightarrow{OC} = \mathbf{c}$.
Find in terms of **a** and **c**

a $\overrightarrow{OB}$ b $\overrightarrow{CA}$ c $\overrightarrow{OM}$ d $\overrightarrow{CM}$

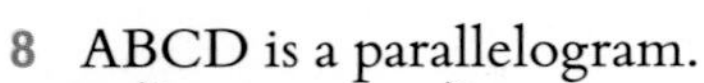

8 ABCD is a parallelogram.
$\overrightarrow{AB} = \mathbf{p}$ and $\overrightarrow{AD} = \mathbf{q}$.
E, F, G and H are the midpoints of the sides.
Find in terms of **p** and **q**

a $\overrightarrow{FG}$ b $\overrightarrow{EH}$ c $\overrightarrow{EF}$ d $\overrightarrow{HG}$

What can you say about quadrilateral EFGH?

9 OABC and BCDE are two congruent parallelograms.
$\overrightarrow{OA} = \mathbf{a}$ and $\overrightarrow{OC} = \mathbf{c}$.
M is the midpoint of BE and N is the midpoint of DE.
Find in terms of **a** and **c**

a $\overrightarrow{AM}$ b $\overrightarrow{AC}$ c $\overrightarrow{DM}$ d $\overrightarrow{MN}$

What can you say about AC and MN?

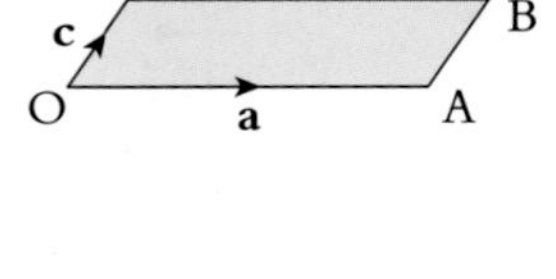

10 ABCD is a trapezium.
BCDE is a parallelogram.
$\overrightarrow{EB} = \mathbf{p}$, $\overrightarrow{AE} = 3\mathbf{p}$ and $\overrightarrow{ED} = \mathbf{q}$.
M is the midpoint of AD.
Find in terms of **p** and **q**

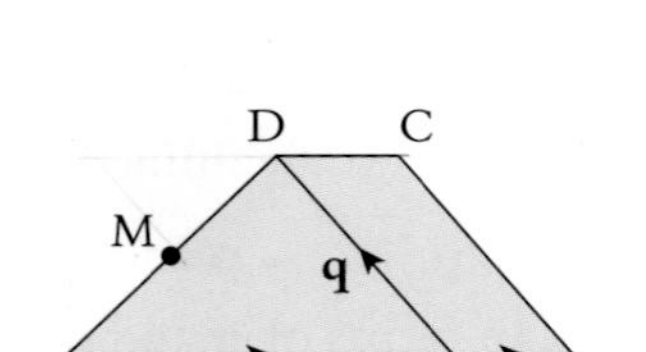

a $\overrightarrow{EC}$ b $\overrightarrow{EM}$ c $\overrightarrow{MC}$ d $\overrightarrow{AM}$

Further vector geometry

Vector geometry can be used to prove that points are **collinear** (lie on a straight line).

If $\overrightarrow{AB} = k\overrightarrow{AC}$, then the points A, B and C are collinear.

(This is because the lines AB and AC must be parallel and the point A lies on both lines.)

Ratios can also be used in vector geometry.

If X is a point on the line AB such that AX : XB = 2 : 3

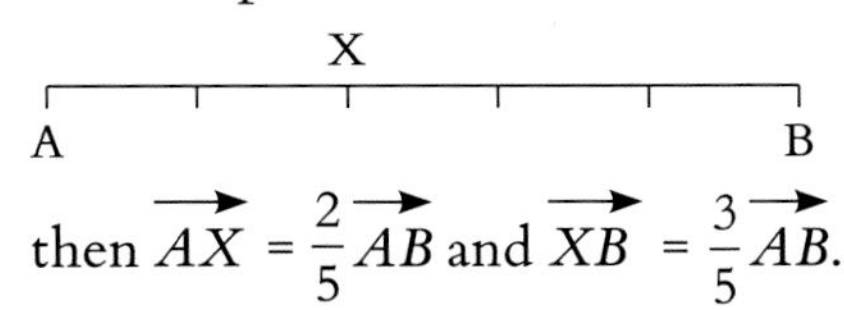

then $\overrightarrow{AX} = \frac{2}{5}\overrightarrow{AB}$ and $\overrightarrow{XB} = \frac{3}{5}\overrightarrow{AB}$.

The next example illustrates these techniques.

EXAMPLE

OPQR is a trapezium.

$\overrightarrow{OP} = 3\mathbf{p}$ and $\overrightarrow{OR} = \mathbf{r}$

$\overrightarrow{RQ} = \frac{1}{3}\overrightarrow{OP}$

X lies on RP such that RX : XP = 1 : 3

R, Q, r, X, O, 3**p**, P

a Find in terms of **p** and **r**

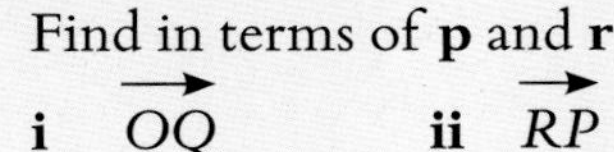

i $\overrightarrow{OQ}$ **ii** $\overrightarrow{RP}$ **iii** $\overrightarrow{RX}$ **iv** $\overrightarrow{OX}$

b What do your answers for $\overrightarrow{OQ}$ and $\overrightarrow{OX}$ tell you about the points O, X and Q?

---

a **i** $\overrightarrow{OQ} = \overrightarrow{OR} + \overrightarrow{RQ}$

$= \mathbf{r} + \mathbf{p}$

ii $\overrightarrow{RP} = \overrightarrow{RO} + \overrightarrow{OP}$

$= -\mathbf{r} + 3\mathbf{p}$

iii $\overrightarrow{RX} = \frac{1}{4}\overrightarrow{RP}$

➡ **NOTE:** $RX : XP = 1 : 3$ means that $\overrightarrow{RX} = \frac{1}{4}\overrightarrow{RP}$.

$= \frac{1}{4}(-\mathbf{r} + 3\mathbf{p})$

$= \frac{3}{4}\mathbf{p} - \frac{1}{4}\mathbf{r}$

iv $\overrightarrow{OX} = \overrightarrow{OR} + \overrightarrow{RX} = \mathbf{r} + \frac{3}{4}\mathbf{p} - \frac{1}{4}\mathbf{r}$

$= \frac{3}{4}\mathbf{r} + \frac{3}{4}\mathbf{p}$

$= \frac{3}{4}(\mathbf{r} + \mathbf{p})$

b $\overrightarrow{OQ} = \mathbf{r} + \mathbf{p}$ and $\overrightarrow{OX} = \frac{3}{4}(\mathbf{r} + \mathbf{p})$

So $\overrightarrow{OX} = \frac{3}{4}\overrightarrow{OQ}$

This means that OX and OQ are parallel.

The point O lies on both of these line segments.

So the points O, X and Q are collinear.

UNIT 9

EXERCISE 9.12

1. OPQ is a triangle and O is the origin.
   $\overrightarrow{OP} = \mathbf{p}$ and $\overrightarrow{OQ} = \mathbf{q}$.
   X is a point on PQ such that PX : XQ = 1 : 2
   Find in terms of **p** and **q**

   a $\overrightarrow{PQ}$ **b** $\overrightarrow{PX}$ **c** the position vector of X.

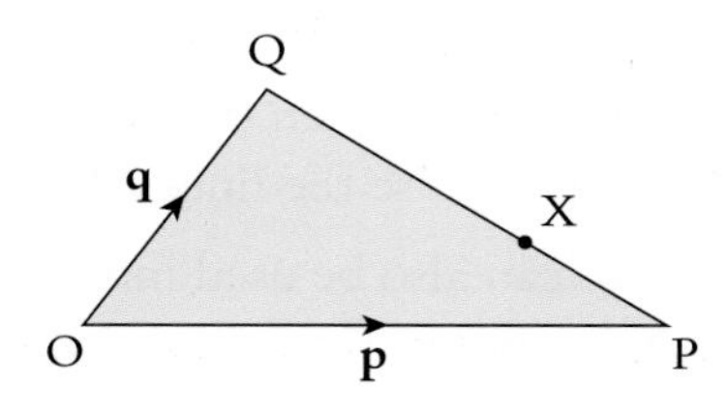

2. OPQ is a triangle and O is the origin.
   $\overrightarrow{OP} = \mathbf{p}$ and $\overrightarrow{OQ} = \mathbf{q}$.
   A is the midpoint on OQ.
   B is a point on PQ such that PB : BQ = 3 : 1
   Find in terms of **p** and **q**

   a $\overrightarrow{PQ}$ **b** $\overrightarrow{PB}$ **c** $\overrightarrow{AB}$ **d** the position vector of B.

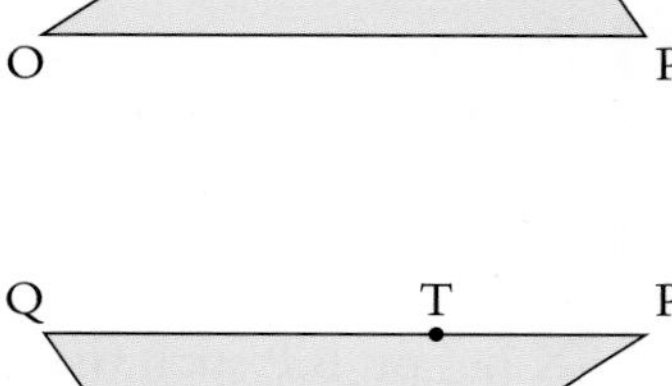

3. OPQ is a triangle and O is the origin.
   $\overrightarrow{OP} = \mathbf{p}$ and $\overrightarrow{OQ} = \mathbf{q}$.
   R is the midpoint on OQ.
   S is a point on OP such that OS : SP = 1 : 2
   T is a point on QP such that QT : TP = 3 : 2
   Find in terms of **p** and **q**

   a $\overrightarrow{OR}$ **b** $\overrightarrow{SR}$ **c** $\overrightarrow{PQ}$ **d** $\overrightarrow{PT}$

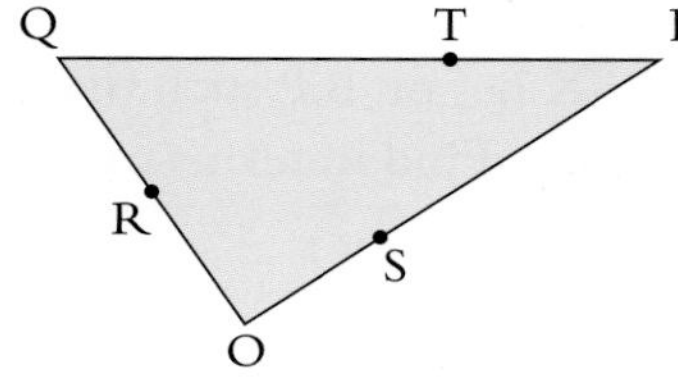

4. OPQR is a parallelogram and O is the origin.
   $\overrightarrow{OP} = \mathbf{p}$ and $\overrightarrow{OR} = \mathbf{r}$.
   X is a point on PQ such that PX : XQ = 1 : 2
   Y is a point on RQ such that RY : YQ = 1 : 3
   Find in terms of **p** and **r**

   a $\overrightarrow{PX}$ **b** $\overrightarrow{RY}$ **c** $\overrightarrow{XY}$

   d the position vector of Y.

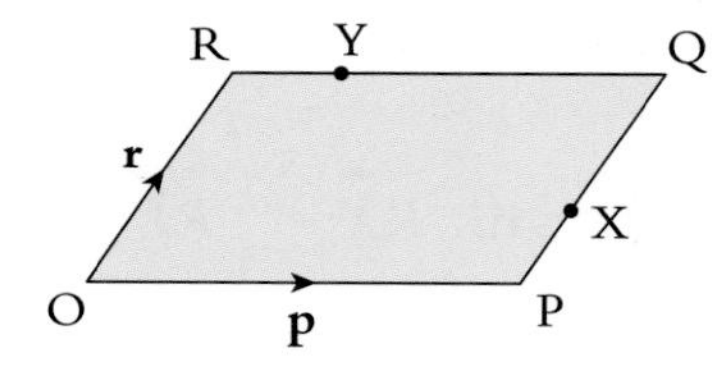

5. OABC is a trapezium.
   $\overrightarrow{OA} = 3\mathbf{a}$, $\overrightarrow{CB} = 2\mathbf{a}$ and $\overrightarrow{OC} = \mathbf{c}$.
   X is a point on AC such that AX : XC = 3 : 2
   Find in terms of **a** and **c**

   a $\overrightarrow{AC}$ **b** $\overrightarrow{AX}$ **c** $\overrightarrow{OX}$ **d** $\overrightarrow{OB}$

   What do your answers for $\overrightarrow{OX}$ and $\overrightarrow{OB}$ tell you about the points O, X and B?

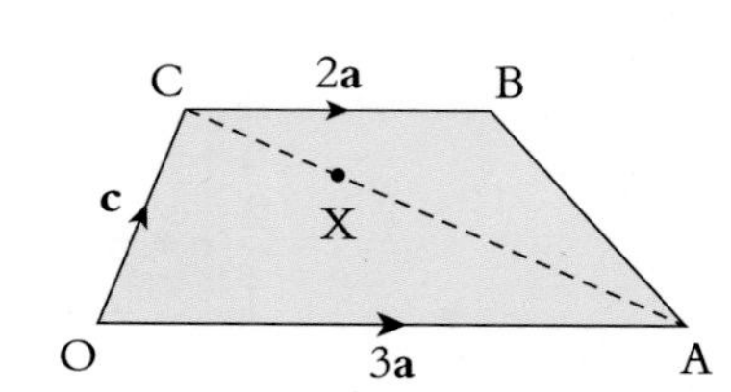

6 OPQR is a parallelogram and O is the origin.
$\overrightarrow{OP} = \mathbf{p}$ and $\overrightarrow{OR} = \mathbf{r}$.
X is a point on PR such that PX : XR = 1 : 3
Find in terms of **p** and **r**

a $\overrightarrow{PR}$ b $\overrightarrow{PX}$ c $\overrightarrow{XQ}$

d the position vector of X.

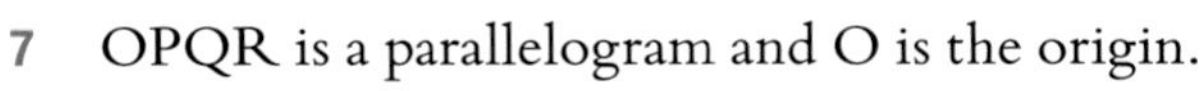

7 OPQR is a parallelogram and O is the origin.
$\overrightarrow{OP} = \mathbf{p}$ and $\overrightarrow{OR} = \mathbf{r}$.
Y is a point on PQ such that PY : YQ = 2 : 3
The line RY is extended to the point X so that
$\overrightarrow{RX} = \frac{5}{3}\overrightarrow{RY}$.
Find in terms of **p** and **r**

a $\overrightarrow{PY}$ b $\overrightarrow{RY}$ c $\overrightarrow{RX}$ d $\overrightarrow{OX}$

Explain why O, P and X are collinear.

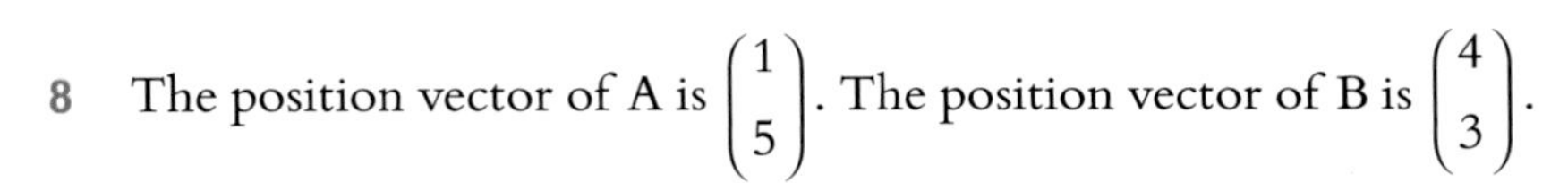

8 The position vector of A is $\begin{pmatrix}1\\5\end{pmatrix}$. The position vector of B is $\begin{pmatrix}4\\3\end{pmatrix}$.

The position vector of C is $\begin{pmatrix}p\\-3\end{pmatrix}$. The position vector of D is $\begin{pmatrix}-7\\q\end{pmatrix}$.

The points A, B, C and D are collinear. Find the values of p and q.

9 $\overrightarrow{OP} = \mathbf{p}$ and $\overrightarrow{OQ} = \mathbf{q}$.
OA : AP = 1 : 2 and OB : BQ = 1 : 2
Prove that $\overrightarrow{PQ} = 3\overrightarrow{AB}$.

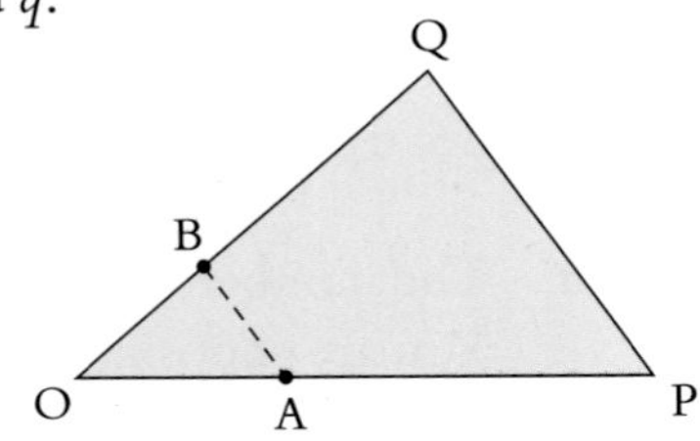

CHALLENGE:

OPQ is a triangle and O is the origin.
$\overrightarrow{OP} = \mathbf{p}$ and $\overrightarrow{OQ} = \mathbf{q}$.
X is the midpoint of OP and Y is the midpoint of OQ.
V is a point on XQ such that
XV : VQ = 1: 2.

a Find the position vector of V.

W is a point on YP such that YW : WP = 1 : 2.

b Find the position vector of W.
What can you say about the points V and W?

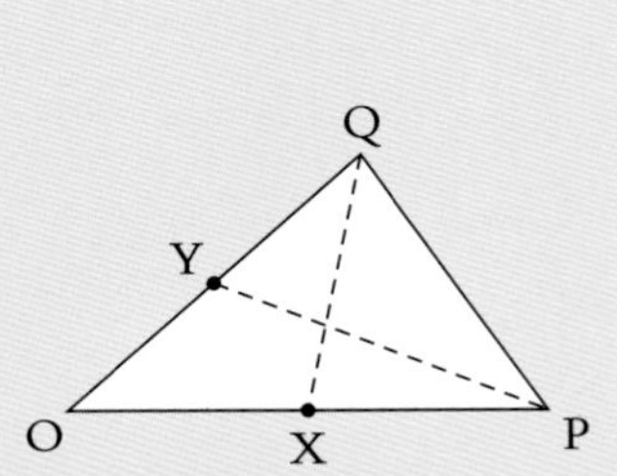

KEY WORDS

vector
magnitude
modulus
resultant vector
scalar
position vector
collinear

Probability 3

THIS SECTION WILL SHOW YOU HOW TO

- Use Venn diagrams to answer probability questions

Calculating probabilities from Venn diagrams

Probabilities can be calculated using information given on a Venn diagram.

EXAMPLE

In a survey, 100 adults are asked whether they like coffee (C) and whether they like tea (T).

The Venn diagram shows the results of the survey.

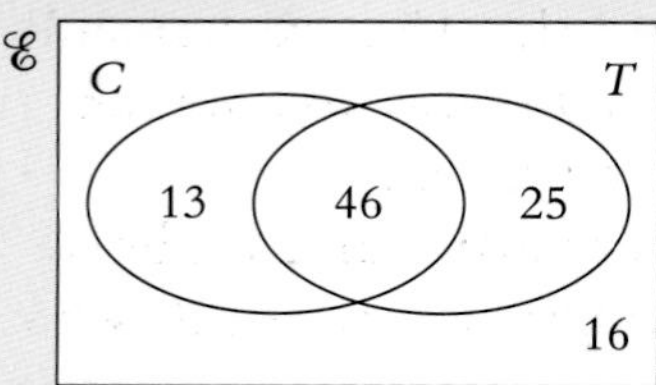

a A person is chosen at random.
Find the probability that the person does not like coffee.

b A person who likes tea is chosen at random.
Find the probability that this person also likes coffee.

c Two people are chosen at random.
Find the probability that they both like coffee.

---

a The number of people who do not like coffee = 25 + 16 = 41

You are choosing out of 100 people.

So, the probability that the person does not like coffee = $\frac{41}{100}$

b The number of people who like tea = 46 + 25 = 71

You are choosing out of 71 people and 46 of these also like coffee.

So, the probability that the person also likes coffee = $\frac{46}{71}$

c The probability the first person likes coffee = $\frac{13+46}{100} = \frac{59}{100}$

One person has been chosen so there are now only 99 to choose from.

Of these 99 people there are now only 58 that like coffee.

The probability the second person likes coffee = $\frac{58}{99}$

So, the probability that they both like coffee = $\frac{59}{100} \times \frac{58}{99} = \frac{1711}{4950} \approx 0.346$ (to 3 s.f.)

EXERCISE 9.13

1 In a survey, 60 people are asked if they own a mobile phone (M) or a computer (C). The Venn diagram shows the results of the survey.

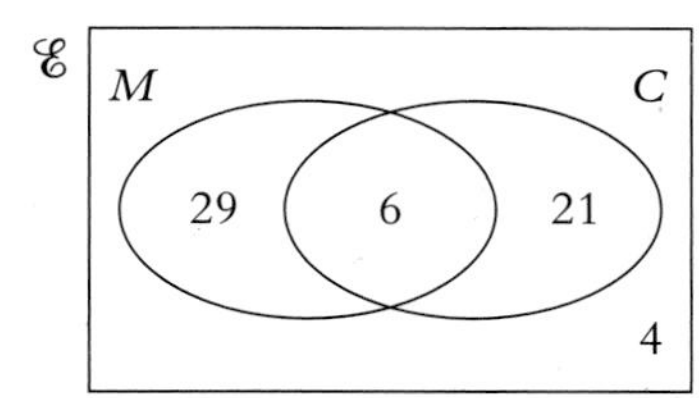

a A person is chosen at random.
   i Find the probability that the person owns a mobile phone.
   ii Find the probability that the person owns both a computer and a mobile phone.
   iii Find the probability that the person does not own a computer.

b A person who owns a computer is chosen at random.
   Find the probability that this person does not own a mobile phone.

c Two people are chosen at random. Find the probability that they both own a computer.

2 There are 100 students in a year group.
50 study art (A), 29 study biology (B) and 13 study both subjects.

a Show this information on a Venn diagram.

b A student is chosen at random.
   i Find the probability that the student studies biology but not art.
   ii Find the probability that the student studies neither biology nor art.

c A student who studies art is chosen at random.
   Find the probability that this student does not study biology.

d Two students are chosen at random. Find the probability that they both study art.

3 In a survey, 150 members of a youth club are asked whether they like singing (S) and whether they like dancing (D).
The Venn diagram shows the results of the survey.

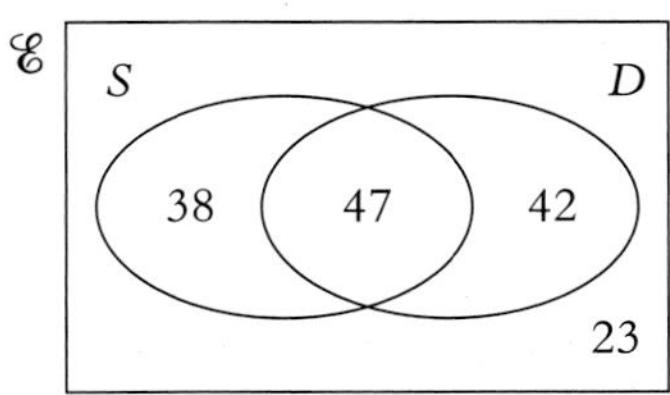

a A member of the youth club is chosen at random.
   i Find the probability that this person likes dancing.
   ii Find the probability that this person does not like singing.
   iii Find the probability that this person likes neither dancing nor singing.

b A member of the youth club who likes dancing is chosen at random.
   Find the probability that this person likes singing.

c Two members of the youth club are chosen at random.
   Find the probability that exactly one of the two members likes dancing.

4 In a survey, 100 students are asked if they like rock climbing (R), windsurfing (W) or swimming (S). The Venn diagram shows the results of the survey.

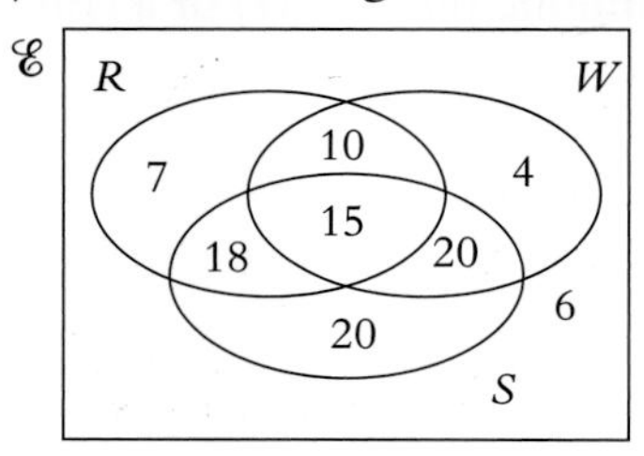

a One student is chosen at random.
  i Find the probability that the student likes swimming.
  ii Find the probability that the student likes rock climbing but not windsurfing.
  iii Find the probability that the student likes all three sports.

b A student who likes rock climbing is chosen at random.
  Find the probability that this student does not like swimming.

c Two students are chosen at random from those who like windsurfing.
  Find the probability that they both like exactly one other sport.

5 There are 35 students in a class.
18 study geography (G), 22 study history and 5 study neither geography nor history.

a Show this information on a Venn diagram.

b A student is chosen at random.
  i Find the probability that the student studies both history and geography.
  ii Find the probability that the student studies history but not geography.

c A student who studies history is chosen at random.
  Find the probability that this student studies geography.

d Two students are chosen at random.
  Find the probability that only one of them studies history.

6 The brakes, steering and lights are tested on 200 cars.
The Venn diagram shows the results of the survey.
B = {cars with faulty brakes} S = {cars with faulty steering} L = {cars with faulty lights}

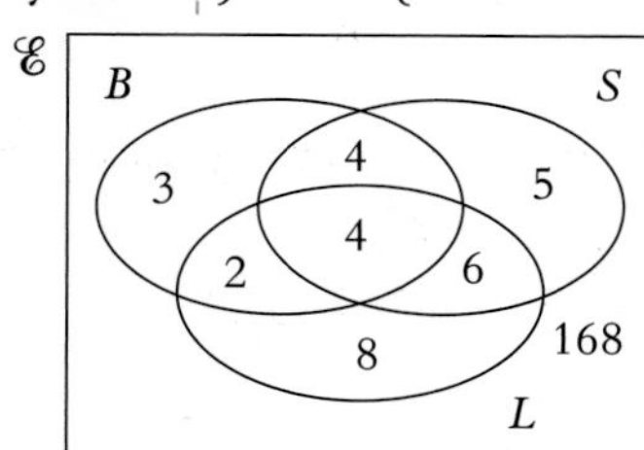

a One of the 200 cars is chosen at random.
  i Find the probability that the car has exactly one of the faults.
  ii Find the probability that the car has exactly two of the faults.
  iii Find the probability that the car has all three of the faults.

b A car with faulty steering is chosen at random.
  Find the probability that this car also has faulty brakes.

c Two cars are chosen at random. Find the probability that they both have faulty lights.

7 $\mathscr{E} = \{1, 2, 3, 4, 5, 6, 7, 8, 9, 10, 11, 12, 13, 14, 15, 16\}$
$A = \{\text{even numbers}\}$ and $B = \{\text{multiples of } 3\}$

a Show these sets on a Venn diagram.

b A number is chosen at random from the set $\mathscr{E}$.
Find the probability that this number is in the set

i B ii A' iii $A \cap B$ iv $A \cup B$ v $(A \cup B)'$

8 In a survey, 50 people are asked if they recycle glass (G), paper (P) or aluminium (A). The Venn diagram shows the results of the survey.

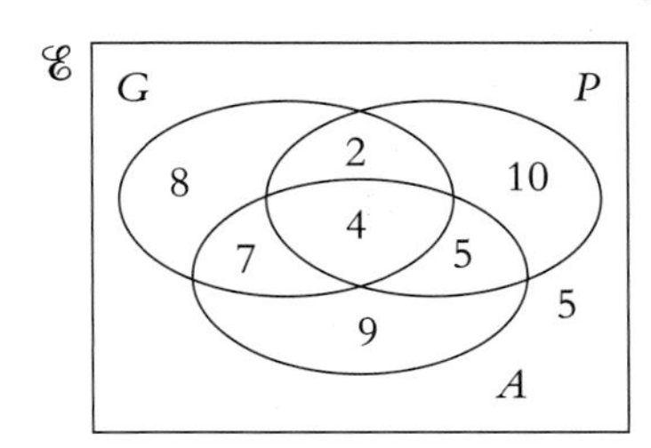

a One of the 50 people is chosen at random.

i Find the probability that they recycle exactly one of the materials.

ii Find the probability that they recycle exactly two of the materials.

iii Find the probability that they recycle all three of the materials.

b A person who recycles aluminium is chosen at random.
Find the probability that this person also recycles paper.

c Two people are chosen at random.
Find the probability that exactly one of them recycles glass.

9 In a survey, a group of 50 students are asked whether they go to a gym (G) or whether they go swimming (S).
16 students go to a gym.
21 students go swimming.
19 students neither go to a gym nor go swimming.

a Show this information on a Venn diagram.

b A student is chosen at random.
Find the probability that the student goes to the gym and goes swimming.

c A student who goes swimming is chosen at random.
Find the probability that this student does not go to the gym.

d Three students are chosen at random.
Find the probability that all three students go to the gym.

Unit 9 Examination-style questions

1 A regular hexagon has sides of length 43 mm, correct to the nearest millimetre.

Calculate the lower bound and the upper bound for the perimeter of this hexagon. [2]

2 A rectangle has sides of length 5.3 cm and 7.5 cm, each correct to 1 decimal place.

Calculate the lower bound and the upper bound for the area of this rectangle. [3]

3 Jack walks a distance of 16.68 km, correct to 2 decimal places.
The time taken to walk this distance is 3.6 hours, correct to 1 decimal place.

Calculate the upper bound for his average speed.
Give your answer in kilometres per hour. [3]

4 Solve the simultaneous equations.
You must show all your working.

$$3x - 4y = 8$$
$$x = 2y - 5$$

[3]

5 Prya buys x apples and y oranges.
She must buy

- at least 5 apples
- at least 7 oranges
- more apples than oranges
- a total of at least 18 apples and oranges.

Exercise 1.8

1 $\frac{1}{2}$ 2 $\frac{5}{7}$ 3 $\frac{3}{4}$
4 $\frac{2}{5}$ 5 $\frac{5}{8}$ 6 $\frac{6}{7}$
7 $\frac{4}{7}$ 8 $\frac{4}{5}$ 9 $\frac{7}{8}$
10 $\frac{3}{11}$ 11 $\frac{17}{20}$ 12 $\frac{7}{12}$
13 $1\frac{3}{5}$ 14 $2\frac{1}{3}$ 15 $2\frac{1}{7}$
16 $1\frac{3}{17}$ 17 $6\frac{1}{2}$ 18 $4\frac{3}{4}$
19 $1\frac{5}{12}$ 20 $3\frac{2}{15}$ 21 $1\frac{7}{8}$
22 $1\frac{1}{21}$ 23 $3\frac{2}{5}$ 24 $3\frac{1}{3}$

Exercise 1.9

1 $3\frac{5}{6}$ 2 $5\frac{9}{20}$ 3 $7\frac{25}{42}$
4 $7\frac{7}{15}$ 5 $4\frac{7}{12}$ 6 $4\frac{7}{30}$
7 $4\frac{7}{40}$ 8 $15\frac{4}{9}$ 9 $2\frac{3}{10}$
10 $\frac{7}{24}$ 11 $3\frac{22}{63}$ 12 $2\frac{11}{40}$
13 $3\frac{11}{40}$ 14 $\frac{17}{30}$ 15 $3\frac{5}{12}$
16 $3\frac{29}{35}$ 17 $-2\frac{3}{10}$ 18 $6\frac{13}{15}$
19 $4\frac{1}{12}$ 20 -1 21 $-1\frac{4}{15}$
22 $1\frac{7}{15}$ 23 Answers as above.
24 $\frac{13}{35}$ 25 $\frac{13}{20}$ 26 $1\frac{9}{10}$
27 $6\frac{31}{35}$ 28 $-\frac{1}{12}$
29 a $\frac{13}{30}, \frac{3}{5}, \frac{2}{3}$ b $\frac{5}{9}, \frac{7}{12}, \frac{11}{36}$
c $\frac{7}{10}, \frac{3}{4}, \frac{4}{5}$ d $\frac{5}{6}, \frac{8}{9}, \frac{11}{12}$

Exercise 1.10

1 $3\frac{2}{3}$ 2 $5\frac{1}{16}$ 3 $3\frac{29}{30}$
4 $19\frac{23}{28}$ 5 $11\frac{5}{18}$ 6 $20\frac{11}{25}$
7 $\frac{9}{14}$ 8 $2\frac{1}{2}$ 9 $1\frac{7}{10}$
10 $3\frac{29}{32}$ 11 $\frac{10}{27}$ 12 $1\frac{1}{20}$
13 $\frac{9}{25}$ 14 $6\frac{1}{4}$ 15 $\frac{27}{8000}$
16 $1\frac{7}{9}$ 17 $\frac{2}{5}$ 18 $1\frac{5}{8}$
19 Answers as above.
20 $1\frac{2}{3}$ 21 $1\frac{1}{2}$ 22 $\frac{15}{56}$
23 $2\frac{3}{11}$ 24 $3\frac{3}{4}$ litres

Exercise 1.11

1 a 5.3 b 8.8 c 6.0
d 30.0 e 0.1 f 8.9
g 4.2 h 16.3 i 644.0
j 7.0
2 a 6.35 b 8.39 c 16.16
d 4.21 e 5.70 f 0.04
g 0.07 h 2.99 i 3.06
j 7.10
3 a 7.222 b 6.162
c 35.686 d 82.009
e 24.789 f 3.000
g 6.009 h 6.667
i 102.103 j 5.010
4 a 3 b 8 c 10
d 50 e 400 f 30 000
g 50 000 h 0.002 i 0.09
j 0.3
5 a 1.7 b 3.1 c 5700
d 61 000 e 16 f 20
g 100 h 0.0063
i 0.058 j 0.00040
6 a 27.3 b 6.51 c 2590
d 149 e 16.7 f 0.349
g 0.0718 h 0.00808
i 10.1 j 40 000
7 a 7 000 000 000
8 a 100 000 000
b 110 000 000
c 106 000 000
9 a 9000 b 8800 c 8850
10 a 0.06 b 0.064 c 0.0640
d 0.1 e 0.06 f 0.064
11 a 40 000 b 40 000
c 40 000 d 40 010
e 40 008.6 f 40 008.63

Exercise 1.12

1 a 60 (62.72) b 16 (16.83)
c 3000 (3555)
d 2500 (2569)
e 80 000 (81290)
f 6 (5.695) g 5 (5.861)
h 5 (4.404) i 16 (16.54)
2 a 10 (10.10) b 2 (2.081)
c 3 (2.564)
3 Accurate answers given in brackets above.
4 \$8 5 \$9500

Exercise 1.13

1 $-x$ 2 $-3y$ 3 $6xy$
4 $-3xy$ 5 $7x - 9y$
6 $5p + 5q$ 7 $5xy - 3x$
8 $7x^2 - 15x$ 9 $7 + 4x$
10 $-7xy$ 11 $-3x^2 + 9$
12 $-6y^2 - 2y$ 13 $8xy - x$
14 $a - 4ab$ 15 $10ab - 6bc$
16 $x^4 + 5x^2 - 5$ 17 $5x^3 - x^2 + x$
18 $-2fg - 4gh$ 19 $4a^2b - 16ab^2$
20 $5cd^2 - 9c^2d$ 21 $6x + \frac{10}{x}$
22 $\frac{9}{x} - \frac{7}{y}$
23 $-2xy^2 + 12xy - 10x^2y + x^2 - y^2$
24 3

Exercise 1.14

1 $7(2x + 3) = 14x + 21$
2 a $4x + 12$ b $6y + 12$
c $4x + 20$ d $6 - 2a$
e $7y - 28$ f $8x - 72$
g $10x + 15$ h $18y + 24$
i $35a + 42$ j $6x + 6y - 3$
k $20a + 25b - 15$
l $24p - 32q - 56r$ m $4x - 1$
n $3x + 2$ o $5y - 2$
3 a $-3x - 12$ b $-2x + 12$
c $-30 + 20x$ d $-16x - 40$
e $-21x + 56$ f $-8x + 72$
g $-5x - 4$ h $-2x + 7$
i $-3x - 8$ j $-18x + 18y - 30$
k $-12p - 16q + 20$
l $-27x^2 - 9x + 6$
4 a $x^2 + 3x$ b $y^2 - 5y$
c $7a - a^2$ d $10x - 6x^2$
e $5y^2 + 40y$ f $6x^2 + 12xy$

UNIT 1

5 a $5x + 30$ b $11x + 32$
c $9y$ d $11x - 74$
e $36x - 4y$ f $2x^2 - 11x$
g $2x^2 + 6x$ h $2x^2 + 4x$
i $38x - 9x^2$
j $12xy - 8x - 15y$
k $-4x^2 + 11xy + 21y^2$

6 $5(2x - 3) - 4(x - 6) =$
$10x - 15 - 4x + 24 = 6x + 11$

7 a x b $2y - 11$
c $23a - 10$ d $36x - 20$
e $2x + 17$ f $11 - 8y$
g $-h - 6g$ h $x^2 - 5x$
i $11p^2 - 44p$
j $63a + 6ab - 51a^2$

8 a $2x + 16$ b $27 - 3x$
c $8 - 6x + 8y$ d $-1 - 5y$
e $22 - 7y$ f $14x - 15$
g $8y + x$ h $11x - 2x^2$

Exercise 1.15

1 a 6 b 9 c 7
d $1\frac{1}{6}$ e $1\frac{1}{2}$ f $13\frac{1}{4}$
g 3 h $-\frac{2}{3}$ i $\frac{3}{4}$
j $1\frac{7}{8}$ k $-2\frac{1}{2}$ l $-2\frac{4}{5}$
m $7\frac{2}{3}$ n $8\frac{1}{2}$ o $\frac{1}{2}$
p $1\frac{5}{6}$ q 7 r $-2\frac{2}{5}$
s 2 t $2\frac{4}{5}$

2 a 2 b 5 c 7
d $3\frac{1}{2}$ e $\frac{1}{8}$ f $3\frac{4}{5}$
g $\frac{1}{5}$ h $-6\frac{1}{2}$ i -5
j 1 k $1\frac{1}{2}$ l 1

3 a 4 b 3 c 4
d 5 e $6\frac{1}{2}$ f -4
g $-\frac{1}{3}$ h $4\frac{1}{2}$ i $1\frac{2}{3}$

4 56, 57, 58

5 20

6 2

7 1.5 cm

8 5

Exercise 1.16

1 a 20 b 60 c 75
d $6\frac{2}{3}$ e 45 f -28
g 25 h $9\frac{1}{3}$ i -14
j 22 k 24 l 20

2 a 7 b 11 c $3\frac{1}{4}$
d $1\frac{1}{2}$ e $-2\frac{1}{2}$ f 5
g $4\frac{1}{2}$ h $1\frac{7}{15}$

3 a 3 b $\frac{1}{2}$ c $1\frac{3}{4}$
d $\frac{3}{10}$ e $\frac{1}{2}$ f 2
g 5 h -2 i $1\frac{1}{2}$
j -2 k $2\frac{1}{2}$ l -4
m 3 n 3 o 1 p 0

4 a 0 b 3 c 4 d 1
e $-\frac{8}{11}$ f -7 g 3
h $-3\frac{7}{8}$

5 $12x + 4 = 12x - 5$ means that $4 = -5$ which is not possible

6 6

Exercise 1.17

1 $P = 1350 + 13n$

2 $Q = 100 - 9n$

3 $C = 4b + 7c$

4 $T = \frac{5r}{100} + \frac{6p}{100} + 4f$

5 $T = 30 + 4n$

6 $T = 45m + 20$

Exercise 1.18

1 -8 2 -23 3 -36
4 11 5 20 6 3
7 4 8 210 9 -2
10 16 11 -48 12 -480
13 -40 14 128 15 12
16 28 17 6 18 8
19 -1 20 -9 21 $-2\frac{1}{2}$
22 3 23 $3\frac{1}{2}$ 24 $-15\frac{2}{3}$
25 0 26 $-5\frac{4}{5}$ 27 -3
28 26

Exercise 1.19

1 a -10 b 5
c -20 d 100

2 a 25 b 9 c 10

3 a 6 b 2.1 c 0.912

4 a 1 b 1.2 c 1.8

5 a 435 b 16

6 a 5 b 8

7 a 7 b 10

8 a 9 b 8

Exercise 1.20

1 a 1 b 4 c $-\frac{1}{3}$
d $\frac{1}{4}$ e -2 f $\frac{2}{3}$
g 0 h $-\frac{3}{2}$ i -1
j 2

2 a 1 b $-\frac{9}{8}$ c $\frac{3}{2}$
d $\frac{1}{3}$ e -4 f $\frac{9}{7}$
g 0 h -11 i 4
j -1 k $\frac{2}{5}$ l $\frac{1}{4}$

3 a 1 b 2 c $\frac{1}{2}$
d 2 e $-\frac{1}{2}$ f infinite(∞)
g 1 h 0

4 6

5 8

6 7

7 8

8 5

9 -3

10 2

11 11.25

Exercise 1.21

1 a $y = 2$ b $y = x$
c $x = 4$ d $y = -x$

2 **a**

x	0	1	2
y	2	3	4

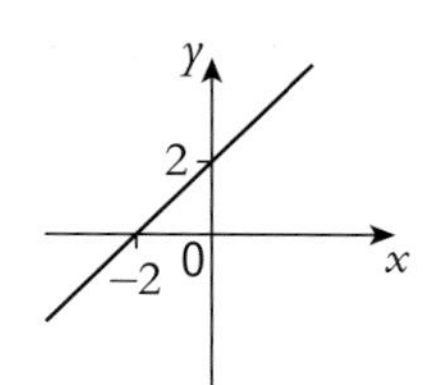

b

x	0	1	2
y	−5	−4	−3

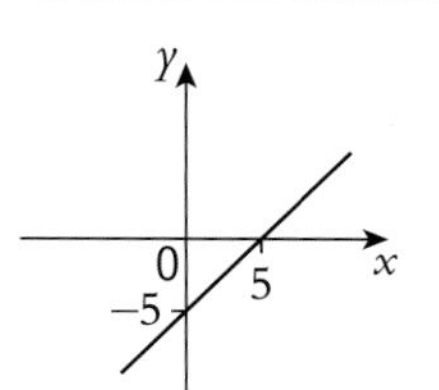

c

x	0	1	2
y	−2	0	2

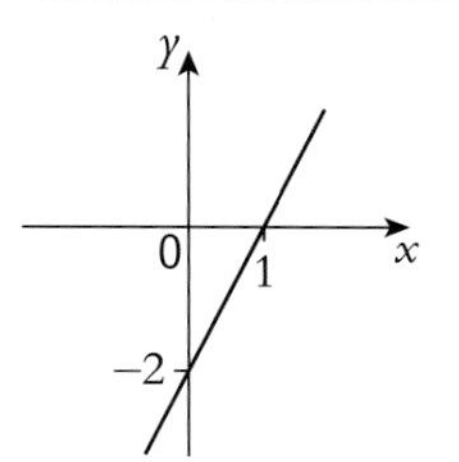

d

x	0	1	2
y	−4	−1	2

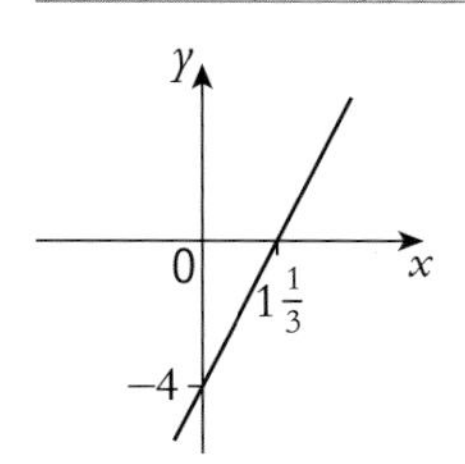

e

x	−2	0	2
y	−2	−1	0

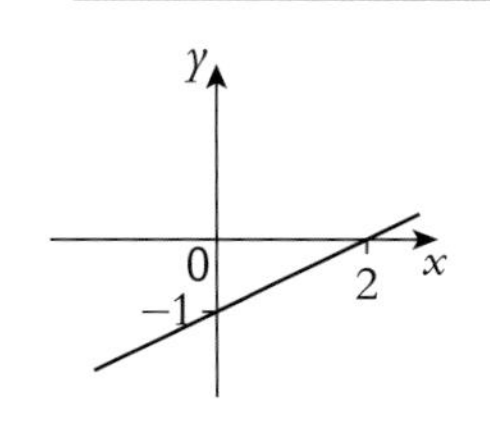

f

x	−3	0	3
y	−5	−4	−3

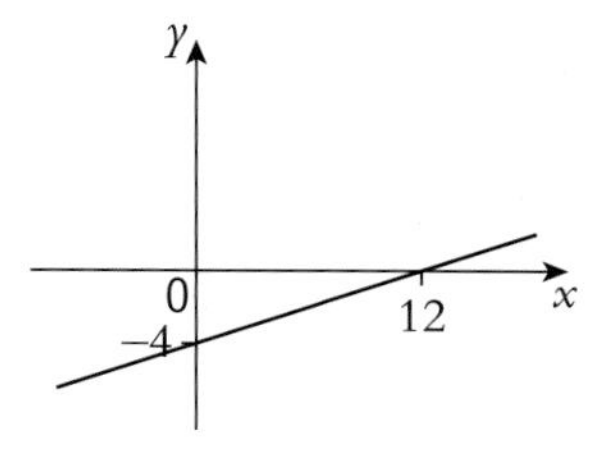

g

x	0	1	2
y	3	1	−1

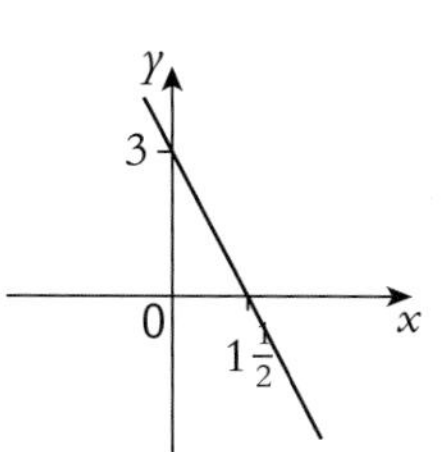

h

x	0	1	2
y	5	2	−1

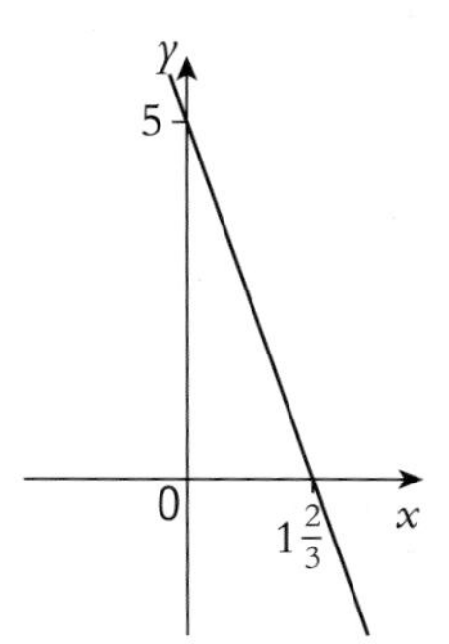

3

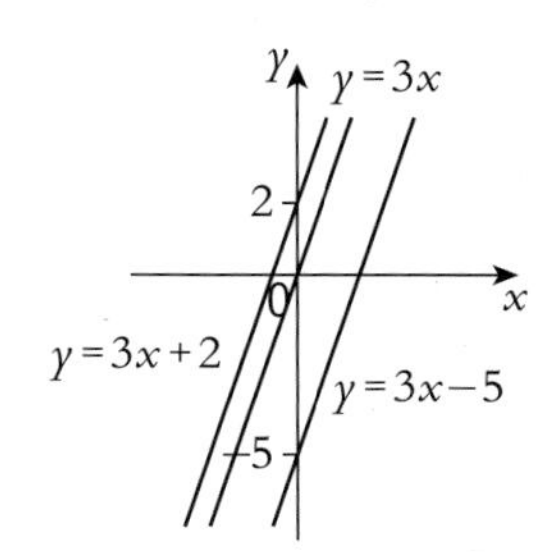

Parallel

4

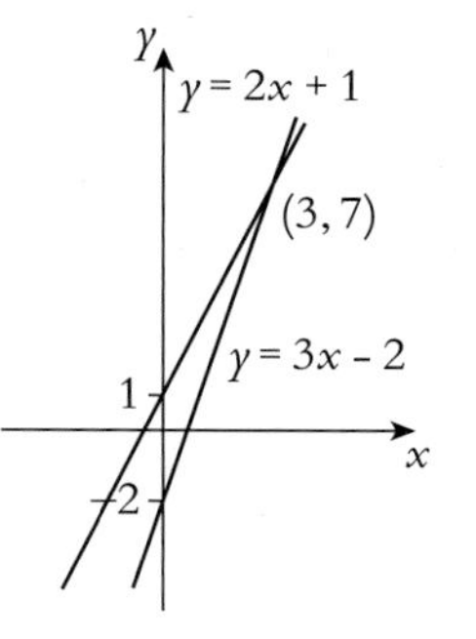

5

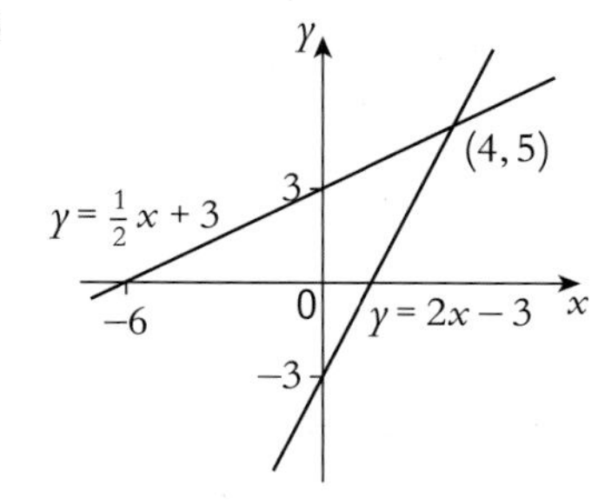

6 2

7 $-4\frac{1}{2}$

8 −14

Exercise 1.22

1

x	0	2
y	4	0

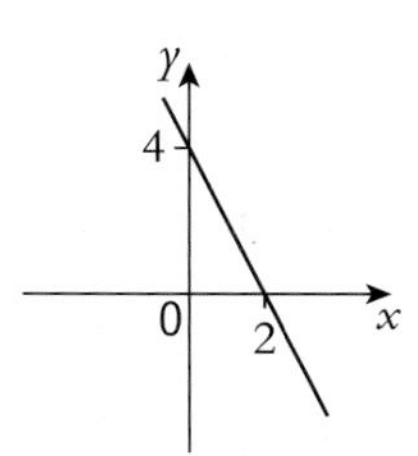

2

x	0	2
y	3	0

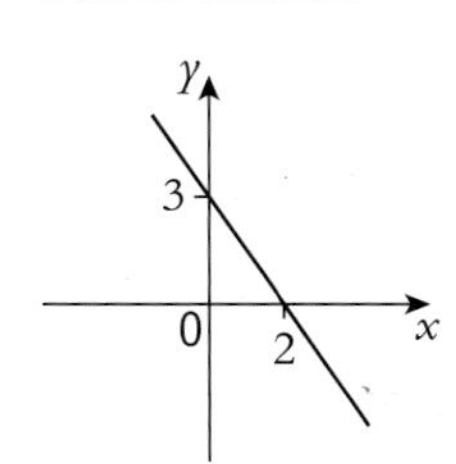

3

x	0	5
y	2	0

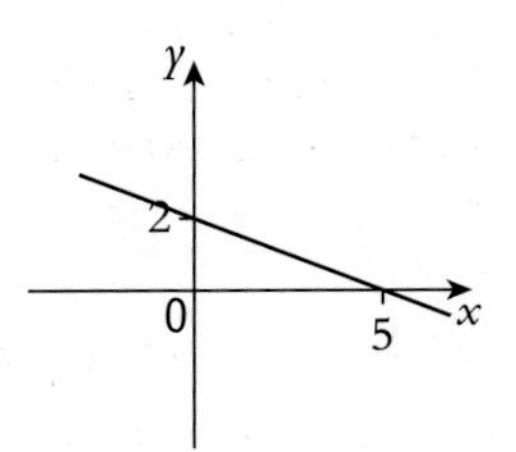

4

x	0	3
y	9	0

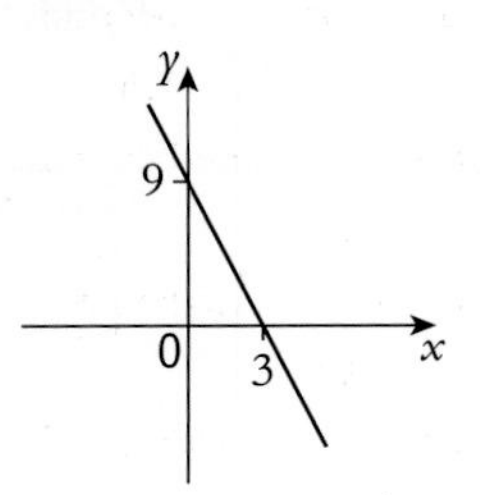

5

x	0	−6
y	−4	0

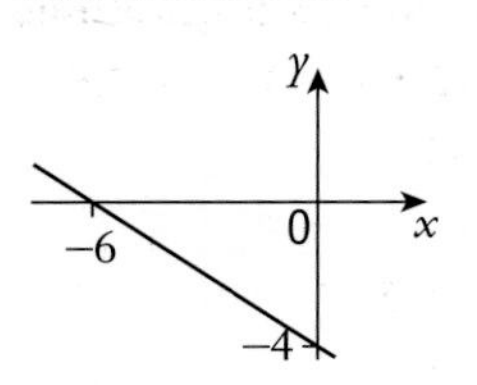

6

x	0	−5
y	−6	0

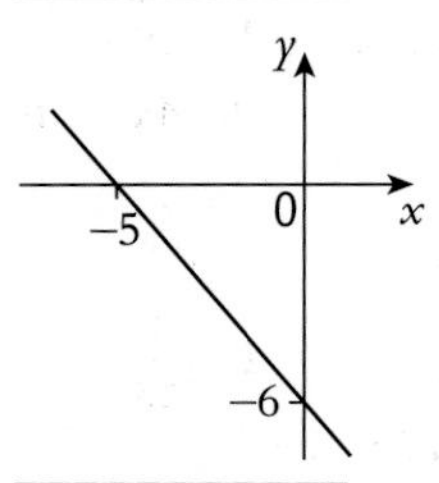

7

x	0	6
y	−3	0

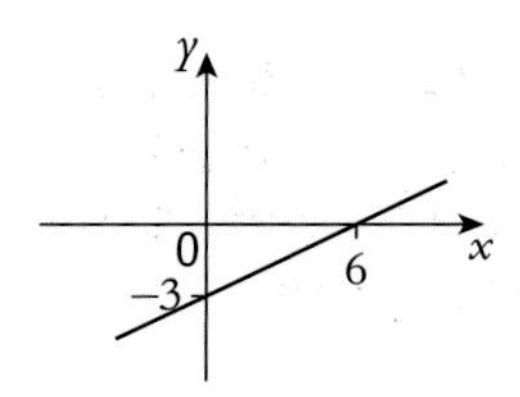

8

x	0	5
y	−4	0

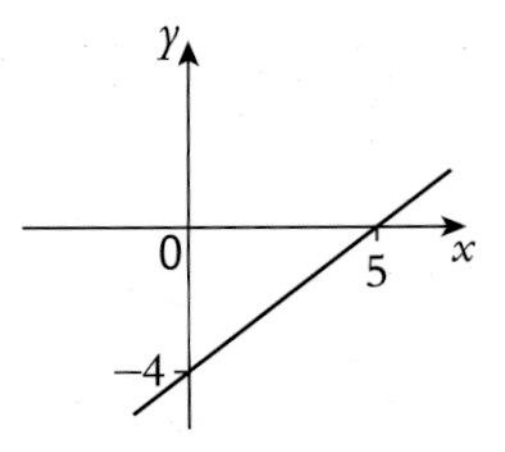

9

x	0	6
y	−4	0

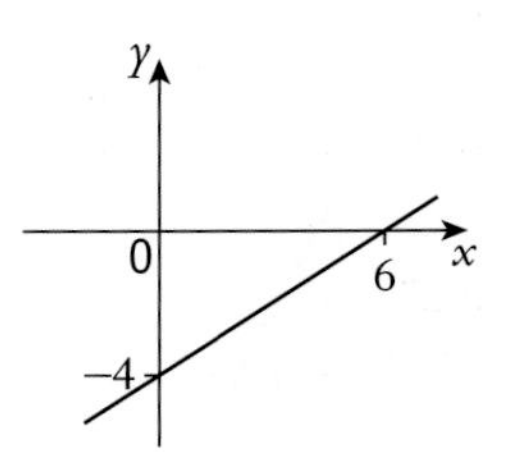

10

x	0	−5
y	15	0

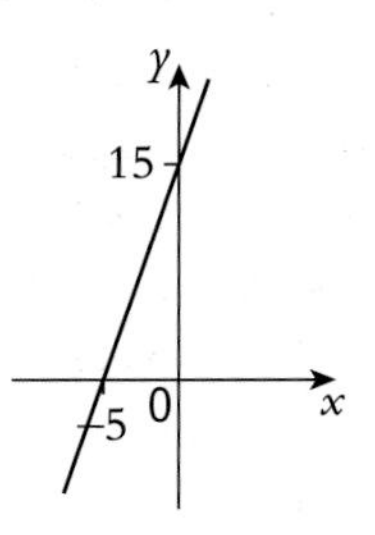

11

x	0	−6
y	10	0

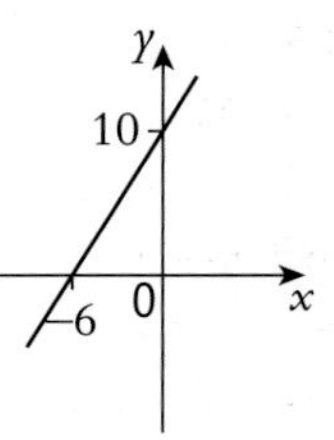

12

x	0	−2
y	12	0

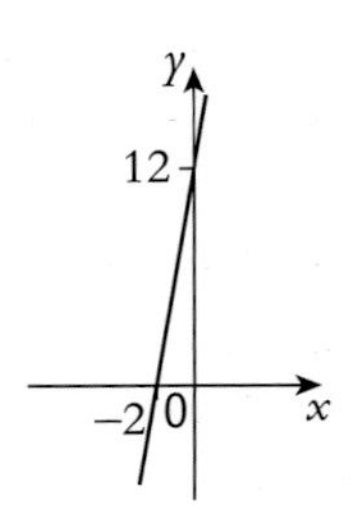

Exercise 1.23

1 $a = 138°$ $b = 42°$ $c = 138°$
2 $a = 150°$
3 $a = 55°$
4 $a = 36°$
5 $a = 49°$
6 $a = 25°$
7 $a = 68°$ $b = 44°$
8 $a = 18°$
9 $a = 50°$ $b = 65°$
10 $a = 40°$
11 $a = 40°$ $b = 35°$
12 $a = 30°$
13 $a = 40°$
14 $a = 120°$
15 $a = 55°$
16 $a = 50°$
17 $a = 40°$ $b = 40°$
18 $a = 45°$
19 $a = 100°$ $b = 130°$
20 $a = 65°$
21 $a = 60°$
22 $a = 60°$ $b = 70°$
23 $a = 73°$ $b = 45°$
24 $a = 109°$
25 33°, 66°, 99°, 162°
26 Several possible proofs, for example

$E\hat{F}B = F\hat{G}D$ (Corresponding angles)
$E\hat{F}B = 180° - B\hat{F}G$
$F\hat{G}D = 180° - F\hat{G}D$ (Angles on a straight line)
$B\hat{F}G + F\hat{G}D = 180°$

Exercise 1.24

1 **a** 4, 4 **b** 2, 2 **c** 0, 2

d 0, 1 **e** 2, 2 **f** 1, 1

2 **a** 1, 1 **b** ∞, ∞ **c** 0, 1

d 2, 2 **e** 2, 2 **f** 8, 8

g 0, 2 **h** 0, 4 **i** 0, 2

j 6, 6 **k** 4, 4 **l** 0, 2

3 **a** 0, 2 **b** 2, 2 **c** 0, 4

4 **a**

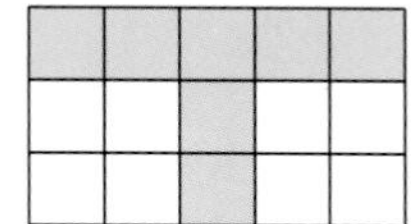

There are other possible answers.

b

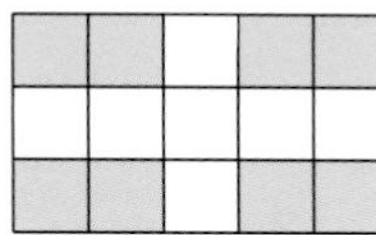

There are other possible answers.

c

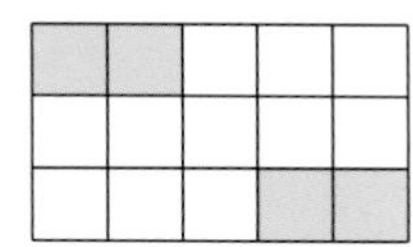

There are other possible answers.

5 **a** 1, 1 **b** 0, 2

c 2, 2 **d** 0, 2

6 The four properties allow you to distinguish all quadrilaterals except a kite and an arrowhead.

Exercise 1.25

1

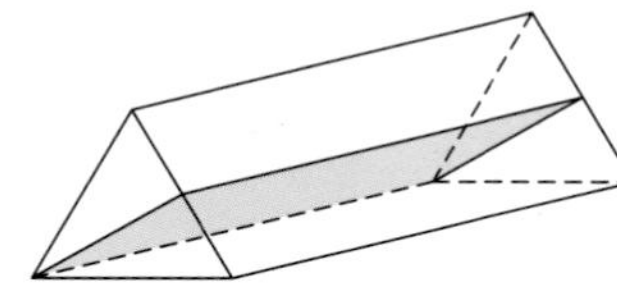

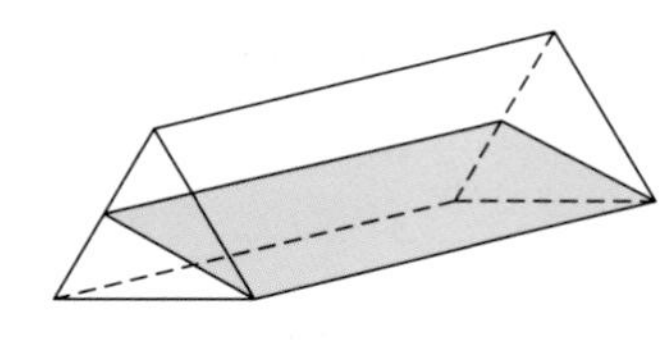

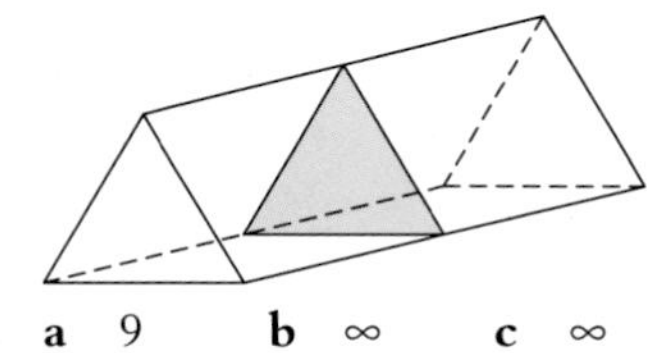

2 **a** 9 **b** ∞ **c** ∞

d ∞ **e** 7

3 **a** 2 **b** 1

c 0 **d** 5

4

2

Exercise 1.26

1 No

Sum of interior angles = $4 \times 180° = 720°$

2 124° **3** 140°

4 12 **5** 20°

6 1440°

7 **a** 117° **b** 58° **c** 41°

d 94° **e** 125° **f** 55°

8 **a** 108° **b** 135° **c** 67.5°

d 22.5° **e** 36° **f** 81°

g 117° **h** 36°

9 18

Exercise 1.27

1 He is incorrect Median = 5

2 **a** mean = 9 median = 10

mode = 10 range = 5

b mean = −2 median = −2

no mode range = 15

c mean = 58 median = 58.5

no mode range = 15

d mean = 3 median = 3

mode = 3 range = 5

e mean = $1\frac{1}{12}$ median = $1\frac{1}{12}$

no mode range = $1\frac{1}{6}$

3 mean = 12 median = 13

mode = 13 range = 9

4 mean = 10.336s

median = 10.305s

mode = 10.28s range = 0.6s

5 **a** Use mediam = 5 or mean = 4.5; mode = 0 or 8 which are extremal.

b Use mode = 7 or median = 7; mean = 11.4 is sensitive to single large value.

c Use median = 8 or mean = 8, though there are no data values near these averages. Mode not defined.

d Use mean = 0 or median = 2 Mode = 3 is extremal.

6 **a** 659.2 kg **b** 7.14 m

7 1.75 m

8 **a** 69 **b** 7 **c** 13 **d** 56

9 68.2 **10** 61 **11** 37.1 cm

Exercise 1.28

1 **a** Incorrect Mode = 1

b Incorrect Median = 1

2 **a** 1 **b** 1 **c** 1.6

3 mode = 1 median = 1

mean = 1.2

4 mode = 1 median = 1

mean = 1.43

5 mode = 7 median = 7

mean = 7.025

6 **a** $x = 25, n = 41$ **b** 5

Examination-style questions

1 $15 - (2 \times 3 + 4) \div 2 = 2.5$

2 100 °C

3 48

4 **a** $2^2 \times 5 \times 7$

b 420

5 53, 59

6 $1\frac{19}{42}$ **7** $2\frac{1}{8}$

8 **a** 302.66 **b** 300

9 100 **10** $4x - 11$

11 −20

12 **a** 1.2 **b** $\frac{6}{7}$

13 **a** $\frac{8}{17}$ **b** $\frac{7}{13}$

14 $T = 170 + 40n$

15 **a** $80x + 910 = 1790$

b 11

16 −617

17 **a** −1.5 **b** (0, 3)

18 −0.5 **19** −3

20 $q = 2p + 23$ **21** −2

22 $x = 107, y = 60$

23 a 4 b 4

24 a 6 b 7

25 20 26 16

27 150

28 a i 7 ii 28

iii 25.5 iv 25.625

b $\frac{26n+205}{n+8}$

29 a 1 b 4

c 3.525 d 3.5

30 8

Unit 2

Excercise 2.1

1 a $\frac{6}{25}$ b $\frac{4}{5}$ c $\frac{29}{50}$

d $\frac{13}{20}$ e $\frac{3}{10}$

2 a 60% b 35% c 72%

d 74% e 87.5%

3 a 0.18 b 0.7 c 0.06

d 0.47 e 0.025

4 a 28% b 80% c 8%

d 120% e 75.5%

5 a \$12 b 312 km

c 135 kg

6 30% of \$60

7 a \$64.96 b \$89.28

c \$190.40 d \$271.40

d \$451 f \$198

8 a \$54.40 b \$272

c \$230 d \$112

e \$482.50 f \$359.28

9 \$735

10 \$883.50

11 \$1277.50

12 \$3510

13 \$18

Excercise 2.2

1 a 2 : 3 b 3 : 4 c 1 : 5

d 8 : 5 e 29 : 36 f 5 : 3

g 8 : 7 h 7 : 11

2 a 1 : 5 b 1 : 20 c 1 : 4

d 1 : 5 000 000 e 25 : 1

f 1 : 50 g 1 : 25 h 5 : 8

i 1 : 4

3 a 7 : 3 b 4 : 3 c 3 : 1

d 1 : 5 e 2 : 5 f 26 : 11

g 7 : 1 h 150 : 1

4 a 2 : 1 b 2 : 5 c 9 : 8

d 3 : 5 e 1 : 4 f 9 : 10

g 3 : 2 h 3 : 2

5 a 1 : 4 : 5 b 3 : 5 : 4

c 1 : 3 : 5 d 8 : 3 : 7

e 21 : 45 : 77 f 1 : 4 : 7

g 8 : 12 : 1 h 10 : 15 : 4

6 a 1 : 4 b $1 : \frac{2}{3}$

c 1 : 0.16 d 1 : 0.625

e 1 : 500 000 f 1 : 400

g 1 : 750 000 h $1 : 333\,333\frac{1}{3}$

7 3 : 1

8 6 : 15 : 11

9 3 : 7

10 1:1.6, 1:1.625, 1:1.615, 1:1.619, 1:1.617, 1:1.618

Tends to a constant ratio 1: 1.61803399....

Excercise 2.3

1 a \$56, \$28 b 105 m, 175 m

c \$1200, \$900

d 187.2 km, 436.8 km

e 112 kg, 70 kg

f \$23.20, \$29

2 a \$70, \$140, \$ 210

b 225 m, 45 m, 90 m

c \$140, \$60, \$20

d 900 km, 750 km, 1200 km

e 8000 kg, 6000 kg, 10 000 kg

f \$269.60, \$377.44, \$431.36

3 \$31.35

4 84°

5 36°

6 $\frac{4}{9}$

7 4 : 3

8 85

9 315

10 1161

11 115 cm

12 15

Excercise 2.4

1 a i 3.5 cm ii 2.6 cm

iii 1.6 cm

b i 70 km ii 52 km

iii 32 km

2 40 km

3 a 15 km b 32.5 km

c 21 km d 4.5 km

4 a 8 km b 5 km

c 7.6 km d 1.4 km

5 a 2 cm b 10 cm

c 5 cm d 8.6 cm

6 3.2 km 7 3 km

8 6 cm

9 27.2 km

10 4.8 cm

11 7.8 cm

Excercise 2.5

1 Incorrect $3^x \times 3^y = 3^{x+y}$

2 a x^9 b b^8 c y^{10}

d c^{12} e a^{11} f p^8

g y^9 h a^7

3 a $15x^6$ b $8b^4$ c $14y^{14}$

d $25c^6$ e $6a^3$ f $70p^8$

g $60y^6$ h $27a^6$

4 a $6x^6y^2$ b $30x^4y^3$

c $10x^2y^3$ d $6x^4y^2$

e $15x^3y^3$ f $35x^3y$

g $4x^4y^2$ h $6a^8b^4$

5 a x^5 b b c a^3

d x^5 e $6x^3$ f $2y^2$

g $5b^7$ h $3c$

6 a $5y^2$ b $12ab$ c $\frac{x}{4}$

d $\frac{y}{2}$ e $2a$ f $4abc$

g a^2b^2c h $5x^2$ i x^5

j y^7 k $6a^7$ l $3x^6$

7 a x^6 b a^{10} c b^{16}

d y^{21} e $3a^6$ f $5x^{20}$

g $8x^9$ h $81y^8$ i $8a^6$

j $9y^8$ k $18x^4y^2$ l $40x^7$

8 **a** $27x^3y^6$ **b** $32a^{10}b^{15}$
c $25x^8y^6$ **d** $1000x^{15}y^{21}$

9 **a** 4 **b** $13x^6$
c $36x^3$ **d** $\frac{ab^3}{9}$
e $4a^4bcd^2$ **f** $32x^{19}y^7$
g $8x^9$ **h** $16x^4y^8$

Excercise 2.6

1 **a** $\frac{1}{3}$ **b** 1 **c** $\frac{1}{8}$ **d** $\frac{1}{27}$
e 1 **f** $\frac{1}{1000}$ **g** $\frac{1}{16}$
h $\frac{1}{7}$ **i** $\frac{1}{10000}$ **j** $\frac{1}{125}$
k $\frac{1}{32}$ **l** $\frac{1}{81}$

2 **a** $\frac{1}{9}$ **b** $-\frac{1}{8}$ **c** $\frac{1}{36}$
d $-\frac{1}{9}$ **e** $\frac{1}{16}$ **f** $-\frac{1}{1000}$

3 **a** $1\frac{1}{2}$ **b** $2\frac{1}{2}$ **c** $1\frac{1}{3}$
d $1\frac{1}{5}$ **e** $3\frac{1}{2}$ **f** $1\frac{1}{4}$

4 **a** $2\frac{1}{4}$ **b** $6\frac{1}{4}$ **c** $1\frac{7}{9}$
d $1\frac{11}{25}$ **e** $12\frac{1}{4}$ **f** $1\frac{9}{16}$
g 8 **h** 1000

5 **a** a^4 **b** b^{-9} **c** c^{-2} **d** d^{-13}
e d^3 **f** x^3 **g** y^4 **h** d^3
i a^5 **j** b **k** 1 **l** d^{-4}

6 **a** x^{-2} **b** x^5 **c** $6x^9$ **d** x^{-5}

7 **a** $8x$ **b** $15x^{-3}$ **c** $6x^3$
d $28x^{-5}$

8 **a** x^{-2} **b** y^{-6} **c** a^{-20} **d** b^{-8}
e a^6 **f** y^2 **g** a^{10} **h** a

9 **a** $9a^{-2}$ **b** $125b^{-3}$ **c** $16c^{-8}$
d $16d^{-4}$ **e** $\frac{1}{4}x^{-4}$ **f** $\frac{1}{27}y^{-6}$
g $\frac{1}{25}x^{-6}$ **h** $\frac{1}{4}x^4$

10 **a** x **b** $\frac{1}{2}x^7$ **c** y^{-1}
d $7x^{-2}$

11 $2a^2 + 3b^{-3}$

Excercise 2.7

1 $x \geq 1$ **2** $x < 2$ **3** $-1 \leq x \leq 2$
4 $-2 \leq x < 3$ **5** $x \leq 6$
6 $x \geq 4$ **7** $x < 7$ **8** $x > 2$
9 $x \geq 1\frac{1}{2}$ **10** $x < 6$ **11** $x \leq -1$
12 $x < 7$ **13** $x \geq 1$ **14** $x < 7\frac{1}{2}$
15 $x \leq -6$ **16** $x < 2$ **17** $x \geq 7$
18 $x < -\frac{1}{5}$ **19** $x > -2\frac{1}{2}$
20 $x \leq -4$ **21** $x \leq 1$ **22** $y > 8$
23 $x \geq 2$ **24** $x < 7$ **25** $x < \frac{1}{2}$
26 $x \geq -3$ **27** $x \leq 4$ **28** $x > 2$
29 $x \geq 3$ **30** $x < -7$ **31** $x \geq 9$
32 $x < 2$ **33** $x \geq 7$ **34** $x \leq 3$
35 $x \leq 2$ **36** $x < 1\frac{1}{2}$ **37** $x \leq 1$
38 $x \geq 18$ **39** $x > 4\frac{1}{2}$
40 $x \leq -3\frac{1}{5}$ **41** $x > 4\frac{2}{5}$
42 $x > 3$ **43** $x \leq -7$
44 $x > -5\frac{1}{2}$ **45** $x \leq -5\frac{1}{2}$
46 $-3 \leq x \leq 5$
$-3, -2, -1, 0, 1, 2, 3, 4, 5$
47 $-3 \leq x < 4$ $-3, -2, -1, 0, 1, 2, 3$
48 $2 < x \leq 9$ $3, 4, 5, 6, 7, 8, 9$
49 $-2 < x \leq 4$ $-1, 0, 1, 2, 3, 4$
50 $2 < x < 4$ 3
51 $-2\frac{1}{2} \leq x < 5\frac{1}{2}$
$-2, -1, 0, 1, 2, 3, 4, 5$
52 $0 < x \leq 4$ $1, 2, 3, 4$
53 $3 \leq x \leq 7$ $3, 4, 5, 6, 7$
54 $3 < x \leq 9$ $4, 5, 6, 7, 8, 9$
55 4 **56** 4
57 $x = 1, y = 1$

Excercise 2.8

1 **a** $\frac{4x}{5}$ **b** $\frac{5x}{7}$ **c** $\frac{x}{2}$
d $\frac{x}{2}$ **e** $\frac{2x}{3}$ **f** $\frac{x}{2}$
g $\frac{2x}{5}$ **h** $\frac{2x}{5}$ **i** $\frac{5x}{7}$

2 $\frac{y}{3} + \frac{2y}{5} = \frac{5y}{15} + \frac{6y}{15} = \frac{11y}{15}$

3 **a** $\frac{3x}{4}$ **b** $\frac{7x}{12}$ **c** $\frac{11x}{15}$
d $\frac{17x}{20}$ **e** $\frac{x}{6}$ **f** $\frac{7x}{15}$
g $\frac{x}{2}$ **h** $\frac{5x}{36}$ **i** $\frac{3x}{5}$

4 **a** $\frac{7x+6}{12}$ **b** $\frac{11x-13}{30}$
c $\frac{17x+13}{35}$ **d** $\frac{37x+13}{35}$
e $\frac{17x+33}{12}$ **f** $\frac{46x+64}{33}$

5 **a** $\frac{21x-2}{12}$ **b** $\frac{19x-1}{40}$
c $\frac{9x-8}{30}$ **d** $\frac{21-37x}{36}$
e $\frac{12x+49}{15}$ **f** $\frac{19x-75}{36}$

6 $\frac{34x+127}{60}$ **7** $\frac{3x}{4} - \frac{2x}{3}$

Excercise 2.9

1 **a** 36 **b** 18 **c** 40
d 12 **e** 40 **f** −36
g 10 **h** 42

2 **a** 15 **b** 4 **c** 4 **d** 6
e −4 **f** $\frac{2}{3}$ **g** 6 **h** −2

3 **a** 2 **b** 7 **c** 5 **d** 3
e −3 **f** 6 **g** 4 **h** 8
i −3 **j** −2 **k** $2\frac{1}{2}$
l $\frac{1}{3}$

4 **a** 7 **b** 1 **c** 3 **d** 6
e 4 **f** 1 **g** 5 **h** 7
i −3 **j** −5 **k** $1\frac{1}{2}$
l $\frac{1}{3}$

5 2

6 **a** $2 \times \left(\frac{x+2}{3} + \frac{4x+7}{5}\right) = 20$

$$5(x+2) + 3(4x+7) = 20 \times 3 \times 5 \div 2$$
$$17x + 31 = 150$$

b $x = 7$; 3, 7

7 $5\frac{1}{2}$

Excercise 2.10

1 **a** 3, 2 **b** 2, −5 **c** $\frac{1}{2}, 3$
d $\frac{2}{3}, -1$ **e** 2, 3 **f** −5, 4
g $-\frac{1}{2}, 7$ **h** $3, \frac{1}{2}$

2 **a** $1\frac{1}{2}, 2$ **b** $2, -\frac{2}{3}$ **c** $\frac{2}{5}, -\frac{3}{5}$
d $-\frac{2}{3}, 1\frac{2}{3}$ **e** $-1\frac{1}{2}, 3$
f $\frac{1}{3}, -2$ **g** $2, -1\frac{1}{2}$
h $-\frac{3}{7}, \frac{2}{7}$

3 **a** **i** $1\frac{1}{2}$ **ii** 3
iii $y = 1\frac{1}{2}x + 3$
b **i** $-\frac{1}{4}$ **ii** 4
iii $y = -\frac{1}{4}x + 4$

UNIT 2

c **i** $\frac{1}{2}$ **ii** -1

iii $y = \frac{1}{2}x - 1$

d **i** $-\frac{5}{3}$ **ii** 5

iii $y = -\frac{5}{3}x + 5$

4 **a** $y = -2$ **b** $y = 2x$

c $y = 3x - 2$ **d** $y = \frac{1}{2}x + 6$

e $3x + 4y = 26$ **f** $2y - x = 8$

5 **a**

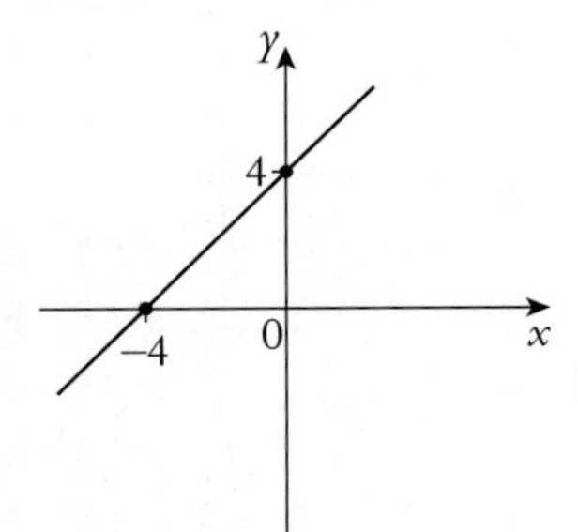

b

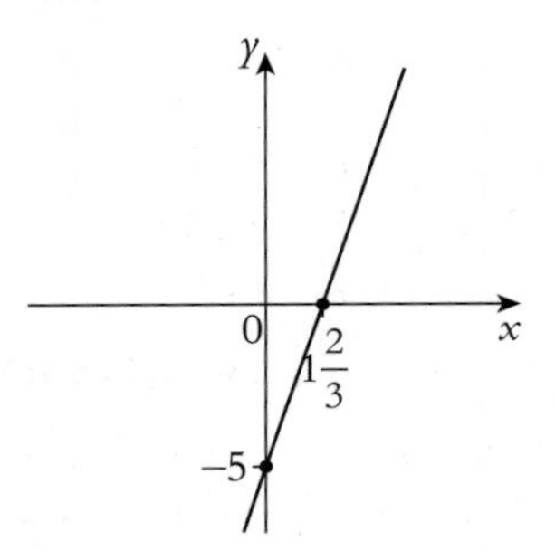

c

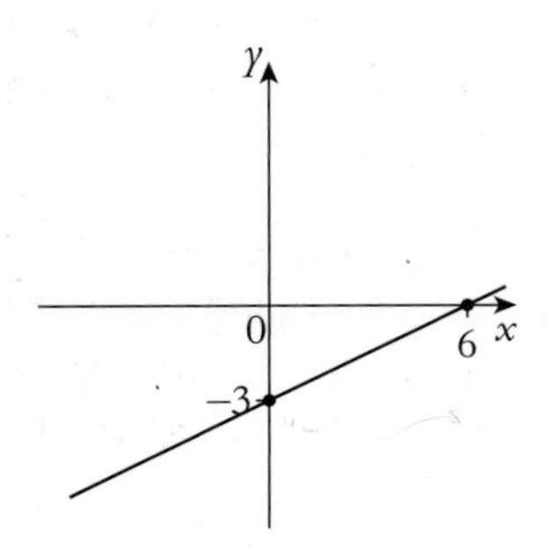

d

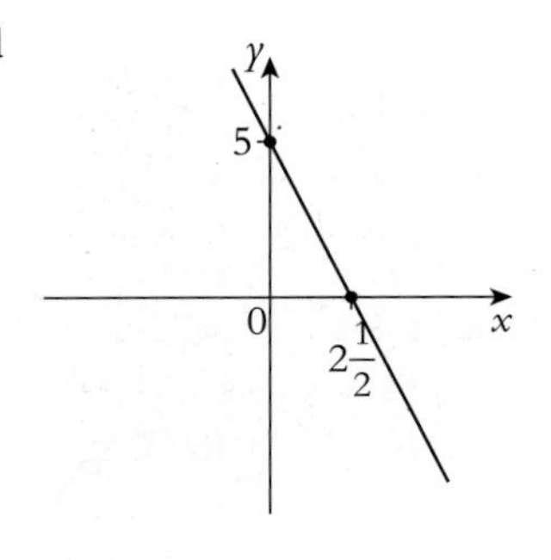

e

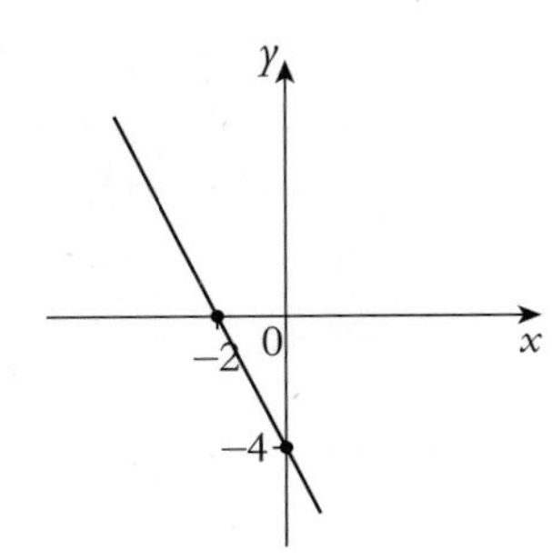

f

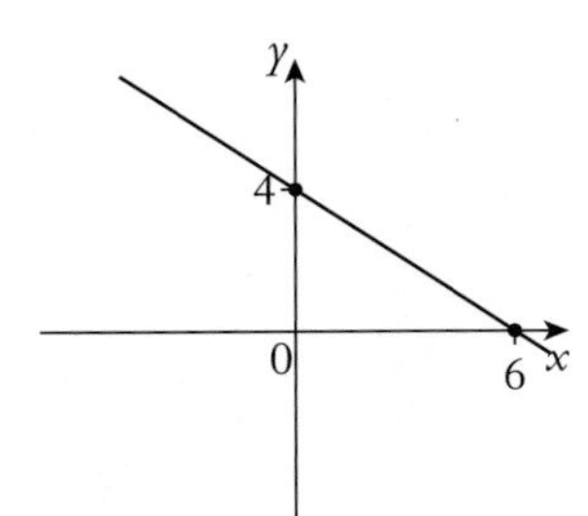

g

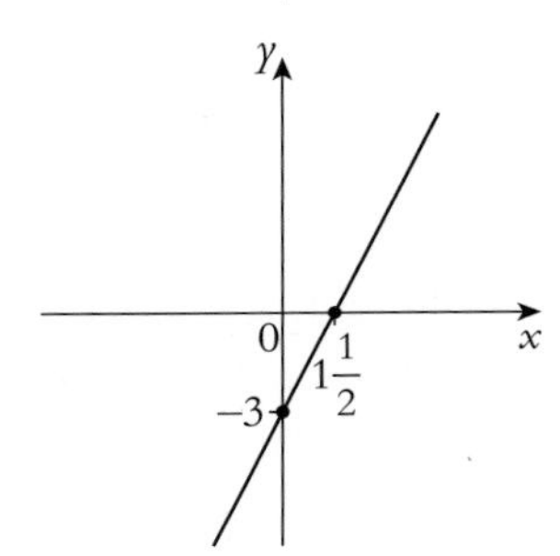

h

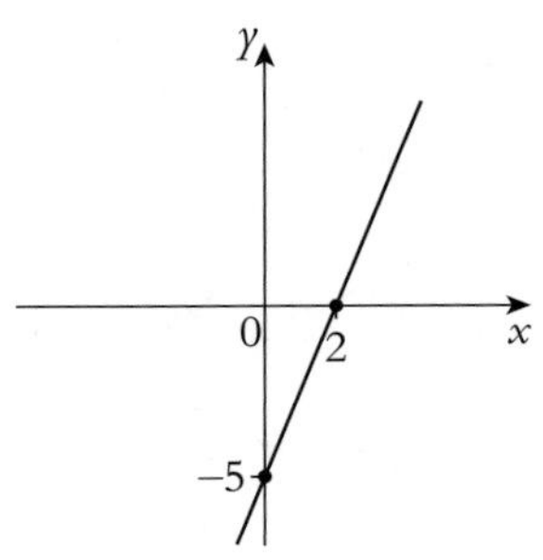

6 **a** E **b** A, E **c** B, C

d D **e** B, D

Excercise 2.11

1 **a** $y = \frac{1}{2}x + 6$ **b** $y = -2x - 3$

c $y = x + 1$ **d** $y = 4x - 4$

e $y = -2x - 2$ **f** $y = \frac{2}{3}x + 5$

g $y = -\frac{3}{5}x - 10$ **h** $y = \frac{2}{5}x + 8$

i $y = \frac{3}{4}x + 5$

2 **a** $y = \frac{1}{2}x + 2$ **b** $y = -3x - 4$

c $y = x + 5$ **d** $y = -2x + 2$

e $y = x - 5$ **f** $y = \frac{2}{3}x - \frac{1}{3}$

g $y = -\frac{1}{5}x + 3\frac{1}{5}$ **h** $y = \frac{1}{4}x - 5$

i $y = \frac{1}{2}x - 1\frac{1}{2}$

Excercise 2.12

1 **a** $(-3, 2)$ **b** $(4\frac{1}{2}, \frac{1}{2})$

c $(2\frac{1}{2}, -1\frac{1}{2})$ **d** $(1\frac{1}{2}, 1\frac{1}{4})$

e $(-2\frac{1}{4}, -1\frac{1}{4})$

2 **a** $(4, 3\frac{1}{2})$ **b** $(-2, 0)$

c $(2\frac{1}{2}, -1)$ **d** $(-4, 5)$

e $(-1\frac{1}{2}, 3\frac{1}{2})$ **f** $(4\frac{1}{2}, -4)$

g $(-3\frac{1}{2}, -4\frac{1}{2})$

h $(-3\frac{3}{4}, -2\frac{1}{4})$ **i** $(2\frac{1}{2}, \frac{3}{4})$

Excercise 2.13

1 **a** $y \le 1$ **b** $x < -2$

c $-1 \le y < 3$ **d** $y \ge x$

e $y > x - 2$ **f** $x + y \le 2$

g $y \ge -\frac{1}{2}x + 1$ **h** $y > 2x - 3$

i $y < -3x + 3$

2 **a**

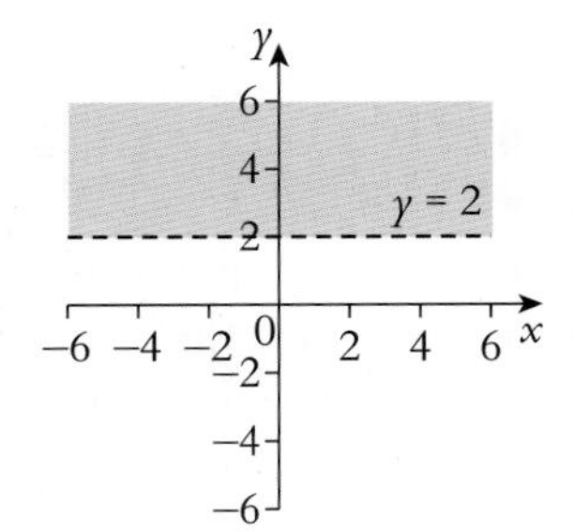

b

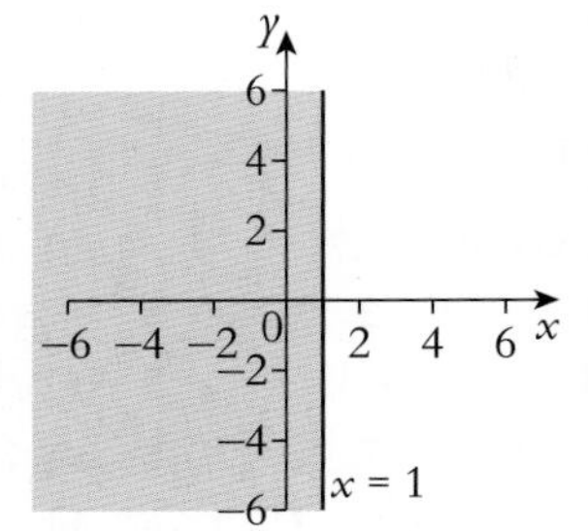

c

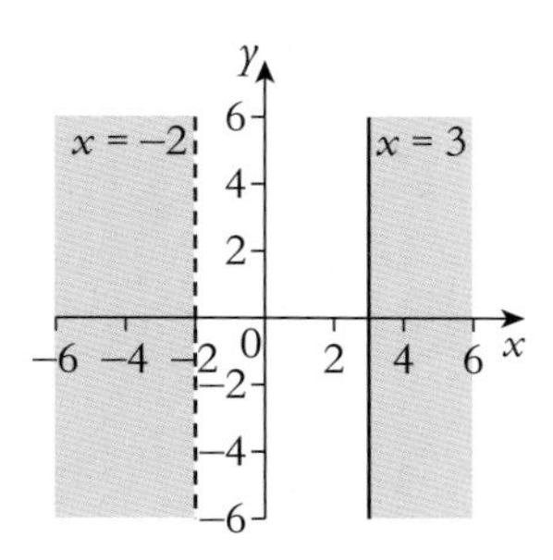

d

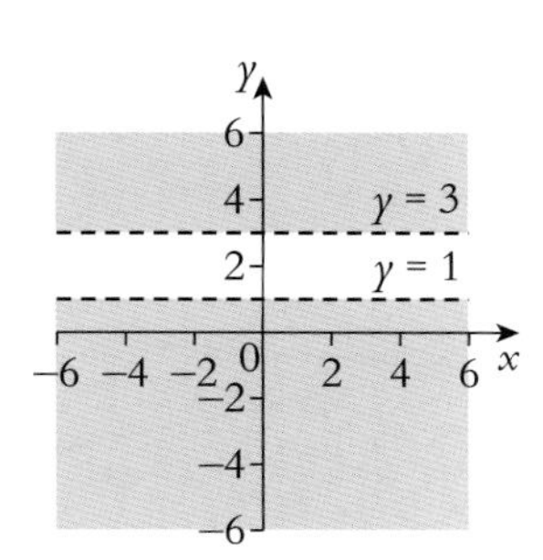

e

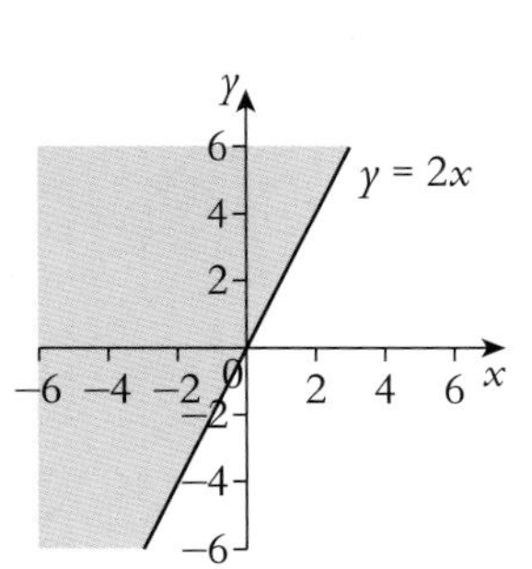

f

$y = \frac{1}{2}x$

g

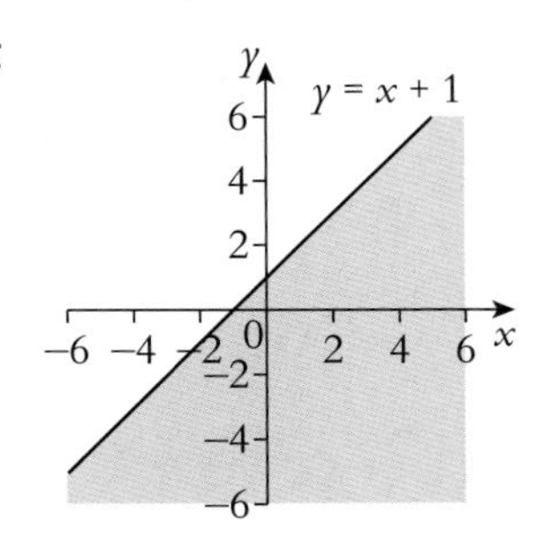

h

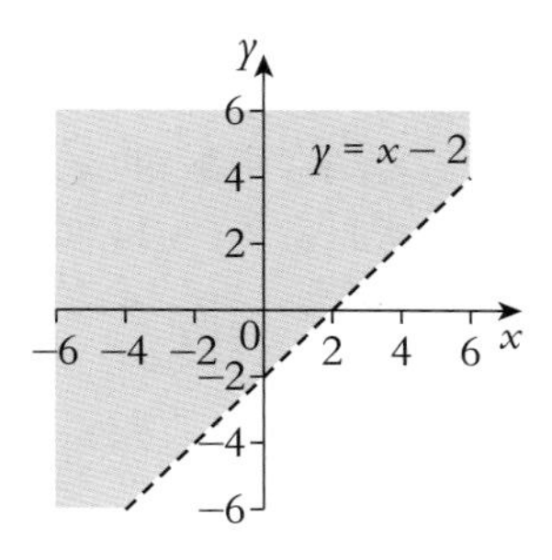

i

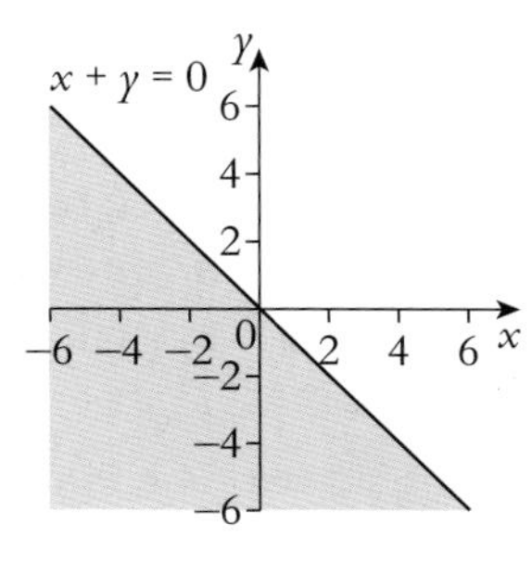

j

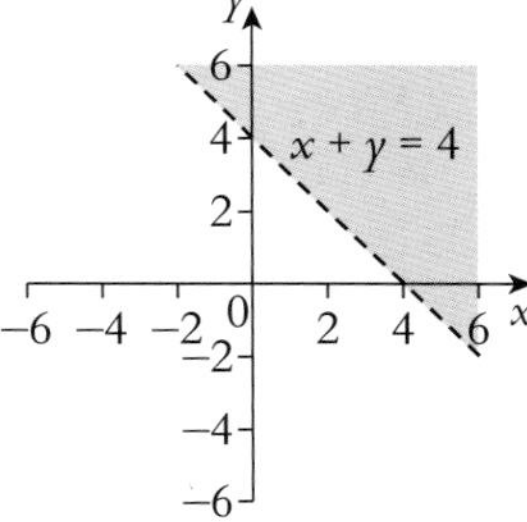
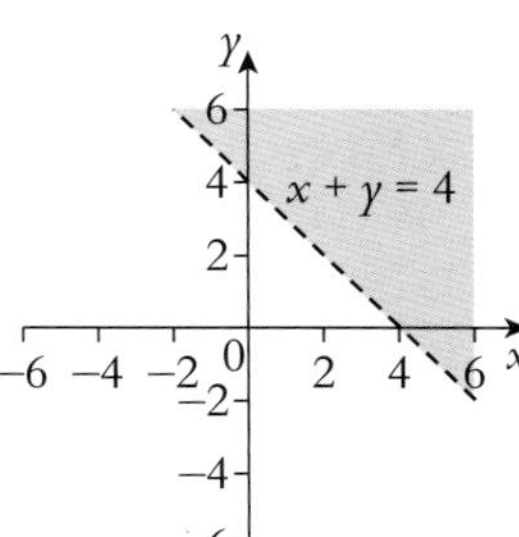

k

$x - 2y = 4$

l

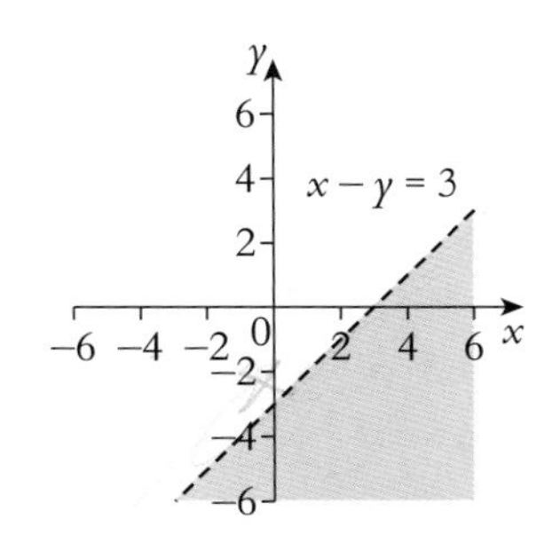

3 a

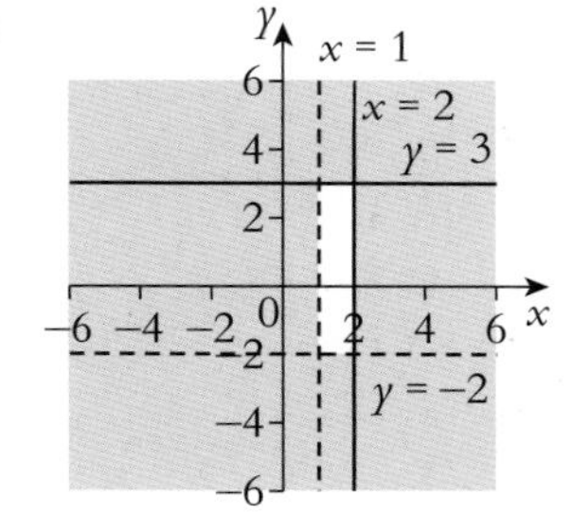

b

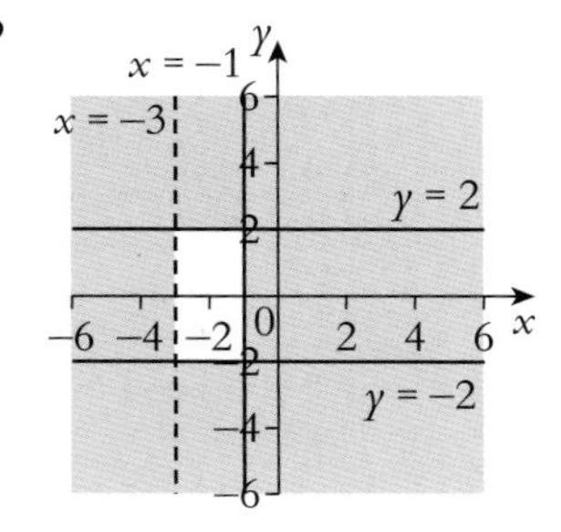

4 a $y \geq 2, \quad y \leq 2x$

b $y \leq 2, \quad y \leq 2x$

c $y \leq 2, \quad y \geq 2x$

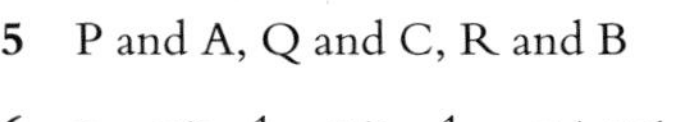

5 P and A, Q and C, R and B

6 a $x > -1, \quad y > -1, \quad x + y \leq 2$

b $x > -1, \quad y > 1, \quad x + y \leq 3$

c $x < -1, \quad y > -1, \quad y < x + 3$

d $x > -1, \quad x + y < 2, \quad y \geq x - 2$

e $x > -2, \quad y \geq -3, \quad y < \frac{1}{2}x + 1$

f $y < 2, \quad y > x, \quad y < -2x - 2$

g $x > -2, \quad y \geq x, \quad x + y < 3$

h $y \leq 4, \quad y > 2x, \quad x + y > 0$

i $x + 2y < 4, \quad y \geq x - 2,$
$y < 3x + 4$

7

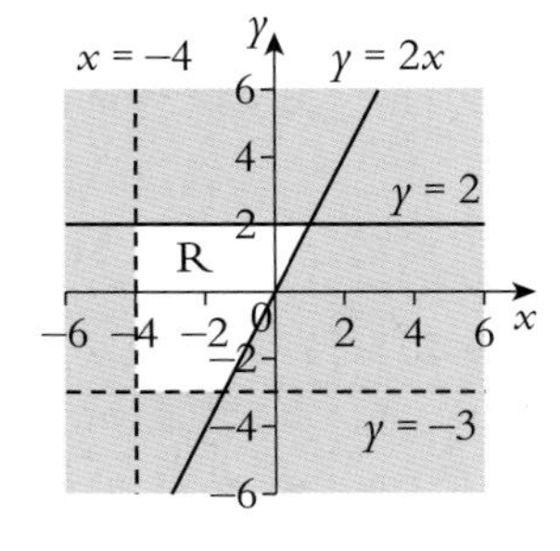

8

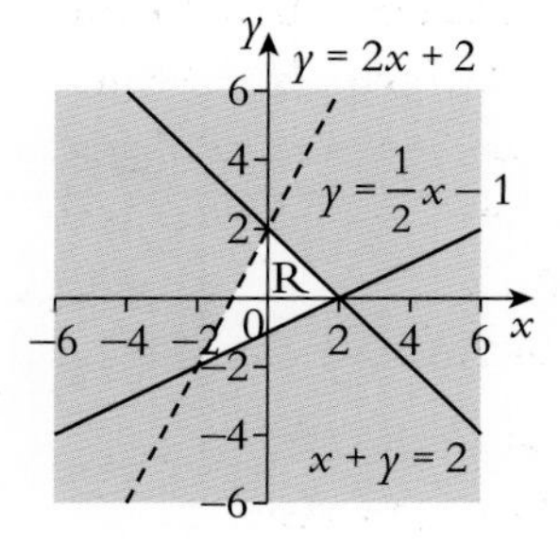

9

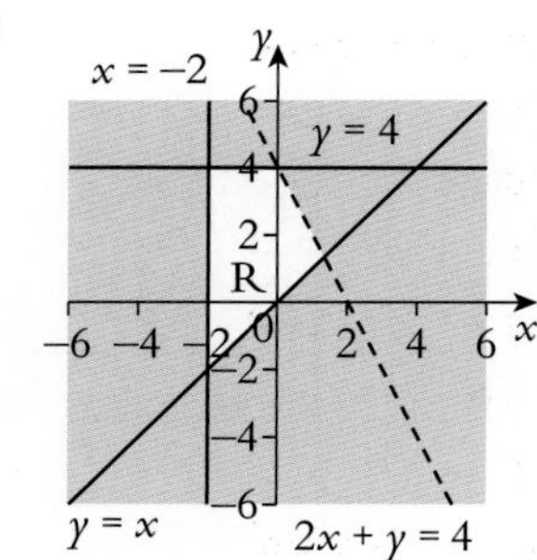

10 **a**

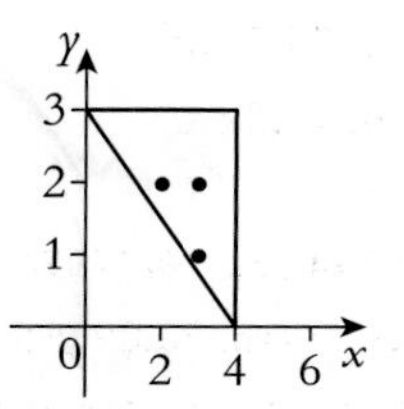

b $x < 4$, $y < 3$, $3x + 4y > 12$

c (2, 2), (3, 1), (3, 2)

Excercise 2.14

1 **a** 36 cm, 62 cm^2
b 24 cm, 30 cm^2
c 36 cm, 84 cm^2
d 42 cm, 84 cm^2

2 **a** 45 cm^2 **b** 40 cm
c 72 cm^2

3 **a** 18 cm^2 **b** 22.5 cm^2
c 45 cm^2 **d** 66.5 cm^2

4 **a** 28 cm^2 **b** 30 cm^2
c 160 cm^2 **d** 49.5 cm^2

5 8 cm

6 8.5

Excercise 2.15

1 **a** 8.06 cm **b** 10.4 cm
c 8.25 cm **d** 6.62 m
e 6.80 m **f** 6.96 cm

2 $x^2 + 4^2 = 6^2$, $x^2 = 20$, $x = 4.47$ cm

3 **a** 6.93 cm **b** 5.66 cm
c 13.3 cm **d** 8.08 cm
e 19.0 cm **f** 14.4 cm

4 38.1 km

5 4.77 m

6 **a** 3.54 cm **b** 8.94 cm
c 2.83 cm

7 7.81 cm

8 8.94 cm

9 10.6 cm

10 **a** 5 cm **b** 22 cm
c 26 cm^2 **d** 10 cm^2

11 **a** 4.33 cm **b** 10.8 cm^2

12 6.63 cm

13 He is incorrect. $4^2 + 5^2 \neq 6^2$

14 Yes. $48^2 + 55^2 = 73^2$

15 5 miles

16 5.29 cm

17 17 cm

18 4.85 cm

19 0.828 cm

Excercise 2.16

1 **a** 5.39 **b** 3.61 **c** 5
d 7.07

2 **a** 8.06 **b** 15.3 **c** 15.6
d 8.54 **e** 11.7 **f** 13.9
g 9.49 **h** 3.16 **i** 4.12

Excercise 2.17

1 **a** 5.39 m **b** 8.31 m
c 12.3 m **d** 12.1 m

2 14.8 cm

3 11.1 cm

Excercise 2.18

Check students' drawings.

5 (5, 5.5)

7 15.2 cm

9 6.95 cm

10 (5, 6)

Excercise 2.19

1

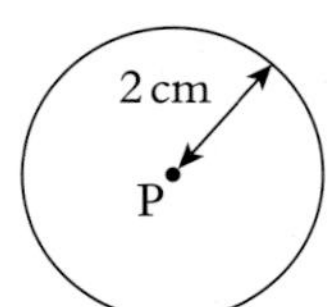

2

3

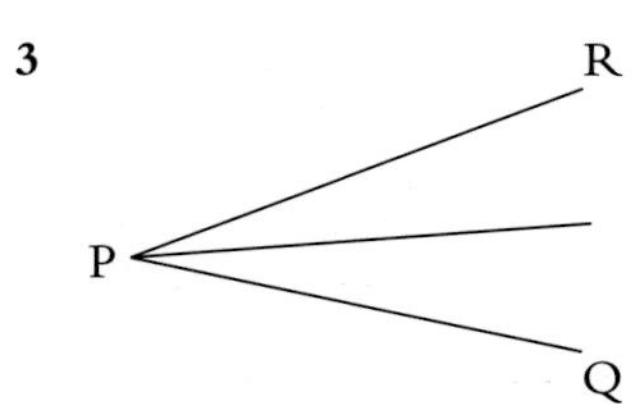

4

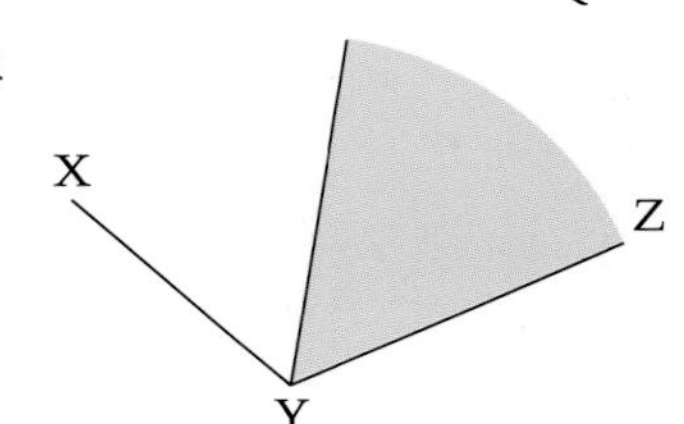

5

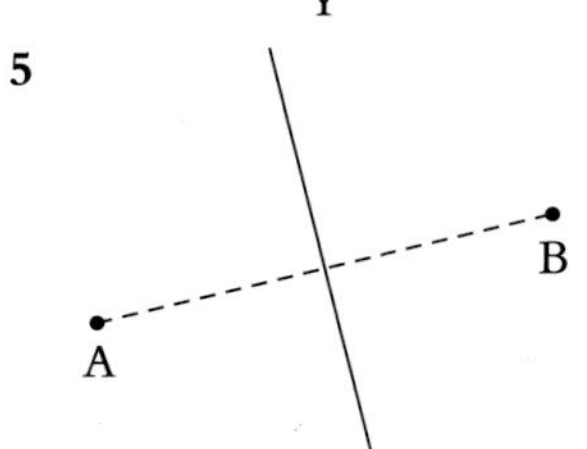

6

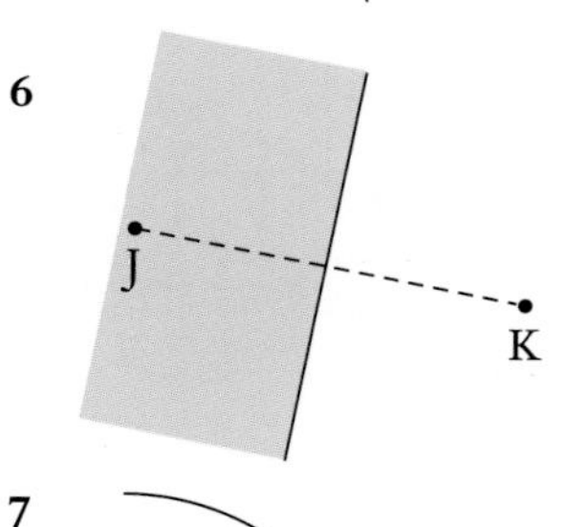

7

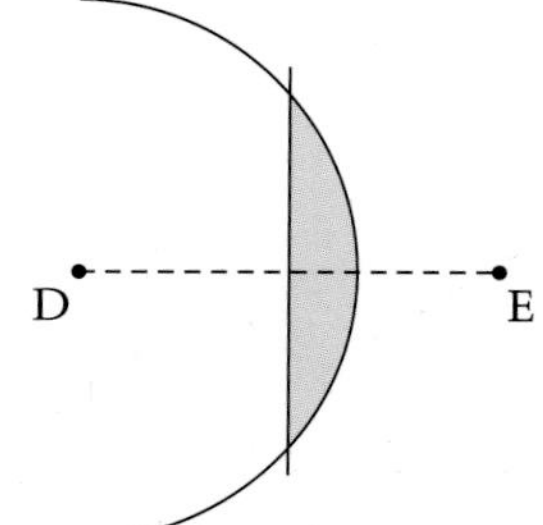

8

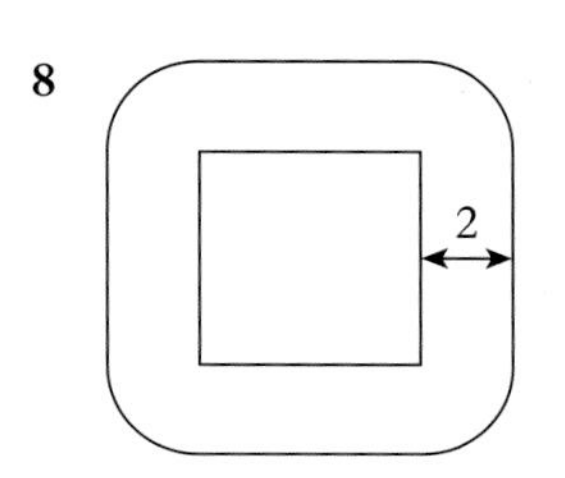

9

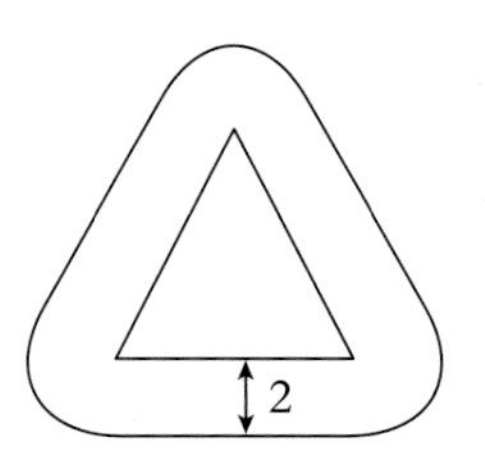

10

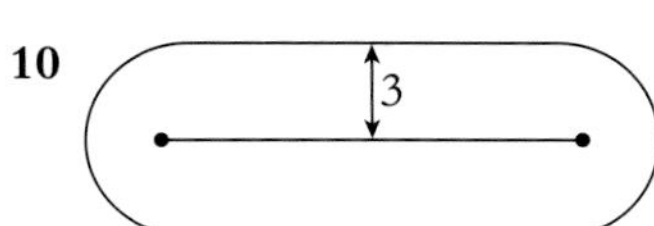

11

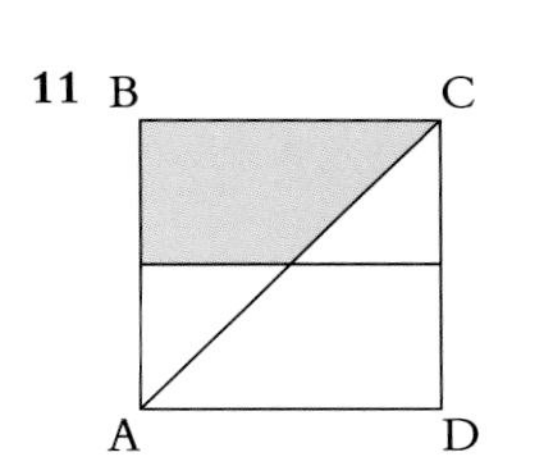

12

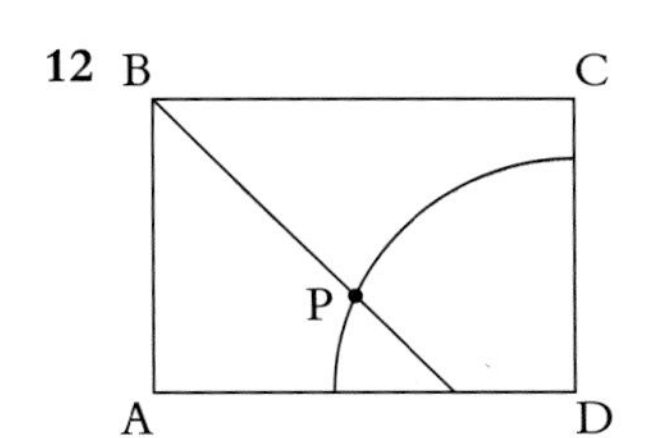

13

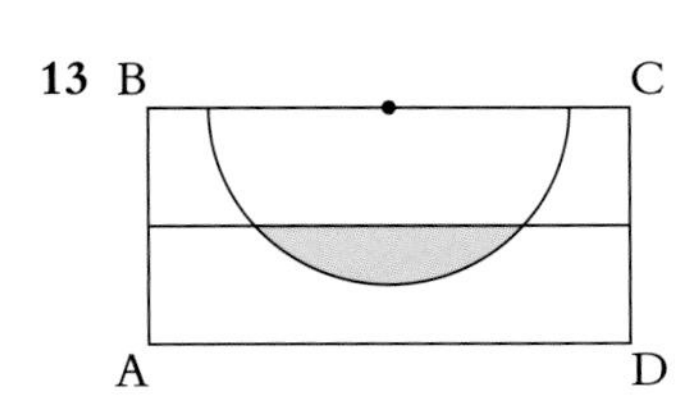

14

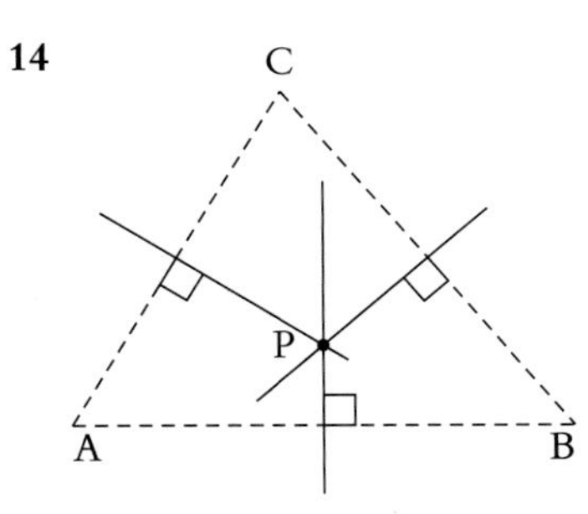

15

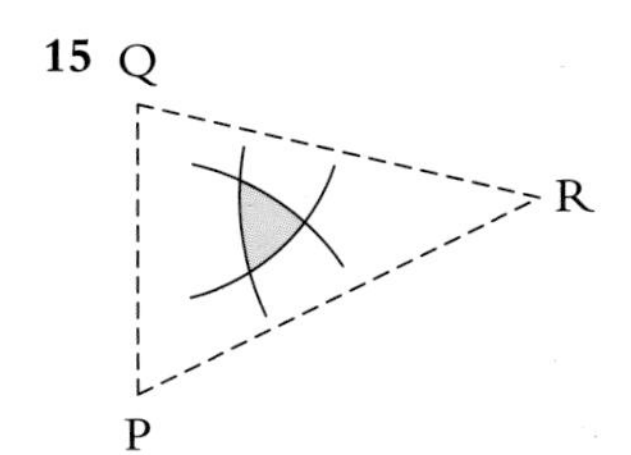

16

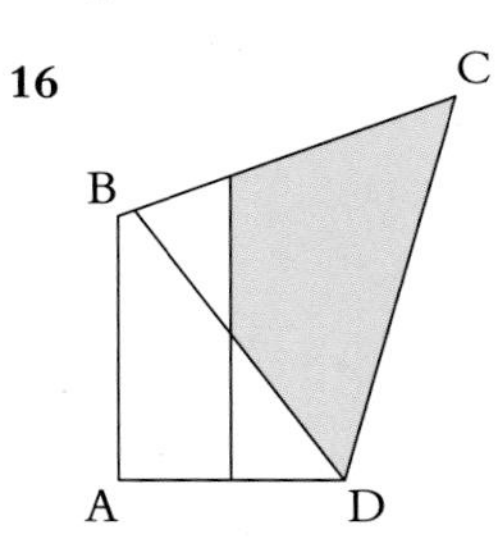

17

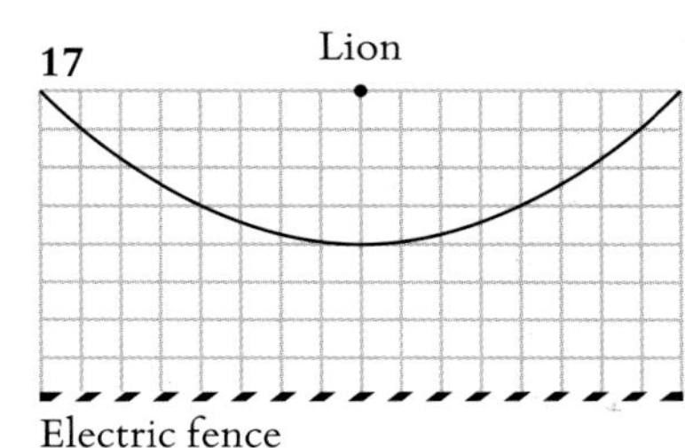

18

$r = 2$, 4, 8, $r = 6$

19 a

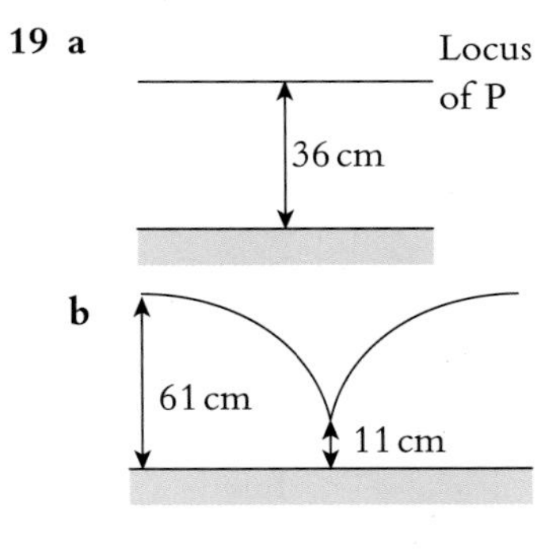

Excercise 2.20

1 a 18.8 cm, 28.3 cm^2
  b 47.1 cm, 177 cm^2
  c 15.1 cm, 18.1 cm^2
  d 23.9 cm, 45.4 cm^2
  e 126 cm, 1260 cm^2
  f 29.5 cm, 69.4 cm^2

2 a 163 cm b 8.17 km

3 a 0.631 cm b 3.96 m

4 a 30.9 cm^2 b 30.9 cm^2
  c 30.9 cm^2 d same

5 50.3 cm^2

6 114 cm^2

Excercise 2.21

1 15.4 cm, 14.1 cm^2

2 25.0 cm, 38.5 cm^2

3 13.4 cm, 9.42 cm^2

4 22.9 cm, 34.8 cm^2

5 23.4 cm, 28.1 cm^2

6 30.3 cm, 44.6 cm^2

7 24.6 cm, 11.4 cm^2

8 40.6 cm, 51.8 cm^2

9 31.4 cm, 21.5 cm^2

10 25.1 cm, 25.1 cm^2

11 14.9 cm, 15.8 cm^2

12 28.6 cm

13 a 1.9544 cm, 10.04878 cm
  b 3.9088 cm, 20.09756 cm
  c When the area quadruples, the lengths double.

14 a 2.523 cm, 9.010 cm
  b 7.569 cm, 27.03 cm
  c When the area is multiplied by 9, the lengths treble.

15 35.5 cm^2

16 7.78 cm

17 4 cm

Excercise 2.22

1 a 3.67 cm b 19.5 cm
  c 6.14 cm d 32.1 cm

2 a 6.68 cm^2 b 22.3 cm^2
  c 127 cm^2 d 159 cm^2

3 a 9.59 cm, 5.59 cm^2
  b 65.1 cm, 61.1 cm^2
  c 40.3 cm, 64.5 cm^2
  d 39.2 cm, 49.7 cm^2

4 76.4°

5 4.09 cm

6 a 40.9 cm b 77.4 cm^2

7 0.161 cm^2

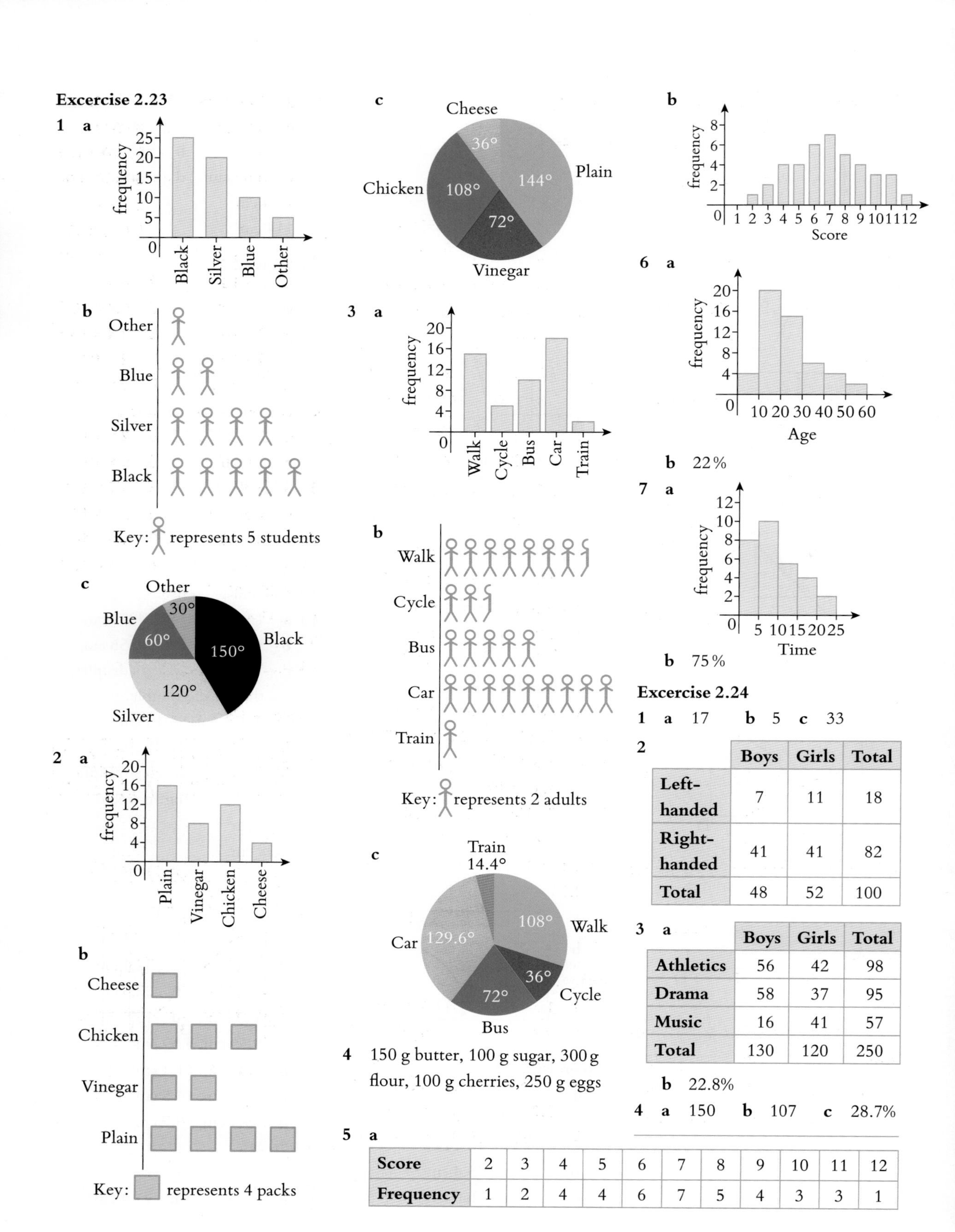

Excercise 2.23

1 a

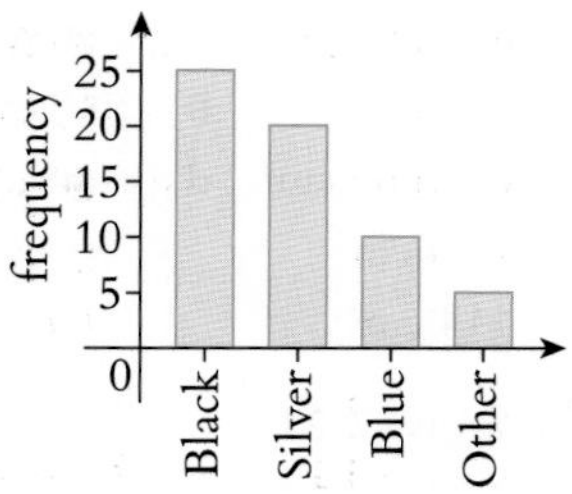

b

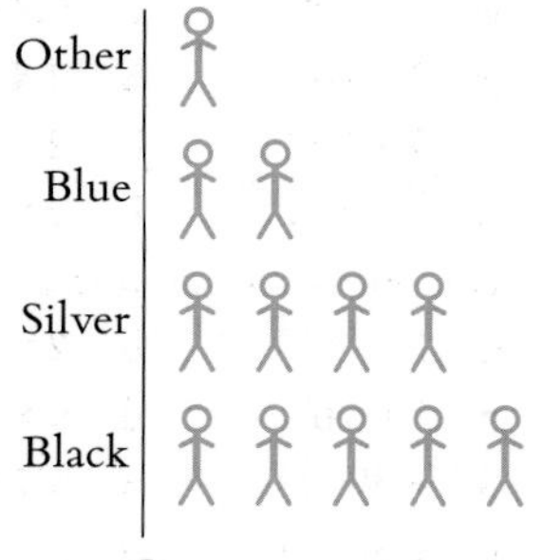

Key: represents 5 students

c

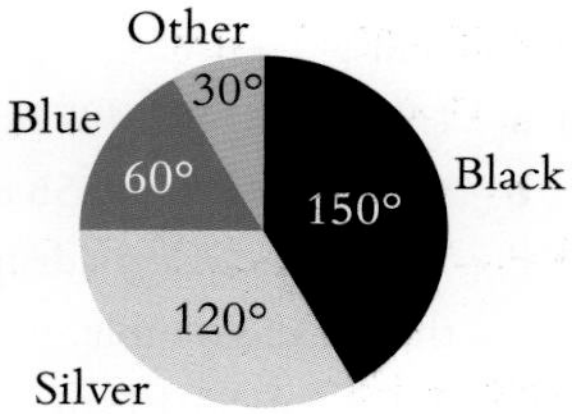

2 a

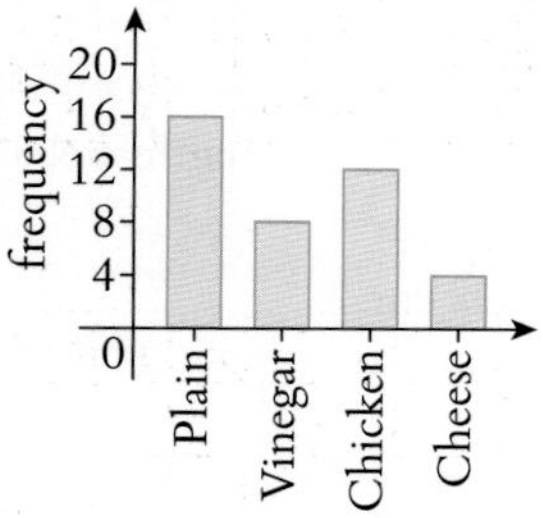

b

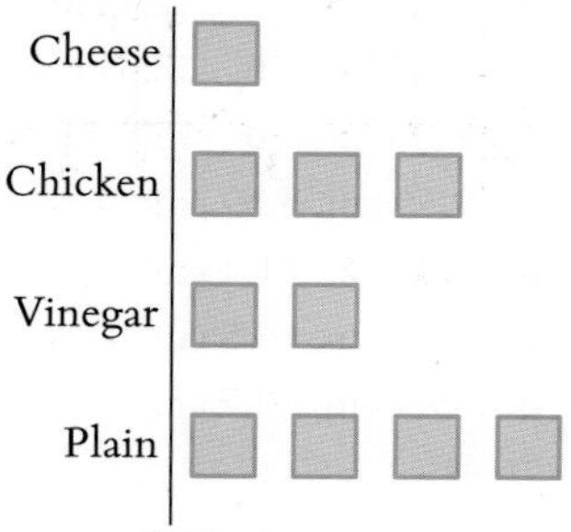

Key: represents 4 packs

c

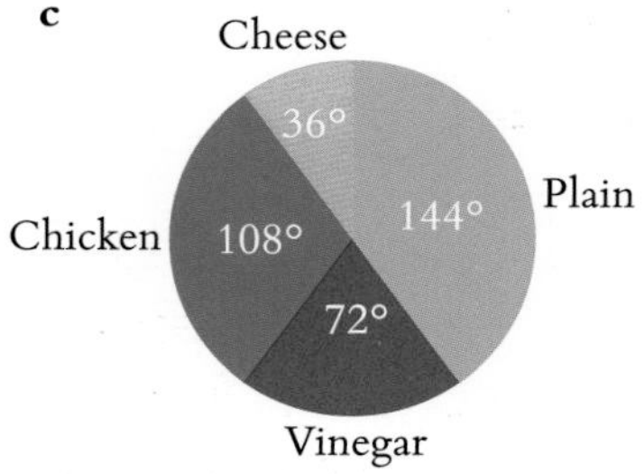

3 a

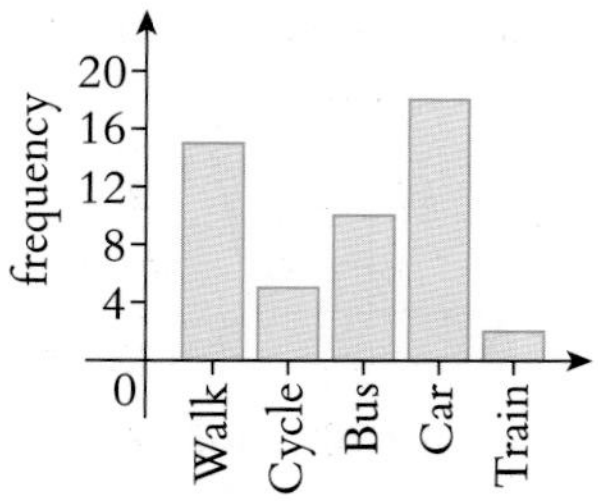

b

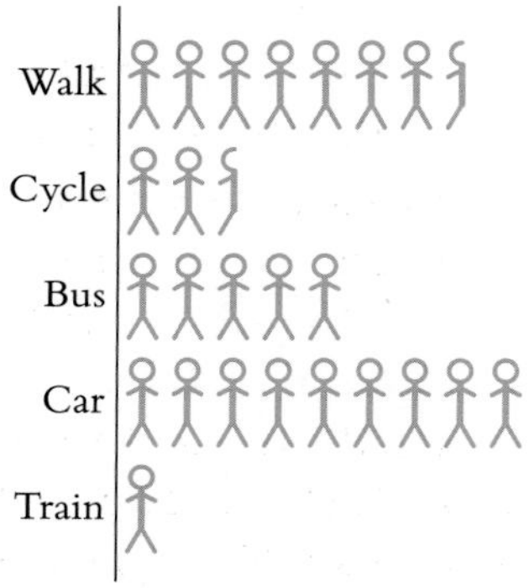

Key: represents 2 adults

c

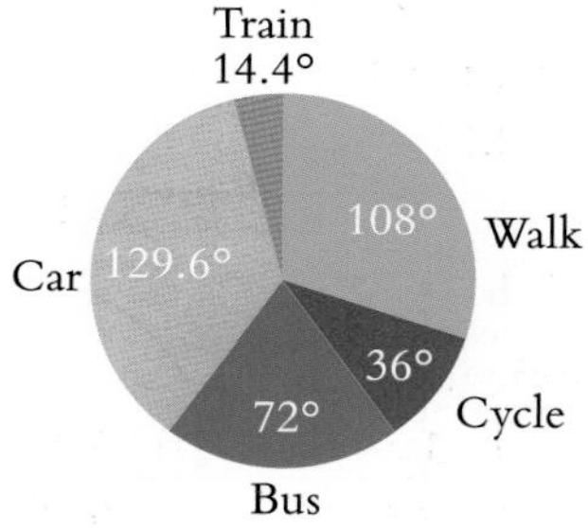

4 150 g butter, 100 g sugar, 300 g flour, 100 g cherries, 250 g eggs

5 a

Score	2	3	4	5	6	7	8	9	10	11	12
Frequency	1	2	4	4	6	7	5	4	3	3	1

b

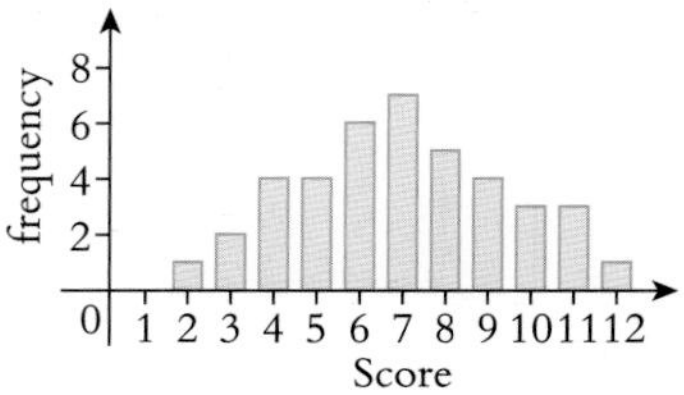

6 a

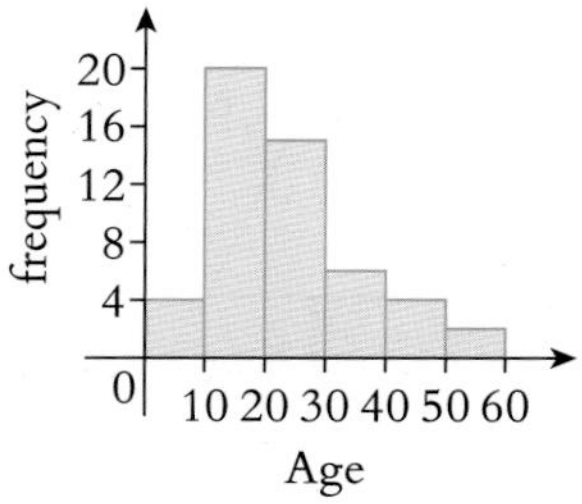

b 22%

7 a

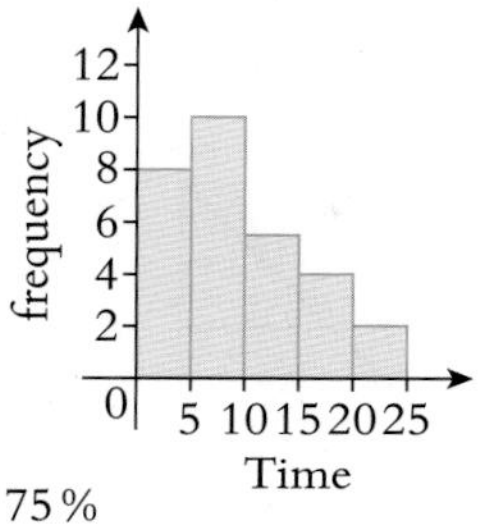

b 75%

Excercise 2.24

1 a 17 **b** 5 **c** 33

2

	Boys	**Girls**	**Total**
Left-handed	7	11	18
Right-handed	41	41	82
Total	48	52	100

3 a

	Boys	**Girls**	**Total**
Athletics	56	42	98
Drama	58	37	95
Music	16	41	57
Total	130	120	250

b 22.8%

4 a 150 **b** 107 **c** 28.7%

Examination-style questions

1 \$15 480 **2** 27 540

3 3.5 **4** \$325, \$175

5 \$1635 **6** 3.2 km

7 a $x^5 \div x^2$ **b** x^{12}

8 a $2\frac{3}{8}$

b i 7 **ii** $32x^2$

9 −4 **10** $x < -2.5$

11 a $x \leq 4\frac{3}{7}$ **b** 1, 2, 3 and 4

12 $\frac{13x}{24}$ **13** $10\frac{1}{4}$

14 $y = 2x - 3$

15 a (1, 3) **b** 13.4 or $6\sqrt{5}$

c $y = -2x + 5$ **d** $y = -2x - 1$

16 9.25 m² **17** 9.5

18

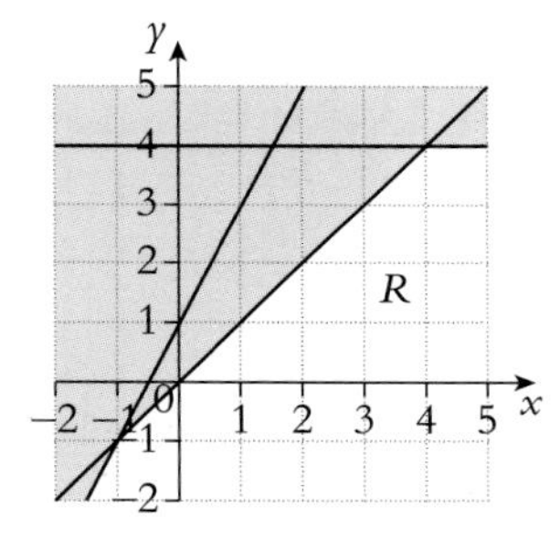

19 a and **b**

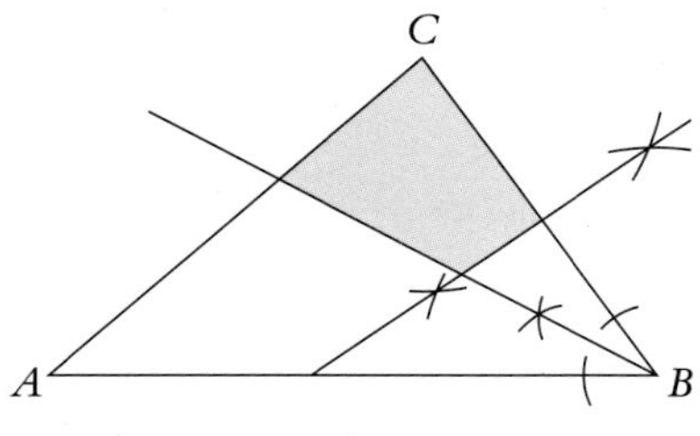

20 7.07 cm **21** 13.2 cm

22 a 78.5 cm² **b** 62.8 cm

23 16.58 cm

24 a Grade B **b** 99°

25 a

	Walked	Cycled	Bus	Total
Boys	8	10	**7**	25
Girls	**17**	7	5	**29**
Total	**25**	**17**	**12**	54

b $33\frac{1}{3}\%$

Unit 3

Exercise 3.1

1 a, c, f, g

2 a 5×10^4 **b** 2×10^2

c 7×10^5 **d** 5.33×10^3

e 8.2×10^7 **f** 3.02×10^3

g 6.66×10^5 **h** 3.636×10^1

3 a 480 000 **b** 370

c 564 000 000 **d** 70 600 000

e 20 000 **f** 1 100 000

g 300.1 **h** 9 500 000 000

4 a 6×10^3 **b** 4.5×10^5

c 3×10^4 **d** 4.1×10^6

e 3.2×10^9 **f** 6.3×10^7

g 4×10^3 **h** 2.5×10^7

i 1.6×10^{13} **j** 5×10^4

k 4.2×10^2 **l** 4.87×10^2

5 Answers as for question **4**.

6 a 6.2×10^9 **b** 5.8×10^9

c 1.1×10^{10} **d** 1.2×10^{18}

e 3×10^1 **f** 3.6×10^{19}

g 3.1×10^1 **h** 9×10^{10}

7 a 30 : 1 **b** 1 : 40 **c** 500 : 1

8 a 9.6×10^4 **b** 2.1×10^4

c 1.46×10^6

9 a 2×10^4 **b** 3×10^2

c 6×10^3 **d** 7×10^4

10 2.13×10^4, 5.8×10^{13}, 2.04×10^{14}, 2.2×10^{14}

11 1.49×10^8 **12** 5 : 12

13 1.2995×10^9

Exercise 3.2

1 a 1.5×10^{-2} **b** 6.7×10^{-4}

c 9×10^{-4} **d** 6.06×10^{-4}

e 3.2×10^6 **f** 5.2×10^{-4}

g 4.4×10^{-1} **h** 5.308×10^7

i 5.7×10^{-8} **j** 6.13×10^{-4}

k 7.002×10^5 **l** 8.9×10^{-4}

2 a 0.000088 **b** 0.00000023

c 0.006 **d** 0.0511

e 99 000 **f** 0.00000104

g 0.00068 **h** 600 000 000

i 0.38 **j** 22

k 408 000 **l** 0.0000095

3 a 6×10^{-2} **b** 7.2×10^{-5}

c 5×10^{-8} **d** 6×10^{-4}

e 3×10^{-9} **f** 6.4×10^{-7}

g 3×10^{12} **h** 2.4×10^{-2}

i 9×10^{-10} **j** 2×10^6

k 2.7×10^3 **l** 2×10^{-8}

4 Answer as for question **3**.

5 a 3×10^{-3} **b** 2×10^{-5}

c 8×10^{-3} **d** 9×10^{-7}

6 a 6.3×10^{-8} **b** 5.7×10^{-8}

c 1.86×10^{-7} **d** 1.8×10^{-16}

e 2×10^1 **f** 3.6×10^{-15}

g 9×10^{-18} **h** 1.5×10^{-6}

7 8.35×10^{-22} grams

8 1.87×10^{23}

9 1.5×10^{-5} seconds

10 2.625×10^{-5} kg

Exercise 3.3

1

x	0	2	4
y	2	4	6

x	0	2	4
y	4	2	0

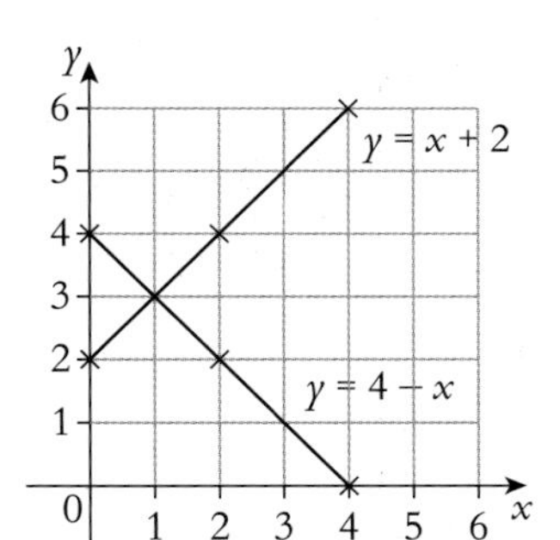

$x = 1, y = 3$

UNIT 3

2

x	0	2	4
y	1	2	3

x	0	4
y	5	0

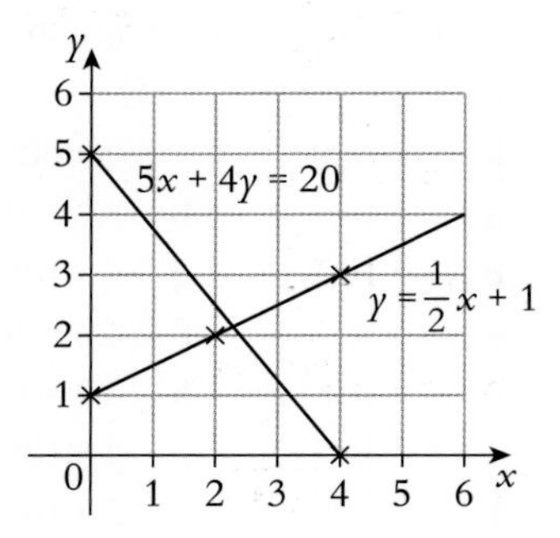

$x = 2\frac{2}{7}, y = 2\frac{1}{7}$

Exercise 3.4

1 a $x = 3, y = 4$
b $x = 5, y = 2$
c $x = 10.5, y = -2$
d $x = 4, y = 4$
e $x = 2, y = -1$
f $x = -3, y = 4$
g $x = -6, y = -2$
h $x = 5, y = -2$

2 a $x = 2, y = 4$
b $x = 9, y = 5$
c $x = 2, y = 10$
d $x = 7, y = 3$
e $x = 5, y = -2$
f $x = -3, y = 4$
g $x = 7, y = -5$
h $x = -2\frac{1}{2}, y = -6$

3 a $x = 6, y = 2$
b $x = 3, y = 3$
c $x = 5, y = -2$
d $x = 1\frac{1}{2}, y = 3$
e $x = -4, y = 2\frac{1}{2}$
f $x = 2, y = 1\frac{1}{3}$
g $x = -1, y = -2\frac{1}{4}$
h $x = -4, y = 3\frac{1}{2}$

4 $x + y = 15, x - y = 4$;
$x = 9\frac{1}{2}, y = 5\frac{1}{2}$

5 a 39, 15 b 15, −7

6 $x = 11, y = 2$

7 $x = 2.92, y = 1.8$

8 \$0.85, \$0.32

9 $1 = -2m + c, 4 = 4m + c$;
$m = \frac{1}{2}, c = 2$

10 \$39.68

Exercise 3.5

1 a $x = 3, y = 5$
b $x = 4, y = 1$
c $x = 6, y = 4$
d $x = 3, y = 7$
e $x = 5, y = -2$
f $x = 10, y = -3$
g $x = 1.625, y = 7.75$
h $x = 3, y = -2.5$

2 a $x = 4, y = 7$
b $x = 3, y = 2$
c $x = 5, y = 3$
d $x = 2, y = 2$
e $x = 8, y = 6$
f $x = 11, y = 4$
g $x = 3, y = 9$
h $x = 1, y = 7$

3 a $x = -\frac{1}{2}, y = 4$
b $x = 3, y = -5$
c $x = -2, y = 7$
d $x = -3, y = -2$
e $x = 2\frac{1}{2}, y = 3$
f $x = 6, y = 1\frac{1}{2}$
g $x = 3\frac{1}{2}, y = -5$
h $x = -9, y = 4$

4 a $x = 4, y = 2$
b $x = 2, y = 1$
c $x = 4, y = 2$
d $x = \frac{5}{7}, y = 4\frac{3}{7}$
e $x = 2, y = -3$
f $x = -1, y = 2$
g $x = -2, y = 1$
h $x = 1, y = 2$

5 a $x = 12, y = 15$
b $x = 20, y = -16$
c $x = -18, y = 10$
d $x = 24, y = -6$

6 $x = 3, y = 7$

7 $x = 4, y = 2.5$

8 17

9 153

10 \$1.85

11 \$0.19

12 $a = 3, b = 5$

13 $a = 2, b = -3$

14 a $x = 3, y = -1$ b 28
c 40

15 a $x = 1\frac{1}{2}, y = 4$ b 20
c 25

16 a $x = 4, y = 3$ b 24
c 27.7

Exercise 3.6

1 a $6(x + y)$ b $5xy(1 - 2y)$
c $2(2x + 3y)$

2 a $4(x + 2)$ b $3(x - 2)$
c $2(4x + 1)$ d $5(2x - 3)$
e $7(y - 2)$ f $3(3y + 4)$
g $4(5 - 2x)$ h $12(3 - y)$
i $7(6 - x)$ j $2(9y - 2)$
k $8(3x + 2)$ l $11(2y - 5)$

3 a $x(x + 5)$ b $x(x - 8)$
c $y(y - 4)$ d $a(a + 9)$
e $4x(x - 2)$ f $3y(y - 3)$
g $2x(5x - 2)$ h $2y(4 - y)$
i $5y(3y + 1)$ j $7x(2 - x)$
k $2x(3 - 2x)$ l $3y(4y - 3)$

4 a $4(2\pi + 1)$ b $3(5 - 3\pi)$
c $\pi r(r + l)$ d $2\pi r(r + h)$

5 a $3xy(2x - 1)$
b $5x^4 y(2y - 1)$
c $8xy(y^2 - 4x^2)$
d $3x^2y^2(5x - 4y^2)$
e $3xy(4x - 5)$
f $7pq(4q - 5p)$
g $3ab(2a + 3)$
h $7x^2y(3x^3 - 4)$
i $7ab^5(8a^3b^2 - 5)$

6 a $3(a + b - c)$
b $5(x + y + x^2)$
c $a(2x - y - 3y^2)$
d $x(x^2 + 4x - 2)$
e $y^2(y^2 + 6y - 6)$

f $pqr(pq - 1 + q)$
g $4ab(5abc^2 - 2ab + 4c^2)$
h $3x^2y(5xy - 9 + 7y)$
i $2p^3q^3r^2(4p^4q^2r + 3p - 2r)$

7 a $(x + 3)(y + 2)$
b $(a + 2)(b - 4)$
c $(p - 5)(q + 3)$
d $(x - 3)(y - 5)$
e $(x + y)^3$
f $(a + b)^2(a + b - 1)$
g $(x - y)^2(1 - x + y)$
h $(x + y)(5 + x + y)$
i $(x - y)^2(x - y - 4)$

8 a $(y + 3)(x + 1)$
b $(5 + x)(y - 2)$
c $(4 + c)(a + b)$
d $(p + 3)(q + 2r)$
e $(3 - y)(x + 4)$
f $(x - 4)(y - 3)$

Exercise 3.7

1 a $x - 2$ b $\frac{x+2}{4}$
c $\frac{3x+4}{5}$

2 a $x + 2$ b $4y + 2$
c $x + 2$ d $3y + 2$
e $\frac{3x-6}{4}$ f $\frac{3x+2}{4}$
g $\frac{3-4y}{2}$ h $\frac{6-4a}{5}$

3 a $x + 9$ b $y - 8$
c $\frac{x-3}{x}$ d $1 - x^2$
e $\frac{2x+y}{z}$ f $\frac{x+y}{z}$
g $\frac{x+y}{y}$ h $\frac{x-2}{y}$

4 a 3 b 5 c $\frac{1}{6}$ d $\frac{1}{5}$
e $\frac{2}{5}$ f $\frac{3x}{7}$ g $\frac{x}{z}$ h $\frac{2}{3}$

5 a x b $\frac{5}{x-5}$ c $\frac{x}{5}$
d $\frac{2y}{5}$ e $\frac{x}{x-1}$ f $\frac{3}{(y-2)^2}$
g x h $\frac{xy}{2}$

6 a $3\frac{1}{3}$ b $4\frac{2}{3}$
c $\frac{x^2}{y^2}$ d $\frac{x+1}{y+1}$
e $\frac{7(x+y)}{y}$ f $\frac{2f}{g}$

7 a $\frac{5xy}{12}$ b $\frac{9}{10}$

Exercise 3.8

1 a $x = \frac{p+q}{3}$ b $x = \frac{8b-a}{2}$
c $x = (y - b)^2$

2 a $x = y - 3$ b $x = 5y + 6$
c $x = 4 - y$ d $x = a + b$
e $x = y + 8$ f $x = 5 - 2y$
g $x = 6y - 8$ h $x = y - 10$

3 a $x = \frac{y}{6}$ b $x = \frac{y}{3}$
c $x = \frac{y-4}{3}$ d $x = \frac{7-5y}{a}$
e $x = 2p$ f $x = 15 + 5d$
g $x = 2y - 16$
h $x = 10a - 4ay$
i $x = \frac{3y}{2}$ j $x = \frac{15y}{4}$
k $x = \frac{7(a+b)}{3}$
l $x = \frac{b(c+d)}{a}$

4 a $x = \frac{d-c}{2}$ b $x = \frac{y+2z}{5}$
c $x = \frac{6-3y}{4}$ d $x = \frac{3b+c+d}{8}$
e $x = \frac{d-b}{a}$ f $x = \frac{c+cd}{b}$
g $x = \frac{a+b+q}{p}$ h $x = \frac{y-8}{y}$

5 a $x = \frac{2}{y+z}$ b $x = \frac{p}{2q+3}$
c $x = \frac{a-3y}{6}$ d $x = \frac{p+qr}{q}$

6 a $x = bc - a$ b $x = 3gh + 2f$
c $x = \frac{cd-by}{a}$ d $x = \frac{p-mr}{q}$
e $x = \frac{p}{q}$ f $x = \frac{g}{h}$
g $x = \frac{p+q}{r}$ h $x = \frac{p-q}{5}$

7 a $x = \pm\sqrt{\frac{b}{a}}$ b $x = \pm\sqrt{\frac{q}{p}}$
c $x = \pm\sqrt{\frac{c}{ab}}$ d $x = \pm\sqrt{\frac{bc}{a}}$
e $x = \pm\sqrt{b-a}$
f $x = \pm\sqrt{d+e}$
g $x = \pm\sqrt{\frac{c-b}{a}}$
h $x = \pm\sqrt{\frac{4r+5q}{p}}$
i $x = \pm\sqrt{4(b+a)}$
j $x = \pm\sqrt{a(d-bc)}$
k $x = \pm\sqrt{g(f-3h)}$
l $x = \pm\sqrt{2(3fg-g)}$

8 a $x = b^2 - a$ b $x = q^2r^2 + p$
c $x = \frac{y+16}{2}$ d $x = \frac{z^2+4y}{4}$

9 They are both correct.

Exercise 3.9

1 a not congruent
b congruent
c congruent
d not congruent
e not congruent
f congruent

2 a SAS b RHS c ASA
d SAS e SSS f RHS

For questions **3** to **5** other answers are possible.

3 SSS 5 SSS 6 RHS

Exercise 3.10

1 B and C
2 B and C
3 a $x = 12, y = 4$
b $x = 3.6, y = 12.5$
c $x = 15, y = 16.5$
d $x = 4, y = 4$
e $x = 8.75$
f $x = 15, y = 17.5$

Exercise 3.11

1 **a** $x = 5, y = 12$
   b $x = 8.4, y = 2.3$

2 $w = 33\frac{1}{3}$

3 **a** 4 **b** 12 **c** 2 **d** $3\frac{1}{3}$

4 54.5 m

5 **a** $x = 8, y = 20$
   b $x = 12, y = 5.5$

6 **a** D = (5, 4), E = (14, 7)
   b $(9\frac{1}{2}, 5\frac{1}{2})$

7 **a** (−3, −1) **b** (0, −3)
   c $(-7\frac{1}{2}, 2)$

8 **a** $x = 4\frac{1}{2}, y = 7\frac{1}{2}$
   b $x = 30.45, y = 20.05$

9 **a** $x = 12, y = 10$
   b $x = 15, y = 7.5$

Exercise 3.12

1 **a**

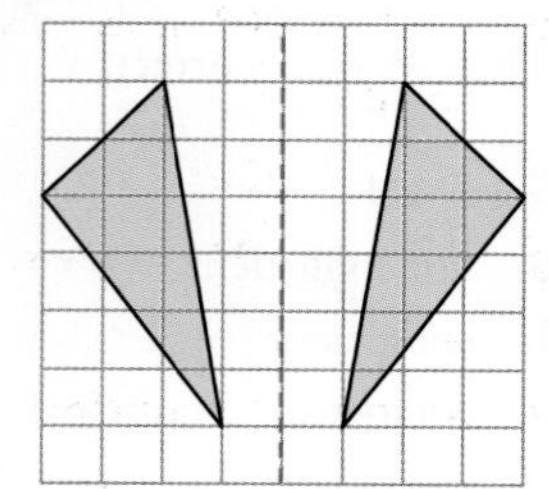

b

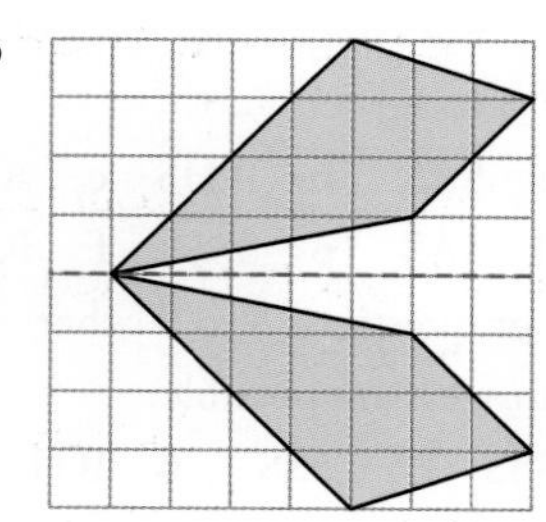

c

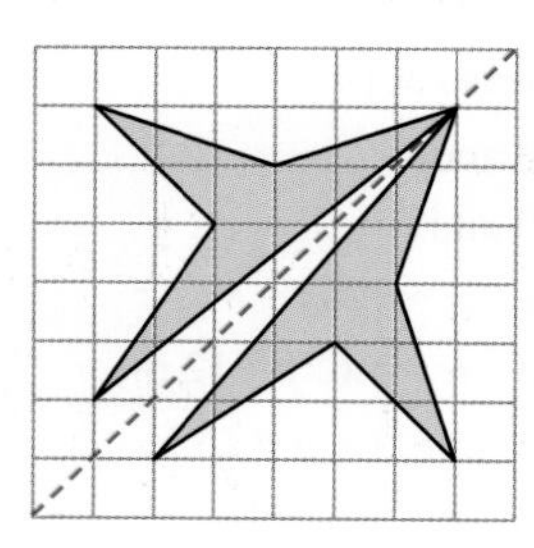

d

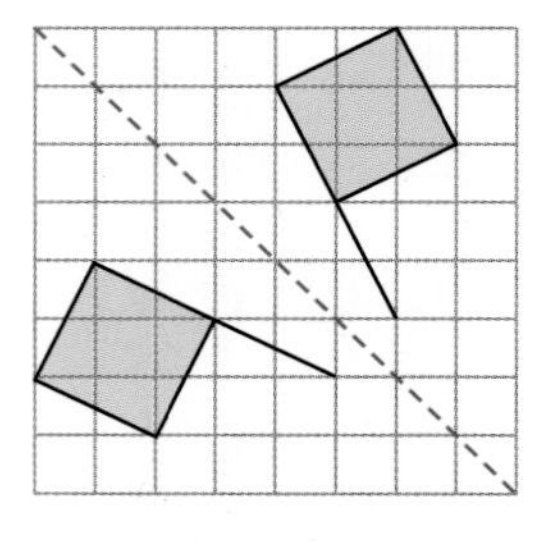

e

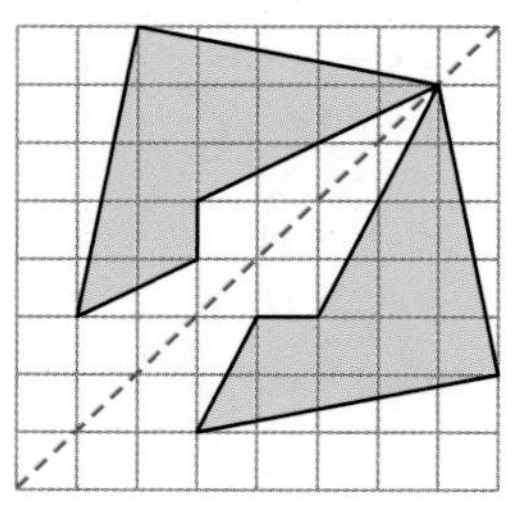

f

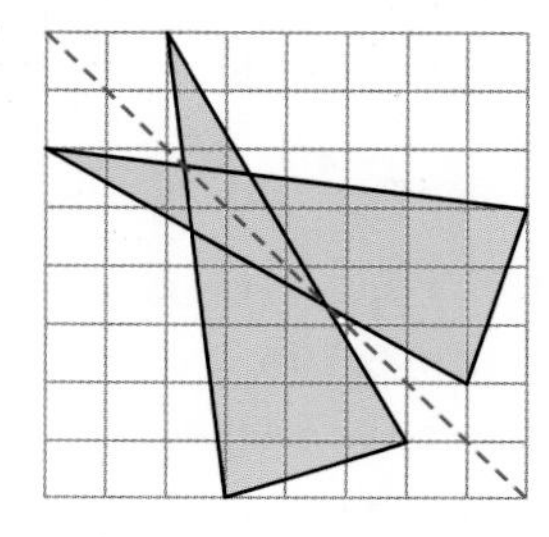

2 **a** $x = 6$ **b** $x = 10.5$
   c $x = 13$ **d** $x = 4$
   e $x = 6$ **f** $x = 11$
   g $x = 11$ **h** $x = 13$

3 **a** $y = 4.5$ **b** $y = 2$
   c $y = 1$ **d** $y = -1$
   e $y = 4$ **f** $y = -1$

4 **a** $x = 0$ **b** $y = 0$
   c $y = 1$

5 **a** $y = x$ **b** $x = 1$
   c $y = x$ **d** $y = 1$
   e $x = 0$

6 **a** $x = -1$ **b** $x = -1$
   c $y = 1$ **d** $y = 1$
   e $y = -x$ **f** $y = x$
   g $y = x$ **h** $y = -x$
   i $y = -x$ **j** $y = -x$

7

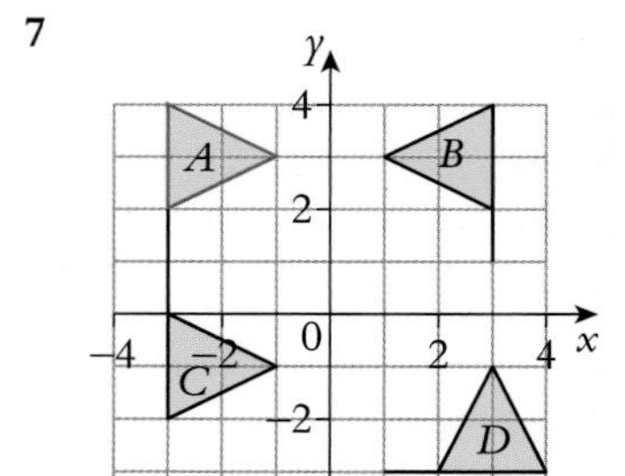

8

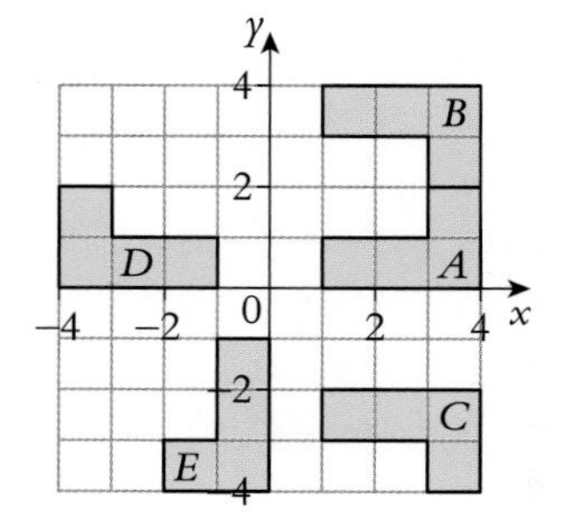

9

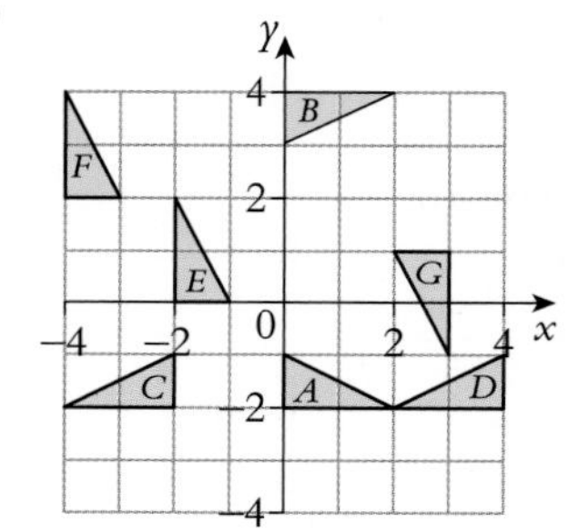

10

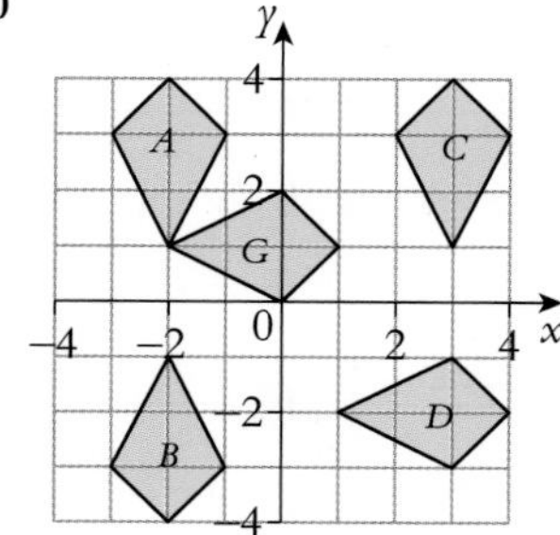

11

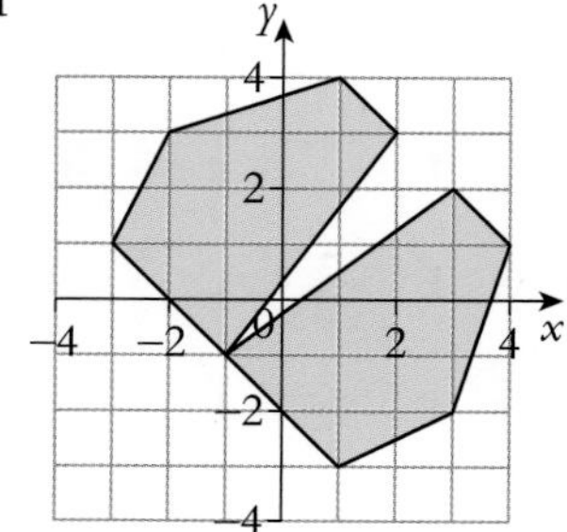

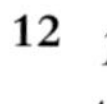

12

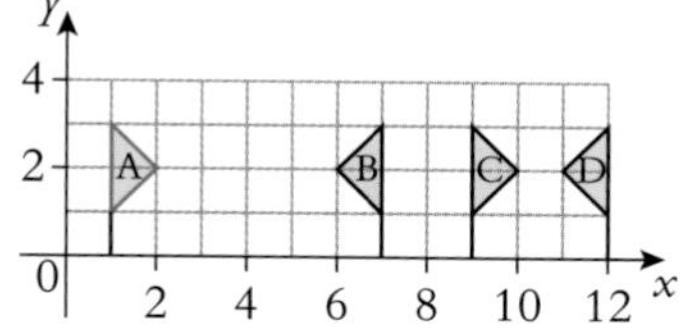

d $x = 6\frac{1}{2}$

13

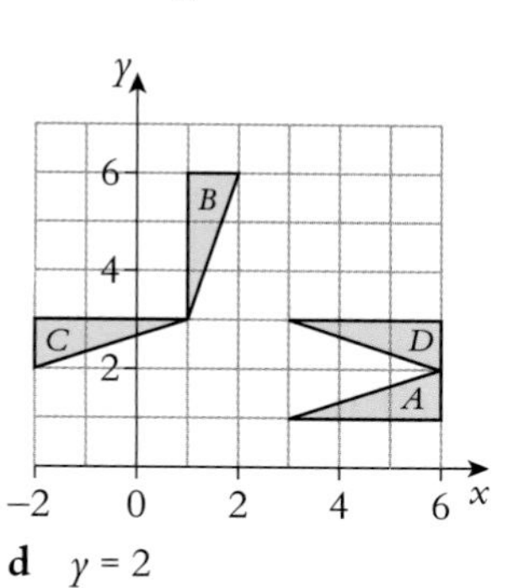

d $y = 2$

Exercise 3.13

1 a

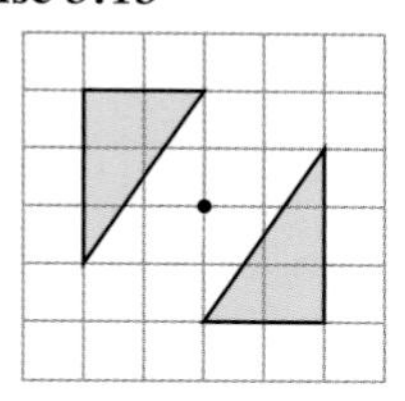

b

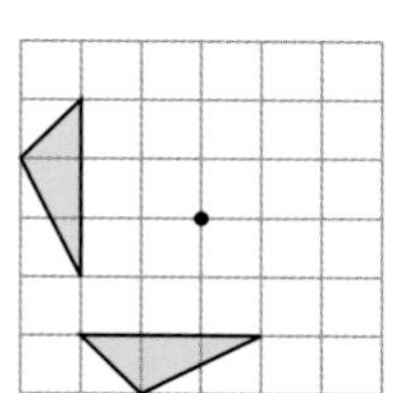

c

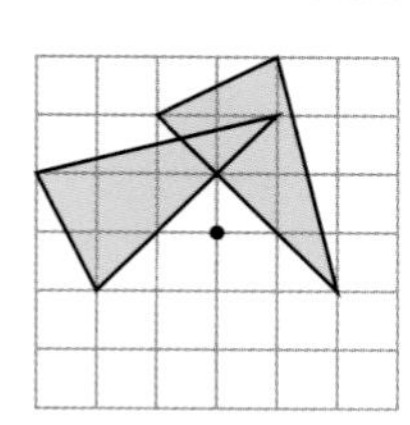

d

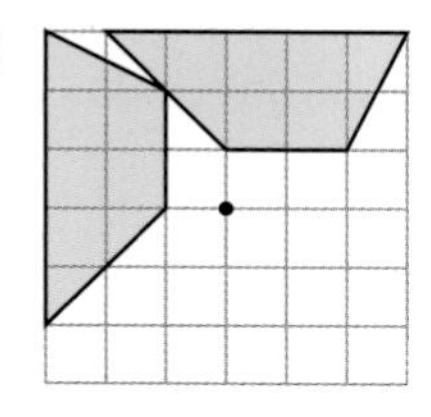

e

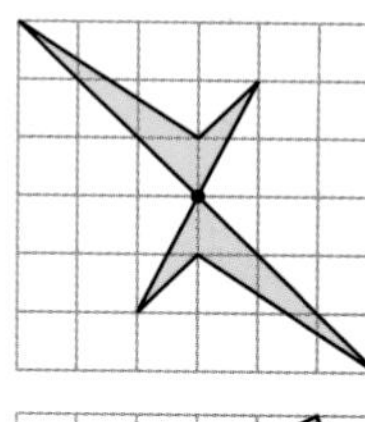

f

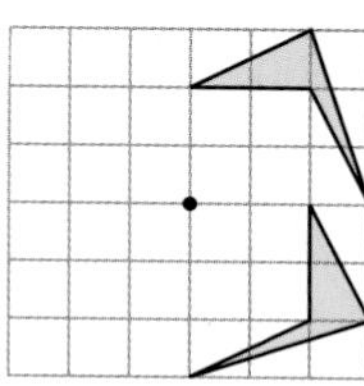

2

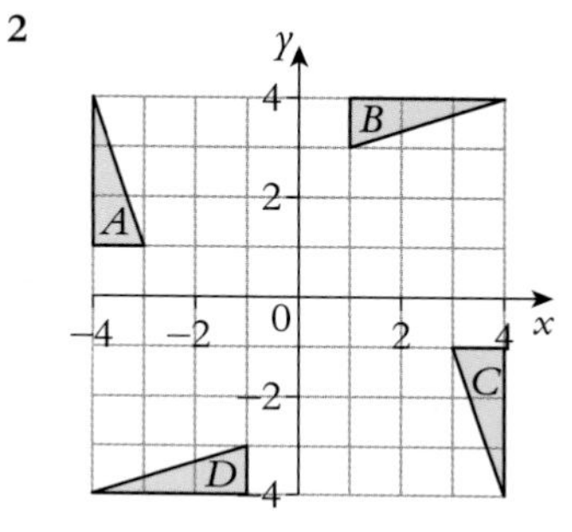

3

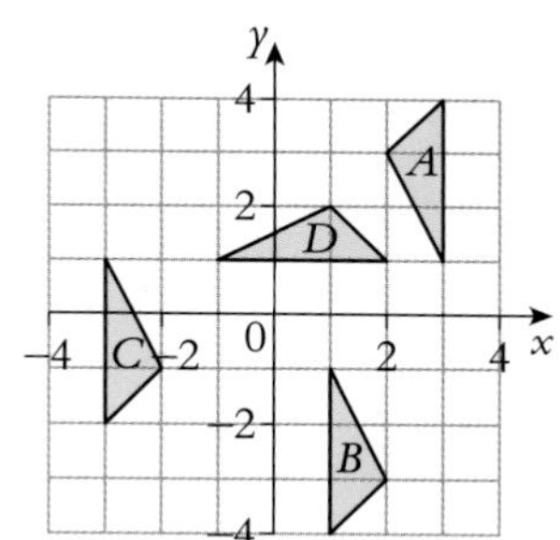

4

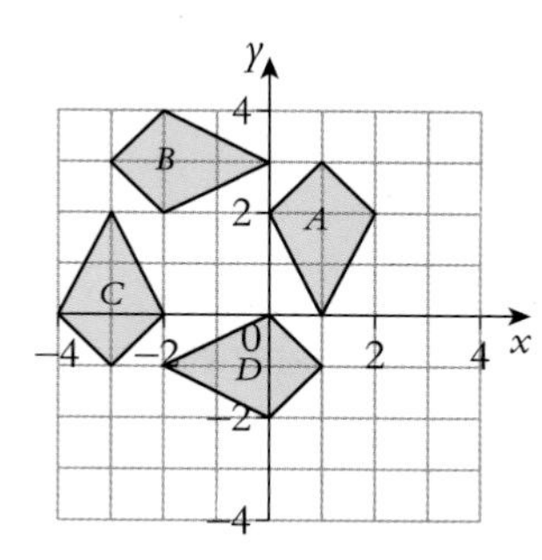

5

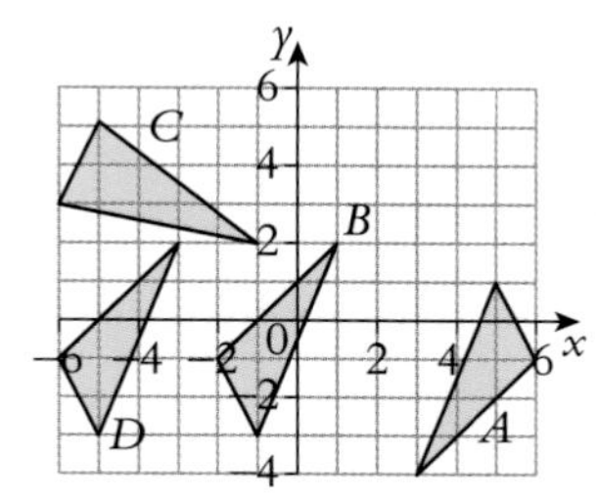

Exercise 3.14

1 **a** rotation, 90° clockwise about (0, 0)

b rotation, 180° about (0, 0)

c rotation, 90° anticlockwise about (0, 3)

d rotation, 90° clockwise about (2, −1)

e rotation, 90° anticlockwise about (2, 3)

f rotation, 180° about $(0, -\frac{1}{2})$

2 **a** rotation, 180° about (3, 3)

b rotation, 180° about (0, 0)

c rotation, 90° anticlockwise about (−1, 2)

d rotation, 90° clockwise about (−1, −1)

e rotation, 90° anticlockwise about (1, −2)

f rotation, 180° about (−1, 2)

g rotation, 90° clockwise about (2, 1)

h rotation, 90° anticlockwise about (1, −1)

i rotation, 90° anticlockwise about (−2, −1)

j rotation, 180° about (0.5, 0.5)

k rotation, 90° anticlockwise about (−1, 2)

l rotation, 90° clockwise about (2, 2)

Exercise 3.15

1 **a** $\begin{pmatrix} 2 \\ 4 \end{pmatrix}$ **b** $\begin{pmatrix} 3 \\ 3 \end{pmatrix}$ **c** $\begin{pmatrix} 5 \\ -1 \end{pmatrix}$

d $\begin{pmatrix} 3 \\ 0 \end{pmatrix}$ **e** $\begin{pmatrix} 0 \\ -3 \end{pmatrix}$ **f** $\begin{pmatrix} -2 \\ 4 \end{pmatrix}$

2 **a**

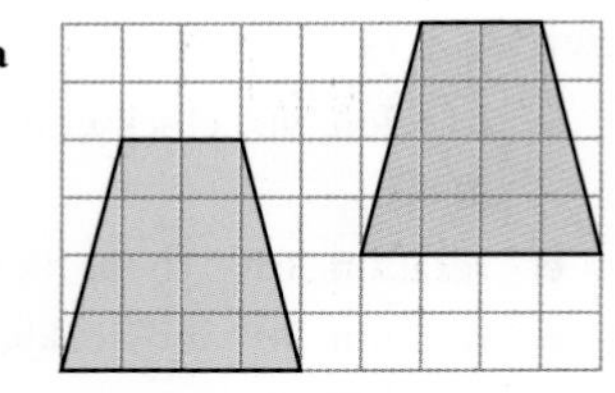

b

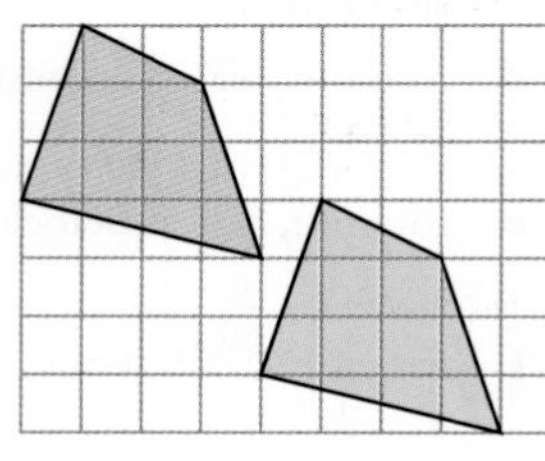

c

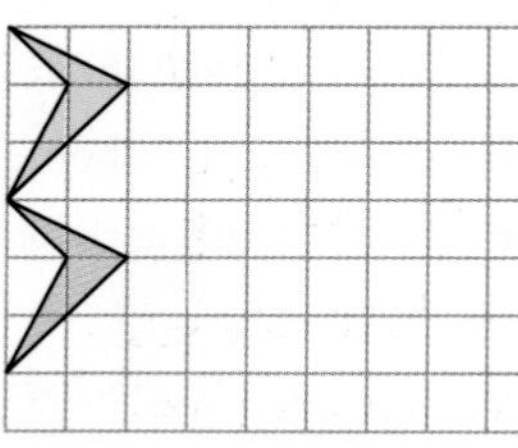

d

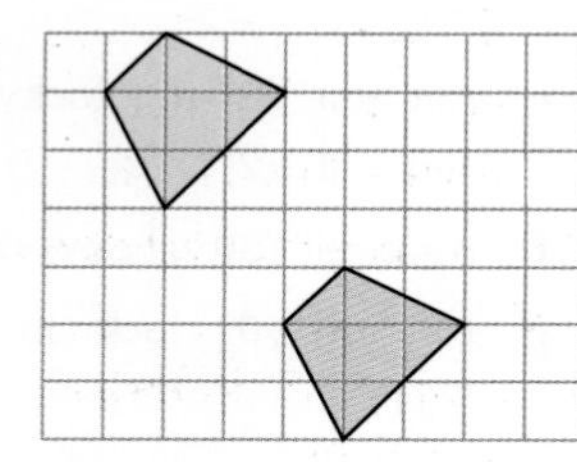

e

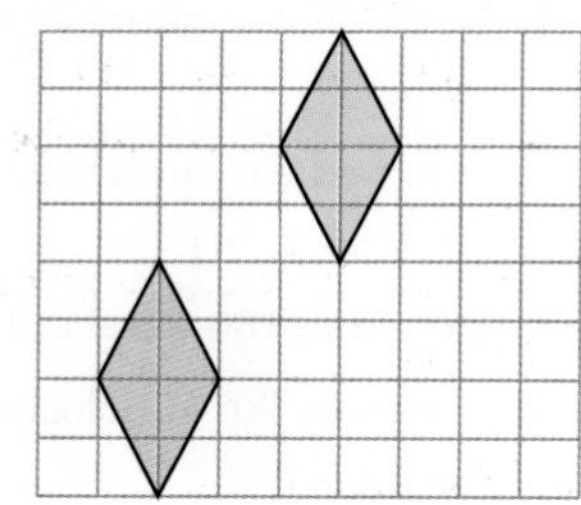

f

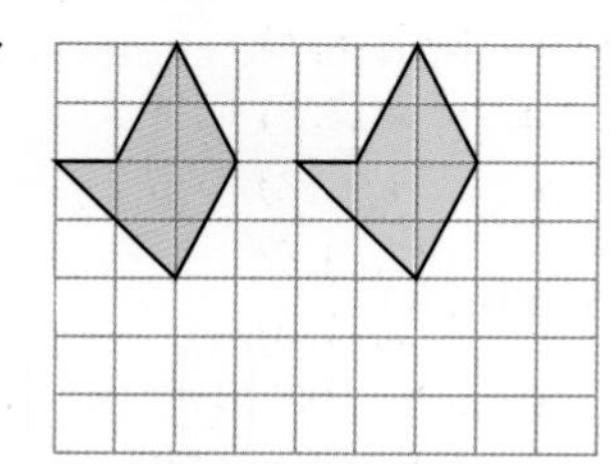

3 **a** (8, 8) **b** (5, −2)
c (−4, −1) **d** (−9, 6)
e (−4, −12) **f** (7, −8)

4 **a** translation $\begin{pmatrix} -4 \\ 1 \end{pmatrix}$

b translation $\begin{pmatrix} -8 \\ -1 \end{pmatrix}$

c translation $\begin{pmatrix} -9 \\ -4 \end{pmatrix}$

d translation $\begin{pmatrix} -5 \\ -5 \end{pmatrix}$

e translation $\begin{pmatrix} 2 \\ -5 \end{pmatrix}$

f translation $\begin{pmatrix} 5 \\ -2 \end{pmatrix}$

g translation $\begin{pmatrix} 5 \\ 1 \end{pmatrix}$

h translation $\begin{pmatrix} 0 \\ -3 \end{pmatrix}$

i translation $\begin{pmatrix} -7 \\ 0 \end{pmatrix}$

j translation $\begin{pmatrix} 9 \\ 0 \end{pmatrix}$

k translation $\begin{pmatrix} 10 \\ -4 \end{pmatrix}$

l translation $\begin{pmatrix} 14 \\ -2 \end{pmatrix}$

5 **a** **b**

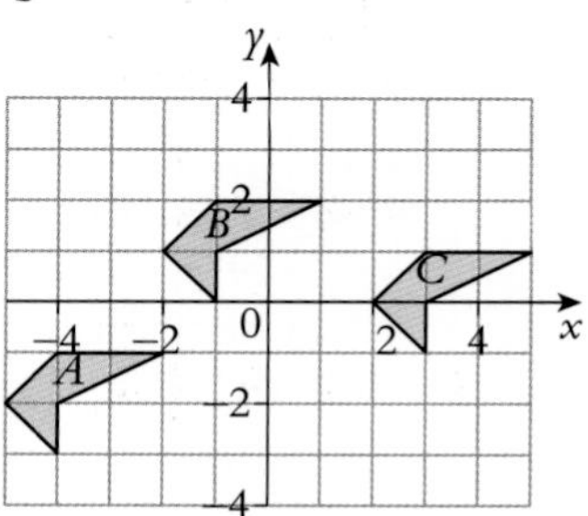

c translation $\begin{pmatrix} 7 \\ 2 \end{pmatrix}$

6 **a** **b**

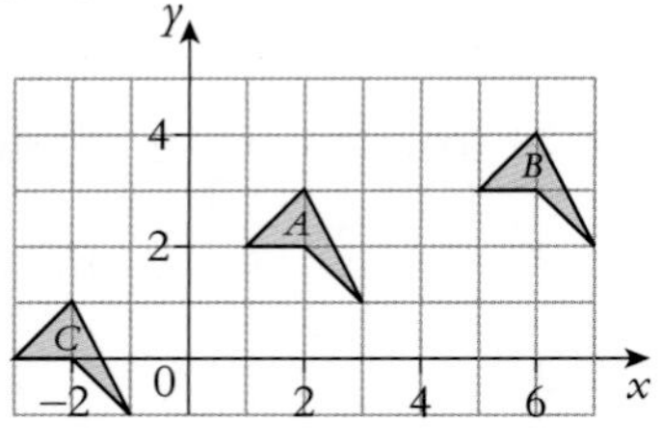

c translation $\begin{pmatrix} 4 \\ -2 \end{pmatrix}$

7 $C = (1, 4)$ $B' = (6, -3)$

8 $A = (4, 1)$ $C = (-1, 3)$
$D' = (-1, -2)$

Exercise 3.16

1 **a** reflection in $x = 3$
b reflection in $y = x$
c reflection in x-axis
d rotation, 180° about (0, 0)
e reflection in $y = -x$
f translation $\begin{pmatrix} -6 \\ -1 \end{pmatrix}$
g rotation, 90° anticlockwise about (1, 1)
h reflection in $y = x$
i rotation, 90° clockwise about (0, 0)
j reflection in $y = -1$
k reflection in y-axis
l reflection in $x = 1$

2 **a** translation $\begin{pmatrix} 1 \\ 6 \end{pmatrix}$
b rotation, 180° about (3.5, 2)
c reflection in $x = 4$
d rotation, 180° about (3, 7)
e rotation, 90° clockwise about (5, 4)
f translation $\begin{pmatrix} 5 \\ 3 \end{pmatrix}$
g translation $\begin{pmatrix} 6 \\ -3 \end{pmatrix}$

h reflection in $y = 5$

i reflection in $x + y = 15$

j translation $\begin{pmatrix} -3 \\ 4 \end{pmatrix}$

k rotation, 90° clockwise about (8, 7)

l rotation, 90° anticlockwise about (4, 1)

Exercise 3.17

1 a b c

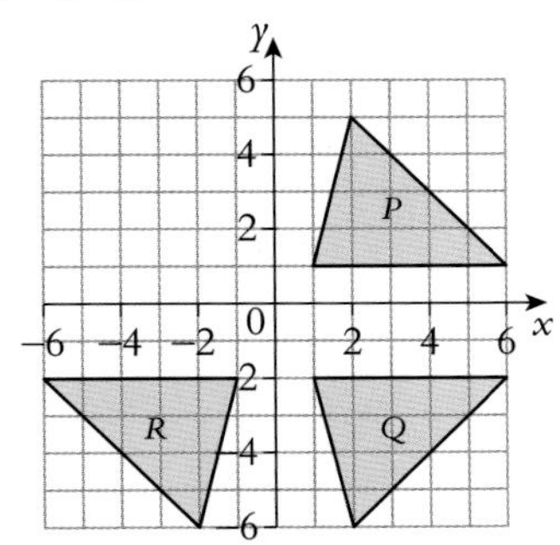

d rotation, 180° about (0, 0)

2 a b c

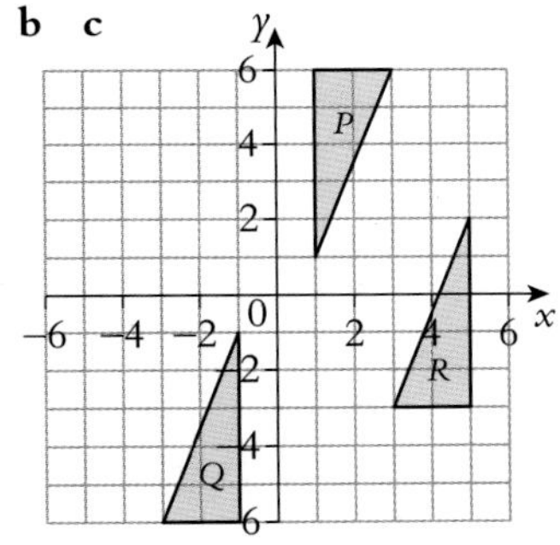

d rotation 180° about (3, 1.5)

3 a b c

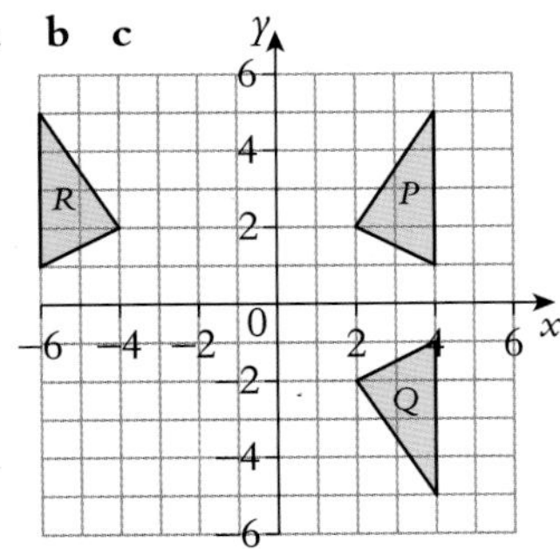

d reflection in $x = -1$

4 a b c

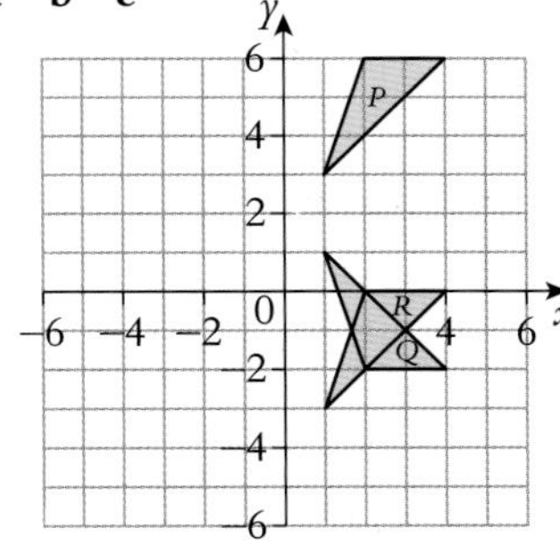

d translation $\begin{pmatrix} 0 \\ -6 \end{pmatrix}$

5 a b c

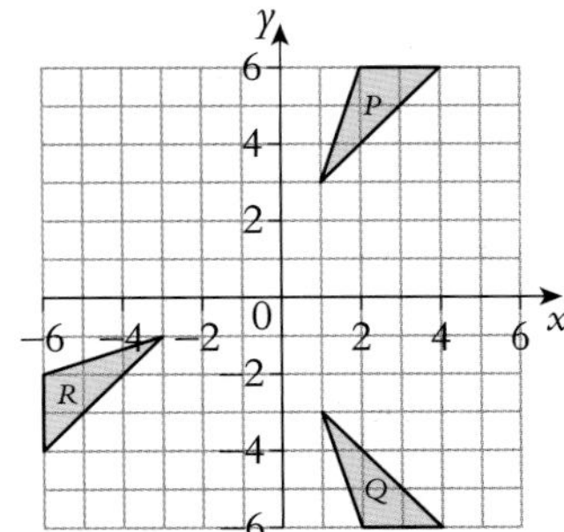

d reflection in $y = -x$

6 a b c

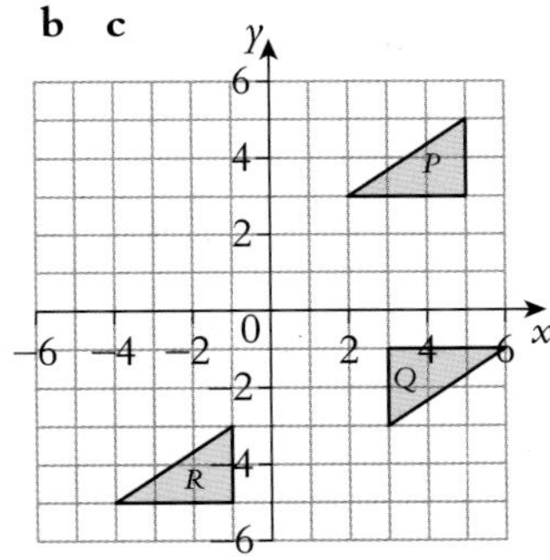

d translation $\begin{pmatrix} -6 \\ -8 \end{pmatrix}$

7 a translation $\begin{pmatrix} 6 \\ 10 \end{pmatrix}$

b translation $\begin{pmatrix} 3 \\ 10 \end{pmatrix}$

8 a rotation, 180° about (3, 2)

b translation $\begin{pmatrix} 0 \\ 6 \end{pmatrix}$

9 a translation $\begin{pmatrix} 4 \\ 2 \end{pmatrix}$

b translation $\begin{pmatrix} -6 \\ 10 \end{pmatrix}$

Exercise 3.18

1 translation $\begin{pmatrix} -2 \\ 4 \end{pmatrix}$

2 translation $\begin{pmatrix} 0 \\ -6 \end{pmatrix}$

3 translation $\begin{pmatrix} 5 \\ -7 \end{pmatrix}$

4 translation $\begin{pmatrix} 8 \\ 0 \end{pmatrix}$

5 reflection in x-axis

6 reflection in y-axis

7 reflection in $y = 3$

8 reflection in $x = -5$

9 reflection in $x + y = 4$

10 rotation, 90° anticlockwise about (0, 0)

11 rotation, 90° clockwise about (0, 0)

12 rotation, 180° about (0, 0)

13 rotation anticlockwise about (5, −2)

14 rotation 90° clockwise about (−4, 7)

Exercise 3.19

1 a SF 2, (6, 4)

b SF 2, (2, 8)

c SF 3, (2, 1)

d SF 2, (0, 0)

e SF 2, (6, 2)

f SF 3, (−0.5, 7)

2 a enlargement, SF 2, centre (8, 3)

b enlargement, SF 2, centre (12, 6)

c enlargement, SF 2, centre (0, 7)

d enlargement, SF 3, centre (13, 4)

e enlargement, SF 2, centre (18, 3)

f enlargement, SF 2, centre (22, 6)

Exercise 3.20

1 a

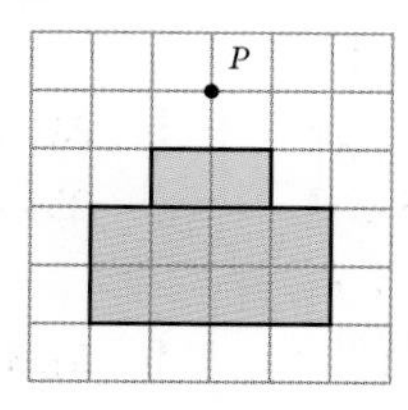

b

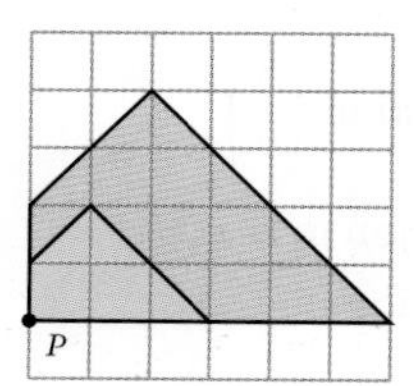

c

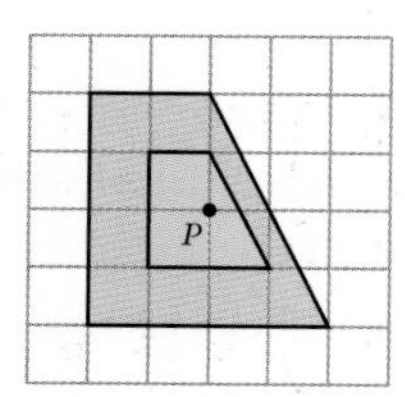

2 a

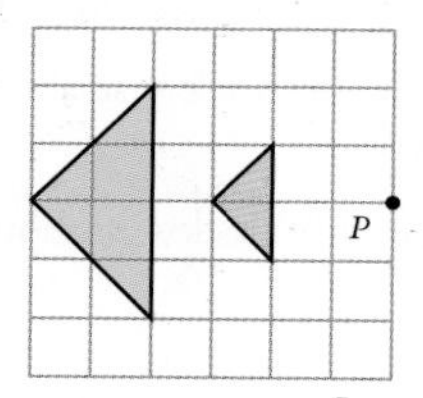

b

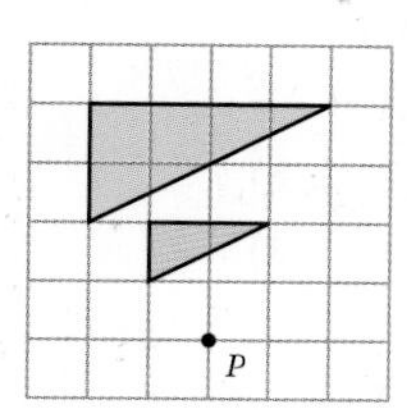

c

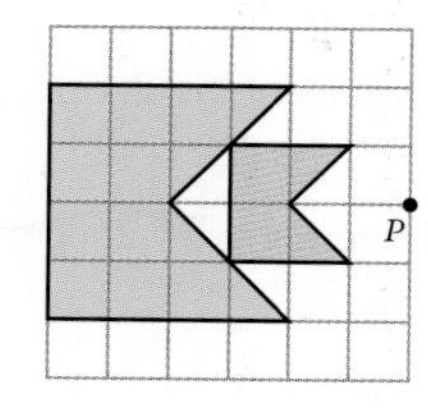

3 a

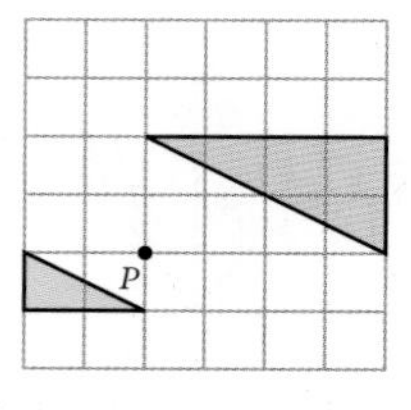

b

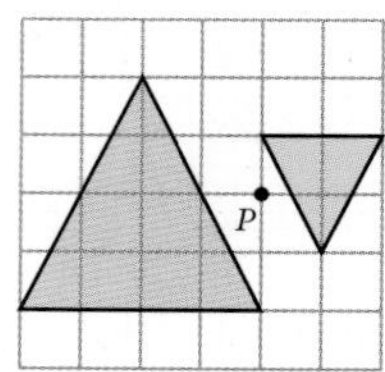

c

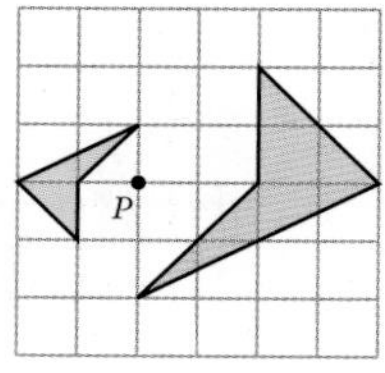

d

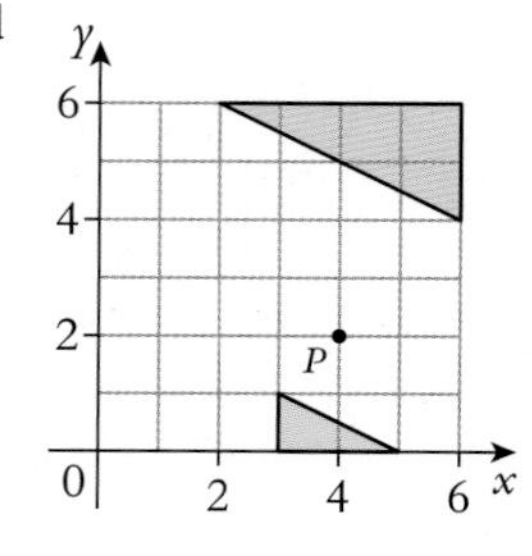

e

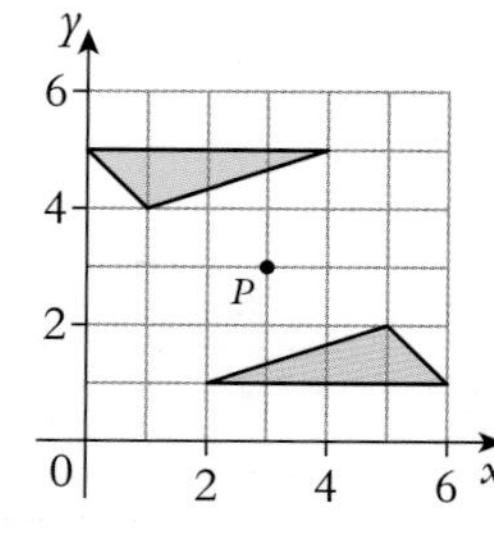

f

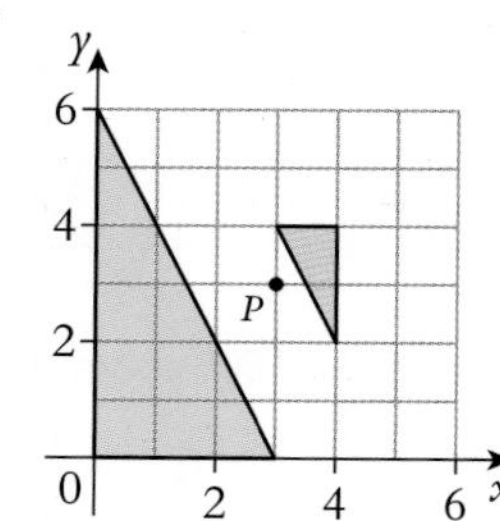

4 a

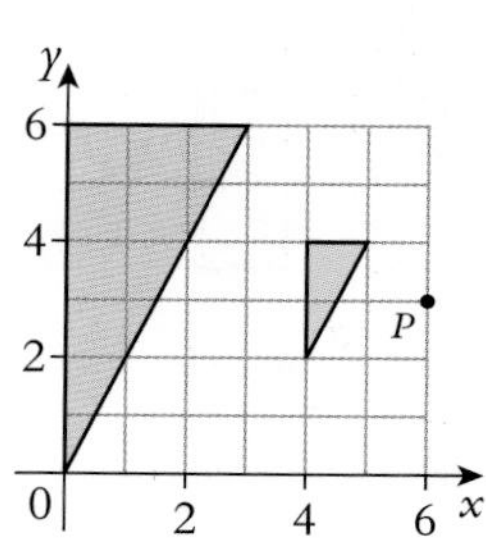

b

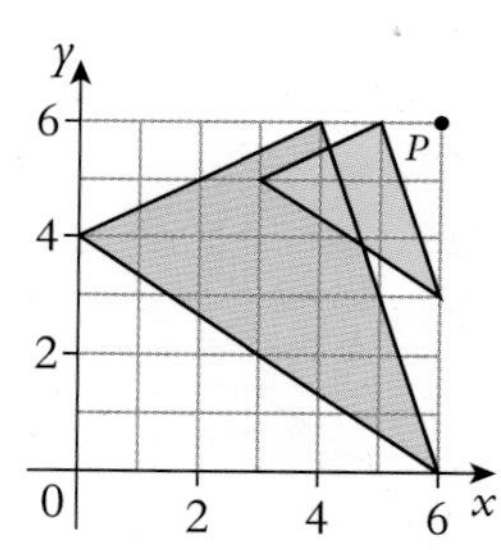

c

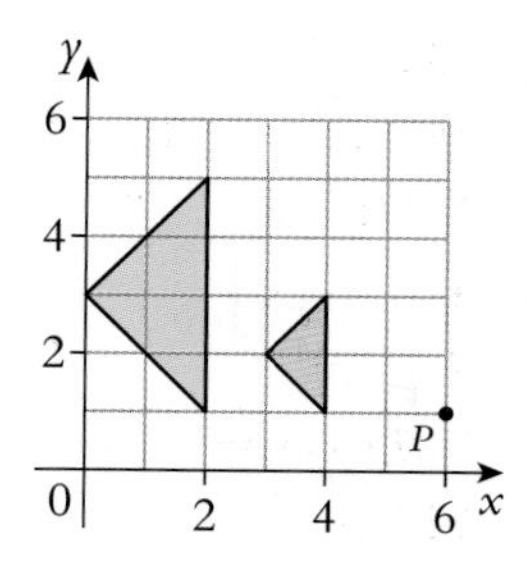

5

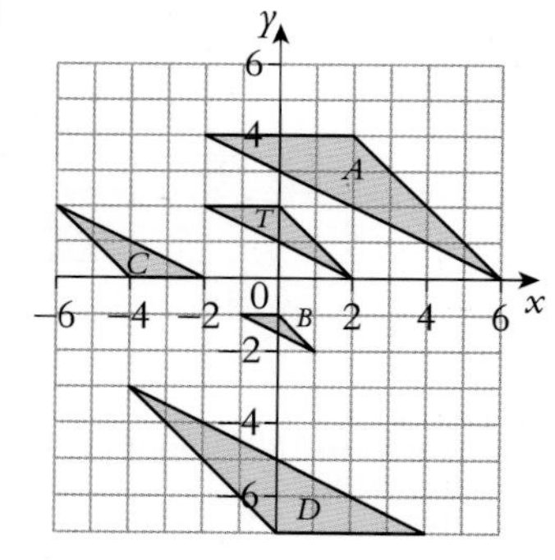

6

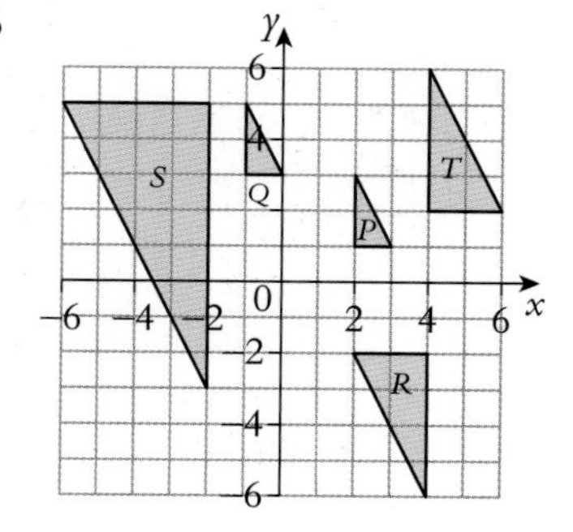

7

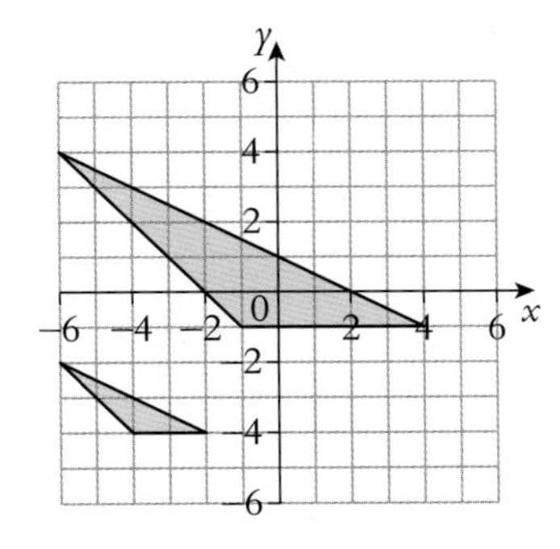

8

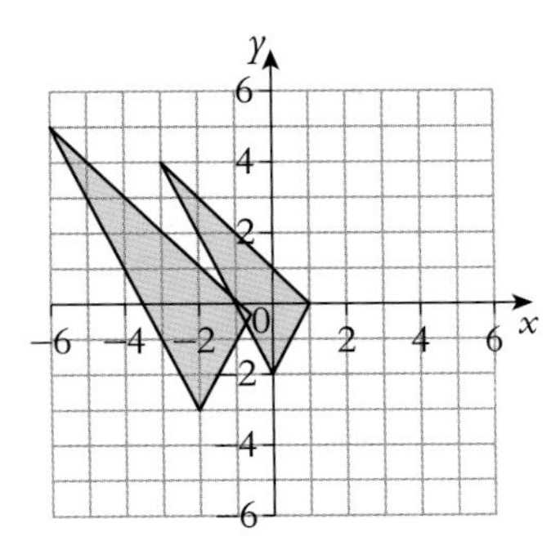

9

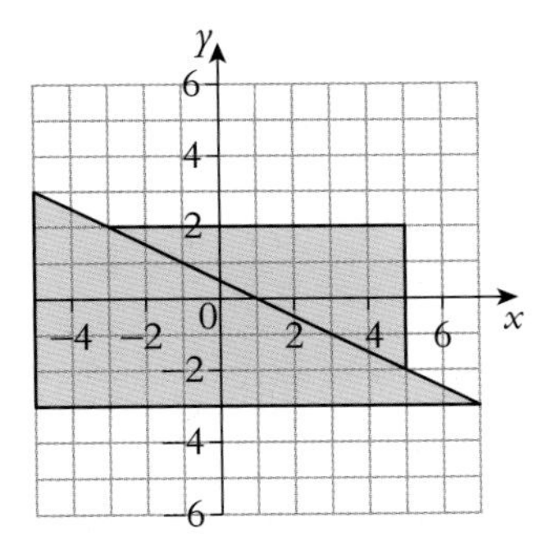

Exercise 3.21

1 a b c

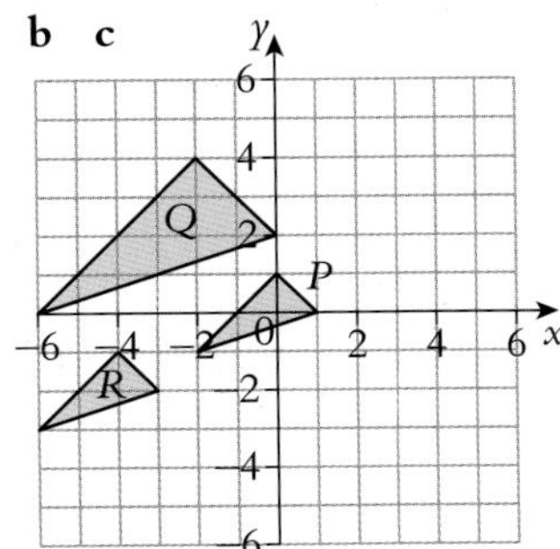

d translation $\begin{pmatrix} -4 \\ -2 \end{pmatrix}$

2 a b c

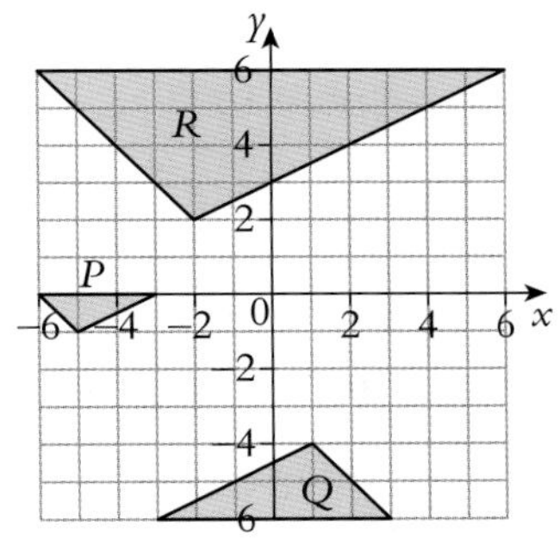

d enlargement, SF 4, centre (−6, −2)

3 a b c

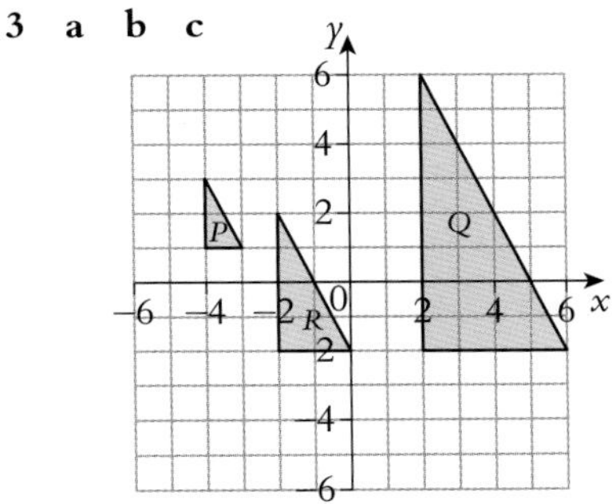

d enlargement, SF 2, centre (−6, 4)

Exercise 3.22

1 a 125 cm^3 b 96 cm^3
c 120 000 cm^3

2 a 180 cm^3 b 192 cm^3
c 16 cm^3

3 a 24 cm^3 b 128 cm^3
c 2240 cm^3 d 90 cm^3
e 450 cm^3 f 540 cm^3

4 a 3.14 m^3 b 226 cm^3
c 236 cm^3

5 a 151 cm^3 b 490 cm^3
c 25100 cm^3

6 20 cm

7 a 7.875 m^3
b 7 875 000 cm^3
c 7875 litres

8 257 cm^3

9 5.64 cm

10 306 cm

11 109.3 cm^3

12 a 166 cm^2 b 4157 cm^3
c 2.7 kg

Exercise 3.23

1 a 216 cm^2 b 184 cm^2
c 928 cm^2

2 a 240 cm^2 b 88 cm^2
c 172 cm^2

3 a 510 cm^2 b 216 cm^2

4 a 101 cm^2 b 254 cm^2
c 176 cm^2

5 a 75.4 cm^2 b 89.5 cm^2
c 50.3 cm^2

6 234 cm^2

Exercise 3.24

1 a $\frac{1}{8}$ b $\frac{1}{2}$ c $\frac{1}{4}$ d $\frac{1}{2}$
e $\frac{1}{4}$ f 1

2 a $\frac{2}{5}$ b $\frac{2}{5}$ c $\frac{4}{5}$ d $\frac{1}{5}$
e $\frac{4}{5}$ f 1

3 a $\frac{1}{10}$ b $\frac{1}{2}$ c $\frac{2}{5}$ d $\frac{9}{10}$
e $\frac{3}{5}$ f $\frac{1}{2}$ g 0 h 1

4 a $\frac{1}{3}$ b $\frac{2}{3}$ c $\frac{4}{9}$ d $\frac{1}{9}$

5 a $\frac{17}{30}$ b $\frac{12}{25}$

Exercise 3.25

1 a

Number	1	2	3
Probability	$\frac{1}{4}$	$\frac{1}{8}$	$\frac{5}{8}$

b i $\frac{3}{8}$ ii $\frac{7}{8}$ iii $\frac{3}{4}$ iv 0

2 a i $\frac{5}{8}$ ii $\frac{3}{8}$
b i $\frac{4}{7}$ ii $\frac{3}{7}$
c i $\frac{5}{7}$ ii $\frac{2}{7}$

3 a $\frac{1}{4}$ b green c $\frac{5}{6}$

4 a 0.2 b 5 c 0.75

Exercise 3.26

1 a

	1	2	3	4	5
1	2	3	4	5	6
2	3	4	5	6	7
3	4	5	6	7	8
4	5	6	7	8	9
5	6	7	8	9	10

b i $\frac{2}{25}$ ii $\frac{2}{25}$ iii 0

iv $\frac{12}{25}$ v $\frac{1}{5}$ vi $\frac{11}{25}$

2 a

	1	2	3	4
1	2	3	4	5
2	3	4	5	6
3	4	5	6	7
4	5	6	7	8

b i $\frac{1}{8}$ ii $\frac{1}{4}$ iii $\frac{1}{2}$ iv $\frac{3}{8}$

3 a

	2	4	6	8
1	2	4	6	8
2	4	8	12	16
3	6	12	18	24

b i $\frac{1}{6}$ ii 0

iii 1 iv $\frac{5}{12}$

Exercise 3.27

1 288 2 145

3 150 4 30

5 492

6 a 0.255, 0.42, 0.325

b 234

7 a 0.18, 0.36, 0.27, 0.19

b 162, 324, 243, 171

Examination-style questions

1 a 5.62×10^8 b 2.53×10^5

2 a 0.000 000 8

b 1.8×10^{-7}

3 $x = 2, y = -1$

4 $x = 1, y = 3$

5 $x = -3.5, y = -2$

6 $x(x - 7)$

7 a $5x^2y(xy - 2)$

b $(2x + 3)(y - 4)$

8 $\frac{5x}{1-3x}$ 9 $a = 5c + 5b$

10 $x = 5y + 2$ 11 $h = \frac{A-\pi r^2}{2\pi r}$

12 $h = \sqrt{\frac{2p-q}{5}}$

13 $P = \frac{9R^2}{4Q}$

14 a 12.5 cm b 7.2 cm

15 $47.5\ \text{m}^3$

16 a $50\ \text{cm}^2$ b $350\ \text{cm}^3$

17 a 8.74 b $1304\ \text{cm}^2$

18 a Rotation, 90° anticlockwise, centre (6, 5)

b Image at (5, 3), (5, 5), (9, 3)

c Image at (10, 8), (18, 8), (18, 4)

19 a Image at (1, 3), (1, 7), (3, 5), (3, 4)

b Image at (2, 8), (6, 8), (5, 6), (4, 6)

c Reflection in $x + y = 9$

20 a $\frac{1}{2}$ b $\frac{5}{6}$ c 0

21 a 0.1 b 5 c 0.55

22 a

	1	2	3	4	5	6
1	1	2	3	4	5	6
2	2	4	6	8	10	12
3	3	6	9	12	15	18
4	4	8	12	16	20	24
5	5	10	15	20	25	30
6	6	12	18	24	30	36

b i $\frac{1}{9}$ ii $\frac{7}{18}$

23 90

Unit 4

Exercise 4.1

1 57.5 % 2 37 % 3 3.55 %

4 10 % 5 60 % 6 5 %

7 1.32 % 8 3.09 % 9 5 %

10 21.4 % 11 3.125 %

12 7.42 % 13 16 %

Exercise 4.2

1 a 1 × 4 b 1 × 1 c 1 × 3

d 1 × 2 e 2 × 2 f 2 × 3

g 2 × 1 h 2 × 4 i 3 × 1

j 3 × 3 k 3 × 2 l 4 × 3

2 a $\begin{pmatrix} 7 & 4 \\ 4 & 13 \end{pmatrix}$ b $\begin{pmatrix} 3 & 4 \\ 2 & 3 \end{pmatrix}$

3 a $(8\ \ 11\ \ 15)$ b $(2\ \ -5\ \ 9)$

4 a $\begin{pmatrix} 12 & 4 & 6 \\ -2 & -2 & 11 \end{pmatrix}$

b $\begin{pmatrix} -2 & -2 & 8 \\ -2 & 2 & 1 \end{pmatrix}$

5 a $\begin{pmatrix} 16 & 7 \\ -16 & 0 \\ 0 & 0 \\ 0 & -7 \end{pmatrix}$

b $\begin{pmatrix} -4 & -3 \\ 0 & 0 \\ 4 & -10 \\ 14 & -1 \end{pmatrix}$

6 a $\begin{pmatrix} 7 & 10 \\ -10 & 2 \end{pmatrix}$

b $\begin{pmatrix} 11 & -1 \\ -1 & -2 \end{pmatrix}$

c $\begin{pmatrix} -1 & 3 \\ -5 & 6 \end{pmatrix}$

d $\begin{pmatrix} -4 & 11 \\ -9 & 4 \end{pmatrix}$

e $\begin{pmatrix} 13 & 8 \\ -8 & -2 \end{pmatrix}$

f $\begin{pmatrix} 1 & 12 \\ -12 & 6 \end{pmatrix}$

g $\begin{pmatrix} 9 & -10 \\ 6 & -2 \end{pmatrix}$

h $\begin{pmatrix} -3 & -6 \\ 2 & 6 \end{pmatrix}$

7 a $\begin{pmatrix} 3 & -7 \\ 0 & 2 \\ 14 & 13 \end{pmatrix}$

b $\begin{pmatrix} 2 & 3 \\ 2 & 6 \\ 10 & -1 \end{pmatrix}$

c $\begin{pmatrix} -1 & -6 \\ 10 & 4 \\ 10 & -4 \end{pmatrix}$

d $\begin{pmatrix} 1 & -10 \\ -2 & -4 \\ 4 & 14 \end{pmatrix}$

e $\begin{pmatrix} 3 & 9 \\ -8 & 2 \\ 0 & 3 \end{pmatrix}$

f $\begin{pmatrix} 4 & -9 \\ -6 & -2 \\ 11 & 22 \end{pmatrix}$

g $\begin{pmatrix} 2 & 11 \\ -2 & 6 \\ 3 & -6 \end{pmatrix}$

h $\begin{pmatrix} -2 & -11 \\ 2 & -6 \\ -3 & 6 \end{pmatrix}$

8 $x = 7, y = -8$

9 $x = -7, y = -5$

Exercise 4.3

1 $(12 \quad -16)$ **2** $\begin{pmatrix} 2 & -\frac{1}{2} \\ 0 & 1 \end{pmatrix}$

3 $\begin{pmatrix} 48 & -24 & -16 \\ 0 & 8 & -8 \end{pmatrix}$

4 $\begin{pmatrix} 6 & -3 & -18 & 15 \\ -3 & 0 & -6 & -3 \end{pmatrix}$

5 a $\begin{pmatrix} 4 & -6 \\ 2 & 2 \end{pmatrix}$ b $\begin{pmatrix} -20 & -30 \\ 0 & 10 \end{pmatrix}$

c $\begin{pmatrix} 15 & 0 \\ -6 & 3 \end{pmatrix}$ d $\begin{pmatrix} -2 & -3 \\ 0 & 1 \end{pmatrix}$

e $\begin{pmatrix} -6 & 9 \\ -3 & -3 \end{pmatrix}$ f $\begin{pmatrix} 19 & -6 \\ -4 & 5 \end{pmatrix}$

g $\begin{pmatrix} -35 & -30 \\ 6 & 7 \end{pmatrix}$ h $\begin{pmatrix} 2 & -9 \\ 2 & 3 \end{pmatrix}$

6 a (17) b (14) c (−72)
d (15) e (8) f (6)
g (44) h (−6) i (18)

7 $x = 9$ **8** $x = 7$

9 $x = -2$

Exercise 4.4

1 a $\begin{pmatrix} 3 & 17 \\ 15 & 13 \end{pmatrix}$ b $\begin{pmatrix} 10 & 2 & 4 \\ 5 & 5 & 3 \end{pmatrix}$

c $\begin{pmatrix} 36 \\ 5 \end{pmatrix}$ d $\begin{pmatrix} 32 & 19 \\ 16 & 8 \end{pmatrix}$

e $\begin{pmatrix} 16 & 4 \\ 11 & 4 \end{pmatrix}$ f (19)

g $\begin{pmatrix} 36 & 0 \\ 10 & 6 \end{pmatrix}$

h $\begin{pmatrix} 19 & 3 & 7 & 11 \\ 11 & 6 & 5 & 4 \\ -9 & 0 & -3 & -6 \\ 34 & 12 & 14 & 16 \end{pmatrix}$

2 a $x = 3, \quad y = 4$
b $2 \times 4 \quad 3 \times 2$ is impossible because the numbers in the middle do not match

3 $\begin{pmatrix} 7 \\ 14 \\ 22 \end{pmatrix}$

Exercise 4.5

1 a 1 b 1 c 1 d −7
e −1 f 2 g 4 h 0

2 $3\frac{1}{2}$ **3** −1

4 a $\begin{pmatrix} -3 & 12 \\ -2 & 6 \end{pmatrix}$ b $\begin{pmatrix} 5 & -4 \\ 4 & -2 \end{pmatrix}$

5 a $\begin{pmatrix} -13 & 8 \\ -7 & -5 \end{pmatrix}$ b $\begin{pmatrix} -14 & -5 \\ 13 & -4 \end{pmatrix}$

c $\begin{pmatrix} 26 & 3 \\ 3 & 5 \end{pmatrix}$ d $\begin{pmatrix} 11 & 0 \\ 0 & 11 \end{pmatrix}$

6 a $\begin{pmatrix} 6 & -3 \\ -9 & 6 \end{pmatrix}$ b $\begin{pmatrix} 10 & 15 \\ 5 & 20 \end{pmatrix}$

c $\begin{pmatrix} 0 & -4 \\ -4 & -2 \end{pmatrix}$ d $\begin{pmatrix} 16 & 12 \\ -4 & 26 \end{pmatrix}$

e $\begin{pmatrix} 3 & 2 \\ -4 & -1 \end{pmatrix}$ f $\begin{pmatrix} -5 & 4 \\ -10 & 7 \end{pmatrix}$

g $\begin{pmatrix} 7 & -4 \\ -12 & 7 \end{pmatrix}$ h $\begin{pmatrix} 7 & 18 \\ 6 & 19 \end{pmatrix}$

i 1 j 5

7 a $\begin{pmatrix} 2 & -2 \\ 2 & 4 \end{pmatrix}$ b $\begin{pmatrix} 20 & -12 \\ 4 & -4 \end{pmatrix}$

c $\begin{pmatrix} -18 & 10 \\ -2 & 8 \end{pmatrix}$ d $\begin{pmatrix} 4 & -2 \\ 7 & -5 \end{pmatrix}$

e $\begin{pmatrix} 2 & -11 \\ 0 & -3 \end{pmatrix}$ f $\begin{pmatrix} 0 & -3 \\ 3 & 3 \end{pmatrix}$

g $\begin{pmatrix} 22 & -12 \\ 4 & -2 \end{pmatrix}$ h $\begin{pmatrix} 16 & -2 \\ 0 & 9 \end{pmatrix}$

i 3 j −2

8 $\begin{pmatrix} -5 & -3 \\ -2 & -1 \end{pmatrix}$

Exercise 4.6

1 a $\begin{pmatrix} 3 & -5 \\ -1 & 2 \end{pmatrix}$ b $\begin{pmatrix} 3 & -4 \\ -2 & 3 \end{pmatrix}$

c $\frac{1}{2}\begin{pmatrix} 2 & -2 \\ -4 & 5 \end{pmatrix}$ d $\begin{pmatrix} -2 & -1 \\ -1 & -1 \end{pmatrix}$

e $\begin{pmatrix} -1 & -2 \\ 2 & 3 \end{pmatrix}$ f $\frac{1}{2}\begin{pmatrix} 1 & 1 \\ 4 & 6 \end{pmatrix}$

g $-\frac{1}{2}\begin{pmatrix} -7 & -4 \\ 3 & 2 \end{pmatrix}$ h $\begin{pmatrix} 4 & 1 \\ -3 & -1 \end{pmatrix}$

i $\frac{1}{2}\begin{pmatrix}-1 & 1\\ 7 & -9\end{pmatrix}$ **j** $\frac{1}{2}\begin{pmatrix}4 & 5\\ 2 & 3\end{pmatrix}$

k $\begin{pmatrix}0.5 & 1\\ 1 & 4\end{pmatrix}$ **l** $\begin{pmatrix}-1 & 2\\ -1 & 1\end{pmatrix}$

2 Determinant = 0

3 $x = -5$

4 $x = 1\frac{1}{2}$

5 $Y = \begin{pmatrix}1 & -3\\ -2 & 7\end{pmatrix}$

6 $X = -\frac{1}{14}\begin{pmatrix}3 & -1\\ -8 & -2\end{pmatrix}$

7 $C = \begin{pmatrix}-1 & 3\\ 2 & -5\end{pmatrix}$

8 a $\begin{pmatrix}18 & 24\\ 4 & 4\end{pmatrix}$ **b** $\begin{pmatrix}13 & 16\\ 4 & 5\end{pmatrix}$

c $\begin{pmatrix}19 & 24\\ 6 & 7\end{pmatrix}$ **d** $\begin{pmatrix}117 & 144\\ 36 & 45\end{pmatrix}$

e $\begin{pmatrix}55 & 68\\ 17 & 21\end{pmatrix}$ **f** -1

g $\begin{pmatrix}-1 & 4\\ 1 & -3\end{pmatrix}$ **h** $\begin{pmatrix}1 & 0\\ 0 & 1\end{pmatrix}$

9 $Y = \begin{pmatrix}2 & -2\\ -3 & -3\end{pmatrix}$

10 $X = \begin{pmatrix}1 & 2\\ -3 & 4\end{pmatrix}$

Exercise 4.7

1 $x^2 + 13x + 30$ **2** $x^2 + 7x + 12$

3 $x^2 + 10x + 21$ **4** $x^2 + 6x + 8$

5 $x^2 + 6x + 5$ **6** $x^2 + 15x + 56$

7 $x^2 + 8x + 12$ **8** $x^2 + 13x + 42$

9 $x^2 + 8x + 16$ **10** $x^2 + 19x + 60$

11 $x^2 + 18x + 72$ **12** $2x^2 + 13x + 15$

13 $6x^2 + 19x + 10$

Exercise 4.8

1 a $x^2 + 4x + 3x + 12$
$= x^2 + 7x + 12$

b $x^2 - 6x + 4x - 24$
$= x^2 - 2x - 24$

c $x^2 - 4x - 5x + 20$
$= x^2 - 9x + 20$

d $x^2 + 4x + 4x + 16 = x^2 + 8x + 16$

2 a $x^2 + 3x + 2$

b $x^2 + 12x + 35$

c $x^2 + 11x + 24$ **d** $x^2 + 6x + 8$

e $x^2 + 9x + 20$ **f** $x^2 + 4x + 3$

g $x^2 + 12x + 27$

h $x^2 + 11x + 30$

i $x^2 + 7x + 12$ **j** $x^2 + 5x + 6$

k $x^2 + 10x + 16$ **l** $x^2 + 5x + 4$

3 a $x^2 + x - 2$ **b** $x^2 + 4x - 12$

c $x^2 + 3x - 18$ **d** $x^2 + 3x - 4$

e $x^2 - 2x - 63$

f $x^2 + 6x - 16$

g $x^2 - 4x - 32$

h $x^2 - 5x - 14$

i $x^2 - 3x - 10$

j $x^2 + 5x - 24$

k $x^2 - 2x - 3$

l $36 - 5x - x^2$

4 a $x^2 - 9$ **b** $x^2 - 64$

c $x^2 - 36$ **d** $x^2 - 100$

e $x^2 - 16$ **f** $x^2 - 81$

g $x^2 - 1$ **h** $144 - x^2$

5 a $x^2 - 7x + 10$

b $x^2 - 5x + 4$

c $x^2 - 11x + 24$

d $x^2 - 8x + 12$

e $x^2 - 14x + 49$

f $x^2 - 9x + 18$

g $x^2 - 9x + 20$

h $x^2 - 11x + 28$

i $x^2 - 13x + 36$

j $x^2 - 2x + 1$

k $x^2 - 6x + 5$

l $4 - 4x + x^2$

6 a $x^2 + 6x + 9$

b $y^2 + 10y + 25$

c $x^2 - 8x + 16$

d $x^2 + 14x + 49$

e $x^2 + 16y + 64$

f $a^2 - 20a + 100$

g $36 - 12x + x^2$

h $4 + 4x + x^2$

7 a $2y^2 - 5y - 3$

b $4x^2 + 5x - 6$

c $5x^2 + 22x + 8$

d $7y^2 + 33y - 10$

e $8y^2 - 22y - 21$

f $10x^2 - 49x - 33$

g $15x^2 - 38x + 24$

h $27y^2 - 96y + 20$

i $24y^2 - 47y + 20$

j $6a^2 - 5a - 21$

k $3 - 13a + 12a^2$

l $30x^2 - 3x - 9$

8 a $4x^2 - 1$ **b** $9x^2 - 4$

c $25x^2 - 49$ **d** $25x^2 - 4$

9 a $4x^2 - 4x + 1$

b $25x^2 + 20x + 4$

c $9x^2 + 12xy + 4y^2$

d $9y^2 - 48y + 64$

10 a $2x^2 + 3x + 19$

b $8x^2 + 12x - 38$

c $x^2 - 13x - 3$

d $14x^2 - 12x + 3$

e $13x^2 - 14x + 26$

f $8x^2 - 8x + 10$

g $32x^2 + 72x + 16$

h $6x^2 - 48x - 73$

Exercise 4.9

1 a -5.5 **b** 2 **c** -2

2 a 35 **b** 23

3 a A and B **b** D and E

c B and E **d** B and C

e B and D **f** A and C

g A and D **h** A and E

4 10

5 a $3x^2 + 9x - 30$

b $2x^2 + 18x - 2$

6 answer given

7 a c^2 **b** $(a + b)^2$

c $2ab$ **d** answer given

b

speed (v km/h)	f	(x)	$(f \times x)$
$30 < v \leq 40$	37	35	$37 \times 35 = 1295$
$40 < v \leq 50$	85	45	$85 \times 45 = 3825$
$50 < v \leq 60$	76	55	$76 \times 55 = 4180$
$60 < v \leq 70$	2	65	$2 \times 65 = 130$
	Total = 200		Total = 9430

mean = 47.15

3 **a** $150 < h \leq 160$ **b** 152.8 cm

4 **a** $3 < t \leq 3.5$ **b** 3.53°C

5 **a** $10.5 < d \leq 11$ **b** 10.955 m

6 **a** $3 < m \leq 4$ **b** 3.12 kg

c 8%

7 **a** 0 to 2 **b** 2.935

8 **a** 5.74 hrs **b** 0.59

9 **a** 1.74 miles **b** 2%

10 **a** $60 < t \leq 90$ **b** 93.6 mins

c $p = 8, q = 72$ **d** 94.2 mins

Examination-style questions

1 98 **2** £21.09 **3** $115.68

4 **a** −19

b $f^{-1}(x) = \frac{x+4}{3}$ **c** 8

5 **a** −4 **b** 11

6 **a** −45 **b** $12x^2 - 26x + 14$

c $g^{-1}(x) = \frac{3-x}{2}$

7 **a** 19 683 **b** $42x^2 - 20x + 28$

c $3^{2x} + 3$

8 **a** $(2x - 5)(2x + 5)$

b $(4x - 1)(x - 6)$

9 **a** $(y - 4)(y + 4)$

b $(y + 8)(y - 2)$

10 $2(a - b)(5a - 5b - 3)$

11 $\frac{x-2}{x-8}$ **12** $\frac{x+8}{2x-1}$

13 **a**

x	−2.5	−2	−1.5	−1	−0.5	0	0.5	1	1.5	2	2.5
y	−1.8	**1**	2.3	**2.5**	1.9	**1**	**3**	0.06	−0.5	1	3.8

b

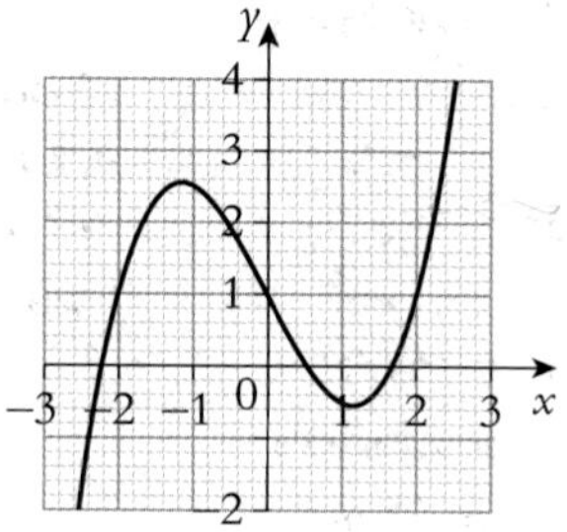

c 3.5 to 4.5

14 408 cm²

15 228 cm²

16 **a** 25.1 cm

b **i** 4 cm

ii 19.6 cm

iii 328 cm³

17 89.8 cm³

18 **a** 328 cm³ **b** 3.456 cm

19 5100 cm² **20** 3.42 cm²

21 148 800 cm³

22 **a** $6.5 < t \leq 7$

b 6.46 minutes

23 **a** $12 < t \leq 15$

b 10.53 hours

c $a = 16, b = 43$

d 10.62 hours

Unit 6

Exercise 6.1

1 3×10^6 **2** $80

3 $20 **4** $300

5 38.2 minutes **6** $16 500

7 $2.32 **8** $3540

9 $8.50 **10** $130

11 $44.80 **12** 21.6°C

13 $5360

Exercise 6.2

1 $60 **2** $756

3 $295 **4** $481.50

5 $530.45 **6** $2153.78

7 $757.70 **8** $63.05

9 3 **10** 15

11 7 years **12** $12 000

13 3365

Exercise 6.3

1 **a** $\frac{1}{3}$ hr **b** $\frac{1}{2}$ hr **c** $\frac{1}{6}$ hr

d $\frac{1}{4}$ hr **e** $\frac{5}{12}$ hr **f** $\frac{8}{15}$ hr

g $\frac{3}{4}$ hr **h** $\frac{1}{15}$ hr **i** $\frac{9}{10}$ hr

j $\frac{2}{5}$ hr **k** $\frac{2}{3}$ hr **l** $\frac{4}{5}$ hr

2 **a** 30 min **b** 3 min **c** 15 min

d 21 min **e** 48 min **f** 42 min

g 6 min **h** 27 min **i** 24 min

j 45 min **k** 25 min **l** 18 min

3 **a** 62.5 km/hr **b** 38 km/hr

c 13 km/hr **d** 48 km/hr

e 18 km/hr **f** 10 km/hr

g 48 km/hr **h** 24 km/hr

i 120 km/hr

4 **a** 20 m/s **b** 25 m/s

c 50 m/s **d** 20 m/s

e 30 m/s **f** 6 m/s

5 **a** 20 m/s **b** $3\frac{1}{3}$ m/s

c 100 m/s **d** 72 km/h

e 126 km/h

f 147.6 km/h

6 20 m/s

7 3×10^8 m/s

8 343 m/s

Exercise 6.4

1 **a** 2 hours **b** 2 hrs 30 mins

c 30 mins

d 2 hours 15 mins

e 2 hours 50 mins

f 2 hours 39 mins

2 **a** 56 km **b** 216 km

c 315 km **d** 25 km

e 10 km **f** 5 km

3 36 km/h
4 12 600 m
5 31.25 s
6 41.4 km/h
7 a 8400 km b 1400 km/h
8 a 23.4 km b $4\frac{1}{3}$ m/s
9 a 520 km b 86.7 km/h
10 30 m/s
11 80

Exercise 6.5

1 a {1, 4, 9, 16, 25, 36, 49}
  b {2, 3, 5, 7, 11, 13, 17, 19, 23}
  c {Jan, Feb, Mar, Apr, May, June, July, Aug, Sept, Oct, Nov, Dec}
  d {1, 2, 3, 4, 6, 12}
  e {5}
  f {2, 4, 6, 8, 12, 14}
  g {21, 23, 25, 27, 29}
  h {151, 157, 163, 167, 173, 179, 181, 191, 193, 197, 199}

2 a {1, 7, 49} b {125}
  c {2, 5}
  d {1, 2, 4, 5, 10, 20}
  e {2, 3}
  f {Jan, June, July}

3 a {square numbers less than 17}
  b {odd numbers}
  c {months of the year beginning with M}
  d {multiples of 4}
  e {cube numbers}
  f {colours}

4 a {prime numbers}
  b {factors of 20}
  c {powers of 2 smaller than 33}
  d {Pythagorean triples}

5 a FALSE b TRUE
  c FALSE d FALSE
  e FALSE f FALSE

6 a FALSE b TRUE
  c FALSE d TRUE

7 a TRUE b FALSE
  c FALSE d TRUE

8 a TRUE b FALSE
  c TRUE d TRUE
  e FALSE f TRUE

Exercise 6.6

1 a

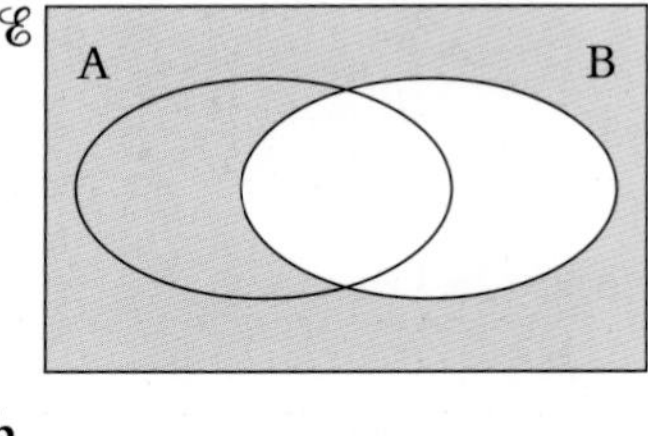

b

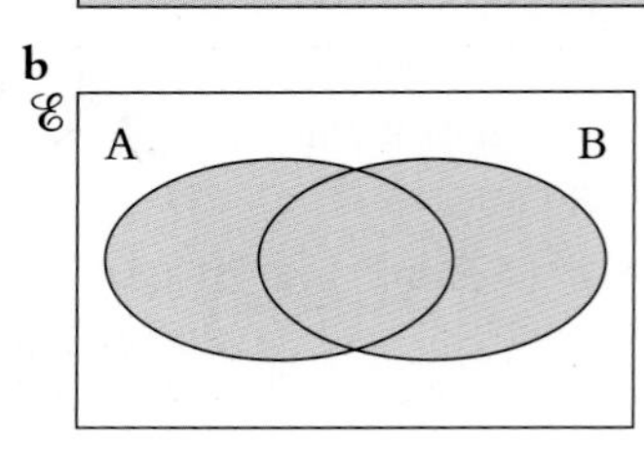

c

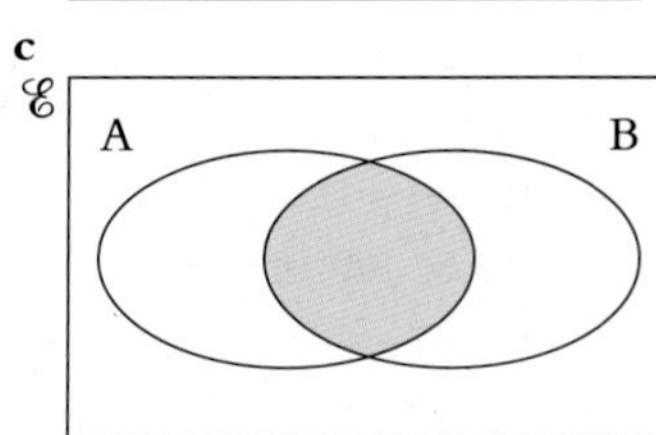

d

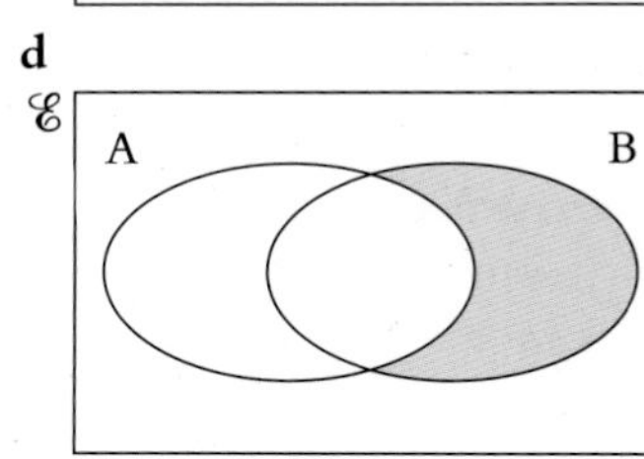

e

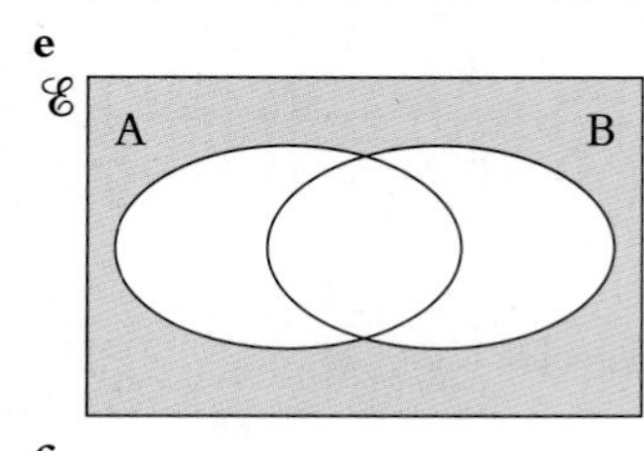

f

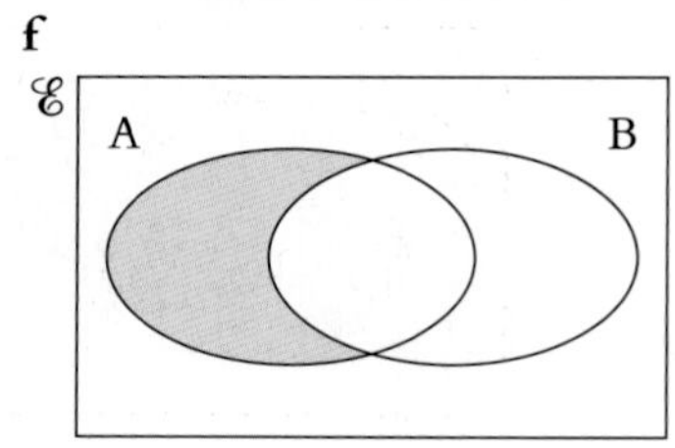

g

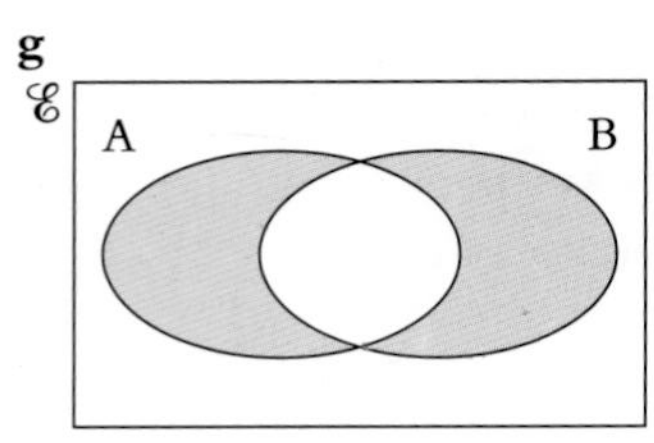

h

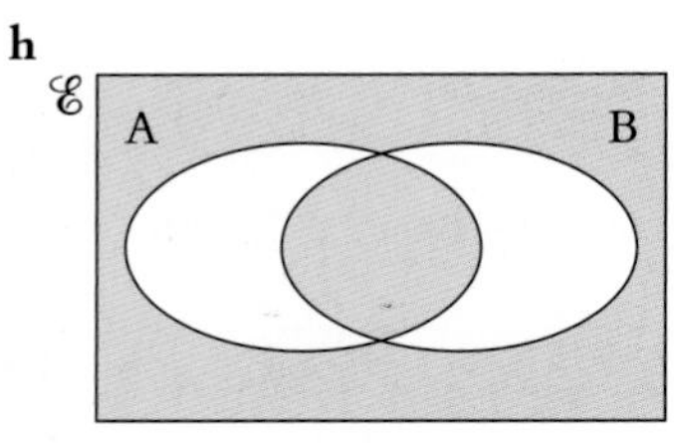

2 a

b

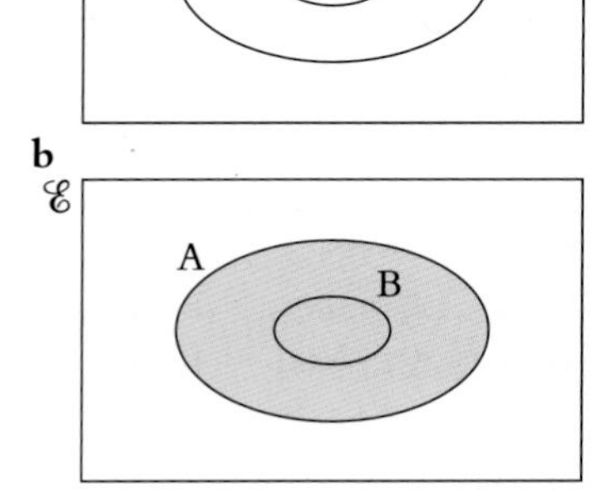

c

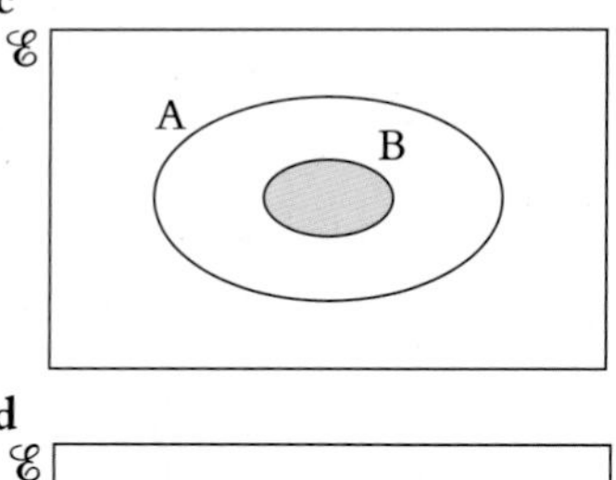

d

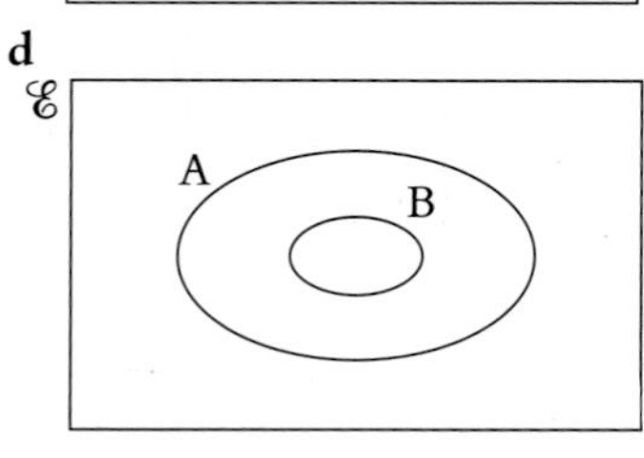

e

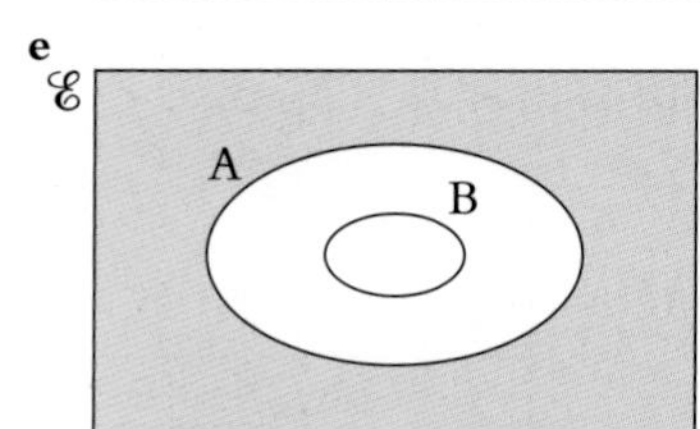

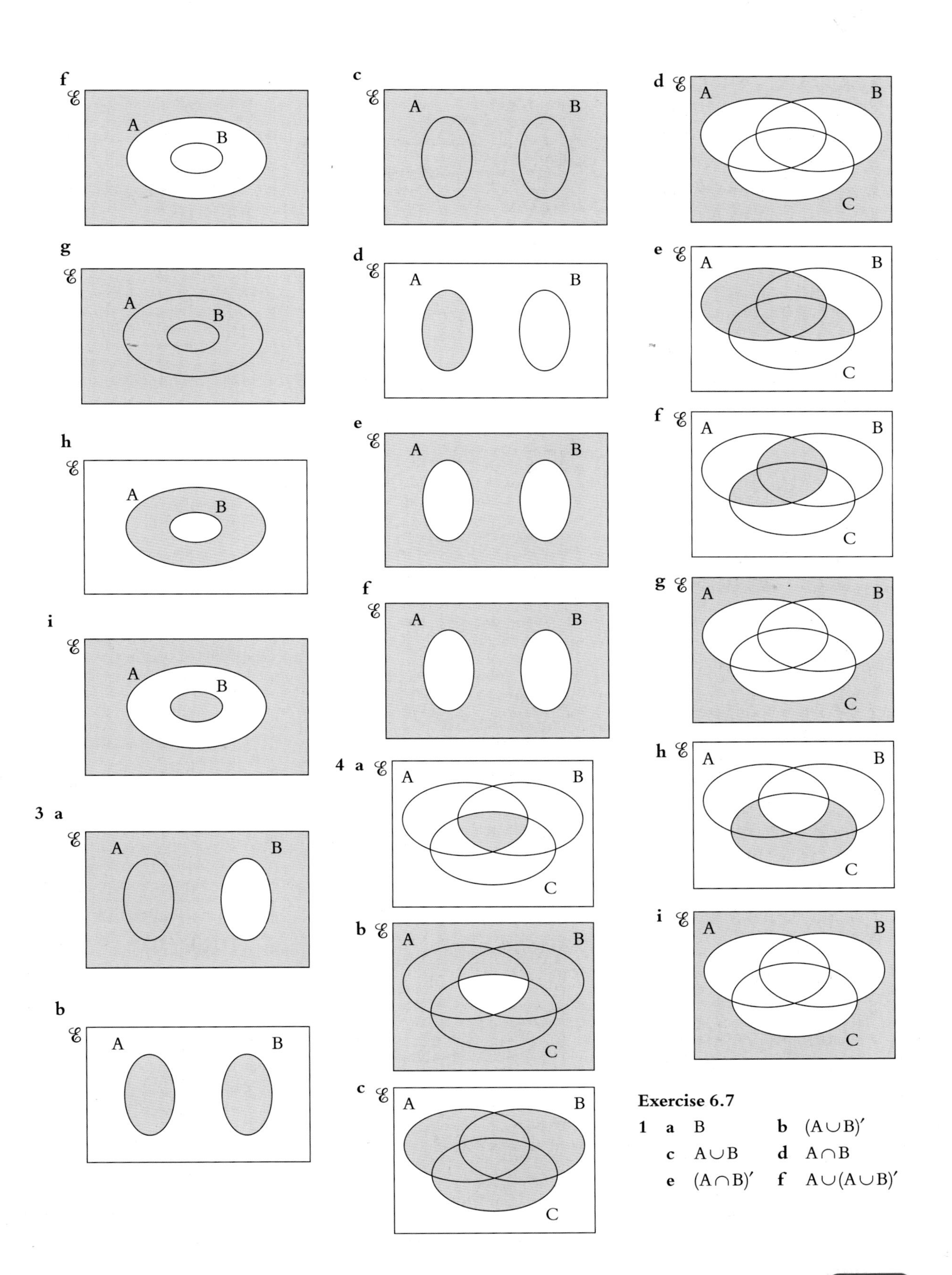

Exercise 6.7

1 **a** B **b** $(A \cup B)'$
c $A \cup B$ **d** $A \cap B$
e $(A \cap B)'$ **f** $A \cup (A \cup B)'$

UNIT 6

2 **a** $A \cup B \cup C$ **b** $A \cap C$
c A **d** $(A \cup B \cup C)'$
e $A \cap (B \cup C)'$ **f** $A \cap C \cap B'$
g $A \cap B'$ **h** $(A \cup C) \cap B'$
i $(A \cup B \cup C) \cap (A \cap B \cap C)'$

3 **a** $B' \cap C$ **b** $A' \cap B \cap C'$
c $C \cup (A \cap B)$

Exercise 6.8

1 **a** {3, 6, 9, 12,...}
b {1, 2, 4, 5, 10, 20}
c {1, 3, 5, 7,...}
d {2, 3, 5, 7, 11, 13, 17, 19}
e {2, 3} **f** {2}
g {−5, 5} **h** {3} **i** ∅

2 **a** {2, 3, 6, 7} **b** {3 , 5, 7, 8, 9}
c {1, 4, 5, 8, 9, 10}
d {1, 2, 4, 6, 10}
e {2, 3, 5, 6, 7, 8, 9}
f {3, 7} **g** {1, 4, 10}
h {2, 6}

3 **a** {1, 3, 4, 8, 9, 10, 13, 14}
b {2, 3, 4, 5, 6, 8, 10, 14}
c {11}
d {1, 3, 4, 7, 9, 11, 12, 13, 14}
e {1, 2, 3, 4, 5, 6, 7, 8, 9, 10, 12, 13, 14}
f {1, 9, 13} **g** {8, 9, 10}
h {7, 9, 12}

4 **a** {1, 2, 4, 5, 10}
b {4, 8} **c** {2, 3, 5, 7}
d ∅ **e** {4} **f** {2, 5}
g ∅ **h** {6, 9}
i

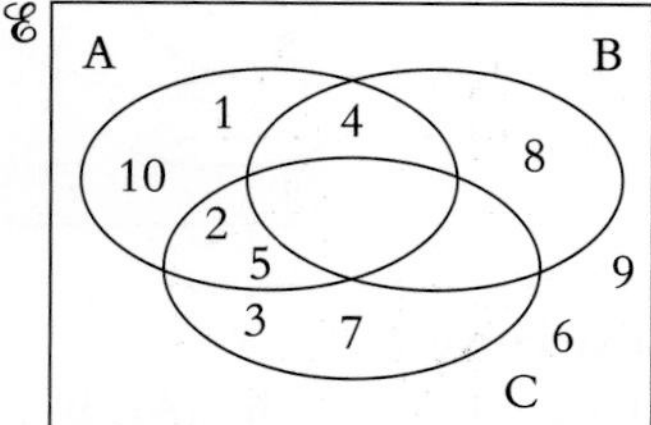

5 **a** {b, d, e, f } **b** {a, e, f, g}
c {c} **d** {a, b, c, d, g}
e {b, d} **f** {a, g}
g {a, b, d, e, f, g}
h {e, f}
i

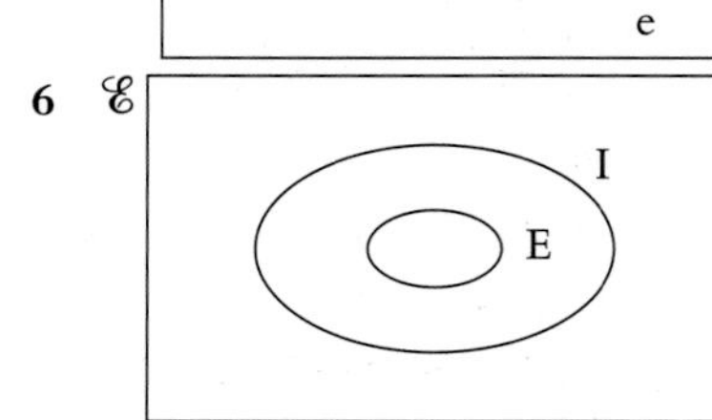

6

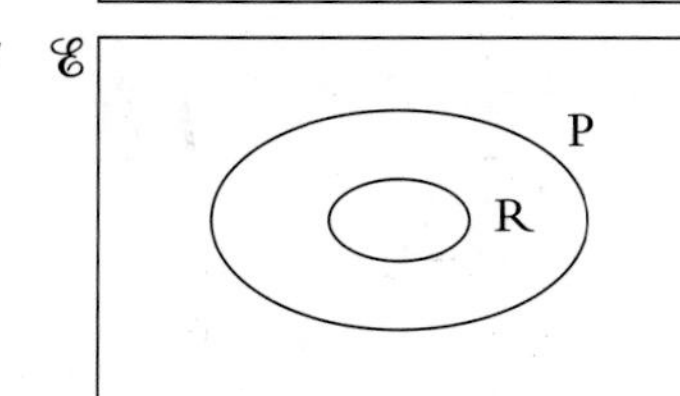

7 ℰ

P
R

Exercise 6.9

1 **a** 14 **b** 19 **c** 8
d 8 **e** 24 **f** 3
g 11 **h** 5

2 28 **3** 19 **4** 55
5 7 **6** 8
7 **a** 17 **b** 10 **c** 8
8 **a** 10 **b** 15
9 **a** 36 **b** 14 **c** 10
10 **a** 17 **b** 22
11 23
12 25
13 35

Exercise 6.10

1

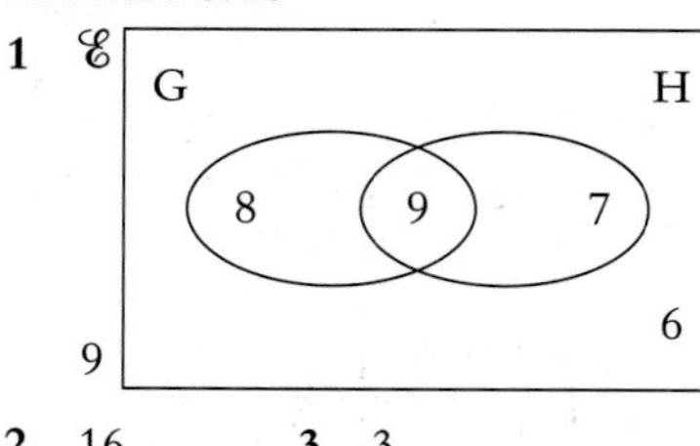

2 16 **3** 3
4 4 **5** 4
6 **a** 2 **b** 2
7 23

Exercise 6.11

1 **a** 3 **b** 8 **c** 3
d 9 **e** 4 **f** 2
g 10 **h** 3 **i** 12
j 13 **k** 20 **l** 20

2 **a** $\frac{1}{3}$ **b** $\frac{1}{2}$ **c** $\frac{1}{3}$
d $\frac{1}{5}$ **e** $\frac{1}{3}$ **f** $\frac{1}{5}$
g $\frac{1}{2}$ **h** $\frac{1}{4}$ **i** $\frac{1}{10}$
j $\frac{1}{6}$ **k** $\frac{1}{11}$ **l** $\frac{1}{2}$

3 **a** 125 **b** 16 **c** 8
d 27 **e** 100 000 **f** 1
g 4 **h** 243
i 100 **j** 2197 **k** 27
l 16

4 **a** $\frac{1}{125}$ **b** $\frac{1}{16}$ **c** $\frac{1}{64}$
d $\frac{1}{8}$ **e** $\frac{1}{1000}$ **f** $\frac{1}{4}$
g $\frac{1}{243}$ **h** $\frac{1}{32}$
i $\frac{1}{100000}$ **j** $\frac{1}{16}$
k $\frac{1}{27}$ **l** 1

5 **a** x^2 **b** x^2 **c** $x^{\frac{1}{2}}$
d x^{-2} **e** x **f** x^2
g x^{-1} **h** x^4 **i** x^3
j x **k** x^{-2} **l** x^2
m x^2 **n** 1 **o** $x^{\frac{5}{6}}$
p $x^{-\frac{1}{6}}$

6 **a** $15x$ **b** 16 **c** $6x^4$
d $3x^{\frac{1}{3}}$ **e** $6x^{1.5}$ **f** $3x^{-1}$
g $2x^{-4.5}$ **h** $4x^{5.5}$

7 **a** $5x^{-3}y^{-2}$ **b** $64x^{18}y^9$
c $5x^3y^{\frac{1}{4}}$ **d** $4x^5y$

Exercise 6.12

1 **a** 5 **b** 2 **c** 1
d 0 **e** 2 **f** 3
g 8 **h** 3 **i** 1
j 3 **k** 4 **l** 4

UNIT 6

2 a -4 b -2 c -7
d -2 e -2 f -3
g -3 h -4 i -1
j -3 k -4 l -1

3 a 3 b 3 c 4
d 2 e 1.5 f 2
g 9 h 1 i 3
j 1 k 4 l 0.5

4 a -2 d -1 c -1.5
d -1 e -4 f -2.5
g -3 h -3

5 a $\frac{1}{5}$ b $\frac{1}{3}$ c $\frac{1}{2}$
d $\frac{1}{2}$ e $\frac{1}{4}$ f $\frac{1}{3}$
g $\frac{1}{3}$ h $\frac{1}{2}$ i -7
j -3 k -1 l -0.5

6 a $\frac{1}{6}$ b 2 c 0
d 1.5 e -1 f 3
g -3 h 2

Exercise 6.13

1 a ± 4 b ± 6 c ± 10
d $\pm\frac{1}{2}$ e ± 5 f ± 3
g ± 6 h ± 11 i $\pm\frac{2}{3}$
j $\pm\frac{3}{4}$ k $\pm\frac{4}{5}$ l $\pm\frac{1}{3}$
m ± 9 n ± 1 o ± 20
p $\pm\frac{3}{2}$

2 No. You cannot find the square root of a negative number.

3 a 0, -6 b 0, 8 c 0, 10
d 0, -2 e 0, 5 f 0, 12
g 0, -2.25 h 0, $\frac{1}{3}$
i 0, -6.5 j 0, 25 k 0, -40
l 0, -3 m 0, 4 n 0, 15
o 0, 11 p 0, 1

4 a -2, -3 b 6, -1 c 2, -3
d 5, 4 e 7, -2 f -3, -9
g 5, -4 h 8, 2 i 3, -10
j 8 k 5 l -1, -3
m 3, -7 n 3 o 8, 1
p 5, -9

5 a 3, $-\frac{1}{2}$ b 1, $\frac{2}{3}$
c 0.6, -1 d 0.5, -7
e 2.5, 0. 5 f $\frac{1}{3}$, $-2\frac{1}{2}$
g 0.8, 0.5 h $\frac{1}{6}$, -5
i $\frac{2}{3}$, $-2\frac{1}{2}$ j 2.5, 1.5
k $\frac{2}{7}$, -5 l $1\frac{1}{4}$, $-\frac{2}{3}$

6 10, 2

7 a 3, 2 b 6, 2 c 3, -6
d -2, -4 e 4 f 7, -1
g 7, 5 h 5, -3 i 3, -6

8 a 6, -1 b 4, -5
c 6, -12 d 4, -3

Exercise 6.14

1 a 7, -6 b 17, -18

2 a 9, -11 b 12, -10

3 15, -18

4 a $(2x + 1)^2 + (2x)^2 = 29^2$
$8x^2 + 4x + 1 = 841$
$2x^2 + x - 210 = 0$
b 20, 21, 29

5 6

6 4

7 11 cm, 14 cm

8 10

9 2 s, 3 s

10 (5, 8), (−3, 0)

Exercise 6.15

1

−5	−4	−3	−2	−1
−1	−1.25	−1.67	−2.5	−5

0	1	2	3	4	5
–	5	2.5	1.67	1.25	1

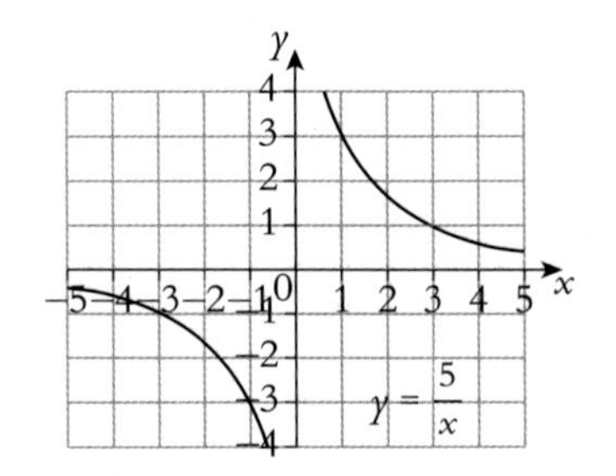

2

−5	−4	−3	−2	−1	0
1	1.25	1.67	2.5	5	–

1	2	3	4	5
−5	−2.5	−1.67	−1.25	−1

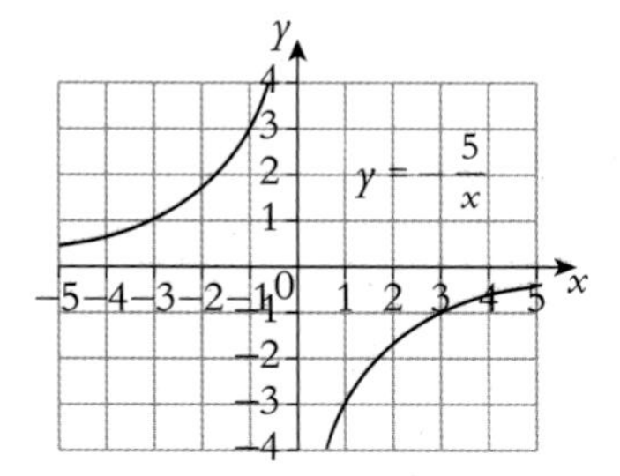

3

−5	−4	−3	−2	−1	0
1.2	1	0.67	0	−2	–

1	2	3	4	5
6	4	3.33	3	2.8

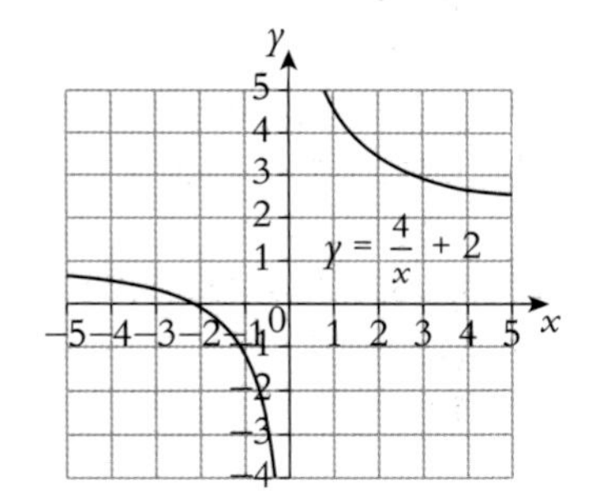

4

−2	−1	0	1	2
−1	−1.25	−1.67	−2.5	−5

3	4	5	6	7	8
–	5	2.5	1.67	1.25	1

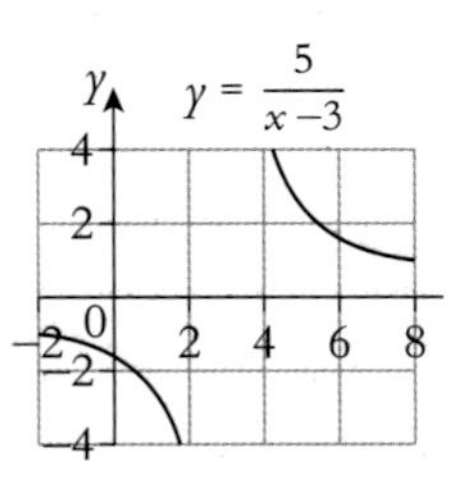

5

−6	−5	−4	−3	−2	−1
−6	−5.2	−4.5	−4	−4	−6

0	1	2	3	4	5	6
–	8	6	6	6.5	7.2	8

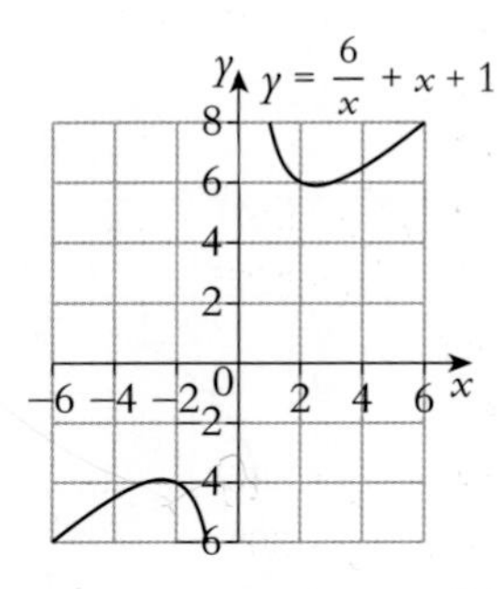

6

−4	−3	−2	−1	0
14.5	7	1	−5	–

1	2	3	4	−4
7	7	11	17.5	14.5

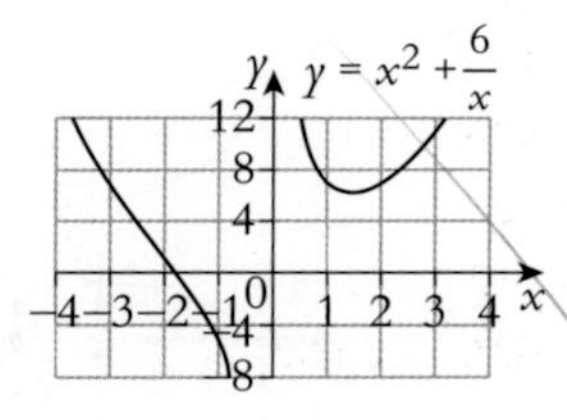

7

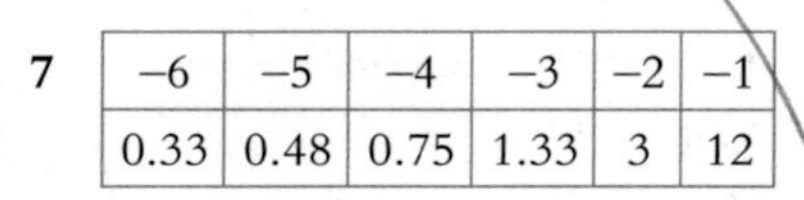

−6	−5	−4	−3	−2	−1
0.33	0.48	0.75	1.33	3	12

0	1	2	3	4	5	6
–	12	3	1.33	0.75	0.48	0.33

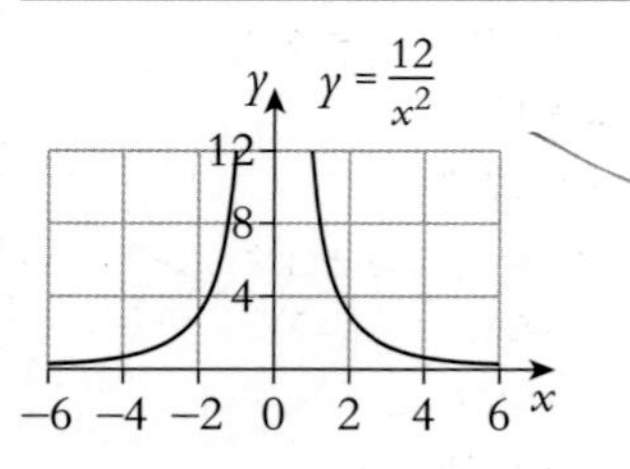

8

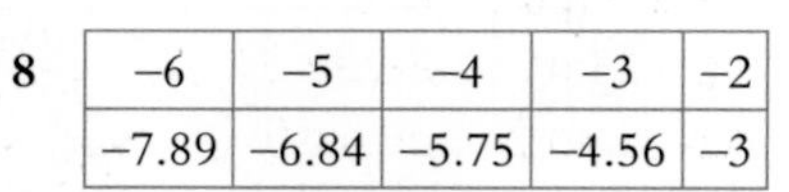

−6	−5	−4	−3	−2
−7.89	−6.84	−5.75	−4.56	−3

−1	0	1	2	3	4	5	6
1	–	3	1	1.44	2.25	3.16	4.11

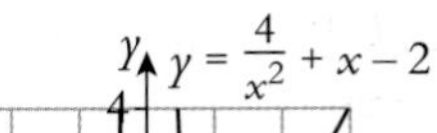

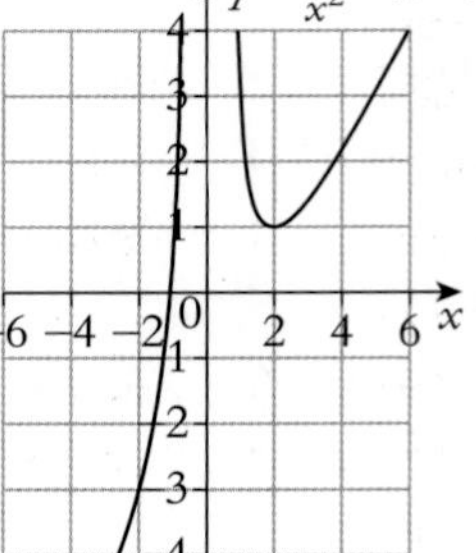

9 **a** answer given

b $T = 2x + y$

c answer given

d

x	5	10	15	20	30	40	50
T	110	70	63.3	65	76.7	92.5	110

e

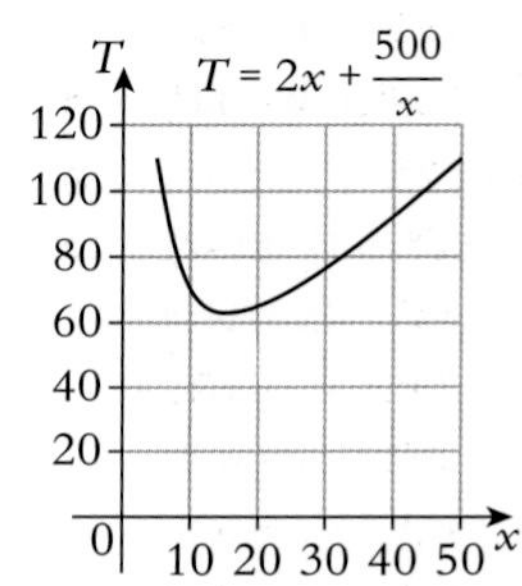

f minimum length = 63.2 m when x =15.8

Exercise 6.16

1 **a**

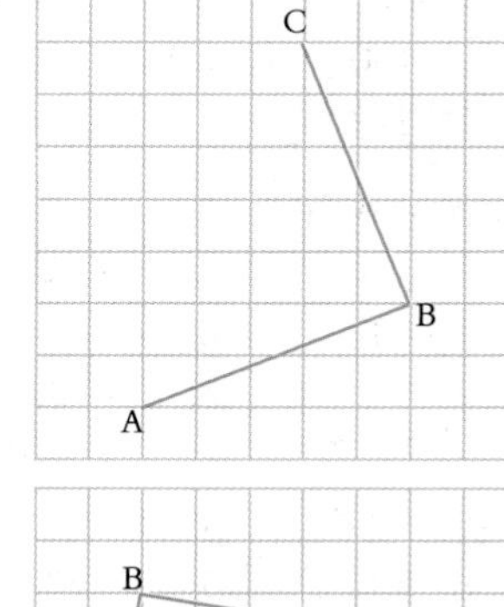

b

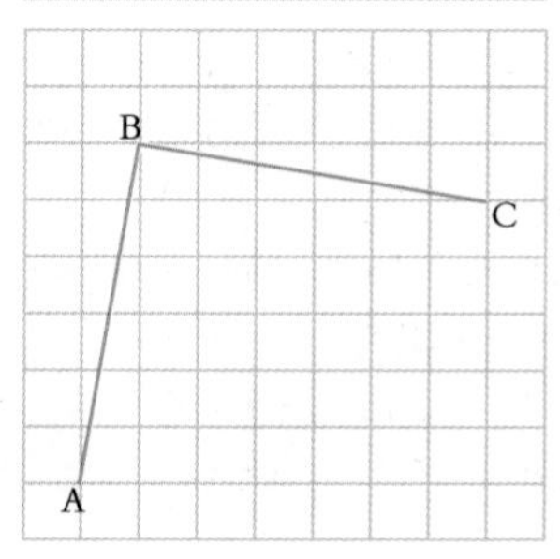

c

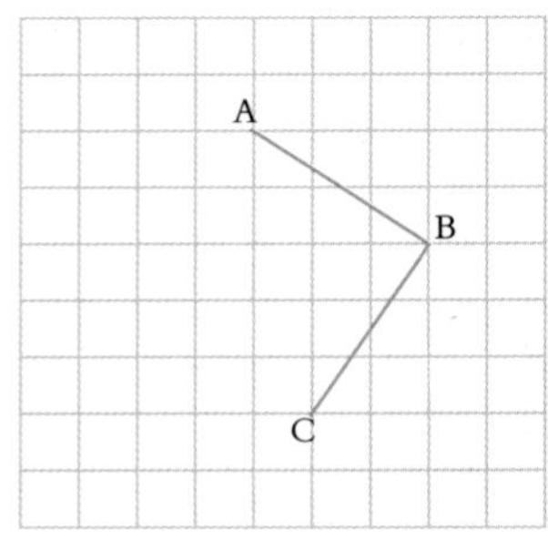

2 **a**

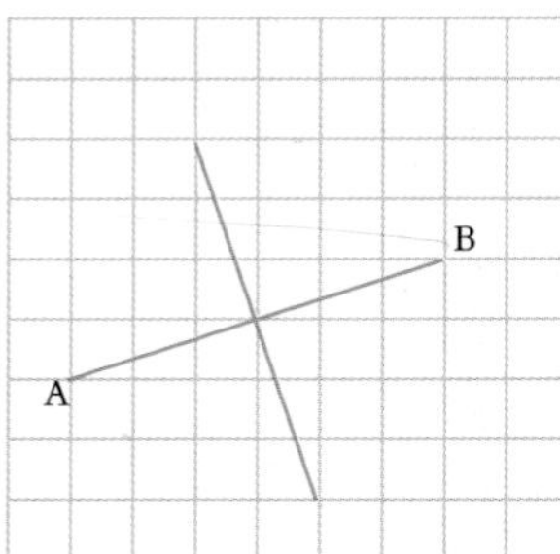

b

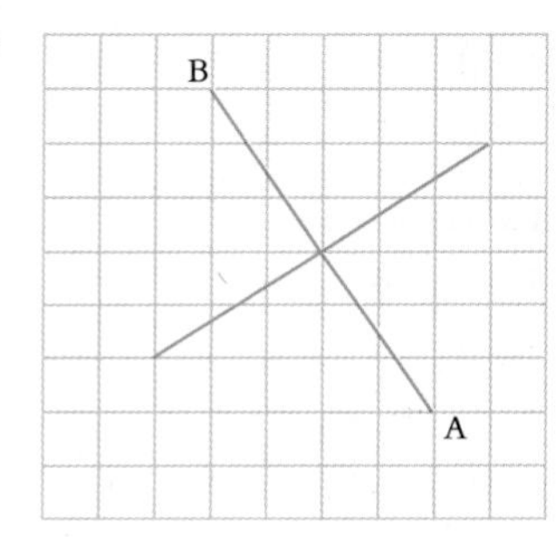

c

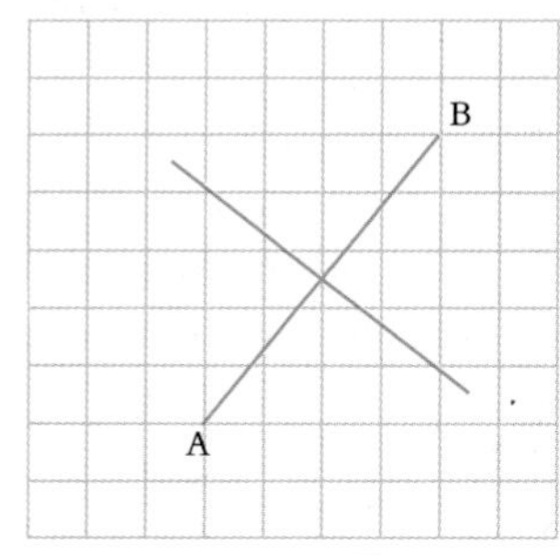

3 **a** $-\frac{1}{5}$ **b** −3 **c** $-\frac{5}{2}$

d $\frac{3}{2}$ **e** $\frac{1}{4}$ **f** $-\frac{7}{5}$

4 **a** neither **b** neither

c neither **d** perpendicular

e perpendicular

f perpendicular

g perpendicular

h neither **i** parallel

5 **a** **i** A, D **ii** C, E

b $y = -\frac{2}{3}x + k$

6 $y = -\frac{1}{3}x + 5y$

7 **a** $y = -\frac{1}{2}x + 5$

b $y = -\frac{1}{5}x + 7$

c $y = -\frac{3}{2}x - 8$

d $y = -\frac{1}{4}x$

8 **a** $\frac{1}{2}$ **b** -2

9 $\frac{1}{3} \times -3 = -1$

10 $-\frac{1}{2} \times 2 = -1$

11 **a** -4 **b** $(0, -4)$

12 $y = -2x$

13 P(−1.5, 0), Q(0, 2.25)

14 **a** (2, 3) **b** $y = \frac{2}{3}x + \frac{5}{3}$

15 (−3, 2)

16 (7, 0)

17 (8, 7), (0, −5)

18 **a** (1, 5) **b** −3

19 **a** $y = 2x - 4$ **b** (2, 0)

20 3 or − 4

21 3 or − 0.5

22 **a** **i** $2x + y = 7$

ii $x - 7y = -4$

b (3, 1)

Exercise 6.17

1 **a** 9.17 cm **b** 18.8 cm

c 63.6° **d** 71.6°

2 **a** 6.63 cm **b** 56.4°

c 33.2 cm^2 **d** 11.5 cm^2

3 **a** 53.1° **b** 106°

c 48 cm^2 **d** 33.4 cm^2

e 14.6 cm^2

4 **a** 8 cm **b** 24 cm

c 13.9 cm **d** 333 cm^2

e 131 cm^2

Exercise 6.18

1 9.17 cm

2 5.29 cm

3 13 cm

4 **a** 9 cm **b** 108 cm^2

5 **a** 1.71 cm **b** 4.70 cm

c 9.40 cm **d** 8.03 cm^2

e 30.5 cm^2 **f** 22.5 cm^2

6 16.0 cm

7 **a** 6 cm **b** 16 cm

c 128 cm^2

Exercise 6.19

1 $x = 240°$, $y = 30°$

2 $x = 100°$, $y = 260°$

3 $x = 110°$, $y = 35°$

4 $x = 286°$, $y = 53°$

5 $x = 76°$, $y = 52°$

6 $x = 100°$, $y = 80°$, $z = 50°$

7 $x = 32°$, $y = 16°$, $z = 16°$

8 $x = 96°$, $y = 84°$, $z = 48°$

9 $x = 43°$, $y = 90°$

10 $x = 30°$, $y = 120°$

11 $x = 50°$, $y = 40°$

12 $x = 90°$, $y = 64°$

13 $x = 54°$, $y = 27°$

14 $x = 74°$

15 $x = 102°$

16 $x = 30°$

17 $x = 20°$, $y = 70°$, $z = 50°$

18 $x = 130°$, $y = 90°$, $z = 65°$

19 $x = 22.5°$

Exercise 6.20

1 $x = 90°$ 2 $x = 62°$

3 $x = 30°$ 4 $x = 84°$

5 $x = 48°$ 6 $x = 52.5°$

7 $x = 70°$ 8 $x = 200°$

9 $x = 170°$ 10 $x = 45°$

11 $x = 30°$ 12 $x = 80°$

13 $x = 16°$, $y = 74°$

14 $x = 115°$ 15 $x = 45°$

16 $x = 28°$, $y = 62°$

17 $x = 88°$, $y = 46°$

18 $x = 50°$, $y = 40°$

Exercise 6.21

1 $x = 50°$

2 $x = 56°$, $y = 100°$

3 $x = 80°$, $y = 105°$

4 $x = 123°$, $y = 56°$

5 $x = 80°$, $y = 110°$

6 $x = 85°$, $y = 95°$

7 $x = 29°$

8 $x = 100°$, $y = 55°$

9 $x = 78°$, $y = 102°$

10 $x = 50°$

11 $x = 66°$

12 $x = 54°$, $y = 126°$

13 $x = 30°$, $y = 30°$

14 $x = 55°$, $y = 45°$

15 $x = 27°$, $y = 35°$

Exercise 6.22

1 $x = 28°$, $y = 62°$

2 $x = 22°$

3 $x = 27°$

4 $x = 17°$

5 $x = 57.5°$

6 $x = 105°$

7 $x = 113°$

8 $x = 68°$, $y = 93°$

9 $x = 9°$

10 $x = 26°$, $y = 64°$

11 $x = 25°$

12 $x = 64°$, $y = 32°$

13 $x = 142°$, $y = 38°$

14 $x = 18°$

15 $x = 68°$, $y = 44°$

16 $x = 130°$

17 $x = 100°$

18 $x = 28°$, $y = 62°$

19 $x = 66°$, $y = 71°$

20 $x = 196°$

21 $x = 130°$, $y = 65°$

22 $x = 70°$, $y = 35°$

23 $x = 40°$, $y = 140°$

24 $x = 70°$, $y = 110°$, $z = 35°$

25 $x = 69°$, $y = 21°$, $z = 21°$

26 $x = 40°$, $y = 43°$

27 $x = 60°$, $y = 50°$, $z = 30°$

28 $x = 60°$, $y = 120°$

29 $x = 22°$

30 $x = 58°$, $y = 52°$

For questions **31–36** explanations are also required.

31 $x = 28°$, $y = 62°$

32 $x = 76°$, $y = 38°$, $z = 14°$

33 $x = 90°$, $y = 60°$, $z = 30°$

34 $x = 56°$

35 $x = 48°$, $y = 24°$

36 $x = 130°$, $y = 90°$

Exercise 6.23

1 a

0.7 H: 0.7 H HH 0.49; 0.3 M HM 0.21

0.3 M: 0.7 H MH 0.21; 0.3 M MM 0.09

b i 0.49 **ii** 0.09
iii 0.42 **iv** 0.91

2 a $\frac{9}{25}$ **b** $\frac{4}{25}$ **c** $\frac{6}{25}$

3 a $\frac{2}{5}$ **b** $\frac{2}{15}$ **c** $\frac{7}{15}$

4 a 0.0001 **b** 0.9801
c 0.0198 **d** 0.0199

5 a

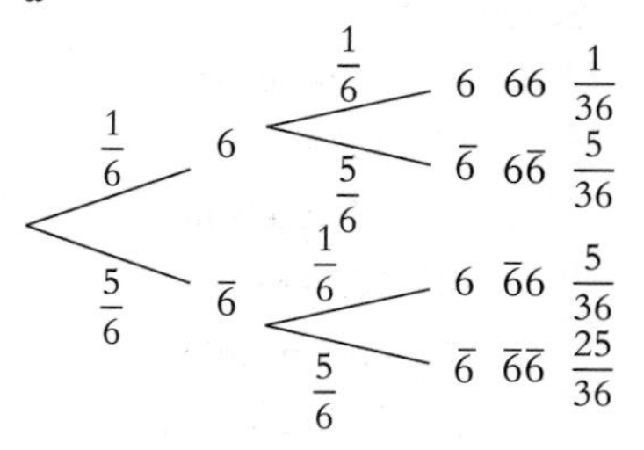

b i $\frac{1}{36}$ **ii** $\frac{25}{36}$
iii $\frac{5}{18}$

6 a $\frac{25}{49}$ **b** $\frac{4}{49}$
c $\frac{29}{49}$ **d** $\frac{45}{49}$

7 a

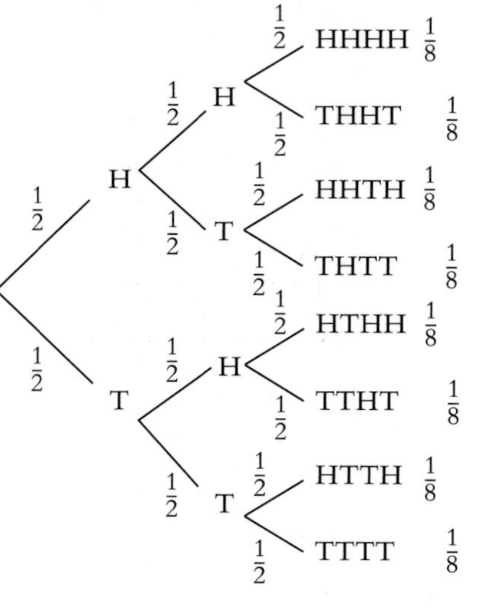

b i $\frac{1}{8}$ **ii** $\frac{1}{8}$
iii $\frac{3}{8}$ **iv** $\frac{1}{2}$

8 a $\frac{8}{27}$ **b** $\frac{1}{27}$
c $\frac{4}{9}$ **d** $\frac{26}{27}$

9 a $\frac{1}{216}$ **b** $\frac{125}{216}$
c $\frac{25}{72}$ **d** $\frac{5}{72}$

10 a 0.512 **b** 0.008 **c** 0.096

11 a

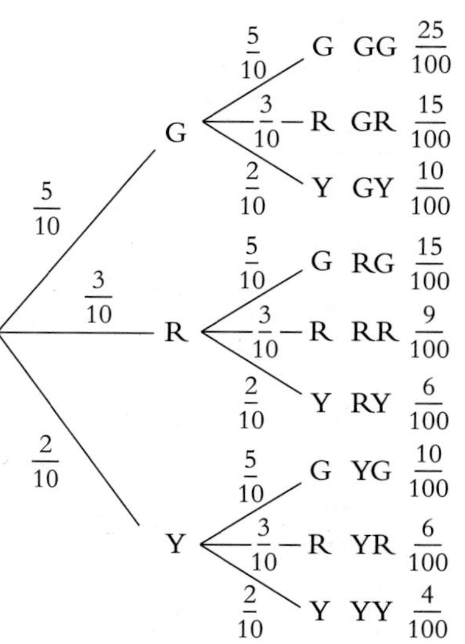

b i $\frac{1}{4}$ **ii** $\frac{9}{100}$ **iii** $\frac{1}{25}$
iv $\frac{19}{50}$ **v** $\frac{31}{50}$

12 a $\frac{1}{64}$ **b** $\frac{1}{27}$
c $\frac{125}{1728}$ **d** $\frac{5}{24}$

Exercise 6.24

1 a $\frac{10}{21}$ **b** $\frac{1}{21}$ **c** $\frac{11}{21}$
d $\frac{10}{21}$

2 a $\frac{1}{2}$ **b** 0 **c** 1

3 a $\frac{2}{5}$ **b** $\frac{1}{15}$ **c** $\frac{8}{15}$
d $\frac{3}{5}$

4 a $\frac{5}{12}$ **b** $\frac{1}{12}$ **c** $\frac{1}{2}$
d $\frac{11}{12}$

5 a $\frac{1}{6}$ **b** $\frac{1}{12}$ **c** $\frac{1}{36}$
d $\frac{5}{18}$ **e** $\frac{13}{18}$

6 a $\frac{1}{3}$ **b** $\frac{1}{4}$ **c** $\frac{5}{12}$
d $\frac{3}{4}$

7 a

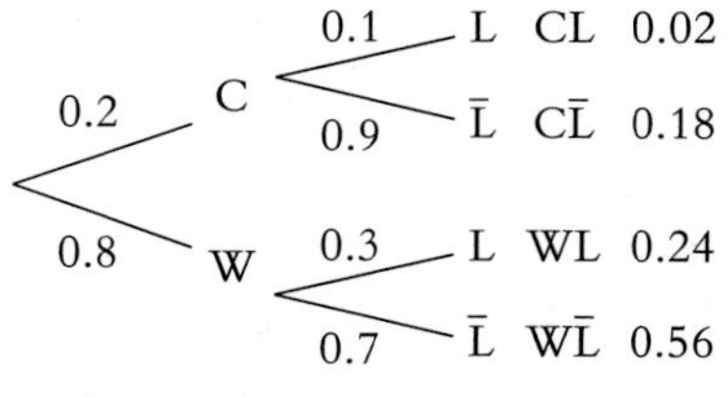

b i 0.26 **ii** 0.74

8 a i

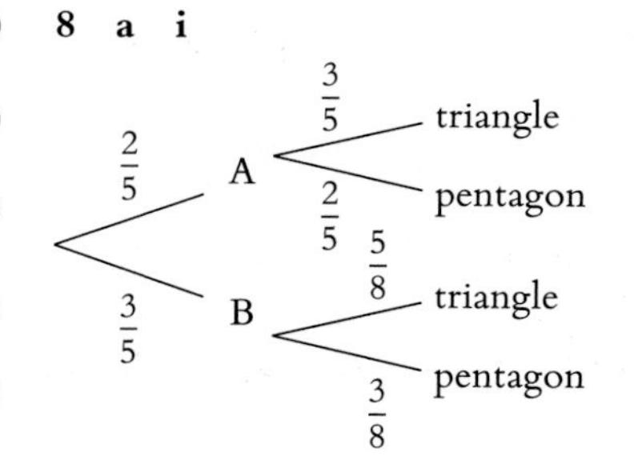

ii $\frac{123}{200}$ iii $\frac{77}{200}$

b i $\frac{71}{210}$ ii $\frac{39}{70}$

Exercise 6.25

1 $\frac{343}{4096}$ **2** $\frac{1}{64}$ **3** $\frac{625}{7776}$

4 $\frac{1}{256}$ **5** $\frac{10}{63}$

Examination-style questions

1 \$645 **2** \$840

3 \$675.90 **4** \$155.03

5 3% **6** 2.08%

7 15 m/s **8** 38.1 km/h

9 30 m/s

10 a i 31 ii 14

b 20

c $\mathscr{E}$

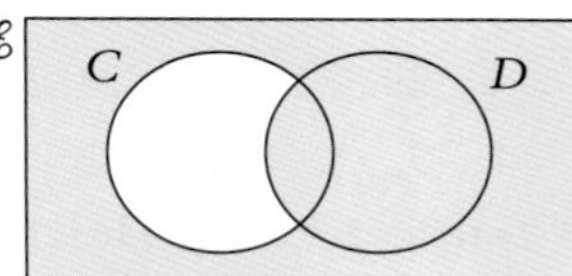

11 a $(A \cap B)' \cap C$

b $(A \cap B) \cap C'$

12 4

13 a $\mathscr{E}$

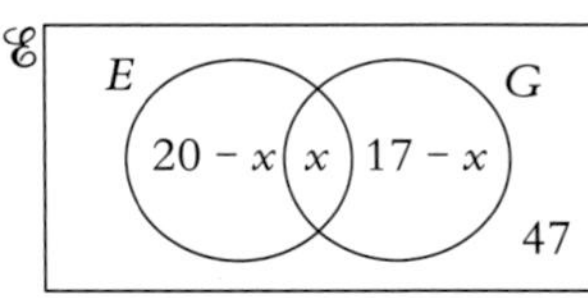

b $20 - x + x + 17 - x = 80$

c 4

14 2.5 **15** −12

16 a 1 **b** p^8

c $8q^{12}$ **d** $\frac{a^3}{216}$

17 a x^7 **b** $20x^3y^{11}$ **c** $9x^4$

18 $\frac{7}{9}$

19 a $(x-5)(x+8)$ **b** −8, 5

20 a $(3x+4)(x-2)$ **b** $-\frac{4}{3}$, 2

21 −1, 2.5

22 b $2(x+9)(2x-19)$ **c** 9.5 cm

23 a $y = \frac{48}{x}$

c

x	1	2	4	6	10	15	20
L	50	**28**	20	**20**	26.8	33.2	**42.4**

d

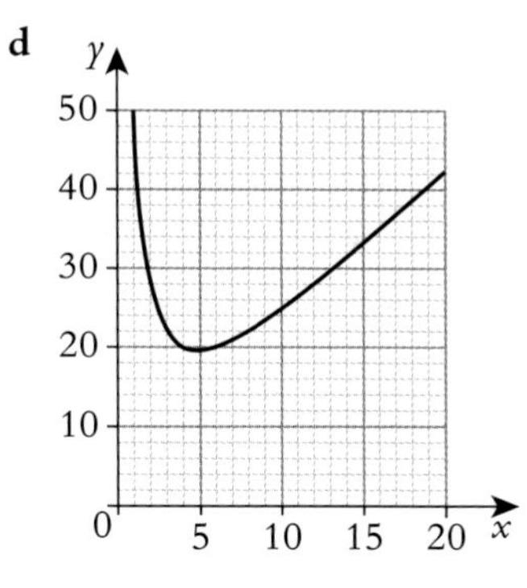

e 19.6 m

24 a $y = \frac{1}{4}x + \frac{23}{4}$

b $y = -4x + 10$

25 $y = \frac{3}{4}x - 2$

26 a $y = -\frac{1}{2}x - \frac{3}{2}$

b $y = 2x - 4$

c (1, −2)

27 $x = 46, y = 38, z = 54$

28 $x = 76, y = 52$

29 a 52° **b** 38°

c 76° **d** 38°

e 28°

30 0.49

31 a 0.14

b i 0.0225 ii 0.111

32 a i

First throw Second throw

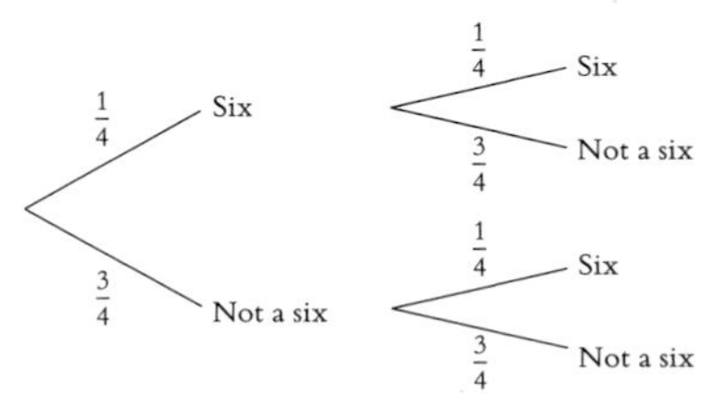

ii $\frac{7}{16}$ **b** $\frac{9}{64}$

33 $\frac{5}{9}$ **34** $\frac{8}{81}$

35 $\frac{3}{28}$

36 b $x = -2\frac{1}{2}$ or $x = 12$

c 35 cm

37 b $x = -\frac{2}{3}$ or $x = 12$

c 16 cm

38 a $\frac{1}{5}$

b i

Box Colour

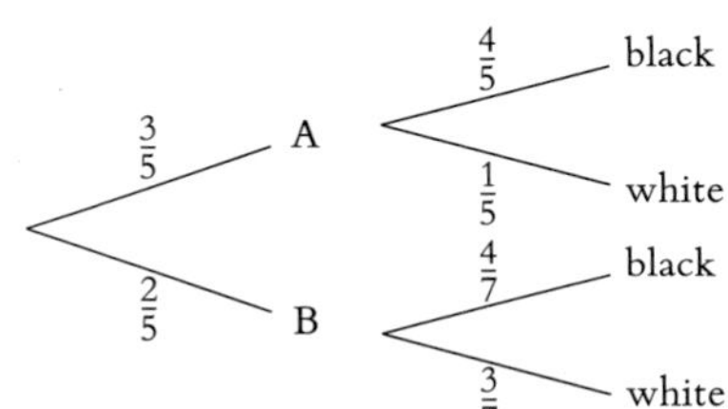

ii $\frac{6}{35}$ iii $\frac{124}{175}$

c i $\frac{27}{56}$ ii $\frac{13}{28}$

Unit 7

Exercise 7.1

1 a i 0.8 km/min

ii 48 km/h

b i 1.7 km/min

ii 102 km/h

c i 0.4 km/min

ii 24 km/h

d 64 km/h

2 a 5 mins

b i 0.08 km/min

ii 4.8 km/h

c i 0.0267 km/min

ii 1.6 km/h

d 2.4 km/h

3 a 1.25 m/s b 1.43 m/s
c 1.33 m/s

4 a 6 km, 10 km
b 5 min, 5 min
c i 0.8 km/min
ii 48 km/h
d i 0.5 km/min
ii 30 km/h

5 a 3 b 18 mins
c between 15 and 20 mins
d 50 km e 50 km/h

6 a answer given b 9:09 am
c

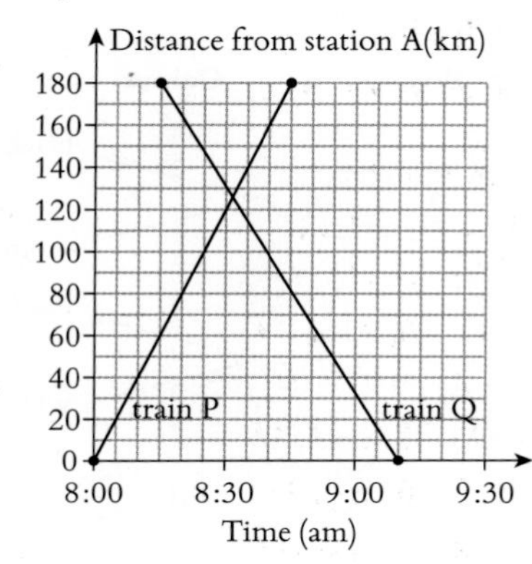

d 8:32 am (approx)

Exercise 7.2

1 a $2\frac{2}{3}$ m/s² b $\frac{2}{7}$ m/s²
c 75 m

2 a 1 m/s² b $\frac{2}{3}$ m/s²
c 700 m

3 a 0.5 m/s² b 0.25 m/s²
c 300 m d 5 m/s

4 a 1.25 m/s² b 250 m
c 1000 m d 20 m/s

5 a 0.25 m/s² b $\frac{1}{3}$ m/s²
c 200 m d 950 m

6 a 1.25 m/s² b −0.5 m/s²
c 950 m d 15.8 m/s

7 a 0.5 m/s² b 2000 m
c 225 m

8 a 0.8 m/s² b 0.4 m/s²
c 1120 m d 2560 m
e 2440 m

9 $9\frac{2}{3}$ km

10 a 32 m/s b 16 m/s

11 a 19.6 m/s b 1.96 m/s²

Exercise 7.3

1 $x = \frac{r}{p-q}$ 2 $x = \frac{b}{a-c}$

3 $x = \frac{c}{a-b}$ 4 $x = \frac{a}{1-b}$

5 $x = \frac{c-a}{1-b}$ 6 $x = \frac{5+b}{a-c}$

7 $x = \frac{3-b}{a+2}$ 8 $x = \frac{-a^2}{a+b}$

9 $x = \frac{3a+4b}{a-b}$ 10 $x = \frac{7b+3d}{3c-7a}$

11 $x = \frac{3a+2}{1-a}$ 12 $x = \frac{b}{c-a}$

13 $x = \frac{2e-1}{d+3e}$ 14 $x = \frac{d-2b}{2a-c}$

15 $x = \frac{3a+4b}{b-a}$ 16 $x = \frac{a^2c^2}{1+b}$

17 $x = \frac{2}{a^2-1}$ 18 $x = \frac{4p}{4-q^2}$

19 $x = \pm\sqrt{\frac{a}{1-b}}$ 20 $x = \sqrt{\frac{c-b}{a+d}}$

21 $x = \frac{fy}{y-f}$ 22 $x = \frac{2a}{a-2}$

23 a $x = \frac{5}{b-a}$ b $x = \frac{b}{a-c}$
c $x = \frac{p}{2r-5q}$ d $x = \frac{6f}{7h-g}$
e $x = \frac{a-b}{3c-2}$ f $x = \frac{a+d}{b+c}$

Exercise 7.4

1 a 23, 27
b 30, 35
c 32, 64
d 14, 15.5
e 8, 5
f 78.125, 195.3125
g $1\frac{1}{3}, \frac{4}{9}$
h 2.5, −0.75
i 13, 21
j 63, 127
k 17, 27
l 100, 1000

2 a 7, 8, 9, 10, 11
b 2, 7, 12, 17, 22
c 6, 10, 14, 18, 22
d 1, 3, 5, 7, 9
e 8, 11, 14, 17, 20
f −2, 0, 2, 4, 6
g 48, 46, 44, 42, 40
h 15, 14, 13, 12, 11
i 22, 19, 16, 13, 10
j 2.5, 3, 3.5, 4, 4.5
k 29.5, 29, 28.5, 28, 27.5
l 3, −1, −5, −9, −13

3 nth term = $2n-1$. He is incorrect.

4 a 13 b 58
c $3n - 2$

5 a $4n + 3$ b $3n - 3$
c $5n + 3$ d $3n - 7$
e $2.5n - 0.5$ f $7n - 15$
g $3n - 2$ h $4n - 5$
i $6n - 8$ j $1.5n - 7$
k $2.5n + 5.5$ l $0.02n + 6$

6 a $23 - 2n$ b $104 - 4n$
c $22 - 7n$ d $20 - 3n$
e $7 - 2n$ f $2.1 - 0.4n$
g $1 - 3n$ h $15 - 2n$
i $2.5 - 0.5n$

7 20th

8 38th

9 25th

10 63rd

Exercise 7.5

1 a 3, 6, 11, 18, 27
b −2, 1, 6, 13, 22

5 **a**

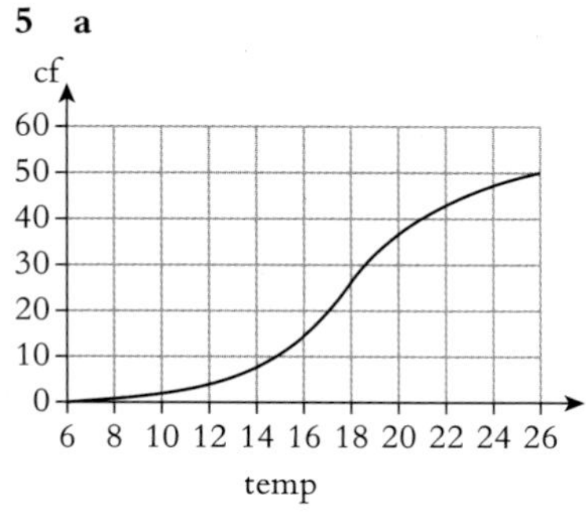

b 20 **c** 26%

6

Speed	Frequency
$0 < v \le 30$	6
$30 < v \le 40$	18
$40 < v \le 50$	26
$50 < v \le 60$	10

7

Mark	Frequency
$0 < m \le 20$	3
$20 < m \le 40$	18
$40 < m \le 60$	37
$60 < m \le 80$	21
$80 < m \le 100$	1

8 **a** 23

b 4

c

Time	cf
$t \le 10$	4
$t \le 20$	16
$t \le 30$	32
$t \le 40$	38
$t \le 50$	40

Time	Frequency
$0 < t \le 10$	4
$10 < t \le 20$	12
$20 < t \le 30$	16
$30 < t \le 40$	6
$40 < t \le 50$	2

d 22.5 minutes

9 **a** 5

b 13%

c

Time	cf
≤ 700	2
≤ 800	12
≤ 900	30
≤ 1000	38
≤ 1100	40

Time	Freq
$600 < t \le 700$	2
$700 < t \le 800$	10
$800 < t \le 900$	18
$900 < t \le 1000$	8
$1000 < t \le 1100$	2

d 845 hours

Exercise 7.14

1 **a** **i** 31 **ii** 22 **iii** 40 **iv** 18 **v** 28

b **i** 57 **ii** 44 **iii** 68 **iv** 24 **v** 52

c **i** 168 **ii** 165.5 **iii** 171 **iv** 5.5 **v** 167

d **i** 47 **ii** 44 **iii** 51 **iv** 7 **v** 45

2 **a**

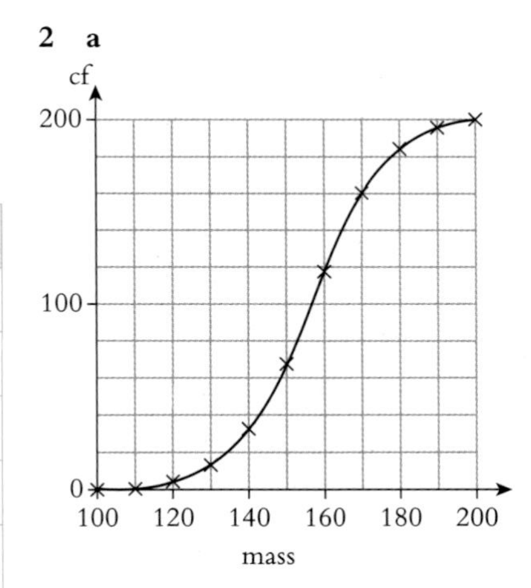

b **i** 156 **ii** 145 **iii** 168 **iv** 23 **v** 164

3 **a**

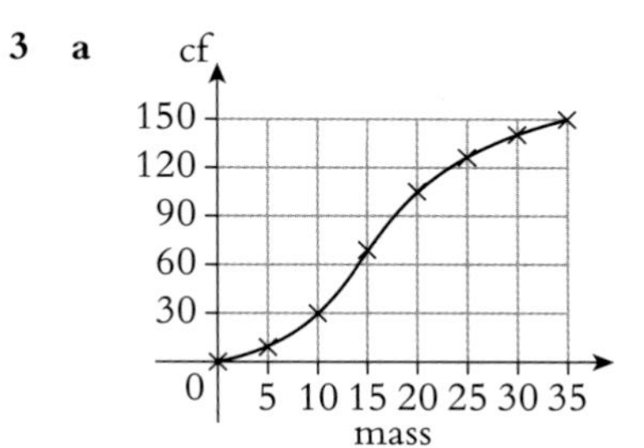

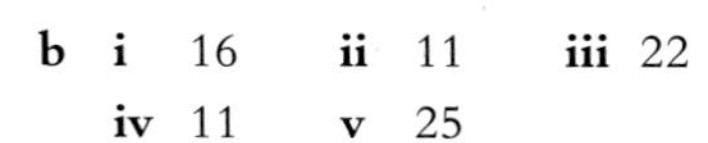

b **i** 16 **ii** 11 **iii** 22 **iv** 11 **v** 25

4 **a**

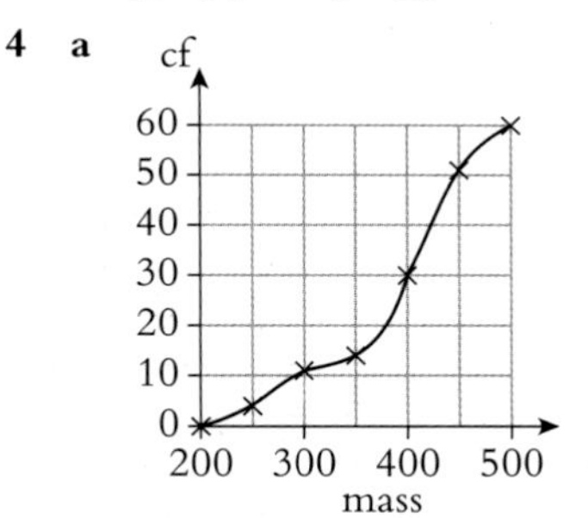

b **i** 400 **ii** 353 **iii** 436 **iv** 83 **v** 377

5 In general, the 10 to 20 year olds react quicker.
The 10 to 20 year olds are less varied.

6 **a**

	Median	LQ	UQ	IQR
Paper 1	46	30	62	32
Paper 2	58	47	67	20

b In general, the marks on paper 2 are higher.
The marks on paper 2 are less varied.

7 **a**

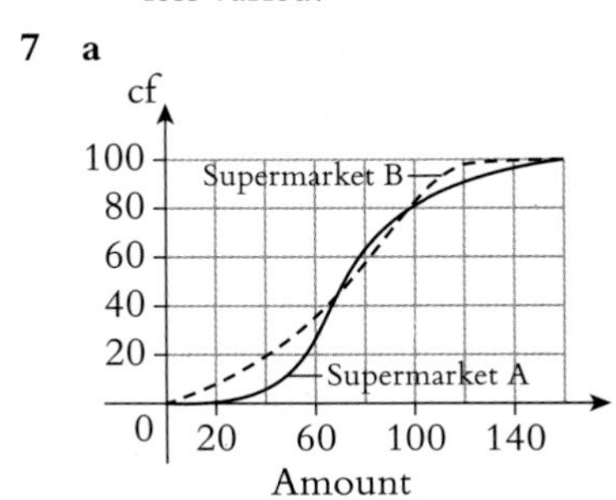

UNIT 7

b

	Median	LQ	UQ	IQR
Supermaket A	75	58	92	34
Supermaket B	71	44	93	49

c In general, shoppers spent more money in Supermarket A.
The amounts spent in Supermarket A are less varied.

Examination-style questions

1 **a** $0.75\,\text{m/s}^2$ **b** 4575 m

2 **a** $0.5\,\text{m/s}^2$ **b** 4525 m
c 22.625 m/s

3 $c = \dfrac{ab}{2b - a}$

4 $p = \dfrac{3}{5r + 4q}$

5 $x = \sqrt{\dfrac{c-d}{a-b}}$

6 3, −1 **7** 45th

8 **a** $7n - 3$ **b** 2^{n+1}

9 **a** $11 - 3n$
b $2n^2 - 3$
c $\dfrac{n(n+1)}{2}\ x^{n+1}$

10 17th

11 $a = 5, b = 3$

12 $p = 3, q = -7, r = 2$

13

Term	1	2	3	4	5		*n*th term
Sequence A	3	5	7	9	**11**		$\mathbf{2n + 1}$
Sequence B	$\frac{2}{9}$	$\frac{3}{16}$	$\frac{4}{25}$	$\frac{5}{36}$	$\mathbf{\frac{6}{49}}$		$\mathbf{\frac{n+1}{(n+2)^2}}$
Sequence C	1	2	8	16	**32**		$\mathbf{2^{n+1}}$
Sequence D	−2	1	6	13	**22**		$\mathbf{n^2 - 3}$
Sequence E	3	1	2	3	**10**		$\mathbf{2^{n+1} - n^2 + 3}$

14 **a**

x	−2	−1.5	−1	−0.5	0	0.5	1	1.5	2	2.5	3
y	2.69	**2.59**	2.44	**2.25**	2	1.66	1.2	0.59	−0.24	**−1.35**	−2.83

b

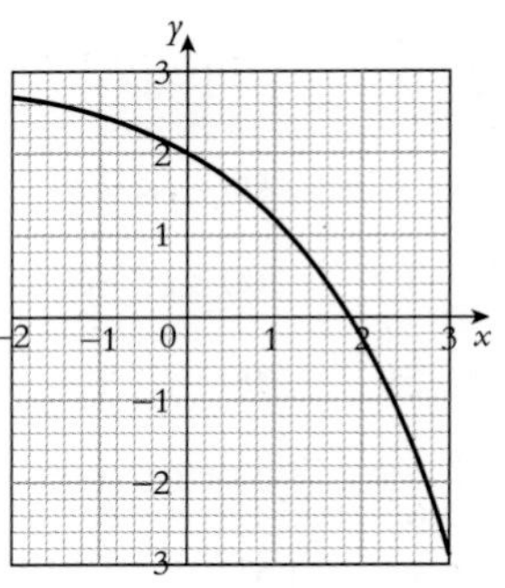

c $x = 1.2$ **d** −1.1 to −0.9

15 $\begin{pmatrix} 0 & 1 \\ -1 & 0 \end{pmatrix}$ **16** $\begin{pmatrix} -2 & 0 \\ 0 & 2 \end{pmatrix}$

17 **a** **i**

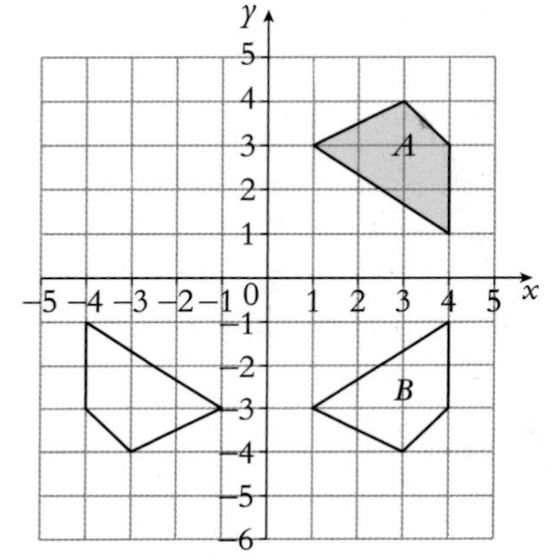

ii Rotation 180° about (0, 0)

b $\begin{pmatrix} 1 & 0 \\ 0 & -1 \end{pmatrix}$

18 **a** **i** Enlargement, scale factor −2, centre (0, 0)
ii $\begin{pmatrix} -2 & 0 \\ 0 & -2 \end{pmatrix}$

b **i** Reflection in the y-axis.
ii $\begin{pmatrix} -1 & 0 \\ 0 & 1 \end{pmatrix}$

c $\begin{pmatrix} 2 & 0 \\ 0 & -2 \end{pmatrix}$

19 **a** **i** enlargement, scale factor 0.5, centre (1, 6)
ii rotation, 90° clockwise, centre (−2, −1)
iii reflection in $y = x$

b $\begin{pmatrix} 0 & 1 \\ 1 & 0 \end{pmatrix}$

20 **a**

Time (t minutes)	$t \le 5.5$	$t \le 6$	$t \le 6.5$	$t \le 7$	$t \le 7.5$	$t \le 8$
Frequency	11	35	**68**	**84**	**96**	100

b

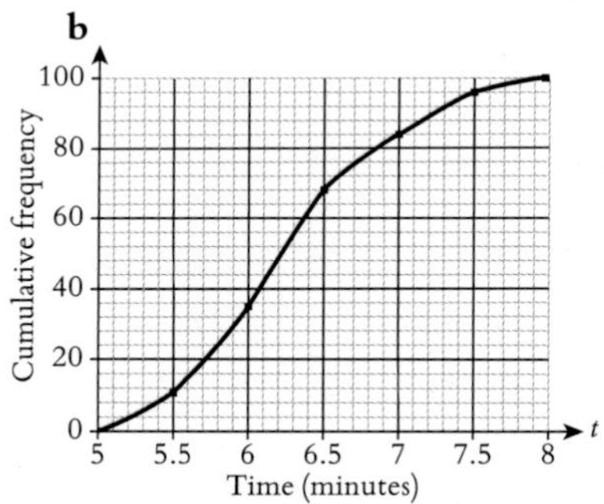

c **i** 6.2
ii 6.7
iii 4

Unit 8

Exercise 8.1

1 **a** $0.\dot{3}$ **b** $0.\dot{5}$ **c** $0.\dot{1}\dot{8}$
d $0.3\dot{1}\dot{8}$ **e** $0.6\dot{1}$ **f** $0.6\dot{5}$
g $0.\dot{1}\dot{5}$ **h** $0.\dot{1}7\dot{1}$
i $0.\dot{2}8571\dot{4}$ **j** $0.\dot{3}8461\dot{5}$

2 **a** $\frac{1}{3}$ **b** $\frac{7}{9}$ **c** $\frac{4}{11}$
d $\frac{68}{99}$ **e** $\frac{14}{99}$ **f** $\frac{145}{999}$

g $\frac{238}{999}$ h $\frac{47}{90}$ i $\frac{11}{18}$

j $\frac{3}{22}$

3 a $\sqrt{36}$ b $\sqrt{40}$ c $\sqrt{64}$

d $\sqrt{49}$ e $\sqrt{\frac{4}{9}}$

4 $\pi, \sqrt{\frac{5}{8}}$

5 $\sqrt{2}, \sqrt{3}, \sqrt{5}, \sqrt{6}$

6 a irrational b rational

7 a 25π cm², irrational

b 100 cm², rational

c $100 - 25\pi$ cm², irrational

Exercise 8.2

1 $890.38

2 $540.56 → $541

3 $5793

4 $330.50

5 $154.47

6 $550

7 $4150

8 3.5%

9 3.25

10 1.89

Exercise 8.3

1 $1774.82

2 $19 683

3 $6333

4 943

5 5.64 million

6 6554

7 24 hours

8 6.35

9 3.25

a $100(\sqrt{2}-1)\%$ (= 4.44% to 3 sf)

b 16

c 268 435 456

Exercise 8.4

1 The value of the square root was rounded too soon. Using $\sqrt{141} = 11.8743...$ gives the correct answers.

2 a 4, −6 b 0.56, −3.56

c 3.19, −2.19

d 2.73, −0.73

e 5.65, 0.35

f 2.53, −5.53

g 1.32, −5.32 h 2.39, 0.28

i 3.39, −0.89

j 1.85, −0.18

k 15.62, −25.62

l 0.54, −2.29

m 2.35, −0.85

n 1.45, −3.45

o 0.68, −3.68

p −1.26, −8.74

3 Impossible to solve because of the square root of a negative number

4 $x = 2.5$. There is only one (repeated) root of the equation.

5 a 2.61, −4.61

b 1.61, −5.61

c 4.45, −0.45

d −0.46, −6.54

e −0.46, −6.54

f $3, -\frac{1}{3}$

g 0.72, −0.61

h 7.52, −2.52

6 a 1.61, −5.61

b 4.46, −2.46

7 a 2.87, −4.87

b 3.48, −0.48

c 2.14, −5.14

Exercise 8.5

1 a 2 b 1 c 0 d 2

e 1 f 2 g 0 h 1

i 2 j 2

2 ±10 3 $k < 2$

Exercise 8.6

1 2.65 2 2.61 3 5.23

4 2.72 5 2.86, 0.14

6 a $A = w(w + 20)$

b 74.26 m, 337 m

7 a answer given

b 5

8 2.30

9 (0.414, −0.586), (−2.414, −3.414)

10 a answer given

b 10 c 12, 96

Exercise 8.7

1 a $\frac{5(x-3)}{x(x-5)}$ b $\frac{11x+27}{(x+3)(x+2)}$

c $\frac{5x+62}{(x-2)(x+7)}$ d $\frac{11x+4}{x(3x+1)}$

e $\frac{17x-24}{(2x+5)(3x-2)}$

f $\frac{-4x-19}{(1-2x)(x-4)}$

g $\frac{2x+5}{(x+2)(x+3)}$

h $\frac{9x+17}{(x+1)(x+3)}$

i $\frac{2+3x}{x(x+1)}$ j $\frac{x+3}{(x+2)(x+4)}$

k $\frac{-3x-1}{(x-3)(x+1)}$ l $\frac{x^2+y^2}{x+y}$

2 a −2.25 b −0.5 c $2, -\frac{1}{3}$

3 a $3, -1\frac{1}{4}$ b $2, -\frac{5}{6}$

c 11, −2 d 5, 4

e $-1, -2\frac{1}{16}$ f $4, 2\frac{5}{7}$

g 2 h $-1\frac{1}{2}, -5\frac{1}{2}$

i 6, 2

Exercise 8.8

1 a $y = 5x$ b 40 c 1.4

2 a $y = 8x$ b 24 c 0.75

3 a $y = 0.5x$ b 1.2 c 10

4 a $y = \frac{1}{3}x$ b $1\frac{1}{3}$ c 10.2

5 a $y = \frac{1}{10}x$ b 0.8 c 70

6

x	2	5	13
y	6	15	39

UNIT 8

7 a $v = 9.8t$ b 29.4 m/s
c 2.5 s

8 a $e = 4.4\,m$ b 3.52 cm
c 0.4 kg

9 a $d = 1.5V$ b 1.8 m
c 0.38 m^3

Exercise 8.9

1 a $y = 2x^2$ b 50 c ±8

2 a $y = 0.5x^2$ b 40.5 c ±8

3 a $y = 4x^3$ b 32 c 4

4 a $y = 2(x + 3)^2$ b 98
c 2, −8

5 a $y = 5(x - 2)^2$ b 180

6 a $y = 6\sqrt{x}$ b 15 c 64

7

x	1	2	5
y	2	16	250

8 a $d = 0.015v^2$ b 24

9 a $E = 12e^2$ b 75

10 a $R = 2v^2$ b 2450

Exercise 8.10

1 a $y = \frac{6}{x}$ b 1 c 4

2 a $y = \frac{24}{x}$ b 6 c 3

3 a $y = \frac{10}{x}$ b 20 c 4

4 a $y = \frac{1}{2x}$ b 0.25 c 0.5

5

x	2	4	6
y	18	9	6

6

x	2.5	4	5
y	40	25	20

7 a $w = \frac{300000}{f}$ b 2000
c 1500

8 a $P = \frac{300}{V}$ b 2 c 120

9 a $t = \frac{24}{n}$ b 8 c 6

Exercise 8.11

1 a B b C c E
d A e D

2 a $y = \frac{36}{x^2}$ b 9 c ±6

3 a $y = \frac{20}{\sqrt{x}}$ b 1.25 c 25

4 a $y = \frac{15}{x+2}$ b 1.5 c 2

5 a $y = \frac{16}{(x+1)^2}$ b 1 c 1, −2

6

x	±1	2	5
y	50	12.5	2

7 a $F = \frac{20}{d^2}$ b 80 c 10

8 a $I = \frac{126000}{d^2}$ b 12.6
c 251

9 a D b A c B d C

Exercise 8.12

1 78.5° 2 45.6°

3 61.6°, 118.4° 4 110.5°

5 11.5°, 168.5° 6 90°

7 180° 8 0°, 180°

9 no solution

10 no solution in range 11 139.5°

12 54.1°, 125.9° 13 73.4°

14 36.9°, 143.1° 15 138.6°

16 53.1°, 126.9°

17 no solution in range

18 no solution 19 60°

20 19.9°, 160.1° 21 30°, 150°

22 60°, 120° 23 41.8°, 138.2°

24 18.4°, 161.6°

Exercise 8.13

1 a 15.3 cm^2 b 44.5 cm^2
c 9.64 cm^2 d 6.68 cm^2
e 15.5 cm^2 f 32.7 cm^2

2 a 21.2 cm^2 b 43.3 cm^2
c 29.1 cm^2

3 a 171 cm^2 b 25.7 cm^2
c 59.4 cm^2

4 a 25.0 cm^2 b 28.4 cm^2
c 20.8 cm^2

5 374 cm^2

6 9.61

7 10.1

8 6.20

9 8.51

10 13.2

11 173 cm^2

12 125 cm^2

13 48.6, 131.4

Exercise 8.14

1 a 1.85 cm^2 b 7.34 cm^2
c 9.83 cm^2 d 11.0 cm^2
e 51.4 cm^2 f 1.33 cm^2

2 22.0

Exercise 8.15

1 a 3.38 b 4.31 c 15.1
d 5.38 e 9.95 f 4.82
g 6.84 h 4.16 i 10.5

2 a 30.9 b 43.7 c 24.2
d 35.8 e 28.0 f 46.4
g 59.1 h 47.6 i 41.2

3 7.01

4 a 5.71 cm b 51.0°

5 25.4°

6 a 9.56 km
b i 110° ii 147°

7 13.1 km

Exercise 8.16

1 34.6, 145.4 2 62.4, 117.6

Exercise 8.17

1 a 3.81 cm b 5.74 cm
c 2.63 cm d 6.03 cm
e 5.99 cm f 5.35 cm
g 5.31 cm h 9.27 cm
i 5.42 cm

2 a 34.8 b 16.1 c 45.5
d 34.1 e 97.2 f 35.6
g 112.9 h 22.3 i 120

Exercise 8.18

1 a 12.8 cm b 7.90 cm

2 a 5 cm b 92.1°

3 a 80° b 7.86 km

4 29.0°

5 104.5°

6 8.61

7 a 85°

b 17.4 cm^2

c 8.24

d 37.2

e 22.8°

8 32.4 cm

9 a 6.48 cm

b 45.2 cm^2

10 a 37.6 cm

b 24.2 cm

c 604 cm^2

11 1020 cm^2

12 a i 5 ii 10 iii 11.2

b 25

13 a answer given b 3.91

c 11.7 cm^2

14 a answer given b 4.74

c 13.9 cm^2

Exercise 8.19

1 a 1, 3.5, 4.5, 3

b

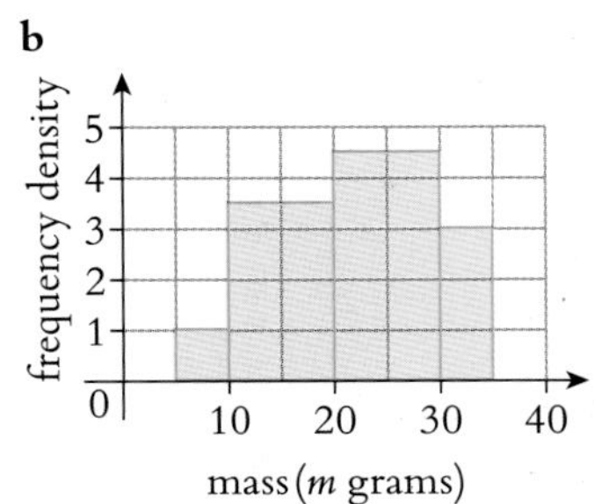

2 a 7, 13, 14, 2

b

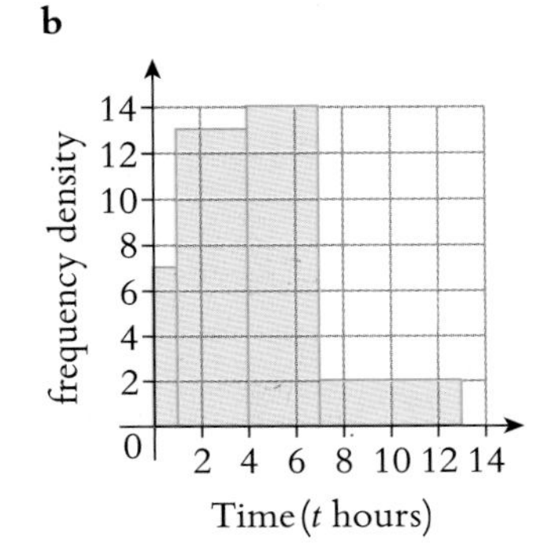

3 a 5, 8, 7, 6, 4

b

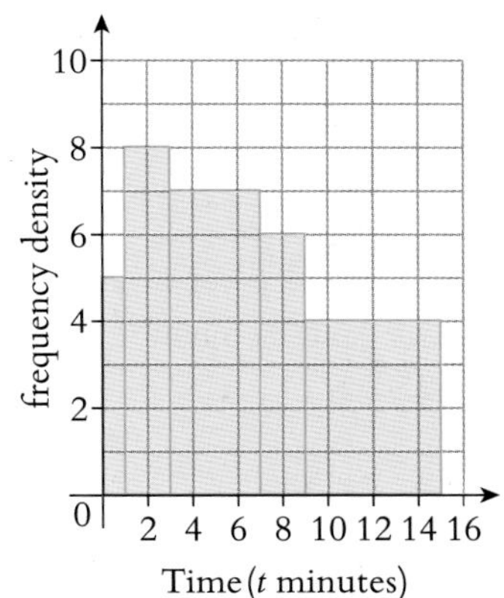

4 a 2.4, 5.8, 12.6, 9, 0.5

b

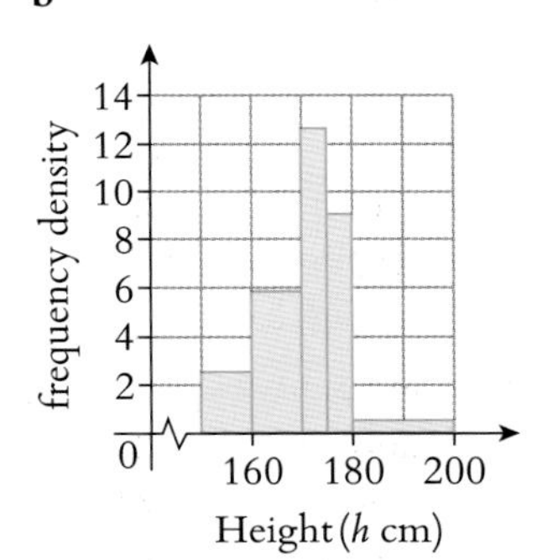

5

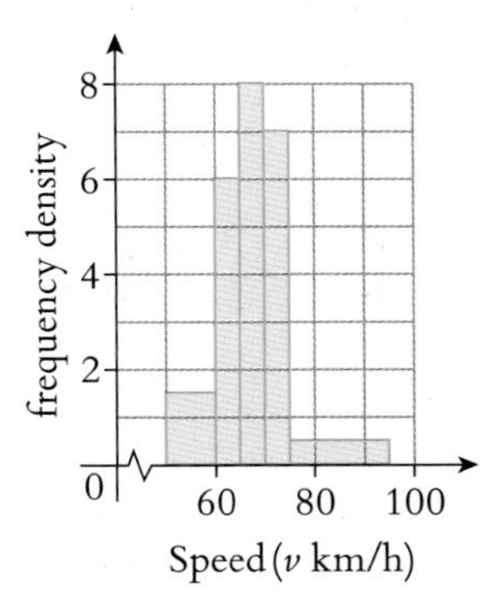

6

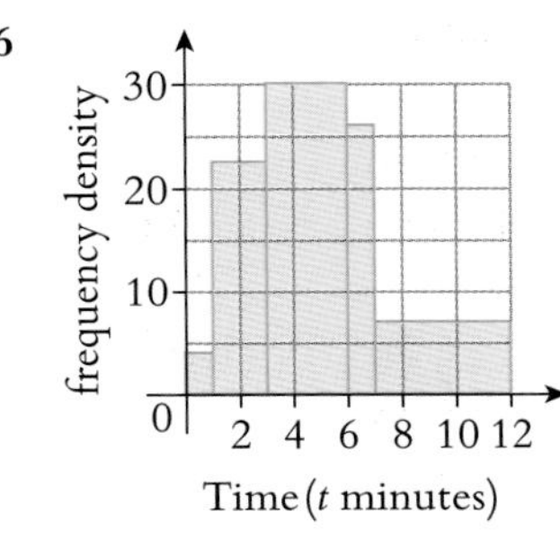

7

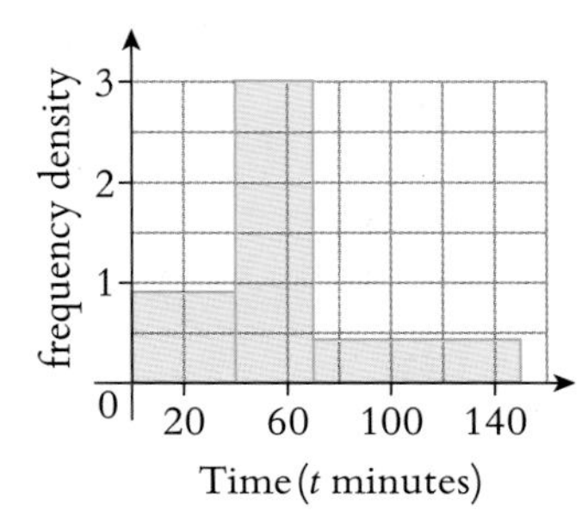

8 a 6, 8, 7, 9, 5

b 5.84 min

9 a 2, 8, 6, 5, 4

b 3.17 kg

10 9 cm, 24 cm, 2 cm

11 8 cm, 10 cm, 2.4 cm

12 a 8, 20, 15, 14, 12

b 19

c 2.19 hours

d $\frac{41}{70}$

13 a 48, 24

b

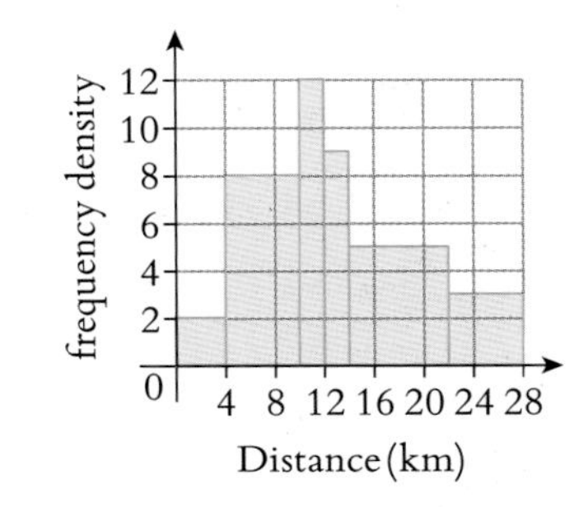

c 12.9 km

Examination-style questions

1 $\frac{5}{9}$

2 $\frac{16}{45}$

3 $\frac{263}{333}$

UNIT 8

4 $\pi, \sqrt{7}$

5 172

6 25

7 1.55

8 a 45 000

b 1.5%

c 28 600

9 −3.61, 1.11

10 −0.53, 1.86

11 a $r^2 + 5r - 75 = 0$

b 6.51

12 $\dfrac{3x+17}{(x-1)(x+3)}$

13 $\dfrac{3x^2-2x-13}{4(x-3)}$

14 −4, 5

15 6

16 $y = \dfrac{250}{x^2}$

17 30, 150

18 12.6 cm²

19 8.81 cm²

20 58.6

21 14.2

22 a i 3.95 km

ii 21.3°

b i 242°

ii 274°

23 a 81.1°

b 67.6°

c 2690 cm²

24

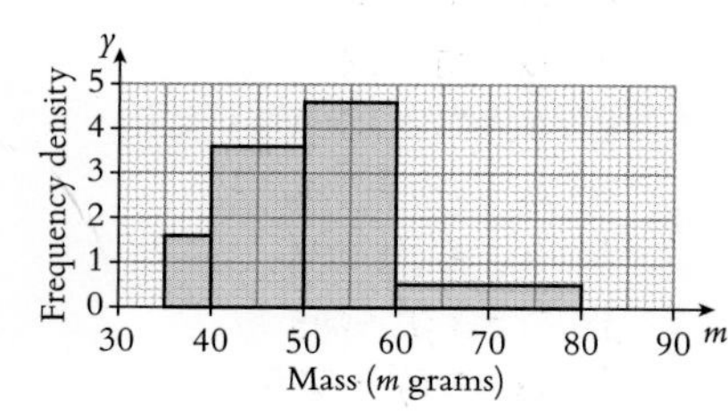

25 a

Mass (x kilograms)	$1.8 < x \le 2$	$2 < x \le 2.5$	$2.5 < x \le 3.5$	$3.5 < x \le 5.1$
Frequency	2	12	18	8

b 2.98 kg

26 a $y = 6$

b ii 1.17 or 6.83

iii 1.85 cm or 30.15 cm

27 a $\dfrac{50}{x}$

b $\dfrac{50}{x+2}$ or $\dfrac{50}{x} - 0.1$

d $x = -32.64$ or $x = 30.64$

e 1 hour 32 minutes

28 693 cm²

Unit 9

Exercise 9.1

1 a 5.5 cm, 6.5 cm

b 31.5 min, 32.5 min

c 31.5 kg, 32.5 kg

d 91.5 mm, 92.5 mm

e 7.65 cm, 7.75 cm

f 2.625 kg, 2.635 kg

g 62.85 g, 62.95 g

h 475.5 s, 476.5 s

i 4.935 m, 4.945 m

j 244.5 m, 245.5 m

2 $249.5 \le x < 250.5$

3 23 kg

4 $18 675, $18 525

5 30 825 cm, 30 675 cm

6 139.5 mm, 136.5 mm

7 66 cm, 65.2 cm

8 a 25.6 cm, 25.2 cm

b 39.8575 cm², 38.5875 cm²

9 a 82.2 cm, 81.8 cm

b 418.6925 cm², 414.5925 cm²

10 28 cm, 24 cm

11 a 107.171875 cm³, 100.544625 cm³

b 135.375 cm², 129.735 cm²

12 a 151.895625 cm³, 143.028375 cm³

b 180.715 cm², 173.995 cm²

13 494.8125 cm³, 352.6875 cm³

14 231

15 5.186249645 cm

16 132.61014 cm²

17 a 11.77765262 cm

b 22.90713754 cm²

18 a 428 cm, 426 cm

b 92 cm, 90 cm

19 214.2664365 cm³

20 1.633802817 m/s²

Exercise 9.2

1 a $x = -2, y = 3$

b $x = 9, y = 4$

c $x = 1, y = -2$

d $x = 8, y = 1\frac{1}{2}$

e $x = \frac{1}{2}, y = 7$

f $x = -2, y = 5$

g $x = 4, y = -3$

h $x = 2\frac{1}{2}, y = 4$

i $x = -\frac{1}{3}, y = 7\frac{2}{3}$

j $x = 2, y = 9$

k $x = -3, y = -3$

l $x = -4, y = 7$

2 a $x = 4, y = 1$

b $x = 16, y = 3$

c $x = -23, y = -78$

d $x = -22, y = -9$

3 a $x = 6, y = -4$

b $x = 3, y = 6$

c $x = -2, y = 1$

d $x = 8, y = -2$

e $x = 1, y = -4$

f $x = 5, y = -2$

g $x = -3, y = -1$

h $x = 8, y = 1\frac{1}{2}$

i $x = -\frac{1}{3}, y = 0$

j $x = -10, y = 9$

k $x = 3, y = 4\frac{1}{2}$

l $x = \frac{2}{3}, y = -2$

4 (2, 4) 5 (−1, −4)

6 34 7 \$3.65

8 42 9 13

10 546

Exercise 9.3

1 a $15x + 3y \le 900$

b $x \ge 40, y \ge 20$

c

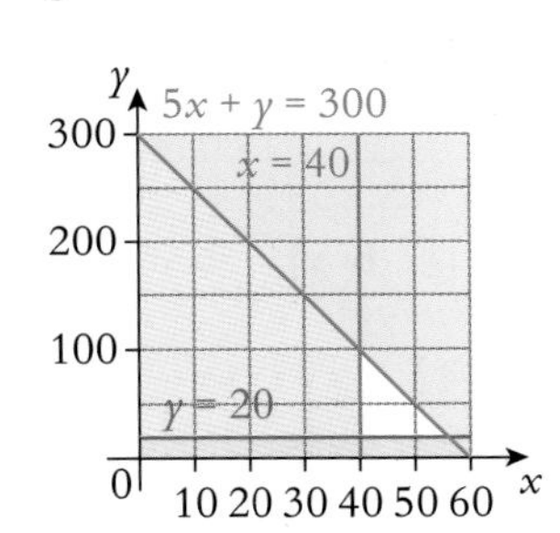

d 56

2 a $x + y \le 15, y \ge 6, y \ge 2x$

b

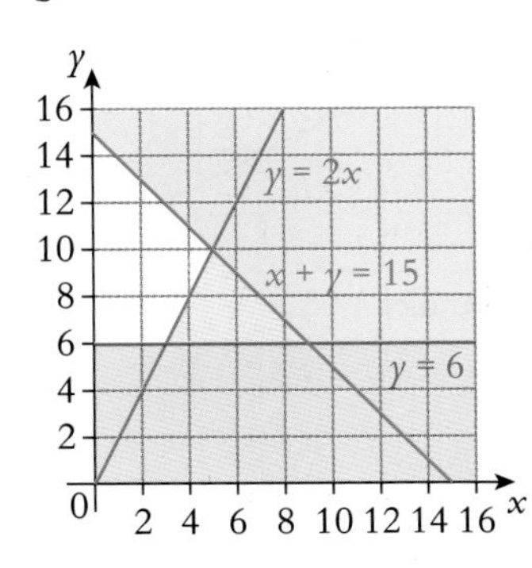

c 5

3 a $400x + 200y \le 9000$

b $y < x, y \ge 10$

c

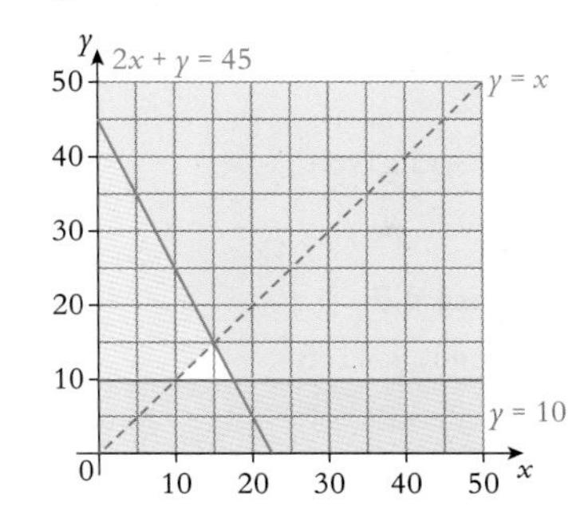

d 17

4 a $16x + 10y \ge 100$

b $x + y \le 8 \quad x \ge 1 \quad y \ge 3$

c

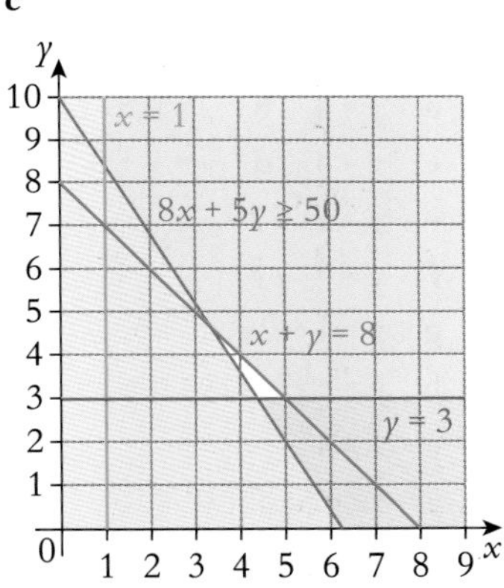

d 5

Exercise 9.4

1 a $4x + 3y \le 48, x + y \le 14,$
$x \ge 5, y \ge 5$

b

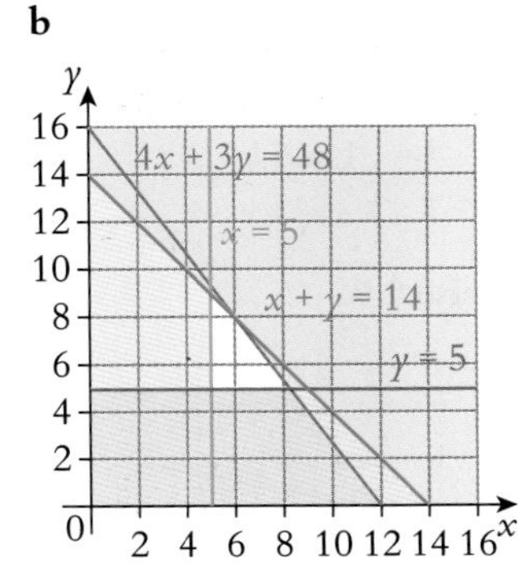

c \$230

2 a $5x + 3y \ge 45, x + y \le 11, y \ge 2$

b

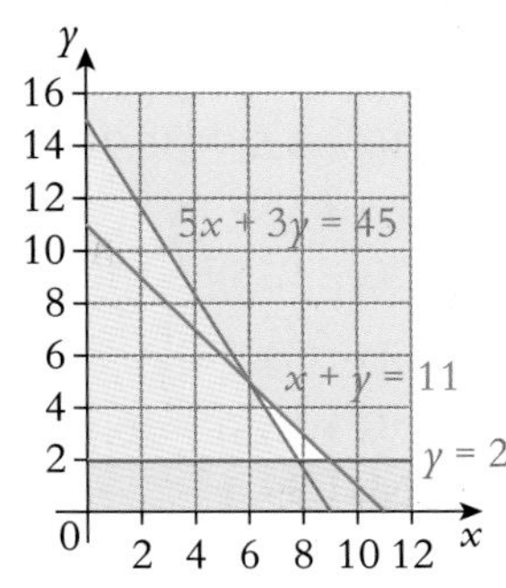

c \$96

3 a $2x + y \le 200, 4x + 5y \le 580,$
$x \ge 50, y \ge 40$

b

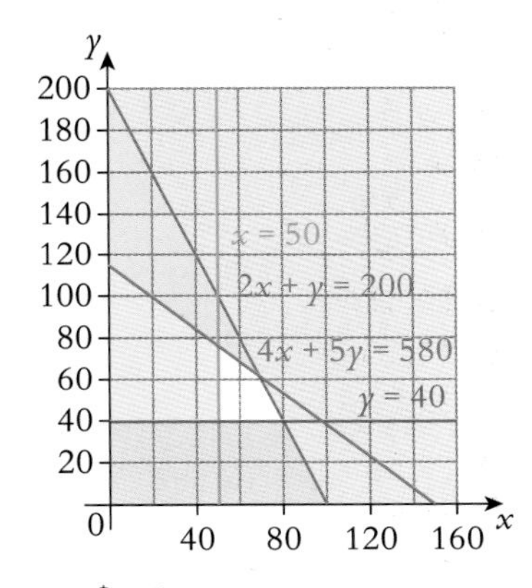

c \$165

Exercise 9.5

1 a $(x - 1)^2 - 1$

b $(x + 2)^2 - 4$

c $(x - 5)^2 - 25$

d $(x + 10)^2 - 100$

e $(x - 8)^2 - 64$

f $(x - 6)^2 - 36$

g $(x + 9)^2 - 81$

h $(x + 18)^2 - 324$

i $(x + 11)^2 - 121$

j $(x + 12)^2 - 144$

k $(x - 1.5)^2 - 2.25$

l $(x + 2.5)^2 - 6.25$

m $(x + 3.5)^2 - 12.25$

n $(x + 4.5)^2 - 20.25$

o $(x - 10)^2 - 100$

p $(x + 0.5)^2 - 0.25$

2 **a** $5(x-2)^2-20$
b $3(x+2)^2-12$
c $4(x+2.5)^2-25$
d $3(x-3.5)^2-36.75$
e $2(x+3)^2-18$
f $2(x-1.25)^2-3.125$
g $3(x-\frac{2}{3})^2-1\frac{1}{3}$
h $2(x+2.25)^2-10.125$

3 **a** $(x-4)^2-1$
b $(x-2)^2-16$
c $(x-3)^2-13$
d $(x+2)^2-7$
e $(x+6)^2-41$
f $(x-4)^2-15$
g $(x+5)^2-28$
h $(x+4.5)^2-22.25$

4 $a=5, b=-21$

5 $p=4, q=-19$

6 $p=2, q=-5$

7 **a** $x=3\pm\sqrt{b-9}$
b $x=-5\pm\sqrt{a+b+25}$

Exercise 9.6

1 **a** −2, −4 **b** −2
c 2, −4 **d** 7.41, 4.59
e 4.41, 1.59
f 0.21, −14.21
g 21.5, −1.53
h 8.16, 1.84
i 3.65, −1.65

2 **a** −1, −2 **b** 4, 3
c 2, −3 **d** 3.62, 1.38
e 2.70, −3.70
f 10.72, 0.28

3 **a** $2, -\frac{1}{3}$ **b** $-\frac{1}{2}, -3$
c −3, −5 **d** 7.74, 0.26
e −0.42, −1.58
f 0.65, −4.65

4 **a** 0.24, −12.24
b 4.61, −2.61

5 **a** 6.90, −2.90
b 3.83, −1.83
c 3.11, −1.61
d 3.30, −0.30
e 1.29, 0.31
f 1.65, −3.65

6 7

7 answer given

8 3

Exercise 9.7

1 −1.41, 1.41, 5

2 −1.2, 3.2

3 **a** −2.45, 2.45 **b** −1.3, 2.3
c −0.7, 2.7

4 **a** 0, 3 **b** −0.45, 4.45
c −1.6, 2.6

5 **a** 2.7 **b** 2.1, 0.2, −2.3
c 0.84

6 **a** $y=4$ **b** $y=0.5$
c $y=4x$ **d** $y=x+2$
e $y=x^2-2$
f $y=\frac{1}{x}+3$
g $y=1-2x$
h $y=\frac{1}{x^2}$

7 $y=x+5$

8 $y=5x-5$

Exercise 9.8

1 **a** −0.8, 4.8 **b** $k=1$
c $k<1$

2 **a** 0.2, 2.8
b $k=2$, $k=3$, $k=4$

3 **a** −2.4, 0.4, 2 **b** $k=3$
c $k<3$

4 **a** −1.9, 0.3, 1.5
b $k=-1$ and $k=3$
c $k>3$ and $k<-1$

Exercise 9.9

1 **a**

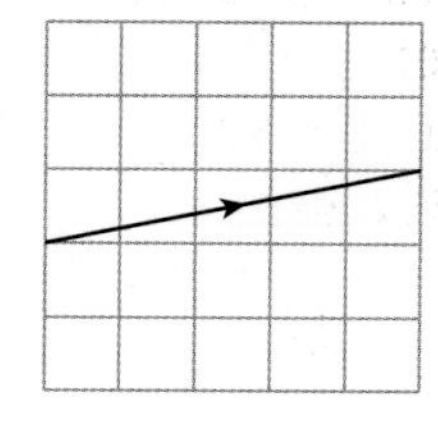

b

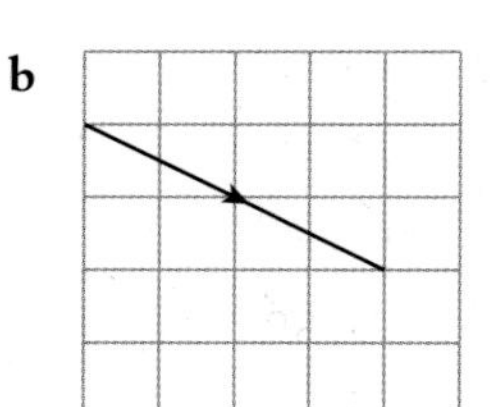

c

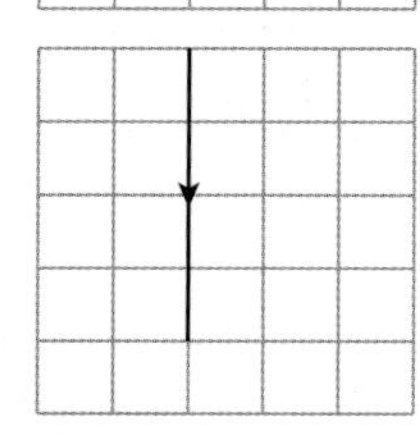

d

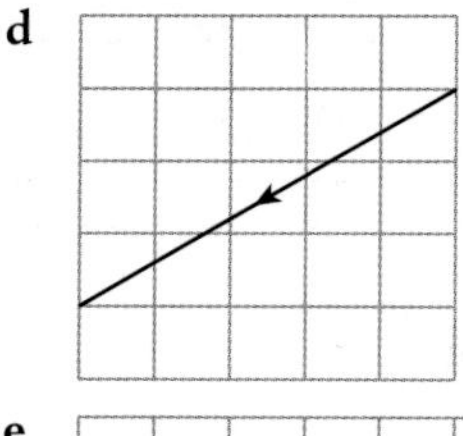

e

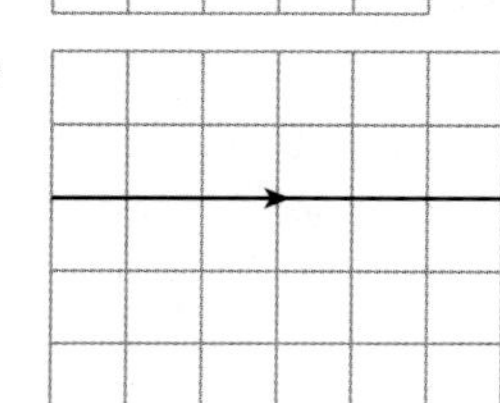

2 **a** $\begin{pmatrix}3\\2\end{pmatrix}$ **b** $\begin{pmatrix}-5\\-1\end{pmatrix}$ **c** $\begin{pmatrix}-2\\4\end{pmatrix}$
d $\begin{pmatrix}-4\\0\end{pmatrix}$ **e** $\begin{pmatrix}-3\\-3\end{pmatrix}$ **f** $\begin{pmatrix}0\\-2\end{pmatrix}$
g $\begin{pmatrix}5\\-3\end{pmatrix}$ **h** $\begin{pmatrix}4\\1\end{pmatrix}$

3

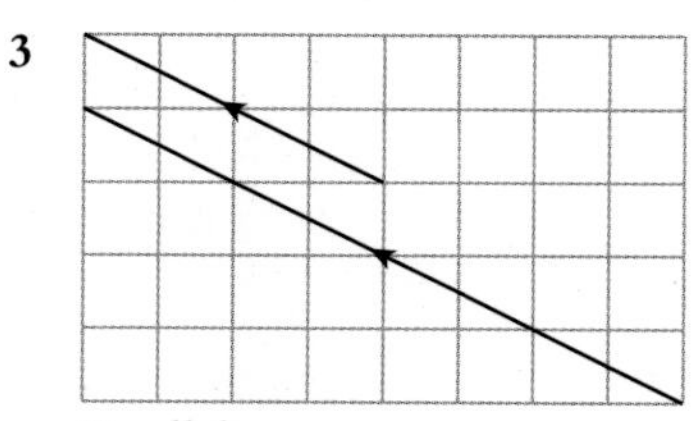

Parallel

4

Perpendicular

5 a 10 b 13 c 4
d 17 e 29

6 a $\overrightarrow{TU}, \overrightarrow{SO}, \overrightarrow{RQ}, \overrightarrow{OP}$
b $\overrightarrow{UP}, \overrightarrow{TO}, \overrightarrow{SR}, \overrightarrow{OQ}$
c $\overrightarrow{OS}, \overrightarrow{UT}, \overrightarrow{QR}, \overrightarrow{PO}$
d $\overrightarrow{QO}, \overrightarrow{OT}, \overrightarrow{PU}, \overrightarrow{RS}$

Exercise 9.10

1 a

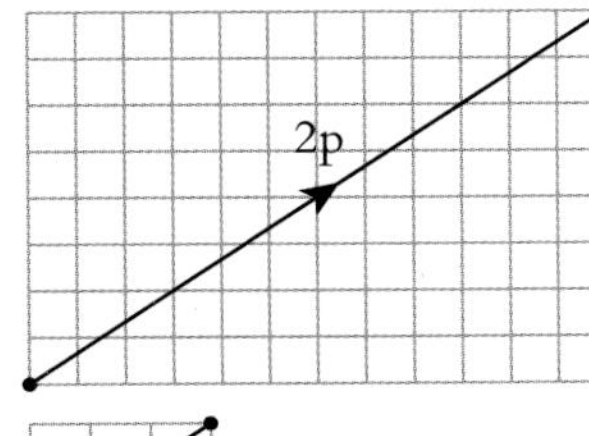

b

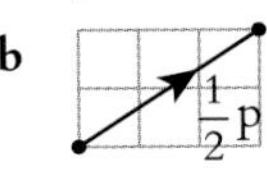

c

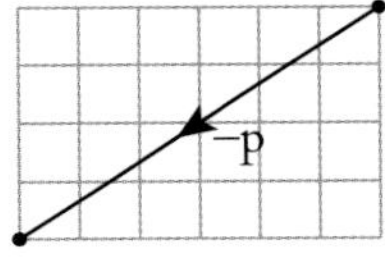

d

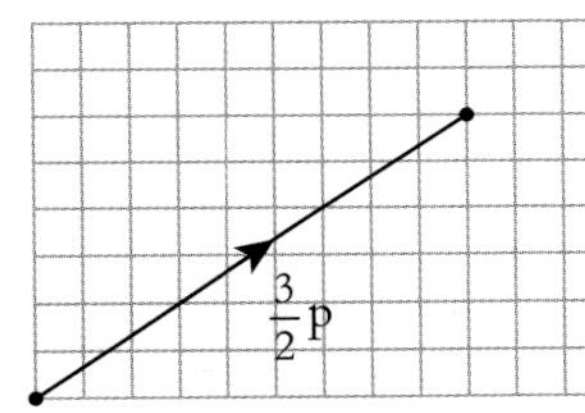

e

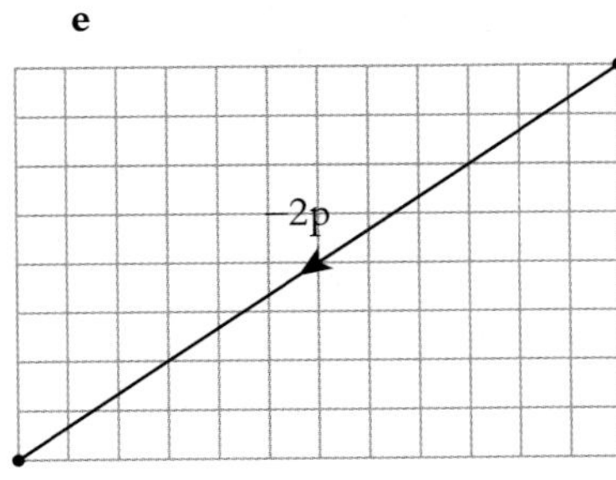

f

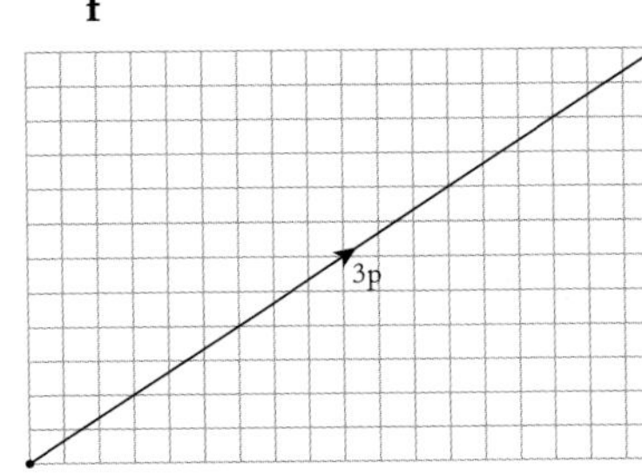

2 a $\begin{pmatrix}-6\\-10\end{pmatrix}$ b $\begin{pmatrix}-5\\-10\end{pmatrix}$ c $\begin{pmatrix}4\\8\end{pmatrix}$
d $\begin{pmatrix}1\\-2\end{pmatrix}$ e $\begin{pmatrix}2\\-1\end{pmatrix}$ f $\begin{pmatrix}-5\\10\end{pmatrix}$

3 a $\begin{pmatrix}4\\3\end{pmatrix}$ b $\begin{pmatrix}6\\-2\end{pmatrix}$ c $\begin{pmatrix}2\\-5\end{pmatrix}$
d $\begin{pmatrix}4\\7\end{pmatrix}$ e $\begin{pmatrix}8\\6\end{pmatrix}$
f $\begin{pmatrix}-2\\-27\end{pmatrix}$ g $\begin{pmatrix}6\\14\end{pmatrix}$ h $\begin{pmatrix}14\\25\end{pmatrix}$
i $\begin{pmatrix}14\\8\end{pmatrix}$ j $\begin{pmatrix}30\\18\end{pmatrix}$ k $\begin{pmatrix}4\\-6\end{pmatrix}$
l $\begin{pmatrix}-3\\-13\end{pmatrix}$

4 C and E, B and H, G and D, A and F

5 a $\begin{pmatrix}4\\1\end{pmatrix}$ b $\begin{pmatrix}2\\3\end{pmatrix}$ c $\begin{pmatrix}-1\\2\end{pmatrix}$
d $\begin{pmatrix}-3\\4\end{pmatrix}$ e $\begin{pmatrix}-3\\0\end{pmatrix}$ f $\begin{pmatrix}-2\\-3\end{pmatrix}$
g $\begin{pmatrix}0\\-1\end{pmatrix}$ h $\begin{pmatrix}1\\-4\end{pmatrix}$

6 a $\begin{pmatrix}2\\-1\end{pmatrix}, \begin{pmatrix}2\\1\end{pmatrix}, \begin{pmatrix}0\\2\end{pmatrix}, \begin{pmatrix}-3\\1\end{pmatrix}, \begin{pmatrix}-1\\-3\end{pmatrix}$
b $\begin{pmatrix}0\\0\end{pmatrix}$ Start and finish at the same place

7 a $\begin{pmatrix}21\\20\end{pmatrix}$ b 29

8 a $\begin{pmatrix}12\\35\end{pmatrix}$ b 37

9 a answer given
b 5.07

10 $p = 6.12, q = -5.14$

11 $p = -28.2, q = 10.3$

12 $x = 7, y = -1$

13 $x = 3, y = -4$

14 a $\begin{pmatrix}8\\1\end{pmatrix}$ b $\begin{pmatrix}1\\3\end{pmatrix}$ c $\begin{pmatrix}1\\3\end{pmatrix}$
d $\begin{pmatrix}2\\6\end{pmatrix}$

15 a $\begin{pmatrix}3\\2\end{pmatrix}$ b $\begin{pmatrix}6\\4\end{pmatrix}$ c (9, 6)

16 a

t	0	1	2	3	4
r	$\begin{pmatrix}3\\5\end{pmatrix}$	$\begin{pmatrix}5\\4\end{pmatrix}$	$\begin{pmatrix}7\\3\end{pmatrix}$	$\begin{pmatrix}9\\2\end{pmatrix}$	$\begin{pmatrix}11\\1\end{pmatrix}$

b

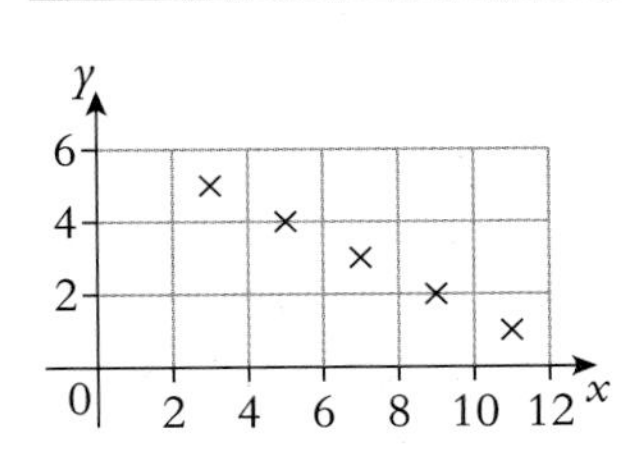

c The points lie in a straight line.

Exercise 9.11

1 a $-\mathbf{q}$ b $2\mathbf{q}$ c $-\mathbf{p} + \mathbf{q}$
d $-2\mathbf{q} + \mathbf{p}$

2 a $-\mathbf{p} + \mathbf{q}$ b $-\frac{1}{2}\mathbf{p} + \frac{1}{2}\mathbf{q}$
c $\frac{1}{2}\mathbf{p} - \frac{1}{2}\mathbf{q}$ d $\frac{1}{2}\mathbf{p} + \frac{1}{2}\mathbf{q}$

3 a $\mathbf{p} + \mathbf{r}$ b $\frac{1}{2}\mathbf{r}$ c $\mathbf{p} + \frac{1}{2}\mathbf{r}$
d $\frac{1}{2}\mathbf{r} - \mathbf{p}$

4 a $\overrightarrow{BA} = -2\mathbf{p}$ b $\mathbf{q} + 2\mathbf{p}$
c $-\mathbf{p} + \mathbf{q}$ d $\mathbf{p} + \mathbf{q}$

5 a $\mathbf{p}$ b $-\mathbf{q}$
c $\mathbf{p} + \mathbf{q}$ d $\mathbf{p} - \mathbf{q}$
e $2\mathbf{q} + \mathbf{p}$ f $-\mathbf{q} - 2\mathbf{p}$

6 a $\frac{1}{2}\mathbf{a}$ b $-\frac{1}{2}\mathbf{c}$
c $-\mathbf{a} + \mathbf{c}$ d $\frac{1}{2}\mathbf{c} - \frac{1}{2}\mathbf{a}$
parallel and AC = 2 MN

7 a $\mathbf{a} + \mathbf{c}$ b $-\mathbf{c} + \mathbf{a}$
c $\frac{1}{2}\mathbf{a} + \frac{1}{2}\mathbf{c}$ d $\frac{1}{2}\mathbf{a} - \frac{1}{2}\mathbf{c}$

8 a $\frac{1}{2}\mathbf{p} + \frac{1}{2}\mathbf{q}$ b $\frac{1}{2}\mathbf{p} + \frac{1}{2}\mathbf{q}$
c $-\frac{1}{2}\mathbf{q} + \frac{1}{2}\mathbf{p}$ d $\frac{1}{2}\mathbf{p} - \frac{1}{2}\mathbf{q}$
parallelogram

9 a $\frac{3}{2}\mathbf{c}$ b $-\mathbf{a}+\mathbf{c}$

c $\mathbf{a}-\frac{1}{2}\mathbf{c}$ d $\frac{1}{2}\mathbf{c}-\frac{1}{2}\mathbf{a}$

parallel and AC = 2 MN

10 a $\mathbf{p}+\mathbf{q}$ b $-\frac{3}{2}\mathbf{p}+\frac{1}{2}\mathbf{q}$

c $\frac{5}{2}\mathbf{p}+\frac{1}{2}\mathbf{q}$ d $\frac{1}{2}(3\mathbf{p}+\mathbf{q})$

Exercise 9.12

1 a $-\mathbf{p}+\mathbf{q}$ b $-\frac{1}{3}\mathbf{p}+\frac{1}{3}\mathbf{q}$

c $\frac{2}{3}\mathbf{p}+\frac{1}{3}\mathbf{q}$

2 a $-\mathbf{p}+\mathbf{q}$ b $-\frac{3}{4}\mathbf{p}+\frac{3}{4}\mathbf{q}$

c $\frac{1}{4}\mathbf{p}+\frac{1}{4}\mathbf{q}$ d $\frac{1}{4}\mathbf{p}+\frac{3}{4}\mathbf{q}$

3 a $\frac{1}{2}\mathbf{q}$ b $-\frac{1}{3}\mathbf{p}+\frac{1}{2}\mathbf{q}$

c $-\mathbf{p}+\mathbf{q}$ d $-\frac{2}{5}\mathbf{p}+\frac{2}{5}\mathbf{q}$

4 a $\frac{1}{3}\mathbf{r}$ b $\frac{1}{4}\mathbf{p}$

c $\frac{2}{3}\mathbf{r}-\frac{3}{4}\mathbf{p}$ d $\mathbf{r}+\frac{1}{4}\mathbf{p}$

5 a $-3\mathbf{a}+\mathbf{c}$ b $-\frac{9}{5}\mathbf{a}+\frac{3}{5}\mathbf{c}$

c $\frac{6}{5}\mathbf{a}+\frac{3}{5}\mathbf{c}$ d $2\mathbf{a}+\mathbf{c}$

collinear

6 a $\mathbf{r}-\mathbf{p}$ b $\frac{1}{4}\mathbf{r}-\frac{1}{4}\mathbf{p}$

c $\frac{1}{4}\mathbf{p}+\frac{3}{4}\mathbf{r}$ d $\frac{3}{4}\mathbf{p}+\frac{1}{4}\mathbf{r}$

7 a $\frac{2}{5}\mathbf{r}$ b $\mathbf{p}-\frac{3}{5}\mathbf{r}$

c $\frac{5}{3}\mathbf{p}-\mathbf{r}$ d $\frac{5}{3}\mathbf{p}$

$\overrightarrow{OX}=\frac{5}{3}\overrightarrow{OP}$

8 $p = 13, q = 10\frac{1}{3}$

9 answer given

Challenge

a $\frac{1}{3}p+\frac{1}{3}q$

b $\frac{1}{3}p+\frac{1}{3}q$

The points coincide.

Exercise 9.13

1 a i $\frac{7}{12}$ ii $\frac{1}{10}$ iii $\frac{11}{20}$

b $\frac{7}{9}$ c $\frac{117}{590}$

2 a

ℰ
A
B
37
13
16
34

b i $\frac{4}{25}$ ii $\frac{17}{50}$

c $\frac{37}{50}$ d $\frac{49}{198}$

3 a i $\frac{89}{150}$ ii $\frac{13}{30}$ iii $\frac{23}{150}$

b $\frac{47}{89}$ c 0.486

4 a i $\frac{73}{100}$ ii $\frac{1}{4}$ iii $\frac{3}{20}$

b $\frac{17}{50}$ c $\frac{145}{392}$

5 a

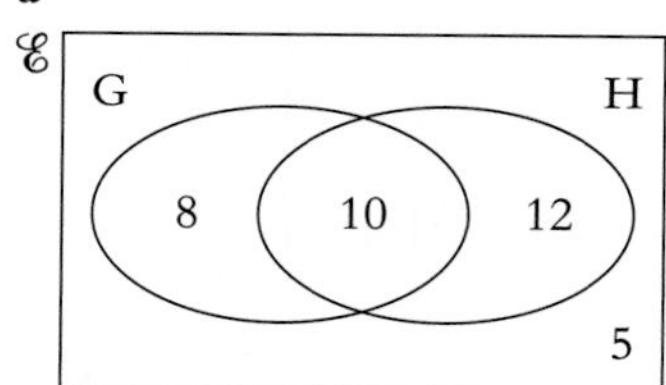

b i $\frac{2}{7}$ ii $\frac{12}{35}$

c $\frac{5}{11}$ d $\frac{286}{595}$

6 a i $\frac{2}{25}$ ii $\frac{3}{50}$ iii $\frac{1}{50}$

b $\frac{8}{19}$ c $\frac{19}{1990}$

7 a

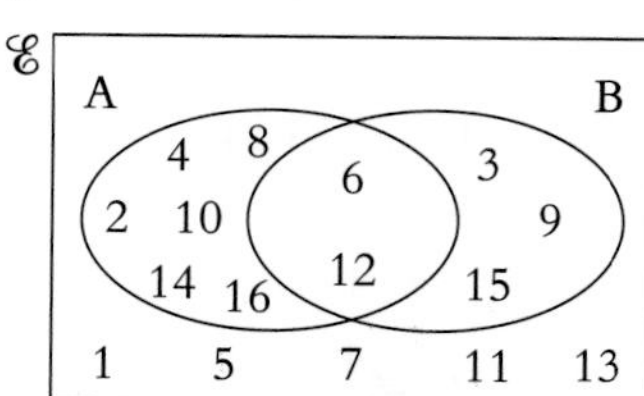

b i $\frac{5}{16}$ ii $\frac{1}{2}$ iii $\frac{1}{8}$

iv $\frac{11}{16}$ v $\frac{5}{16}$

8 a i $\frac{27}{50}$ ii $\frac{7}{25}$ iii $\frac{2}{25}$

b $\frac{9}{25}$ c $\frac{87}{175}$

9 a

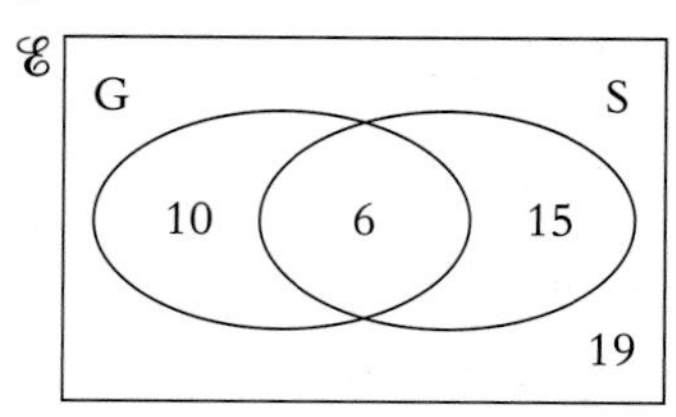

b $\frac{3}{25}$ c $\frac{5}{7}$ d $\frac{1}{35}$

Examination-style questions

1 LB = 255 mm, UB = 261 mm

2 LB = 39.1125 cm², UB = 40.3925 cm²

3 4.7 km/h

4 $x = 18, y = 11.5$

5 a $x \geq 5, y \geq 7, x > y, x + y \geq 18$

b

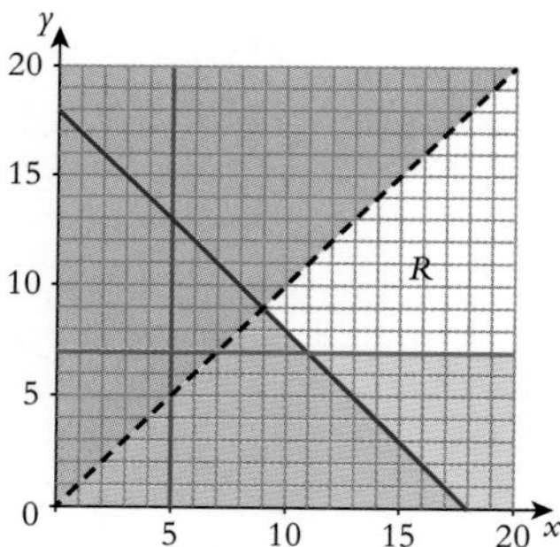

c $12.20

6 $a = -4, b = -6$

7 $p = 2, q = 7$

8 a $e = 2.25, f = -7.125$

b -7.125

9 a $x = -0.2, x = 4.2$

b i $y = x - 2$

ii $x = 1.4, x = 3.6$

10 a $-4.8, -0.2, 3.0$

b $k = 10$

c $k < 10$

11 a $\begin{pmatrix} 14 \\ -4 \end{pmatrix}$

b $\begin{pmatrix} -7 \\ 2 \end{pmatrix}$

c $\begin{pmatrix} 3 \\ -5 \end{pmatrix}$

d 5

12 $\frac{1}{3}\mathbf{a} + \frac{2}{3}\mathbf{b}$

13 a i $\mathbf{a} - \mathbf{b}$

ii $\frac{3}{5}(\mathbf{a} - \mathbf{b})$

iii $\frac{3}{5}\mathbf{a} + \frac{2}{5}\mathbf{b}$

iv $\frac{3}{4}\mathbf{a} - \mathbf{b}$

b $\frac{3k}{5}\mathbf{a} + \frac{2k}{5}\mathbf{b}$

c $\frac{3m}{4}\mathbf{a} + (1 - m)\mathbf{b}$

d $k = \frac{5}{6},\ m = \frac{2}{3}$

14 a i $-3\mathbf{a} + 4\mathbf{b}$

ii $3\mathbf{a} - 2\mathbf{b}$

b i $\frac{12k}{7}\mathbf{a} - 4k\mathbf{b}$

ii $\frac{12\lambda}{7}\mathbf{a} + (2 - 4\lambda)\mathbf{b}$

c 0.7

15 a $\frac{10}{19}$

b $\frac{11}{87}$

c $\frac{442}{1015}$

16 a i $\frac{19}{50}$

ii $\frac{13}{25}$

b $\frac{1}{2}$

c i $\frac{496}{1225}$

ii $\frac{5763}{1225}$

d $\frac{87}{196}$

UNIT 9

Index